Maritime Welding

Level 1

Trainee Guide

Pearson

Boston Columbus Indianapolis New York San Francisco Amsterdam
Cape Town Dubai London Madrid Milan Munich Paris Montreal Toronto Delhi
Mexico City Sao Paulo Sydney Hong Kong Seoul Singapore Taipei Tokyo

NCCER

President and Chief Executive Officer: Don Whyte
President: Boyd Worsham
Vice President: Steve Greene
Chief Operations Officer: Katrina Kersch
Electrical Project Manager: Chris Wilson
Senior Development Manager: Mark Thomas
Senior Production Manager: Tim Davis
Art Manager: Kelly Sadler
Managing Editor: Natalie Richoux
Desktop Publishing Manager: James McKay
Production Specialists: Gene Page, Eric Caraballoso
Editors: Graham Hack, Jordan Hutchinson

Writing and development services provided by Topaz Publications, Liverpool, NY

Lead Writer/Project Manager: Troy Staton
Desktop Publisher: Joanne Hart
Art Director: Alison Richmond
Permissions Editor: Tonia Burke
Writers: Troy Staton, Carol Herbert, Thomas Burke

Pearson Education, Inc.

Director of Alliance/Partnership Management: Kelly Trakalo
Content Producer: Alexandrina B. Wolf
Assistant Content Producer: Alma Dabral
Digital Content Producer: Jose Carchi
Senior Marketing Manager: Brian Hoehl
Composition: NCCER
Printer/Binder: LSC Communications
Cover Printer: LSC Communications
Text Fonts: Palatino and Univers

Credits and acknowledgments for content borrowed from other sources and reproduced, with permission, in this textbook appear at the end of each module.

9 2023

Pearson

Perfect bound ISBN13: 978-0-13-560371-0
ISBN10: 0-13-560371-4

Preface

Welding is a high-tech industry that can take you to places all over the world. Welds are everywhere — just look around. From ladders to aircraft carriers; from NASCAR to national defense; from the laboratory to sales and repair; the welding trade impacts virtually every industry.

Few career choices offer as many options for employment and opportunities for growth. According to the Department of Labor, "the basic skills of welding are similar across industries, so welders can easily shift from one industry to another, depending on where they are needed most." Welders will continue to be in high demand since there is a continual need to rebuild aging infrastructure such as highways and buildings. In addition, the creation of new power generation facilities will yield new job prospects for trained and skilled welders.

Of course, one of the places to find a prosperous career as a welder is in the maritime industry. Shipyards and repair facilities are plentiful in coastal areas and on major rivers and lake systems. Maritime welding can be done on land or sea and above or underneath the surface of the water. It includes the work on ships, boats, and off-shore oil rigs. It includes the installation and repair of metal structures and components. .

Maritime Welding

This *Maritime Welding* curriculum is a product of collaboration between NCCER and the National Maritime Education Council (NMEC). Several maritime welding subject matter experts determined which of NCCER's *Welding* modules are suitable to maritime industry training. This edition of *Maritime Welding* can be used for welding training and can be supplemented by instructors with information specific to the maritime industry.

We invite you to visit the NCCER website at **www.nccer.org** for the latest releases, training information, newsletter, and much more. You can also reference the Pearson product catalog online at **www.nccer.org**. Your feedback is welcome. You may e-mail your comments to **curriculum@nccer.org** or send general comments and inquiries to **info@nccer.org**.

NMEC

The National Maritime Education Council (NMEC) is a multi-regional, industry-centered organization working to address critical workforce issues facing its member yards and the maritime industry. Established in 2012 to provide a centralized voice for the industry as it relates to workforce development, NMEC's primary mission is to lead the maritime industry in the development, promotion, and implementation of a national workforce development system that includes standardized craft training, portable credentials, and journey-level assessments.

Through its partnership with NCCER, NMEC is working in collaboration with its member yards to:

- Establish a system of career awareness and recruiting that will allow the industry to speak with one voice and deliver a consistent message to students, student influencers, the underemployed, women, and returning military.
- Establish and implement an industry-wide, comprehensive, standardized training and credentialing system for maritime craft professionals.
- Grow a pipeline of skilled craft professionals trained and credentialed through an industry-driven, uniform process.
- Improve industry outcomes by developing a better skilled, safer, more productive workforce.

NCCER Standardized Curricula

NCCER is a not-for-profit 501(c)(3) education foundation established in 1995 by the world's largest and most progressive construction companies and national construction associations. It was founded to address the severe workforce shortage facing the industry and to develop a standardized training process and curricula. Today, NCCER is supported by hundreds of leading construction and maintenance companies, manufacturers, and national associations. The NCCER Standardized Curricula was developed by NCCER in partnership with Pearson Education, Inc., the world's largest educational publisher.

Some features of the NCCER Standardized Curricula are as follows:

- An industry-proven record of success
- Curricula developed by the industry for the industry
- National standardization providing portability of learned job skills and educational credits
- Compliance with Office of Apprenticeship requirements for related classroom training (*CFR 29:29*)
- Well-illustrated, up-to-date, and practical information

NCCER also maintains a Registry that provides transcripts, certificates, and wallet cards to individuals who have successfully completed a level of training within a craft in the NCCER Standardized Curricula. *Training programs must be delivered by an NCCER Accredited Training Sponsor in order to receive these credentials.*

Special Features

In an effort to provide a comprehensive user-friendly training resource, we have incorporated many different features for your use. Whether you are a visual or hands-on learner, this book will provide you with the proper tools to get started in the welding industry.

Introduction Page

This page is found at the beginning of each module and lists the Objectives, Performance Tasks, and Trade Terms, for that module. The Objectives list the skills and knowledge you will need in order to complete the module successfully. The Performance Tasks give you an opportunity to apply your knowledge to real-world tasks. The list of Trade Terms identifies important terms you will need to know by the end of the module.

29105-15
BASE METAL PREPARATION

Objectives

When you have completed this module, you will be able to do the following:

1. Identify safety practices related to preparing base metals and describe basic cleaning procedures.
   a. Identify safety practices related to preparing base metals.
   b. Describe the basic properties and types of carbon and stainless steel.
   c. Describe basic metal cleaning procedures and concerns.
2. Identify and describe basic weld joint design and types of welds.
   a. Identify and describe the loads that are routinely placed on weld joints.
   b. Identify and describe the various types of weld joints.
   c. Describe a welding procedure specification (WPS) and the information it provides.
3. Describe how to prepare joints for welding.
   a. Describe how to mechanically prepare joints for welding.
   b. Describe how to thermally prepare joints for welding.

Performance Tasks

Under the supervision of your instructor, you should be able to do the following:

1. Mechanically or hand grind a bevel on the edge of a ¼"- to ¾"-thick mild steel plate (6 to 20 mm metric plate) at 22½ degrees.
2. Thermally bevel the edge of a ¼"- to ¾"-thick mild steel plate (6 to 20 mm metric plate) at 22½ degrees.

Trade Terms

Annealed	Ferritic	Oxide	Tempering
Austenitic	Ferrous	Piping porosity	Tensile strength
Backing	Load	Porosity	Weathering steel
Base metal	Malleable	Quench	Welding procedure specification (WPS)
Castings	Martensitic	Root face	Weldment
Distortion	Melt-through	Root opening	Wrought
Ductile	Nonferrous metal	Surfacing	

Industry Recognized Credentials

If you are training through an NCCER-accredited sponsor, you may be eligible for credentials from NCCER's Registry. The ID number for this module is 29105-15. Note that this module may have been used in other NCCER curricula and may apply to other level completions. Contact NCCER's Registry at 888.622.3720 or go to **www.nccer.org** for more information.

Notes, Cautions, and Warnings

Safety features are set off from the main text in highlighted boxes and organized into three categories based on the potential danger of the issue being addressed. Notes simply provide additional information on the topic area. Cautions alert you of a danger that does not present potential injury but may cause damage to equipment. Warnings stress a potentially dangerous situation that may cause injury to you or a co-worker.

NOTE
Some companies have physical activity requirements that must be met by apprentices. These requirements vary from company to company.

CAUTION
Loosening regulator adjusting screws closes the regulators and prevents damage to the regulator diaphragms when the cylinder valves are opened.

WARNING!
Never apply heat directly to a cylinder or regulator. This can cause excessive pressure, resulting in an explosion.

Special Features

Features present technical tips and professional practices from the construction industry. These features often include real-life scenarios similar to those you might encounter on the job site.

Commonly Available Carbon Electrodes

The carbon electrodes that are usually readily available from general welding suppliers are the copper-coated round and rectangular electrodes used with direct current electrode positive (DCEP) current. Other types and styles, including jointed,

Going Green

Going Green looks at ways to preserve the environment, save energy, and make good choices regarding the health of the planet. Through the introduction of new construction practices and products, you will see how the "greening of the world" has already taken root.

GOING GREEN

Cooling Water

Welders involved in metal-cutting activities often allow the water used in the process to simply flow into the ground or down the drain. To better protect the environment, water used for cooling the plasma arc cutting equipment should be contained whenever

Did You Know?

The Did You Know? features introduce historical tidbits or modern information about the construction industry. Interesting and sometimes surprising facts about construction are also presented.

Did You Know?

PHMSA

The US Department of Transportation (DOT) established the Pipeline and Hazardous Materials Safety Administration (PHMSA) in 2004 to develop and enforce regulations for the safe, reliable, and environmentally sound operation of the 2.6 million

Case History

Case History features emphasize the importance of safety by citing examples of the costly (and often devastating) consequences of ignoring safety rules and regulations.

> **Case History**
>
> **Mexico City Refinery**
>
> In Mexico City in 1984, a weld on a liquid propane gas (LPG) tank cracked. The escaping gas cloud drifted toward a nearby residential area until it encountered an ignition source. When the gas ignited, the fireball quickly burned its way back to the leak source. The resulting explosion destroyed the refinery and many residential neighborhoods, leaving 503 people dead and more than 4,000 injured.
>
> **The Bottom Line:** Welds must be made cor-

Color Illustrations and Photographs

Full-color illustrations and photographs are used throughout each module to provide vivid detail. These figures highlight important concepts from the text and provide clarity for complex instructions. Each figure is denoted in the text in *italic type* for easy reference.

Figure 15 Worker in a confined space.

Step-by-Step Instructions

Step-by-step instructions are used throughout to guide you through technical procedures and tasks from start to finish. These steps show you not only how to perform a task but also how to do it safely and efficiently.

> Follow these steps to make an overhead fillet weld:
>
> *Step 1* Tack two plates together to form a T-joint for the fillet weld coupon.
>
> *Step 2* Tack-weld the coupon so it is in the overhead position.

Trade Terms

Each module presents a list of Trade Terms that are discussed within the text, defined in the Glossary at the end of the module, and reinforced with a Trade Terms Quiz. These terms are denoted in the text with **blue bold type** upon their first occurrence. To make searches for key information easier, a comprehensive Glossary of Trade Terms from all modules is found at the back of this book.

> Metals are classified into two basic groups: **ferrous** metals, which are composed mainly of iron, and **nonferrous metals**, which contain very little or no iron. Ferrous metals include all steel, cast iron, **wrought** iron, **malleable** iron, and **ductile**

Section Review

The Section Review features helpful additional resources and review questions related to the objectives in each section of the module.

> **Additional Resources**
>
> *2014 Technical Training Guide*. Current Edition. Cleveland, OH. USA: The Lincoln Electric Company. www.lincolnelectric.com
>
> *Welding Handbook*. Current Edition. Miami, FL: The American Welding Society.
>
> **2.0.0 Section Review**
>
> 1. Which of the following can be quenched without concern?
>   a. Weld test coupons
>   b. Functional welds
>   c. Practice coupons
>   d. Welds done in the 3G position
> 2. On an open-root V-groove weld in the 4G position, the electrode work angle should be ____.
>   a. 30 degrees
>   b. 45 degrees
>   c. 90 degrees
>   d. varied as needed to tie in

Review Questions

Review Questions are provided to reinforce the knowledge you have gained. This makes them a useful tool for measuring what you have learned.

> **Review Questions**
>
> 1. The A-CAC process uses a high-current electric arc between the base metal and the end of a(n) ____.
>   a. carbon-graphite electrode to melt the base metal
>   b. high-frequency pilot arc to melt the base metal
>   c. plasma jet to melt the base metal
>   d. idle arc to melt the base metal
> 2. An A-CAC torch requires compressed air to blow the ____.
>   a. carbon dioxide away from the cut
>   b. molten metal away from the cut
>   c. nitrogen away from the cut
>   d. carbon arc away from the cut
> 3. When cutting and gouging with A-CAC equipment, the jet orifices of the torch head(s) should be located ____.
>   a. beside the electrode
>   b. above the electrode
>   c. ahead of the electrode
>   d. between the electrode and the base metal surface
> 4. To prevent overheating, the electrode tip to the V-groove head should be no farther than ____.
>   a. 1" (2 cm)
>   b. 2" (5 cm)
>   c. 4" (10 cm)
>   d. 6" (15 cm)
> 5. Light-duty gouging or cutting is typically done with a ____.
>   a. 10A welding machine
>   b. 200A welding machine
>   c. 600A welding machine
>   d. 1,200A welding machine
> 6. Heavy-duty cutting can require a torch airflow rate up to ____.
>   a. 20 cfm (566 lpm)
>   b. 35 cfm (991 lpm)
>   c. 40 cfm (1,133 lpm)
>   d. 50 cfm (1,416 lpm)
> 7. Compressed air is supplied to the torch through a hose attached to a fitting on the ____.
>   a. welding machine
>   b. end of the torch cable
>   c. air jet orifices
>   d. workpiece
> 8. Electrodes designed for A-CAC contain a rare-earth, ____.
>   a. magnet metal
>   b. undercurrent ingredient
>   c. amplifier ingredient
>   d. arc-stabilizing metal
> 9. The most common style of A-CAC electrode used for manual cutting or gouging is the ____.
>   a. flat electrode
>   b. round electrode
>   c. half-round electrode
>   d. round-jointed electrode
> 10. When performing A-CAC cutting or gouging, the minimum groove width is controlled by the ____.
>   a. size of the torch
>   b. size of the electrode
>   c. speed of travel
>   d. angle of the electrode holder
> 11. Because A-CAC activities tend to blow molten metal considerable distances, it is important to station a(n) ____.
>   a. observer in the A-CAC work area
>   b. slag trap in the A-CAC work area
>   c. oxygen monitor in the A-CAC work area
>   d. fire watch in the A-CAC work area
> 12. To wash a projection from a metal surface using A-CAC equipment, the electrode is typically used at a push angle of about ____.
>   a. 30 degrees to the base metal surface
>   b. 45 degrees to the base metal surface
>   c. 55 degrees to the base metal surface
>   d. 90 degrees to the base metal surface

NCCER Standardized Curricula

NCCER's training programs comprise more than 80 construction, maintenance, pipeline, and utility areas and include skills assessments, safety training, and management education.

Boilermaking
Cabinetmaking
Carpentry
Concrete Finishing
Construction Craft Laborer
Construction Technology
Core Curriculum: Introductory Craft Skills
Drywall
Electrical
Electronic Systems Technician
Heating, Ventilating, and Air Conditioning
Heavy Equipment Operations
Highway/Heavy Construction
Hydroblasting
Industrial Coating and Lining Application Specialist
Industrial Maintenance Electrical and Instrumentation Technician
Industrial Maintenance Mechanic
Instrumentation
Insulating
Ironworking
Masonry
Millwright
Mobile Crane Operations
Painting
Painting, Industrial
Pipefitting
Pipelayer
Plumbing
Reinforcing Ironwork
Rigging
Scaffolding
Sheet Metal
Signal Person
Site Layout
Sprinkler Fitting
Tower Crane Operator
Welding

Maritime

Maritime Industry Fundamentals
Maritime Pipefitting
Maritime Structural Fitter

Green/Sustainable Construction

Building Auditor
Fundamentals of Weatherization
Introduction to Weatherization
Sustainable Construction Supervisor
Weatherization Crew Chief
Weatherization Technician
Your Role in the Green Environment

Energy

Alternative Energy
Introduction to the Power Industry
Introduction to Solar Photovoltaics
Introduction to Wind Energy
Power Industry Fundamentals
Power Generation Maintenance Electrician
Power Generation I&C Maintenance Technician
Power Generation Maintenance Mechanic
Power Line Worker
Power Line Worker: Distribution
Power Line Worker: Substation
Power Line Worker: Transmission
Solar Photovoltaic Systems Installer
Wind Turbine Maintenance Technician

Pipeline

Control Center Operations, Liquid
Corrosion Control
Electrical and Instrumentation
Field Operations, Liquid
Field Operations, Gas
Maintenance
Mechanical

Safety

Field Safety
Safety Orientation
Safety Technology

Supplemental Titles

Applied Construction Math
Tools for Success

Management

Fundamentals of Crew Leadership
Project Management
Project Supervision

Spanish Titles

Acabado de concreto: nivel uno
Aislamiento: nivel uno
Albañilería: nivel uno
Andamios
Carpintería: Formas para carpintería, nivel tres
Currículo básico: habilidades introductorias del oficio
Electricidad: nivel uno
Herrería: nivel uno
Herrería de refuerzo: nivel uno
Instalación de rociadores: nivel uno
Instalación de tuberías: nivel uno
Instrumentación: nivel uno, nivel dos, nivel tres, nivel cuatro
Orientación de seguridad
Paneles de yeso: nivel uno
Seguridad de campo

Acknowledgments

This curriculum was revised as a result of the farsightedness and leadership of the following sponsors:

Austal USA
Bollinger Shipyards
Claddagh Enterprises
Central Louisiana Technical College
Continental Maritime
Exelon Generation
Gulf States, Inc., Michigan Division
Ingalls Shipbuilding
KBR Industrial Services
Lee College
Lincoln Electric
Northeast Community College
Northland Pioneer College
Robins & Morton
Tecnico Corporation
TIC-The Industrial Company
Toledo Refining Co. LLC
Zachry Industrial, Inc.

This curriculum would not exist were it not for the dedication and unselfish energy of those volunteers who served on the Authoring Team. A sincere thanks is extended to the following:

Tom Ashley
Paul Bennett
Gerald Bickerstaff
Curtis Casey
Bill Cherry
Gerald Guthrie
Rod Hellyer
Frank Johnson
John Knapp
Jason Langford
Jerry McDaniel
Tim Riley
Kyle Savoie
Jason Scales
Dan Sterry
Holley Thomas
Jerry Trainor
Adam Webb

NCCER Partners

American Fire Sprinkler Association
Associated Builders and Contractors, Inc.
Associated General Contractors of America
Association for Career and Technical Education
Association for Skilled and Technical Sciences
Construction Industry Institute
Construction Users Roundtable
Construction Workforce Development Center
Design Build Institute of America
GSSC – Gulf States Shipbuilders Consortium
ISN
Manufacturing Institute
Mason Contractors Association of America
Merit Contractors Association of Canada
NACE International
National Association of Minority Contractors
National Association of Women in Construction
National Insulation Association
National Technical Honor Society
NAWIC Education Foundation
North American Crane Bureau
North American Technician Excellence
Pearson
Pearson Qualifications International
Prov
SkillsUSA®
Steel Erectors Association of America
U.S. Army Corps of Engineers
University of Florida, M. E. Rinker School of Building Construction
Women Construction Owners & Executives, USA

NCCER Business Partners

Contents

Module Nine

SMAW – Beads and Fillet Welds

Describes the preparation and setup of arc welding equipment and the process of striking an arc. Explains how to detect and correct arc blow. Describes how to make stringer, weave, overlapping beads, and fillet welds. (Module ID 29109-15; 100 hours)

Module Ten

Joint Fit-Up and Alignment

Describes job code specifications. Explains how to use fit-up gauges and measuring devices to check fit-up and alignment and use plate and pipe fit-up and alignment tools to properly prepare joints. Explains how to check for joint misalignment and poor fit. (Module ID 29110-15; 5 hours)

Glossary

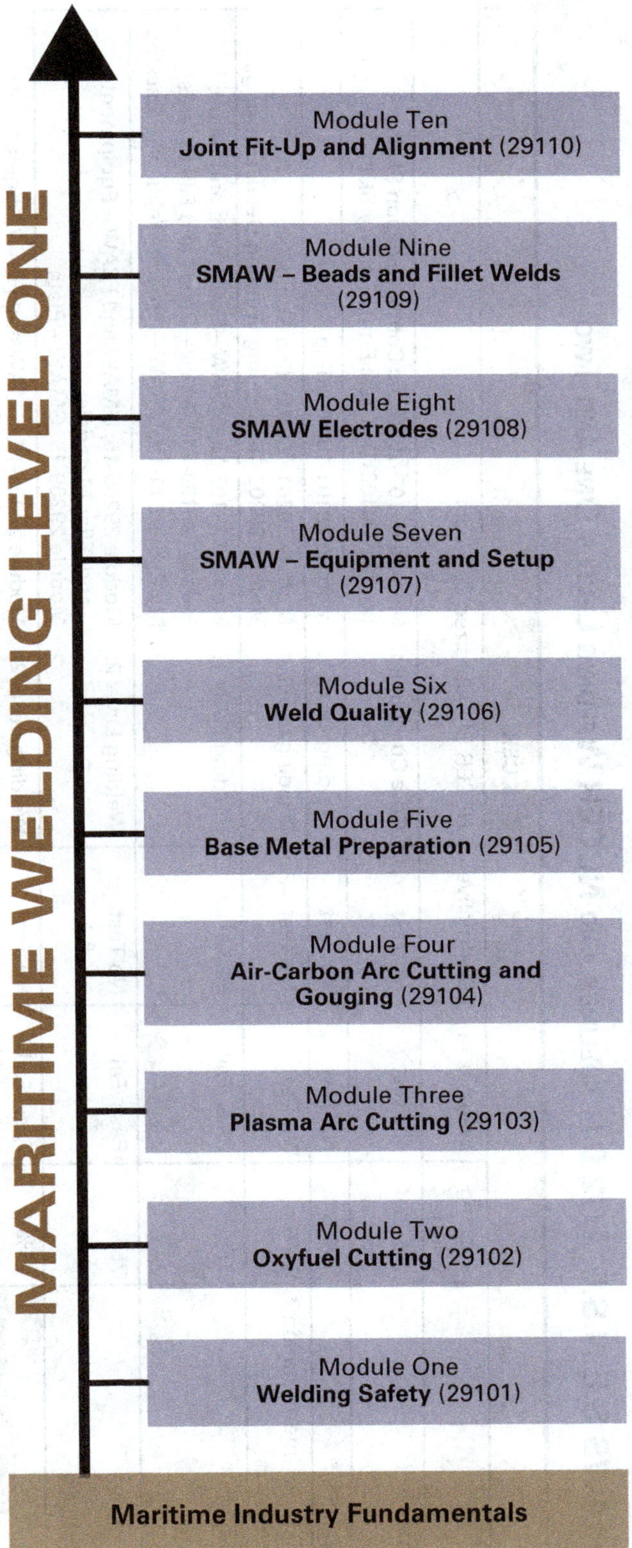

This course map shows all of the modules in *Maritime Welding Level One*. The suggested training order begins at the bottom and proceeds up. Skill levels increase as you advance on the course map. The local Training Program Sponsor may adjust the training order.

Correlations Chart: AWS S.E.N.S.E. EG2.0 Guidelines and NCCER Welding Levels One and Two

SENSE Guidelines				NCCER
Modules	Passing Score	Visual	Destructive	NCCER Standardized Curricula
1 – Occupational Orientation *	No Test	No Test	No Test	Core Curriculum: Module 00107-15, Basic Communication Skills; Module 00108-15, Basic Employability Skills
2 – Safety and Health of Welders *	100%	No Test	No Test	Welding Level 1: Module 29101-15, Welding Safety
3 – Drawing and Welding Symbol Interpretation *	75%	No Test	No Test	Welding Level 2: Module 29201-15, Welding Symbols; Module 29202-15, Reading Welding Detail Drawings
4 – Shielded Metal Arc Welding	75%	Pass /Fail	Pass /Fail	Welding Level 1: Module 29107-15, SMAW—Equipment and Setup; Module 29109-15, SMAW—Beads and Fillet Welds; Module 29111-15, SMAW—Groove Welds with Backing
5 – Gas Metal Arc Welding	75%	Pass /Fail	No Test	Welding Level 2: Module 29205-15, GMAW and FCAW – Equipment and Filler Materials; Module 29209-15, GMAW – Plate
6 – Flux Cored Arc Welding	75%	Pass /Fail	No Test	Welding Level 2: Module 29205-15, GMAW and FCAW – Equipment and Filler Materials; Module 29210-15, FCAW – Plate
7 – Gas Tungsten Arc Welding	75%	Pass /Fail	No Test	Welding Level 2: Module 29207-15, GTAW – Equipment and Filler Materials; Module 29208-15, GTAW – Plate
8 – Thermal Cutting Process * †	75%	No Test	No Test	
Unit 1 Manual Oxyfuel Gas Cutting (OFC) †		Pass /Fail	No Test	Welding Level 1: Module 29102-15, Oxyfuel Cutting
Unit 2 Mechanized Oxyfuel Gas Cutting (OFC) †		Optional	No Test	Welding Level 1: Module 29105-15, Base Metal Preparation
Unit 3 Manual Plasma Arc Cutting—PAC †		Pass /Fail	No Test	Welding Level 1: Module 29103-15, Plasma Arc Cutting
Unit 4 Manual Air Carbon Arc Cutting †		Optional	No Test	Welding Level 1: Module 29104-15, Air-Carbon Arc Cutting and Gouging
9 – Welding Inspection and Testing *	75%	Pass /Fail	No Test	Welding Level 1: Module 29106-15, Weld Quality

* Required module for Level I Entry Welder Completion (plus one welding process module)

† Completion of Units 1 and 3 minimum

Welding Safety

Overview

Qualified welders are in high demand all over the world. Typical welding tasks may include working in heavy construction; joining pipe sections for oil and natural gas pipelines; building ships; and working in a variety of industrial environments such as power plants, refineries, chemical plants, and manufacturing facilities. This module introduces trainees to the all-important topic of safety in the welding trade.

Module 29101

Trainees with successful module completions may be eligible for credentialing through the NCCER Registry. To learn more, go to **www.nccer.org** or contact us at **1.888.622.3720**. Our website has information on the latest product releases and training, as well as online versions of our *Cornerstone* magazine and Pearson's product catalog.

Your feedback is welcome. You may email your comments to **curriculum@nccer.org**, send general comments and inquiries to **info@nccer.org**, or fill in the User Update form at the back of this module.

This information is general in nature and intended for training purposes only. Actual performance of activities described in this manual requires compliance with all applicable operating, service, maintenance, and safety procedures under the direction of qualified personnel. References in this manual to patented or proprietary devices do not constitute a recommendation of their use.

29101 V5

29101
WELDING SAFETY

Objectives

When you have completed this module, you will be able to do the following:

1. Describe basic welding processes, the welding trade, and training/apprenticeship programs.
   a. Describe basic welding processes and the welding trade.
   b. Describe NCCER standardized training and explain apprenticeship programs.
2. Identify and describe personal protective equipment (PPE) related to the welding trade.
   a. Identify and describe body, foot, and hand protective gear.
   b. Identify and describe ear, eye, face, and head protective gear.
3. Identify and describe welding safety practices related to specific hazards or environments.
   a. Describe the importance of welding safety and identify factors related to accidents.
   b. Describe basic welding safety practices related to the general work area.
   c. Describe hot work permits and fire watch requirements.
   d. Describe confined spaces and their related safety practices.
   e. Identify safety practices related to welding equipment.
   f. Identify and describe respiratory hazards, respiratory safety equipment, and ways to ventilate welding work areas.
   g. Explain the purpose of the SDS/MSDS and how it is used.

Performance Tasks

This is a knowledge-based module; there are no Performance Tasks.

Trade Terms

Arc burn
Bonded
Carbon steel
Electric arc
Electrically grounded
Electrode
Flash burn
Flux
Fume plume
Galvanized steel
Immediate danger to life and health (IDLH)
Purge gas
Shielding gas
Stainless steel
Ultraviolet (UV) radiation
Ventricular fibrillation

Industry Recognized Credentials

If you are training through an NCCER-accredited sponsor, you may be eligible for credentials from NCCER's Registry. The ID number for this module is 29101. Note that this module may have been used in other NCCER curricula and may apply to other level completions. Contact NCCER's Registry at 888.622.3720 or go to **www.nccer.org** for more information.

Contents

Topics to be presented in this module include:

Figures and Tables

Section One

1.0.0 The Welding Trade

Objective

Describe basic welding processes, the welding trade, and training/apprenticeship programs.

a. Describe basic welding processes and the welding trade.
b. Describe NCCER standardized training and explain apprenticeship programs.

Trade Terms

Carbon steel: An alloy of iron combining iron and a small amount of carbon (usually less than 1 percent) to provide hardness.

Electric arc: The flow of an electrical current across an air gap or gaseous space.

Electrode: The point from which a welding arc is produced. In most welding processes, the electrode is a filler metal which is consumed in the process and becomes part of the weld.

Flux: A material used to dissolve or prevent the formation of oxides and other undesirable substances on a weld joint.

Shielding gas: A gas such as argon, helium, or carbon dioxide used to protect the welding electrode wire from contamination in gas metal arc welding (GMAW) and flux core arc welding (FCAW).

Stainless steel: An iron-based alloy usually containing at least 11 percent chromium.

This module is an extension of the Core Curriculum *Basic Safety* module. Its purpose is to identify general safety considerations that apply to various aspects of welding, along with the steps needed to avoid job-related deaths and injuries. Specific equipment and process hazards are covered in more detail in modules that deal with the use of the equipment or process in a particular application.

1.1.0 Welding Work

Welding is the process of joining similar metals using a high-intensity **electric arc** at temperatures from 6,000°F (3,316°C) to 10,000°F (5,538°C). The arc melts and fuses the base metals. In most cases, a filler metal is melted along with the base metal in order to strengthen the joint. The finished joint is as strong as, or even stronger than, the base metals that have been joined. Welding is commonly done on **carbon steel**, **stainless steel**, and aluminum.

Welders are key contributors to a number of construction and industrial crafts, including building and bridge construction; ship and oil rig construction and repair; heavy industry; and pipeline work. Welding is a craft that requires extensive knowledge and skill as well as a commitment to performing high-quality work and, most importantly, a strong commitment to safety.

There is a lot more to welding than simply striking an electric arc on base metals. Welders must know the characteristics of the metals being welded and the special materials and techniques that apply to the different metals. They must also know how to properly handle and store base metals and filler metals; how to prepare metals for welding; and how to interpret welding symbols and specifications.

The most common type of welding is shielded metal arc welding (SMAW). In the SMAW method, a **flux**-coated, consumable **electrode** is used to place the weld (*Figure 1*). SMAW is used primarily for welding iron and steel, including stainless steel.

Gas metal arc welding (GMAW), sometimes called MIG, or metal inert gas welding, uses a consumable wire that is fed through the welding gun to make the weld. It also uses a **shielding gas**. GMAW is used on steel and other metals. It is widely used in industrial applications, especially in the manufacture of automobiles where robotic welding is common. Flux core arc welding (FCAW) is very similar to GMAW. The main differences are that the FCAW filler metal is a tubular flux-cored wire and shielding gas may or may not be used.

Gas tungsten arc welding (GTAW), sometimes called TIG (tungsten inert gas) welding, uses a non-consumable tungsten electrode to produce the weld. GTAW is commonly used in welding stainless steel and light metals such as aluminum and magnesium.

In addition to welding, welders are often required to cut steel using oxyfuel, air carbon-arc, or plasma-arc equipment (*Figure 2*). Oxyfuel cutting is the most common method. In the oxyfuel method, oxygen and a fuel gas are mixed in a cutting torch to produce a high-intensity flame that can cut through thick steel. Acetylene is the most common fuel gas because it produces the most intense heat. Propane, MAPP® gas, and butane are also used for cutting work. Plasma-arc cutting relies on a jet of superheated gas that is passed through an electric arc. It is used to cut steel as well as other metals. Carbon-arc cutting uses

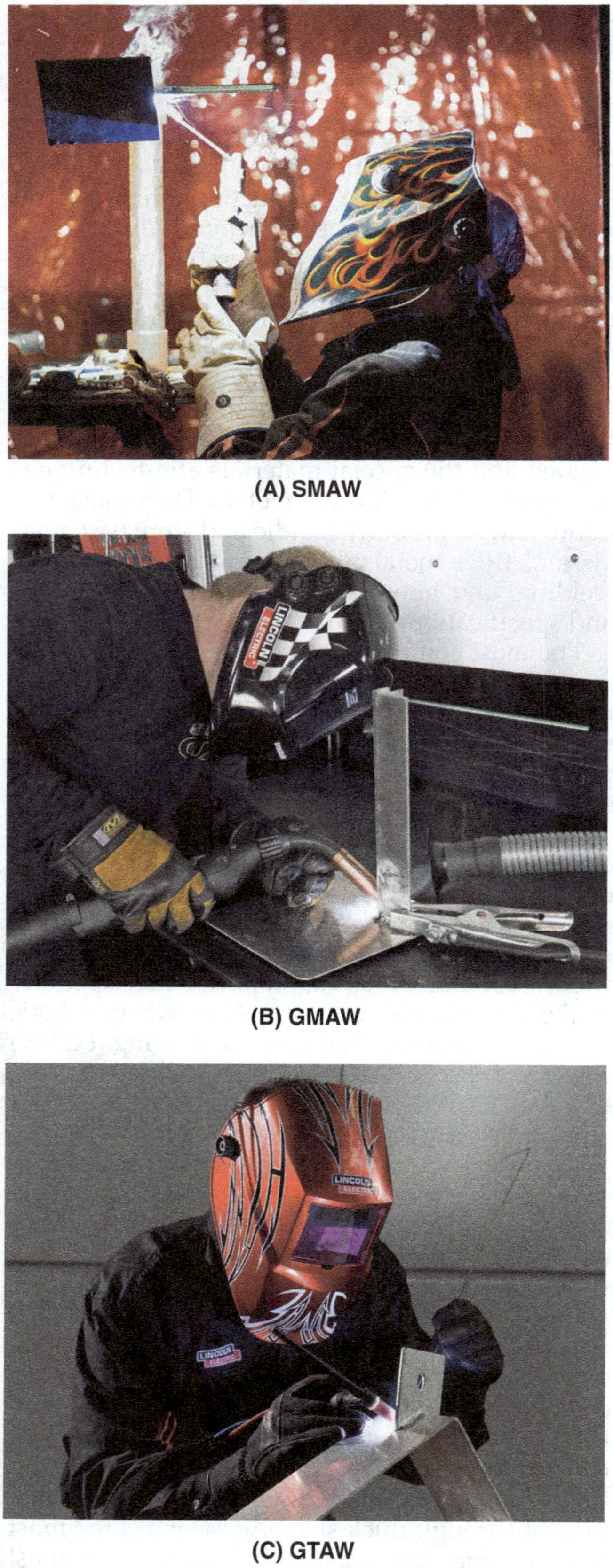

Figure 1 Welding methods.

an electric arc formed between a carbon electrode and the base metal to melt the base metal. Carbon-arc cutting is often used in preparing a base metal for welding.

Welders work on critical structures such as bridges, high-rise buildings, ships, pipelines, and automobiles. As you can imagine, defective welds in any of these applications can lead to a disaster. For that reason, welding work is subjected to intense quality control inspections. The more critical the weld, the more intense the inspection is. Inspectors use a variety of methods to examine welds, including X-ray inspection.

In order to qualify as a welder, a person must undergo extensive training that requires many hours of practice. The training provided by this program will prepare you to weld plate and pipe in a variety of positions and to cut carbon steel plate and pipe using common cutting equipment and techniques.

There are numerous opportunities for welders in heavy construction, manufacturing, ship building, and auto racing. Once you have completed your training and have become a journey-level welder, you can advance in the trade in a variety of ways (see *Figure 3*). The more training and education you complete, the greater the opportunities.

1.2.0 NCCER Standardized Training

NCCER is a not-for-profit education foundation established by the nation's leading construction companies. NCCER was created to provide the industry with standardized construction education materials, the NCCER Standardized Curricula (the welding modules are part of this series), and a system for tracking and recognizing students' training accomplishments—the NCCER Registry.

NCCER also offers accreditation, instructor certification, and craft assessments. NCCER is committed to developing and maintaining a training process that is internationally recognized, standardized, portable, and competency-based.

Working in partnership with industry and academia, NCCER has developed a system for program accreditation that is similar to those found in institutions of higher learning. NCCER's accreditation process ensures that students receive quality training based on uniform standards and criteria. These standards are outlined in NCCER's *Accreditation Guidelines* and must be adhered to by NCCER Accredited Training Sponsors.

Figure 2 Cutting methods.

More than 603 training centers across the United States and eight other countries are proud to be NCCER Accredited Training Sponsors/ Accredited Assessment Centers. Millions of craft professionals and construction managers have received quality construction education through NCCER's network of Accredited Training Sponsors. Every year the number of NCCER Accredited Training Sponsors increases significantly.

A craft instructor is a journey-level craft professional or career and technical educator trained and certified to teach the NCCER Standardized Curricula. This network of certified instructors ensures that NCCER training programs meet the standards of instruction set by the industry. There are more than 5,919 master trainers and 61,698 craft instructors within the NCCER instructor network. More information is available at **www.nccer.org**.

The Welding curriculum developed by NCCER provides participants with portable, industry-recognized credentials. These credentials include transcripts, certificates, and wallet cards that are tracked through NCCER's Registry.

This curriculum provides trainees with industry-driven training and education using a competency-based learning approach. This means that trainees must show the instructor that they possess the knowledge and skills needed to safely perform the hands-on tasks that are covered in each module.

When the instructor is satisfied that a trainee has demonstrated through written exams and hands-on performance testing that he or she has the required knowledge and skills for a given module, that information is sent to NCCER and kept in the Registry. The NCCER Registry can then confirm training and skills for workers as they move from state-to-state, company-to-company, or position-to-position.

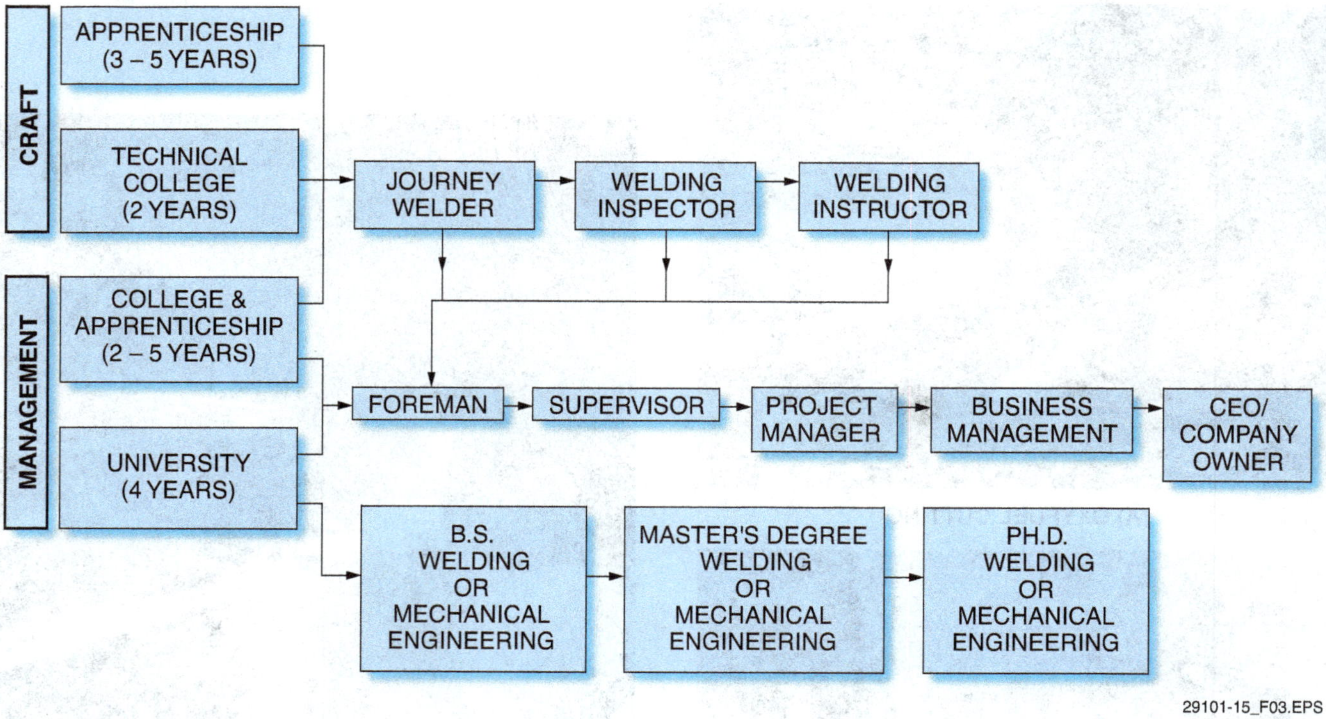

Figure 3 Opportunities in the welding trade.

1.2.1 Apprenticeship Training

Apprentice training goes back thousands of years. Its basic principles have not changed over time. First, it is a means for a person entering the craft to learn from those who have mastered the craft. Second, it focuses on learning by doing; real skills versus theory. Some theory is presented in the classroom. However, it is always presented in a way that helps the apprentice understand the purpose behind the skill that is to be learned.

The US Department of Labor's Office of Apprenticeship sets the minimum standards for training programs across the country. These programs rely on mandatory classroom instruction and on-the-job learning (OJL). They require at least 144 hours of classroom instruction per year and 2,000 hours of OJL per year. In a typical four-year apprenticeship program, trainees spend 576 hours in classroom instruction and 8,000 hours in OJL before receiving certificates issued by registered apprenticeship programs.

NCCER uses the Department of Labor apprenticeship standards as a foundation for comprehensive training that provide trainees with in-depth classroom and OJL experience.

All apprenticeship standards prescribe certain work-related or on-the-job learning. This OJL is broken down into specific tasks in which the apprentice receives hands-on training. In addition, a specified number of hours is required in each task. In a competency-based program, it may be possible to shorten the time required by testing out of specific tasks through a series of written and performance tests.

In a traditional program, the required OJL may be acquired in increments of 2,000 hours per year. Layoffs or illness may affect the duration. The apprentice must log all work time and turn the log in to the apprenticeship committee so that accurate records are maintained.

The classroom instruction and work-related training will not always run concurrently due to such reasons as layoffs, the type of work needed to be done in the field, and other factors. Apprentices with special job experience or coursework may obtain credit toward their classroom requirements. This reduces the total time required in the classroom while maintaining the total 8,000-hour on-the-job training requirement. These special cases depend on the type of program and the regulations and standards under which it operates.

Informal OJL provided by employers is usually less thorough than that provided through a formal apprenticeship program. The degree of training and supervision in this type of program often depends on the size of the employing firm. A small contractor may provide training in only one area, while a large company may be able to provide training in several areas.

For those entering an apprenticeship program, a high school or technical school education is desirable. Courses in shop, mechanical drawing, and

Did You Know?

An Ancient Apprenticeship Contract

Historical records include a contract between a young Greek named Heracles and a weaver. In exchange for food, a tunic (a long shirt), and 20 holidays a year, Heracles worked for the weaver for five years. After two and a half years, he was paid 12 drachmas. In his fifth year, his pay was scheduled to double.

In addition to learning to become a journeyman weaver, Heracles most likely worked seven long days each week and had to clean the shop, build fires, pick up supplies, make deliveries, and do whatever other jobs the master weaver needed to have done. It's also likely that any damage he caused came out of his promised pay.

So how much pay is 12 drachmas? There is no way of knowing for certain, but the word *drachma* means handful—not much pay for almost three years of hard work. Yet Heracles was willing to put in the time to become a skilled master in his chosen trade. As a master weaver, Heracles could look forward to a better income and a better life.

general mathematics are helpful. Manual dexterity, good physical condition, and quick reflexes are important. The ability to solve problems quickly and accurately and to work closely with others is essential. You must also have high respect for safety and its place in the construction environment.

NOTE

Some companies have physical activity requirements that must be met by apprentices. These requirements vary from company to company.

1.2.2 Youth Training and Apprenticeship Programs

Youth apprenticeship programs are available that allow students to begin their apprenticeship or craft training while still in high school. A student entering the program in the 11th grade may complete as much as one year of the training program by high school graduation. In cooperation with local construction industry employers, students may be able to work in the craft and earn money while still in school. Upon graduation, students can enter the industry at a higher level and with more pay than someone just starting in a training program.

Students participating in apprenticeship training may enter the second level or year of the program wherever it is offered. They may also have the option of applying credits at two-year or four-year colleges that offer degree or certificate programs in their selected field of study.

Did You Know?

Apprenticeships in the United States

In 2012, the US Department of Labor reported that there were more than 440,000 apprentices receiving registered apprenticeship training in 37,000 programs in the United States.

Additional Resources

Arc Welding Safety E205. Latest Edition. Cleveland, OH: Lincoln Electric.

Safety Topics in Welding, Cutting, and Brazing. **www.osha.gov**.

AWS F32M/F32, Ventilation Guide for Weld Fumes. Latest Edition. Miami, FL: American Welding Society.

1.0.0 Section Review

1. Flux core arc welding (FCAW) is similar to _____.
   a. SMAW
   b. GMAW
   c. GTAW
   d. carbon-arc

2. Before graduation, a high school student entering an apprenticeship program in the 11th grade may complete _____.
   a. as much as one year of the program
   b. nearly two years of the program
   c. more than two years of the program
   d. three years of the program

Section Two

2.0.0 Personal Protective Equipment

Objective

Identify and describe personal protective equipment (PPE) related to the welding trade.

a. Identify and describe body, foot, and hand protective gear.
b. Identify and describe ear, eye, face, and head protective gear.

Trade Terms

Arc burn: Burn to the skin produced by brief exposure to intense radiant heat and ultraviolet light.

Flash burn: Burns to the eyes sometimes called welder's flash; caused by exposure to intense radiant heat and ultraviolet light.

Ultraviolet (UV) radiation: Invisible rays capable of causing burns. UV rays from the sun are the cause of sunburn.

Each person in the shop environment wears general work clothing. Extra protection is needed for each person that is in direct contact with hot materials. Depending on the specific job and conditions, protective equipment can include the following:

- Body protection
- Foot protection
- Hand protection
- Ear protection
- Eye, face, and head protection

2.1.0 Body, Foot, and Hand Protection

Welding produces hazards from molten metal and sparks. For that reason, welders and anyone working in the welding environment must have proper protection.

2.1.1 Body Protection

Basic clothing for welders, as shown in *Figure 4*, should offer protection from flying sparks, heat, and ultraviolet (UV) radiation from electric arcs. Shirts should be made of a tight-weave fabric, have long sleeves and pocket flaps, and be worn with the collar buttoned. Pants must not have cuffs and should fit so they hang straight down the leg, touching the shoe tops without creases. Cuffs and creases can catch sparks, which can cause fires. Never wear polyester or other synthetic fibers; sparks will melt these materials, causing serious burns. Wear wool and cotton, which are more resistant to sparks. Cotton clothing, if worn, should be treated with a fire-retardant.

29101-15_F04.EPS

Figure 4 Welder with proper protective clothing.

It is important to follow the manufacturer's recommendations for laundering fire-resistant clothing. Fabric softeners may reduce the effectiveness of the treatment, increasing the flammability. Some fire-resistant clothing may be ruined by repeated washings or incorrect washing. When purchasing fire-resistant clothing, look for fire-resistant labeling on the garment itself, rather than on the packaging alone.

Additional protective coverings are required for the following:

- Cutting operations
- Out-of-position or overhead welding/cutting
- Welding operations that produce spattering molten metal and sparks or large amounts of heat

Figure 5 shows a welder wearing protective clothing and equipment. Protective coverings, usually made of leather because of its durability, are often referred to as leathers. However, because leather is heavy and hot to wear, lightweight fire- and heat-resistant clothing is also available. *Figure 6* shows a variety of leather and fire-resistant items that can be worn in combinations to provide various degrees of covering.

Full jackets offer the most protection. A cape is cooler, but offers only shoulder, chest, and arm protection. Leather pants, in combination with a jacket, will protect a welder's lap and legs. If the weather is hot and full leather pants are uncomfortable, leggings, sometimes called chaps, are available. They are strapped to the legs, leaving the back open. However, an apron alone is cooler and will protect a welder's lap and most of the leg area when squatting, sitting, or bending over a table. In some cases, if the welding is done at or below waist level, only full- or half-sleeve arm covers are required in combination with an apron. The welding position and type of welding operation typically dictate the appropriate personal protective equipment (PPE), in conjunction with company or site policies.

29101-15_F05.EPS

Figure 5 Personal protective equipment (PPE).

> **Terms**
>
> Don't confuse the terms *fire-retardant* and *fireproof*. A product impregnated with a fire-retardant delays ignition. A fireproof product is not supposed to burn. Clothing treated with a fire-retardant may lose some of its protective ability if washed, so it may need retreating after washing.

> **WARNING!**
>
> Sparks of molten metal from welding and metal cutting can reach temperatures greater than 9,900°F (5,500°C) and can travel a significant distance. They can land on the clothing or exposed skin of the welder or others in the immediate area, potentially resulting in burns. Sparks can also land on flammable material a significant distance away, resulting in a fire. For this reason, welders and nearby workers must wear the proper protective clothing. Flammable materials nearby must be covered with nonflammable blankets.

2.1.2 Foot Protection

The Occupational Safety and Health Administration (OSHA) requires that protective footwear be worn when working where falling, rolling, or sharp objects pose a danger of foot injuries and where feet are exposed to electrical hazards. For welders, sparks and droplets of molten metal provide additional hazards. Footwear with leather-reinforced soles or innersoles of flexible metal are recommended. High-top safety shoes or boots should be worn. Ensure that the pant leg covers the tongue and lace area of the footwear. If the tongue and lace area is exposed, wear leather spats (as shown in *Figure 6*) under the pants or leggings and over the front top of the footwear to protect them from sparks or falling molten metal. Spats will prevent sparks from burning through the front of lace-up boots. Protective footwear must comply with *ANSI Z49.1*. Sneakers, tennis shoes, and similar types of footwear must never be worn on the job site.

Figure 6 Leather and fire-resistant protective coverings.

2.1.3 Hand Protection

Gloves are the primary type of hand protection. Gloves must be selected on the basis of the hazards involved in doing the work. Gauntlet-type welding gloves must be worn when welding or cutting to protect against UV rays from an electric arc and heat from any thermal welding/cutting process. The most common glove is the standard leather welding glove shown in *Figure 7*. For heavy-duty welding or cutting, special heat-resistant or heat-reflective gloves are also available.

WARNING!

Don't weld or cut metal with gloves that have been used to handle petroleum products, especially when oxygen is being used. Also consider that the transfer of petroleum residue to the welding rod can affect the weld quality.

2.2.0 Protection Above the Neck

Special attention must be given to eye and ear safety when welding or cutting. Severe burns to the eyes can result from viewing the welding arc without the proper protection. Ears can be damaged by noise and sparks. The face, head, and neck must also be protected from sparks and heat.

2.2.1 Hearing Protection

Welding areas can be very noisy. If overhead work is being performed, hot sparks can cause burns to the ears and ear canals unless a leather hood is used. Such burns are extremely painful, so it is important to wear ear plugs when welding (*Figure 8*). Disposable earplugs are the most common form of hearing protection used in welding. These devices usually have an outer layer of pliable foam and a core layer of acoustical fiber that filters out harmful noise, yet allows you to hear normal conversation. To use disposable earplugs, simply roll each plug into a cone and insert the tapered end into the ear canal while pulling up on the upper portion of your ear. The earplugs will expand, filling the ear canal and creating a proper fit.

WARNING!

Plain cotton placed in the ear is not acceptable protection. Only approved ear protection is to be used.

2.2.2 Eye, Face, and Head Protection

The heat and light produced by cutting or welding operations can damage the skin and eyes. Injury to the eyes may result in permanent loss of vision. Oxyfuel cutting and welding can cause eye fatigue and mild burns to the skin because of the infrared heat radiated by the process. Welding or cutting operations involving an electric arc of any kind produce UV light. The UV light can cause a severe **arc burn** to exposed skin and **flash burns** to unprotected eyes. (The sun also produces UV light, some of which is not filtered out by the Earth's atmosphere and can also dam-

Figure 7 Welding gloves.

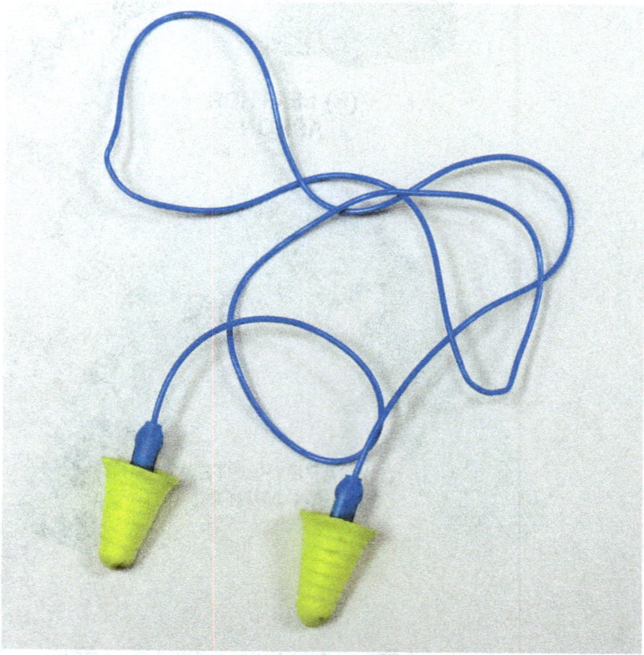

Figure 8 Ear plugs.

age the eyes.) A flash burn can very easily cause permanent damage to the retina of the eye. Welders should never view an electric arc directly or indirectly without wearing a properly tinted lens designed for electric arc use.

> **NOTE**
>
> Goggles are normally used with a leather soft cap to protect the face and head from UV rays.

As specified in *ANSI/AWS Z49.1*, protective spectacles with side shields, arc goggles, or other approved eye protection must be worn by welders in addition to a helmet or face shield with a filtered lens (*Figure 9*). The spectacles or goggles may have either clear or filtered lenses, depending on the amount of exposure to welding or cutting radiation (see *Table 1*). For overhead oxyfuel cutting operations, a leather hood may be used in place of the face shield to provide protection from sparks and molten metal. Goggles can be paired with a leather soft-cap for cutting and burning. The combination face shield with flip-up lenses can be used at some sites with a lens that is dark enough for the application.

For electric arc operations, a welding helmet with a properly tinted lens (shades 9 to 14) must be worn over safety glasses to provide proper protection. Many varieties of helmets are available; typical styles are shown in *Figure 10*. Some of the helmets are available with additional side-view lenses in a lighter tint so that welders can sense, by peripheral vision, any activities occurring beside them. A tinted lens that flips up to reveal a clear safety lens is a good feature. The clear lens enables the welder to continue using the helmet for protection during chipping and grinding.

Most welding and cutting tasks require the use of safety glasses, chemical-resistant goggles, dust goggles, or face shields. Always check your company's policy for the welding or cutting process being used to find out what type of eye protection is needed.

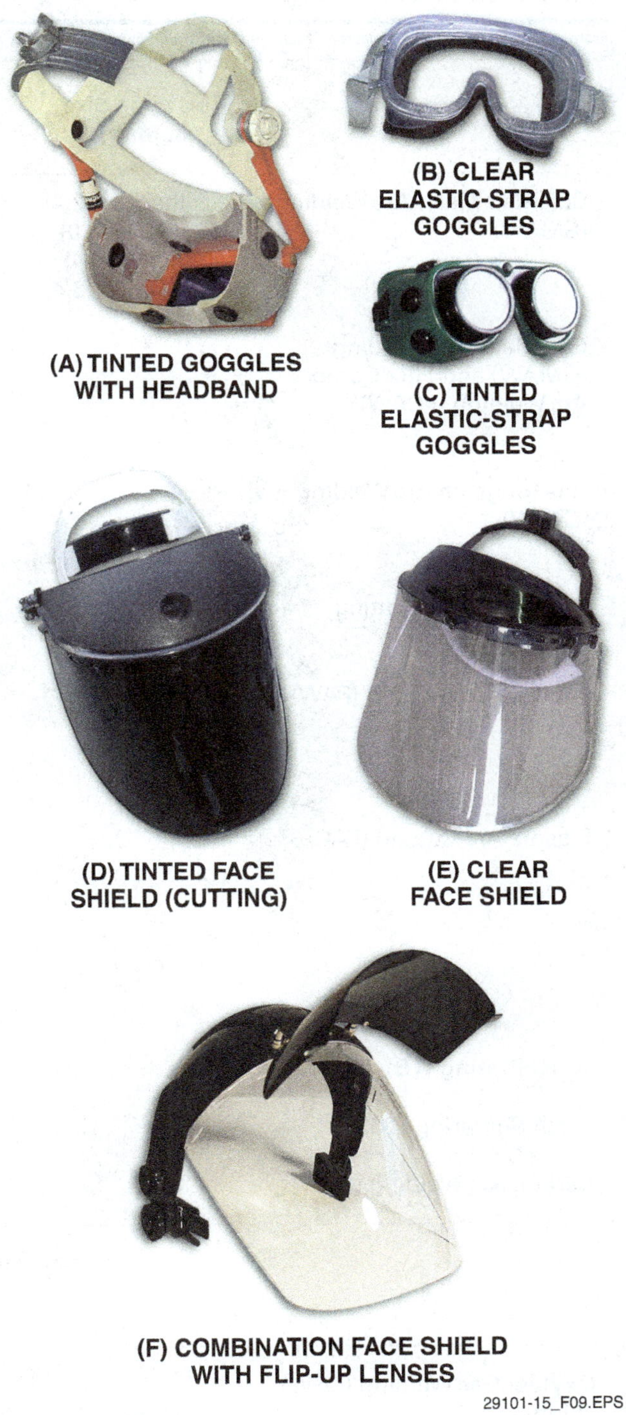

Figure 9 Oxyfuel welding/cutting goggles and face shield combinations.

Table 1 Guide for Shade Numbers

From *AWS F2.2 Lens Shade Selector.* Shade numbers are given as a guide only and may be varied to suit individual needs.

Process	Electrode Size in (mm)	Arc Current (Amperes)	Minimum Protective Shade	Suggested* Shade No. (Comfort)
Shielded Metal Arc Welding (SMAW)	Less than 3/32 (2.4)	Less than 60	7	–
	3/32–5/32 (2.4–4.0)	60–160	8	10
	5/32–1/4 (4.0–6.4)	160–250	10	12
	More than 1/4 (6.4)	250–550	11	14
Gas Metal Arc Welding (GMAW) and Flux Cored Arc Welding (FCAW)		Less than 60	7	–
		60–160	10	11
		160–250	10	12
		250–500	10	14
Gas Tungsten Arc Welding (GTAW)		Less than 50	8	10
		50–150	8	12
		150–500	10	14
Air Carbon Arc Cutting (CAC-A)	(Light)	Less than 500	10	12
	(Heavy)	500–1000	11	14
Plasma Arc Welding (PAW)		Less than 20	6	6 to 8
		20–100	8	10
		100–400	10	12
		400–800	11	14
Plasma Arc Cutting (PAC)		Less than 20	4	4
		20–40	5	5
		40–60	6	6
		60–80	8	8
		80–300	8	9
		300–400	9	12
		400–800	10	14
Torch Brazing (TB)		–	–	3 or 4
Torch Soldering (TS)		–	–	2
Carbon Arc Welding (CAW)		–	–	14

Process	Plate Thickness in	Plate Thickness mm	Suggested* Shade No. (Comfort)
Oxyfuel Gas Welding (OFW)			
Light	Under 1/8	Under 3	4 or 5
Medium	1/8 to 1/2	3 to 13	5 or 6
Heavy	Over 1/2	Over 13	6 or 8
Oxygen Cutting (OC)			
Light	Under 1	Under 25	3 or 4
Medium	1 to 6	25 to 150	4 or 5
Heavy	Over 6	Over 150	5 or 6

*As a rule of thumb, start with a shade that is too dark to see the weld zone. Then go to a lighter shade which gives sufficient view of the weld zone without going below the minimum. In oxyfuel gas welding, cutting, or brazing where the torch and/or the flux produces a high yellow light, it is desirable to use a filter lens that absorbs the yellow or sodium line of the visible light spectrum.

29105-15_T01.EPS

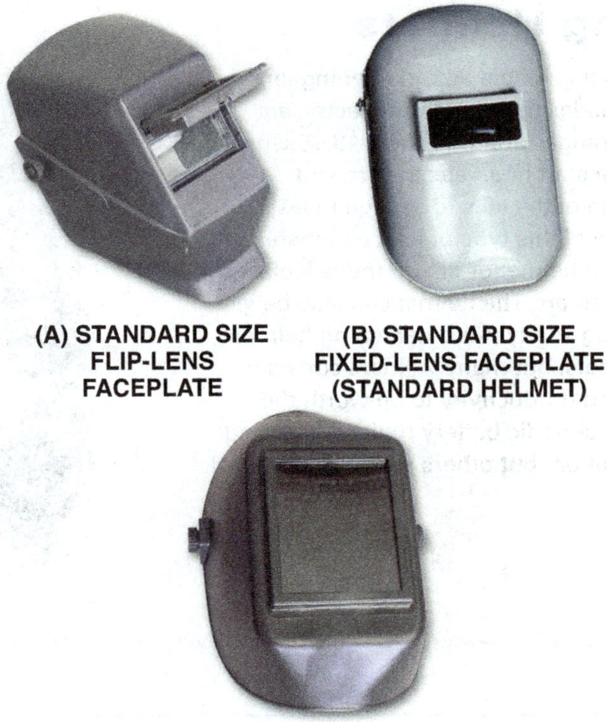

Figure 10 Arc-welding helmets.

Lightweight Welding Gloves

Soft, flexible, leather gloves are used for light-duty work and for operations involving GMAW, GTAW, brazing, soldering, and oxyfuel welding where free hand movement is required. These types of welding produce less heat and spatter than others. Lightweight gloves provide more flexibility for the required motions and pulling the trigger of the welding torch.

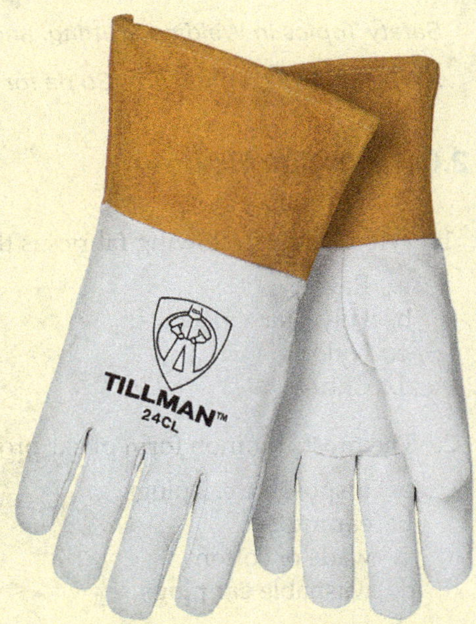

29101-15_SA01.EPS

Auto-Darkening Helmets

Helmets are available with an adjustable auto-darkening lens that automatically changes shades the instant an electric arc starts to occur. Typical shade ranges are 5-14 and 9-13. A sensor just above the lens detects the start of an electric arc and electronically causes the shade of the lens to change in less than ¹⁄₂₅,₀₀₀ of a second. The use of this type of lens eliminates the need for the welder to raise the helmet or flip up the faceplate to view the work before striking an arc. The helmet can also be left down for chipping and grinding safety. Auto-darkening helmets cost more than standard helmets, but many welders consider the improvement in quality and productivity to be worth the additional cost. Some require periodic battery replacement and testing to ensure proper operation, but others are solar-powered.

29101-15_SA02.EPS

Additional Resources

ANSI/AWS Z49.1, Safety in Welding, Cutting, and Allied Processes. Miami, FL: American Welding Society.

Arc Welding Safety E205, Latest Edition. Cleveland, OH: Lincoln Electric.

Personal Protective Equipment (PPE) for Welding and Cutting. Miami, FL: American Welding Society.

Safety Topics in Welding, Cutting, and Brazing. **www.osha.gov**.

AWS F32M/F32, Ventilation Guide for Weld Fumes. Latest Edition. Miami, FL: American Welding Society.

2.0.0 Section Review

1. Which of the following fabrics is the best choice for welding clothing?
   a. Rayon
   b. Polyester
   c. Nylon
   d. Cotton

2. The most common form of ear protection is ____ .
   a. disposable ear plugs
   b. ear muffs
   c. wads of cotton
   d. washable ear plugs

SECTION THREE

3.0.0 WELDING SAFETY PRACTICES

Objective

Identify and describe welding safety practices related to specific hazards or environments.

a. Describe the importance of welding safety and identify factors related to accidents.
b. Describe basic welding safety practices related to the general work area.
c. Describe hot work permits and fire watch requirements.
d. Describe confined spaces and their related safety practices.
e. Identify safety practices related to welding equipment.
f. Identify and describe respiratory hazards, respiratory safety equipment, and ways to ventilate welding work areas.
g. Explain the purpose of the SDS/MSDS and how it is used.

Trade Terms

Bonded: The permanent joining of metallic parts to form an electrically conductive path that will assure electrical continuity and the capacity to safely conduct any current likely to be imposed on it.

Electrically grounded: Connected to Earth or to some conducting body that serves in place of the earth.

Fume plume: The fumes, gases, and particles from the consumables, base metal, and base metal coating during the welding process.

Galvanized steel: Carbon steel dipped in zinc to inhibit corrosion.

Immediate danger to life and health (IDLH): As defined by OSHA, this term refers to an atmosphere that poses an immediate threat to life, would cause irreversible adverse health effects, or would impair an individual's ability to escape from a dangerous atmosphere.

Purge gas: An inert gas, such as nitrogen, used to drive oxygen away from a weld site.

Ventricular fibrillation: Irregular heart rhythm characterized by a rapid fluttering heartbeat.

There are a number of special hazards associated with welding, including intense heat, flame, and dangers of exposure to a high-intensity electric arc. Metal cutting brings additional hazards to the workplace, including explosive and flammable gases and the intense flame used to cut metal. These conditions create the need for specialized safety practices that must be followed by those performing the work, as well as those in the vicinity of the work.

3.1.0 Welding Safety Hazards

The welding industry is fortunate to have a well-developed standard for safety: *American National Standards Institute (ANSI) Z49.1, Safety in Welding, Cutting, and Allied Processes.* Everyone involved with welding, cutting, and related processes should be familiar with this document. The document discusses personal protective equipment, ventilation, and fire prevention, as well as welding in confined spaces. It contains precautionary information that is important to welders, supervisors, and managers. The document also details specific welding process safety procedures.

There are many causes of accidents. They can usually be divided into two broad categories: personal and physical. Workers must be aware of the factors that can cause accidents so they can understand the consequences of their actions on the job site.

3.1.1 Personal Factors That Cause Accidents

Accidents can often be traced to personal factors such as poor health, lack of experience, and the improper use of alcohol and medications. Something as simple as excessive use of colognes, after-shaves, or perfumes that contain alcohol can represent a burn hazard. A good general rule is: whenever you feel burning or excessive heat, immediately stop work and check it out.

People who are ill or injured may not be able to concentrate on their work. In addition, they may be physically weakened and unable to handle strenuous work. A person doing work that is inherently dangerous is more likely to have an accident when sick or injured. Mental stress can also play a role in accidents; concentration is the issue. An employee who is worried about a serious personal problem is likely to be distracted and unable to focus on work. Workers need to realize that they endanger themselves and those around them when they are not 100 percent effective. As a worker, you need to be able to recognize when others may not be up to par and take appropriate precautions.

Age and inexperience often play a role in accidents. Insurance company studies show that a person who lacks experience is more likely to take risks that cause accidents. Sometimes accidents occur because someone hasn't had enough experience to learn how to avoid them. In fact, insurance company statistics show that more accidents occur in the under-18 age group than in any other. An inexperienced person often is not able to foresee the outcome of an action, or lacks the knowledge to know what works and what doesn't.

People who consume alcohol or use illegal drugs while working risk their lives and the lives of their co-workers. A drug or alcohol hangover is nearly as bad. Legal prescription and over-the-counter drugs can also cause problems. Some cold remedies and cough medicines contain alcohol or other substances that will make people drowsy. Alcohol and drugs affect a person's coordination, alertness, and decision-making ability. Never use these substances while you are working, and don't work with people who use them on the job.

3.1.2 Physical Factors That Cause Accidents

Accidents are often caused by conditions at the job site. For example, foot and vehicle traffic congestion increases at starting and quitting times. As quitting time approaches, workers are more likely to hurry, possibly taking risks they might not otherwise take. People often slow down right after a meal, so workers may be less attentive after lunch than at other times.

People who learned to put their toys away when they were young probably keep a neat, safe work site as adults. The others are likely to have tools, equipment, scrap material, and other items lying around their work site that could trip people. If you keep a messy work site, eventually you or someone else will be injured because of it. Put away your tools, dispose of scrap materials, and secure your equipment and cables when not in use. If you have flammable or hazardous material at your work site, keep it properly contained and covered. When you're done with it, return it to the designated location for disposal or storage.

Damaged or defective welding equipment is dangerous. For example, a frayed or damaged welding cable could give someone a lethal shock or start a fire. Any damaged or defective equipment should be so tagged and either repaired or replaced. All equipment should be periodically inspected for damage. Make sure that periodic maintenance procedures recommended by the manufacturer are performed on schedule.

3.2.0 Work Area Safety

An important factor in area safety is good housekeeping. The work area should be picked up and swept clean. The floors and workbenches should be free of dirt, scrap metal, grease, oil, and anything that is not essential to accomplishing the given task. Collections of steel, welding electrode stubs, wire, hoses, and cables are difficult to work around and easy to trip over. An electrode caddy can be used to hold the electrodes and stubs. Hooks can be made to hold hoses and cables, and scrap steel should be thrown into scrap bins.

The ideal welding shop should have bare concrete floors and **electrically grounded** and **bonded** bare metal walls and ceilings to reduce the possibility of fire. Never weld or cut over wood floors, as this increases the possibility of fire.

It is important to keep flammable liquids as well as rags, wood scraps, piles of paper, and other combustibles out of the welding area.

If you must weld in an enclosed building, make every effort to eliminate anything that could trap a spark. Sparks can smolder for hours and then burst into flames. Regardless of where you are welding, be sure to have a fire extinguisher nearby. Also, keep a bucket of water handy to cool off hot metal (other than a test coupon or functional weld) and quickly douse small fires. If a piece of hot metal must be left unattended, use soapstone to write the word HOT on it before leaving. This procedure can also be used to warn people of hot tables, vises, firebricks, and tools.

Perform cutting activities in the workshop in a properly prepared cutting area. Be aware that welding and cutting sparks can ignite any type of flammable vapors. Prevent hot droplets from landing directly on the torch hoses, as they could melt a hole in the hose.

Whenever welding must be done outside a welding booth, use portable screens to protect other personnel from the bright glare of the welding arc (*Figure 11*). The portable screen also prevents drafts of air from interfering with the stability of the arc. Welding blankets (*Figure 12*) should be used to cover material and equipment that could be burned or otherwise damaged.

The most common welding accident is burned hands and arms. Keep first-aid equipment nearby to treat burns in the work area. Eye injuries can also occur if you are careless. Post emergency phone numbers in a prominent location.

29101-15_F11.EPS

Figure 11 A typical welding screen.

3,000°F (1,620°C) INTERMITTENT, 1,500°F (820°C) CONTINUOUS SILICON DIOXIDE CLOTH

29101-15_F12.EPS

Figure 12 A welding blanket.

The following are some work-area reminders:

- Eliminate tripping hazards by coiling cables and keeping clamps and other tools off the floor.
- Don't get entangled in cables, loose wires, or clothing while you work. You must be able to move freely, especially should your clothing ignite or some other accident occur.
- Clean up oil, grease, or other agents that may ignite and splatter off surfaces while welding.
- Shut off the welding machine and disconnect the power before performing any service or maintenance. A circuit breaker or power disconnect switch should be provided for each machine when hard-wired. Others may be equipped with only a cord and plug.
- Keep the floor free of electrodes and electrode stubs. They could easily cause a slip or fall. All used electrodes should be disposed of in a proper metal container. Throwing electrodes on the floor can result in disciplinary action or even dismissal, due to the safety hazard and poor housekeeping attitude it represents.
- Always remove a SMAW electrode from the electrode holder before laying it down. Leaving a stub in the electrode holder allows it to easily arc to nearby surfaces.
- Work in a dry area, booth, or other shielded area whenever possible.
- Make sure there are no open doors or windows through which sparks may travel to flammable materials.
- Don't coil electrical leads over your body while working on scaffolding.

Conditions on a job site can change quickly. When performing welding or cutting work, it is a good idea to stop occasionally and check your surroundings.

3.3.0 Hot Work Permits and Fire Watches

A hot work permit (*Figure 13*) is an official authorization from the site manager to perform work that may pose a fire hazard. The purpose of the permit is to ensure that a survey process takes place to confirm that the area is safe for cutting or welding, while also establishing accountability for the safety of the job. The permit includes information such as the time, location, and type of work being done. The hot work permit system promotes the development of standard fire safety guidelines. Hot work permits and fire watches are required by OSHA. Permits also help managers keep records of who is working where and at what time. This information is essential in the event of an emergency or at other times when personnel need to be evacuated.

During a fire watch, a person other than the welder or cutting operator must constantly scan the work area for fires. Fire watch personnel must have ready access to fire extinguishers and alarms and know how to use them. When a person is standing fire watch, it should be their only duty.

Cutting operations must never be performed without a fire watch. Whenever oxyfuel cutting equipment is used, there is a great danger of fire. Hot work permits and fire watches are used to minimize this danger. Most sites require the use of hot work permits and fire watches. When they are violated, severe penalties are imposed.

HOT WORK PERMIT

For Cutting, Welding, or Soldering with Portable Gas or ARC Equipment

(References: 1997 Uniform Fire Code Article 49 & National Fire Protection Association Standard NFPA 51B.)

Job Date___________ Start Time___________ Expiration___________ WO #___________

Name of Applicant_______________________ Company___________ Phone___________

Supervisor_______________________ Phone___________

Location / Description of work _______________________

IS FIRE WATCH REQUIRED?

1. _______ (yes or no) Are combustible materials in building construction closer than 35 feet to the point of operation?
2. _______ (yes or no) Are combustibles more than 35 feet away but would be easily ignited by sparks?
3. _______ (yes or no) Are wall or floor openings within a 35 foot radius exposing combustible material in adjacent areas, including concealed spaces in floors or walls?
4. _______ (yes or no) Are combustible materials adjacent to the other side of metal partitions, walls, ceilings, or roofs which could be ignited by conduction or radiation?
5. _______ (yes or no) Does the work necessitate disabling a fire detection, suppression, or alarm system component?

YES to any of the above indicates that a qualified fire watch is required.

Fire Watcher Name(s) _______________________ Phone___________

NOTIFICATIONS

Notify the following groups at least 72 hours prior to work and 30 minutes after work is completed. Write in names of persons contacted.

Notify in person OR by phone ONLY if question #5 above is answered "yes":

- Facilities Management Fire Alarm Supervisor

Notify by phone or in person: (If by phone, write down name of person and send them a completed copy of this permit.)

- Facilities Management Fire Protection Group
- Environmental Health & Safety Industrial Hygiene Group

SIGNATURES REQUIRED

University Project Manager_______________________ Date ___________ Phone___________

I understand and will abide by the conditions described in this permit. I will implement the necessary precautions which are outlined on both sides of this permit form. Thirty minutes after each hot work session, I will reinspect work areas and adjacent areas to which spark and heat might have spread to verify that they are fire safe, and contact Facilities Management Alarm Technicians to have any disabled fire protection systems reactivated.

_______________________ _______________________ Date ___________ Phone___________

Permit Applicant Company or Department

1/17/15

29101-15_F13.EPS

Figure 13 Hot work permit.

> **WARNING!**
>
> Never perform any type of heating, cutting, or welding until you have obtained a hot work permit and established a fire watch. If you are unsure of the procedure, check with your supervisor. Violation of hot work permit and fire watch procedures can result in serious injury or death. OSHA requires that the fire watch be maintained for one-half hour after welding or cutting is finished.

3.4.0 Confined Spaces

A confined space refers to a relatively small or restricted space, such as a storage tank, boiler, or pressure vessel, or small compartments, such as underground utility vaults, small rooms, or the unventilated corners of a room. *Figure 14* shows an example of a confined space entry permit typically found on a job site.

OSHA 29 *CFR* 1910.146 defines a confined space as a space that:

- Is large enough and so configured that an employee can bodily enter and perform assigned work
- Has a limited or restricted means of entry or exit; for example, tanks, vessels, silos, storage bins, hoppers, vaults, and pits
- Is not designed for continuous employee occupancy

OSHA 29 *CFR* 1910.146 further defines a permit-required confined space as a space that:

- Contains or has the potential to contain a hazardous atmosphere
- Contains a material that has the potential for engulfing an entrant
- Has an internal configuration such that an entrant could be trapped or asphyxiated by inwardly converging walls or by a floor that slopes downward and tapers to a smaller cross-section
- Contains any other recognized serious safety or health hazard

For safe working conditions, the oxygen level in a confined space atmosphere must range between 19.5 percent and 23.5 percent by volume, with 21.5 percent being considered the normal level (*Table 2*). Oxygen concentrations below 19.5 percent by volume are considered deficient.

> **WARNING!**
>
> If too much oxygen is ventilated into a confined space, it can be absorbed by the welder's clothing and ignite. If too little oxygen is present, it can lead to the welder's death in minutes. For this reason, make sure confined spaces are ventilated properly for cutting or welding purposes and *never* use oxygen in confined spaces for ventilation purposes.

Before operations are started, the wheels of heavy portable equipment must be securely blocked to prevent accidental movement. Where a welder must enter a confined space through a manhole or other opening, all means must be provided for quickly removing the worker in case of emergency (*Figure 15*). When safety belts and lifelines are used for this purpose, they must be attached to the welder's body so that he or she

Oxyfuel Gas Welding Safety Precautions

- Always light the oxyfuel gas torch flame using an approved torch lighter to avoid burning your fingers.
- Never point the torch tip at anyone when lighting it or using it.
- Never point the torch at the cylinders, regulators, hoses, or anything else that may be damaged and cause a fire or explosion.
- Never lay a lighted torch down on the bench or workpiece, and do not hang it up while it is lighted. If the torch is not in the welder's hands, it must be off.
- To prevent possible fires and explosions, check valves and flashback arrestors must be installed in all oxyfuel gas welding and cutting outfits.
- When cutting with oxyfuel gas equipment, clear the area of all combustible materials.
- Skin contact with liquid oxygen can cause frostbite. Be careful when handling liquid oxygen.
- Never use oxygen as a substitute for compressed air.
- All oxygen cylinders with leaky valves or safety fuse plugs and discs should be set aside and marked for the attention of the supplier. Do not tamper with or attempt to repair oxygen cylinder valves. Do not use a hammer or wrench to open the valves.

Master Card No.____________

1. Work Description

Area____________ Equipment Location____________

Work to be done:

2. Gas Test		Results	Recheck	Recheck
Required	☐ Instrument Check			
☐ Yes	☐ Oxygen % 20.8 Min.			
☐ No	☐ Combustible % LFL			
		Date/Time/Sig.	Date/Time/Sig.	Date/Time/Sig.

3. Special Instructions: ☐ Check issuer before beginning work ☐ None

4. Hazardous Materials: ☐ None What did the line / equipment last contain?

5. Special Protection Required: ☐ None ☐ Forced Air Ventilation

☐ Avoid Skin Contact ☐ Gloves __________ ☐ Suit __________

☐ Goggles or Face Shield ☐ Respirator __________ ☐ Safety Harness

☐ Self-Contained Breathing Equipment ☐ Hoseline Breathing Equipment

☐ Other, Specify __________ ☐ Standby - Name: __________

SAMPLE

6. Fire Protection Required: ☐ None ☐ Portable Fire Extinguisher ☐ Fire Hose and Nozzle

☐ Fire Watch ☐ Other, specify:

7. Condition of Area and Equipment

Required Yes	No	THESE KEY POINTS MUST BE CHECKED
		a. Lines disconnected & blinded or where disconnecting is not possible, blinds installed? (Includes drains, vents and instrument leads) and appropriate valves locked out?
		b. Equipment cleaned, washed, purged, ventilated?
		c. Low voltage or GFCI-protected electrical equipment provided?
		d. Explosion-proof electrical equipment provided?
		e. Life lines required to be attached to safety harnesses?

Comments

rev.6/10/15

29101-15_F14A.EPS

Figure 14A Confined space entry permit (sheet 1 of 2).

8. Approval	Date	Time	Permit Authorization Area Supervisor	Permit Acceptance Maint./Contractor Supervisor/Engineer	Date	Time
Issued by						
Endorsed by						
Endorsed by						
Endorsed by						

9. Individual Review

I have been instructed on proper Safety Procedures and proper Confined Space Entry Procedures. I have signed in on the appropriate Master Card and have affixed personal locks on energy isolation devices as appropriate.

SAMPLE

Signature of all personnel covered by this permit.

Forward to Production Superintendent 7 days after completion of work.

rev.6/10/15

29101-15_F14B.EPS

Figure 14B Confined space entry permit (sheet 2 of 2).

Table 2 Effects of an Increase or Decrease in Oxygen Levels

Oxygen Level	Effects
> 23.5%	Easy ignition of flammable material such as clothes
19.5% – 21.5%*	Normal
17%	Deterioration of night vision, increased breathing volume, accelerated heartbeat
14% – 16%	Very poor muscular coordination, rapid fatigue, intermittent respiration
6% – 10%	Nausea, vomiting, inability to perform, unconsciousness
< 6%	Spasmodic breathing, convulsive movements, and death in minutes

*There is a safety zone of 2%. Above 23.5% is considered enriched.

cannot be jammed in a small exit opening. An attendant with a pre-planned rescue procedure must be stationed outside to observe the welder at all times and must be capable of putting the rescue operations into effect.

When welding or cutting operations are suspended for any substantial period of time, such as during lunch or overnight, all electrodes must be removed from the holders, and the holders must be carefully located so that accidental contact cannot occur. The welding machines must also be disconnected from the power source.

In order to eliminate the possibility of gas escaping through leaks or improperly closed valves when gas welding or cutting, the gas and oxygen supply valves must be closed, the regulators

29101-15_F15.EPS

Figure 15 Worker in a confined space.

released, the gas and oxygen lines bled, and the valves on the torch shut off when the equipment will not be used for a substantial period of time. Where practical, the torch and hose must also be removed from the confined space. After welding operations are completed, the welder must mark the hot metal or provide some other means of warning other workers.

> **WARNING!**
> Nitrogen and argon are used as backing and **purge gases** for some welding applications. These gases will displace all the oxygen in the space where they are used.

3.5.0 Welding Equipment Safety

Anyone working with welding equipment must be aware of the equipment-related hazards. Of particular concern are the hazards associated with the voltages and electrical currents produced by welding machines. There are additional hazards related to falling equipment and equipment powered by gasoline engines. *Figure 16* shows some examples of safety notices typically found in operator manuals for welding units.

3.5.1 Electrical Hazards

Electric shock from welding and cutting equipment can result in death, injury, and severe burns. Injury can also result if a fall occurs because of the shock. The amount of current that passes through the human body determines the outcome of an electric shock. The higher the voltage, the greater the chance for a fatal shock. Electric current flows along the path of least resistance to return to its source. If you come in contact with a live conductor, you become a load in the circuit and the current will flow though you. *Figure 17* shows how much resistance the human body presents under various circumstances and how this converts into amps or milliamps when the voltage is 110V. Note

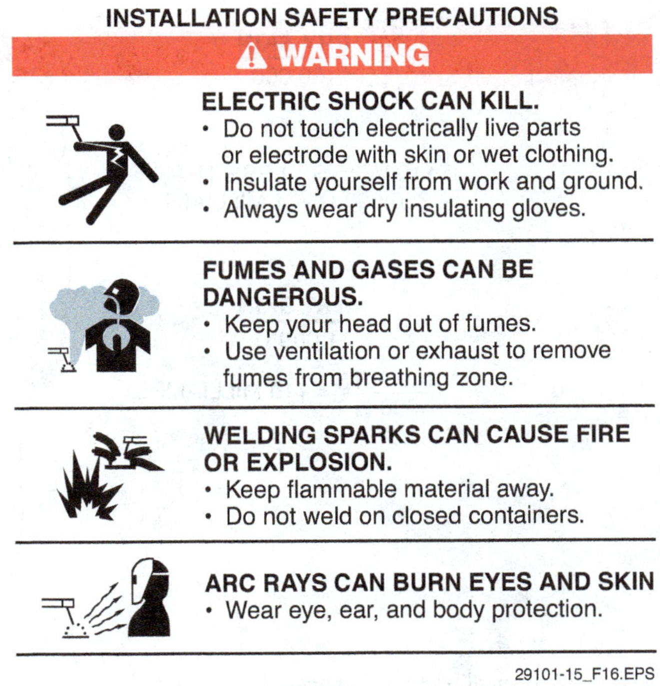

Figure 16 Safety hazard notices.

that the potential for shock increases dramatically if the skin is damp. A cut will also reduce resistance and increase shock potential. Currents of less than 1 amp can severely injure and even kill a person. Contact with a live electrical circuit can also cause severe burns. Under the right conditions, even a small current can cause serious injury or death. The Construction Employers Association has issued the following guidelines regarding the seriousness of electric shock:

- *Greater than 3 milliamps (mA)* – Indirect accident
- *Greater than 10 mA* – Muscle contraction
- *Greater than 30 mA* – Lung paralysis, usually temporary
- *Greater than 50 mA* – Possible **ventricular fibrillation**
- *100 mA to 4A* – Certain ventricular fibrillation
- *Greater than 4A* – Heart paralysis, severe burns

Did You Know?

Oxygen-Deficient Air

People don't realize that entering an oxygen-deficient area can result in immediate death. Even a single breath of oxygen-deficient air can prevent muscles from responding and prevent you from escaping. Persons with physical problems such as heart ailments are at the greatest risk. At a refinery in Corpus Christi, Texas, while the welder was at lunch, a helper entered a pipe in which argon had been used as a backing gas. When the welder returned, he found the helper dead.

Nitrogen is lighter than air and, if allowed, will float upward. Argon, on the other hand, is heavier than air and will settle in low-lying areas. When removing nitrogen from a confined space, open the top to allow it to float out. When removing argon from a confined space, provide some way for it to escape out the bottom or push/pull it out using air and mechanical equipment.

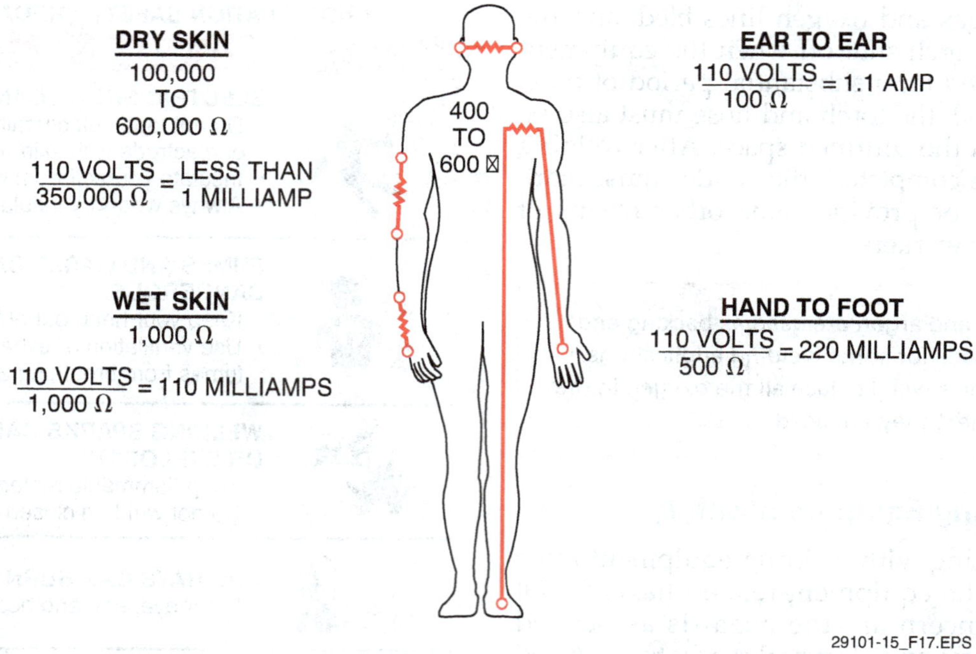

Figure 17 Typical body resistances and current.

Each welder and operator of the equipment must be trained and able to recognize the dangers associated with each particular type of equipment to avoid injuries, fatalities, and other electrical accidents.

It is a good idea for all involved to know first-aid procedures, such as cardiopulmonary resuscitation (CPR) for treating electric shock. Treat electrical burns like any other type of burn by applying ice or a cold compress over the burned area. Prevent infection and cover with a clean, dry dressing. Seek medical attention if necessary.

Welding-related electrical hazards come from two sources. The first, which we will call primary hazards, come from the source voltage that powers the welding machine. This voltage is usually 230 or 460 volts. These are potentials that create the possibility of serious, even fatal, shocks. It is important to know that this voltage exists inside the welding machine any time the machine is plugged in. The only way to disable it is to unplug the machine or turn off the power at the disconnect switch. The welding machine must also be properly grounded. This is done by connecting the ground terminal on the machine to a known ground.

> **WARNING!**
> Be sure to follow applicable lockout/tagout procedures when disabling the power to the unit. Also, have a qualified electrician install the welding machine so that it will be correctly wired and grounded. If the welding power supply does not work properly, have a qualified technician repair it. Never open the unit yourself.

Secondary hazards exist in the welding process itself. This type of shock can occur if a person touches part of the electrode circuit, such as a bare spot on the electrode cable, and touches the metal work material with another part of the body at the same time. Such a condition can occur if the welder is leaning against the work while welding. Here are some things to remember that will help keep you safe:

- The voltage is highest at the electrode when you are not welding. Due to the high current flow through the leads, the voltage drops somewhat during welding.
- The jaws and screws on the electrode holder, as well as the stick electrode itself, are electrically energized. Make sure they do not touch your skin or clothing, especially if the clothing is damp or wet.

- Never touch an electrode to a grounded conductive surface because these surfaces will become electrically energized. The electrode and work circuit is electrically energized when the welding machine switch is on.
- Never operate arc-welding equipment on a wet or damp floor.
- Wear dry gloves that are in good condition when welding. Keep an extra pair on hand.
- Keep dry insulation between your body, including your arms and legs, and the metal being welded or a ground such as a metal floor or wet earth. Use plywood or a rubber mat between you and the ground or floor.
- Keep the welding cables and electrode holder in good condition.

If you must work in conditions where you might come into contact with the work while welding, use one of the following types of machines:

- Semiautomatic DC constant-voltage welder
- DC manual (stick) welder
- AC welder with reduced voltage control
- Use a ground fault circuit interrupter (GFCI) for power tools that plug in

3.5.2 Gasoline-Powered Units

Gasoline- or diesel-powered welding machines are often used on sites where there is no direct access to 230- or 460-volt electrical outlets. In fact, this type of unit is very common. They range from small, portable units that look like the portable generators you would see at hardware stores, to trailer-mounted units like the one shown in *Figure 18*. In addition to the electrical hazards previously covered, gasoline- and diesel-powered units have their own particular hazards:

Shock Prevention

- Floors must be dry at all times. Use wooden platforms or rubberized carpet/floor coverings or other insulated material.
- Only qualified electricians may work on electric arc welding machine power connections.
- Workbenches must be grounded.
- With the power off, cable connections must be checked for cracked insulators, loose contacts, and other signs of damage.

- Gasoline-powered units produce deadly carbon monoxide just like automobile engines. They must never be operated indoors or in an enclosed space unless the exhaust is vented to the outdoors.
- Gasoline is highly flammable, so it must be kept in approved containers and stored away from hot work.
- When refueling a generator, turn off any welding or cutting flames. Allow the engine to cool to prevent the gasoline from contacting hot engine parts.

3.6.0 Fumes and Gases

The gases, fumes, and dust produced by welding and cutting processes can be hazardous. Adequate ventilation must be provided to prevent workers from breathing these products. There are three methods available to protect personnel against fumes and gases:

- *Natural ventilation* – The movement of air through the workplace caused by natural forces is often enough to remove fumes and gases.
- *Mechanical ventilation* – If natural ventilation is not adequate, portable or fixed fans can be used to provide the necessary ventilation.
- *Source extraction* – This method uses a mechanical device to capture welding fumes at or near the arc.

4-WHEEL TRAILER WITH MULTI-PROCESS WELDER

29101-15_F18.EPS

Figure 18 Gasoline-powered welding machine.

Regardless of the ventilation method used, it is important to avoid breathing in the welding fumes. This means keeping your head up and away from the **fume plume**, which is the term used to describe the fumes, gases, and particles created by the welding process. The heated fumes and smoke generally rise straight up or move towards the ventilation system opening, where a negative pressure exists. Observe the column of smoke and position yourself to avoid it. A small fan may also be used to divert the smoke, but take care to keep the fan from blowing directly on the work area. The gases from flux and the shielding gases controlled by the welding machine must be present at the welding arc in order to protect the molten metal from the surrounding air. A fan directed at the weld will disturb the process and create flaws.

In addition to the fume plume itself, some consumables and base metals contain toxic materials that require special ventilation. These materials include the following:

- Barium
- Cadmium (found in some steel and fasteners)
- Chromium (found in stainless steel)
- Base metal coatings such as paint
- Cobalt
- Copper
- Manganese
- Nickel
- Silica
- Zinc (found in galvanized metals)

Always check to determine what material you are welding or cutting. Some of the materials listed here produce fumes that are known to cause cancer or can lead to serious respiratory illness. Long-term exposure can even result in death. Some consumables are designated as special ventilation products as a result of the hazards they represent. When welding or cutting these materials, wear approved respiratory equipment and/or ensure that source extraction is used.

Chromium 6, also known as hexavalent chrome, is used in stainless steel. It gives off toxic fumes when heated. Manganese is another hazardous metal used in stainless steel. In addition, it is used in some welding rods, so welders have a greater risk of exposure. Exposure to manganese fumes and dust have been linked to Parkinson's Disease.

Galvanized steel, which is very common, is dipped in zinc to inhibit rust. When flame-cut or welded, it can produce a condition known as zinc chills, which causes flu-like symptoms. For occasional short-term exposure to fumes from zinc- or copper-coated materials, an approved respirator must be used.

In many cases, natural ventilation provides the required level of protection. However, it may be necessary to have a specialist determine if the ventilation is adequate. Some guidelines used in making this determination are as follows:

- The welding area must contain at least 10,000 cubic feet. This represents an area of 22' × 22' × 22' for each welder.
- The ceiling height must be no less than 16 feet.
- Cross-ventilation must not be blocked by partitions, equipment, or other barriers.
- The area must not be designated a confined space.

3.6.1 Source Extraction Equipment

Source extraction, also called local exhaust, can be provided by portable, mobile, or stationary systems such as those shown in *Figure 19*. As a general rule, the farther the extraction hose is from the arc, the greater the volume of air required. There are two classes of source extraction equipment: low vacuum/high volume and high vacuum/low volume.

Low vacuum/high volume equipment can be used in situations where the extraction hose is 6 to 15 inches (≈15 to 38 cm) from the arc (*Figure 20*). It uses extraction hoses, or hose and tubing combinations, with a diameter of 6 to 8 inches (≈15 to 20 cm) to accommodate the higher air volume of 560 to 860 cubic feet per minute (≈16 to 24 cubic meters per minute). The mobile unit shown in *Figure 19* is an example of a low vacuum/high volume system.

High vacuum/low volume systems are designed for close-proximity work and are effective in difficult to reach or confined work spaces. In this case, close proximity means the inlet opening needs to be 2" to 4" (≈5 to 10 cm) from the fume source. There are two types of high vacuum/low volume systems. The fume extraction gun type has a fume extraction nozzle built into the welding gun. Suction nozzles—the second type—are positioned near the weld, usually within 4 inches (10 cm) of the work. Adequate ventilation can be a problem in tight or cramped working quarters. To ensure adequate room ventilation in such cases, local exhaust ventilation should be used to capture fumes. Positioning is very important with these systems, to ensure the immediate area around the weld is not disturbed.

The exhaust hood should be kept close enough to the source of the fumes to ensure their capture.

Figure 19 Source extraction equipment.

29101-15_F20.EPS

Figure 20 Low vacuum/high volume fume extraction.

Welders should recognize that fumes of any type, regardless of their source, should not be inhaled. The best way to avoid problems is to provide adequate ventilation and keep your face out of the fume plume. Respirators must be used. *Figure 21* shows a welder using both a fume extraction system and a powered air-purifying respirator unit for increased safety.

29101-15_F21.EPS

Figure 21 Personal filtration and an extraction system combined.

WARNING!

Fumes and gases can be dangerous. Overexposure can cause nausea, headaches, dizziness, metal fume fever, and severe toxic effects that can be fatal. Studies have demonstrated irritation to eyes, skin, and the respiratory system and even more severe complications. In confined spaces, fumes and gases may cause asphyxiation.

3.6.2 Respirators

Special metals require the use of respirators to protect welders from harmful fumes. Respirator selection is based on the type and concentration of the contaminants. Respirators are grouped into three main types based on how they work to protect the wearer from contaminants. The types are the following:

- Air-purifying respirators
- Supplied-air respirators (SARs)
- Self-contained breathing apparatus (SCBA)

If respirators are used, your employer must offer worker training on respirator fitting and use. Medical screenings should also be provided.

Air-Purifying Respirators – Air-purifying respirators provide the lowest level of protection. They are made for use only in atmospheres that have enough oxygen to sustain life (at least 19.5 percent). Air-purifying respirators use special filters and cartridges to remove specific gases, vapors, and particles from the air. The respirator cartridges contain charcoal, which absorbs certain toxic vapors and gases. When the wearer detects any taste or smell, the charcoal's absorption capacity has been reached and the cartridge can no longer remove the contaminant. The respirator filters remove particles such as dust, mists, and metal fumes by trapping them in the filter material. Filters should be changed when it becomes difficult to breathe. Depending on the contaminants, cartridges can be used alone or in combination with a filter/pre-filter and filter cover. Air-purifying respirators should be used for protection only against the types of contaminants listed on the filters and cartridges and on the National Institute for Occupational Safety and Health (NIOSH) approval label affixed to each respirator carton and replacement filter/cartridge carton. Respirator manufacturers typically classify air-purifying respirators into four groups:

- No-maintenance
- Low-maintenance
- Reusable
- Powered air-purifying respirators (PAPRs)

No-maintenance and low-maintenance respirators are typically used for residential or light commercial work that does not call for constant and heavy respirator use. No-maintenance respirators are typically half-mask respirators with permanently attached cartridges or filters. The entire respirator is discarded when the cartridges or filters are spent. Low-maintenance respirators generally are also half-mask respirators that use replaceable cartridges and filters. However, they are not designed for constant use.

Reusable respirators (*Figure 22*) are made in half-mask and full facepiece styles. These respirators require the replacement of cartridges, filters, and respirator parts. Their use also requires a complete respirator maintenance program.

Powered air-purifying respirators (PAPRs) are made in half-mask, full facepiece, and hood styles. They use battery-operated blowers to pull outside air through the cartridges and filters attached to the respirator. The blower motors can be either mask- or belt-mounted. Depending on the cartridges used, they can filter particulates, dusts, fumes, and mists, along with certain gases and vapors. PAPRs like the one shown in *Figure 23* have a belt-mounted, powered air-purifier unit connected to the mask by a breathing tube. The hood style of PAPR can be seen in *Figure 21*. Many models also have an audible and visual alarm that is activated when airflow falls below the required minimum level. This feature gives an immediate indication of a loaded filter or low battery-charge condition. Units with the blower mounted in the mask do not use the breathing tube and belt-mounted package. Since they are more compact, they typically have less filtration capacity.

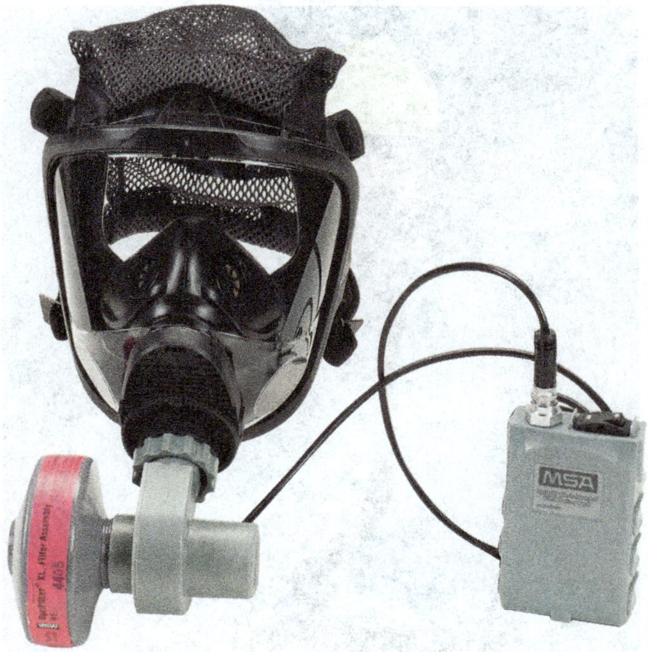

29101-15_F23.EPS

Figure 23 Full facepiece powered air-purifying respirator (PAPR).

Supplied-Air Respirators – Supplied-air respirators (*Figure 24*) provide a supply of breathable air for extended periods of time via a high-pressure hose that is connected to an external source, such as a compressor, compressed-air cylinder, or pump. They provide a higher level of protection in atmospheres where air-purifying respirators are

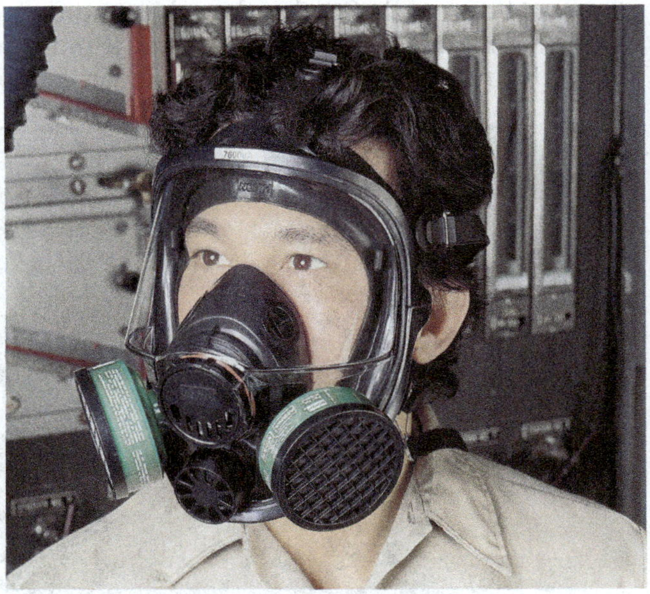

(A) FULL FACEPIECE MASK

(B) HALF MASK

29101-15_F22.EPS

Figure 22 Full-face and half-face air-purifying respirators.

29101-15_F24.EPS

Figure 24 Supplied-air respirator.

not capable of capturing the hazardous fumes. Supplied-air respirators are typically used in toxic atmospheres. Some can be used in atmospheres that are classified as **immediately dangerous to life and health (IDLH)** as long as they are equipped with an air cylinder for emergency escape. An atmosphere is considered IDLH if it poses an immediate hazard to life or produces immediate, irreversible, adverse effects on health. There are two types of supplied-air respirators: continuous-flow and pressure-demand. The unit shown in *Figure 24* is a continuous-flow model.

The continuous-flow supplied-air respirator provides air to the user in a constant stream. One or two hoses may be used to deliver the air from the air source to the facepiece. Unless the compressor or pump is especially designed to filter the air or a portable air-filtering system is used, the unit must be located where there is breathable air (Grade D or better as described in *Compressed Gas Association [CGA] Commodity Specification G-7.1*). Continuous-flow respirators are made with tight-fitting half-masks or full facepieces. They are also made with hoods. The flow of air to the user may be adjusted either at the air source (fixed flow) or on the unit's regulator (adjustable flow). Pressure-demand supplied-air respirators are similar to the continuous-flow type except that they supply air to the user's facepiece via a pressure-demand valve as the user inhales and fresh air is required. They typically have a two-position exhalation valve that allows the worker to switch between pressure-demand and negative-pressure modes to permit entry into, movement within, and exit from a work area.

Self-Contained Breathing Apparatus (SCBA) – SCBAs (*Figure 25*) can be used in oxygen-deficient atmospheres (below 19.5 percent oxygen); in poorly ventilated or confined spaces; and in IDLH atmospheres. These respirators provide a supply of air for 30 to 60 minutes from a compressed-air cylinder worn on the user's back.

An emergency escape breathing apparatus (EEBA) is a smaller version of an SCBA cylinder. EEBA units are used for escape from hazardous environments and generally provide only a five- to ten-minute supply of air. This type is usually kept in strategic locations where they can be used in the event of an emergency. In facilities that require that they be available, all workers must become familiar with their exact location.

3.6.3 Respiratory Program

In the United States, OSHA regulations, along with your company's procedures, must be followed when selecting the proper type of respirator for a particular job (*Figure 26*). Respirator selection is based on the contaminant present and its concentration level. This determination should be made by a trained safety specialist. The respirator must be properly fitted and used in accordance with the manufacturer's instructions. It must be worn during all times of exposure. Regardless of

29101-15_F25.EPS

Figure 25 Self-contained breathing apparatus (SCBA).

BEFORE USING A RESPIRATOR YOU MUST DETERMINE THE FOLLOWING:

1. THE TYPE OF CONTAMINANT(S) FOR WHICH THE RESPIRATOR IS BEING SELECTED
2. THE CONCENTRATION LEVEL OF THAT CONTAMINANT
3. WHETHER THE RESPIRATOR CAN BE PROPERLY FITTED ON THE WEARER'S FACE

YOU MUST READ AND UNDERSTAND ALL RESPIRATOR INSTRUCTIONS, WARNINGS, AND USE LIMITATIONS CONTAINED ON EACH PACKAGE BEFORE USE.

29101-15_F26.EPS

Figure 26 Use the right respirator for the job.

the kind of respirator needed, OSHA regulations require employers to have a respirator protection program consisting of the following:

- Standard operating procedures for selection and use
- Employee training
- Regular cleaning and disinfecting
- Sanitary storage
- Regular inspection
- Annual fit testing
- Pulmonary function testing

As an employee, you are responsible for wearing respiratory protection when needed. Both vapors and fumes can be eliminated in certain concentrations by the use of air-purifying devices, as long as oxygen levels are acceptable. Examples of fumes are smoke billowing from a fire, or the fumes generated when welding. Always check the cartridge on your respirator to make sure it is the correct type to use for the air conditions and contaminants found on the job site.

When selecting a respirator to wear while working with specific materials, first identify the hazardous ingredients contained in the material and their exposure levels, then choose the proper respirator to protect yourself at these levels. Always read the product's safety data sheet (SDS) or material safety data sheet (MSDS). It identifies the hazardous ingredients and should list the type of respirator and cartridge recommended for use with the product.

Limitations that apply to all half-mask (air-purifying) respirators are as follows:

- These respirators do not completely eliminate exposure to contaminants, but they will reduce the level of exposure to below-hazardous levels.
- These respirators do not supply oxygen and must not be used in areas where the oxygen level is below 19.5 percent.
- These respirators must not be used in areas where chemicals provide weak warning signs of their presence, such as a lack of taste or a unique odor.

If your breathing becomes difficult; if you become dizzy or nauseated; if you smell or taste the chemical; or if you have other noticeable effects, leave the area immediately, return to a fresh air area, and seek assistance.

All respirators are useless unless properly fit-tested to each individual. To obtain the best protection from your respirator, you must perform positive and negative fit checks each time you wear it. These fit checks must be done until you have obtained a good face seal.

To perform the positive fit check, do the following:

Step 1 Adjust the facepiece for the best fit, then adjust the head and neck straps to ensure good fit and comfort.

WARNING!

Do not overtighten the head and neck straps. Tighten them only enough to stop leakage. Overtightening can cause facepiece distortion and dangerous leaks.

Step 2 Block the exhalation valve with your hand or other material.

Step 3 Breathe out into the mask to increase the pressure inside.

Step 4 Check for air leakage around the edges of the facepiece.

Step 5 If the facepiece puffs out slightly for a few seconds, a good face seal has been obtained.

To perform a negative fit check, do the following:

Step 1 Block the inhalation valve with your hand or other material.

Step 2 Attempt to inhale.

Step 3 Check for air leakage around the edges of the facepiece.

Step 4 If the facepiece caves in slightly for a few seconds, a good face seal has been obtained.

3.7.0 Safety Data Sheets

Safety data sheets (SDSs) and material safety data sheets (MSDSs) are fact sheets prepared by the chemical manufacturer or importer. All manufacturers of potentially hazardous materials must provide detailed information regarding possible hazards resulting from the use of their product. An SDS/MSDS must be made available to anyone using the product or anyone working in the area where the product is in use. Each product used on a construction site must have an SDS/MSDS readily available for examination by workers. *Figure 27* shows a sample MSDS storage box where the information is available on site. An SDS/MSDS describes the substance, its hazards, safe handling, first aid, and emergency spill procedures.

The SDS replaces the MSDS that has been used for a number of years. The primary difference in the two is the arrangement of the entries; both provide the same basic information and have the same function. The Hazard Communication (HAZCOM) standard requires a new, internationally standardized format for the SDS. The sections of the SDS form include the following:

- *Section 1, Product identification, manufacturer information, recommended uses and restrictions*
- *Section 2, Hazard identification*
- *Section 3, Composition/information on ingredients*
- *Section 4, First aid measures*
- *Section 5, Firefighting information*
- *Section 6, Accidental release measures*
- *Section 7, Handling and storage*
- *Section 8, Exposure controls/personal protection*
- *Section 9, Physical and chemical properties*
- *Section 10, Stability and reactivity*
- *Section 11, Toxicological properties*
- *Section 12, Ecological properties*
- *Section 13, Disposal considerations*
- *Section 14, Transport information*
- *Section 15, Regulatory information*
- *Section 16, Other information*

29101-15_F27.EPS

Figure 27 SDS/MSDS storage box.

Respirator Inspection, Care, and Maintenance

A respirator must be clean and in good condition, and all of its parts must be in place for it to give you proper protection. Respirators must be cleaned every day. Failure to do so will limit their effectiveness and offer little or no protection. For example, suppose you wore the respirator yesterday and did not clean it. The bacteria from breathing into the respirator, plus the airborne contaminants that managed to enter the facepiece, will have made the inside of your respirator unsanitary. Continued use may cause you more harm than good. Remember, only a clean and complete respirator will provide you with the necessary protection. Follow these guidelines:

- Inspect the condition of your respirator before and after each use.
- Do not wear a respirator if the facepiece is distorted or if it is worn and cracked. You will not be able to get a proper face seal.
- Do not wear a respirator if any part of it is missing. Replace worn straps or missing parts before using.
- Do not expose respirators to excessive heat or cold, chemicals, or sunlight.
- Clean and wash your respirator after every time you use it. Remove the cartridge and filter, hand wash the respirator using mild soap and a soft brush, and let it air dry overnight.
- Sanitize your respirator each week. Remove the cartridge and filter, then soak the respirator in a sanitizing solution for at least two minutes. Thoroughly rinse with warm water and let it air dry overnight.
- Store the clean and sanitized respirator in its resealable plastic bag. Do not store the respirator face down. This will cause distortion of the facepiece.

The SDS format is gradually replacing the MSDS and is required for all new products. During the transition, you may find that the data sheets for many existing products are still in the MSDS format.

Figure 28 shows a sample SDS for solvent cement used with PVC pipe products. The most important things to look for on an SDS are the specific hazards, personal protection, handling procedures, and first aid information. Most SDSs have a 24-hour emergency-response number.

Look for the information you would need to use the cement described on the SDS form in *Figure 28*. First locate the hazards. Section 2 shows that the adhesive is flammable and an eye and skin irritant that can cause respiratory irritation, dizziness, and drowsiness.

Next, find out how to minimize these hazards. Section 7 gives general handling and storage information. It indicates that ventilation is needed to reduce hazardous vapors. This can be as simple as a fan in an open window. If ventilation is not enough, respiratory protection is needed. Section 8 tells you how to protect your eyes and skin.

Section 4 lists the first-aid measures for eye contact, skin contact, or inhalation. Section 5 explains fire hazards and firefighting measures. You now have the information you need in case of an emergency.

WARNING!

The SDSs must be kept in the work area and be readily accessible to all workers. Hazardous chemical producers are required to provide an SDS/MSDS in the official language(s) of each country or market where it is being produced and sold. Employers should ensure that workers are capable of reading and understanding the contents of an SDS/MSDS, and take steps to provide the information in a language they can understand. The company's safety officer or competent person should always review the SDS/MSDS before the hazardous material is used.

All SDSs/MSDSs must be kept on site. Ask your supervisor to tell you where they are located and to point out the sections that relate to your job. The health and safety of you and your co-workers depends on it. Companies often post these sheets on a bulletin board or put them in a convenient place near the work area.

At this writing, the MSDS is still in use for many products. The layout of the MSDS may vary somewhat between products and/or manufacturers. This is one of the reasons for the change to the SDS format. *Figure 29* shows a portion of an MSDS for Excalibur 7018-1 welding electrodes. As a welder, it is important to be familiar with the safety precautions related to the consumable products in use. Remember that some consumables contain potentially hazardous substances.

WELD-ON

GHS SAFETY DATA SHEET

WELD-ON® 705™ Low VOC Cements for PVC Plastic Pipe

Date Revised: **DEC 2011**
Supersedes: **FEB 2010**

SECTION I - PRODUCT AND COMPANY IDENTIFICATION

PRODUCT NAME: **WELD-ON® 705™ Low VOC Cements for PVC Plastic Pipe**
PRODUCT USE: Low VOC Solvent Cement for PVC Plastic Pipe
SUPPLIER:
MANUFACTURER: IPS Corporation
17109 South Main Street, Carson, CA 90248-3127
P.O. Box 379, Gardena, CA 90247-0379
Tel. 1-310-898-3300

EMERGENCY: Transportation: CHEMTEL Tel. 800.255-3924, 813-248-0585 (International)
Medical: Tel. 800.451.8346, 760.602.8703 3E Company (International)

SECTION 2 - HAZARDS IDENTIFICATION

GHS CLASSIFICATION:

Health		Environmental		Physical	
Acute Toxicity:	Category 4	Acute Toxicity:	None Known	Flammable Liquid	Category 2
Skin Irritation:	Category 3	Chronic Toxicity:	None Known		
Skin Sensitization:	NO				
Eye:	Category 2B				

GHS LABEL:  OR  **Signal Word: Danger**

WHMIS CLASSIFICATION: CLASS B, DIVISION 2

Hazard Statements	Precautionary Statements
H225: Highly flammable liquid and vapor	P210: Keep away from heat/sparks/open flames/hot surfaces – No smoking
H319: Causes serious eye irritation	P261: Avoid breathing dust/fume/gas/mist/vapors/spray
H332: Harmful if inhaled	P280: Wear protective gloves/protective clothing/eye protection/face protection
H335: May cause respiratory irritation	P304+P340: IF INHALED: Remove victim to fresh air and keep at rest in a position comfortable for breathing
H336: May cause drowsiness or dizziness	P403+P233: Store in a well ventilated place. Keep container tightly closed
EUH019: May form explosive peroxides	P501: Dispose of contents/container in accordance with local regulation

SECTION 3 - COMPOSITION/INFORMATION ON INGREDIENTS

	CAS#	EINECS #	REACH Pre-registration Number	CONCENTRATION % by Weight
Tetrahydrofuran (THF)	109-99-9	203-726-8	05-2116297729-22-0000	25 - 50
Methyl Ethyl Ketone (MEK)	78-93-3	201-159-0	05-2116297728-24-0000	5 - 36
Cyclohexanone	108-94-1	203-631-1	05-2116297718-25-0000	15 - 30

All of the constituents of this adhesive product are listed on the TSCA inventory of chemical substances maintained by the US EPA, or are exempt from that listing.
* Indicates this chemical is subject to the reporting requirements of Section 313 of the Emergency Planning and Community Right-to-Know Act of 1986 (40CFR372).
# indicates that this chemical is found on Proposition 65's List of chemicals known to the State of California to cause cancer or reproductive toxicity.

SECTION 4 - FIRST AID MEASURES

Contact with eyes: Flush eyes immediately with plenty of water for 15 minutes and seek medical advice immediately.
Skin contact: Remove contaminated clothing and shoes. Wash skin thoroughly with soap and water. If irritation develops, seek medical advice.
Inhalation: Remove to fresh air. If breathing is stopped, give artificial respiration. If breathing is difficult, give oxygen. Seek medical advice.
Ingestion: Rinse mouth with water. Give 1 or 2 glasses of water or milk to dilute. Do not induce vomiting. Seek medical advice immediately.

SECTION 5 - FIREFIGHTING MEASURES

Suitable Extinguishing Media: Dry chemical powder, carbon dioxide gas, foam, Halon, water fog.
Unsuitable Extinguishing Media: Water spray or stream.
Exposure Hazards: Inhalation and dermal contact
Combustion Products: Oxides of carbon, hydrogen chloride and smoke
Protection for Firefighters: Self-contained breathing apparatus or full-face positive pressure airline masks.

	HMIS	NFPA	
			0-Minimal
Health	2	2	1-Slight
Flammability	3	3	2-Moderate
Reactivity	0	0	3-Serious
PPE	B		4-Severe

SECTION 6 - ACCIDENTAL RELEASE MEASURES

Personal precautions: Keep away from heat, sparks and open flame.
Provide sufficient ventilation, use explosion-proof exhaust ventilation equipment or wear suitable respiratory protective equipment.
Prevent contact with skin or eyes (see section 8).
Environmental Precautions: Prevent product or liquids contaminated with product from entering sewers, drains, soil or open water course.
Methods for Cleaning up: Clean up with sand or other inert absorbent material. Transfer to a closable steel vessel.
Materials not to be used for clean up: Aluminum or plastic containers

SECTION 7 - HANDLING AND STORAGE

Handling: Avoid breathing of vapor, avoid contact with eyes, skin and clothing.
Keep away from ignition sources, use only electrically grounded handling equipment and ensure adequate ventilation/fume exhaust hoods.
Do not eat, drink or smoke while handling.
Storage: Store in ventilated room or shade below 44°C (110°F) and away from direct sunlight.
Keep away from ignition sources and incompatible materials: caustics, ammonia, inorganic acids, chlorinated compounds, strong oxidizers and isocyanates.
Follow all precautionary information on container label, product bulletins and solvent cementing literature.

SECTION 8 - PRECAUTIONS TO CONTROL EXPOSURE / PERSONAL PROTECTION

EXPOSURE LIMITS:

Component	ACGIH TLV	ACGIH STEL	OSHA PEL	OSHA STEL:
Tetrahydrofuran (THF)	50 ppm	100 ppm	200 ppm	
Methyl Ethyl Ketone (MEK)	200 ppm	300 ppm	200 ppm	
Cyclohexanone	20 ppm	50 ppm	50 ppm	

Engineering Controls: Use local exhaust as needed.
Monitoring: Maintain breathing zone airborne concentrations below exposure limits.
Personal Protective Equipment (PPE):
Eye Protection: Avoid contact with eyes, wear splash-proof chemical goggles, face shield, safety glasses (spectacles) with brow guards and side shields, etc. as may be appropriate for the exposure.
Skin Protection: Prevent contact with the skin as much as possible. Butyl rubber gloves should be used for frequent immersion.
Use of solvent-resistant gloves or solvent-resistant barrier cream should provide adequate protection when normal adhesive application practices and procedures are used for making structural bonds.
Respiratory Protection: Prevent inhalation of the solvents. Use in a well-ventilated room. Open doors and/or windows to ensure airflow and air changes. Use local exhaust ventilation to remove airborne contaminants from employee breathing zone and to keep contaminants below levels listed above.
With normal use, the Exposure Limit Value will not usually be reached. When limits approached, use respiratory protection equipment.

29101-15_F28A.EPS

Figure 28A Solvent Cement SDS (sheet 1 of 2).

WELD-ON

GHS SAFETY DATA SHEET

WELD-ON® 705™ Low VOC Cements for PVC Plastic Pipe

Date Revised: **DEC 2011**
Supersedes: **FEB 2010**

SECTION 9 - PHYSICAL AND CHEMICAL PROPERTIES

Appearance:	Clear or gray, medium syrupy liquid		
Odor:	Ketone	**Odor Threshold:**	0.88 ppm (Cyclohexanone)
pH:	Not Applicable		
Melting/Freezing Point:	-108.5°C (-163.3°F) Based on first melting component: THF	**Boiling Range:**	66°C (151°F) to 156°C (313°F)
Boiling Point:	66°C (151°F) Based on first boiling component: THF	**Evaporation Rate:**	> 1.0 (BUAC = 1)
Flash Point:	-20°C (-4°F) TCC based on THF	**Flammability:**	Category 2
Specific Gravity:	0.9611 @23°C (73°F)	**Flammability Limits:**	**LEL:** 1.1% based on Cyclohexanone
Solubility:	Solvent portion soluble in water. Resin portion separates out.		**UEL:** 11.8% based on THF
Partition Coefficient n-octanol/water:	Not Available	**Vapor Pressure:**	129 mm Hg @ 20°C (68°F)based on THF
Auto-ignition Temperature:	321°C (610°F) based on THF	**Vapor Density:**	>2 (Air = 1)
Decomposition Temperature:	Not Applicable	**Other Data: Viscosity:**	Medium bodied
VOC Content:	When applied as directed, per SCAQMD Rule 1168, Test Method 316A,VOC content is: ≤ 510 g/l.		

SECTION 10 - STABILITY AND REACTIVITY

Stability: Stable
Hazardous decomposition products: None in normal use. When forced to burn, this product gives off oxides of carbon, hydrogen chloride and smoke.
Conditions to avoid: Keep away from heat, sparks, open flame and other ignition sources.
Incompatible Materials: Oxidizers, strong acids and bases, amines, ammonia

SECTION 11 - TOXICOLOGICAL INFORMATION

Likely Routes of Exposure: Inhalation, Eye and Skin Contact
Acute symptoms and effects:
Inhalation: Severe overexposure may result in nausea, dizziness, headache. Can cause drowsiness, irritation of eyes and nasal passages.
Eye Contact: Vapors slightly uncomfortable. Overexposure may result in severe eye injury with corneal or conjunctival inflammation on contact with the liquid.
Skin Contact: Liquid contact may remove natural skin oils resulting in skin irritation. Dermatitis may occur with prolonged contact.
Ingestion: May cause nausea, vomiting, diarrhea and mental sluggishness.
Chronic (long-term) effects: None known to humans

Toxicity:	LD_{50}	LC_{50}
Tetrahydrofuran (THF)	Oral: 2842 mg/kg (rat)	Inhalation 3 hrs. 21,000 mg/m^3 (rat)
Methyl Ethyl Ketone (MEK)	Oral: 2737 mg/kg (rat), Dermal: 6480 mg/kg (rabbit)	Inhalation 8 hrs. 23,500 mg/m^3 (rat)
Cyclohexanone	Oral: 1535 mg/kg (rat), Dermal: 948 mg/kg (rabbit)	Inhalation 4 hrs. 8,000 PPM (rat)

Reproductive Effects	Teratogenicity	Mutagenicity	Embryotoxicity	Sensitization to Product	Synergistic Products
Not Established	Not Established	Not Established	Not Established	Not Established	Not Established

SECTION 12 - ECOLOGICAL INFORMATION

Ecotoxicity: None Known
Mobility: In normal use, emission of volatile organic compounds (VOC's) to the air takes place, typically at a rate of ≤ 510 g/l.
Degradability: Biodegradable
Bioaccumulation: Minimal to none.

SECTION 13 - WASTE DISPOSAL CONSIDERATIONS

Follow local and national regulations. Consult disposal expert.

SECTION 14 - TRANSPORT INFORMATION

Proper Shipping Name: Adhesives
Hazard Class: 3
Secondary Risk: None
Identification Number: UN 1133
Packing Group: PG II
Label Required: Class 3 Flammable Liquid
Marine Pollutant: NO

EXCEPTION for Ground Shipping
DOT Limited Quantity: Up to 5L per inner packaging, 30 kg gross weight per package.
Consumer Commodity: Depending on packaging, these quantities may qualify under DOT as "ORM-D".

TDG INFORMATION	
TDG CLASS:	FLAMMABLE LIQUID 3
SHIPPING NAME:	ADHESIVES
UN NUMBER/PACKING GROUP:	UN 1133, PG II

SECTION 15 - REGULATORY INFORMATION

Precautionary Label Information: Highly Flammable, Irritant
Symbols: F, Xi

Ingredient Listings: USA TSCA, Europe EINECS, Canada DSL, Australia AICS, Korea ECL/TCCL, Japan MITI (ENCS)

Risk Phrases:
R11: Highly flammable.
R20: Harmful by inhalation.
R36/37: Irritating to eyes and respiratory system.
R66: Repeated exposure may cause skin dryness or cracking
R67: Vapors may cause drowsiness and dizziness

Safety Phrases:
S9: Keep container in a well-ventilated place.
S16: Keep away from sources of ignition - No smoking.
S25: Avoid contact with eyes.
S26: In case of contact with eyes, rinse immediately with plenty of water and seek medical advice.
S33: Take precautionary measures against static discharges.
S46: If swallowed, seek medical advise immediately and show this container or label.

SECTION 16 - OTHER INFORMATION

Specification Information:
Department issuing data sheet: IPS, Safety Health & Environmental Affairs
E-mail address: <EHSinfo@ipscorp.com>

All ingredients are compliant with the requirements of the European Directive on RoHS (Restriction of Hazardous Substances).

Training necessary: Yes, training in practices and procedures contained in product literature.
Reissue date / reason for reissue: 12/14/2011 / Updated GHS Standard Format
Intended Use of Product: Solvent Cement for PVC Plastic Pipe

This product is intended for use by skilled individuals at their own risk. The information contained herein is based on data considered accurate based on current state of knowledge and experience. However, no warranty is expressed or implied regarding the accuracy of this data or the results to be obtained from the use thereof.

29101-15_F28B.EPS

Figure 28B Solvent Cement SDS (sheet 2 of 2).

LINCOLN ELECTRIC

Date: 1/21/2013	MSDS No.: US-M292
Trade Name:	Excalibur 7018-1 MR
Sizes:	All
Supersedes:	8/29/2011

MATERIAL SAFETY DATA SHEET

For Welding Consumables and Related Products

Conforms to Hazard Communication Standard 29CFR 1910.1200 Rev. October 1988

SECTION I - IDENTIFICATION

Manufacturer/ Supplier: The Lincoln Electric Company
22801 St. Clair Avenue
Cleveland, OH 44117-1199
(216) 481-8100

Product Type: Covered Electrode

Classification: AWS E7018-1H4R

SECTION II - HAZARDOUS MATERIAL (1)

IMPORTANT!

This section covers the materials from which this product is manufactured. The fumes and gases produced during welding with the normal use of this product are covered by Section V; see it for industrial hygiene information.
CAS Number shown is representative for the ingredients listed. All ingredients listed may not be present in all sizes.
(1) The term "hazardous" in "Hazardous Materials" should be interpreted as a term required and defined in the Hazards Communication Standard and does not necessarily imply the existence of any hazard. All materials are listed on the TSCA inventory.

Ingredients:	CAS No.	Wt.%	TLV mg/m^3	PEL mg/m^3
Iron	7439-89-6	15	10*	15*
Limestone and/or calcium carbonate	1317-65-3	10	10*	15
Titanium dioxides	13463-67-7	< 5	10	15
Fluorides (as F)	7789-75-5	< 5	2.5	2.5
Silicates and other binders	1344-09-8	< 5	10*	15*
Manganese and/or manganese alloys and compounds (as Mn)*****	7439-96-5	< 5	0.2	5 (c)
Mineral silicates	1332-58-7	< 5	5**	5**
Silicon and/or silicon alloys and compounds (as Si)	7440-21-3	1	10*	15*
Zirconium alloys and compounds (as Zr)	12004-83-0	1	5	5
Cellulose and other carbohydrates	65996-61-4	0.5	10*	15*
Quartz	14808-60-7	< 0.5	#0.025**	#0.1**
Molybdenum alloys (as Mo)	7439-98-7	< 0.5	10	10
Lithium compounds (as Li)	554-13-2	< 0.5	10*	15*
Carbon steel core wire	7439-89-6	55	10*	15*

Supplemental Information:

(*) Not listed. The OSHA PEL for nuisance particles is 15 milligrams per cubic meter. The ACGIH guideline for total particulate is 10 milligrams per cubic meter. PEL value for iron oxide is 10 milligrams per cubic meter. TLV value for iron oxides is 5 milligrams per cubic meter.

(**) As respirable dust.

(*****) Subject to the reporting requirements of Sections 311, 312, and 313 of the Emergency Planning and Community Right-to-Know Act of 1986 and of 40CFR 370 and 372.

(c) Value is for manganese fume. Present PEL is 5 milligrams per cubic meter (ceiling value). Values proposed by OSHA in 1989 were 1.0 milligrams per cubic meter TWA and 3.0 milligrams per cubic meter STEL (Short Term Exposure Limit).

(#) Crystalline silica (quartz) is on the IARC (International Agency for Research on Cancer) and NTP (National Toxicology Program) lists as posing a carcinogenic risk to humans.

SECTION III - HAZARD DATA

Non Flammable; Welding arc and sparks can ignite combustibles and flammable products. See Z49.1 referenced in Section VI. Product is inert, no special handling or spill procedures required. Not regulated by DOT.

Rev 1/13

(CONTINUED ON SIDE TWO)

29101-15_F29A.EPS

Figure 29A Welding electrode MSDS. (sheet 1 of 2)

Product: Excalibur 7018-1 MR

Date: 1/21/2013

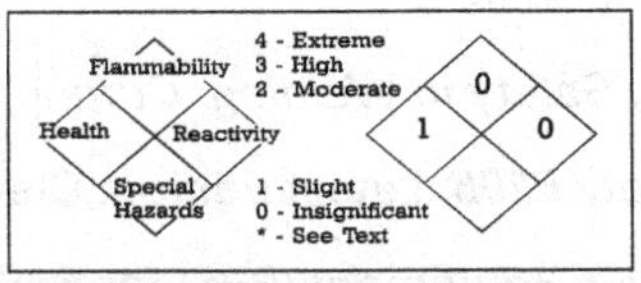

SECTION IV - HEALTH HAZARD DATA

Threshold Limit Value: The ACGIH recommended general limit for Welding Fume NOS - (Not Otherwise Specified) is 5 mg/m^3. ACGIH-1999 preface states that the TLV-TWA should be used as guides in the control of health hazards and should not be used as fine lines between safe and dangerous concentrations. See Section V for specific fume constituents which may modify this TLV. Threshold Limit Values are figures published by the American Conference of Government Industrial Hygienists. Units are milligrams per cubic meter of air.

Effects of Overexposure: Electric arc welding may create one or more of the following health hazards:
Fumes and Gases can be dangerous to your health. Common entry is by inhalation. Other possible routes are skin contact and ingestion.

Short-term (acute) overexposure to welding fumes may result in discomfort such as metal fume fever, dizziness, nausea, or dryness or irritation of nose, throat, or eyes. May aggravate pre-existing respiratory problems (e.g. asthma, emphysema).

Long-term (chronic) overexposure to welding fumes can lead to siderosis (iron deposits in lung) and may affect pulmonary function. Manganese overexposure can affect the central nervous system, resulting in impaired speech and movement. Bronchitis and some lung fibrosis have been reported. Repeated exposure to fluorides may cause excessive calcification of the bone and calcification of ligaments of the ribs, pelvis and spinal column. May cause skin rash. Titanium dioxide is listed by the IARC (International Agency for Research on Cancer) as a Group 2B carcinogen (possibly carcinogenic to humans based on animal studies). Respiratory exposure to the crystalline silica present in this welding electrode is not anticipated during normal use. Respiratory overexposure to airborne crystalline silica is known to cause silicosis, a form of disabling pulmonary fibrosis which can be progressive and may lead to death. Crystalline silica is on the IARC (International Agency for Research on Cancer) and NTP (National Toxicology Program) lists as posing a cancer risk to humans. WARNING: This product, when used for welding or cutting, produces fumes or gases which contain chemicals known to the State of California to cause birth defects and, in some cases, cancer. (California Health & Safety Code Section 25249.5 et seq.)

Arc Rays can injure eyes and burn skin. *Skin cancer has been reported.*
Electric Shock can kill. If welding must be performed in damp locations or with wet clothing, on metal structures or when in cramped positions such as sitting, kneeling or lying, or if there is a high risk of unavoidable or accidental contact with workpiece, use the following equipment: Semiautomatic DC Welder, DC Manual (Stick) Welder, or AC Welder with Reduced Voltage Control.

Emergency and First Aid Procedures: Call for medical aid. Employ first aid techniques recommended by the American Red Cross.
IF BREATHING IS DIFFICULT give oxygen. IF NOT BREATHING employ CPR (Cardiopulmonary Resuscitation) techniques.
IN CASE OF ELECTRICAL SHOCK, turn off power and follow recommended treatment. In all cases call a physician.

SECTION V - REACTIVITY DATA

Hazardous Decomposition Products: Welding fumes and gases cannot be classified simply. The composition and quantity of both are dependent upon the metal being welded, the process, procedure and electrodes used.

Other conditions which also influence the composition and quantity of the fumes and gases to which workers may be exposed include: coatings on the metal being welded (such as paint, plating, or galvanizing), the number of welders and the volume of the worker area, the quality and amount of ventilation, the position of the welder's head with respect to the fume plume, as well as the presence of contaminants in the atmosphere (such as chlorinated hydrocarbon vapors from cleaning and degreasing activities.)

When the electrode is consumed, the fume and gas decomposition products generated are different in percent and form from the ingredients listed in Section II. Decomposition products of normal operation include those originating from the volatilization, reaction, or oxidation of the materials shown in Section II, plus those from the base metal and coating, etc., as noted above.

Reasonably expected fume constituents of this product would include: Primarily iron oxide and fluorides; secondarily complex oxides of manganese, potassium, silicon, sodium, and titanium.

Maximum fume exposure guideline for this product (based on manganese content) is 4.0 milligrams per cubic meter.

Gaseous reaction products may include carbon monoxide and carbon dioxide. Ozone and nitrogen oxides may be formed by the radiation from the arc.

Determine the composition and quantity of fumes and gases to which workers are exposed by taking an air sample from inside the welder's helmet if worn or in the worker's breathing zone. Improve ventilation if exposures are not below limits. See ANSI/AWS F1.1, F1.2, F1.3 and F1.5, available from the American Welding Society, 550 N.W. LeJeune Road, Miami, FL 33126.

SECTION VI AND VII
CONTROL MEASURES AND PRECAUTIONS FOR SAFE HANDLING AND USE

Read and understand the manufacturer's instruction and the precautionary label on the product. Request Lincoln Safety Publication E205. See American National Standard Z49.1, "Safety In Welding, Cutting and Allied Processes" published by the American Welding Society, 550 N.W. LeJeune Road, Miami, FL, 33126 (both available for free download at http://www.lincolnelectric.com/community/safety/) and OSHA Publication 2206 (29CFR1910), U.S. Government Printing Office, Superintendent of Documents, P.O. Box 371954, Pittsburgh, PA 15250-7954 for more details on many of the following:

Ventilation: Use enough ventilation, local exhaust at the arc, or both to keep the fumes and gases from the worker's breathing zone and the general area. Train the welder to keep his head out of the fumes. *Keep exposure as low as possible.*

Respiratory Protection: Use respirable fume respirator or air supplied respirator when welding in confined space or general work area when local exhaust or ventilation does not keep exposure below TLV.

Eye Protection: Wear helmet or use face shield with filter lens shade number 12 or darker. Shield others by providing screens and flash gog gles.

Protective Clothing: Wear hand, head, and body protection which help to prevent injury from radiation, sparks and electrical shock. See Z49.1.
At a minimum this includes welder's gloves and a protective face shield, and may include arm protectors, aprons, hats, shoulder protection, as well as dark substantial clothing. Train the welder not to permit electrically live parts or electrodes to contact skin . . . or clothing or gloves if they are wet. Insulate from work and ground.

Disposal Information: Discard any product, residue, disposable container, or liner as ordinary waste in an environmentally acceptable manner according to Federal, State and Local Regulations unless otherwise noted. No applicable ecological information available.

29101-15_F29B.EPS

Figure 29B Welding electrode MSDS. (sheet 2 of 2)

Additional Resources

ANSI/AWS Z49.1, Safety in Welding, Cutting, and Allied Processes. Miami, FL: American Welding Society.

Arc Welding Safety E205. Latest Edition. Cleveland, OH: Lincoln Electric.

Personal Protective Equipment (PPE) for Welding and Cutting. Miami, FL. American Welding Society.

Safety Topics in Welding, Cutting, and Brazing. **www.osha.gov**.

AWS F32M/F32, Ventilation Guide for Weld Fumes. Latest Edition. Miami, FL: American Welding Society.

3.0.0 Section Review

1. The AWS standard for welding and cutting safety is ____.
   a. *Z49.1*
   b. *CFR* 1929
   c. *CFR* 1910
   d. *Z45.5*

2. The primary purpose of a welding screen is to ____.
   a. keep non-welders from watching welders at work
   b. protect others from the glare of the welding arc
   c. prevent insects from affecting the welding arc
   d. keep nearby material from catching fire

3. When a worker is assigned the responsibility of providing a fire watch, the worker ____.
   a. cannot do another task
   b. can retrieve additional consumables for the welder from the shop
   c. can also weld or cut to remain productive
   d. may conduct weld testing

4. Argon is used as a ____.
   a. breathing gas
   b. cutting fuel gas
   c. solvent
   d. purge gas

5. Which of the following is a *correct* statement about electrical hazards associated with welding?
   a. The jaws of the electrode holder do not carry electrical current.
   b. GFCIs should not be used when welding.
   c. Electrical shocks can cause severe burns.
   d. Welding machines typically operate on 120-volt house current.

6. Which of these metals is *least* likely to produce toxic fumes when heated?
   a. Manganese
   b. Hexavalent chrome
   c. Carbon steel
   d. Galvanized steel

7. In which section of an SDS are hazards identified?
   a. Section 1
   b. Section 2
   c. Section 5
   d. Section 12

SUMMARY

Welders join metal using a high-intensity electric arc to melt and fuse the base metals. Welders may work on critical structures such as pipelines, high-rise buildings, ships, and bridges, so the trade requires a high level of skill and concentration, which can only be achieved through extensive training and practice. Welders also are called upon to cut and gouge metal using equipment that produces a high-intensity flame. The most common cutting method is oxyfuel cutting, in which oxygen is mixed with a fuel gas such as acetylene to produce the cutting flame.

Safety is everyone's responsibility. Proper clothing, footwear, and eye protection are required for safe welding and cutting. Workers who fail to comply with safety rules are subject to dismissal. All welding shops must have established plans for dealing with accidents. Take the time to learn the proper procedures for accident response and reporting before you need to respond to an emergency.

Accidents are very harmful for both employees and employers, and they are often caused by poor behavior and unsafe conditions. Most accidents can be prevented. By knowing and avoiding the behaviors that cause accidents and by keeping working conditions safe, it may be possible to avoid injuries and reduce hazards.

Trying to bluff your way through a job you do not understand is asking for trouble. Even if you think you know the correct procedures, a review may bring out an important part of the job that you may have forgotten. Do not be afraid to ask questions. The responses you receive may help new or less experienced co-workers get answers to questions they may be embarassed to ask.

Materials, fumes, welding radiation exposure, and storing and handling cylinders present particular hazards for the welder. Developing an attitude of safety is an excellent way for every worker to avoid or reduce all of these hazards. Practicing good safety attitudes means that you do the following:

- Report all unsafe conditions and behaviors immediately.
- Keep work areas clean and orderly at all times.
- Immediately report all accidents and injuries, no matter how minor.
- Be certain you completely understand the instructions given before starting work.
- Know how and where medical help may be obtained.
- Wear the required protective devices when working in a hazardous operation area.
- Do not use alcohol or drugs. If you are ill and must take prescribed medication, notify your supervisor immediately.

Review Questions

1. The most common type of welding is ____ .
   a. GTAW
   b. GMAW
   c. FCAW
   d. SMAW

2. A competency-based curriculum is one in which ____ .
   a. the focus is on classroom training
   b. students must demonstrate knowledge and skills
   c. instructors must have four-year degrees
   d. all the training is done on the job

3. Youth apprenticeship programs are available that allow students to begin their apprenticeship or craft training ____ .
   a. in middle school
   b. in high school
   c. instead of their traditional subjects
   d. instead of English classes

4. Body protection while welding is best provided by wearing ____ .
   a. pants with cuffs
   b. polyester clothing
   c. cotton clothing
   d. loose-fitting shirts

5. Which of the following is a *correct* statement?
   a. It is not necessary to wear other eye protection if you are wearing a welding helmet.
   b. The same shade of lens can be used for almost all welding and cutting tasks.
   c. The recommended shade for electric arc operations is 9–14.
   d. Safety glasses with side shields should not be used because they block peripheral vision.

6. Accidents are more likely for employees in the age group ____ .
   a. 18 and under
   b. 19 to 25
   c. 26 to 45
   d. 60 and over

7. A good work area safety practice is ____ .
   a. welding near an open window
   b. welding over wooden floors
   c. using cardboard to control welding/grinding sparks
   d. writing HOT on hot metal before leaving it unattended

8. A hot work permit ____ .
   a. authorizes the performance of work potentially posing a fire hazard
   b. promotes development of standard fire safety guidelines
   c. records unsafe conditions at a job site
   d. helps the manager keep records of hazardous spaces

9. Cutting operations should never be performed without which of these items in the area?
   a. A bucket of sand
   b. A bucket of water
   c. A fire watch
   d. A fire hose

10. A confined space is one that ____ .
    a. has a flammable atmosphere
    b. has unrestricted means of entry or exit
    c. is designed for continuous employee occupancy
    d. is large enough that a worker can enter and work

11. An atmosphere is considered oxygen deficient when the oxygen level is below ____ .
    a. 19.5 percent
    b. 20 percent
    c. 21.5 percent
    d. 23.5 percent

12. The potential for electric shock ____ .
    a. decreases when the skin is damp
    b. remains the same when the skin is damp
    c. increases when the skin is damp
    d. decreases when the skin is cut

13. A potentially toxic coating that is used in galvanized metal is ____ .
    a. zinc
    b. chromium
    c. barium
    d. manganese

14. To make sure a respirator provides proper protection, what must be done each time it is worn?
    a. Only a positive fit check must be performed.
    b. Positive and negative fit checks must be performed.
    c. Only a negative fit check must be performed.
    d. The head and neck straps must be tightened as much as possible

15. An SDS is a form used to ____ .
    a. file a worker's compensation claim with an insurance company
    b. list the contents, hazards, and precautions that pertain to a chemical or material
    c. record unsafe conditions that exist at a job site
    d. recommend changes to an employer safety program

Trade Terms Quiz

Fill in the blank with the correct term that you learned from your study of this module.

1. An electric circuit that is connected to the earth is said to be _______ .
2. Helium or argon is used as a(n) _______ in some welding and cutting operations.
3. The cloud of gases and other materials that rises during the welding process is called the _______ .
4. An irregular heart rhythm characterized by a fluttering heartbeat is called _______.
5. A condition that poses an immediate or delayed threat to life or health is referred to as _______ .
6. When metal parts are joined together to form an electrically conductive path, they are said to be _______ .
7. When skin is exposed to intense ultraviolet light, it can cause a(n) _______ .
8. When carbon is added to steel to increase hardness, the result is _______ .
9. Steel dipped in zinc is referred to as _______ .
10. If your eyes are exposed to intense ultraviolet light, the injury is referred to as a(n) _______ .
11. When nitrogen is used to drive oxygen away from a weld site, it is called a(n) _______ .
12. Chromium, manganese, and other materials are added to steel to produce _______ .
13. The flow of an electrical current across an air gap is called a(n) _______ .
14. The point from which a welding arc is produced is called a(n) _______ .
15. A(n) _______ is a material used to dissolve or prevent the formation of oxides and other undesirable substances on a weld joint.
16. The invisible rays produced by arc welding that can cause burns to the eye are _______ .

Trade Terms

Arc burn
Bonded
Carbon steel
Electric arc
Electrically grounded
Electrode
Flash burn
Flux
Fume plume
Galvanized steel
Immediate danger to life and health (IDLH)
Purge gas
Shielding gas
Stainless steel
Ultraviolet (UV) radiation
Ventricular fibrillation

Cornerstone of Craftsmanship

Bill D. Cherry

Zachry Industrial, Inc.
Manager—Weld Testing

How did you choose a career in the construction industry?
I liked looking at a finished product that resulted from my efforts and knowing that it made an impact on people's lives. Most, if not all, of the structures I welded are still in use today. This gives me a strong sense of accomplishment.

Who inspired you to enter the industry?
I was inspired by the father of a school friend who owned a private construction company.

How important are education and training in construction?
They help me maintain my current position and also to teach the trades I have experience with.

What kinds of work have you done in your career?
After graduating from high school, I started out in the road-boring business. This is where I learned how to weld. My next job was at a coal-fired power plant in San Antonio, Texas where I was exposed to many different welding processes and procedures. After that, I was given the opportunity to learn welding inspection and training. I was able to achieve Certified Welding Instructor (CWI), Certified Welding Educator (CWE), and NCCER Master Welding Instructor certifications. I have also trained students to become welders since that time.

Tell us about your present job.
I primarily test welders for our construction and maintenance jobsites. I also have been training welders for 25 years.

What factors have contributed most to your success?
I attribute my success to one person, who gave me the opportunity to advance my career. Without that one individual, I might still be in an unfulfilling career. One person can make a big difference in your life. I started as a welding helper at Zachary Industrial. I eventually became a certified welder, and then a welding inspector, and now manager of weld testing.

What types of training have you been through?
The majority of my welding skills were self-taught. I did receive formal training to become a Certified Welding Instructor (CWI) and have been through the NCCER Instructor Certification Training Program (ICTP). Of course, on-the-job training also played a major role.

Would you suggest construction as a career to others? Why?
Absolutely. Obviously it is not for everyone, but it can be a very rewarding and lucrative career.

What advice would you give to those new to the field?
My advice to those new to the field is never stop learning and always strive to better yourself. Never just sit back and be content with what you are currently doing. The opportunities are always there. You have to actively pursue them.

Tell us an interesting career-related fact or accomplishment.
I hold the only known certification from NCCER as a Master Welding Instructor. Also, I have done many different types of construction. I have administered more than 10,000 weld tests to date. Another thing that means a lot to me is having former students stop by my shop just to say hello and let me know how well they are doing in the welding field.

How do you define craftsmanship?
I define it as the ability to do quality work the first time around.

Trade Terms Introduced in This Module

Arc burn: Burn to the skin produced by brief exposure to intense radiant heat and ultraviolet light.

Bonded: The permanent joining of metallic parts to form an electrically conductive path that will assure electrical continuity and the capacity to safely conduct any current likely to be imposed on it.

Carbon steel: An alloy of iron combining iron and usually less than 1 percent carbon to provide hardness.

Electric arc: The flow of an electrical current across an air gap or gaseous space.

Electrically grounded: Connected to Earth or to some conducting body that serves in place of the earth.

Electrode: The point from which a welding arc is produced.

Flash burn: Burns to the eyes sometimes called welder's flash; caused by exposure to intense radiant heat and ultraviolet light.

Flux: A material used to dissolve or prevent the formation of oxides and other undesirable substances on a weld joint.

Fume plume: The fumes, gases, and particles from the consumables, base metal, and base metal coating during the welding process.

Galvanized steel: Carbon steel dipped in zinc to inhibit corrosion.

Immediate danger to life and health (IDLH): As defined by OSHA, this term refers to an atmosphere that poses an immediate threat to life, would cause irreversible adverse health effects, or would impair an individual's ability to escape from a dangerous atmosphere.

Purge gas: An inert gas, such as nitrogen, used to drive oxygen away from a weld site.

Shielding gas: A gas such as argon, helium, or carbon dioxide used to protect the welding electrode wire from contamination in GMAW and FCAW welding.

Stainless steel: An iron-based alloy usually containing at least 11 percent chromium.

Ultraviolet (UV) radiation: Invisible rays capable of causing burns. UV rays from the sun are the causes of sunburn.

Ventricular fibrillation: Irregular heart rhythm characterized by a rapid fluttering heartbeat.

Additional Resources

This module presents thorough resources for task training. The following resource material is suggested for further study.

ANSI/AWS Z49.1, Safety in Welding, Cutting, and Allied Processes. Miami, FL: American Welding Society.

Arc Welding Safety E205. Latest Edition. Cleveland, OH: Lincoln Electric.

Personal Protective Equipment (PPE) for Welding and Cutting. Miami, FL: American Welding Society.

Safety Topics in Welding, Cutting, and Brazing. **www.osha.gov**.

AWS F32M/F32, Ventilation Guide for Weld Fumes. Latest Edition. Miami, FL: American Welding Society.

Figure Credits

Justin Poland, Robins and Morton, Module Opener, Figure 1A

The Lincoln Electric Company, Cleveland, OH, USA, Figures 1B, 1C, 2A, 2B, 4, 5, 6A–6C, 18–21, 29, SA02

Zachry Industrial, Inc., Figure 2C

Topaz Publications, Inc., Figures 6D–6F, 7, 8, 9A–9E, 10, 12

John Tillman Company, SA01

AWS F2.2:2001, Lens Shade Selector, reproduced with permission from the American Welding Society (AWS), Miami, FL, USA, Table 1

Sellstrom Manfacturing, Figures 9F, 11

North Safety Products USA, Figure 22A

MSA The Safety Company, Figures 22B, 23–25

Photo courtesy of Justrite Mfg. Co. LLC, Figure 27

Weld-On Adhesives, Inc., a division of IPS Corporation, Figure 28

Section Review Answer Key

Answer	Section Reference	Objective
Section One		
1. b	1.1.0	1a
2. a	1.2.2	1b
Section Two		
1. d	2.1.1	2a
2. a	2.2.1	2b
Section Three		
1. a	3.1.0	3a
2. b	3.2.0	3b
3. a	3.3.0	3c
4. d	3.4.0	3d
5. c	3.5.1	3e
6. c	3.6.0	3f
7. b	3.7.0	3g

NCCER CURRICULA — USER UPDATE

NCCER makes every effort to keep its textbooks up-to-date and free of technical errors. We appreciate your help in this process. If you find an error, a typographical mistake, or an inaccuracy in NCCER's curricula, please fill out this form (or a photocopy), or complete the online form at **www.nccer.org/olf**. Be sure to include the exact module ID number, page number, a detailed description, and your recommended correction. Your input will be brought to the attention of the Authoring Team. Thank you for your assistance.

Instructors – If you have an idea for improving this textbook, or have found that additional materials were necessary to teach this module effectively, please let us know so that we may present your suggestions to the Authoring Team.

NCCER Product Development and Revision

13614 Progress Blvd., Alachua, FL 32615

Email: curriculum@nccer.org

Online: www.nccer.org/olf

❑ Trainee Guide ❑ Lesson Plans ❑ Exam ❑ PowerPoints Other ____________________

Craft / Level: Copyright Date:

Module ID Number / Title:

Section Number(s):

Description:

Recommended Correction:

Your Name:

Address:

Email: Phone:

Oxyfuel Cutting

Overview

Oxyfuel cutting is a method for cutting metal that uses an intense flame produced by burning a mixture of a fuel gas and pure oxygen. It is a versatile metal cutting method that has many uses on job sites. Because of the flammable gases and open flame involved, there is a danger of fire and explosion when oxyfuel equipment is used. However, these risks can be minimized when the operator is well-trained and knowledgeable about the function and operation of each part of an oxyfuel cutting outfit.

Module 29102

Trainees with successful module completions may be eligible for credentialing through the NCCER Registry. To learn more, go to **www.nccer.org** or contact us at **1.888.622.3720**. Our website has information on the latest product releases and training, as well as online versions of our *Cornerstone* magazine and Pearson's product catalog.

Your feedback is welcome. You may email your comments to **curriculum@nccer.org**, send general comments and inquiries to **info@nccer.org**, or fill in the User Update form at the back of this module.

This information is general in nature and intended for training purposes only. Actual performance of activities described in this manual requires compliance with all applicable operating, service, maintenance, and safety procedures under the direction of qualified personnel. References in this manual to patented or proprietary devices do not constitute a recommendation of their use.

29102 V5

29102

OXYFUEL CUTTING

Objectives

When you have completed this module, you will be able to do the following:

1. Describe oxyfuel cutting and identify related safe work practices.
   a. Describe basic oxyfuel cutting.
   b. Identify safe work practices related to oxyfuel cutting.
2. Identify and describe oxyfuel cutting equipment and consumables.
   a. Identify and describe various gases and cylinders used for oxyfuel cutting.
   b. Identify and describe hoses and various types of regulators.
   c. Identify and describe cutting torches and tips.
   d. Identify and describe other miscellaneous oxyfuel cutting accessories.
   e. Identify and describe specialized cutting equipment.
3. Explain how to set up, light, and shut down oxyfuel equipment.
   a. Explain how to properly prepare a torch set for operation.
   b. Explain how to leak test oxyfuel equipment.
   c. Explain how to light the torch and adjust for the proper flame.
   d. Explain how to properly shut down oxyfuel cutting equipment.
4. Explain how to perform various oxyfuel cutting procedures.
   a. Identify the appearance of both good and inferior cuts and their causes.
   b. Explain how to cut both thick and thin steel.
   c. Explain how to bevel, wash, and gouge.
   d. Explain how to make straight and bevel cuts with portable oxyfuel cutting machines.

Performance Tasks

Under the supervision of your instructor, you should be able to do the following:

1. Set up oxyfuel cutting equipment.
2. Light and adjust an oxyfuel torch.
3. Shut down oxyfuel cutting equipment.
4. Disassemble oxyfuel cutting equipment.
5. Change empty gas cylinders.
6. Cut shapes from various thicknesses of steel, emphasizing:
   - Straight line cutting
   - Square shape cutting
   - Piercing
   - Beveling
   - Cutting slots
7. Perform washing.
8. Perform gouging.
9. Use a track burner to cut straight lines and bevels.

Trade Terms

Backfire
Carburizing flame
Drag lines
Dross
Ferrous metals
Flashback
Gouging
Kerf
Neutral flame
Oxidizing flame
Pierce
Soapstone
Washing

Industry Recognized Credentials

If you are training through an NCCER-accredited sponsor, you may be eligible for credentials from NCCER's Registry. The ID number for this module is 29102. Note that this module may have been used in other NCCER curricula and may apply to other level completions. Contact NCCER's Registry at 888.622.3720 or go to **www.nccer.org** for more information.

Contents

Topics to be presented in this module include:

Contents (continued)

Figures and Tables

Figures and Tables (continued)

SECTION ONE

1.0.0 OXYFUEL CUTTING BASICS

Objective

Describe oxyfuel cutting and identify related safe work practices.

a. Describe basic oxyfuel cutting.
b. Identify safe work practices related to oxyfuel cutting.

Trade Terms

Dross: The material (oxidized and molten metal) that is expelled from the kerf when cutting using a thermal process. It is sometimes called slag.

Ferrous metals: Metals containing iron.

Soapstone: Soft, white stone used to mark metal.

In order to perform oxyfuel cutting in a safe and effective manner, it is critical to understand the basic principles of oxyfuel cutting and be thoroughly familiar with safe work practices associated with the process.

1.1.0 The Oxyfuel Cutting Process

Oxyfuel cutting, also called flame cutting or burning, is a process that uses the flame and oxygen from a cutting torch to cut ferrous metals. The flame is produced by burning a fuel gas mixed with pure oxygen. The flame heats the metal to be cut to the kindling temperature (a cherry-red color); then a stream of high-pressure pure oxygen is directed from the torch at the metal's surface. This causes the metal to instantaneously oxidize or burn away. The cutting process results in oxides that mix with molten iron and produce dross, which is blown from the cut by the jet of cutting oxygen. This oxidation process, which takes place during the cutting operation, is similar to an accelerated rusting process. *Figure 1* shows an operator performing oxyfuel cutting.

The oxyfuel cutting process is usually performed only on ferrous metals such as straight carbon steels, which oxidize rapidly. This process can be used to quickly cut, trim, and shape ferrous metals, including the hardest steel.

Oxyfuel cutting can be used for certain metal alloys, such as stainless steel; however, the process requires higher preheat temperatures (white

29102-15_F01.EPS

Figure 1 Oxyfuel cutting.

heat) and about 20 percent more oxygen for cutting. In addition, sacrificial steel plate or rod may have to be placed on top of the cut to help maintain the burning process. Other methods, such as carbon arc cutting, powder cutting, inert gas cutting, and plasma arc cutting, are much more practical for cutting steel alloys and nonferrous metals. Some of these methods are covered in other modules.

1.2.0 Oxyfuel Safety Summary

Cutting activities present unique hazards depending upon the material being cut and the fuel used to power the equipment. The proper safety equipment and precautions must be used when working with oxyfuel equipment because of the potential danger from the high-pressure flammable gases and high temperatures used. The following is a summary of safety procedures and practices that must be observed while cutting or welding. Keep in mind that this is just a summary. Complete safety coverage is provided in the *Welding Safety* module. Trainees who have not completed that module should do so before continuing.

1.2.1 Protective Clothing and Equipment

Oxyfuel cutting produces intense light and heat. It can also produce flying sparks and toxic fumes. To avoid injury, operators must wear appropriate personal protective equipment (PPE) when performing oxyfuel cutting operations. The following is a list of safety practices related to protective clothing and equipment that operators should follow:

- Always use safety glasses with a full face shield or a helmet (*Figure* 2). The glasses, face shield, or helmet lens must have the proper light-reducing tint for the type of cutting to be performed.
- Wear proper protective leather and/or flame-retardant clothing along with welding gloves that protect from flying sparks and molten metal as well as heat.
- Wear high-top safety shoes or boots. Make sure that the tongue and lace area of the footwear will be covered by a pant leg. If the tongue and lace area is exposed or the footwear must be protected from burn marks, wear leather spats under the pants or chaps and over the top of the footwear.
- Wear a 100-percent cotton cap with no mesh material included in its construction. The bill of the cap points to the rear. If a hard hat is required for the environment, use one that allows the attachment of rear deflector material and a face shield. A hard hat with a rear deflector is generally preferred when working overhead, and may be required by some employers and job sites.

> **WARNING!**
> Do not wear a cap with a button in the middle. The conductive metal button beneath the fabric represents a safety hazard.

- Wear a face shield over safety glasses for cutting. Either the face shield or the lenses of the safety glasses must be an approved shade for the application. A welding hood equipped with a properly tinted lens is also acceptable. A shade 3 to 6 filter is recommended, depending on the thickness of the metal being cut, as required by *ANSI Z49.1*.
- Wear earplugs to protect ear canals from sparks. Wear hearing protection to protect against the consistent sound of the torch.

> **WARNING!**
> Ear protection is essential to protect ears from the noise of the torch. Other personal protective equipment (PPE) must be worn to protect the operator from hot metal and slag.

Figure 2 Personal protective equipment (PPE) for oxyfuel cutting.

- Cutting operations involving materials or coatings containing cadmium, mercury, lead, zinc, chromium, and beryllium result in toxic fumes. For long-term cutting of such materials, always wear an approved full face, supplied-air respirator (SAR) that uses breathing air supplied externally of the work area. For occasional, very short-term exposure to zinc or copper fumes, a high-efficiency particulate arresting (HEPA)-rated or metal-fume filter may be used on a standard respirator.

1.2.2 Fire/Explosion Prevention

Most welding environment fires occur during oxyfuel gas welding or cutting. To minimize fire and explosion hazards, all cutting should be done in designated areas of the shop if possible. These areas should be made safe for cutting operations with concrete floors, arc filter screens, protective drapes, and fire extinguishers. A welding blanket (*Figure 3*) can be used to protect items in the area that would otherwise be damaged. No combustibles should be stored nearby. The work area should be kept neat and clean, and any metal scrap or dross must be cold before disposal.

Operators should be well-trained in the function and operation of each part of an oxyfuel gas welding or cutting station. In addition, it is often required that at least one fire watch be posted with an extinguisher to watch for possible fires.

The following list contains other steps that operators should follow to help prevent fires and explosions.

- Never carry matches or gas-filled lighters. Sparks can cause the matches to ignite or the lighter to explode, resulting in serious injury.
- Always comply with any site requirement for a hot-work permit and/or a fire watch.
- Never use oxygen to blow off clothing. The oxygen can remain trapped in the fabric for a time. If a spark hits the clothing during this time, the clothing can burn rapidly and violently out of control.
- Never release a large amount of oxygen or use oxygen in place of compressed air. Its presence around flammable materials or sparks can cause rapid and uncontrolled combustion. Keep pure oxygen away from oil, grease, and other petroleum products.
- Make sure that any flammable material in the work area is moved or shielded by a fire-resistant covering.

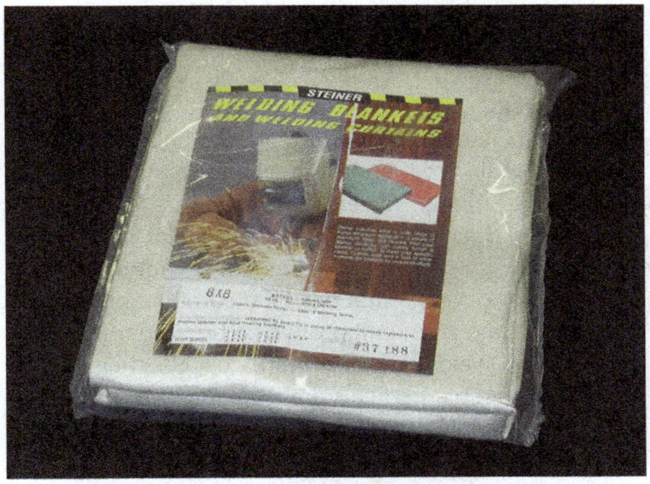

3,000°F (1,649°C) INTERMITTENT, 1,500°F (816°C) CONTINUOUS SILICON DIOXIDE CLOTH

29102-15_F03.EPS

Figure 3 Welding blanket.

- Approved fire extinguishers must be available before attempting any heating, welding, or cutting operations. Make sure the extinguisher is charged, the inspection tag is valid, and any individual that may be required to operate it knows how to do so.
- Never release a large amount of fuel gas, especially acetylene. Propylene and propane tend to concentrate in and along low areas and can ignite at a considerable distance from the release point. Acetylene is lighter than air but is even more dangerous. When mixed with air or oxygen, it will explode at much lower concentrations than any other fuel gas.
- To prevent fires, maintain a neat and clean work area, and make sure that any metal scrap or slag is cold before disposal.

Before cutting containers such as tanks or barrels, check to see if they have contained any explosive, hazardous, or flammable materials, including petroleum products, citrus products, or chemicals that decompose into toxic fumes when heated. Proper procedures for cutting or welding hazardous containers are described in the *American Welding Society (AWS) F4.1, Safe Practices for the Preparation of Containers and Piping for Welding and Cutting,* and *ANSI Z49.1*. As a standard practice, always clean and then fill any tanks or barrels with water, or purge them with a flow of inert gas such as nitrogen to displace any oxygen.

> **WARNING!**
>
> Welding or cutting must never be performed on drums, barrels, tanks, vessels, or other containers until they have been emptied and cleaned thoroughly, eliminating all flammable materials and all substances (such as detergents, solvents, greases, tars, or acids) that might produce flammable, toxic, or explosive vapors when heated. Do not assume that a container that has held combustibles is clean and safe until proven so by proper tests. Do not weld in places where dust or other combustible particles are suspended in air or where explosive vapors are present.

Containers must be cleaned by steam cleaning, flushing with water, or washing with detergent until all traces of the material have been removed.

> **WARNING!**
>
> Clean containers only in well-ventilated areas. Vapors can accumulate during cleaning, causing explosions or injury.

After cleaning the container, fill it with water (*Figure 4*) or a purge gas, such as carbon dioxide, argon, or nitrogen to displace the explosive fumes. Air, which contains oxygen, is displaced from inside the container by the water or inert gas. Without oxygen, combustion cannot take place.

A water-filled vessel is the best alternative. When using water, position the container to minimize the air space. When using an inert gas, provide a vent hole so the inert gas can push the air and other vapors out to the atmosphere. Keep in mind that these precautions do not guarantee the absence of flammable materials inside. For that reason, these types of activities should not be done without adequate supervision and the use of proper testing methods.

NOTE: ANSI Z49.1 AND AWS STANDARDS SHOULD BE FOLLOWED.

29102-15_F04.EPS

Figure 4 Eliminating/minimizing oxygen in a container.

1.2.3 Work Area Ventilation

Cutting operations should always be performed in a well-ventilated area. This greatly reduces the risks associated with toxic fumes. This rule is especially true for confined spaces. Confined-space procedures must be followed before cutting begins. Proper ventilation must be provided before any cutting procedures take place, but never use oxygen for ventilation purposes.

Work area safety rules apply to all workers in the area. In a typical work area, there might be numerous workers performing various tasks. Always remain aware of other personnel in the area and take the necessary precautions to ensure their safety as well as your own.

1.2.4 Cylinder Handling and Storage

Operators must be aware of the hazards involved in the use of fuel gas, oxygen, and shielding gas cylinders and know how to store these cylinders are stored safely. One basic rule is that only compressed gas cylinders containing the correct gas for the process should be used. Regulators must be correct for the gas and pressure and must function properly. All hoses, fittings, and other parts must be suitable and maintained in good condition.

Any cylinder that leaks, bad valves, or damaged threads must be identified and reported to the supplier. Use a piece of soapstone or a tag to write the problem on the cylinder. If closing the cylinder valve cannot stop the leak, move the cylinder outdoors to a safe location, away from any source of ignition, and notify the supplier. Post a warning sign and then slowly release the pressure.

In its gaseous form, acetylene is extremely unstable and explodes easily. For this reason it must remain at pressures below 15 pounds per square inch (psi) or 103 kilopascals (kPa). If an acetylene cylinder is tipped over, stand it upright and wait at least an hour before using it. If the cylinder has spent a significant period of time laying on its side,

> **Oxygen Consumption**
>
> Two-thirds of the oxygen in a neutral flame comes from the air you breathe inside the work space. This could be an issue in a heavily occupied, tight work space.

it is best to allow at least two hours to ensure all of the liquid has drained down to the base of the cylinder. Acetylene cylinders contain liquid acetone. If the liquid is withdrawn from a cylinder, it will foul the safety check valves and regulators and decrease the stability of the acetylene stored in the cylinder. For this reason, acetylene must never be withdrawn at a per-hour rate that exceeds one-seventh of the volume of the cylinder(s) in use. Acetylene cylinders in use should be opened no more than one and one-half turns and, preferably, no more than three-fourths of a turn.

Other precautions associated with gas cylinders include the following:

- Keep cylinders in the upright position and securely chained to an undercarriage or fixed support so that they cannot be knocked over accidentally. Even though they are more stable, cylinders attached to a gas distribution manifold should be chained or otherwise confined, as should cylinders stored in a special room used only for cylinder storage.
- Keep cylinders that contain combustible gases in one area of the building for safety. Cylinders must be at a safe distance from arc welding or cutting operations and any other source of heat, sparks, or flame.
- Cylinder storage areas must be located away from halls, stairwells, and exits so that in case of an emergency they will not block an escape route. Storage areas should also be located away from heat, radiators, furnaces, and welding sparks. The location of storage areas should be where unauthorized people cannot tamper with the cylinders. A warning sign that reads "Danger—No Smoking, Matches, or Open Lights", or similar wording, should be posted in the storage area.
- Oxygen and fuel gas cylinders or other flammable materials must be stored separately. The storage areas must be separated by 20 feet (6.1 m) or by a wall 5 feet (1.5 m) high with at least a 30-minute burn rating. The purpose of the distance or wall is to keep the heat of a small fire from causing the oxygen cylinder safety valve to release. If the safety valve releases the oxygen, a small fire could quickly become an inferno.
- Inert gas cylinders may be stored separately or with either fuel cylinders or oxygen cylinders. Empty cylinders must be stored separately from full cylinders, although they may be stored in the same room or area. All cylinders must be stored vertically and have the protective caps screwed on firmly.
- Never allow a welding electrode, electrode holder, or any other electrically energized parts to come in contact with the cylinder.
- When opening a cylinder valve, operators should stand with the valve stem between themselves and the regulator.
- If a cylinder is frozen to the ground, use warm water (not boiling) to loosen it.

Cylinders equipped with a valve protection cap must have the cap in place unless the cylinder is in use. The protection cap prevents the valve from being broken off if the cylinder is knocked over. If the valve of a full high-pressure cylinder (such as argon or oxygen) is broken off, the cylinder can take flight like a missile if it has not been secured properly. Never lift a cylinder by the safety cap or valve. The valve can easily break off or be damaged. When moving cylinders, the valve protection cap must be replaced, especially if the cylinders are mounted on a truck or trailer. Cylinders must never be dropped or handled roughly.

WARNING!

Using a wrench inserted through the cap can open the valve. If the cap is stuck, use a strap wrench to remove, or call the gas supplier.

Additional Resources

ANSI Z49.1, Safety in Welding, Cutting, and Allied Processes. Miami, FL: American Welding Society.

Uniweld Products, Inc. Numerous videos are available at **www.uniweld.com/en/uniweld-videos**. Last accessed: November 30, 2014.

The Harris Products Group, a division of Lincoln Electric. Numerous videos are available at **www.harrisproductsgroup.com/en/Expert-Advice/videos.aspx**. Last accessed: November 30, 2014.

1.0.0 Section Review

1. The oxyfuel cutting process creates oxides that mix with molten iron and produce dross, which is blown from the cut by a jet of ____ .
   a. pure nitrogen
   b. pure oxygen
   c. carbon dioxide
   d. cooling water

2. Carbon dioxide, argon, and nitrogen are all gases that are suitable for ____ .
   a. mixing with a fuel gas to enable oxyfuel cutting
   b. oxidizing dross to create reusable slag
   c. filling and pressurizing confined spaces
   d. purging explosive fumes from containers

Section Two

2.0.0 Oxyfuel Cutting Consumables and Equipment

Objective

Identify and describe oxyfuel cutting equipment and consumables.

a. Identify and describe various gases and cylinders used for oxyfuel cutting.
b. Identify and describe hoses and various types of regulators.
c. Identify and describe cutting torches and tips.
d. Identify and describe other miscellaneous oxyfuel cutting accessories.
e. Identify and describe specialized cutting equipment.

Trade Terms

Backfire: A loud snap or pop as a torch flame is extinguished.

Flashback: The flame burning back into the tip, torch, hose, or regulator, causing a high-pitched whistling or hissing sound.

Gouging: The process of cutting a groove into a surface.

Kerf: The gap produced by a cutting process.

Washing: A term used to describe the process of cutting out bolts, rivets, previously welded pieces, or other projections from the metal surface.

The equipment used to perform oxyfuel cutting includes oxygen and fuel gas cylinders, oxygen and fuel gas regulators, hoses, and a cutting torch. A typical movable oxyfuel (oxy-acetylene) cutting outfit is shown in *Figure 5*.

2.1.0 Cutting Gases

Many oxyfuel cutting outfits use oxygen and acetylene. However, other fuel gases including natural gas and liquefied gases are also used with oxygen for cutting. This section examines some of the more common cutting gases.

Figure 5 Typical oxyfuel welding/cutting outfit.

2.1.1 Oxygen

Oxygen (O_2) is a colorless, odorless, tasteless gas that supports combustion. It is not considered a fuel gas, but is necessary for combustion. Combined with burning material, pure oxygen causes a fire to flare and burn out of control. When mixed with fuel gases, oxygen produces the high-temperature flame required to flame-cut metals.

Oxygen is stored at more than 2,000 psi (13,790 kPa) in hollow steel cylinders. The cylinders come in a variety of sizes based on different international standards. Specific details about the cylinders are typically regulated by the government agency for the country in which the cylinders are transported, such as the US Department of Transportation (DOT), Transport Canada (TC), and the Department for Transport (DfT) in Europe. *Figure 6* shows high-pressure oxygen cylinder markings and capacities in cubic feet based on US DOT specifications. The smallest standard cylinder holds about 85 cubic feet (2.4 cubic meters) of oxygen, and the largest ultra-high-pressure cylinder holds about 485 cubic feet (13.7 cubic meters). The most common oxygen cylinder size used for cutting operations is the 227-cubic foot (6.4-cubic meter) cylinder. It is more than 4' (1.2 m) tall and 9" (10.2 cm) in diameter. Regardless of their size and capacity, oxygen cylinders must be tested every 10 years in the United States. Other locations may have a different testing standard.

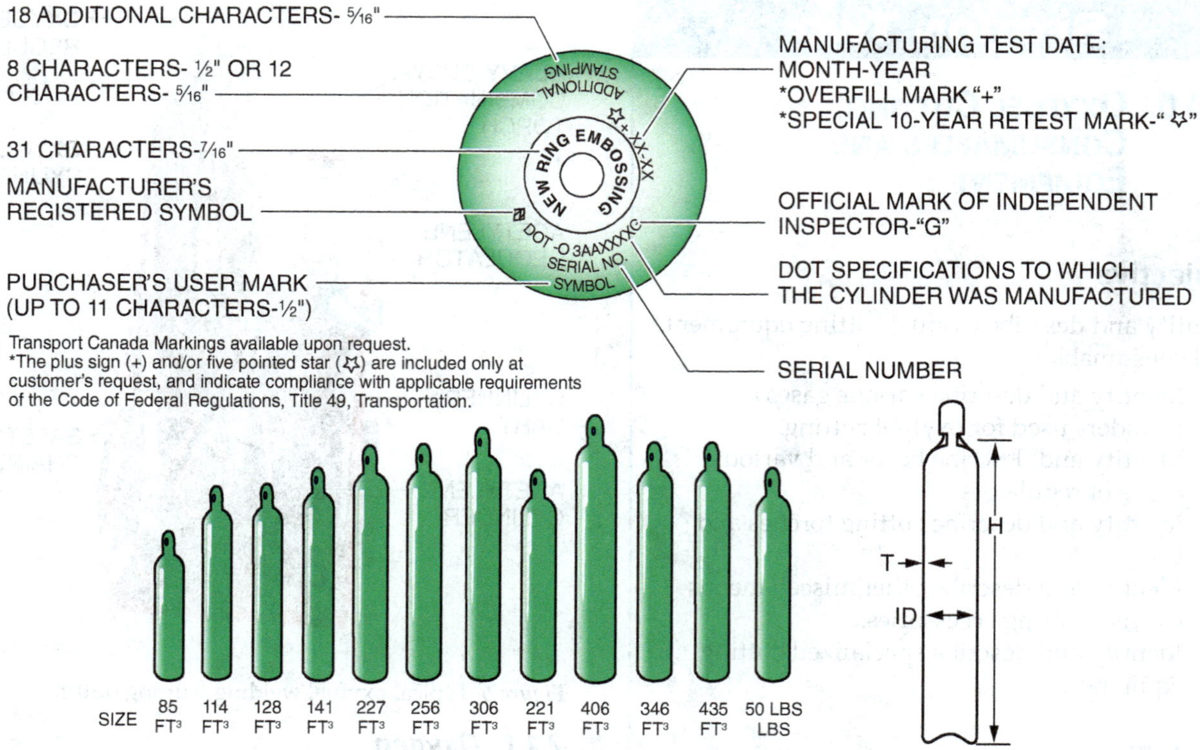

HIGH PRESSURE CYLINDER MARKINGS

DOT SPECIFICATIONS	O_2 CAPACITY (FT^3) AT RATED SERVICE PRESSURE	O_2 CAPACITY (FT^3) AT 10% OVERCHARGE	WATER CAPACITY (IN^3) MINIMUM	WATER CAPACITY (IN^3) MAXIMUM	NOMINAL DIMENSIONS (IN) AVG. INSIDE DIAMETER "ID"	NOMINAL DIMENSIONS (IN) HEIGHT "H"	NOMINAL DIMENSIONS (IN) MINIMUM WALL "T"	NOMINAL WEIGHT (LB)	PRESSURE (PSI) SERVICE	PRESSURE (PSI) TEST
STANDARD HIGH PRESSURE CYLINDERS[1]										
3AA2015	85	93	960	1040	6.625	32.50	0.144	48	2015	3360
3AA2015	114	125	1320	1355	6.625	43.00	0.144	61	2015	3360
3AA2265	128	140	1320	1355	6.625	43.00	0.162	62	2265	3775
3AA2015	141	155	1630	1690	7.000	46.00	0.150	70	2015	3360
3AA2015	227	250	2640	2710	8.625	51.00	0.184	116	2015	3360
3AA2265	256	281	2640	2710	8.625	51.00	0.208	117	2265	3775
3AA2400	306	336	2995	3060	8.813	55.00	0.226	140	2400	4000
3AA2400	405	444	3960	4040	10.060	56.00	0.258	181	2400	4000
ULTRALIGHT® HIGH PRESSURE CYLINDERS[1]										
E-9370-3280	365	NA	2640	2710	8.625	51.00	0.211	122	3280	4920
E-9370-3330	442	NA	3181	3220	8.813	57.50	0.219	147	3330	4995
ULTRA HIGH PRESSURE CYLINDERS[2]										
3AA3600	347[3]	374	2640	2690	8.500	51.00	0.336	170	3600	6000
3AA6000	434[3]	458	2285	2360	8.147	51.00	0.568	267	6000	10000
E-10869-4500	435[3]	NA	2750	2890	8.813	51.00	0.260	148	4500	6750
E-10869-4500	485[3]	NA	3058	3210	8.813	56.00	0.260	158	4500	6750

1. Regulators normally permit filling these cylinders with 10% overcharge, provided certain other requirements are met.
2. Under no circumstances are these cylinders to be filled to a pressure exceeding the marked service pressure at 70°F.
3. Nitrogen capacity at 70°F.

All cylinders normally furnished with ¾" NGT internal threads, unless otherwise specified.
Nominal weights include neck ring but exclude valve and cap, add 2 lbs. (.91 kg) for cap and 1½ lb. (.8 kg) for valve.
Cap adds approximately 5 in. (127 mm) to height.
Cylinder capacities are approximately 5 in. (127 mm) to height.
Cylinder capacities are approximately at 70°F. (21°C).

29102-15_F06.EPS

Figure 6 High-pressure oxygen cylinder markings and sizes.

Oxygen cylinders have bronze cylinder valves on top (*Figure 7*). The cylinder valve controls the flow of oxygen out of the cylinder. A safety plug on the side of the cylinder valve allows oxygen in the cylinder to escape if the pressure in the cylinder rises too high. Although the escaping oxygen presents a hazard, the risk of explosion represents an even more significant hazard. Oxygen cylinders are usually equipped with Compressed Gas Association (CGA) Valve Type 540 valves for service up to 3,000 pounds per square inch gauge (psig), or 20,684 kPa. Some cylinders are equipped with CGA Valve Type 577 valves for up to 4,000 psig (27,579 kPa) oxygen service or CGA Valve Type 701 valves for up to 5,500 psig (34,473 kPa) service. Each CGA valve type is for a specific type of gas and pressure rating. Use care when handling oxygen cylinders because oxygen is stored at such high pressures. When it is not in use, always cover the cylinder valve with the protective steel safety cap and tighten it securely (*Figure 8*).

WARNING!

Do not remove the protective cap unless the cylinder is secured. If the cylinder falls over and the valve assembly breaks off, the cylinder will be propelled like a rocket, causing severe injury or death to anyone in its path.

2.1.2 Acetylene

Acetylene gas (C_2H_2), a compound of carbon and hydrogen, is lighter than air. It is formed by dissolving calcium carbide in water. It has a strong, distinctive, garlic-like odor, which is added to the gas intentionally so that it can be detected. In its

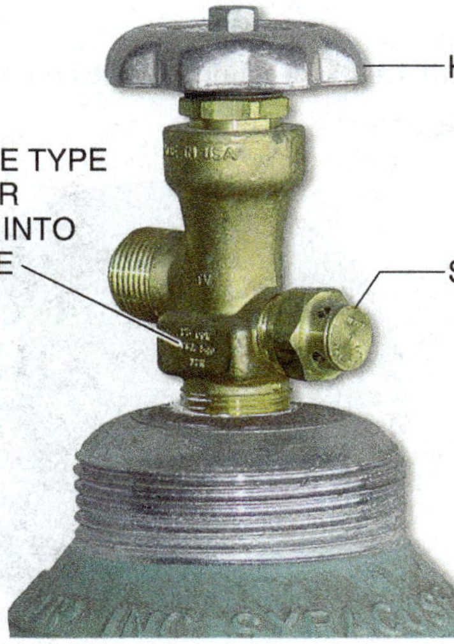

Figure 7 Oxygen cylinder valve.

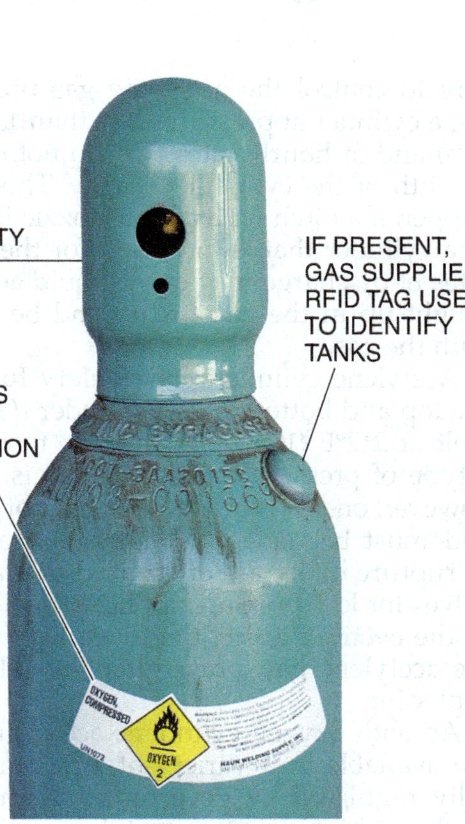

Figure 8 Oxygen cylinder with standard safety cap.

gaseous form, acetylene is extremely unstable and explodes easily. Because of this instability, it cannot be compressed at pressures of more than 15 psi (103 kPa) when in its gaseous form. At higher pressures, acetylene gas breaks down chemically, producing heat and pressure that could result in a violent explosion. When combined with oxygen, acetylene creates a flame that burns hotter than 5,500°F (3,037°C), one of the hottest gas flames. Acetylene can be used for flame cutting, welding, heating, flame hardening, and stress relieving.

Because of the explosive nature of acetylene gas, it cannot be stored above 15 psi (103 kPa) in a hollow cylinder. To solve this problem, acetylene cylinders are specially constructed to store acetylene. The acetylene cylinder is filled with a porous material that creates a solid cylinder, instead of a hollow cylinder as used for all other common gases. The porous material is soaked with liquid acetone, which absorbs the acetylene gas, stabilizing it and allowing for storage at pressures above 15 psi (103 kPa).

Because of the liquid acetone inside the cylinder, acetylene cylinders must always be used in an upright position. If the cylinder is tipped over, stand the cylinder upright and wait at least one hour before using it. If liquid acetone is withdrawn from a cylinder, it will foul the safety check valves and regulators. It will also cause extremely unstable torch operation. Always take care to control the acetylene gas pressure leaving a cylinder at pressures less than 15 psig (103.4 kPa) and at hourly rates that do not exceed one-seventh of the cylinder capacity. This can easily happen if a torch with a large nozzle is connected to a cylinder that is too small for the task. High rates of discharge may cause liquid acetone to be caught up in the gas stream and be carried out with the gas.

Acetylene cylinders have safety fuse plugs in the top and bottom of the cylinder (*Figure 9*) that melt at 212°F (100°C). A fuse, or fusible, plug is a type of pressure relief device. It is not a valve however; once it is activated, it cannot be reclosed and must be replaced. Fuse plugs, also known as rupture disks, are often used in place of relief valves for low-pressure applications like this one. In the event of a fire, the fuse plugs will release the acetylene gas, preventing the cylinder from exploding.

As with oxygen cylinders, acetylene cylinders are available in a variety of sizes and are typically regulated by government agencies. *Figure 10* shows high-pressure acetylene cylinder markings and capacities in cubic feet based on US DOT specifications. The smallest standard cylinder holds about 10 cubic feet (0.28 cubic meters) of gas. The largest standard cylinder holds about 420 cubic feet (11.9 cubic meters) of gas. A cylinder that holds about 850 cubic feet (24.1 cubic meters) is also available. Like oxygen cylinders, acetylene cylinders used in the United States must be tested every 10 years.

Acetylene cylinders are usually equipped with a CGA 510 brass cylinder valve. The valve controls the flow of acetylene from the cylinder into a regulator. Some acetylene cylinders are equipped with an alternate CGA 300 valve. Some obsolete valves still in use require a special long-handled wrench with a square socket end to operate the valve.

Figure 9 Standard acetylene cylinder valve and fuse plugs.

Lifetime Cylinder Management

Gas cylinders may be fitted with radio-frequency identification (RFID) tags so that they can be readily identified and tracked. The RFID tag is electronically scanned and a coded number is matched against the records for identification and tracking. This aids in quickly determining the identity of the purchaser or user of the cylinder, where it has been, and for determining the testing and maintenance records of the cylinder. The RFID tag may appear to be a button like the one shown here. However, it can also be concealed in the walls of the cylinder neck and covered for protection, or have a tight-fitting collar that snaps around the cylinder valve neck.

Alternate High-Pressure Cylinder Valve Cap

High-pressure cylinders can be equipped with a clamshell cap that can be closed to protect the cylinder valve with or without a regulator installed on the valve. This enables safe movement of the cylinder after the cylinder valve is closed. This type of cap is usually secured to the cylinder body cap threads when it is installed so that it cannot be removed. When the clamshell is closed, it can also be padlocked to prevent unauthorized operation of the cylinder valve.

CLAMSHELL OPEN TO ALLOW CYLINDER VALVE OPERATION

CLAMSHELL CLOSED FOR MOVEMENT OR PADLOCKED TO PREVENT OPERATION OF CYLINDER VALVE

CLAMSHELL CLOSED FOR TRANSPORT

29102-15_SA02.EPS

Alternate Acetylene Cylinder Safety Cap

Acetylene cylinders can be equipped with a ring guard cap that protects the cylinder valve with or without a regulator installed on the valve. This enables safe movement of the cylinder after the cylinder valve is closed. This type of cap is usually secured to the cylinder body cap threads when it is installed so that it cannot be removed. Some other types of cylinders, such as propane cylinders, may also be fitted with this type of guard.

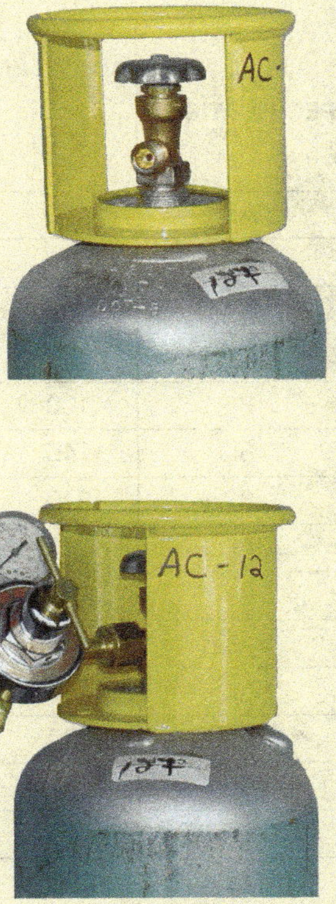

29102-15_SA03.EPS

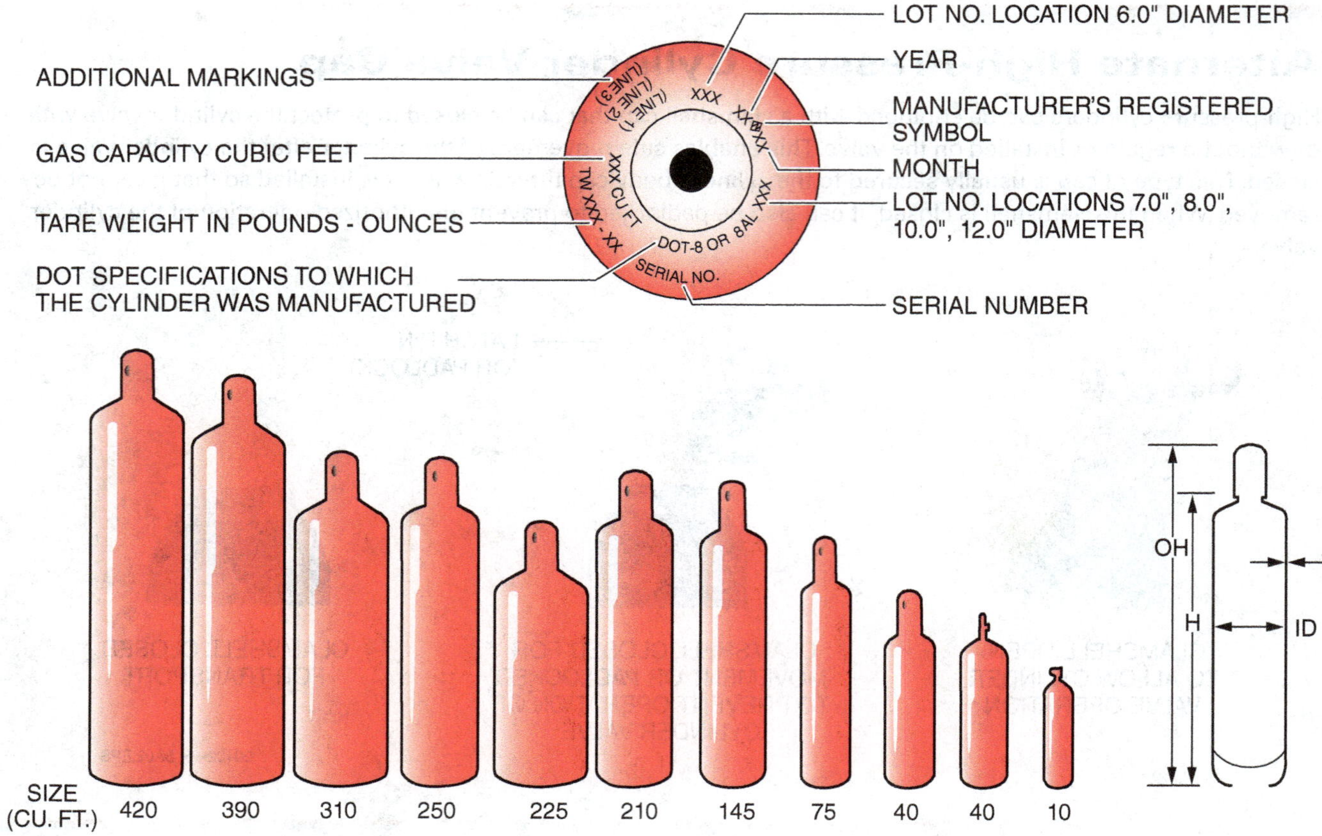

ACETYLENE CYLINDER MARKINGS

DOT SPECIFICATIONS	CAPACITY ACETYLENE (FT³)	CAPACITY MIN. WATER (IN³)	CAPACITY MIN. WATER (LB.)	NOMINAL DIMENSIONS (IN) AVG. INSIDE DIAMETER "ID"	NOMINAL DIMENSIONS (IN) HEIGHT W/OUT VALVE OR CAP "H"	NOMINAL DIMENSIONS (IN) HEIGHT W/VALVE AND CAP "OH"	NOMINAL DIMENSIONS (IN) MINIMUM WALL "T"	ACETONE (LB - OZ)	APPROXIATE TARE WEIGHT WITH VALVE WITHOUT CAP (LB)
8 AL[1]	10	125	4.5	3.83	13.1375	14.75	0.0650	1-6	8
8[1]	40	466	16.8	6.00	19.8000	23.31	0.0870	5-7	25
8[2]	40	466	16.8	6.00	19.8000	28.30	0.0870	5-7	28
8[3]	75	855	30.8	7.00	25.5000	31.25	0.0890	9-8	45
8	100	1055	38.0	7.00	30.7500	36.50	0.0890	12-2	55
8	145	1527	55.0	8.00	34.2500	40.00	0.1020	18-10	76
8	210	2194	79.0	10.00	32.2500	38.00	0.0940	25-13	105
8AL	225	2630	94.7	12.00	27.5000	32.75	0.1280	29-6	110
8	250	2606	93.8	10.00	38.0000	43.75	0.0940	30-12	115
8AL	310	3240	116.7	12.00	32.7500	38.50	0.1120	39-5	140
8AL	390	4151	150.0	12.00	41.0000	46.75	0.1120	49-14	170
8AL	420	4375	157.5	12.00	43.2500	49.00	0.1120	51-14	187
8	60	666	24.0	7.00	25.79 OH		0.0890	7-11	40
8	130	1480	53.3	8.00	36.00 OH		0.1020	17-2	75
8AL	390	4215	151.8	12.00	46.00 OH		0.1120	49-14	180

1. Tapped for 3/8" valve but are not equipped with valve protection caps.
2. Includes valve protection cap.
3. Can be tared to hold 60 ft³ (1.7 m³) of acetylene gas.

Standard tapping (except cylinders tapped for 3/8") 3/4"-14 NGT.

Weight includes saturation gas, filler, paint, solvent, valve, fuse plugs.
Does not include cap of 2 lb. (.91 kg.)
Cylinder capacities are based upon commercially pure acetylene gas at 250 psi (17.5 kg/cm²), and 70°F (15°C).

29102-15_F10.EPS

Figure 10 Acetylene cylinder markings and sizes.

The smallest standard acetylene cylinder, which holds 10 cubic feet (0.28 cubic meters), is equipped with a CGA 200 small series valve, and 40 cubic foot (1.13 cubic meter) cylinders use a CGA 520 small series valve. As with oxygen cylinders, place a protective valve cap on the acetylene cylinders during transport (*Figure 11*).

WARNING!

Do not remove the protective cap unless the cylinder is secured. If the cylinder falls over and the nozzle breaks off, the cylinder will release highly explosive gas.

2.1.3 Liquefied Fuel Gases

Many fuel gases other than acetylene are used for cutting. They include natural gas and liquefied fuel gases such as propylene and propane. Their flames are not as hot as acetylene, but they have higher British thermal unit (Btu) ratings and are cheaper and safer to use. Job site policies typically determine which fuel gas to use. *Table 1* compares the flame temperatures of oxygen mixed with various fuel gases.

Propylene mixtures are hydrocarbon-based gases that are stable and shock-resistant, making them relatively safe to use. They are purchased under trade names such as High Purity Gas (HPG™), Apachi™, and Prestolene™. These gases and others have distinctive odors to make leak detection easier. They burn at temperatures around 5,193°F (2,867°C), hotter than natural gas and propane. Propylene gases are used for flame cutting, scarfing, heating, stress relieving, brazing, and soldering.

Propane is also known as liquefied petroleum (LP) gas. It is stable and shock-resistant, and it has a distinctive odor for easy leak detection. It burns at 4,580°F (2,526°C), which is the lowest temperature of any common fuel gas. Propane has a slight tendency toward backfire and flashback and is used quite extensively for cutting procedures.

Figure 11 Acetylene cylinder with standard valve safety cap.

Table 1 Flame Temperatures of Oxygen With Various Fuel Gases

Type of Gas	Flame Temperature
Acetylene	More than 5,500°F (3,038°C)
Propylene	5,130°F (2,832°C)
Natural gas	4,600°F (2,538°C)
Propane	4,580°F (2,527°C)

Natural gas is delivered by pipeline rather than by cylinders. Manifolds must be available on site for the connection of regulators and hoses. It burns at about 4,600°F (2,537°C). Natural gas is relatively stable and shock-resistant and has a slight tendency toward backfire and flashback. Because of its recognizable odor, leaks are easily detectable. Natural gas is used primarily for cutting on job sites with permanent cutting stations.

Liquefied fuel gases are shipped in hollow steel cylinders (*Figure 12*). When empty, they are much lighter than acetylene cylinders.

The hollow steel cylinders for liquefied fuel gases come in various sizes. They can hold from 30 to 225 pounds (13.6 to 102.1 kilograms) of fuel gas. As the cylinder valve is opened, the

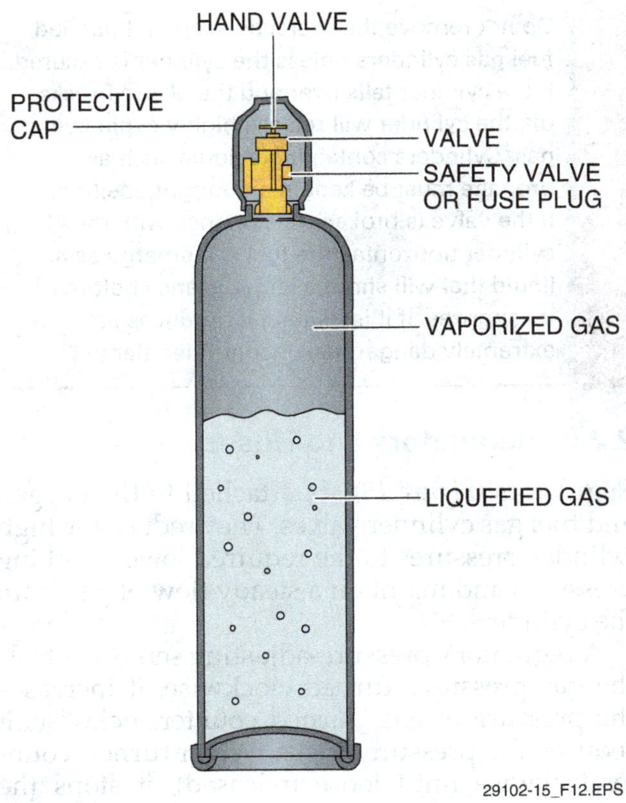

Figure 12 Liquefied fuel gas cylinder.

vaporized gas is withdrawn from the cylinder. The remaining liquefied gas absorbs heat and releases additional vaporized gas. The pressure of the vaporized gas varies with the outside temperature. The colder the outside temperature, the lower the vaporized gas pressure will be. If high volumes of gas are removed from a liquefied fuel gas cylinder, the pressure drops, and the temperature of the cylinder will also drop. A ring of frost can form around the base of the cylinder due to the cooling effect as the liquid vaporizes and absorbs heat. If high withdrawal rates continue, the regulator may also start to ice up. If high withdrawal rates are required, special regulators with electric heaters should be used.

WARNING!

Never apply heat directly to a cylinder or regulator. This can cause excessive pressure, resulting in an explosion.

The pressure inside a liquefied fuel gas cylinder is not an indicator of how full or empty the cylinder is. The weight of a cylinder determines how much liquefied gas is left. Liquefied fuel gas cylinders are equipped with CGA 510, 350, or 695 valves, depending on the fuel and storage pressures.

WARNING!

Do not remove the protective cap on liquefied fuel gas cylinders unless the cylinder is secured. If the cylinder falls over and the nozzle breaks off, the cylinder will release highly explosive gas. Cylinders containing a liquid such as propane must be kept in an upright position. If the valve is broken or is opened with the cylinder horizontal, the fuel can emerge as a liquid that will shoot a long distance before it vaporizes. If it is ignited, it produces an extremely dangerous, uncontrolled flame.

2.2.0 Regulators and Hoses

Regulators (*Figure 13*) are attached to the oxygen and fuel gas cylinder valves. They reduce the high cylinder pressures to the required lower working pressures and maintain a steady flow of gas from the cylinder.

A regulator's pressure-adjusting screw controls the gas pressure. Turned clockwise, it increases the pressure of gas. Turned counterclockwise, it reduces the pressure of gas. When turned counterclockwise until loose (released), it stops the flow of gas. When the adjusting screw feels very loose, it is an indication that the end of the screw is no longer in contact with the regulating spring. However, the regulator adjusting screw should never be considered a shut-off valve. When shut-off is desired, use the cylinder valve. The regulator can remain at its set position, allowing any gas remaining between the cylinder valve and the regulator to escape.

Handling and Storing Liquefied Gas Cylinders

Liquefied fuel gas cylinders have a safety valve built into the valve at the top of the cylinder. The safety valve releases gas if the pressure begins to rise. Use care when handling fuel gas cylinders because the gas in cylinders is stored at significant pressures. Cylinders should never be dropped or hit with heavy objects, and they should always be stored in an upright position. When not in use, the cylinder valve must always be covered with the protective steel cap.

Most regulators contain two gauges. The high-pressure or cylinder-pressure gauge indicates the actual cylinder pressure (upstream); the low-pressure gauge indicates the pressure of the gas leaving the regulator (downstream).

Oxygen regulators differ from fuel gas regulators. Oxygen regulators may have green markings and always have right-hand threads on all connections. The oxygen regulator's high-pressure gauge generally reads up to 3,000 psi (20,684 kPa) and includes a second scale that shows the amount of oxygen in the cylinder in terms of cubic feet. The low-pressure or working-pressure gauge may read 100 psi (689 kPa) or higher.

Fuel gas regulators may have red markings and usually have left-hand threads on all the connections. As a reminder that the regulator has left-hand threads, a V-notch may be cut into the corners of the fitting nut. These notches are visible on the fitting nut of the fuel gas regulator shown in *Figure 13*. The fuel gas regulator's high-pressure gauge usually reads up to 400 psi (2,758 kPa). The low-pressure or working-pressure gauge may read up to 40 psi (276 kPa). Acetylene gauges, however, are always red-lined at 15 psi (103 kPa) as a reminder that acetylene pressure should not be increased beyond that point.

Single-stage and two-stage regulators will be discussed in the following sections.

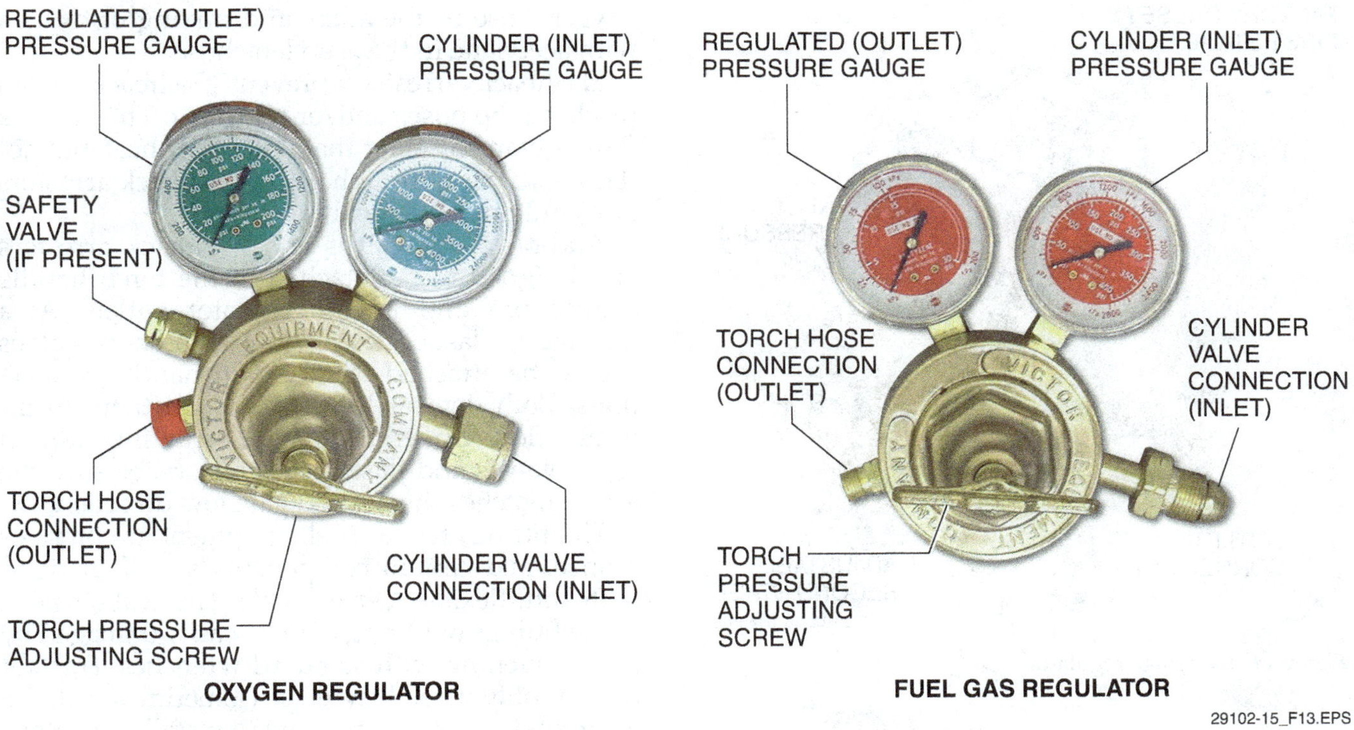

Figure 13 Oxygen and acetylene regulators.

> **WARNING!**
>
> To prevent injury and damage to regulators, always follow these guidelines:
>
> - Never subject regulators to jarring or shaking, as this can damage the equipment beyond repair.
> - Always check that the adjusting screw is fully released before the cylinder valve is turned on and when the welding has been completed.
> - Always open cylinder valves slowly and stand with the valve stem between you and the regulator.
> - Never use oil to lubricate a regulator. This can result in an explosion when the regulator is in use.
> - Never use fuel gas regulators on oxygen cylinders, or oxygen regulators on fuel gas cylinders.
> - Never work with a defective regulator. If it is not working properly, shut off the gas supply and have the regulator repaired by someone who is qualified to work on it.
> - Never use large wrenches, pipe wrenches, pliers, or slip-joint pliers to install or remove regulators.

2.2.1 Single-Stage Regulators

Single-stage, spring-compensated regulators reduce pressure in one step. As gas is drawn from the cylinder, the internal pressure of the cylinder decreases. A single-stage, spring-compensated regulator is unable to automatically adjust for this decrease in internal cylinder pressure. Therefore, it becomes necessary to adjust the spring pressure periodically to modify the output gas pressure as the gas in the cylinder is consumed. These regulators are the most commonly used because of their low cost and high flow rates.

2.2.2 Two-Stage Regulators

The two-stage, pressure-compensated regulator reduces pressure in two steps. It first reduces the input pressure from the cylinder to a predetermined intermediate pressure. This intermediate pressure is then adjusted by the pressure-adjusting screw. With this type of regulator, the delivery pressure to the torch remains constant, and no readjustment is necessary as the gas in the cylinder is consumed. Standard two-stage regulators (*Figure 14*) are more expensive than single-stage regulators and have lower flow rates. There are also heavy-duty types with higher flow rates that are usually preferred for thick material and/or continuous-duty cutting operations.

2.2.3 Check Valves and Flashback Arrestors

Check valves and flashback arrestors (*Figure 15*) are safety devices for regulators, hoses, and torches. Check valves allow gas to flow in one direction only. Flashback arrestors stop fire from being able to travel backwards through the hose.

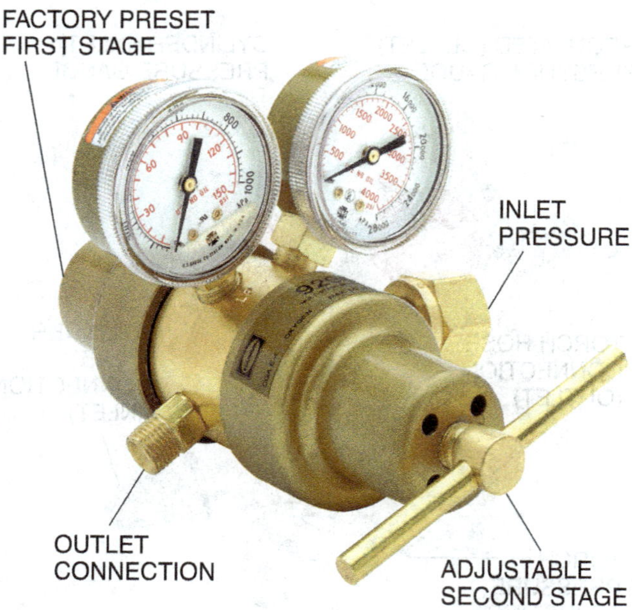

Figure 14 Two-stage regulator.

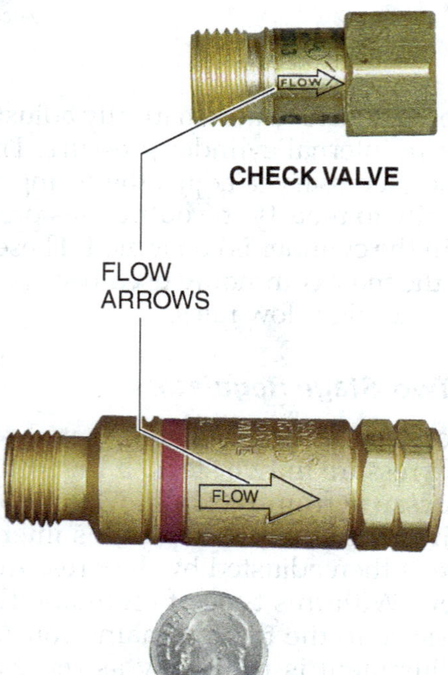

Figure 15 Add-on check valve and flashback arrestor.

Check valves consist of a ball and spring that open inside a cylinder. The valve allows gas to move in one direction but closes if the gas attempts to flow in the opposite direction. When a torch is first pressurized or when it is being shut off, back-pressure check valves prevent the entry and mixing of acetylene with oxygen in the oxygen hose or the entry and mixing of oxygen with acetylene in the acetylene hose.

Flashback arrestors prevent flashbacks from reaching the hoses and/or regulator. They have a flame-retarding filter that will allow heat, but not flames, to pass through. Most flashback arrestors also contain a check valve.

Add-on check valves and flashback arrestors are designed to be attached to the torch handle connections and to the regulator outlets. At a minimum, flashback arrestors with check valves should be attached to the torch handle connections. Both devices have arrows on them to indicate flow direction. When installing add-on check valves and flashback arrestors, be sure the arrow matches the desired gas flow direction.

The fittings for oxyfuel equipment are brass or bronze, and certain components are often fitted with soft, flexible, O-ring seals. The seal surfaces of the fittings or O-rings can be easily damaged by over-tightening with standard wrenches. For that reason, only a torch wrench (sometimes called a gang wrench) should be used to install regulators, hose connections, check valves, flashback arrestors, torches, and torch tips. Longer wrenches that provide more leverage should be avoided.

The universal torch wrench shown in *Figure 16* is equipped with various size wrench cutouts for use with a variety of equipment and standard CGA components. The length of a torch wrench is limited to reduce the chance of damage to fittings because of excessive torque. In some cases,

Serious Cutting

The cutting power of oxyfuel equipment can be very surprising. This 6" (15.2 cm) carbon steel block is no match for a worker with the right torch.

29102-15_SA04.EPS

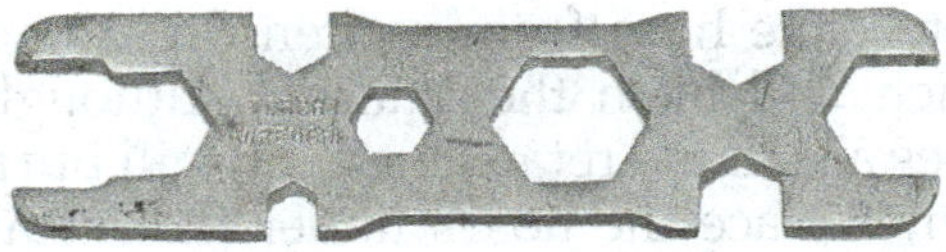

29102-15_F16.EPS

Figure 16 Universal torch wrench.

manufacturers specify only hand-tightening for certain fitting connections of a torch set (tips or cutting/welding attachments, for example). In any event, follow the manufacturer's specific instructions when connecting the components of a torch set.

2.2.4 Gas Distribution Manifolds

In some applications, cutting gases are used on a large scale. Instead of using a cylinder for each operator, a large bank of cylinders connected to a common manifold is used. *Figure 17* is a representation of such an arrangement. High-pressure hoses called pigtails are used to connect the gas supply tanks to the manifold. Regulators are provided at both the source and the workstation hookups.

Figure 18 shows a manifold setup that might be used in a pipe fabrication shop. Operators would tie-in with their hoses at the drops on the manifolds. Each of these manifolds has four drops. Each drop would be provided with a pressure gauge like the one shown. One function of this gauge is to provide an indication if there is a leak

29102-15_F18.EPS

Figure 18 Operator hookups for cutting gases.

in a hose or in the torch. Hose length must be considered when using a manifold distribution system because the pressure drop increases with

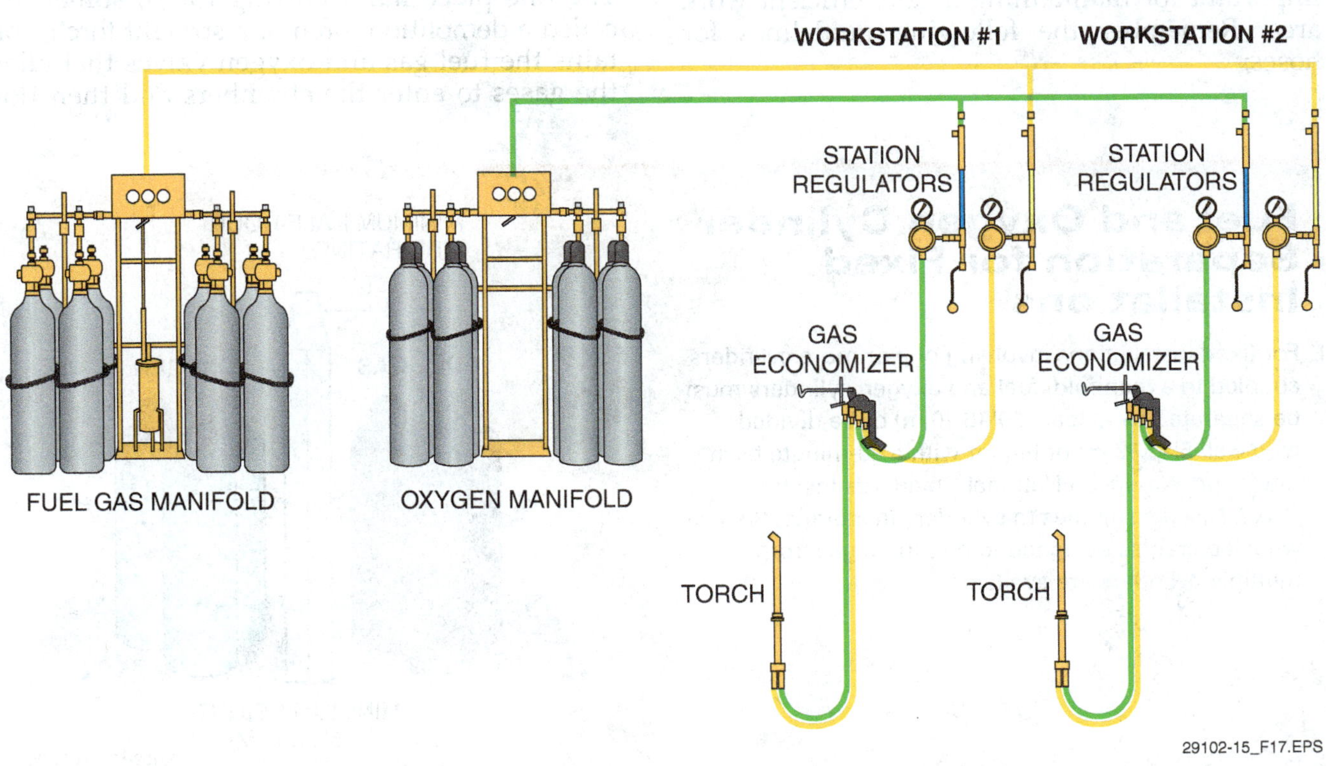

29102-15_F17.EPS

Figure 17 Gas distribution to stations through a manifold.

the length of the hoses. Hoses with a larger inside diameter are needed for long runs.

OSHA provides the following safety precautions specifically for manifold systems:

- Fuel gas and oxygen manifolds must bear the name of the substance they contain.
- Fuel gas and oxygen manifolds must not be placed in confined spaces; they must be placed in safe, well ventilated, and accessible locations.
- Hose connections must be designed so that they cannot be interchanged between fuel gas and oxygen manifolds and supply header connections. Adapters may not be used to interchange hoses.
- Hose connections must be kept free of grease and oil.
- Manifold and header hose connections must be capped when not in use.
- Nothing may be placed on a manifold that will damage the manifold or interfere with the quick closing of the valves.

2.2.5 Hoses

Hoses transport gases from the regulators to the torch. Oxygen hoses are usually green with right-hand threaded connections. Hoses for fuel gas are usually red and have left-hand threaded connections. The fuel gas connection fittings are grooved as a reminder that they have left-hand threads.

Proper care and maintenance of the hose is important for maintaining a safe, efficient work area. Remember the following guidelines for hoses:

- Protect the hose from molten dross or sparks, which will burn the exterior. Although some hoses are flame retardant, they will burn.
- Do not place the hoses under the metal being cut. If the hot metal falls on the hose, the hose will be damaged. Keep hoses as far away from the cutting activity as possible.
- Frequently inspect and replace hoses that show signs of cuts, burns, worn areas, cracks, or damaged fittings. The hoses are tough and durable, but not indestructible.
- Never use pipe-fitting compounds or lubricants around hose connections. These compounds often contain oil or grease, which ignite and burn or explode in the presence of oxygen.

Propane and propylene require hoses designed for the mixture of hydrocarbons present in these fuels. Ensure that any hoses used are appropriate for the fuel gas. Hose and fuel gas providers can help provide the correct hoses.

2.3.0 Torches and Tips

Cutting torches mix oxygen and fuel gas for the torch flame and control the stream of oxygen necessary for the cutting jet. Depending on the job site, either a one-piece or a combination cutting torch may be used.

2.3.1 One-Piece Hand Cutting Torch

The one-piece hand cutting torch, sometimes called a demolition torch or a straight torch, contains the fuel gas and oxygen valves that allow the gases to enter the chambers and then flow

Fuel and Oxygen Cylinder Separation for Fixed Installations

For fixed installations involving one or more cylinders coupled to a manifold, fuel and oxygen cylinders must be separated by at least 20' (6.10 m) or be divided by a wall 5' (1.52 m) or higher with a 30-minute burn rating, per American National Standards Institute *Z49.1.* This also applies to cylinders in storage. Special wheeled cradles designed to distribute gas from multiple cylinders are available.

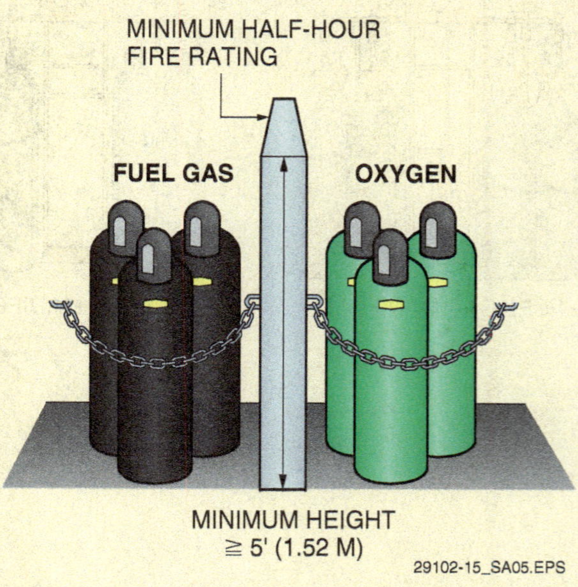

into the part of the torch where they are mixed. The main body of the torch is called the handle. The torch valves control the fuel gas and oxygen flow needed for preheating the metal to be cut. The cutting oxygen lever, which is spring-loaded, controls the jet of cutting oxygen. Hose connections are located at the end of the torch body behind the valves.

Figure 19 shows a three-tube, one-piece hand cutting torch in which the preheat fuel and oxygen are mixed in the tip. These torches are designed for heavy-duty cutting and little else. They have long supply tubes from the torch handle to the torch head to reduce radiated heat to the operator's hands. Cutting torches are generally available with sufficient capacity to cut steel up to 12" (≈30 cm) thick. Larger-capacity torches, with the ability to cut steel up to 36" (≈90 cm) thick, can also be obtained. Torches with this kind of capacity require a significant oxygen and fuel gas supply to perform.

Two different types of oxyfuel cutting torches are in general use. The positive-pressure torch (*Figure 19*) is designed for use with fuel supplied through a regulator from pressurized fuel storage cylinders. The injector torch is designed to use a vacuum created by the oxygen flow to draw the necessary amount of fuel from a very-low-pressure fuel source, such as a natural gas line or acetylene generator. The injector torch, when used, is most often found in continuous-duty, high-volume manufacturing applications. Both types may employ one of two different fuel-mixing methods:

- Torch-handle, or supply-tube, mixing
- Torch-head or tip mixing

The two methods can normally be distinguished by the number of supply tubes from the torch handle to the torch head. Torches that use three tubes from the handle to the head mix the preheat fuel and oxygen at the torch head or tip. This method tends to help eliminate any flashback damage to the torch head supply tubes and torch handle. One tube carries fuel gas to the head. The other two tubes carry oxygen; one carries the oxygen for the preheat flame while the other carries the oxygen for cutting.

The cutting torch with two tubes usually mixes the preheat fuel and oxygen in a mixing chamber in the torch body or in one of the supply tubes (*Figure 20*). Injector torches usually have the injector located in one of the supply tubes, and the mixing occurs in the tube between the injector and the torch head. Some older torches that have only two visible tubes are actually three-tube torches that mix the preheat fuel and oxygen in the torch head or tip. This is accomplished by using a separate preheat fuel tube inside a larger preheat oxygen tube.

2.3.2 Combination Torch

The combination torch consists of a cutting torch attachment that fits onto a welding torch handle. These torches are normally used in light-duty or medium-duty applications. Fuel gas and oxygen valves are on the torch handle. The cutting attachment has a cutting oxygen lever and another oxygen valve to control the preheat flame. When the cutting attachment is screwed onto the torch handle, the torch handle oxygen valve is opened all the way, and the preheat oxygen is controlled by an oxygen valve on the cutting attachment. When the cutting attachment is removed, brazing and heating tips can be screwed onto the torch handle. *Figure 21* shows a two-tube combination torch in which preheat mixing is accomplished in a supply tube. These torches are usually positive-pressure torches with mixing occurring in the attachment body, supply tube, head, or tip. These torches may be equipped with built-in flashback arrestors and check valves.

2.3.3 Cutting Torch Tips

Cutting torch tips, or nozzles, fit into the cutting torch and are either screwed in or secured with a tip nut. There are one- and two-piece cutting tips (*Figure 22*).

One-piece cutting tips are made from a solid piece of copper. Two-piece cutting tips have a separate external sleeve and internal section.

Torch manufacturers supply literature explaining the appropriate torch tips and gas pressures for various applications. *Table 2* shows a sample cutting tip chart that lists recommended tip sizes and gas pressures for use with acetylene fuel gas and a specific manufacturer's torch and tips.

CAUTION

Do not use the cutting tip chart from one manufacturer for the cutting tips of another manufacturer. The gas flow rate of the tips may be different, resulting in excessive flow rates. Different gas pressures may also be required. The cutting torch tip to be used depends on the base metal thickness and fuel gas being used. Special-purpose tips are also available for use in such operations as gouging and grooving.

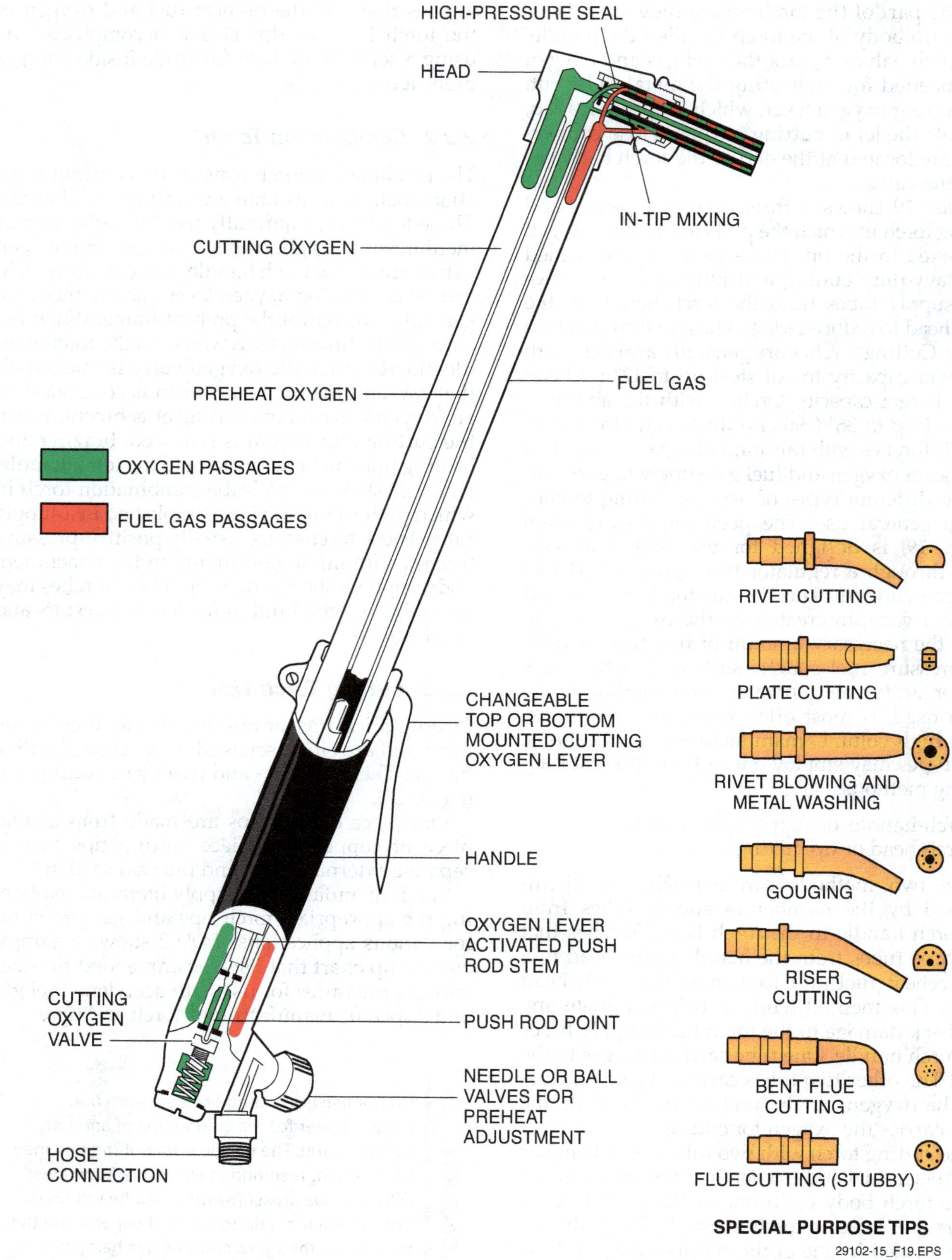

Figure 19 Heavy-duty three-tube one-piece positive-pressure hand cutting torch.

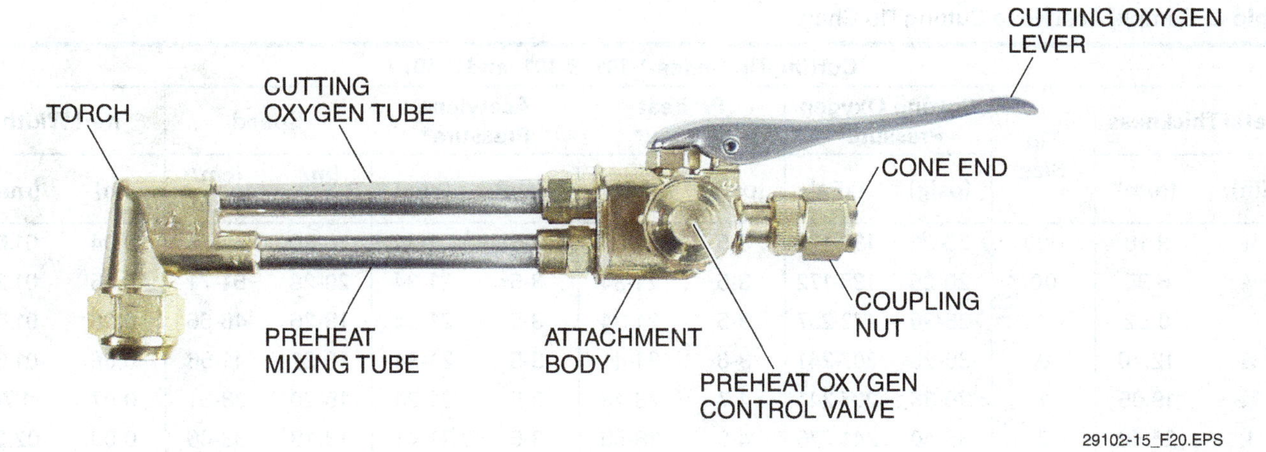

Figure 20 Cutting torch attachment.

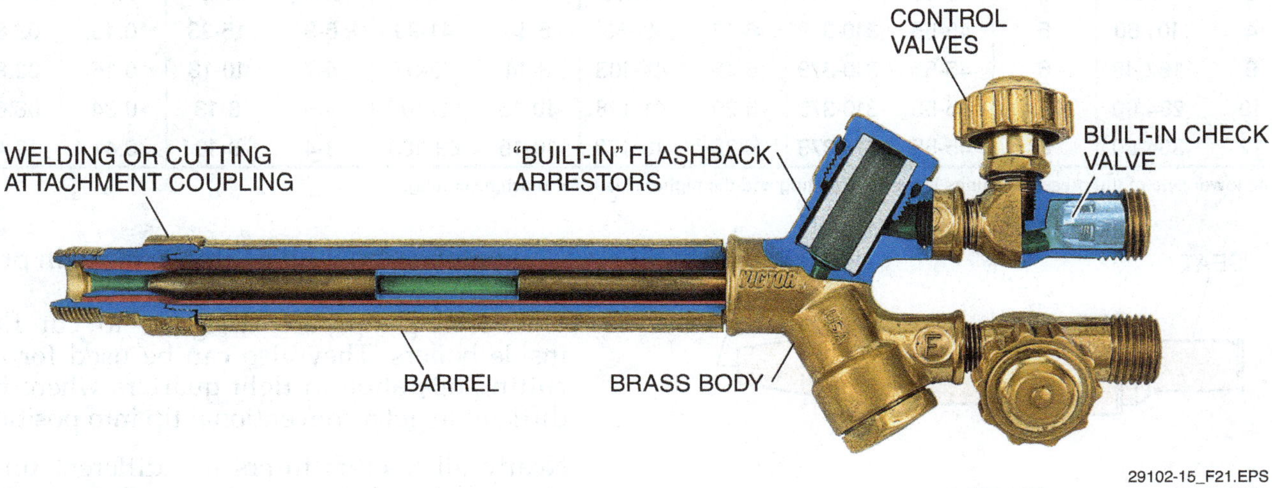

Figure 21 Combination cutting torch

The cutting torch tip to be used depends on the base metal thickness and fuel gas being used. Special-purpose tips are also available for such operations as gouging and grooving.

One-piece torch tips are generally used with acetylene cutting because of the high temperatures involved. They can have four, six, or eight preheat holes in addition to the single cutting hole. *Figure 23* shows the arrangement of typical acetylene torch cutting tips.

Tips used with liquefied fuel gases must have at least six preheat holes (*Figure 24*). Because fuel gases burn at lower temperatures than acetylene, more holes are necessary for preheating. Tips used with liquefied fuel gases can be one- or two-piece cutting tips. *Figure 25* shows a typical two-piece cutting tip used with liquefied fuel gases.

Special-purpose tips are available for special cutting jobs such as cutting sheet metal, rivets, risers, and flues, as well as washing and gouging. *Figure 26* shows special-purpose torch cutting tips, which are described as follows.

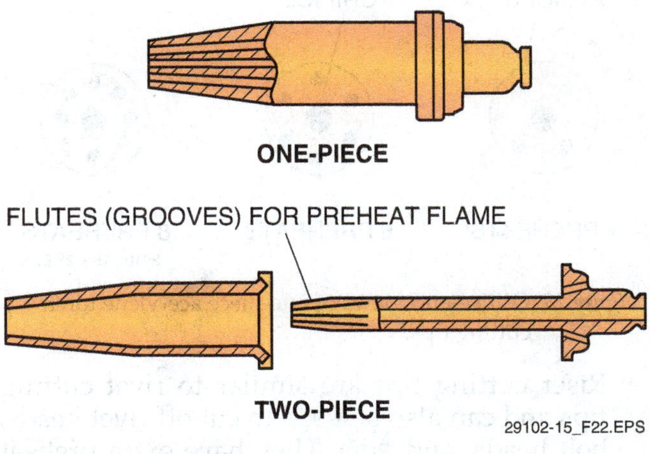

Figure 22 One- and two-piece cutting tips.

- The sheet metal cutting tip has only one preheat hole. This minimizes the heat and prevents distortion in the sheet metal. These tips are normally used with a motorized carriage, but can also be used for hand cutting.
- Rivet cutting tips are used to cut off rivet heads, bolt heads, and nuts.

Table 2 Sample Acetylene Cutting Tip Chart

Cutting Tip Series 1-101, 3-101, and 5-101													
Metal Thickness		Tip Size	Cutting Oxygen Pressure*		Preheat Oxygen*		Acetylene Pressure*		Speed		Kerf Width	
(in)	(mm)		(psig)	(kPa)	(psig)	(kPa)	(psig)	(kPa)	(in/ min)	(cm/ min)	(in)	(mm)
⅛	3.18	000	20-25	138-172	3-5	21-34	3-5	21-34	20-30	51-76	0.04	01.02
¼	6.35	00	20-25	138-172	3-5	21-34	3-5	21-34	20-28	51-71	0.05	01.27
⅜	9.52	0	25-30	172-207	3-5	21-34	3-5	21-34	18-26	46-66	0.06	01.52
½	12.70	0	30-35	207-241	3-6	21-41	3-5	21-34	16-22	41-56	0.06	01.52
¾	19.05	1	30-35	207-241	4-7	28-48	3-5	21-34	15-20	38-51	0.07	1.78
1	25.40	2	35-40	241-276	4-8	28-55	3-6	21-41	13-18	33-46	0.09	02.29
2	50.80	3	40-45	276-310	5-10	34-69	4-8	28-55	10-12	25-30	0.11	02.79
3	76.20	4	40-50	276-345	5-10	34-69	5-11	34-76	8-10	20-25	0.12	03.05
4	101.60	5	45-55	310-379	6-12	41-83	6-13	41-90	6-9	15-23	0.15	03.81
6	152.40	6	45-55	310-379	6-15	41-103	8-14	55-97	4-7	10-18	0.15	03.81
10	254.00	7	45-55	310-379	6-20	41-138	10-15	69-103	3-5	8-13	0.34	08.64
12	304.80	8	45-55	310-379	7-25	48-172	10-15	69-103	3-4	8-10	0.41	10.41

*The lower side of the pressure listings is for hand cutting and the higher side is for machine cutting.

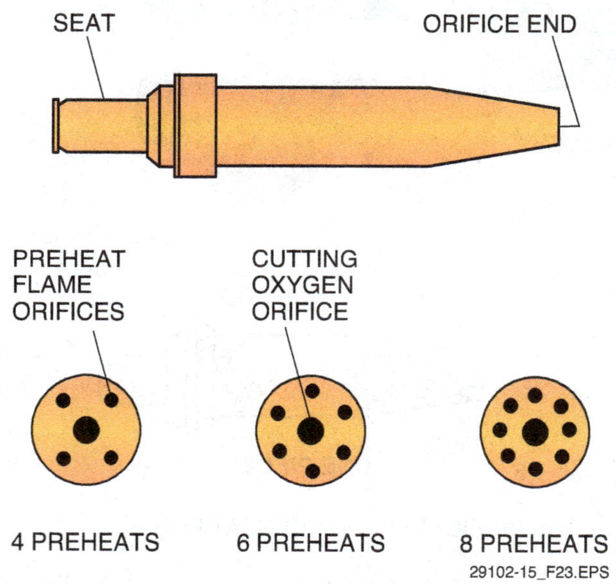

Figure 23 Orifice-end views of one-piece acetylene torch cutting tips.

- Riser cutting tips are similar to rivet cutting tips and can also be used to cut off rivet heads, bolt heads, and nuts. They have extra preheat holes to cut risers, flanges, or angle legs faster. They can be used for any operation that requires a cut close to and parallel to another surface, such as in removing a metal backing.
- Rivet blowing and metal washing tips are heavy-duty tips designed to withstand high heat. They are used for coarse cutting and for removing such items as clips, angles, and brackets.
- Gouging tips are used to groove metal in preparation for welding.
- Flue cutting tips are designed to cut flues inside boilers. They also can be used for any cutting operation in tight quarters where it is difficult to get a conventional tip into position.

Nearly all manufacturers use different tip-to-torch mounting designs, sealing surfaces, and diameters. In addition, tip sizes and flow rates are usually not the same between manufacturers even though the model number designations may be the same. This makes it impossible to safely interchange cutting tips between torches from different manufacturers. Even though some tips from different manufacturers may appear to be the same, do not interchange them. The sealing surfaces are very precise, and serious leaks may occur that

Acetylene Flow Rates

Manufacturers provide listings of the maximum fuel flow rate for each acetylene tip size in addition to recommended acetylene pressures. When selecting a tip, make sure that its maximum flow rate (in cubic feet or cubic meters per hour) does not exceed one-seventh of the total fuel capacity for the acetylene cylinder in use. Multiple cylinders must be manifolded together if the flow rate exceeds the cylinder(s) in use in order to prevent withdrawal of acetone along with acetylene.

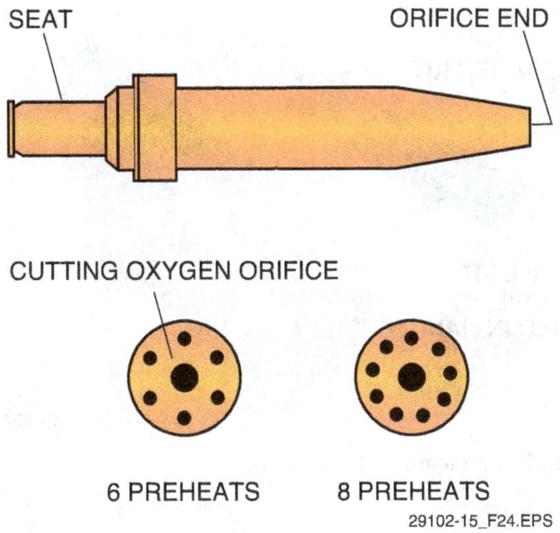

Figure 24 Orifice-end view of one-piece cutting tip for liquefied fuel gases.

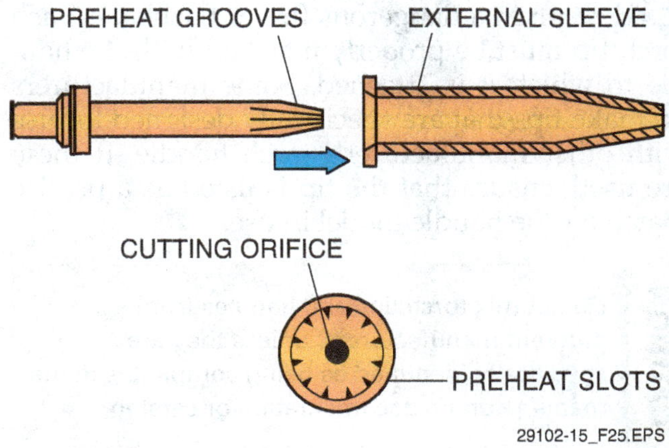

Figure 25 Orifice-end view of two-piece cutting tip for liquefied fuel gases.

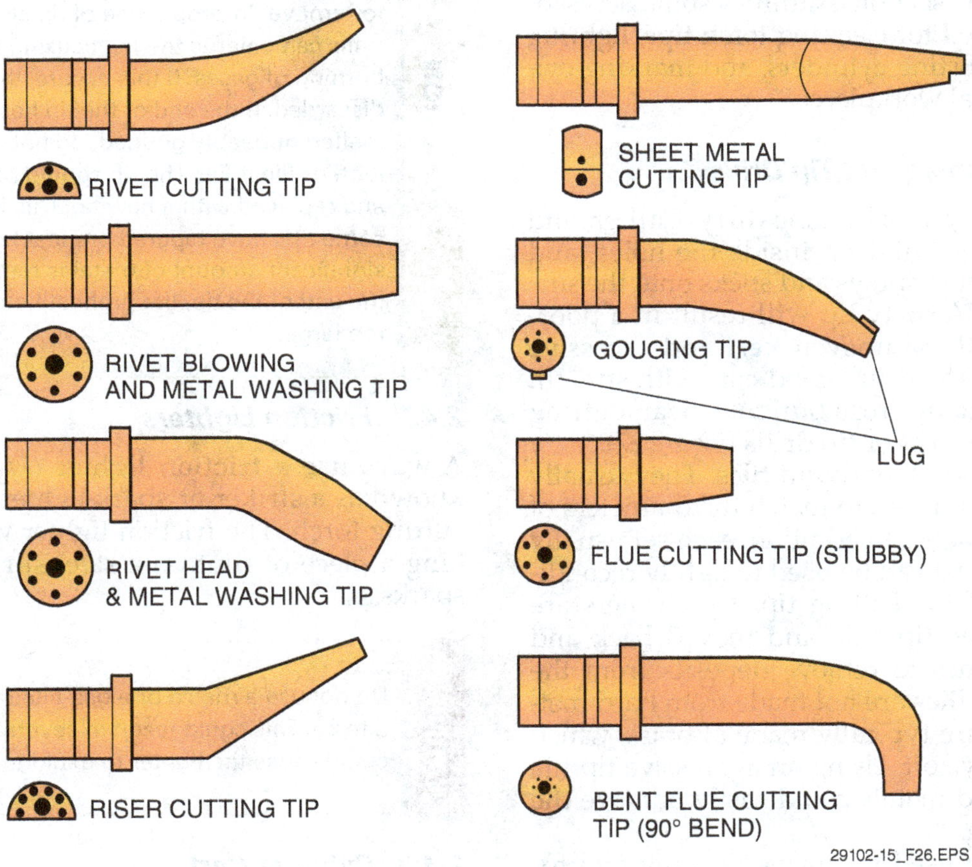

Figure 26 Special-purpose torch cutting tips.

could result in a dangerous fire or flashback. Each torch tip must be properly matched with the handle to which it is attached. Some manufacturers do make tips that are specifically designed for use with other manufacturer's torch handle. If these are used, ensure that the tip is listed as a precise match for the handle model in use.

CAUTION

Do not mix torch tips and handles from different manufacturers, unless they are specifically identified as being compatible in the manufacturer's documentation or catalog.

2.4.0 Other Accessories

In addition to the components that make up the actual oxyfuel cutting equipment, there are numerous accessories that are used along with the equipment. This section examines some accessories that are used for cleaning torch tips, lighting torches, transporting cylinders, and marking cylinders and metal workpieces.

2.4.1 Tip Cleaners and Tip Drills

With use, cutting tips become dirty. Carbon and other impurities build up inside the holes, and molten metal often sprays and sticks onto the surface of the tip. A dirty tip will result in a poor-quality cut with an uneven **kerf** and excessive dross buildup. To ensure good cuts with straight kerfs and minimal dross buildup, clean cutting tips with tip cleaners or tip drills (*Figure 27*).

Tip cleaners are tiny round files. They usually come in a set with files to match the diameters of the various tip holes. In addition, each set usually includes a file that can be used to lightly recondition the face of the cutting tip. Tip cleaners are inserted into the tip hole and moved back and forth a few times to remove deposits from the hole. The small files are not made from hard metals. Torch tips are typically made of brass, which is also relatively soft. Using an aggressive tip file made from hard metals could easily damage the precise opening.

Tip drills are used for major cleaning and for holes that are plugged. Tip drills are tiny drill bits that are sized to match the diameters of tip holes. The drill fits into a drill handle for use. The handle is held, and the drill bit is turned carefully inside the hole to remove debris. They are more brittle than tip cleaners, making them more difficult to use. If a torch tip is properly cared for, tip drills are rarely needed.

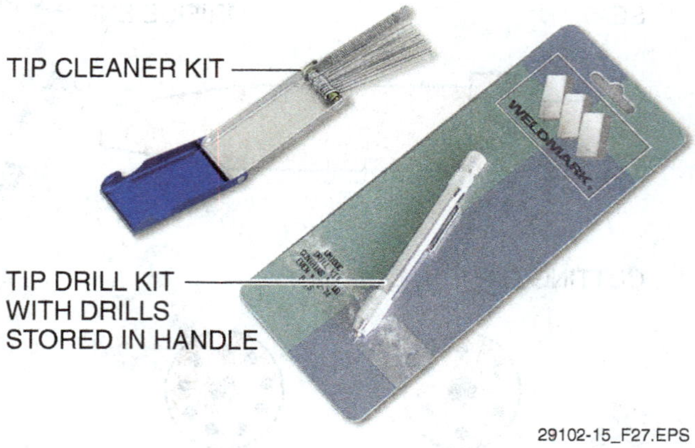

Figure 27 Tip cleaner and drill kits.

CAUTION

Tip cleaners and tip drills are brittle. Care must be taken to prevent these devices from breaking off inside a hole. Broken tip cleaners are difficult to remove. Improper use of tip cleaners or tip drills can enlarge the tip, causing improper burning of gases. If this occurs, the tip must be discarded. If the end of the tip has been partially melted or deeply gouged, do not attempt to cut it off or file it flat. The tip should be discarded and replaced with a new tip. This is because some tips have tapered preheat holes, and if a significant amount of metal is removed from the end of the tip, the preheat holes will become too large.

2.4.2 Friction Lighters

Always use a friction lighter (*Figure 28*), also known as a striker or spark-lighter, to ignite the cutting torch. The friction lighter works by rubbing a piece of flint on a steel surface to create sparks.

WARNING!

Do not use a match or a gas-filled lighter to light a torch. This could result in severe burns and/or could cause the lighter to explode.

2.4.3 Cylinder Cart

The cylinder cart, or bottle cart, is a modified hand truck that has been equipped with chains or straps to hold cylinders firmly in place. Bottle carts help ensure the safe transportation of gas cylinders. *Figure 29* shows two cylinder carts used for oxyfuel cylinders. Some carts are equipped with tool trays or boxes as well as rod holders.

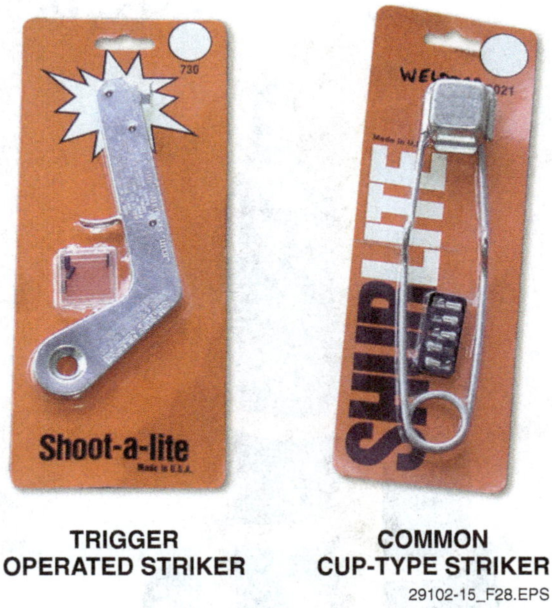

Figure 28 Friction lighters.

2.4.4 Soapstone Markers

Because of the heat involved in cutting operations, along with the tinted lenses that are required, ordinary pen or pencil marking for cutting lines or welding locations is not effective. The oldest and most common material used for marking is soapstone in the form of sticks or cylinders (*Figure 30*). Soapstone is soft and feels greasy and slippery. It is actually steatite, a dense, impure form of talc that is heat resistant. It also shows up well through a tinted lens under the illumination of an electric arc or gas cutting flame. Some welders prefer to use silver-graphite pencils (*Figure 30*) for marking dark materials and red-graphite pencils for aluminum or other bright metals. Graphite is also highly heat resistant. A few manufacturers also market heat-resistant paint/dye markers for cutting and welding.

2.5.0 Specialized Cutting Equipment

In addition to the common hand cutting torches, other types of equipment are used in oxyfuel cutting applications. This equipment includes mechanical guides used with a hand cutting torch, various types of motorized cutting machines, and oxygen lances. All of the motorized units use special straight body machine cutting or welding torches with a gear rack attached to the torch body to set the tip distance from the work.

2.5.1 Mechanical Guides

On long, circular, or irregular cuts, it is very difficult to control and maintain an even kerf with a hand cutting torch. Mechanical guides can help maintain an accurate and smooth kerf along the cutting line. For straight line or curved cuts, use a one- or two-wheeled accessory that clamps on the torch tip in a fixed position. The wheeled accessory maintains the proper tip distance while the tip is guided by hand along the cutting line.

The torch tip fits through and is secured to a rotating mount between two small metal wheels. The wheel heights are adjustable so that the tip distance from the work can be set. The radius of

Cup-Type Striker

When using a cup-type striker to ignite a welding torch, hold the cup of the striker slightly below and to the side of the tip, parallel with the fuel gas stream from the tip. This prevents the ignited gas from deflecting back from the cup and reduces the amount of carbon soot in the cup. Note that the flint in a striker can be replaced.

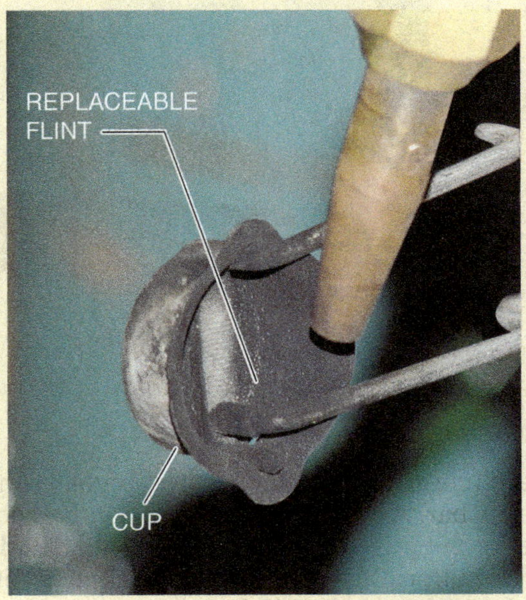

Figure 29 Cylinder carts.

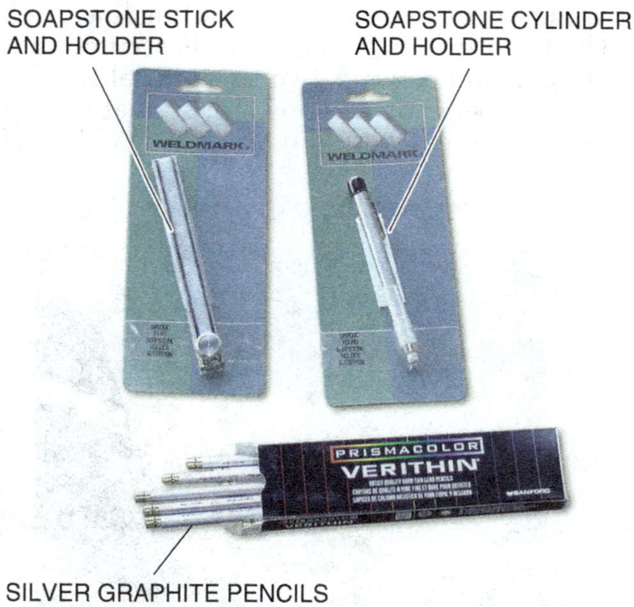

Figure 30 Soapstone and graphite markers.

the circle is set by moving the pivot point on a radius bar. After a starting hole is cut (if needed), the torch tip is placed through and secured to the circle cutter rotating mount. Then the pivot point is placed in a drilled hole or a magnetic holder at the center of the circle. When the cut is restarted, the torch can be moved in a circle around the cut, guided by the circle cutter. The magnetic guide is used for straight line cuts. The magnets hold it securely in place during cutting operations.

When large work with an irregular pattern must be cut, a template is often used. The torch is drawn around the edges of the template to trace the pattern as the cut is made. If multiple copies must be cut, a metal pattern held in place and designed to allow space for tip distance from the pattern is usually used. For a one- or two-time

Sharpening Soapstone Sticks

The most effective way to sharpen a soapstone stick marker is to shave it on one side with a file. By leaving one side flat, accurate lines can be drawn very close to a straightedge or a pattern.

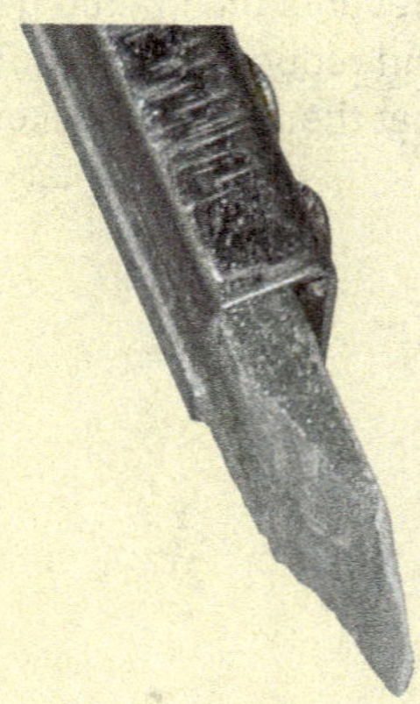

copy, a heavily weighted Masonite or aluminum template that is spaced off the workpiece could be carefully used and discarded.

2.5.2 Motor-Driven Equipment

A variety of fixed and portable motorized cutting equipment is available for straight and curved cutting/welding. The computer-controlled plate cutting machine (*Figure 31*) is a fixed-location machine used in industrial manufacturing applications. The computer-controlled machine can be programmed to **pierce** and then cut any pattern from flat metal stock. There is also an optical pattern-tracing machine that follows lines on a drawing using a light beam and an optical detector. They both can be rigged to cut multiple items using multiple torches operated in parallel. Both units have a motor-driven gantry that travels the length of a table and a transverse motor-driven torch head that moves back and forth across the table. Both units are also equipped to use oxyfuel or plasma cutting torches.

Other types of pattern-tracing machines use metal templates that are clamped in the machine. A follower wheel traces the pattern from the template. The pattern size can be increased or decreased by electrical or mechanical linkage to a moveable arm holding one or more cutting torches that cut the pattern from flat metal stock.

Portable track cutting machines, or track burners, can be used in the field for straight or curved cutting and beveling. *Figures 32* and *33* show units driven by a variable-speed motor. The unit shown in *Figure 32* is available with track extensions for any length of straight cutting or beveling, along with a circle cutting attachment. Some models use a single, centered track.

Machine oxyfuel gas cutters or track burners are basic guidance systems driven by a variable speed electric motor to enable the operator to cut or bevel straight lines at any desired speed. The device (*Figure 33*) is usually mounted on a track or used with a circle-cutting attachment to enable the operator to cut various circle diameters up to 96" (≈244 cm). It consists of a heavy-duty tractor unit fitted with an adjustable torch mount and gas hose attachments. It is also equipped with an On/Off switch, a Low/High speed range switch, a Forward/Reverse directional switch, and a speed-adjusting dial calibrated in inches (or metric equivalent) per minute.

The device shown in *Figure 33* offers the following operational features:

- Makes straight-line cuts of any length
- Makes circle cuts up to 96" (≈244 cm) in diameter
- Makes bevel or chamfer cuts
- Has an infinitely variable cutting speed from 1" to 110" (2.5 cm to 279 cm) per minute
- Has dual speed and directional controls to enable operation of the machine from either end

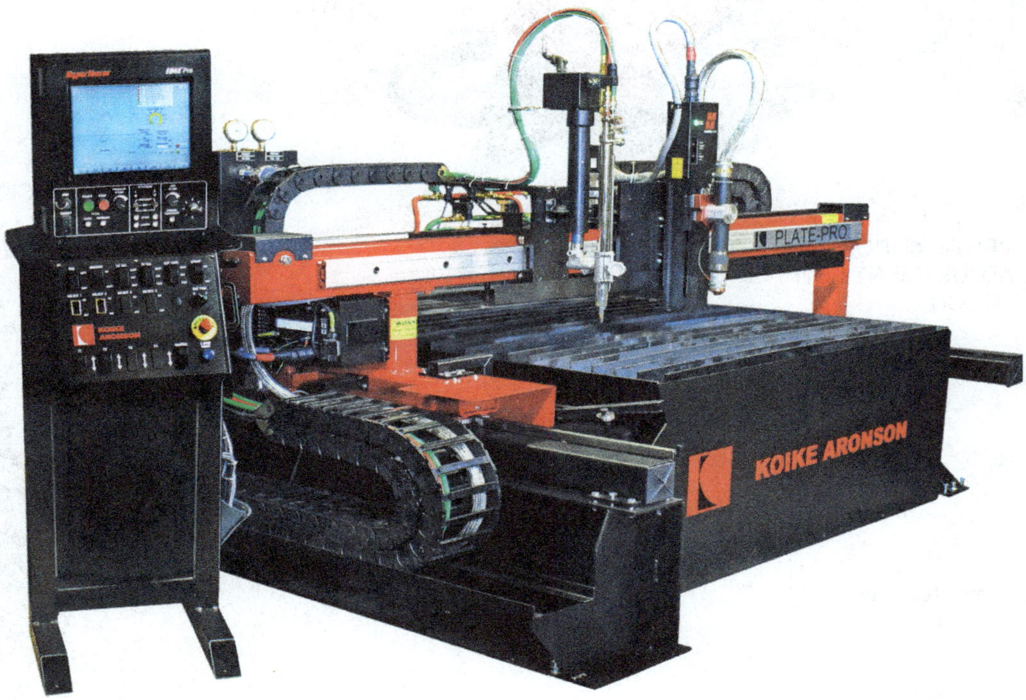

29102-15_F31.EPS

Figure 31 Computer-controlled plate cutting machine.

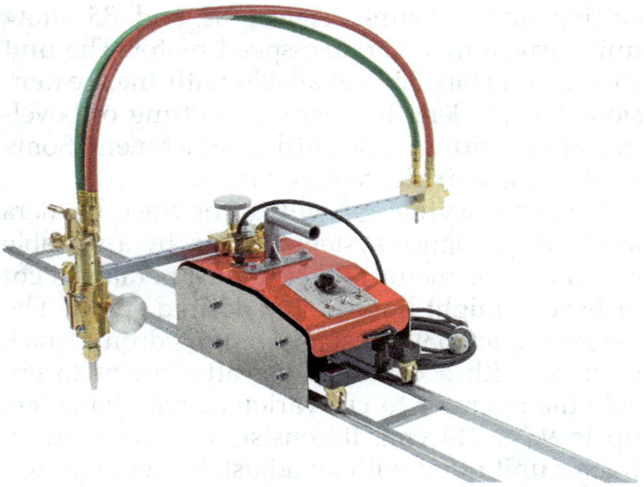
29102-15_F32.EPS

Figure 32 Track burner with oxyfuel machine torch.

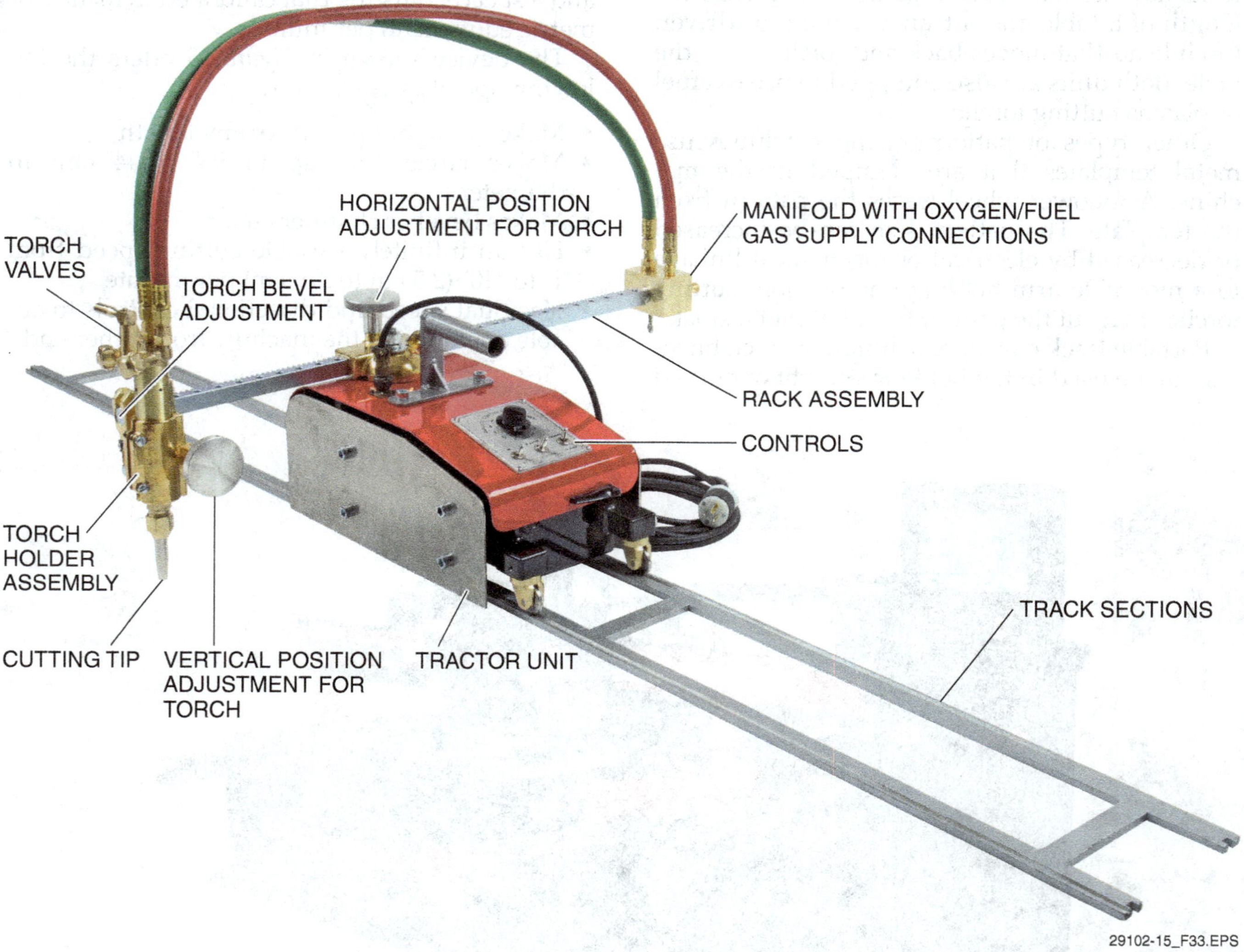

Figure 33 Track burner features.

Figure 34 shows the location of the following controls:

- *Power on/power off* – Turn the machine on and off by toggling the On/Off toggle switch.
- *Speed range control* – Set the machine's speed range by toggling the switch up or down. The use of two speed ranges allows for more precise speed control.
- *Directional control* – Set the machine's direction.
- *Speed control*– Turn the large knob to adjust the cutting speed based on the percentage of the selected speed range.

A portable, motor-driven band track or hand-cranked ring gear cutter/beveler can be set up in the field for cutting and beveling pipe with oxyfuel or plasma machine torches (*Figures 35*

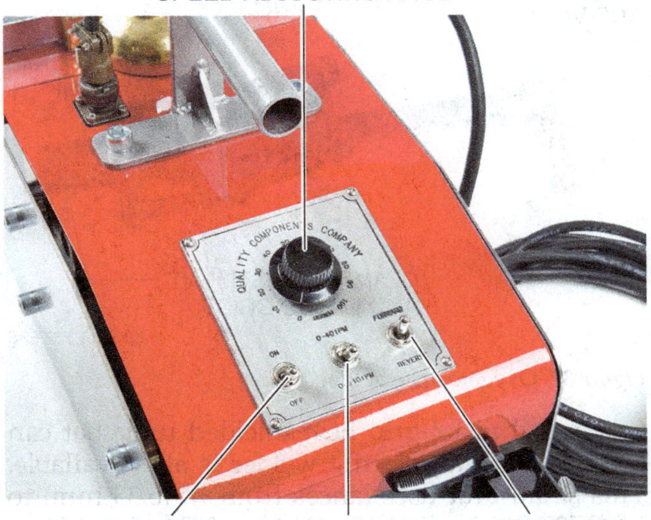

Figure 34 Track burner controls.

Figure 35 Band-track pipe cutter/beveler.

and *36*). The stainless steel band track cutter uses a chain and motor sprocket drive to rotate the machine cutting torch around the pipe a full 360 degrees. The all-aluminum ring gear type of cutter/beveler is positioned on the pipe, and then the saddle is clamped in place. In operation, the ring gear and the cutting torch rotate at different rates around the saddle for a full 360-degree cut.

2.5.3 Exothermic Oxygen Lances

Exothermic (combustible) oxygen lances are a special oxyfuel cutting tool usually used in heavy industrial applications and demolition work. The lance is a steel pipe that contains magnesium- and aluminum-cored powder or rods (fuel). In operation, the lance is clamped into a holder (*Figure 37*) that seals the lance to a fitting that supplies oxygen to it through a hose at pressures of 75 to 80 psi (517 to 552 kPa). With the oxygen turned on, the end of the lance is ignited with an acetylene torch or flare. As long as the oxygen is applied, the lance will burn and consume itself. The oxygen-fed flame of the burning magnesium, aluminum, and steel pipe creates temperatures approaching 10,000°F (5,538°C). At this temperature, the lance will rapidly cut or pierce any material, including steel, metal alloys, and cast iron, even under water. The lances for the holder shown in *Figure 37* are 10' (≈3 m) long and range in size from ⅜" (9.5 mm) to 1" (25.4 mm) in diameter. The larger sizes can be coupled to obtain a longer lance.

Shop-Made Straight-Line Cutting Guide

A simple solution for straight-line cutting is to clamp a piece of angle iron to the work and use a band clamp around the cutting torch tip to maintain the cutting tip distance from the work. When the cut is started, the band clamp rests on the top of the vertical leg of the angle iron, and the torch is drawn along the length of the angle iron at the correct cutting speed.

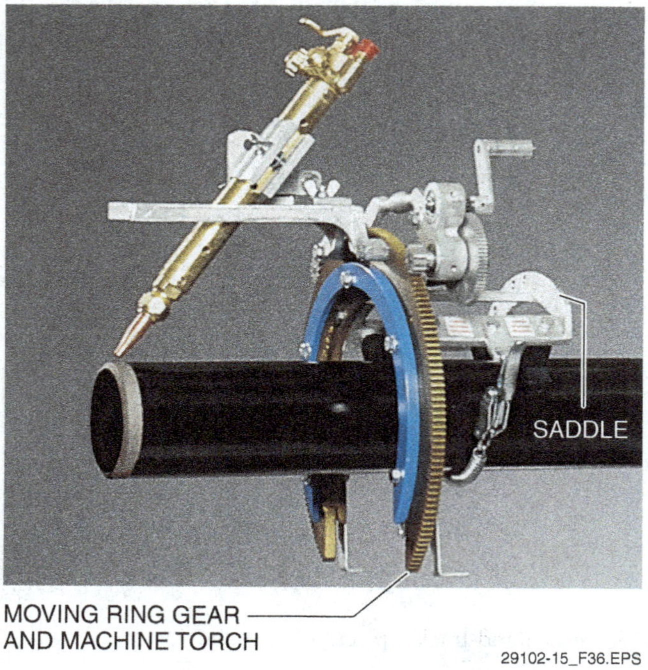

Figure 36 Ring gear pipe cutter/beveler.

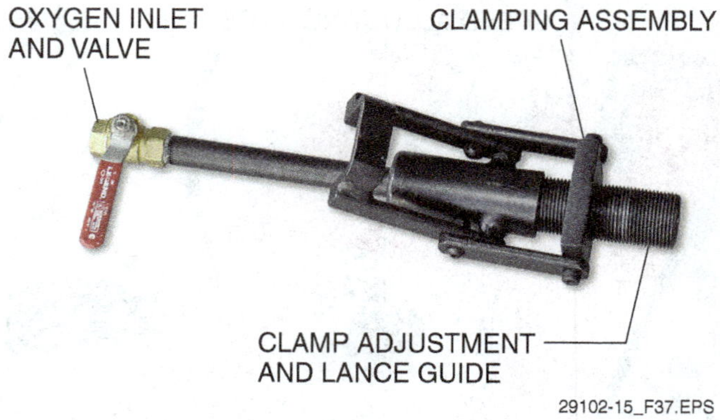

Figure 37 Oxygen lance holder.

A small pistol-grip heat-shielded unit that can be used with an electric welder is also available. This small unit uses lances from ¼" (6.35 mm) to ⅜" (9.52 mm) in diameter that are 22" to 36" (55.9 to 91.4 cm) long and that cut very rapidly at a maximum burning time of 60 to 70 seconds. The small unit is primarily used to burn out large frozen pins and frozen headless bolts or rivets. Like a large lance, it can be used to cut any material, including concrete-lined pipe. Both units are relatively inexpensive and can be set up in the field with only an oxygen cylinder, hose, and ignition device.

Additional Resources

ANSI Z49.1, Safety in Welding, Cutting, and Allied Processes. Miami, FL: American Welding Society.

Uniweld Products, Inc. Numerous videos are available at **www.uniweld.com/en/uniweld-videos**. Last accessed: November 30, 2014.

The Harris Products Group, a division of Lincoln Electric. Numerous videos are available at **www.harrisproductsgroup.com/en/Expert-Advice/videos.aspx**. Last accessed: November 30, 2014.

2.0.0 Section Review

1. A lighter-than-air compound of carbon and hydrogen that is formed by dissolving calcium carbide in water is ____ .
   a. argon
   b. acetylene
   c. propane
   d. propylene

2. Safety devices used on regulators, hoses, and torches to prevent reverse gas flow and protect against fires and explosions are ____ .
   a. check valves and flashback arrestors
   b. orifice plates and torch tips
   c. HEPA filters and flashback valves
   d. distribution manifolds and friction strikers

3. What basic type of oxyfuel cutting torch is designed for use with fuel that is supplied through a regulator from a pressurized fuel storage cylinder?
   a. Motor-controlled torch
   b. Vacuum-injector torch
   c. Neutral-flame torch
   d. Positive-pressure torch

4. The most common material used for marking cutting lines on metal is a form of heat-resistant talc called ____ .
   a. soapstone
   b. Masonite
   c. graphite
   d. flint

5. A type of oxyfuel cutting tool that uses a steel pipe containing magnesium- and aluminum-cored powder or rods for heavy industrial cutting applications and demolition work is a(n) ____ .
   a. circle cutting accessory
   b. plasma machine torch
   c. exothermic oxygen lance
   d. ring gear beveler

SECTION THREE

3.0.0 OXYFUEL EQUIPMENT SETUP AND SHUTDOWN

Objective

Explain how to set up, light, and shut down oxyfuel equipment.

a. Explain how to properly prepare a torch set for operation.
b. Explain how to leak test oxyfuel equipment.
c. Explain how to light the torch and adjust for the proper flame.
d. Explain how to properly shut down oxyfuel cutting equipment.

Performance Tasks

1. Set up oxyfuel equipment.
2. Light and adjust an oxyfuel torch.
3. Shut down oxyfuel cutting equipment.
4. Disassemble oxyfuel cutting equipment.
5. Change empty gas cylinders.

Trade Terms

Carburizing flame: A flame burning with an excess amount of fuel; also called a reducing flame.

Neutral flame: A flame burning with correct proportions of fuel gas and oxygen.

Oxidizing flame: A flame burning with an excess amount of oxygen.

Operators should be trained and tested in the correct methods of safely preparing, starting, testing for leaks, and shutting down an oxyfuel cutting station. This part of the module examines procedures for setting up oxyfuel equipment, leak testing the equipment, lighting and adjusting the torch, and shutting down the equipment after use.

3.1.0 Setting Up Oxyfuel Equipment

When setting up oxyfuel equipment, follow procedures to ensure that the equipment operates properly and safely. The following sections explain the procedures for setting up oxyfuel equipment.

3.1.1 Transporting and Securing Cylinders

Cylinders should be transported to the workstation in an upright position on an appropriate hand truck or bottle cart. Once the cylinders are at the workstation, they must be secured with chain to a fixed support so that they cannot be knocked over accidentally. Leaving the cylinders in a proper cylinder cart is common; removal of the cylinders from the cart is not required. Then the protective cap from each cylinder can be removed and the outlet nozzles inspected to make sure that the seat and threads are not damaged. Place the protective caps where they will not be lost and where they will be readily available for reinstallation.

WARNING!

Always handle cylinders with care. They are under high pressure and should never be dropped, knocked over, rolled, or exposed to heat in excess of 140°F (60°C). When moving cylinders, always be certain that the valve caps are in place. Use a cylinder cage to lift cylinders. Never use a sling or electromagnet for cylinder lifting.

3.1.2 Cracking Cylinder Valves

To crack open a cylinder valve, start by ensuring that the cylinder is fully secured. Then crack open the cylinder valve momentarily to remove any dirt from the valve opening (*Figure 38*).

WARNING!

Operators should always stand with the valve stem between themselves and the regulator when opening valves to avoid injury from dirt that may be lodged in the valve. If a cloth is used during the cleaning process, it must not have any oil or grease on it. Oil or grease mixed with compressed oxygen can cause an explosion.

Hoisting Cylinders

Never attempt to lift a cylinder using the holes in a safety cap. Always use a lifting cage. Make sure that the cylinder is secured in the cage. Cages designed for storing and lifting cylinders are available in various sizes. The model shown here includes a partition to separate oxygen cylinders from fuel gas cylinders.

29102-15_SA09.EPS

CAUTION

Use care not to over-tighten connections. The brass connections used will strip if over-tightened. Repair or replace equipment that does not seal properly. If the torch valves leak, try lightly tightening the packing nut at the valves. If a cutting attachment or torch tip leaks, disassemble and check the sealing surfaces and/or O-rings for damage. If the equipment or torch tip is new and the sealing surfaces do not appear damaged, over-tighten the connection slightly to seat the sealing surfaces; then loosen and retighten normally. If the leaks persist or the equipment is visibly damaged, replace the torch, cutting attachment (if used), or cutting tip as necessary.

3.1.3 Attaching Regulators

To attach the regulators, first check that the regulator is closed (adjustment screw is backed out and loose/turns with no resistance).

Check the regulator fittings to ensure that they are free of oil and grease (*Figure 39*).

Connect and tighten the oxygen regulator to the oxygen cylinder using a torch wrench (*Figure 40*).

29102-15_F38.EPS

Figure 38 Cracking a cylinder valve.

WARNING!

Do not work with a regulator that shows signs of damage, such as cracked gauges, bent thumbscrews, or worn threads. Set it aside for repairs. Operators should never attempt to repair regulators. When tightening the connections, always use a torch wrench.

Figure 39 Checking connection fittings.

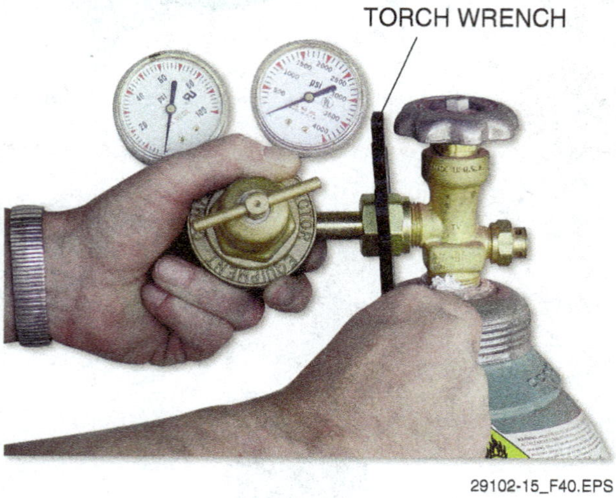

Figure 40 Tightening regulator connection.

Figure 41 Cleaning the regulator.

Figure 42 Attaching a flashback arrestor.

Connect and tighten the fuel gas regulator to the fuel gas cylinder. Remember that most fuel gas fittings have left-hand threads. Next, clean the outlet connection of the regulator. Crack the cylinder valve slightly and turn the pressure adjustment screw clockwise until you feel some resistance and the regulator begins to allow gas to pass through and expel any debris. Shut the cylinder valve and close the regulator (*Figure 41*).

3.1.4 Installing Flashback Arrestors or Check Valves

The installation of flashback arrestors or check valves is important and easy to accomplish if they are not already installed. Attach a flashback arrestor and/or check valve to the hose connection on the oxygen regulator or torch handle (*Figure 42*) and tighten with a torch wrench. Then attach and tighten a flashback arrestor and/or check valve to the fuel gas regulator or torch handle. Once again, remember that fuel gas fittings have left-hand threads.

> **WARNING!**
>
> At least one flashback arrestor must be used with each hose. The absence of a flashback arrestor could result in flashback. Flashback arrestors can be attached either to the regulator, the torch, or both; however, flashback arrestors installed at the torch handle are preferred if only one is being used.

3.1.5 Connecting Hoses to Regulators

New hoses contain talc and possibly loose bits of rubber. These materials must be blown out of the hoses before the torch is connected. If they are not blown out, they will clog the tiny torch needle valves or tip openings.

To connect the hoses to the regulators, first inspect both the oxygen and fuel gas hoses for any damage, burns, cuts, or fraying. Replace any damaged hoses. Then connect the oxygen hose to the oxygen regulator flashback arrestor or check valve (*Figure 43*), and connect the fuel gas hose to the fuel gas regulator flashback arrestor and/or check valve. Complete the installation by opening the cylinder valves and regulators and purging the hoses until they are clear.

3.1.6 Attaching Hoses to the Torch

To attach the hoses to the torch, first attach flashback arrestors to the oxygen and fuel gas hose connections on the torch body—unless the torch has built-in flashback arrestors and check valves. Attach and tighten the oxygen hose to the oxygen fitting on the flashback arrestor or torch (*Figure 44*). Then attach and tighten the hose to the fuel gas fitting on the flashback arrestor or torch.

3.1.7 Connecting Cutting Attachments (Combination Torch Only)

If cutting attachments are being connected to a combination torch, be sure to check the torch manufacturer's instructions for the correct installation method. Then connect the attachment and tighten by hand as required.

3.1.8 Installing Cutting Tips

Before installing a cutting tip in a cutting torch, first identify the thickness of the material to be cut. Then identify the proper size cutting tip from the manufacturer's recommended tip size chart for the fuel being used.

WARNING!

If acetylene fuel is being used, make sure that the maximum fuel flow rate per hour of the tip does not exceed one-seventh of the fuel cylinder capacity. If a purplish flame is observed when the torch is operating, the fuel rate is too high and acetone is being withdrawn from the acetylene cylinder along with the acetylene gas.

Once the cutting tip has been selected, inspect the cutting tip sealing surfaces and orifices for damage or plugged holes. If the sealing surfaces are damaged, discard the tip. If the orifices are plugged, clean them with a tip cleaner or drill.

Check the torch manufacturer's instructions for the correct method of installing cutting tips. Then install the cutting tip and secure it with a torch wrench or by hand as required (*Figure 45*).

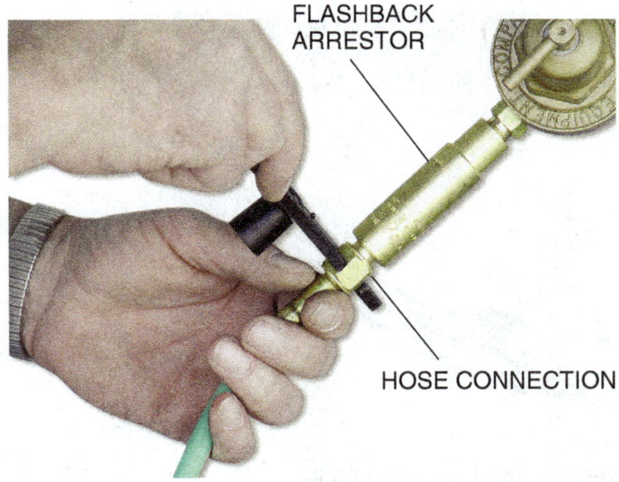

Figure 43 Connecting hose to regulator flashback arrestor.

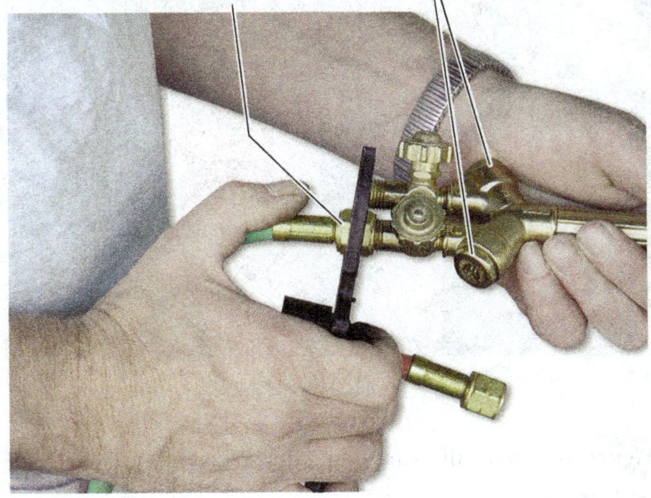

Figure 44 Connecting hoses to torch body.

3.1.9 Closing Torch Valves and Loosening Regulator Adjusting Screws

Closing the torch valves and loosening the regulator adjusting screws (*Figure 46*) are done before opening either cylinder valve. First, check the fuel and oxygen valves on the torch to be sure they are closed. Then check both the oxygen and fuel gas regulator adjusting screws to be sure they are loose (backed out).

CAUTION

Loosening regulator adjusting screws closes the regulators and prevents damage to the regulator diaphragms when the cylinder valves are opened.

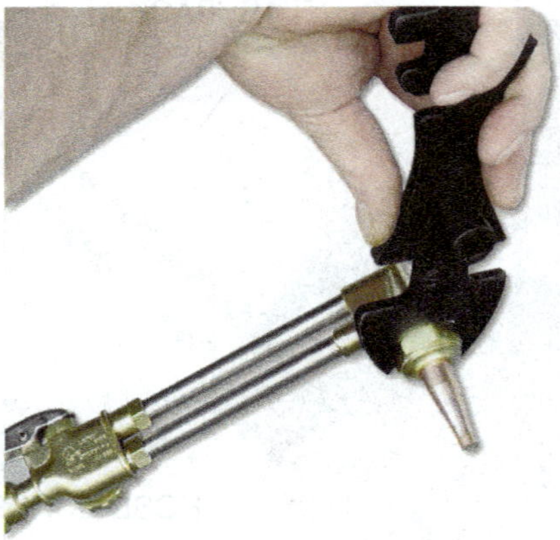

Figure 45 Installing a cutting tip.

Figure 46 Torch valves and regulator adjusting screws.

3.1.10 Opening Cylinder Valves

To open cylinder valves (*Figure 47*), stand on the opposite side of the cylinder from the regulator and crack open the oxygen cylinder valve until the pressure on the regulator gauge rises and stops. The pressure in the cylinder and the pressure at the inlet of the regulator are now equal. Now open the oxygen cylinder valve all the way.

> **WARNING!**
>
> When opening a cylinder valve, keep the cylinder valve stem between you and the regulator. Never stand directly in front of or behind a regulator. The regulator adjusting screw can blow out, causing serious injury. Always open the cylinder valve gradually. Quick openings can damage a regulator or gauge or even cause a gauge to explode.

Figure 47 Cylinder valve and gauges.

Oxygen cylinder valves must be opened all the way until the valve seats at the top. Seating the valve at the fully open position prevents high-pressure leaks at the valve stem.

Once the oxygen cylinder valve is open, slowly open the fuel gas cylinder valve until the cylinder pressure gauge indicates the cylinder pressure, but no more than one and a half turns. This allows it to be quickly closed in case of a fire. This is especially important with acetylene.

3.1.11 Purging the Torch and Setting the Working Pressures

After the oxygen and fuel gas tank valves have been opened, the torch valves are opened to purge the torch and set the working pressures on the regulators.

Fully open the oxygen valve on the torch. Then depress and hold the cutting oxygen lever. Turn the oxygen regulator adjusting screw clockwise until the working pressure gauge shows the correct working pressure with the gas flowing. Allow the gas to flow for five to ten seconds to purge the torch and hoses of air or fuel gas. Then release the cutting lever and close the oxygen valve.

Note that, on single-stage regulators, the working pressure shown on the gauge will likely rise when the cutting lever is released. You will see the pressure gauge fall somewhat each time the lever is opened; this is normal operation. Set the pressure on the oxygen regulator while the gas is flowing to ensure it will be as desired when working.

Open the fuel valve on the torch about one eighth of a turn. Turn the fuel regulator adjusting screw clockwise until the working pressure gauge shows the correct fuel gas working pressure with the gas flowing. Allow the gas to flow for five to ten seconds to purge the hoses and torch of air. Then close the torch fuel valve. If acetylene is used, check that the acetylene static pressure does not rise above 15 psig (103 kPa). If it does, immediately open the torch fuel valve and reduce the regulator output pressure as needed. Because of its instability, acetylene cannot be used at pressures of more than 15 psig (103kPa) when in gaseous form. At higher pressures, acetylene gas breaks down chemically, producing heat and pressure that could result in a violent explosion inside the hose.

WARNING!

The working pressure gauge readings on single-stage regulators will rise after the torch valves are turned off. This is normal. However, if acetylene is being used as the fuel gas, make sure that the static pressure does not rise above 15 psig (103 kPa). Make sure that equipment is purged and leak tested in a well-ventilated area to avoid creating an explosive concentration of gases.

3.2.0 Testing for Leaks

Equipment must be tested for leaks immediately after it is set up and periodically thereafter. The torch should be checked for leaks before each use. Leaks could cause a fire or explosion. To test for leaks, apply a commercially prepared leak-testing formula (*Figure 48*) or a solution of detergent and water to each potential leak point. If bubbles form, a leak is present. Be aware though, that solutions made from common household soaps are not usually as effective as commercially prepared leak detection solutions.

WARNING!

If a detergent is used for leak testing, make sure the detergent contains no oil. In the presence of oxygen, oil can cause fires or explosions.

There are numerous leak points to test, including the following:

- Oxygen cylinder valve
- Fuel gas cylinder valve
- Oxygen regulator and regulator inlet and outlet connections
- Fuel gas regulator and regulator inlet and outlet connections

Portable Oxyacetylene Equipment

Most oxyfuel equipment used on large job sites and in shops is very heavy and is usually transported using a special hand truck. This type of equipment is typically used when extensive cutting must be accomplished. However, for small tasks in unusual locations, such as a commercial building rooftop, portable equipment that can be hand carried by one person is often used.

29102-15_SA10.EPS

APPLYING THE SOLUTION

REACTION TO A LEAK

29102-15_F48.EPS

Figure 48 Leak detection fluid use.

- Hose connections at the regulators, check valves/flashback arrestors, and torch
- Torch valves and cutting oxygen lever valve
- Cutting attachment connection (if used)
- Cutting tip

If there is a leak at the fuel gas cylinder valve stem, attempt to stop it by tightening the packing gland at the base of the stem. If this does not stop the leak, mark and remove the cylinder from service and notify the supplier. For other leaks, tighten the connections slightly with a wrench. If this does not stop the leak, turn off the gas pressure, open all connections, and inspect the fitting for damage.

WARNING!

Do not use Teflon® tape or pipe dope on these fittings since they do not provide a good seal. Make sure that equipment is purged and leak tested in a well-ventilated area to avoid creating an explosive concentration of gases.

3.2.1 Initial and Periodic Leak Testing

Initial and periodic leak testing is performed during initial equipment setup and periodically thereafter. First, set the equipment to the correct working pressures with the torch valves turned off. Then, using a leak-test solution, check for leaks at the cylinder valves, regulator relief ports, and regulator gauge connections (*Figure 49*). Also, check for leaks at hose connections, regulator connections, and check valve/flame arrestor connections up to the torch.

Fuel Cylinder Wrench

If the fuel cylinder is equipped with a valve requiring a T-wrench, always leave the wrench in place on the valve so that the fuel can be quickly turned off. This type of valve is obsolete but still in use.

29102-15_SA11.EPS

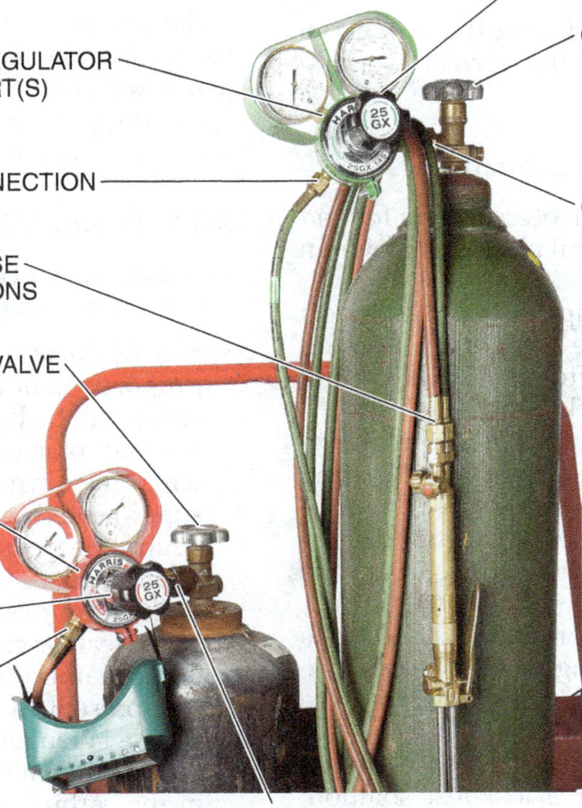

Figure 49 Typical initial and periodic leak-test points.

3.2.2 Leak-Down Testing of Regulators, Hoses, and Torch

Before the torch is ignited for use, the regulators, hoses, and torch should be quickly tested for leaks. First, set the equipment to the correct working pressures with the torch valves turned off. Then loosen both regulator adjusting screws. Check the working pressure gauges after a minute or two to see if the pressure drops. If the pressure drops, check the hose connection and regulators for leaks; otherwise, proceed with the test.

Place a thumb or finger over the cutting tip orifices and press tightly to block them (*Figure 50*). Turn on the torch oxygen valve and then depress and hold the cutting oxygen lever down. After the gauge pressure drops slightly, observe the oxygen working pressure gauge for a minute to see if the pressure continues to drop. If the pressure keeps dropping, perform the leak test described in the following section to determine the source of the leak. If the pressure does not change, close the torch oxygen valve and release the pressure at the cutting tip.

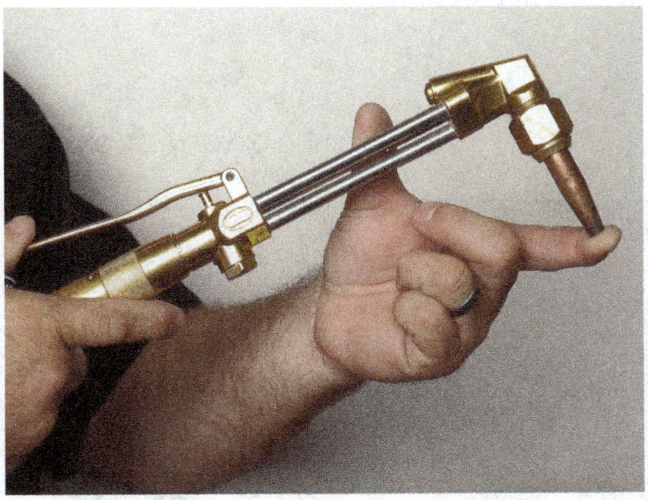

Figure 50 Blocking cutting tip for a leak test.

Next, block the tip again and turn on the torch fuel valve. After the gauge pressure drops slightly, carefully observe the fuel working pressure gauge for a minute. If the pressure continues to drop, perform the leak test described in the following section to determine the source of the leak. If the pressure does not change, close the torch fuel valve and release the pressure at the cutting tip.

If no leaks are apparent during the leak-down test, set the equipment to the correct working pressures.

3.2.3 Full Leak Testing of a Torch

Performing a full leak test involves testing for and isolating torch leaks in several places, as shown in *Figure 51*.

Start by setting the equipment to the correct working pressures with the torch valves turned off. Then, place a thumb or finger over the cutting tip orifices and press to block the orifices.

Turn on the torch oxygen valve and then depress the cutting oxygen lever. With the cutting tip blocked, check for leaks using a leak-test solution at the torch oxygen valve, cutting oxygen lever valve, cutting attachment connection to the handle (if used), preheat oxygen valve (if present), and cutting tip seal at the torch head. If no leaks are found, release the cutting oxygen lever, close the torch oxygen valve, and release the pressure at the cutting tip.

Next, with the cutting tip again blocked, open the torch fuel valve. Using a leak-test solution, check for leaks at the torch fuel valve, cutting attachment connection to the handle (if used), and cutting tip seal at the torch head. If no leaks are detected, close the torch fuel valve and release the pressure at the cutting tip.

3.3.0 Controlling the Oxyfuel Torch Flame

To be able to safely use a cutting torch, the operator must understand the flame and be able to adjust it and react to unsatisfactory conditions. The following sections will explain the oxyfuel flame and how to control it safely.

3.3.1 Oxyfuel Flames

There are three types of oxyfuel flames: **neutral flame**, **carburizing flame**, and **oxidizing flame**.

- *Neutral flame* – A neutral flame burns proper proportions of oxygen and fuel gas. The inner cones will be light blue in color, surrounded by a darker blue outer flame envelope that results when the oxygen in the air combines with the super-heated gases from the inner cone. A neutral flame is used for all but special cutting applications.
- *Carburizing flame* – A carburizing flame has a white feather created by excess fuel. The length of the feather depends on the amount of excess fuel present in the flame. The outer flame envelope is longer than that of the neutral flame, and it is much brighter in color. The excess fuel in the carburizing flame (especially acetylene) produces large amounts of carbon. The carbon will combine with red-hot or molten metal, making the metal hard and brittle. The carburizing flame is cooler than a neutral flame and is never used for cutting. It is used for some special heating applications.

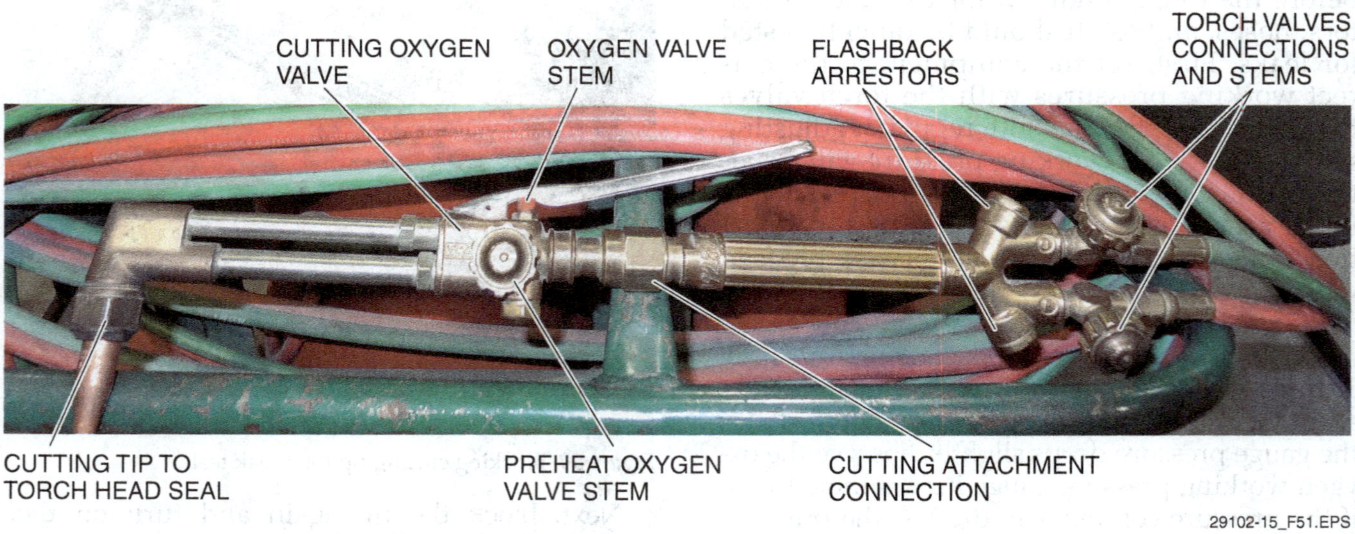

Figure 51 Torch leak-test points.

- *Oxidizing flame* – An oxidizing flame has an excess of oxygen. The inner cones are shorter, much bluer in color, and more pointed than a neutral flame. The outer flame envelope is very short and often fans out at the ends. An oxidizing flame is the hottest flame. A slightly oxidizing flame is recommended with some special fuel gases, but in most cases it is not used. The excess oxygen in the flame can combine with many metals, forming a hard, brittle, low-strength oxide. However, the preheat flames of a properly adjusted cutting torch will be slightly oxidizing when the cutting oxygen is shut off.

Figure 52 shows the various flames that occur at a cutting tip for both acetylene and LP gas.

> ## Always Leak Test a Cutting Torch
>
> Always take the time to perform a leak test on regulators, hoses, and the torch before using them the first time each day. A leak test should also be performed on a torch after tips have been changed or after converting the torch from one operation to another. Leaks, especially fuel gas leaks, can cause a fire or explosion after a cutting operation begins. A dangerous fire occurring at or near the torch may not be immediately noticed by the operator due to limited visibility through the tinted lenses.

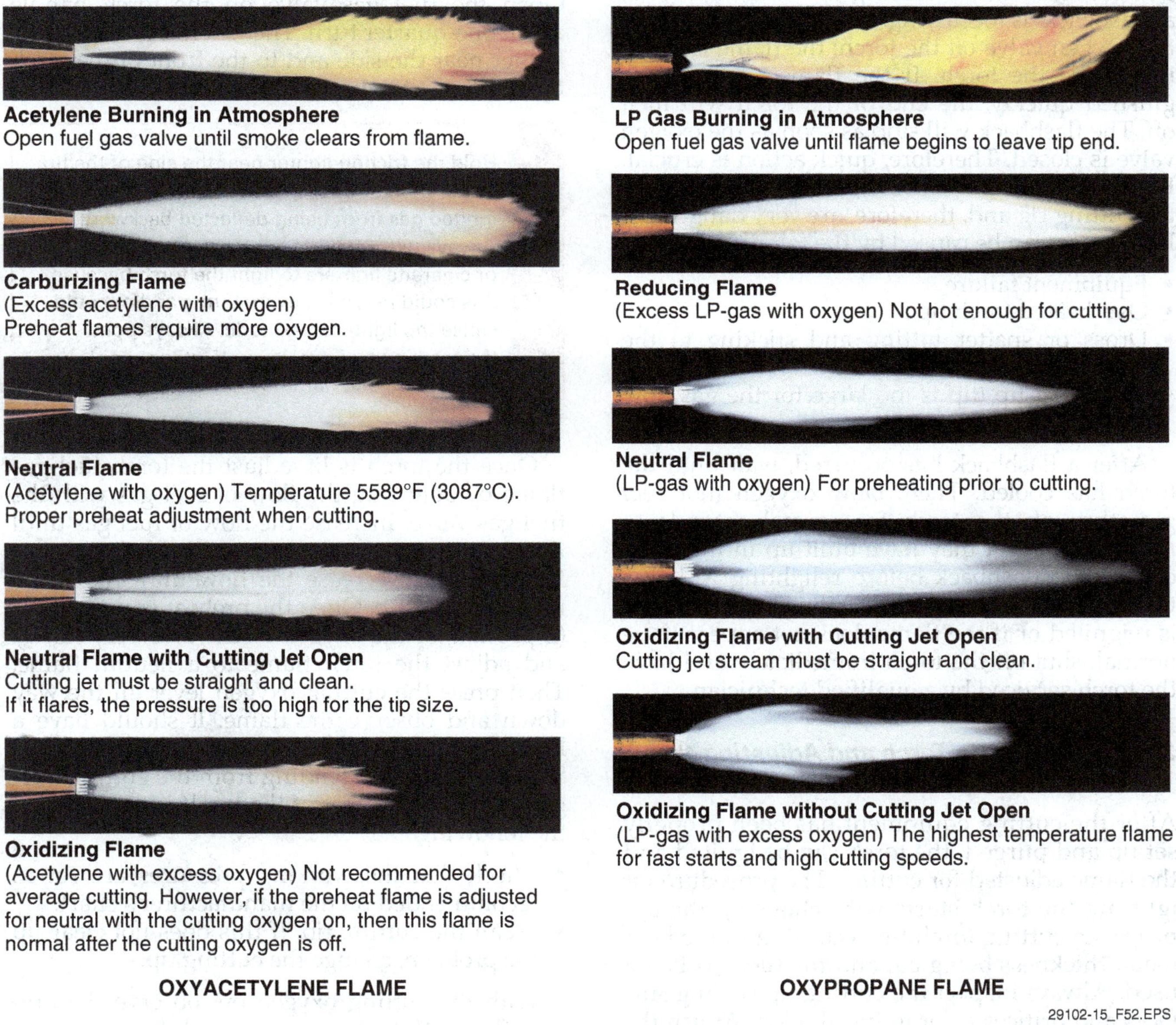

Figure 52 Acetylene and LP (propane) gas flames.

3.3.2 Backfires and Flashbacks

When the torch flame goes out with a loud pop or snap, a backfire has occurred. Backfires are usually caused when the tip or nozzle touches the work surface or when a bit of hot dross briefly interrupts the flame. When a backfire occurs, relight the torch immediately. Sometimes the torch even relights itself. If a backfire recurs without the tip making contact with the base metal, shut off the torch and find the cause. Possible causes are the following:

- Improper operating pressures
- A loose torch tip
- Dirt in the torch tip seat or a bad seat

When the flame goes out and burns back inside the torch with a hissing or whistling sound, a flashback is occurring. Immediately shut off the oxygen valve on the torch; the flame is burning inside the torch. If the flame is not extinguished quickly, the end of the torch will melt off. The flashback will stop as soon as the oxygen valve is closed. Therefore, quick action is crucial. Flashbacks can cause fires and explosions within the cutting rig and, therefore, are very dangerous. Flashbacks can be caused by the following:

- Equipment failure
- Overheated torch tip
- Dross or spatter hitting and sticking to the torch tip
- Oversized tip (tip is too large for the gas flow rate being used)

After a flashback has occurred, wait until the torch has cooled. Then, blow oxygen (not fuel gas) through the torch for several seconds to remove soot that may have built up in the torch during the flashback before relighting it. If the torch makes a hissing or whistling sound after it is reignited or if the flame does not appear to be normal, shut off the torch immediately and have the torch serviced by a qualified technician.

3.3.3 Igniting the Torch and Adjusting the Flame

After the cutting equipment has been properly set up and purged, the torch can be ignited and the flame adjusted for cutting. The procedure for igniting the torch starts with choosing the appropriate cutting torch tip according to the base metal thickness being cut and the fuel gas being used. Always inspect the cutting tip sealing surfaces and orifices prior to installation. Attach the tip to the cutting torch by placing it on the end of the torch and tightening the nut. Some manufacturers recommend tightening the nut with a torch wrench, while others recommend tightening the nut by hand. Check the manufacturer's documentation for the equipment in use to ensure the tip is being installed correctly.

NOTE

Refer to the manufacturer's charts. Depending on the tip selected, the oxygen and fuel gas pressure may have to be adjusted.

Prior to opening the oxygen and fuel gas valves, be sure to put on the proper PPE. Also ensure that you are not depressing the oxygen cutting lever. If present, close the preheat oxygen valve and open the torch oxygen valve fully. Open the fuel gas valve on the torch handle about one-quarter turn. Then, holding the friction lighter near the side and to the front of the torch tip, ignite the torch.

WARNING!

Hold the friction lighter near the side of the tip, rather than directly in front of it, to prevent the ignited gas from being deflected backwards. Always use a friction lighter. Never use matches or cigarette lighters to light the torch because this could result in severe burns and/or could cause the lighter to explode. Always point the torch away from yourself, other people, equipment, and flammable material.

Once the torch is lit, adjust the torch fuel gas flame by adjusting the flow of fuel gas with the fuel gas valve. Increase the flow of fuel gas until the flame stops smoking or pulls slightly away from the tip. Decrease the flow until the flame returns to the tip. Open the preheat oxygen valve (if present) or the oxygen torch valve very slowly and adjust the torch flame to a neutral flame. Then press the cutting oxygen lever all the way down and observe the flame. It should have a long, thin, high-pressure oxygen cutting jet up to 8" (≈20 cm) long, extending from the cutting oxygen hole in the center of the tip. If it does not, do the following:

- Check that the working pressures are set as recommended on the manufacturer's chart.
- Clean the cutting tip. If this does not clear up the problem, change the cutting tip.

With the cutting oxygen on, observe the preheat flame. If it has changed slightly to a carburizing flame, increase the preheat oxygen until the flame is neutral. After this adjustment, the preheat flame will change slightly to an acceptable oxidizing flame when the cutting oxygen is shut off.

3.3.4 Shutting Off the Torch

Shutting off the torch itself is done by releasing the cutting oxygen lever and then closing the torch or preheat oxygen valves. After that, quickly close the torch fuel gas valve to extinguish the flame.

> **WARNING!**
> Always turn off the oxygen flow first to prevent a possible flashback into the torch.

3.4.0 Shutting Down Oxyfuel Cutting Equipment

When a cutting job is completed and the oxyfuel equipment is no longer needed, it must be shut down. *Figure 53* identifies the order in which various pieces of the oxyfuel cutting equipment are shut down.

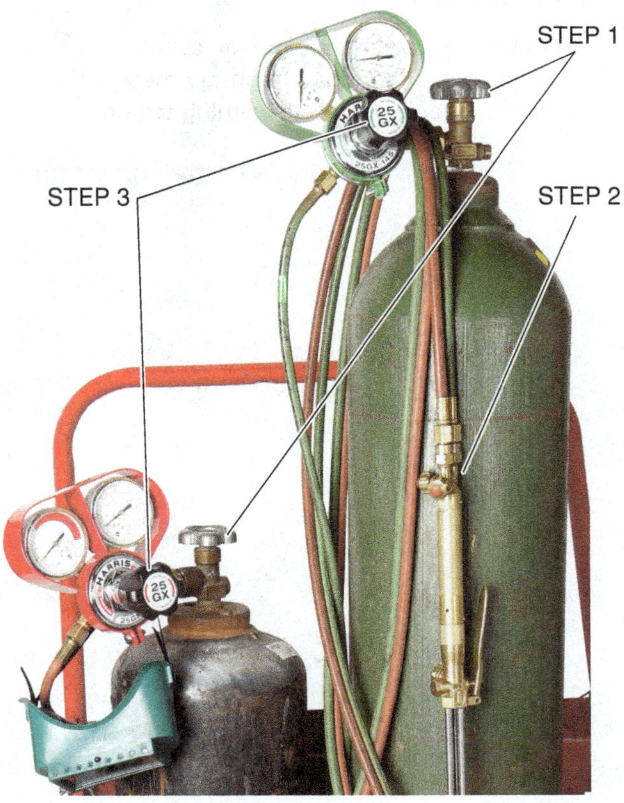

Figure 53 Shutting down oxyfuel cutting equipment.

Step 1 To begin the shutdown, first close the fuel gas and oxygen cylinder valves. Leave the regulators at their present setting.

Step 2 Open the fuel gas valve on the torch to allow all remaining gas to escape, and then close it. Next, open the oxygen valve on the torch to allow all remaining gas to escape, and then close it. These actions relieve the gas pressure in the hose and regulators, all the way back to the cylinder valve. Do not proceed to the next step until all pressure is released and all regulator gauges — both inlet and outlet — read zero.

Step 3 Turn the fuel gas and oxygen regulator adjusting screws counterclockwise to back them out, until they are loose.

Step 4 Coil and secure the hose and torch to prevent damage.

3.4.1 Disassembling Oxyfuel Equipment

In some situations, it may be necessary to disassemble the oxyfuel equipment after it has been used. Before any disassembly takes place, make sure that the equipment has been properly shut down. This includes checking that the cylinder valves are closed and all pressure gauges read zero.

Remove both hoses from the torch assembly and then detach the hoses from the regulators. Remove both regulators from the cylinders and reinstall the protective caps on the cylinders. The cylinders should now be returned to their proper storage place.

> **WARNING!**
> Always transport and store gas cylinders in the upright position. Be sure they are properly secured (chained) and capped. Regardless of whether the cylinders are empty or full, never store fuel gas cylinders and oxygen cylinders together without providing the required separation distance or fire-rated partition.

Obtaining Maximum Fuel Flow

Increasing the fuel flow until the flame pulls away from the tip and then decreasing the flow until the flame returns to the tip sets the maximum fuel flow for the tip size in use.

3.4.2 Changing Cylinders

Empty is a relative term when discussing gas cylinders. These cylinders should never be completely emptied because reverse flow could occur. Oxygen tanks, for example, should never get below the required working pressure, or about 25 psi (172 kPa). Once the cylinder pressure drops near the working pressure value, the torch will stop performing properly and the cylinder will have to be replaced. As a result, some residual pressure is typically present. Follow these procedures to change a cylinder:

> **WARNING!**
>
> When moving cylinders, always be certain that they are in the upright position and the valve caps are secured in place. Never use a sling or electromagnet to lift cylinders. To lift cylinders, use a cylinder cage.

Step 1 Begin by making sure that the equipment has been properly shut down. This includes checking that the cylinder valves are closed and all pressure gauges read zero.

Step 2 Remove the regulator from the empty cylinder and replace the protective cap on the cylinder.

Step 3 Mark MT (empty) and the date (or the accepted site notation for indicating an empty cylinder) near the top of the cylinder using soapstone (*Figure 54*).

29102-15_F54.EPS

Figure 54 Typical empty cylinder marking.

Step 4 Transport the empty cylinder from the workstation to the storage area. Place the empty cylinder in the empty cylinder section of the storage area for the type of gas it contained.

> **Marking and Tagging Cylinders**
>
> Do not use permanent markers on cylinders; use soapstone or another temporary marker. If a cylinder is defective, place a warning tag on it.

Additional Resources

ANSI Z49.1, Safety in Welding, Cutting, and Allied Processes. Miami, FL: American Welding Society.

Uniweld Products, Inc. Numerous videos are available at **www.uniweld.com/en/uniweld-videos**. Last accessed: November 30, 2014.

The Harris Products Group, a division of Lincoln Electric. Numerous videos are available at **www.harrisproductsgroup.com/en/Expert-Advice/videos.aspx**. Last accessed: November 30, 2014.

3.0.0 Section Review

1. Gas cylinders can be lifted to height ____ .
   a. with a strong electromagnet
   b. using a cylinder cage
   c. with a sling routed through the openings in the cap
   d. with a cable routed through the openings in the cap

2. Immediately after setup and periodically thereafter, oxyfuel equipment must be tested for ____ .
   a. purging
   b. carburizing
   c. flashbacks
   d. leaks

3. An oxidizing torch flame is one that has a(n) ____ .
   a. excess of fuel gas
   b. proper oxygen/fuel gas mix
   c. excess of oxygen
   d. abnormally low temperature

4. When oxyfuel cutting equipment is being shut down, how should the fuel gas and oxygen regulator adjusting screws be positioned after bleeding the remaining gas pressure in the hoses?
   a. Both screws tight
   b. Fuel gas screw tight; oxygen screw loose
   c. Fuel gas screw loose; oxygen screw tight
   d. Both screws loose

Section Four

4.0.0 Performing Cutting Procedures

Objective

Explain how to perform various oxyfuel cutting procedures.

a. Identify the appearance of both good and inferior cuts and their causes.
b. Explain how to cut both thick and thin steel.
c. Explain how to bevel, wash, and gouge.
d. Explain how to make straight and bevel cuts with portable oxyfuel cutting machines.

Performance Tasks

6. Cut shapes from various thicknesses of steel, emphasizing:
   - Straight line cutting
   - Square shape cutting
   - Piercing
   - Beveling
   - Cutting slots
7. Perform washing.
8. Perform gouging.
9. Use a track burner to cut straight lines and bevels.

Trade Terms

Drag lines: The lines on the edge of the material that result from the travel of the cutting oxygen stream into, through, and out of the metal.

The following sections explain how to recognize good and bad cuts, how to prepare for cutting operations, and how to perform straight-line cutting, piercing, bevel cutting, washing, and gouging.

4.1.0 Preparing for Oxyfuel Cutting with a Hand Cutting Torch

Before metal can be cut, the equipment must be set up and the metal prepared. One important step is to properly lay out the cut by marking it with soapstone or punch marks. The few minutes this takes will result in a quality job, reflecting craftsmanship and pride. The following procedures describe how to prepare to make a cut.

Prepare the metal to be cut by cleaning any rust, scale, or other foreign matter from the surface. If possible, position the work so that it can be cut comfortably. Mark the lines to be cut with soapstone or a scriber. Then select the correct cutting torch tip according to the thickness of the metal to be cut, the type of cut to be made, the amount of preheat needed, and the type of fuel gas to be used. Ignite the torch and use the procedures outlined in the following sections for performing specific types of cutting operations.

4.1.1 *Inspecting the Cut*

Before attempting to make a cut, operators must be able to recognize good and bad cuts and know what causes bad cuts. This is explained in the following list and illustrated in *Figure 55*:

- A good cut features a square top edge that is sharp and straight, not ragged. The bottom edge can have some dross adhering to it but not an excessive amount. What dross there is should be easily removable with a chipping hammer. The **drag lines** should be near vertical and not very pronounced.
- When preheat is insufficient, bad gouging results at the bottom of the cut because of slow travel speed.
- Too much preheat will result in the top surface melting over the cut, an irregular cut edge, and an excessive amount of dross.
- When the cutting oxygen pressure is too low, the top edge will melt over because of the resulting slow cutting speed.
- Using cutting oxygen pressure that is too high will cause the operator to lose control of the cut, resulting in an uneven kerf.
- A travel speed that is too slow results in bad gouging at the bottom of the cut and irregular drag lines.
- When the travel speed is too fast, there will be gouging at the bottom of the cut, a pronounced break in the drag line, and an irregular kerf.
- A torch that is held or moved unsteadily across the metal being cut can result in a wavy and irregular kerf.
- When a cut is lost and then not restarted carefully, bad gouges will result at the point where the cut is restarted.

A square kerf face with minimal notching not exceeding 1⁄16" (1.6 mm) deep is expected and, in fact, required in the Performance Accreditation Tasks for this module.

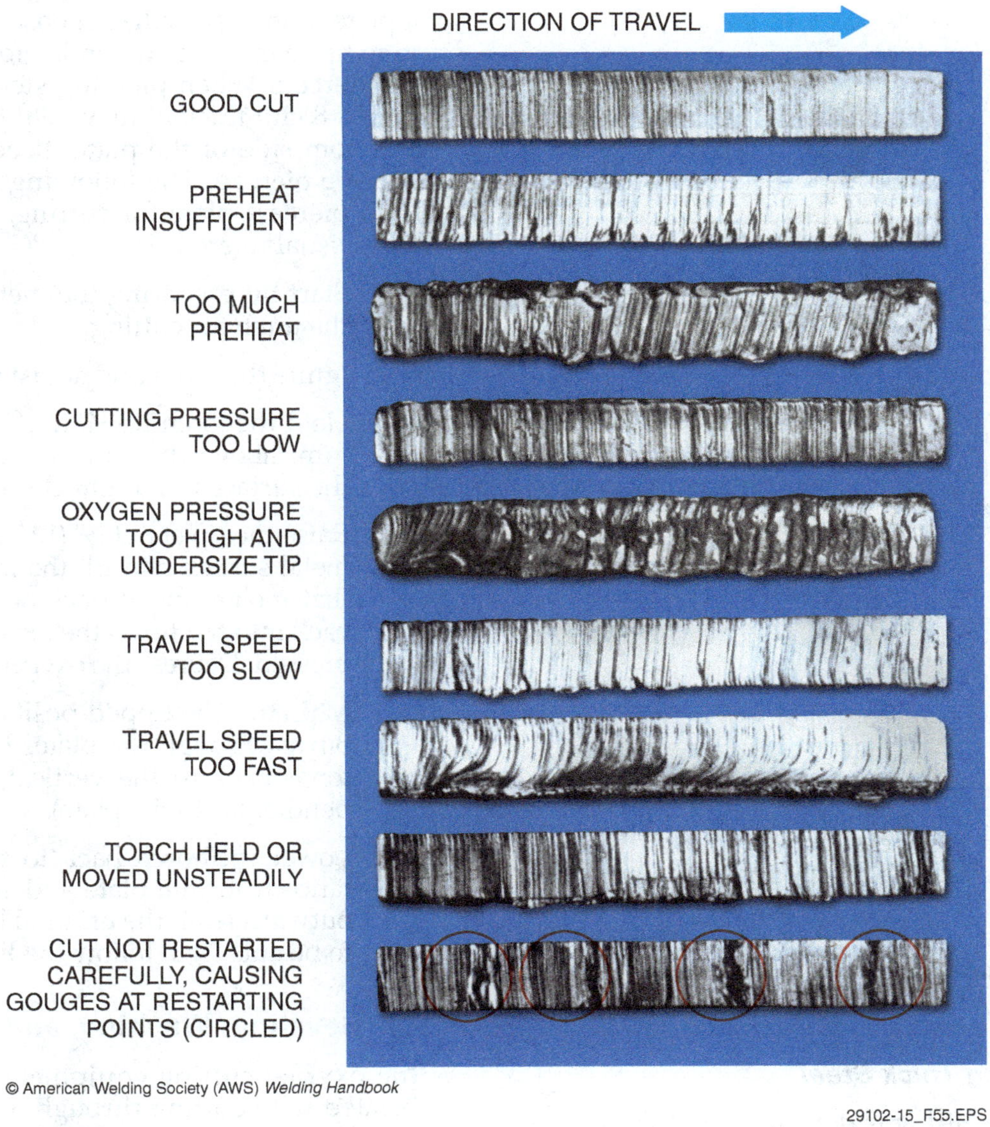

Figure 55 Examples of good and bad cuts.

The tasks in the sections that follow are designed to develop skills with a cutting torch. Each task should be practiced until there is thorough familiarity with the procedure. After each task is completed, it should be taken to the instructor for evaluation. Do not proceed to the next task until the instructor says to continue.

4.2.0 Cutting Steel

The effectiveness of cutting steel with an oxyfuel cutting outfit depends on factors such as the thickness of the steel, the cutting tip that is being used, and the skill of the operator.

4.2.1 Cutting Thin Steel

Thin steel is considered material ³⁄₁₆" (≈5 mm) thick or less. A major concern when cutting thin steel is distortion caused by the heat of the torch and the cutting process. To minimize distortion, move as quickly as possible without losing the cut.

To begin the process for cutting thin steel, first prepare the metal surface. Then light the torch and hold it so that the tip is pointing in the direction the torch is traveling at a 15- to 20-degree angle. Make sure that a preheat orifice and the cutting orifice are centered on the line of travel next to the metal (*Figure 56*).

> **CAUTION**
>
> Holding the tip upright (perpendicular to the metal) when cutting thin steel will overheat the metal, causing distortion. Maintain the 15 to 20-degree push angle as shown.

Preheat the metal to a dull red. Use care not to overheat thin steel because this will cause distortion. The edge of the tip can be lightly rested on

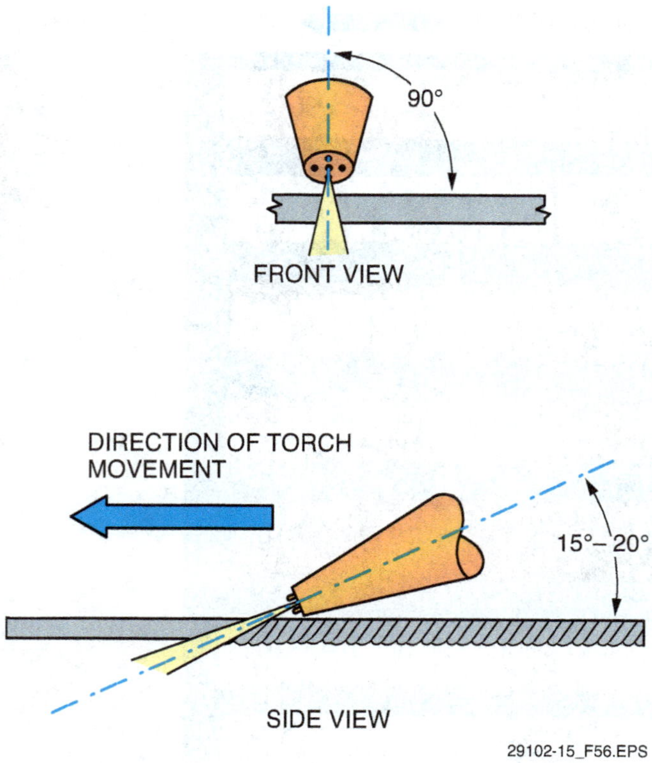

Figure 56 Cutting thin steel.

the surface of the metal being cut and then slid along the surface when making the cut in thin metal. Press the cutting oxygen lever to start the cut, and then move quickly along the line. To minimize distortion, move as quickly as possible without losing the cut.

4.2.2 Cutting Thick Steel

Most oxyfuel cutting is done on steel that is more than 3⁄16" (≈5 mm) thick. Whenever heat is applied to metal, distortion is a problem, but as the steel gets thicker, it becomes less of a problem.

To cut thick steel with a cutting torch, start by preparing the metal surface. Then light the torch and adjust the torch flame. Follow the number sequence shown in *Figure 57* to perform the cut.

The torch can be moved from either right to left or left to right. Choose the direction that allows the best visibility of the cut. When cutting begins, the tips of the preheat flame should be held 1⁄16" to 1⁄8" (1.6 to 3.2 mm) above the workpiece. For steel up to 3⁄8" (10 mm) thick, the first and third procedures can usually be omitted.

4.2.3 Piercing a Plate

Before holes or slots can be cut in a plate, the plate must be pierced. Piercing puts a small hole through the metal where the cut can be started. Because more preheat is necessary on the surface of a plate than at the edge, choose the next-larger cutting tip than is recommended for the thickness to be pierced. When piercing steel that is more than 3" (≈8 cm) thick, it may help to first preheat the bottom side of the plate directly under the spot to be pierced. The following steps describe how to pierce a plate for cutting. *Figure 58* provides a visual reference.

Step 1 Start by preparing the metal surface and the torch for cutting.

Step 2 Ignite the torch and adjust the flame.

Step 3 Hold the torch tip 1⁄4" to 5⁄16" (6.4 mm to 7.9 mm) above the spot to be pierced until the surface is a bright cherry red.

Step 4 Raise the tip about 1⁄2" (12.7 mm) above the metal surface and tilt the torch slightly so that molten metal does not blow directly back into the tip as the oxygen lever is depressed. Depress the oxygen lever.

Step 5 Maintain the tipped position until a hole burns through the plate. Then rotate the torch back to the vertical position (perpendicular to the plate).

Step 6 Lower the torch back to the initial distance from the plate and continue to cut outward from the original hole to the line to be cut. Then follow the line.

4.3.0 Beveling, Washing, and Gouging

While oxyfuel cutting equipment is commonly associated with cutting through metal plate, it is also well suited for cutting angles in the edge of steel plate, removing bolts and rivets, and cutting groves in metal surfaces.

4.3.1 Cutting Bevels

Bevel cutting is often performed to prepare the edge of steel plate for welding. The procedure for bevel cutting is illustrated in *Figure 59.*

Step 1 Prepare the metal surface and the torch.

Step 2 Ignite the torch and adjust the flame.

Step 3 Hold the torch so that the tip faces the metal at the desired bevel angle. Using a piece of angle iron as a cutting guide as shown in *Figure 59* will result in a 45-degree bevel angle. Angle iron can be used as a guide for any angle, as long as the operator consciously maintains the torch at the proper bevel angle.

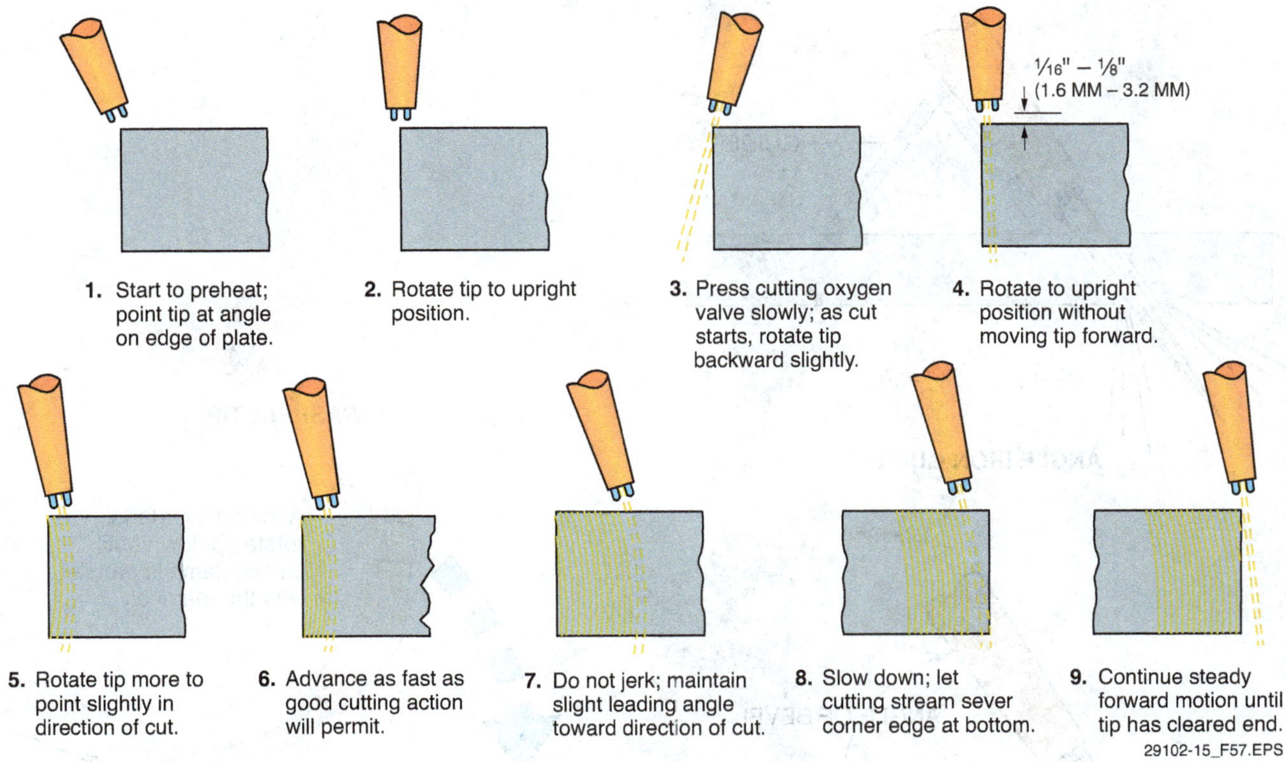

Figure 57 Flame cutting with a hand torch.

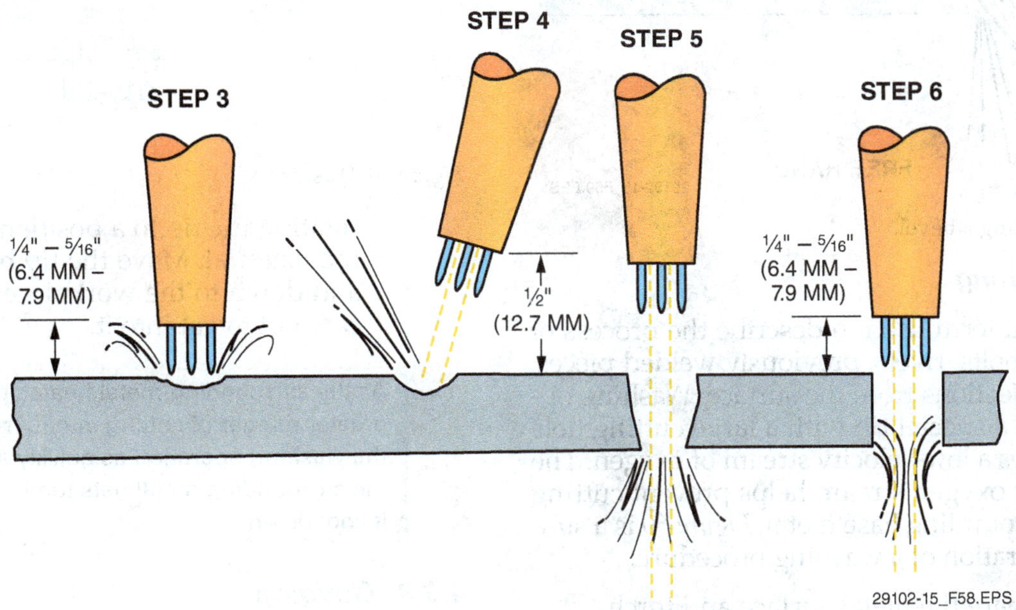

Figure 58 Steps for piercing steel.

Step 4 Preheat the edge to a bright cherry red.

Step 5 Press the cutting oxygen lever to start the cut.

Step 6 As cutting begins, move the torch tip at a steady rate along the line to be cut. Pay particular attention to the torch angle to ensure it creates a uniform bevel along the entire length of the cut.

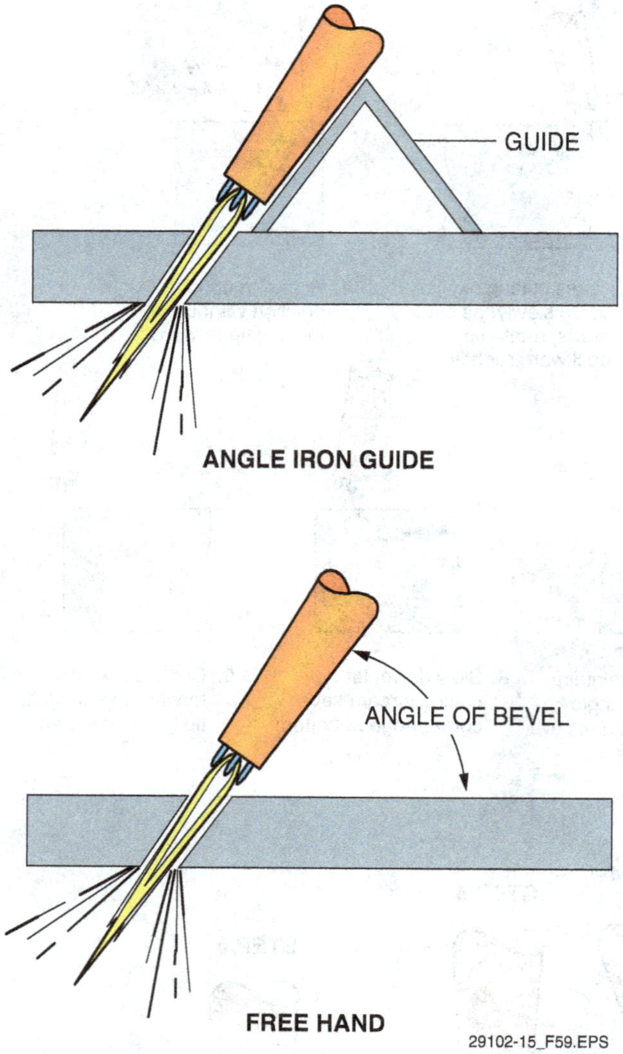

Figure 59 Cutting a bevel.

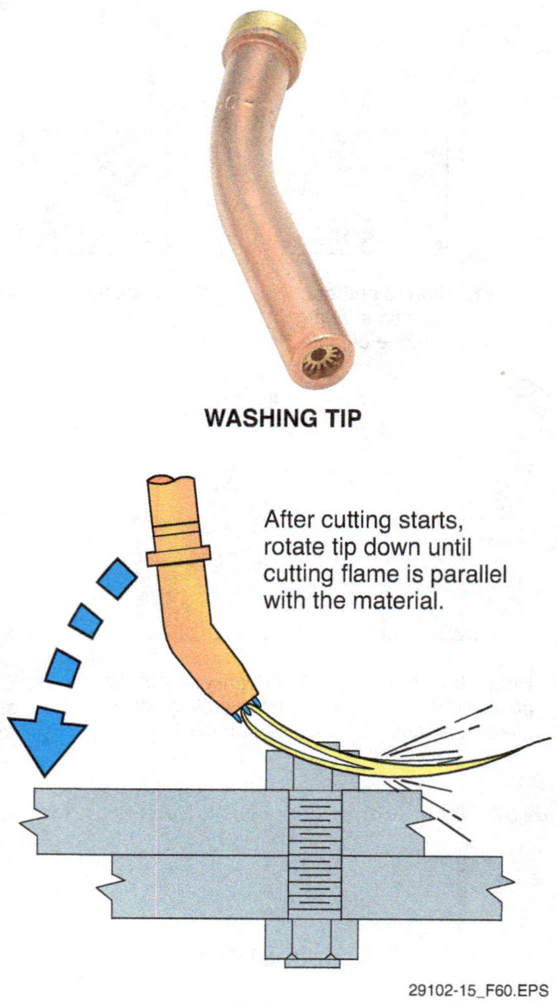

Figure 60 Washing.

4.3.2 Washing

Washing is a term used to describe the process of cutting out bolts, rivets, previously welded pieces, or other projections from the surface. Washing operations use a special tip with a large cutting hole that produces a low-velocity stream of oxygen. The low-velocity oxygen stream helps prevent cutting into the surrounding base metal. *Figure 60* is a simplified illustration of a washing procedure.

Step 1 Prepare the metal surface and torch.

Step 2 Ignite the torch and adjust the flame.

Step 3 Preheat the metal to be cut until it is a bright cherry red.

Step 4 Rotate the cutting torch tip to roughly a 55-degree angle to the metal surface.

Step 5 At the top of the material, press the cutting oxygen lever to begin cutting the material to be removed. Continue moving back and forth across the material while rotating the tip to a position parallel with the material. Move the tip back and forth and down to the workpiece surface. Take care not to cut into it.

CAUTION

As the surrounding metal heats up, there is a greater danger of cutting into it. Try to complete the washing operation as quickly as possible. If the surrounding metal gets too hot, stop and let it cool down.

4.3.3 Gouging

Gouging (*Figure 61*) is the process of cutting a groove into a surface. Gouging operations use a special curved tip that produces a low-velocity stream of oxygen that curves up, allowing the operator to control the depth and width of the groove. It is an effective means to gouge out cracks or weld defects for welding. Gouging tips can also be used to remove steel backing from welds or to wash off bolt or rivet heads. However, gouging tips are not as effective as washing tips for removing the shank of a bolt or rivet.

Figure 61 Gouging.

The travel speed and torch angle are very important when gouging. If the travel speed or torch angle is incorrect, the gouge will be irregular and there will be a buildup of dross inside the gouge. Practice until the gouge is clean and even, with a consistent depth.

Step 1 Prepare the metal surface and torch.

Step 2 Ignite the torch and adjust the flame.

Step 3 Holding the torch so that the preheat holes are pointed directly at the metal, preheat the surface until it becomes a bright cherry red.

Step 4 When the steel has been heated to a bright cherry red, slowly roll the torch away from the metal so that the holes are at an angle that will cut the gouge to the correct depth. While rolling the torch away, depress the cutting oxygen lever gradually.

Step 5 Continue to move the cutting torch along the line of the gouge while rocking it back and forth to create a gouge of the required depth and width.

4.4.0 Operating Oxyfuel Track Burners

Oxyfuel track burners, such as the one shown in *Figure 62*, provide a convenient way for operators to make straight cuts, curved cuts, and beveled cuts in the field. They can also enhance precision and uniformity in cuts. When a number of cuts are needed, track burners significantly increase productivity. Most models are portable, allowing them to be set up on the job site.

4.4.1 Torch Adjustment

The rack assembly on the track burner permits the torch holder assembly to move toward or away from the tractor unit. The torch holder allows vertical positioning of the torch. The torch bevel adjustment allows torch positioning at any commonly required angle. After adjusting the torch to the desired position, tighten all clamping screws to prevent the torch from making any unexpected movements.

4.4.2 Straight-Line Cutting

The following provides some basic information regarding the set-up of track burners. Although most track burners have a great deal in common, be sure to follow the manufacturer's operating procedures for the system in use.

To perform straight-line cutting with an oxyfuel track burner, first place the machine track on the workpiece and line it up before placing the machine on the track. Be sure the track is long enough for the cut to be made. If not, install additional track. Connect track sections carefully and ensure it is properly supported. When properly connected, the machine should travel smoothly from one track section to the next. If the cut is long, the track may have to be clamped at both ends beyond the cut to keep the track from moving during the cutting process.

WARNING!

Many cutting machines are not designed to detect the end of their track or workpiece. Take care that an unattended machine does not fall from an elevated workpiece while in operation.

Once the track has been positioned, place the machine on the track. Be sure that the supply gas hoses and the power lines are long enough and free to move with the machine so that it can complete the cut properly. Move the machine to the approximate point where the cut will start. Then set the Low/High speed switch to the desired cutting speed or speed range. Set the On/Off switch to the Off position. Next, plug the power cord into an appropriate power supply outlet. Ensure that all clamping screws are properly tightened. Now ignite and properly adjust the torch, and preheat the start of the cut. Set the Forward/Reverse switch to the desired direction of travel. Simultaneously turn on the cutting oxygen and rotate the cutting speed control knob to the desired rate of travel. When the cut is completed, stop the machine and shut off the torch.

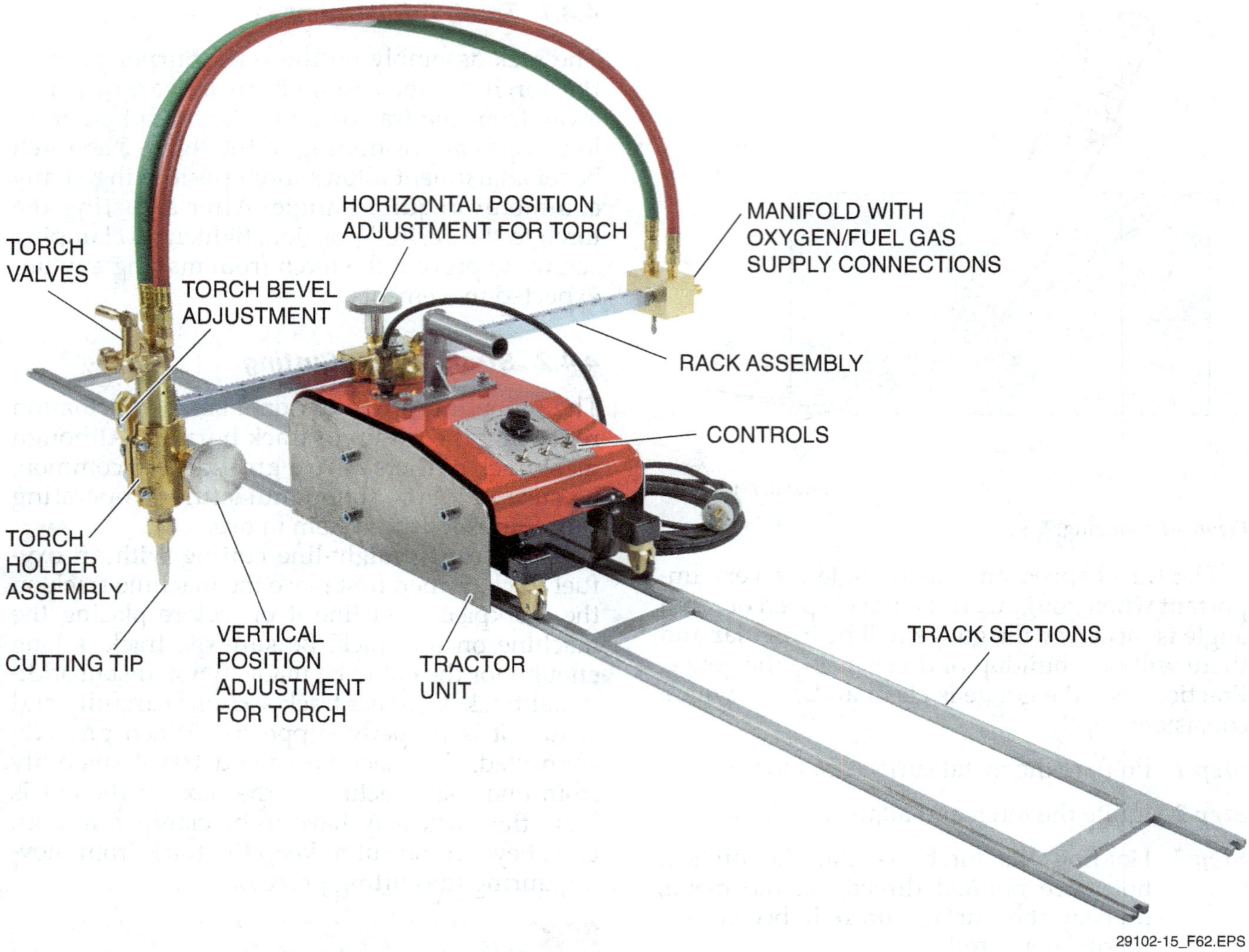

Figure 62 Portable oxyfuel track burner.

4.4.3 Bevel Cutting

Bevel cutting can also be performed with a portable oxyfuel track burner. As before, first place the machine track on the workpiece and line it up before placing the machine on the track. The track must be long enough for the cut to be made. If it is not, install additional track. Connect track sections carefully. Extend the track on both sides of the cut and support the track. When properly connected, the machine should travel smoothly from one track section to the next. If the cut is long, the track may have to be clamped at both ends beyond the cut to keep the track from moving during the cut.

With the track properly positioned, place the machine on the track. Make sure the supply gas hoses and the power lines are long enough and free to move with the machine so that it can complete the cut properly. Loosen the bevel adjusting knob, set the torch angle to the desired bevel angle, and then tighten the bevel adjusting knob.

Move the machine to the approximate point where the cut will start. Then set the Low/High speed switch to the desired cutting speed. Set the On/Off switch to the Off position. Next, plug the power cord into a 115 alternating current (AC), 60 Hertz (Hz) power outlet. Ensure that all clamping screws are properly tightened. Now ignite and properly adjust the torch, and preheat the start of the cut. Set the Forward/Reverse switch to the desired direction of travel. Simultaneously turn on the cutting oxygen and rotate the cutting speed control knob to the desired rate. When the cut is completed, stop the machine and shut off the torch.

Additional Resources

ANSI Z49.1, Safety in Welding, Cutting, and Allied Processes. Miami, FL: American Welding Society.

Uniweld Products, Inc. Numerous videos are available at **www.uniweld.com/en/uniweld-videos**. Last accessed: November 30, 2014.

The Harris Products Group, a division of Lincoln Electric. Numerous videos are available at **www.harrisproductsgroup.com/en/Expert-Advice/videos.aspx**. Last accessed: November 30, 2014.

4.0.0 Section Review

1. An oxyfuel cut that has gouging at the bottom, a pronounced break in the drag line, and an irregular kerf is most likely to occur when the ____.
   a. travel speed is too fast
   b. preheat is insufficient
   c. cutting oxygen pressure is too low
   d. cutting oxygen pressure is too high

2. The process of burning a small hole through the metal where an oxyfuel cut can be started is called ____.
   a. gouging
   b. beveling
   c. piercing
   d. washing

3. A washing operation uses a special torch tip with a large cutting hole that produces a ____.
   a. high-velocity stream of oxygen
   b. grooved cut in the base metal surface
   c. cleansing layer of dross in the cut
   d. low-velocity stream of oxygen

4. On a portable oxyfuel track burner, the cutting torch can be angled in a plane that is perpendicular to the track by using the torch ____.
   a. clamping screws
   b. bevel adjustment
   c. kerf positioner
   d. holder assembly

SUMMARY

Oxyfuel cutting has many uses on many different job sites. It can be used to cut metal plate and shapes to size, prepare joints for welding, clean metals or welds, and disassemble structures. Oxyfuel cutting equipment can range in size from small, portable sets to large, automated, fixed-position machines. In all cases, high pressures and flammable gases are involved. For that reason, there is always the danger of fire and explosion when using oxyfuel equipment. These risks can be minimized when the operator is well trained and knowledgeable. By understanding the safety precautions and equipment fundamentals presented in this module, operators will be better prepared to use oxyfuel cutting equipment and cutting techniques in their workplaces.

Review Questions

1. Oxyfuel cutting is best suited for use on ____.
   a. steel alloys
   b. ferrous metals
   c. nonferrous metals
   d. stainless steel

2. The stream of high-pressure cutting oxygen that is directed from an oxyfuel torch causes the metal to instantaneously ____ .
   a. solidify
   b. de-kerf
   c. magnitize
   d. oxidize

3. The recommended range of tinting for either the face shield or the safety glass lenses used during an oxyfuel cutting operation is ____ .
   a. 1 to 2
   b. 3 to 6
   c. 7 to 8
   d. 9 to 10

4. Most welding environment fires occur during ____ .
   a. oxyfuel gas equipment transporting
   b. carbon arc welding activities
   c. oxyfuel gas welding or cutting
   d. acetylene tank installation

5. When preparing a tank or vessel that might have contained flammable materials to be cut with an oxyfuel cutting torch, the best approach is to ____ .
   a. fill it with water
   b. purge it with oxygen
   c. fill it with air
   d. purge it with acetylene

6. When pure oxygen is combined with a fuel gas, it produces a ____ .
   a. high-pressure jet of cutting oxygen
   b. non-explosive, non-flammable vapor
   c. colorless, odorless, and tasteless gas
   d. high-temperature flame for cutting

7. The most common size of oxygen cylinder used in oxyfuel cutting applications is ____ .
   a. 85 cubic feet (2.4 cubic meters)
   b. 227 cubic feet (6.4 cubic meters)
   c. 350 cubic feet (9.9 cubic meters)
   d. 485 cubic feet (13.7 cubic meters)

8. If an acetylene cylinder is found lying on its side, what should be done before it is used?
   a. Contact the supplier for instructions.
   b. Release a small amount of the gas to atmosphere.
   c. Stand it upright and wait at least one hour.
   d. Add a liquid stabilizer to the gas.

9. The maximum hourly rate at which acetylene gas can be withdrawn from a cylinder is ____ .
   a. one half of the cylinder's capacity
   b. one third of the cylinder's capacity
   c. one fifth of the cylinder's capacity
   d. one seventh of the cylinder's capacity

10. The fuel gas that burns with the lowest flame temperature is ____ .
    a. propane
    b. propylene
    c. acetylene
    d. butane

11. Most gas pressure regulators contain two gauges—one that indicates the cylinder pressure and one that indicates the ____ .
    a. ideal cylinder pressure for the conditions
    b. pressure of the gas at the regulator outlet
    c. pressure in the accompanying cylinder
    d. maximum pressure for the cylinder

12. Oxyfuel cutting equipment that has left-hand threads and a V-notch in the nut are most likely to be ____ .
    a. oxygen fittings
    b. purging gas fittings
    c. aftermarket fittings
    d. fuel gas fittings

13. The type of cutting torch that uses a vacuum created by oxygen flow to draw in fuel from a very-low-pressure fuel source is a(n) ____ .
 a. siphon torch
 b. inert torch
 c. injector torch
 d. suspension torch

14. A friction lighter produces sparks when its steel surface is rubbed with a piece of ____ .
 a. flint
 b. soapstone
 c. graphite
 d. steatite

15. A computer-controlled plate cutting machine and an optical pattern-tracing machine are examples of ____ .
 a. portable oxyfuel cutters
 b. exothermic machines
 c. manual guide cutters
 d. fixed-location machines

16. Cracking the cylinder valves during the setup of oxyfuel equipment is a way of ____ .
 a. leak testing the valve regulators
 b. equalizing the cylinder pressures
 c. removing dirt from the valves
 d. venting residual gas from the cylinders

17. Before opening cylinder valves, verify that the adjusting screws on the oxygen and fuel gas regulators have been ____ .
 a. tightened
 b. closed
 c. loosened
 d. purged

18. Oxyfuel cutting equipment is typically leak tested using a(n) ____ .
 a. supervisor's sense of smell
 b. solution that produces bubbles
 c. lit match or a candle
 d. ultrasonic detector

19. An oxyfuel cutting flame that has an excess of fuel is called a(n) ____ .
 a. hot lean flame
 b. neutral flame
 c. oxidizing flame
 d. carburizing flame

20. When disassembling oxyfuel equipment, verify that all pressure gauges are ____ .
 a. reading zero
 b. open and showing atmospheric pressure
 c. chained securely to each other
 d. marked MT for storage

21. Gas cylinders should never be completely emptied because of the risk of ____ .
 a. reverse flow
 b. valve cracking
 c. disproportional mixing
 d. cylinder implosion

22. When a cut has been made with oxyfuel cutting equipment, the drag lines of the cut should be close to ____ .
 a. thirty degrees with minimal notching
 b. forty-five degrees with a wavy kerf
 c. horizontal and notched at the kerf
 d. vertical and not very pronounced

23. A major concern when cutting thin steel with an oxyfuel cutting torch is ____ .
 a. flashback
 b. sparking
 c. distortion
 d. backfire

24. An oxyfuel cutting process that is often used for cutting off bolts, rivets, and other projections is called ____ .
 a. beveling
 b. gouging
 c. piercing
 d. washing

25. What part of an oxyfuel track burner allows the vertical positioning of the torch to be controlled?
 a. The clamping screws
 b. The torch holder
 c. The kerf regulator
 d. The bevel adjustment

Trade Terms Quiz

Fill in the blank with the correct term that you learned from your study of this module.

1. The gap produced by a cutting process is called a(n) ________.
2. A loud snap or pop that can be heard as a torch flame is extinguished is called a(n) ________.
3. The soft, white material that is commonly used to mark metal is ________.
4. Metals that contain iron are called ________.
5. Creating a groove in the surface of a workpiece is a process called ________.
6. A flame that is burning with too much fuel is called a(n) ________.
7. A flame that is burning with too much oxygen is called a(n) ________.
8. When correct portions of fuel gas and oxygen are fed to a flame, the flame is said to be a(n) ________.
9. When a thermal process is used for cutting, the material expelled from the kerf is called ________.
10. Cutting off projections such as bolts, rivets, and previous welded pieces is a process referred to as ________.
11. The name used to describe what occurs when an oxyfuel cutting torch penetrates a metal plate is ________.
12. When a flame burns back into the tip of a torch and causes a high-pitched whistling sound, the condition is called ________.
13. The lines on the edge of a cut that result from the cutting oxygen streaming into, through, and out of the metal are called ________.

Trade Terms

Backfire
Carburizing flame
Drag lines
Dross
Ferrous metals
Flashback
Gouging
Kerf
Neutral flame
Oxidizing flame
Pierce
Soapstone
Washing

Appendix

Performance Accreditation Tasks

The American Welding Society (AWS) School Excelling through National Skills Standards Education (SENSE) program is a comprehensive set of minimum Standards and Guidelines for Welding Education programs. The following performance accreditation is aligned with and designed around the SENSE program.

The Performance Accreditation Tasks (PATs) correspond to and support the learning objectives in *AWS EG2.0, Guide for the Training and Qualification of Welding Personnel: Entry-Level Welder.*

Note that in order to satisfy all learning objectives in *AWS EG2.0*, the instructor must also use the PATs contained in the second level of the NCCER Welding curriculum.

PATs 1 and 2 correspond to *AWS EG2.0, Module 8 – Thermal Cutting Processes, Unit 1 – Manual OFC Principles*, Key Indicators 5, 6, and 7.

PAT 3 corresponds to *AWS EG2.0, Module 8 – Thermal Cutting Processes, Unit 1 – Manual OFC Principles*, Key Indicators 3 and 4.

PATs provide specific acceptable criteria for performance and help to ensure a true competency-based welding program for students.

The following tasks are designed to test your competency with an oxyfuel cutting torch. Do not perform these cutting tasks until directed to do so by your instructor.

SETTING UP, IGNITING, ADJUSTING, AND SHUTTING DOWN OXYFUEL EQUIPMENT

Using oxyfuel equipment that has been completely disassembled, demonstrate how to:

- Set up oxyfuel equipment
- Ignite and adjust the flame
  - Carburizing
  - Neutral
  - Oxidizing
- Shut off the torch
- Shut down the oxyfuel equipment

Criteria for Acceptance:

- Set up the oxyfuel equipment in the correct sequence ____________
- Demonstrate that there are no leaks ____________
- Properly adjust all three flames ____________
- Shut off the torch in the correct sequence ____________
- Shut down the oxyfuel equipment ____________

CUTTING A SHAPE

Using a carbon steel plate, lay out and cut the shape and holes shown in the figure. If available, use a machine track cutter to straight cut the longer dimension.

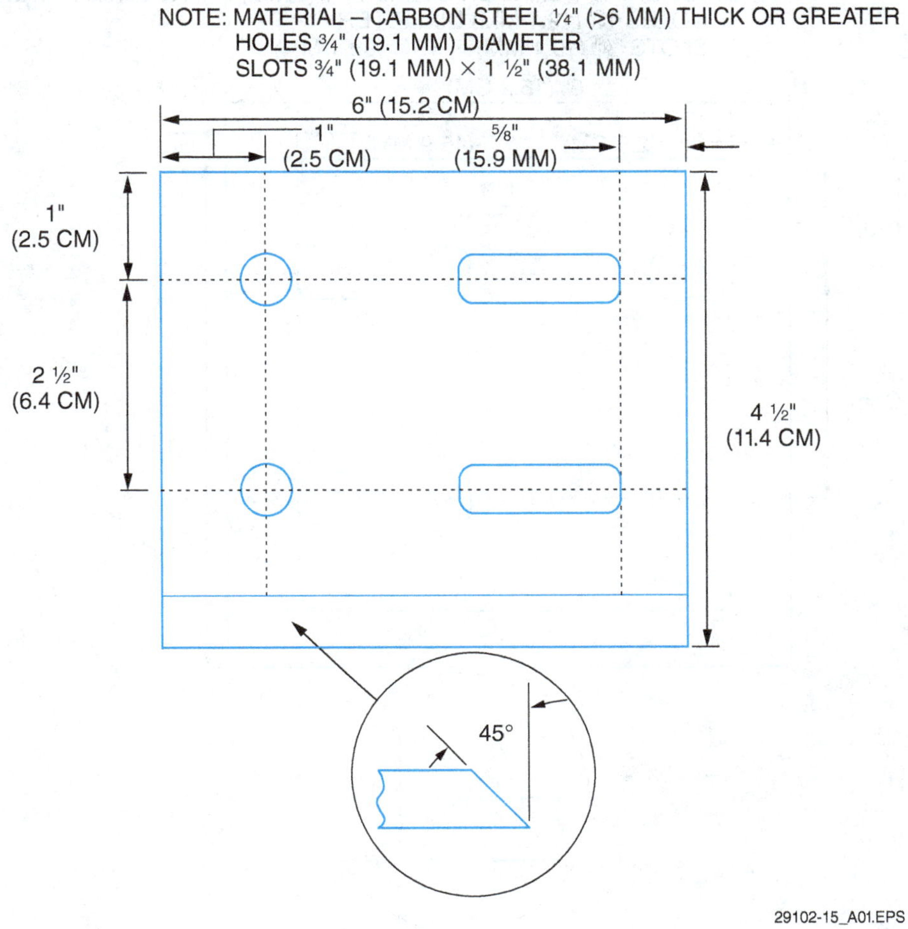

Criteria for Acceptance:

- Perform this task in the flat position (1G) ________
- Outside dimensions ±⅛" (3.2 mm) ________
- Inside dimensions (holes and slots) ±⅛" (3.2 mm) ________
- Square ±5 degrees ________
- Minimal amount of dross sticking to plate which can be easily removed ________
- Square kerf face with minimal notching not exceeding 1/16" (1.6 mm) deep ________

CUTTING A SHAPE

Using a carbon steel plate, lay out and cut the shape and holes shown in the figure. If available, use a machine track cutter to bevel and straight cut the longer dimension.

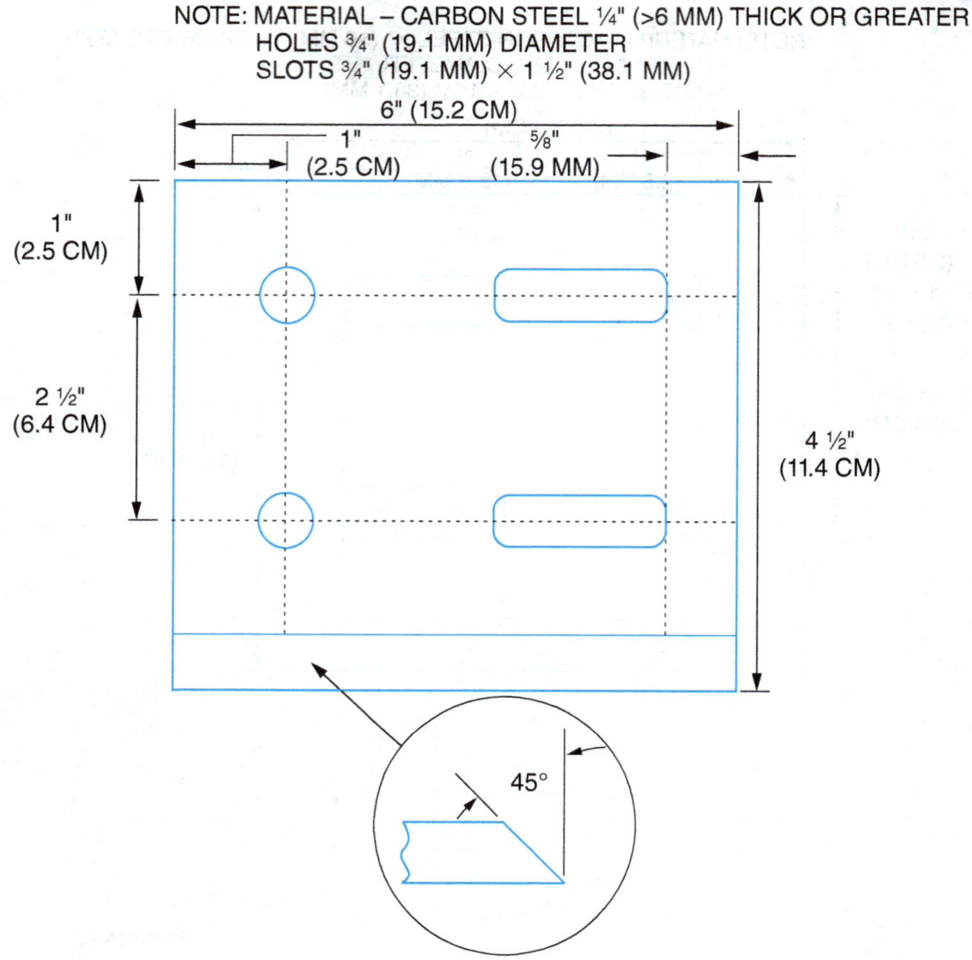

29102-15_A02.EPS

Criteria for Acceptance:

- Perform this task in the horizontal position (2G) ________
- Outside dimensions ±⅛" (3.2 mm) ________
- Inside dimensions (holes and slots) ±⅛" (3.2 mm) ________
- Square ±5 degrees ________
- Bevel ±2 degrees ________
- Minimal amount of dross sticking to plate which can be easily removed ________
- Square kerf face with minimal notching not exceeding 1/16" (1.6 mm) deep ________

Trade Terms Introduced in This Module

Backfire: A loud snap or pop as a torch flame is extinguished.

Carburizing flame: A flame burning with an excess amount of fuel; also called a reducing flame.

Drag lines: The lines on the edge of the material that result from the travel of the cutting oxygen stream into, through, and out of the metal.

Dross: The material (oxidized and molten metal) that is expelled from the kerf when cutting using a thermal process. It is sometimes called slag.

Ferrous metals: Metals containing iron.

Flashback: The flame burning back into the tip, torch, hose, or regulator, causing a high-pitched whistling or hissing sound.

Gouging: The process of cutting a groove into a surface.

Kerf: The gap produced by a cutting process.

Neutral flame: A flame burning with correct proportions of fuel gas and oxygen.

Oxidizing flame: A flame burning with an excess amount of oxygen.

Pierce: To penetrate through metal plate with an oxyfuel cutting torch.

Soapstone: Soft, white stone used to mark metal.

Washing: A term used to describe the process of cutting out bolts, rivets, previously welded pieces, or other projections from the metal surface.

Additional Resources

This module presents thorough resources for task training. The following resource material is suggested for further study.

ANSI Z49.1, Safety in Welding, Cutting, and Allied Processes. Miami, FL: American Welding Society.

Plasma Cutters Handbook: Choosing Plasma Cutters, Shop Safety, Basic Operation, Cutting Procedures, ANSI Z49.1, Safety in Welding, Cutting, and Allied Processes. Miami, FL: American Welding Society.

The Harris Products Group, a division of Lincoln Electric. Numerous videos are available at **www.harrisproductsgroup.com/en/Expert-Advice/videos.aspx**. Last accessed: November 30, 2014.

Uniweld Products, Inc. Numerous videos are available at **www.uniweld.com/en/uniweld-videos**. Last accessed: November 30, 2014.

Figure Credits

The Lincoln Electric Company, Cleveland, OH, USA, Module Opener, Figures 1, 2, 5, 14, 32–34, 48–50, 53, 60 (photo), 62, SA04

Topaz Publications, Inc., Figures 3, 7–9, 11, 13, 15, 16, 18, 27, 28, 29A, 30, 37–45, SA02, SA03, SA06–SA08, SA11

Courtesy of Uniweld Products, Figure 20, SA10

Victor Technologies, Figure 21

Vestil Manufacturing, Figure 29B

Koike Aronson, Inc. – Worldwide manufacturer of cutting, welding and positioning equipment, Figure 31

Courtesy of H & M Pipe Beveling Machine Company, Inc., Figures 35, 36

Zachry Industrial, Inc., Figures 46, 47, 51

Courtesy of Smith Equipment, Figure 52

© American Welding Society (AWS) *Welding Handbook* 1991, Welding Processes Volume No. 2, Edition No. 8, Miami: American Welding Society, Figure 55

Xerafy, SA01

Courtesy of Saf-T-Cart, SA09

Section Review Answer Key

Answer	Section Reference	Objective
Section One		
1. b	1.1.0	1a
2. d	1.2.2	1b
Section Two		
1. b	2.1.2	2a
2. a	2.2.3	2b
3. d	2.3.1	2c
4. a	2.4.4	2d
5. c	2.5.3	2e
Section Three		
1. b	3.1.1	3a
2. d	3.2.0	3b
3. c	3.3.1	3c
4. d	3.4.0	3d
Section Four		
1. a	4.1.1	4a
2. c	4.2.3	4b
3. d	4.3.2	4c
4. b	4.4.1	4d

NCCER CURRICULA — USER UPDATE

NCCER makes every effort to keep its textbooks up-to-date and free of technical errors. We appreciate your help in this process. If you find an error, a typographical mistake, or an inaccuracy in NCCER's curricula, please fill out this form (or a photocopy), or complete the online form at **www.nccer.org/olf**. Be sure to include the exact module ID number, page number, a detailed description, and your recommended correction. Your input will be brought to the attention of the Authoring Team. Thank you for your assistance.

Instructors – If you have an idea for improving this textbook, or have found that additional materials were necessary to teach this module effectively, please let us know so that we may present your suggestions to the Authoring Team.

NCCER Product Development and Revision

13614 Progress Blvd., Alachua, FL 32615

Email: curriculum@nccer.org

Online: www.nccer.org/olf

❑ Trainee Guide ❑ Lesson Plans ❑ Exam ❑ PowerPoints Other ____________

Craft / Level: Copyright Date:

Module ID Number / Title:

Section Number(s):

Description:

Recommended Correction:

Your Name:

Address:

Email: Phone:

Plasma Arc Cutting

Overview

Welders must be familiar with the task of plasma arc cutting. This module presents information related to the plasma arc cutting process, including safety procedures, setup, gas types, flow rates, and equipment.

Module 29103

Trainees with successful module completions may be eligible for credentialing through the NCCER Registry. To learn more, go to **www.nccer.org** or contact us at **1.888.622.3720**. Our website has information on the latest product releases and training, as well as online versions of our *Cornerstone* magazine and Pearson's product catalog.

Your feedback is welcome. You may email your comments to **curriculum@nccer.org**, send general comments and inquiries to **info@nccer.org**, or fill in the User Update form at the back of this module.

This information is general in nature and intended for training purposes only. Actual performance of activities described in this manual requires compliance with all applicable operating, service, maintenance, and safety procedures under the direction of qualified personnel. References in this manual to patented or proprietary devices do not constitute a recommendation of their use.

29103 V5

29103
Plasma Arc Cutting

Objectives

When you have completed this module, you will be able to do the following:

1. Explain plasma arc cutting processes and identify related safety precautions.
   a. Describe the plasma arc cutting processes.
   b. Identify safety practices related to plasma arc cutting.
2. Identify and describe plasma arc cutting equipment.
   a. Identify and describe plasma arc power units.
   b. Identify and describe plasma arc torches and accessories.
   c. Identify and describe plasma arc cutting gases and gas control devices.
3. Describe how to set up, safely operate, and care for plasma arc cutting equipment.
   a. Describe how to set up plasma arc cutting equipment and the adjacent work area.
   b. Describe how to safely operate plasma arc cutting equipment.
   c. Describe how to care for plasma arc cutting equipment.

Performance Tasks

Under the supervision of your instructor, you should be able to do the following:

1. Set up plasma arc cutting equipment.
2. Set the amperage and gas pressures or flow rates for the type and thickness of metal to be cut using plasma arc equipment.
3. Square-cut metal using plasma arc equipment.
4. Bevel-cut metal using plasma arc equipment.
5. Pierce and cut slots in metal using plasma arc equipment.
6. Dismantle and store the equipment.

Trade Terms

Amperage
Dross
Duty cycle
Phase
Plasma
Potential
Solenoid valve

Industry Recognized Credentials

If you are training through an NCCER-accredited sponsor, you may be eligible for credentials from NCCER's Registry. The ID number for this module is 29103. Note that this module may have been used in other NCCER curricula and may apply to other level completions. Contact NCCER's Registry at 888.622.3720 or go to **www.nccer.org** for more information.

Contents

Topics to be presented in this module include:

Figures and Tables

SECTION ONE

1.0.0 PLASMA ARC CUTTING PROCESSES

Objective

Explain plasma arc cutting processes and identify related safety precautions.

a. Describe plasma arc cutting processes.
b. Identify safety practices related to plasma arc cutting.

Trade Terms

Dross: A waste byproduct of molten metal.

Plasma: A fourth state of matter (not solid, liquid, or gas) created by heating a gas to such a high temperature that it boils electrons off the gas molecules (ionization).

Plasma arc cutting (PAC) uses a jet of **plasma** to pierce, cut, and gouge metal. The plasma is created by superheating gas in an electric arc. The temperature of the plasma jet is hot enough to melt almost any metal. The process can be set up for manual or mechanized operation (*Figure 1*) and, in most cases, is faster and more efficient than any other cutting method. PAC offers the following advantages:

- Cuts both ferrous and nonferrous metals
- Produces minimal **dross**
- Creates a minimal heat-affected zone
- Causes little or no distortion
- Possesses high cutting speeds
- Produces a very narrow kerf

The plasma arc cutting setup (*Figure 2*) requires an electrical power supply and a cutting gas, which can be compressed air. Depending on the equipment, a separate shielding or cooling gas may also be required. PAC torches convert the cutting gas into plasma by passing the gas through an electric arc inside the torch. The torch has a relatively small opening (orifice), which constricts the arc and high-pressure gas flow. This results in the plasma stream exiting the torch as a supersonic jet hotter than any flame. Depending on the current flow, the plasma cutting jet may reach temperatures of 30,000°F (16,650°C) or higher. When the jet contacts metal, it transfers its tremendous heat, causing the metal to melt instantly. The high-velocity jet blasts the molten metal away, forming a hole, groove, or gouge. Be advised that this high-velocity jet blast can carry molten material 40 feet (12 m) away from the workpiece.

(A) MANUAL PLASMA ARC CUTTING

(B) MECHANIZED PLASMA ARC CUTTING

29103-15_F01.EPS

Figure 1 Plasma arc cutting (PAC).

1.1.0 Plasma Arc Cutting Process

There are two types of plasma arc cutting processes: transferred arc and non-transferred arc.

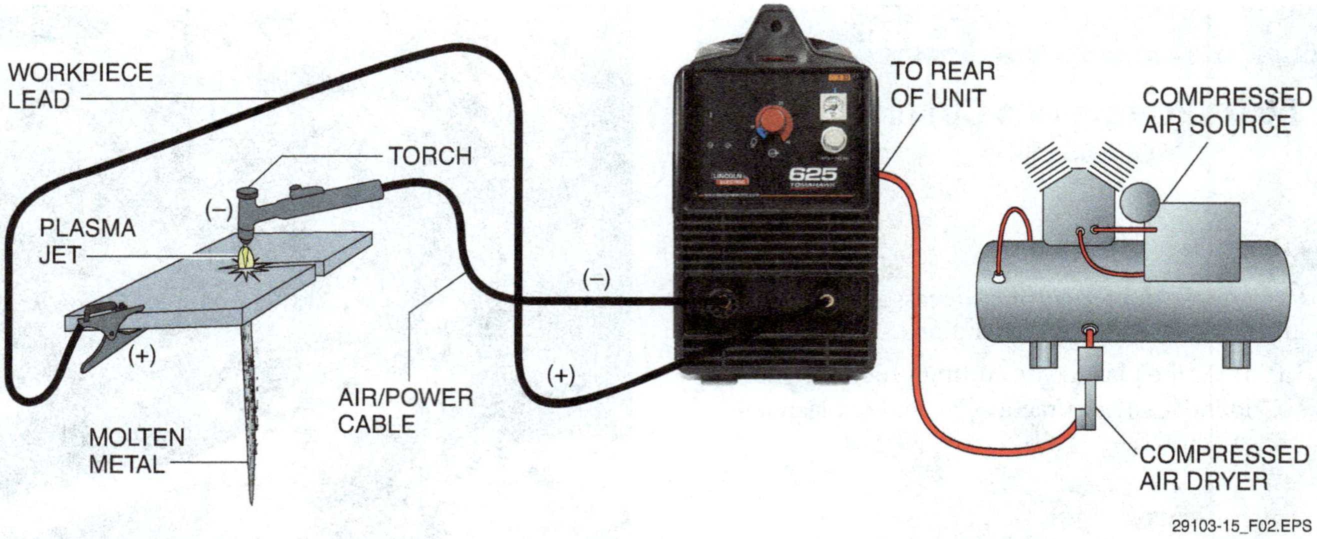

Figure 2 Plasma arc cutting setup diagram.

The transferred arc process is the most common and is used to cut materials that are electrically conductive, such as metals. The non-transferred arc process is less common and is used to cut materials that are not electrically conductive, such as ceramics and concrete.

1.1.1 Transferred Arc Process

In the transferred-arc process, the workpiece is part of the electrical circuit. The initial arc is established between the electrode and the workpiece via an arc transfer process. A number of different arc-initiation methods are used for the transfer process. These include contact starting, discharge-starting to the workpiece, and discharge-starting using a pilot arc. Of these methods, discharge-starting using a pilot arc is the most common. In this method, a low-voltage, low-current, high-frequency (HF) arc is established between the electrode and the torch nozzle inside the torch. This small arc between the torch electrode and torch nozzle is called a pilot arc. The pilot arc is used to ionize the air jet, resulting in plasma generation. When the torch is near the workpiece, a higher-voltage, current-cutting arc is transferred and established between the torch electrode and the work via the plasma. The higher-voltage, current-cutting arc maintains the plasma jet and the pilot arc is then shut off. For the transferred-arc process to work, one lead of the power supply (the positive lead) must be connected to the metal to be cut. *Figure 3* shows a schematic of the transferred-arc process.

1.1.2 Non-Transferred Arc Process

In the non-transferred arc process, the entire arc is established within the torch, between the torch electrode and the torch nozzle. The material being cut is not electrically connected to the arc circuit, since it is typically non-conductive material. The non-transferred arc process is not normally found in facilities that cut only metal. *Figure 4* shows a schematic of the non-transferred arc process.

1.2.0 Safety Practices

The following is a summary of safety procedures and practices that must be observed while cutting. Keep in mind that this is a summary only. Complete safety coverage is provided in the *Welding Safety* module. If you have not completed that module, do so before continuing. Above all, be sure to wear appropriate protective clothing and equipment when welding or cutting.

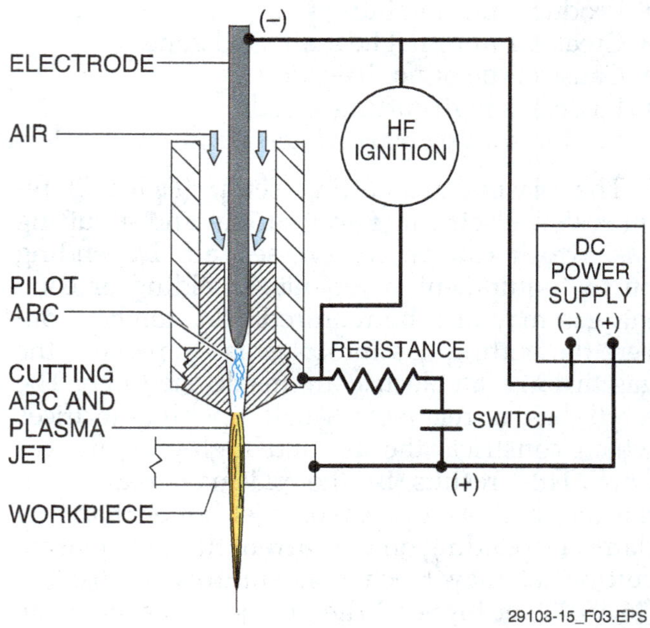

Figure 3 Schematic of the transferred-arc process.

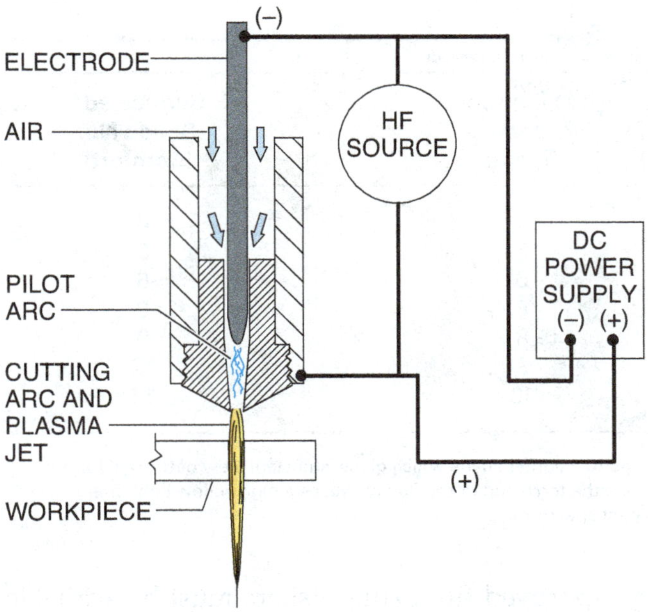

Figure 4 Schematic of the non-transferred arc process.

A plasma arc cutting process that uses air or nitrogen and is set up correctly is generally safer than oxyfuel processes. There is no chance of flashbacks and other dangers caused by flammable gases such as those used in oxyfuel processes. However, the voltages used, the extremely high temperatures of the cutting jet, and the intense ultraviolet radiation of the arc are hazardous. In addition to the safety practices provided in the following section, the following practices should also be observed:

- Make sure that all sources of energy to the PAC equipment are turned off before disassembling a plasma torch to replace any consumable components. Open-circuit DC voltages as high as 400 volts may be present between the torch components and the workpiece when they are connected.
- The extremely high temperatures of the plasma cutting jet can quickly and easily cut through gloves and flesh, resulting in severe injuries.

WARNING!

Working with energized PAC equipment is extremely hazardous. Always disconnect the power supply and all other sources of energy before servicing or otherwise working with plasma cutting equipment. Never place your hands under the workpiece near the cutting jet. Even with protective clothing, you can be severely injured.

1.2.1 Protective Clothing and Equipment

Welding activities can cause injuries unless you wear all of the protective clothing and equipment that is designed specifically for the welding industry. The following information includes important safety guidelines about protective clothing and equipment:

- Always use safety glasses with a full face shield or a helmet. The glasses, face shield, or helmet lens must have the proper light-reducing tint for the type of welding or cutting being performed. Use corrective lenses if needed; some companies will not allow contact lenses to be worn during cutting or welding activities. Never view an electric arc directly or indirectly without using a properly tinted lens.
- Wear proper protective leather and/or flame-retardant clothing along with welding gloves that will protect you from flying sparks and molten metal.
- Wear high-top safety shoes or boots. Make sure that the tongue and lace area of the footwear will be covered by a pant leg or other protective material. Boots with no laces at all are generally preferred for both welding and cutting activities. If the tongue and lace area is exposed, or if the footwear must be protected from burn marks, wear leather spats under your pants or chaps and over the top front area of the footwear.
- Wear a 100-percent cotton cap with no mesh material included in its construction. The bill of the cap points to the rear. If a hard hat is required for the environment, use one that allows the attachment of rear deflector material and a face shield. A hard hat with a rear deflector is generally preferred when working overhead, and may be required by some employers and job sites.

WARNING!

Do not wear a cap with a button in the middle. The conductive metal button beneath the fabric represents a safety hazard.

- Wear a face shield over snug-fitting cutting goggles or safety glasses for gas welding or cutting. Either the face shield or the lenses of the welding goggles must be an approved shade for the application. A welding hood equipped with a properly tinted lens is also acceptable. See *Table 1* for a list of lens shades based on arc current.

Table 1 Plasma Arc Cutting – Guide to Lens Shade Numbers

From *AWS F2.2 Lens Shade Selector*. Shade numbers are given as a guide only and may be varied to suit individual needs.

Process	Arc Current (Amperes)	Minimum Protective Shade	Suggested* Shade No. (Comfort)
Plasma Arc Cutting (PAC)	Less than 20	4	4
	20–40	5	5
	40–60	6	6
	60–80	8	8
	80–300	8	9
	300–400	9	12
	400–800	10	14

*As a rule of thumb, start with a shade that is too dark to see the weld zone. Then go to a lighter shade which gives sufficient view of the weld zone without going below the minimum. In oxyfuel gas welding, cutting, or brazing where the torch and/or the flux produces a high yellow light, it is desirable to use a filter lens that absorbs the yellow or sodium line of the visible light spectrum.

29103-15_T01.EPS

- Wear earplugs to protect ear canals from sparks. Wear hearing protection to protect against the consistent sound of the torch.

WARNING!

Ear protection is essential to protect ears from the noise of the torch. Other personal protective equipment (PPE) must be worn to protect the operator from hot metal and slag.

1.2.2 Fire/Explosion Prevention

Welding activities usually involve the use of fire or extreme heat to melt metal. Whenever fire is used, it must be controlled and contained. Welding or cutting activities are often performed on vessels that may once have contained flammable or explosive materials. Residues from those materials can catch fire or explode when a welder begins work on such a vessel. The following are fire and explosion prevention guidelines associated with welding:

- Never carry matches or gas-filled lighters in your pockets. Sparks can cause the matches to ignite or the lighter to explode, causing serious injury.
- Always comply with all site and/or employer requirements for a hot-work permit and a fire watch.
- Never use oxygen to blow dirt or dust off clothing. The oxygen can remain trapped in the fabric for a time. If a spark hits the oxygen in the fabric, the clothing can burn rapidly and violently.
- Make sure that any flammable material in the work area is moved or shielded by a fire-resistant covering.
- Approved fire extinguishers must be available before attempting any heating, welding, or cutting operations. Make sure the extinguisher is charged, the inspection tag is valid, and any individual that may be required to operate it knows how to do so.
- Never release a large amount of oxygen or use oxygen in place of compressed air. The presence of oxygen around flammable materials or sparks can cause rapid and uncontrolled combustion. Keep oxygen away from oil, grease, and other petroleum products.
- Never release a large amount of fuel gas, especially acetylene. Methane and propane are heavier than air and tend to migrate to and concentrate in low areas. As a result, they can ignite at a considerable distance from the release point. Acetylene is lighter than air but is even more dangerous than methane; when mixed with air or oxygen, it will explode at much lower concentrations than any other common fuel gas.
- To prevent fires, maintain a neat and clean work area, and make sure that any metal scrap or slag is cold before disposal.

Before cutting or welding containers such as tanks or barrels, find out if they contained any explosive, hazardous, or flammable materials, including petroleum products, citrus products, or chemicals that decompose into toxic fumes when heated. Proper procedures for cutting or welding hazardous containers are described in the *American Welding Society (AWS) F4.1, Safe Practices for the Preparation of Containers and Piping for Welding and Cutting,* and *ANSI Z49.1.* As a standard practice, always clean and then fill any tanks or barrels with water, or purge them with a flow of inert gas such as nitrogen to displace any oxygen.

> **WARNING!**
>
> Welding or cutting must never be performed on drums, barrels, tanks, vessels, or other containers until they have been emptied and cleaned thoroughly, eliminating all flammable materials and all substances (such as detergents, solvents, greases, tars, or acids) that might produce flammable, toxic, or explosive vapors when heated. Do not assume that a container that has held combustibles is clean and safe until proven so by proper tests. Do not weld in places where dust or other combustible particles are suspended in air or where explosive vapors are present.

Containers must be cleaned by steam cleaning, flushing with water, or washing with detergent until all traces of the material have been removed.

> **WARNING!**
>
> Clean containers only in well-ventilated areas. Vapors can accumulate during cleaning, causing explosions or injury.

After cleaning the container, fill it with water or a purging gas, such as carbon dioxide, argon, or nitrogen to displace the explosive fumes. Air, which contains oxygen, is displaced from inside the container by the water or inert gas. Without oxygen, combustion cannot take place.

A water-filled vessel is the best alternative. When using water, position the container to minimize the air space. When using an inert gas, provide a vent hole so the inert gas can push the air and other vapors out to the atmosphere. Keep in mind, though, that even these precautions do not guarantee the absence of flammable materials inside. For that reason, these types of activities should not be done without proper supervision and the use of proper testing methods.

1.2.3 Work Area Ventilation

Vapors and fumes tend to rise in the air from their sources. Welders often have to work above the welding area where the fumes are being created. Welding fumes can cause personal injuries. Good work area ventilation helps to remove the vapors and to protect the welder. The following is a list of work area ventilation guidelines to consider before and during welding activities:

- Make sure confined-space procedures are followed before conducting any welding or cutting in the confined space.
- Always perform cutting or welding operations in a well-ventilated area. Cutting or welding operations involving materials such as coatings that contain cadmium, lead, zinc, or chromium will result in toxic fumes. For cutting or welding of such materials, always wear an approved respirator with an appropriate filter as directed by your employer.
- Make sure confined spaces are ventilated properly for cutting or welding purposes.
- Never use pure oxygen from cylinders for the purpose of ventilation or breathing air.

Career Opportunities

The growth of the automobile racing industry has created career opportunities for workers who are skilled at welding and cutting metal. In addition to fabrication of race cars at the shop, there is a demand for at-the-track repair work. Lincoln Electric offers a Motorsports Welding School to address the special needs of the racing community.

29103-15_SA01.EPS

Additional Resources

AWS F3.2M/F3.2, Ventilation Guide for Weld Fume. Latest Edition. Miami, FL: American Welding Society.

Plasma Cutters Handbook: Choosing Plasma Cutters, Shop Safety, Basic Operation, Cutting Procedures, Advanced Cutting Tips, CNC Plasma Cutters, Troubleshooting, and Sample Projects. Eddie Paul. New York, NY: Penguin Group.

1.0.0 Section Review

1. The plasma arc cutting process type most commonly used is the ____ .
   a. electrical process
   b. non-transferred arc process
   c. oxyfuel pilot process
   d. transferred arc process

2. When operating plasma arc cutting equipment, it is acceptable to wear a hat made of ____ .
   a. polyester
   b. nylon
   c. 100-percent cotton
   d. wool

SECTION TWO

2.0.0 PLASMA ARC CUTTING EQUIPMENT

Objective

Identify and describe plasma arc cutting equipment.

a. Identify and describe plasma arc power units.
b. Identify and describe plasma arc torches and accessories.
c. Identify and describe plasma arc cutting gases and gas control devices.

Trade Terms

Amperage: The unit of measure used for the intensity of an electric current; often abbreviated to *amp* or *amps.*

Duty cycle: The percentage of time a plasma arc cutting machine can cut without overheating within a ten-minute period.

Phase: In a three-phase power supply system, the sine waves of the voltage on each of the three separate conductors is displaced from each of the others by 120 degrees, although all conductors are carrying alternating current at the same frequency and are synchronized. Each of the three sine waves is referred to as a phase of the power source.

Potential: The relative electrical voltage difference between two points of reference; the electromotive force or difference in potential between two points, expressed in volts.

Solenoid valve:: A valve used to control the flow of gases or liquids that is opened or closed by the action of an energized electromagnet.

There are many different manufacturers of PAC equipment. The following sections provide general information common to most equipment types. For information on the operation of specific equipment at your site, always refer to the manufacturer's documentation.

Basic plasma arc cutting equipment includes the following items:

- A power unit, where the process is controlled
- Plasma arc cutting torch with torch cable
- Workpiece lead assembly
- Plasma, shielding and cooling gases, and gas control components

2.1.0 PAC Power Units

Plasma-arc cutting is performed using direct-current, electrode-negative (DCEN) power. This means that the electrode of the torch has a negative **potential** compared to the workpiece lead. The power units are designed specifically for plasma arc cutting, and they are available in many sizes and configurations. The smallest and simplest PAC units are designed for light-duty manual cutting of sheet metal and light gauge plate. They typically plug into a single-**phase** 115VAC or 230VAC outlet. Typical maximum cutting **amperages** for these light-duty units range from 14 to 30 amperes. Their **duty cycle** is usually in the 35 percent range. The simplest units use filtered, compressed air for plasma and cooling gas. They have a high-frequency generator for easy arc initiation. Some of these small air units contain their own air compressor, to supply air for both cooling and plasma generation.

Commercial-grade portable units are available that can operate on various single-phase or three-phase power supplies. These are rated for cutting material up to 1¼" (3.2 cm) thick at duty cycles of 50 to 60 percent. At maximum current and operating at a reduced duty cycles, they are capable of severing material up to 1½" (3.8 cm) thick. These units normally require an external air or gas supply for the torch. The size of the unit needed is determined by the type and thickness of the metal to be cut.

Console controls are relatively simple. They usually include most of the following controls and indicators:

- Power On/Off switch
- Power On and/or unit-ready indicator
- Air or gas On/Off switch or setting (for monitoring and adjusting torch air or gas flow, or for purging lines)
- Output current control
- Power unit trouble indicator(s)

Figure 5 shows a typical portable commercial-grade plasma arc cutting unit.

The Hazards of PAC

Ultraviolet radiation, hot particle matter, and noise are all hazards associated with PAC, but they are manageable with the proper equipment. Devices such as water tables and water mufflers use water to control such hazards in mechanized plasma cutting applications.

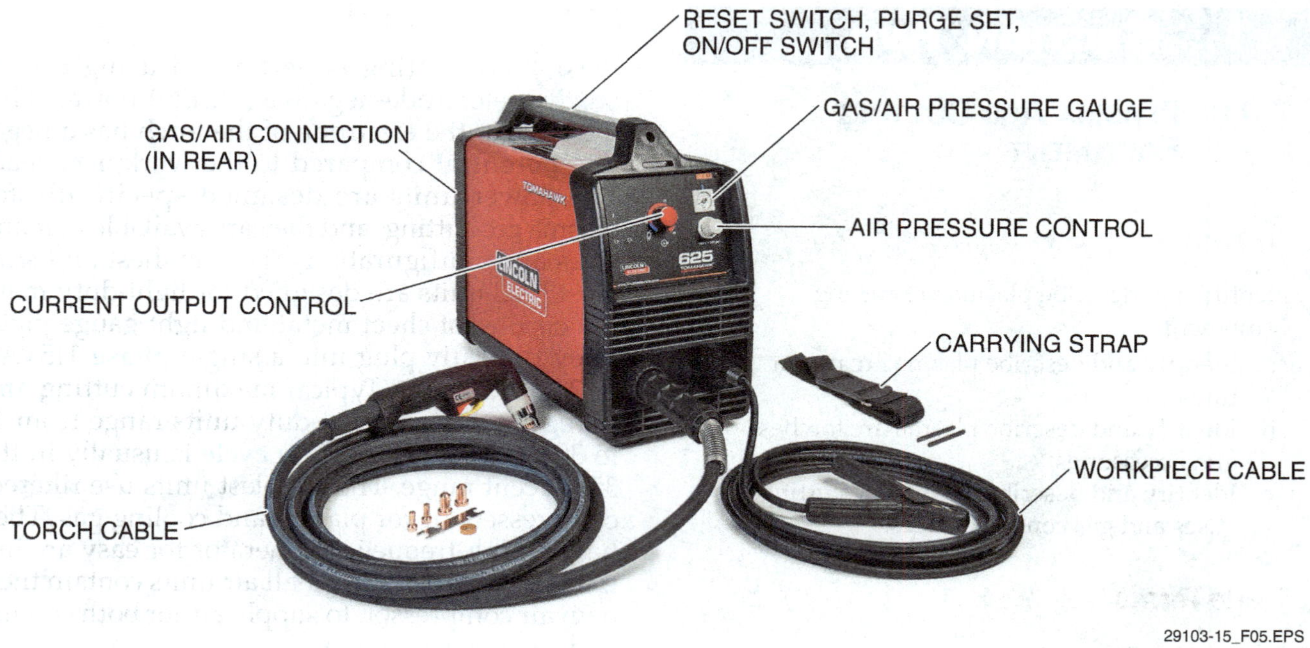

Figure 5 Commercial-grade plasma arc cutting unit.

Larger industrial units are used for heavy-duty manual cutting and general-duty mechanized cutting. These units usually operate on a 230/460VAC three-phase power supply, producing DC output currents up to 750 amperes. Their duty cycle is usually 100 percent, meaning that they are capable of constant operation. They can cut stainless steel and aluminum up to 2½" (6.4 cm) thick and carbon steel up to 3" (7.6 cm) thick. Some units are water-cooled, often containing their own closed-loop torch cooling system.

The larger PAC units are often dual-flow, mechanized units that operate with several types of gases. One gas is used for cutting, and the other is used as a shield around the cutting area, to improve the quality of the cut. Some units may use only one of the gases for the pilot arc.

Although gouging is not the primary use for plasma arc torches, some commercial and industrial units may be used for gouging when fitted with a gouging torch, or a torch with gouging components. Plasma arc gouging is sometimes referred to using the nonstandard acronym PAG.

Controls on the larger units typically include a power On/Off switch, output current control and ammeter, several gas flow controls and gauges, gas selector controls, local/remote switch, and open-circuit voltmeter. A leather cable cover accessory is often used to protect the torch cable from damage from the resulting hot or molten metal. The covers are typically designed with snaps or Velcro® to secure it around the torch cable.

2.1.1 Workpiece Clamp

The workpiece clamp (*Figure 6*) provides the connection between the end of the workpiece lead and the workpiece. The area on the workpiece where the clamp is being attached must be cleaned well enough for good metal-to-metal contact. Workpiece clamps are mechanically connected to the welding lead, and they come in a variety of shapes and sizes. Workpiece clamps are rated by the current they can carry without overheating. If a workpiece clamp needs to be replaced, be sure to select one that is rated at or above the output capacity of the PAC power unit.

When PAC will be used to cut equipment containing electrical or electronic components, batteries, bearings, or seals, the workpiece clamp must be properly located to prevent damage to associated components. If the cutting current passes through any type of bearing, seal, valve, contacting surface, or electrical or electronic component, it could cause severe damage to the item from arcing and overheating.

Cutting Ratings

Manufacturers may rate the cutting capability of their equipment in different ways. Some may rate it at the listed duty cycle, and others may rate it at the maximum cutting (severing) capability at a much reduced duty cycle. When selecting a power supply, make sure that you understand the manufacturer's rating system.

Control Interlocks

Interlocks may be incorporated into the controls of PAC systems. An interlock common in PAC systems monitors the air supply pressure. If the plasma torch is operated without an adequate supply of air, internal arcing may damage the torch. For this reason, a gas pressure safety switch is included in the circuit. Adequate gas pressure must be present before the torch will operate. The safety switch also shuts down the torch if there is a gas supply failure during cutting. Another interlock is an over-temperature switch that shuts down the unit if the duty cycle is exceeded and the unit overheats. Some units are also equipped with a retaining cup switch that shuts down the unit if the retaining cup holding the tip, nozzle, and swirl ring in the torch is loose.

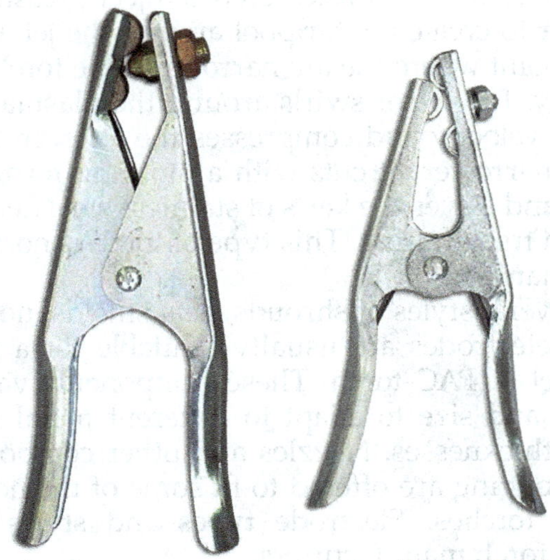

29103-15_F06.EPS

Figure 6 Typical workpiece clamps.

Carefully evaluate the cutting task and position the workpiece clamp so that the cutting current will not pass through any sensitive components or across gaps. If in doubt, ask your supervisor for assistance before proceeding.

CAUTION

To prevent damage from arcing across gapped or contacting surfaces, make sure the path from the workpiece clamp to the torch arc does not pass through these surfaces. Cutting current can severely damage bearings, seals, valves, or contacting surfaces. Position the workpiece clamp to prevent the cutting current from passing through them.

Disconnect the grounded battery lead on any mobile equipment that is being repaired using the plasma arc cutting process. This prevents the cutting current from causing battery explosion or battery damage. Never operate PAC equipment near a battery.

Do not cut near batteries. A spark could cause a battery to explode, showering the area with battery acid. Always remove batteries from the work area. Also, workpiece leads must never be connected to pipes carrying flammable, volatile, or corrosive materials. The cutting current could cause overheating or sparks, resulting in an explosion or fire.

CAUTION

The cutting current can destroy electronic or electrical equipment. Have an electrician check the equipment and, if necessary, isolate the system before cutting.

2.2.0 PAC Torches

The most common plasma arc cutting torch uses a pilot arc and the transferred arc process. A high-frequency pilot arc is first started between the electrode and the nozzle. A small amount of gas is injected into the arc chamber where it is heated to a plasma and escapes through the nozzle as a fine jet. However, the torch current circuit is designed so that the full voltage, or potential, and current are not available in the torch between the internal electrode and nozzle. The base metal to be cut is connected to the power unit by the workpiece clamp to make a complete circuit. This gives the workpiece a higher potential difference than the nozzle. When the torch is moved near the workpiece, the higher-current arc extends (transfers) to the workpiece because of its higher potential. The control unit automatically increases the amperage and gas flow(s) to produce the longer cutting arc, and disables the pilot arc. *Figure 7* shows the result of the plasma transfer process that takes place between the torch and the workpiece.

GOING GREEN

Cooling Water

Welders involved in metal-cutting activities often allow the water used in the process to simply flow into the ground or down the drain. To better protect the environment, water used for cooling the plasma arc cutting equipment should be contained whenever possible and sent to a waste treatment facility.

29103-15_F07.EPS

Figure 7 A transferred plasma cutting arc.

Plasma arc torches may be handheld for manual cutting or machine mounted for mechanized cutting. The heaviest duty torches are usually mechanized and water-cooled. Handheld torches are usually light-duty to heavy-duty, with heavy-duty models typically being water-cooled. Some torches are equipped with separate passages and orifices for cooling and shielding gases. Torches have a button or lever on the torch handle to start the pilot arc. A foot pedal can also be used to start the arc. When a foot pedal is used, the switch on the control panel is set for remote operation.

The typical PAC torch usually contains a replaceable retaining cup, a copper electrode with a hafnium or thoriated tungsten insert, a gas swirl ring, and a nozzle. Nozzles have various-sized holes for cutting metals of different thicknesses. The torch also contains coolant passages for gas or water cooling. *Figure 8* shows a typical handheld PAC torch and its consumable components.

Auto-Voltage Adjustments

Some newer PAC units sense the input voltage and frequency when they are connected to a power source. The units continuously and automatically reconfigure themselves to operate properly with the connected power. This eliminates improper initial power connections and provides continuous adjustment to changing input voltage so that the PAC arc current remains stable during cutting operations. This feature also increases the flexibility of the unit to operate in various locations.

Mechanized PAC torches are designed to be operated on automated carriers. They can usually be used in the same carriers designed for oxyfuel cutting heads. However, any carrier used must be capable of moving at the much greater speeds required for effective plasma arc cutting.

Water-injection PAC torches inject pressurized water to create a whirlpool around the jet, below the point where the arc narrows in the torch's arc cavity. The water swirls around the plasma jet at high velocity and compresses the jet even more. The narrower jet cuts with a more uniform melt rate and leaves the kerfs of stainless steel and titanium free of dross. This type of torch is normally mechanized.

Several styles of shrouds, heat shields, nozzles, and electrodes are usually available for a given model of PAC torch. These components vary in style and size to adapt to different metal types and thicknesses. Nozzles and other components for gouging are offered to fit some of the heavier duty torches. Electrode types and styles vary with torch manufacturers.

The most common parts that are changed in a torch are the electrode and the nozzle. The electrode is changed when it becomes contaminated. The nozzle is changed when it becomes worn and no longer true to its original shape or size, or when metal of a different type or thickness needs to be cut. When performing low-amperage cutting, it is possible to drag the torch on the workpiece using a drag shield or standoff post guide without causing damage or sacrificing the cut. Cutting at current levels above 60 amps or for an extended time usually requires a standoff guide, which attaches to the cup and holds the tip at the required distance from the workpiece. *Figure 9* shows two common styles. The standoff distance, or the distance between the torch tip and the workpiece, should be kept between 1⁄16" and 1⁄4" (1.6 to 6.4 mm) depending on the thickness of the workpiece.

2.3.0 PAC Gases and Controls

All PAC units require one or more types of gases. The specific type of gases required and the controls necessary to set and adjust them vary with the type, size, and manufacturer of the PAC equipment.

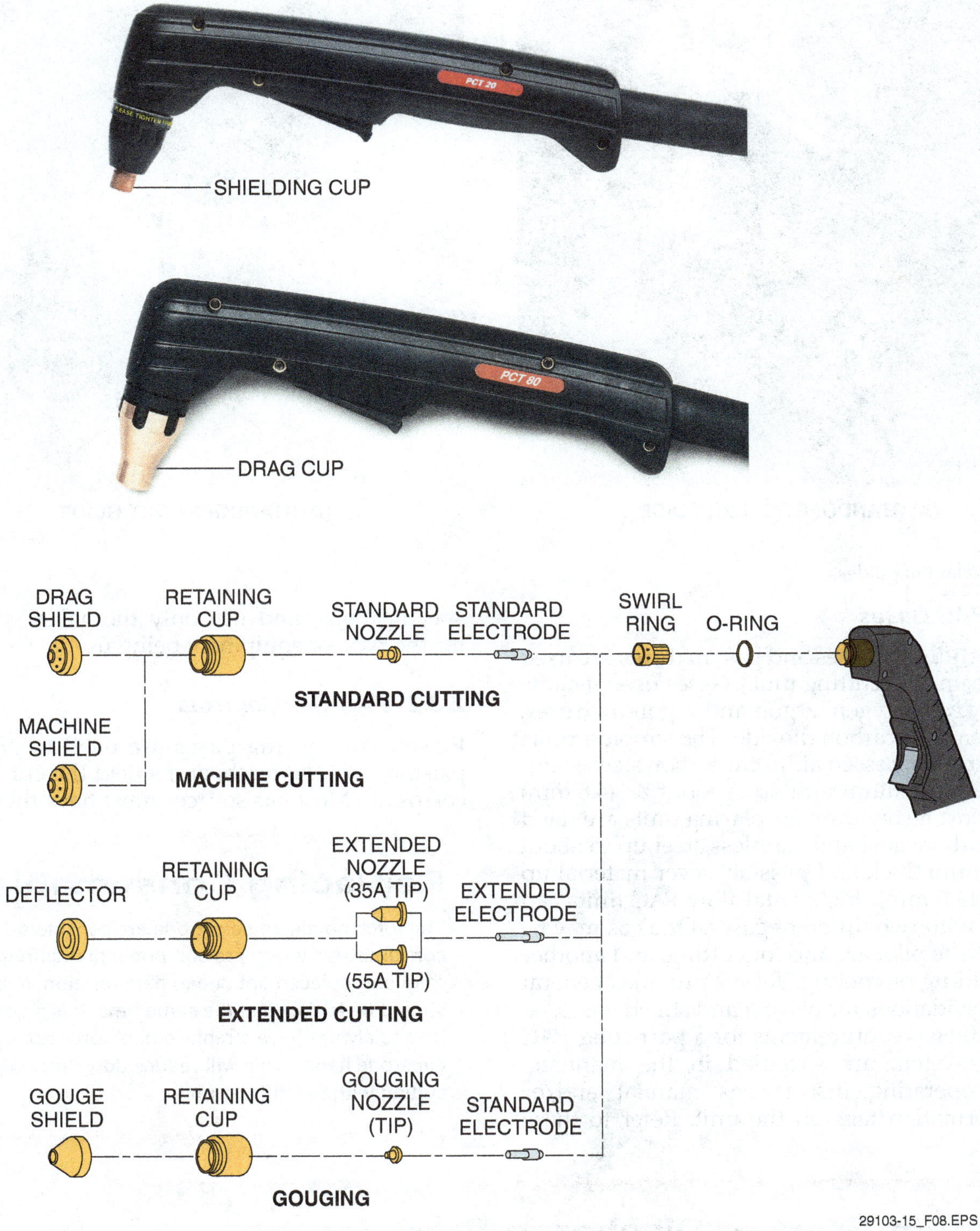

Figure 8 Handheld plasma torches.

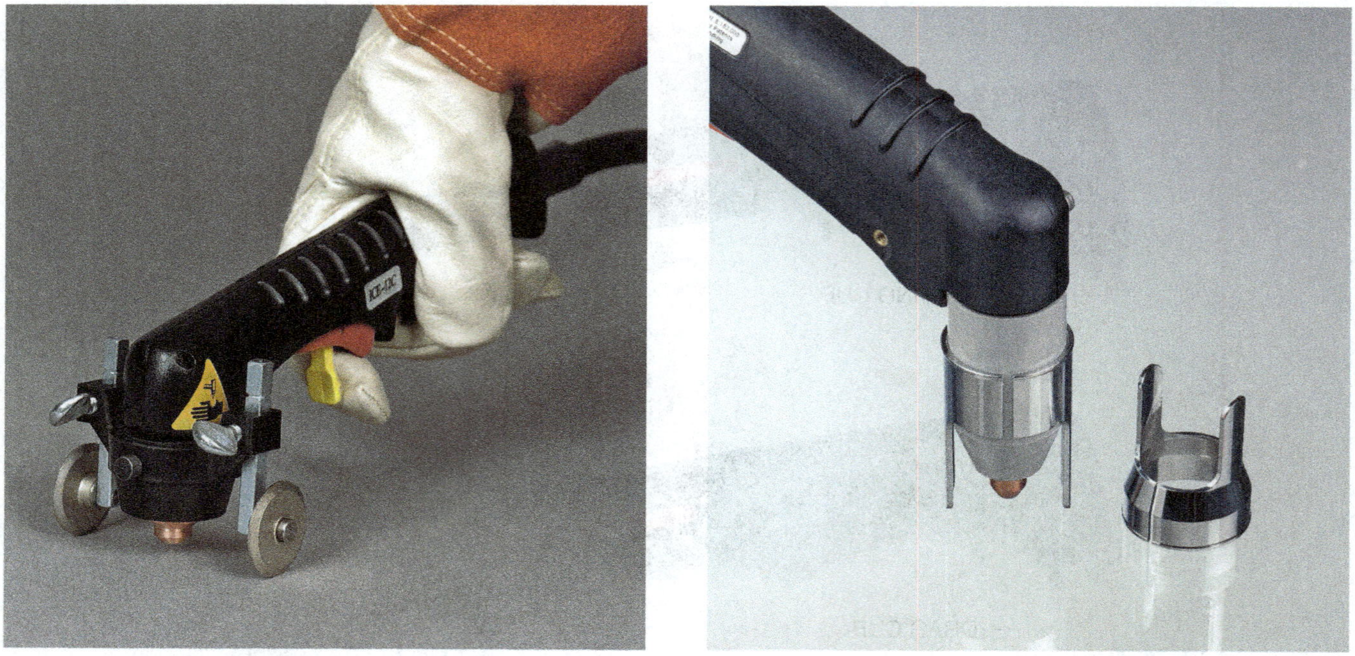

(A) STANDOFF ROLLER GUIDE

(B) STANDOFF POST GUIDE

29103-15_F09.EPS

Figure 9 Standoff guides.

2.3.1 PAC Gases

Several different gases and gas mixtures are used with plasma arc cutting units. Gases used include air, nitrogen, oxygen, argon and argon mixtures, hydrogen, and carbon dioxide. The simplest units use clean compressed air to cut carbon steel, stainless steel, and aluminum up to about 3⁄16" (4.8 mm) thick. Most heavy-duty air plasma units are rated to cut carbon steel and stainless steel up to about 1¼" (32 mm) thick and possibly sever material up to 1¾" (44.5 mm) thick. Dual-flow PAC units can operate with two different gases. One gas may be used for the pilot arc and for cutting, and another for shielding or cooling. *Table 2* provides general recommendations for plasma and shield gases.

Specific gas requirements for a particular PAC unit or system are specified in the manufacturer's operating instructions manual and/or the information tags on the unit. Refer to these specifications, and use only the gases specified for the type of equipment being used.

2.3.2 PAC Gas Controls

Plasma arc cutting gases are used to generate plasma, cool the torch, and shield the cut against corrosion. Most gas sources must have their pres-

Replacing Consumables

The torch nozzle and electrode are considered consumables, which are components requiring frequent replacement due to deterioration, and both should be replaced at the same time. It is a good idea to always have a replacement torch nozzle and electrode handy. This will reduce downtime when cutting with the plasma unit.

Contact-Start or Discharge-Start Torches

High-frequency pilot arc starting of a PAC torch can interfere with nearby electronic circuits, including computers and microprocessor controllers. To eliminate the use of a high-frequency start, some newer PAC torches use an internal contact or capacitor-discharge pilot arc starting system. One such contact-start system has a spring-loaded electrode that is in contact with the nozzle when the torch is off. When a trigger is pressed, the PAC unit senses the shorted electrode via high-current flow and opens the gas solenoid. Gas pressure then forces the electrode away from the nozzle. This action draws an arc that establishes the plasma flow. The arc instantly transfers to the workpiece because of the higher voltage difference between the electrode and the workpiece.

Table 2 Recommended Plasma/Shield Gas Combinations

Material	Air/Air	O_2/Air	N_2/CO_2	N_2/Air	$H35/N_2$
Carbon Steel	Most Economical Good Cut Quality Good Speed Good Gouging Good Weldability	Best Cut Quality Maximum Cut Speed Best Weldability	Some Dross Long Electrode Life	Not Recommended	Best Gouging Long Electrode Life Some Dross
Stainless Steel	Most Economical Good Speed Some Dross	Not Recommended	Good Cut Quality Good Gouging Minimal Dross Long Electrode Life	Long Electrode Life Lowest Shield Gas Cost	Best Cut Quality Best Gouging Minimal Dross Long Electrode Life Cuts Thicker Material
Aluminum	Most Economical Good Speed Some Dross	Not Recommended	Good Cut Quality Good Gouging Minimal Dross Long Electrode Life	Not Recommended	Best Cut Quality Best Gouging Minimal Dross Long Electrode Life Cuts Thicker Material

O_2 = Oxygen
N_2 = Nitrogen
CO_2 = Carbon Dioxide
H35 = Mixture of 35% Hydrogen and 65% Oxygen

29103-15_T02.EPS

sures reduced before they can be used. The type of regulator used to control the gas pressure depends on the type of PAC equipment being used and the base metal being cut.

When a compressed-air supply is being used, a heavy-duty air filter and dryer is required to eliminate oil and moisture from the pressurized air. The high-pressure shop air must be reduced by an external regulator, or internally by the PAC, to the required working pressure. Also, depending on the equipment being used, the gas must be provided at a given flow rate for the task at hand, generally measured in cubic feet of gas per hour (cfh). Always check to be sure that the air supply will provide the flow rate required for the PAC equipment being used. If the airflow rate is too low, it will cause poor cuts and overheating of the torch. Refer to the manufacturer's specifications for the required flow rate for the equipment being used. To conserve consumables, make sure to have air dryers in the compressed air system as well. Moisture in the compressed air can damage the equipment.

Gases other than a local compressed air source are normally supplied in pressurized cylinders at 1,500 to 2,000 psi (103 to 138 bars). The high cylinder pressures are reduced by pressure regulators (*Figure 10*) to a lower working pressure. A gas hose connected to the outlet of the regulator delivers the gas to the PAC unit. Most heavy-duty systems require two high-pressure cylinders and two pressure regulators. One cylinder contains the cutting gas and the other a shielding or cooling gas.

GOING GREEN

Fumes and Gases

The various gases used in PAC activities, as well as the fumes generated from these activities, are often released into the atmosphere without much thought about their effect on the environment. To help protect the environment, efforts are needed to better contain the release of PAC gases and fumes. Whenever possible, PAC work should be performed using some type of capture and recovery system so that the fumes and gases are not released into the atmosphere.

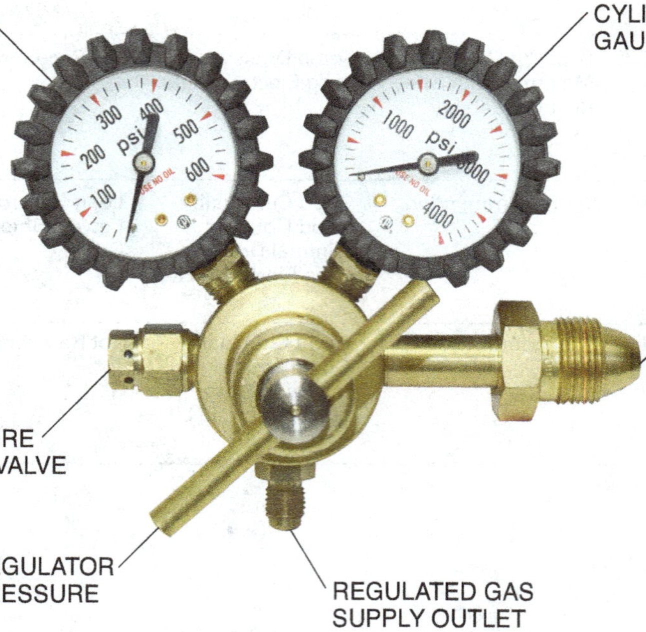

Figure 10 Gas cylinder pressure regulator.

The PAC process was originally developed to cut nonferrous metals using inert gases. Advances in technology led to using air and/or oxygen as the plasma gas so steel could be cut using PAC.

Some systems are able to manage two gas sources to enable the user to manage different applications. *Figure 11* shows a unit commonly known as a dual-flow PAC with two internal gas pressure regulators.

Gas flow to the torch is usually controlled by a **solenoid valve**. Solenoid valves are usually mounted inside the unit and are controlled by a pushbutton on the torch. There is one solenoid valve for each gas source.

Many smaller PAC units come equipped with a built-in air compressor. These light-duty machines are portable and run on 110V power, which makes them ideal for use at remote job sites. *Figure 12* shows a light-duty unit.

Figure 11 Industrial dual-flow PAC unit.

Plasma Torch Cutting Guides

Many manufacturers furnish cutting guide accessories that can be used on several models of their torches. A circle-cutting guide kit can also be used as a freehand standoff roller guide.

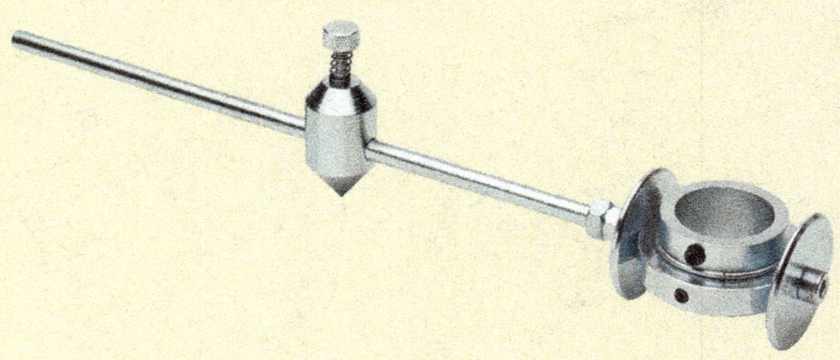

29103-15_F12.EPS

Figure 12 Light-duty PAC unit in use.

Regulator Selection

When selecting a regulator for PAC systems, first determine the required gas pressure from the PAC unit specifications. The desired pressure should be around the center of the gas pressure regulators adjustment range for the best accuracy. For example, if the desired gas pressure to the PAC is 50 psig (345 kPa), a regulator with a 0 to 100 adjustment range would be an ideal choice. However, note that a match between the regulator and the equipment is rarely this perfect. The idea is to avoid operating at the low or high end of the regulator's range, as these areas are where the least amount of accuracy and control occur in a pressure regulator.

Additional Resources

Plasma Cutters Handbook: Choosing Plasma Cutters, Shop Safety, Basic Operation, Cutting Procedures, Advanced Cutting Tips, CNC Plasma Cutters, Troubleshooting, and Sample Projects. Eddie Paul. New York, NY: Penguin Group.

2.0.0 Section Review

1. The duty-cycle range for a light-duty unit is typically ____ .
   a. 35 percent
   b. 50 percent
   c. 75 percent
   d. 90 percent

2. The typical plasma arc cutting torch usually contains a replaceable retaining cup, comprised of a copper electrode with a(n) ____ .
   a. magnesium tungsten insert
   b. aluminum or thoriated tungsten insert
   c. steel tungsten alloy insert
   d. hafnium or thoriated tungsten insert

3. Some types of PAC torches inject water to create a whirlpool of water around the jet to ____ .
   a. cut with a more uniform melt rate
   b. ensure the cut will have a wider kerf
   c. allow the torch to be used on a carrier
   d. minimize the amount of dross from the cut

SECTION THREE

3.0.0 PLASMA ARC CUTTING EQUIPMENT OPERATION AND CARE

Objective

Describe how to set up, safely operate, and care for plasma arc cutting equipment.

a. Describe how to set up plasma arc cutting equipment and the adjacent work area.
b. Describe how to safely operate plasma arc cutting equipment.
c. Describe how to care for plasma arc cutting equipment.

Performance Tasks

1. Set up plasma arc cutting equipment.
2. Set the amperage and gas pressures or flow rates for the type and thickness of metal to be cut.
3. Square-cut metal using plasma arc cutting equipment.
4. Bevel-cut metal using plasma arc equipment.
5. Pierce and cut slots in metal using plasma arc cutting equipment.
6. Dismantle and store the equipment.

The following section explains how to correctly set up PAC equipment and use it for cutting. Various types of cutting techniques are also presented.

3.1.0 Setting Up PAC Equipment

PAC equipment is relatively easy to set up because it is supplied as a complete system. However, the following are some things to consider before cutting is attempted:

- The unit must have the rated power and duty cycle to cut the intended material and thickness.
- The required power characteristics of the PAC unit, including the phase, voltage, and amperage must be provided.
- The appropriate nozzle must be installed in the torch for the task at hand.
- The required gas or gases must be available and at the required pressures and flow rates.
- Spare consumable components must be on hand.

3.1.1 Preparing the Work Area for PAC

The PAC control unit and workpiece must be located close enough to each other for the torch cable to comfortably reach the workpiece and still provide maneuverability. If possible, the workpiece should be located at a comfortable height and position for the torch operator. The cutting area should be well ventilated and cleared of all combustible material.

Plasma arc cutting or gouging can spray molten metal for considerable distances, possibly as far as 40 ft (12 m). It is important that everything combustible is removed from the range of the sprayed metal. If necessary, flame-resistant shields or curtains should be erected to protect workers and equipment from ultraviolet arc rays and metal splatter. Some employers and/or job sites may require a hot-work permit to be completed and a fire watch to be posted.

WARNING!

PAC produces fumes and gases that can be harmful to your health. The composition and rate of generation of fumes and gases depend on factors such as arc current, cutting speed, material being cut, and gases used. Various paints and coatings can also create or contribute to the production of harmful fumes. The fume and gas byproducts usually consist of ozone, oxides of the metal being cut, and oxides of nitrogen. These fumes must be removed from the work area through a capture and exhaust or recovery system. Check all applicable local codes; some codes may require that exhausted fumes be filtered before being vented to the atmosphere.

Eye and ear protection are essential when performing PAC as a safeguard against noise and to prevent metal spray from entering the operator's eyes and ears. If the site ventilation is not adequate to keep the smoke and fumes away from the operator, a respirator may also be required.

WARNING!

Ear protection is essential to protect hearing. Other PPE must be worn to protect the operator from hot metal and slag.

3.1.2 Setting the Correct Cutting Amperage

The cutting amperage depends on the type of equipment being used, the type and thickness of the material being cut, and the type of gas being used (*Figure 13*). Light-gauge sheet steel may require as little as 7 amps of current, while 2" (5.1 cm) aluminum plate may require 250 amps. Always refer to the manufacturer's recommendations to identify the correct amperage for the specific equipment and job. Using the incorrect amperage will result in a poor-quality cut. If the amperage is too high, severe damage to the torch may also occur.

3.1.3 Setting Gas Parameters and Installing Gas Cylinders

Several different gases may be used on the same system for cutting the same or different types and thicknesses of metals. In many cases, the same gas or mixture will not be used for both thick and thin metals of the same type. Refer to the manufacturer's instructions or the unit data plate for specific gas recommendations.

Some gases used in cutting can be dangerous if accidentally released. Welders routinely have to remove empty cylinders and replace them with filled cylinders, increasing the potential for a release.

When connecting a gas cylinder to the PAC unit, follow these steps:

Step 1 After identifying the correct gas type, place it adjacent to the unit and secure it in an upright position.

Step 2 Remove the protective valve cap.

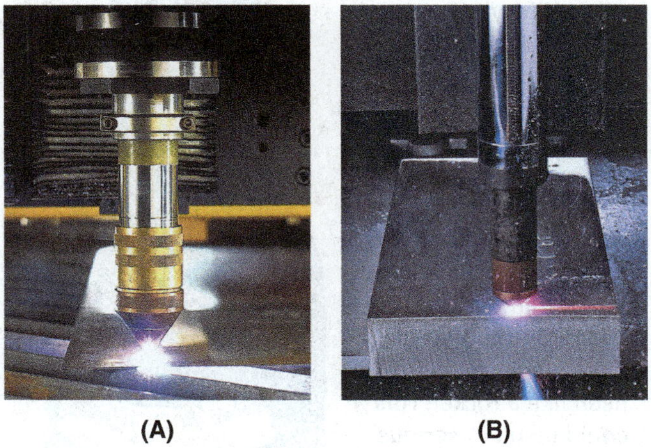

(A) (B)

29103-15_F13.EPS

Figure 13 Gas and amperage settings vary according to material type and thickness.

WARNING!

If oxygen is being used, be sure any cloth used on or near the fittings or regulators is free of oil or grease. Oil or grease mixed with oxygen can explode, causing personal injury or death.

Step 3 While standing to one side of the cylinder valve outlet, and with safety glasses in place, quickly crack open the valve a small amount to blow out any dirt or debris. Reclose immediately to prevent unnecessary gas loss.

WARNING!

To avoid personal injury, always stand with the cylinder between you and the attached regulator when opening valves.

Step 4 Connect the regulator to the cylinder valve and tighten with the proper wrench. Do not over-tighten. Connect the hose from the PAC unit to the regulator outlet.

Step 5 Open the cylinder valve very slowly at first, then open fully. Observe the regulator inlet gauge to determine the pressure remaining in the cylinder.

Step 6 Slowly adjust the regulator to the desired output pressure. The pressure usually has to fall into a range to satisfy the requirements of the PAC unit. The PAC unit regulates the final pressure to the torch.

CAUTION

Do not use pliers or pipe wrenches to tighten the regulator's connections. Use an adjustable wrench or correctly sized open-end wrench. Also, be careful not to over-tighten the connections because they are made of soft bronze and can be damaged easily.

3.2.0 Operating PAC Equipment

PAC equipment is relatively simple to operate. The cutting unit's operating instructions provide information for properly setting gas pressure and flow rates, output current, torch standoff distance, and torch travel speed (for mechanized cutting) for various metal types and thicknesses.

To operate the PAC equipment, follow these general steps. Note that this is a general reference only. Refer to the manufacturer's documentation for specific instructions on the PAC equipment being used.

Step 1 Make sure that the capacity of the PAC power unit matches or exceeds the work to be done. Also ensure that the power source for the unit is sufficient, based on the unit data plate.

Step 2 Identify the location of the primary disconnect for the electrical source to be used in case of emergency.

Step 3 Check to make sure that the required gas(es) is connected. If a pressurized cylinder is used, ensure that it has adequate pressure for the work to be done. If compressed air will be used, ensure that it is also connected and at the appropriate pressure.

WARNING!

Some stainless steel contains chromium 6, also known as hexavalent chrome. This metal will give off toxic fumes when heated. Always use proper respiratory protection when cutting this type of stainless steel.

Step 4 Plug the unit in (if required), energize the PAC unit power source, and turn the unit on.

Step 5 In accordance with the PAC manufacturer's instructions, adjust the gas regulator pressure and/or flow rate to the values recommended for the type and thickness of metal being cut.

Step 6 Set the output current to the value recommended in the manufacturer's instructions for the type and thickness of metal being cut.

Step 7 Attach the work lead clamp to the workpiece to be cut.

CAUTION

If cutting on machinery or other assembled equipment, be sure to position the work lead clamp so that the cutting current will not pass through seals, bearings, or other contacting surfaces that could be damaged from heat or arcing. Also isolate any electrical components from the PAC system/workpiece circuit.

Step 9 Prepare the work area as previously described.

WARNING!

Be advised that various gases are extremely flammable, and PAC processes operate at extreme temperatures. The appropriate fire suppression equipment must be stationed in close proximity for emergency use.

Step 10 Put on the appropriate PPE.

Step 11 Hold the torch at the recommended standoff distance above the point where the cut is to begin, lower your hood, and press the torch arc-start button or foot pedal. As soon as cutting begins, move the torch at a 0-degree travel angle with the recommended speed along the cutting line.

3.2.1 Square-Cutting Metal

Square edges can be cut with PAC by holding the torch at 90 degrees (also referred to as a 0-degree work angle) to the metal surface while it is advanced smoothly, with no side-to-side movement. A straightedge or metal angle can be used to guide and steady the torch. The straightedge can be aligned and quickly clamped down at each end. When cutting thicker metals, the cut does taper slightly as it narrows with depth. This can

Safety Caps

Safety caps should be kept with cylinders at all times. Immediately after removing a cylinder's regulator assembly, reinstall the safety cap over the cylinder valve. This protects the cylinder valve if the cylinder should fall over. If the cylinder valve is suddenly sheared off, the cylinder may propel itself like a rocket. This could result in serious injury or death.

29103-15_SA03.EPS

result in cuts where one or both faces are not quite square with the plate face. The angle of the face is referred to as the cut angle. To compensate for the cut angle on thick materials, the torch should be slightly angled, leaning toward the scrap portion of the workpiece, to produce a square edge on the desired piece. All the taper will then be occur on the scrap side (*Figure* 14[*A*]).

3.2.2 Bevel-Cutting Metal

Bevel cuts are made with the same technique used for square cuts, except that the torch is held at the required bevel angle. A length of angle iron, laid on its open face and shimmed as necessary, can be used to support the torch's side at the required angle. *Figure* 14(*B*) shows how to use a metal angle to hold a bevel angle.

3.2.3 Piercing and Slot-Cutting in Metal

PAC can be used to pierce and cut slots in metal in any position. Full, fire-resistant body protection, including ear protection, is recommended when piercing or cutting metal in situations where out-of-position orientations could cause the metal splash to fall on the torch operator.

To pierce very thin metal, hold the torch directly over the point to be pierced at a 0-degree work angle, and press the arc button. As soon as the jet passes completely through the workpiece, move the torch head in a smooth circle (or other pattern) to produce the desired diameter or shape for the hole. When piercing thicker materials, rotate the torch to a 10-degree work angle. This will help to prevent metal from splashing directly up into the torch. When the jet passes through the workpiece, rotate the torch back to a 0-degree work angle and complete the cut.

When cutting slots or patterns, a template can be used to precisely guide the torch tip.

Gas Pressure and Flow Rates

PAC unit gas pressures and/or flow rates vary with equipment design, gas type, and cut depth (metal thickness). Typical gas flow rates vary from 15 cfh for nitrogen when cutting 0.1" (2.5 mm) aluminum, carbon steel, or stainless steel, up to 62 cfh of argon and 31 cfh of nitrogen when cutting 3" (7.6 cm) carbon steel, or 2½" (6.4 cm) aluminum or stainless steel. Refer to the manufacturer's instructions for the correct gas pressure and flow rates for each situation.

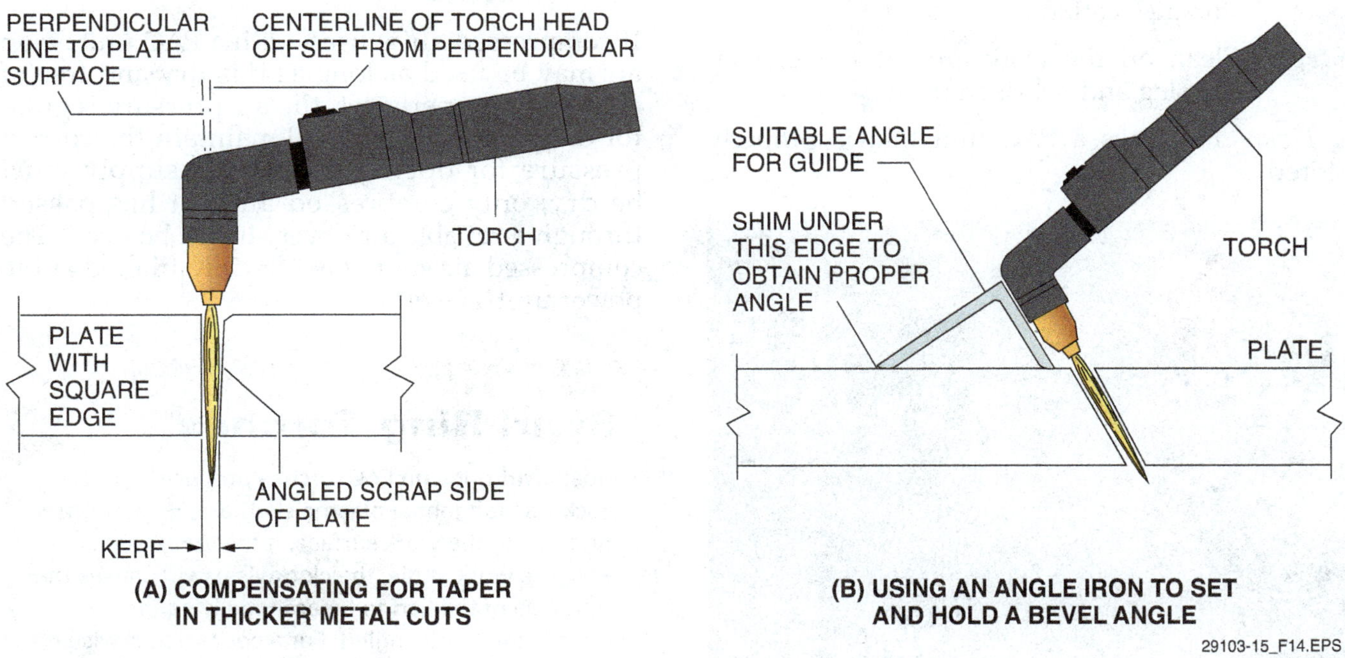

Figure 14 Compensating for taper and holding bevel angle in thicker metal cuts.

3.2.4 *Proper Equipment Storage and Housekeeping*

Proper equipment storage and housekeeping are essential for work efficiency and safety. When finished using the PAC equipment, follow these steps to store the equipment and maintain good housekeeping:

Step 1 Turn off the power source.

Step 2 Turn off all gas sources, bleed pressure slowly and carefully, and disconnect the lines from the PAC unit. When using cylinders, be sure to close the bottle valve, remove the regulator assembly, and install the protective cap before moving the cylinder.

> **WARNING!**
> If cylinders are to be removed, be sure to remove the pressure regulators and replace the cylinder caps before releasing and moving the cylinders.

Step 3 Unplug and coil the power unit cord and all gas lines.

Step 4 Coil the torch cable and the workpiece lead.

Step 5 Return the PAC equipment to its proper storage location.

Step 6 Clean off the work area and sweep up any slag and debris in the area.

Figure 15 shows a PAC unit that is properly stored.

29103-15_F15.EPS

Figure 15 Properly stored PAC unit.

Starting the Torch on a Thick Workpiece

Similar to the oxyfuel cutting of a thick workpiece, it is easier to begin a manual cut at the edge by rotating the PAC torch slightly so that the plasma cutting jet begins cutting at the top edge only. The torch is then rotated vertically so that the cutting jet slices into the remaining thickness before the torch is moved along the cutting line.

3.3.0 Care of Plasma Arc Cutting Equipment

As with all equipment, regular maintenance should be performed according to the manufacturer's recommendations. The cutting equipment should not be altered or modified in any way, except for those changes identified in the manufacturer's literature. The plasma machine requires relatively little maintenance, but its consumables must be inspected regularly and replaced as needed. Minor external repairs usually involve these basic parts only.

3.3.1 *Understanding Basic Operating Concerns*

If compressed air is used with a PAC unit, shop air may be used as long as it is dry and free of oil. When properly set, the air pressure regulator on the power unit will maintain the correct pressure for operation. The air supply must be dry; only compressed air that has passed through a reliable air dryer should be used. The compressed air supply is typically filtered at the power unit as well.

Swirl Ring Torches

Most swirl rings on PAC torches produce a clockwise tangential plasma swirl as viewed from the torch to the work surface. If the torch is at a 0-degree work angle, the clockwise swirl causes the right side of the kerf to be essentially square and the left side to be angled. For a counterclockwise swirl, the opposite is true.

> **WARNING!**
>
> Always switch off the machine power, unplug the power cord, and disconnect gas sources before changing consumable parts or performing any repairs on the power unit. If the power unit is directly connected to a line disconnect switch, move the line disconnect switch to the Off position. All sources of energy must be disconnected from the unit before maintenance is performed.

Your ability to change and replace parts will depend on your knowledge of the procedures contained in the manufacturer's literature for the equipment in question.

Perform the following procedures, and always abide by the manufacturer's guidelines for removal and replacement of worn or damaged parts:

- Check the shield for any signs of wear.
- The shield should be clean and clear of metal debris.
- Unscrew the shield and check the inside for wear. If the holes are blocked, attempt to clean them using a torch tip cleaner.
- If the shield cannot be cleaned, replace it.
- Inspect the retaining cup for damage and replace as necessary.
- Examine the nozzle for wear or damage. If the hole in the nozzle is worn or oval-shaped, replace the nozzle.
- Remove the electrode and check the tip for pitting. If pits exist and are more than 1⁄16" (1.6 mm) deep, replace the electrode.
- Inspect the swirl ring, which should be clean, and the holes on the sides free of clogs. If the swirl ring is damaged, replace it.
- Check the O-ring. Apply a small amount of lubricant (provided in a repair kit) to the O-ring before reinstalling it. The O-ring should make a firm seal when all parts are assembled and tight.

3.3.2 Moisture

A significant concern when using PAC equipment is that moisture may enter the unit through the air supply. This can cause the torch to sputter or hiss. If your PAC equipment experiences moisture problems, first drain the air filter/moisture separator bowl. The filter element may also need to be replaced. Refer to the manufacturer's literature for cleaning and replacement of the filter element.

3.3.3 Cooling Filter

Many power units contain an air filter housing to ensure the air supply is clean. Refer to the manufacturer's literature and clean the filter element periodically as required. An air filter that is neglected may cause the PAC unit to overheat and shut down unexpectedly.

Additional Resources

Plasma Cutters Handbook: Choosing Plasma Cutters, Shop Safety, Basic Operation, Cutting Procedures, Advanced Cutting Tips, CNC Plasma Cutters, Troubleshooting, and Sample Projects. Eddie Paul. New York, NY: Penguin Group.

3.0.0 Section Review

1. To remove, attach, or tighten the regulator's connection to a gas cylinder, you would use ____.
   a. no tools; this connection is to be hand tightened only
   b. a correctly sized wrench with an added extension for leverage to apply maximum force
   c. a properly sized pipe wrench adjusted to fit the regulator fitting
   d. an adjustable wrench or correctly sized open-end wrench

2. To pierce very thin metal with a plasma torch, you should ____.
   a. hold the torch directly over the point to be pierced at a 0-degree work angle
   b. start from the bottom, piercing the metal from below
   c. hold the torch directly over the point to be pierced at 10-degree work angle
   d. hold the torch directly over the point to be pierced at a 45-degree work angle

3. If the PAC equipment exhibits a moisture problem, you should first ____.
   a. attach a vacuum pump to the air filter/moisture separator bowl assembly
   b. replace the filter element and air filter/moisture separator bowl assembly
   c. drain the air filter/moisture separator bowl
   d. stop using air and operate the system using argon for at least 12 hours of operation

Summary

Plasma arc cutting is useful for cutting and piercing many types of metals. Unlike oxyfuel cutting, it can be used to cut aluminum, magnesium, copper, nickel, and stainless steel. PAC is usually much faster than oxyfuel cutting, and it produces little or no distortion or alteration zone changes. The cuts are fast and clean, with little or no dross, and there is no chance of carbon inclusions, as can happen with carbon arc cutting. However, because the torch jet produces metal spray, good site fire-prevention practices and the consistent use of appropriate PPE are essential.

With an understanding of metals and the knowledge to correctly set up a PAC unit for the task at hand, workers can quickly become competent in the use of PAC equipment to produce excellent cuts.

Review Questions

1. Plasma arc cutting uses a jet of plasma that can reach temperatures as high as ____ .
   a. 10,000°F (5,537°C)
   b. 20,000°F (11,093°C)
   c. 25,000°F (13,871°C)
   d. 30,000°F (16,650°C)
2. What are the two primary types of plasma arc cutting processes?
   a. Transferred arc and non-transferred arc
   b. Transferred arc and phased arc
   c. Phased arc and non-transferred arc
   d. Oxyfuel pilot arc and non-transferred arc
3. In the transferred arc process, the pilot arc is used to ____ .
   a. cut the metal
   b. ionize the air jet
   c. establish contact with the metal
   d. keep the torch heated
4. In a non-transferred arc process, the material being cut is ____ .
   a. adjacent to the arc circuit
   b. dependent on the arc circuit
   c. electrically connected to the arc circuit
   d. not electrically connected to the arc circuit
5. Which of the following statements regarding PAC safety is true?
   a. If a cap or hat is worn, it should be constructed of 100-percent polyester or nylon.
   b. The electric arc can be safely observed without tinted lenses, but it is not recommended for an extended period.
   c. Boots with no laces at all are generally preferred for both welding and cutting activities.
   d. Flammable materials can remain near the cutting activity, due to the cool nature of the plasma arc.
6. Light-duty PAC equipment typically requires a ____ .
   a. 115VAC, single-phase power source
   b. 220VAC, three-phase power source
   c. 277VAC, single-phase power source
   d. 440VAC, three-phase power source
7. Some heavy-duty PAC systems ____ .
   a. require no torch cooling
   b. are water-cooled
   c. use very little current
   d. weigh less than 5 pounds (2.3 kg)
8. The workpiece clamp is rated by the ____ .
   a. voltage that it can carry without overheating
   b. capacitance that it can carry without overheating
   c. amperage that it can carry without overheating
   d. gas volume that it uses at maximum load
9. Severe damage from arcing and overheating can be caused by the PAC current passing through a ____ .
   a. workpiece
   b. cable
   c. cutting table
   d. bearing
10. Mechanized PAC torches are designed to be ____ .
    a. handheld
    b. mounted in a vise
    c. used at a slow speed
    d. mounted on automated carriers
11. Gases other than compressed air are normally supplied in pressurized cylinders at ____ .
    a. 200 psi (14 bars) to 400 psi (28 bars)
    b. 800 psi (55 bars) to 1,000 psi (69 bars)
    c. 1,000 psi (69 bars) to 1,500 psi (103 bars)
    d. 1,500 psi (103 bars) to 2,000 psi (138 bars)
12. If site ventilation is not adequate to keep fumes away from the operator, a(n) ____ .
    a. fan must be used
    b. respirator may be required
    c. fire watch must be used
    d. airline must be used
13. Light-gauge sheet steel may be cut using a cutting amperage as low as ____ .
    a. 7 amps
    b. 9 amps
    c. 15 amps
    d. 20 amps

14. Square edges can be cut in thin materials by holding the torch at ____.
    a. 30 degrees to the metal surface
    b. 45 degrees to the metal surface
    c. 75 degrees to the metal surface
    d. 90 degrees to the metal surface

15. When inspecting an electrode before use, it should be replaced if pits exist and are deeper than ____.
    a. 1⁄32" (0.8 mm)
    b. 1⁄16" (1.6 mm)
    c. 1⁄8" (3.2 mm)
    d. 3⁄16" (4.8 mm)

Trade Terms Quiz

Fill in the blank with the correct term that you learned from your study of this module.

1. A fourth state of matter created by heating a gas to such a high temperature that it boils electrons off the gas molecules is called _______ .
2. The relative electrical voltage difference between two points of reference is called _______ .
3. A(n) _______ can control the flow of gases and liquids, and is operated by an electromagnet.
4. The percentage of time a PAC unit can cut without overheating within a 10-minute period is called its _______ .
5. Each of the three sine waves of a power supply that uses three conductors, all carrying alternating current, is referred to as a(n) _______ .
6. The unit of measure for electrical current is _______ .
7. A waste byproduct of molten metal resulting from plasma cutting processes is known as _______ .

Trade Terms

Amperage
Dross
Duty cycle
Phase
Plasma
Potential
Solenoid valve

Performance Accreditation Tasks

The American Welding Society (AWS) School Excelling through National Skills Standards Education (SENSE) program is a comprehensive set of minimum Standards and Guidelines for Welding Education programs. The following performance accreditation tasks are aligned with and designed around the SENSE program.

The Performance Accreditation Tasks (PATs) correspond to and support the learning objectives in *AWS EG2.0, Guide for the Training and Qualification of Welding Personnel: Entry-Level Welder*.

Note that in order to satisfy all learning objectives in *AWS EG2.0*, the instructor must also use the PATs contained in the second level of the NCCER Welding curriculum.

PATs 1 and 2 correspond to *AWS EG2.0, Module 8 – Thermal Cutting Processes, Unit 3 – Manual Plasma Arc Cutting (PAC)*, Key Indicators 3, 4, and 5.

PATs provide specific acceptable criteria for performance and help to ensure a true competency-based welding program for students.

The following tasks are designed to develop your competency with a plasma arc cutting torch. Practice each task until you are thoroughly familiar with the procedure.

As you complete each task, take it to your instructor for evaluation. Do not proceed to the next task until instructed to do so by your instructor.

PLASMA ARC CUTTING

Using electrically conductive material, lay out and cut the shape shown in the figure in the horizontal position. When it is finished, cut your initials into the shape.

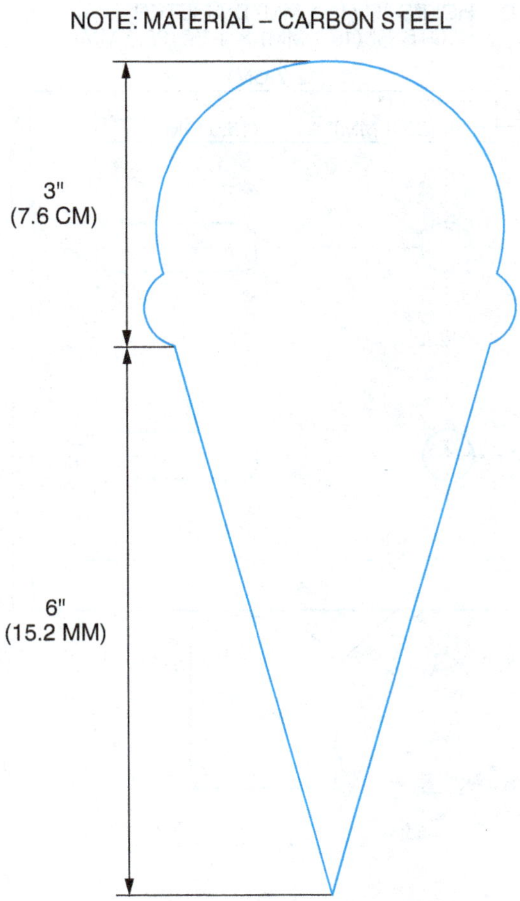

29103-15_A02.EPS

Criteria for Acceptance:

- Dimensions ±1⁄16" ________
- Minimal amount of dross sticking to plate which can be easily removed ________

PLASMA ARC CUTTING

Using electrically conductive material, lay out and cut the shape shown in the figure in the flat position.

NOTE: MATERIAL – CARBON STEEL
HOLES ¾" (19.1 MM) DIAMETER
SLOTS ¾" (19.1 MM) × 1 ½" (38.1 MM)

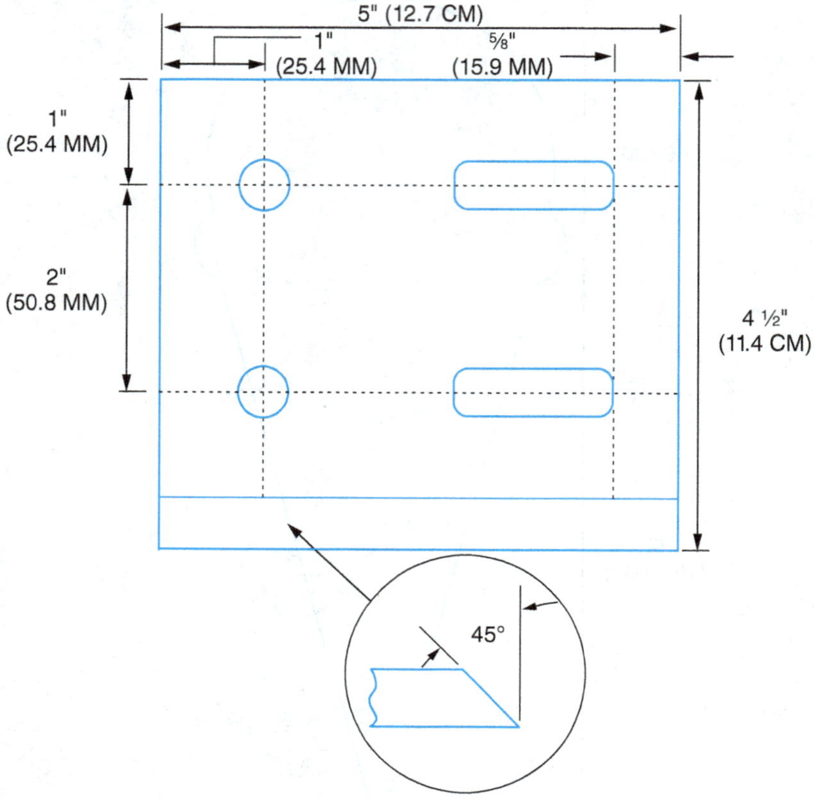

29103-15_A01.EPS

Criteria for Acceptance:

- Outside dimensions ±1/16" ______
- Inside (holes and slots) dimensions ±1/16" ______
- Square ±2° ______
- Bevel ±2° ______
- Minimal amount of dross sticking to plate which can be easily removed ______
- Square kerf face with minimal notching not exceeding ±1/16" deep ______

Cornerstone of Craftsmanship

Curtis Casey

Northland Pioneer College
Emeritus Welding Faculty

Curtis received his welding training in the US Navy, where he was originally slated to be trained as a submarine power plant operator. However, those jobs were filled when he joined, so he chose advanced training as a welder. Curtis welded on SSBN class submarines for six years, working on high-tensile hulls, pressure pipe, and nuclear components. After the Navy, a new opportunity opened up at Bechtel Power Corporation, where he became a Certified Welding Inspector. For the next 15 years, Curtis traveled around the United States helping to build nuclear power plants. He also found the time to earn a Welding Technology degree on the G.I. Bill and to take on supervisory responsibilities. Not long after joining the faculty at Northland Pioneer College, Curtis was appointed chairperson of the Welding Department, which now serves hundreds of students each year.

How did you choose a career in the construction industry?
I took the opportunity to train while in the US Navy's nuclear welding program after graduating from high school. There I got a chance to weld on submarines and other naval vessels.

Who inspired you to enter the industry?
My stepfather inspired me to enlist in the US Navy where I received advanced training in welding. He was a career sailor who always had great stories about seeing the world.

What types of training have you been through?
In the Navy, I received training in the metals trades to be a hull maintenance technician. I finished welding school, learning to weld on both common and exotic materials. After the military I went to college while I worked as a welding inspector in nuclear power plants and got a chance to teach weld inspection as an adjunct professor while attending college. From then on, I was hooked on training and found the opportunity to teach welding full time at Northland Pioneer College. Currently I am a welding consultant wearing many hats within the industry from inspecting to auditing to teaching and consulting.

How important are education and training in construction?
Education is an ongoing process; life-long learning is another way to say it. Education and training are the tools we fit ourselves with to be able to create and mold our careers. Without them our direction in life could easily be aimless and unproductive.

How important are NCCER credentials to your career?
NCCER credentialing gives me an edge over many of my colleagues and peers because of the clout the organization has within the industry. Many companies now are requiring these credentials just to get your foot in their door.

How has training/construction impacted your life?
Simply put, it has given me great sense of satisfaction knowing that I can design and build on ideas that come to my mind. I can take a project from inception to creation because of my training.

What kinds of work have you done in your career?
I started my adult working life as a welder, then worked as a welding inspector, a field welding engineer, a welding business owner, a welding instructor, welding department chair, welding curriculum developer, and welding consultant. I think welding sums it up pretty well. I still maintain my AWS Certified Welding Instructor (CWI) and Certified Welding Educator (CWE) certifications, and I'm actively involved with my students in SkillsUSA, a national student leadership organization.

Tell us about your present job.
I'm currently consulting in the welding arena while occasionally inspecting and auditing other NCCER training and assessment centers.

What do you enjoy most about your job?
My passion has been teaching welding at Northland Pioneer College for the past 15 years, and now in semi-retirement. I love the versatility in what I do. One day I could be doing R & D at a fabrication shop, and the next I could be inspecting welds or auditing programs. It is refreshing to see the newest technology being put to use at certain companies; in this job I see it firsthand.

What factors have contributed most to your success?
Work ethics! I learned at an early age to be responsible for my own successes and failures. I pitched papers on a paper route when I was 11 years old. That taught me to be dependable and gave me exposure to the business world and having to deal with customers, good and bad.

Would you suggest construction as a career to others? Why?
If they were interested in seeing an idea come to fruition from design to completion and if they were willing to get their hands dirty, then I would recommend construction. One can receive great satisfaction and get a sense of worth when they take part in the creation of something that people can actually use in their lives.

What advice would you give to those new to the welding field?
Try it out through a job shadowing program to see if you like it. Then if you dream about it that night, you probably have a passion for it. Then simply follow your passion! Get the right training, surround yourself with like-minded people, keep a good attitude, network a lot, and go to work!

Tell us an interesting career-related fact or accomplishment.
I have had the chance to teach close to 1,000 people the craft that has given me my career, and have seen many former students return to share their success in the industry. That's an emotional paycheck worth the effort.

How do you define craftsmanship?
Developing a good balance between technical and artistic skills.

Trade Terms Introduced in This Module

Amperage: The unit of measure used for the intensity of an electric current; often abbreviated to *amp* or *amps*.

Dross: A waste byproduct of molten metal.

Duty cycle:: The percentage of time a plasma arc cutting machine can cut without overheating within a ten-minute period.

Phase: In a three-phase power supply system, the sine waves of the voltage on each of the three separate conductors is displaced from each of the others by 120 degrees, although all conductors are carrying alternating current at the same frequency and are synchronized. Each of the three sine waves is referred to as a phase of the power source.

Plasma: A fourth state of matter (not solid, liquid, or gas) created by heating a gas to such a high temperature that it boils electrons off the gas molecules (ionization).

Potential: The relative electrical voltage difference between two points of reference; the electromotive force or difference in potential between two points, expressed in volts.

Solenoid valve: A valve used to control the flow of gases or liquids that is opened or closed by the action of an energized electromagnet.

Additional Resources

This module presents thorough resources for task training. The following resource material is suggested for further study.

Plasma Cutters Handbook: Choosing Plasma Cutters, Shop Safety, Basic Operation, Cutting Procedures, Advanced Cutting Tips, CNC Plasma Cutters, Troubleshooting, and Sample Projects. Eddie Paul. New York, NY: Penguin Group.

AWS F3.2M/F3.2, Ventilation Guide for Weld Fume. Latest Edition. Miami, FL: American Welding Society.

Figure Credits

The Lincoln Electric Company, Cleveland, OH, USA, Module Opener, Figures 1A, 2 (photo), 5, 8 (photos), 11, 12, 15, SA01

Courtesy of BUG-O Systems Inc., Figure 1B

AWS F2.2:2001, Lens Shade Selector, reproduced with permission from the American Welding Society (AWS), Miami, FL, USA, Table 1

Topaz Publications, Inc., Figures 6, SA03

ESAB Welding and Cutting Products, Figures 7, 13A

Courtesy of Miller Electric Mfg. Co., Figure 9A

Victor Technologies, Figure 9B, 13B

Courtesy of Uniweld Products, Figure 10

The Eastwood Company, SA02

Section Review Answer Key

Answer	Section Reference	Objective
Section One		
1. d	1.1.0	1a
2. c	1.2.1	1b
Section Two		
1. a	2.1.0	2a
2. d	2.2.0	2b
3. a	2.2.0	2c
Section Three		
1. d	3.1.3	3a
2. a	3.2.3	3b
3. c	3.3.2	3c

NCCER CURRICULA — USER UPDATE

NCCER makes every effort to keep its textbooks up-to-date and free of technical errors. We appreciate your help in this process. If you find an error, a typographical mistake, or an inaccuracy in NCCER's curricula, please fill out this form (or a photocopy), or complete the online form at **www.nccer.org/olf**. Be sure to include the exact module ID number, page number, a detailed description, and your recommended correction. Your input will be brought to the attention of the Authoring Team. Thank you for your assistance.

Instructors – If you have an idea for improving this textbook, or have found that additional materials were necessary to teach this module effectively, please let us know so that we may present your suggestions to the Authoring Team.

NCCER Product Development and Revision

13614 Progress Blvd., Alachua, FL 32615

Email: curriculum@nccer.org

Online: www.nccer.org/olf

❑ Trainee Guide ❑ Lesson Plans ❑ Exam ❑ PowerPoints Other ____________________

Craft / Level: Copyright Date:

Module ID Number / Title:

Section Number(s):

Description:

Recommended Correction:

Your Name:

Address:

Email: Phone:

Air-Carbon Arc Cutting and Gouging

Overview

This module presents information related to air-carbon arc cutting (A-CAC) and gouging. This method is a useful process in which metal materials to be cut and gouged are melted by the heat of a carbon arc. In combination with a jet of compressed air, the molten material is pushed away from the cutting and gouging action.

Module 29104

Trainees with successful module completions may be eligible for credentialing through the NCCER Registry. To learn more, go to **www.nccer.org** or contact us at **1.888.622.3720**. Our website has information on the latest product releases and training, as well as online versions of our *Cornerstone* magazine and Pearson's product catalog.

Your feedback is welcome. You may email your comments to **curriculum@nccer.org**, send general comments and inquiries to **info@nccer.org**, or fill in the User Update form at the back of this module.

This information is general in nature and intended for training purposes only. Actual performance of activities described in this manual requires compliance with all applicable operating, service, maintenance, and safety procedures under the direction of qualified personnel. References in this manual to patented or proprietary devices do not constitute a recommendation of their use.

29104 V5

29104
Air-Carbon Arc Cutting and Gouging

Objectives

When you have completed this module, you will be able to do the following:

1. Define air-carbon arc cutting and identify the related equipment and consumables.
   a. Define air-carbon arc cutting.
   b. Identify and describe air-carbon arc cutting equipment.
   c. Identify and describe various types of electrodes.
   d. Identify safety practices related to air-carbon arc cutting.
2. Describe how to set up, safely operate, and care for air-carbon arc cutting equipment.
   a. Describe how to prepare the equipment and work area for air-carbon arc cutting.
   b. Describe how to wash and gouge metals.
   c. Describe how to care for air-carbon arc cutting equipment.

Performance Tasks

Under the supervision of your instructor, you should be able to do the following:

1. Select and install air-carbon arc cutting electrodes.
2. Prepare the work area and air-carbon arc cutting equipment for safe operation.
3. Use air-carbon arc cutting equipment for washing.
4. Use air-carbon arc cutting equipment for gouging.
5. Perform storage and housekeeping activities for air-carbon arc cutting equipment.

Trade Terms

Carbon-graphite electrode
Concentric cable system
Oxidize
Polarity

Industry Recognized Credentials

If you are training through an NCCER-accredited sponsor, you may be eligible for credentials from NCCER's Registry. The ID number for this module is 29104. Note that this module may have been used in other NCCER curricula and may apply to other level completions. Contact NCCER's Registry at 888.622.3720 or go to **www.nccer.org** for more information.

Contents

Topics to be presented in this module include:

Figures and Tables

This page is intentionally left blank.

SECTION ONE

1.0.0 AIR-CARBON ARC CUTTING

Objectives

Define air-carbon arc cutting and identify the related equipment and consumables.

a. Define air-carbon arc cutting.
b. Identify and describe air-carbon arc cutting equipment.
c. Identify and describe various types of electrodes.
d. Identify safety practices related to air-carbon arc cutting.

Performance Task

1. Select and install air-carbon arc cutting electrodes.

Trade Terms

Carbon-graphite electrode: An electrode composed of a mixture of soft amorphous carbon and hard graphite carbon that may be coated with copper.

Oxidize: To combine with oxygen, such as in burning (rapid oxidation) or rusting (slow oxidation).

Air-carbon arc cutting (A-CAC) is an arc-based cutting and gouging process, unlike oxyfuel processes. The base metal is melted with an electric arc formed between a carbon electrode and the base metal. A jet of compressed air passing through the electrode holder and its head blows the molten metal away. The process can be used in all positions on any metal that will conduct electricity, such as carbon steel, stainless steel, cast iron, and copper, as well as their alloys. It is primarily used to gouge weld grooves, to back-gouge welds, to gouge out defective welds, and to prepare cracks for welding. Carbon electrode arcs produce cleaner cuts than metal electrode arc cutting methods because no foreign metals are introduced at the arc. In addition, A-CAC is often used in washing and beveling operations. *Figure 1(A)* shows a groove being cut by the A-CAC process. *Figure 1(B)* shows the washing method using the A-CAC process.

This module will explain A-CAC equipment, its operation, and its use for cutting and gouging metals.

1.1.0 A-CAC Process and Equipment

Common applications of air-carbon arc cutting include the following:

- Cutting or gouging out defective metal
- Removing defective welds
- Creating a groove for a weld

The A-CAC process uses a high-current electric arc between the base metal and the end of a **carbon-graphite electrode** to melt the base metal. The currents used with this process are generally much higher than those required for the shielded metal arc welding (SMAW) process.

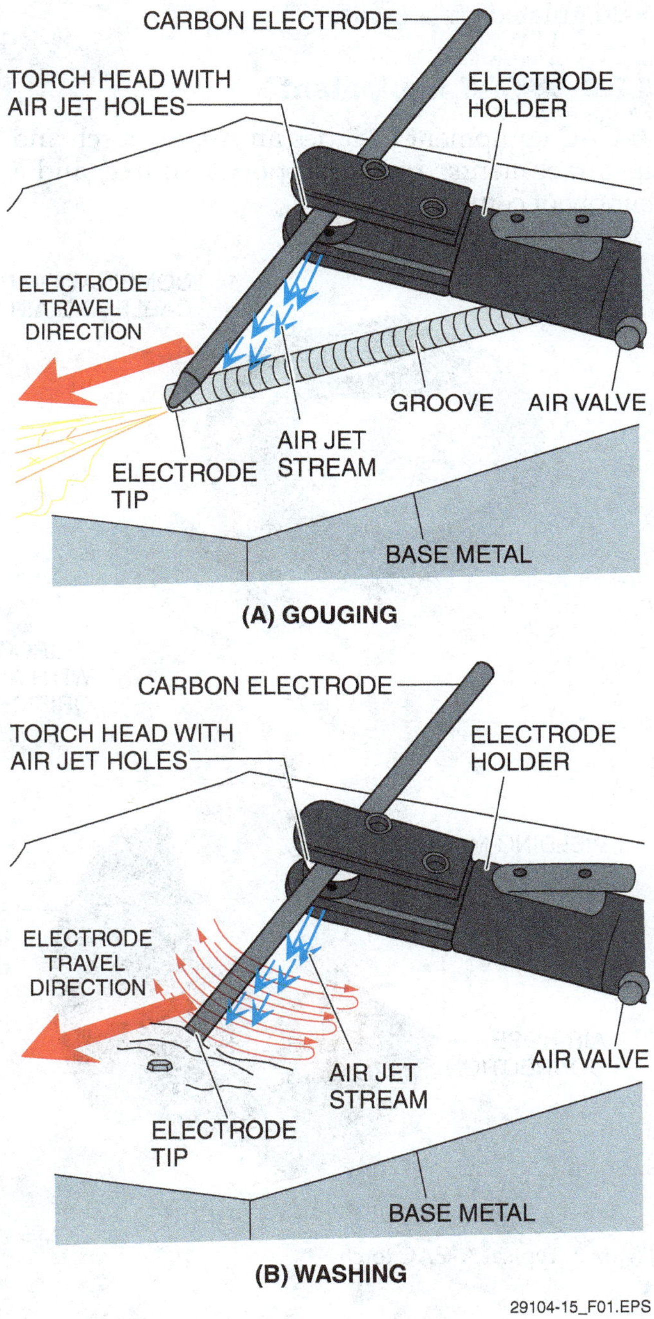

Figure 1 Gouging and washing with A-CAC process.

Air-jet orifices are located in the special torch heads. When gouging or beveling, the orifices direct jets of compressed air under the A-CAC electrode, between the electrode and the base metal, to blow the molten metal pool away from the cut. When the electrode is moved along the surface of the base metal, a groove or bevel can be made.

A groove's depth and width are controlled by the following:

- Electrode size
- Arc amperage
- Electrode angle to the base metal
- Electrode advancement rate
- Electrode movement
- Regulated air pressure

1.2.0 A-CAC Equipment

A-CAC equipment includes an A-CAC torch and its attachments, a welding power source, and a supply of compressed air.

1.2.1 A-CAC Torch and Attachments

Typical A-CAC equipment consists of a carbon-graphite electrode gripped in a specialized electrode holder. The electrode holder is commonly referred to as a torch, torch handle, or torch holder. The torch handle is equipped with V-groove heads containing air jet holes, and a combination power cable and compressed air hose is attached to the handle. *Figure 2* shows a typical A-CAC torch. *Figure 3* is a simplified diagram of an A-CAC setup.

The electrode is gripped by the V-groove head. The electrode tip should be within 6" (15 cm) of the V-groove head to prevent overheating. The contacts are mounted in the lower and upper jaws of the electrode holder. The jaws are spring-loaded to clamp onto the electrode. Depending on the torch design, the contact area of one head may be flat and the other V-grooved, or both may be V-grooved. If both are V-grooved, one or both heads will contain air-jet orifices.

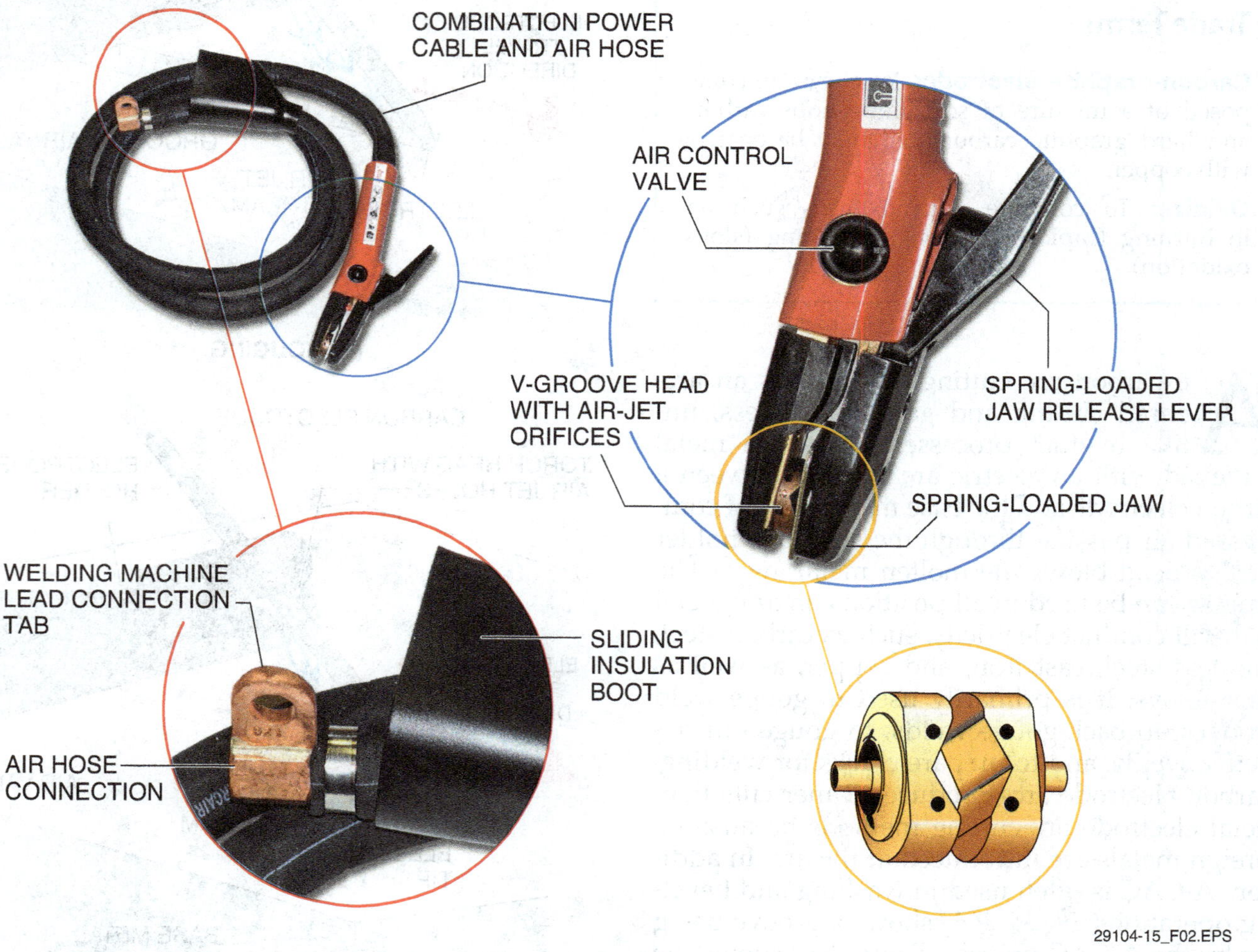

Figure 2 Typical A-CAC torch.

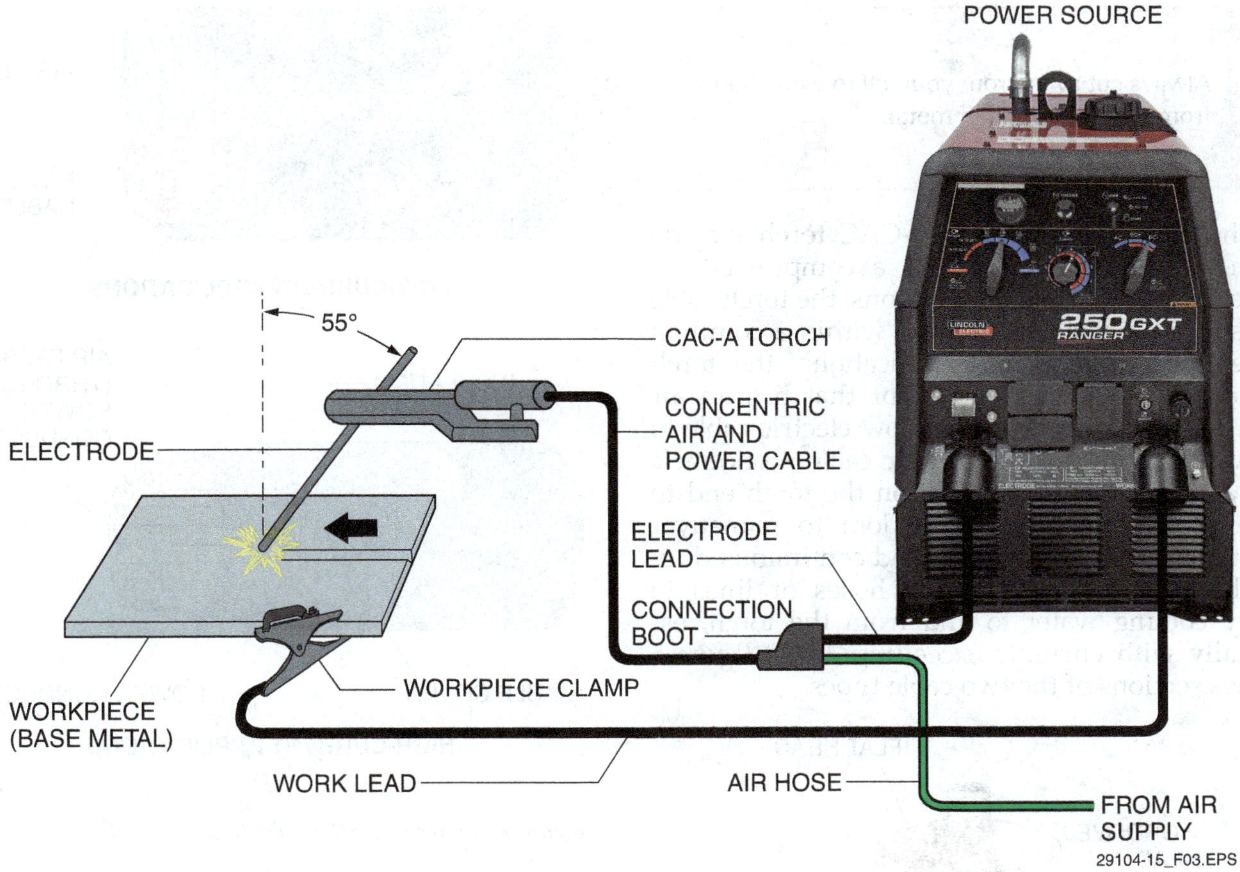

Figure 3 A-CAC torch connected to welding machine.

The air-jet orifices direct jets of air parallel to the electrode to blow the molten metal away. If required, heads may be replaced or head positions exchanged to alter orifice positions for different operations. When cutting and gouging, the jets must be located under the electrode (between the electrode and the base metal surface). For washing, the jet orifices should be located on one or both sides of the electrode. This helps prevent cutting into the base metal surface. *Figure 4* shows different head types and arrangements for a heavy-duty torch. These light- and heavy-duty heads are available in different sizes to accommodate ⅛" (3.2 mm) to 1" (2.5 cm) electrodes.

A-CAC Cutting

Since A-CAC cutting is not dependent on oxidation, it can cut metals that normally cannot be cut with oxyacetylene cutting methods.

At least one V-groove is needed to position the electrode in the center of the head. The V-grooved head provides about twice as much electrical contact surface as the flat head. A flat head is usually fixed, but a V-grooved head is free to rotate in its mount inside the torch handle. This allows the electrode and grooved head to be swiveled within the torch handle for the best or most comfortable work angle. Pressing the lever on the top of the torch handle raises the upper jaw and head, releasing the electrode for repositioning or replacement.

Air passages within the torch handle supply the head orifices with compressed air. A push-push air control valve extends through the torch handle. Pushing it from one side of the torch handle opens the valve and air flows to the head orifices; pushing it back from the other side closes the valve. Air-carbon arc cutting air pressure should be regulated to the lowest air pressure that cleanly blows away the molten metal, usually a range of 40 to 100 pounds (2.8 to 6.9 bars) per square inch gauge (psig).

WARNING!

Always cut away from yourself to avoid injury from sprays of molten metal.

The connections of the A-CAC torch include both an electrical cable and a compressed air hose. For low-current applications, the torch cable consists of a cable and a hose within a common cover. For high-current applications, the torch cable is a concentric conductor that has an air conduit in the center of a hollow electric cable to provide cooling for the electric cable. Concentric cables may include a swivel on the torch end to give the operator more freedom to maneuver the torch. Some heavy-duty and continuous-duty torches also have additional hoses or lines to carry cooling water to and from the torch, especially with currents exceeding 300A. *Figure 5* shows sections of the two cable types.

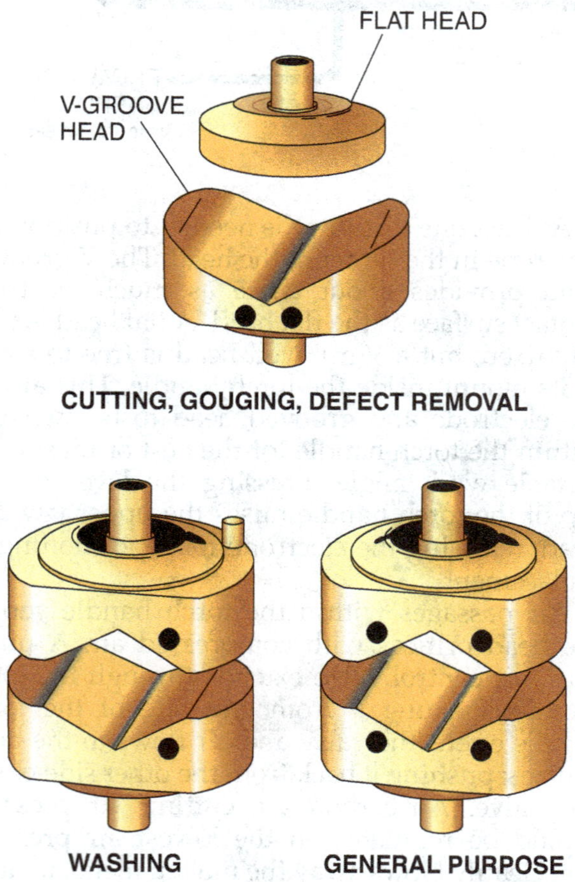

Figure 4 Head types and arrangements.

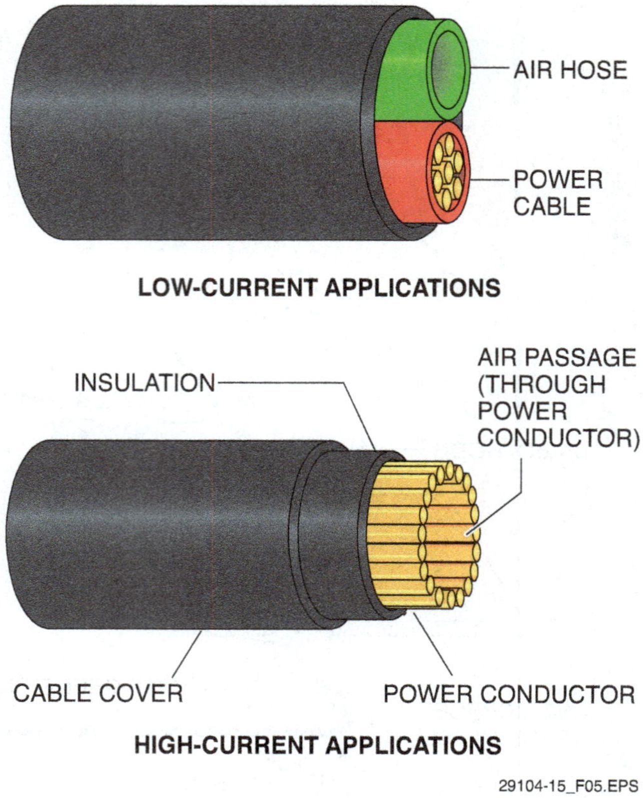

Figure 5 Torch cable configurations.

1.2.2 A-CAC Cutting Power Supply Options

Either AC or DC constant current welding machines may be used to supply power to A-CAC torches, but DC welding machines are usually preferred. The type of welding current used (AC or DC) depends upon the base metal to be cut or gouged and the electrode type used.

Foundry Torch

The heads shown in *Figure 4* are designed for interchangeable use in a heavy-duty foundry torch, such as the one shown here. This allows the torch to be used for defect removal or pad washing, or as a general fine removal or piercing tool.

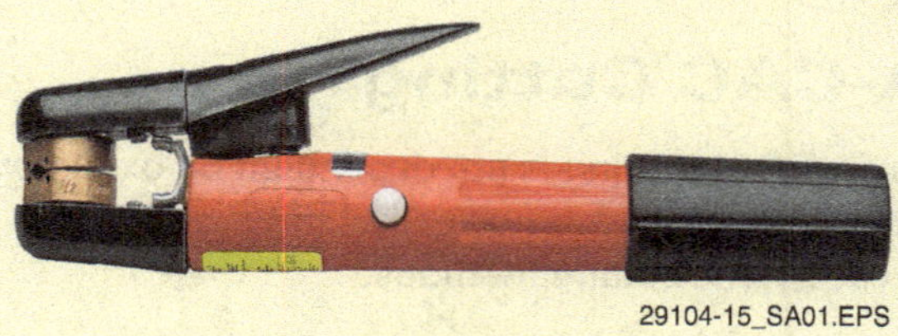

Light-duty gouging and cutting can be done with a 200 amp machine. Medium-duty gouging and cutting will require a 600 amp machine. Heavy-duty cutting may require 1,200- to 2,000-amp equipment.

> **NOTE**
> Avoid using constant voltage (CV) welding machines with A-CAC processes because of their voltage and duty cycle limitations.

Table 1 lists common base metals along with the recommended welding current and polarity to be used with each metal.

1.2.3 A-CAC Torch Air Supply

An A-CAC torch requires regulated compressed air to blow the molten metal from the cut. Compressed air is supplied to the torch through a hose attached to a fitting on the end of the torch cable. While air pressure and volume settings can be over a wide range, they must be high enough that all molten metal will be blown from the cut and dross will not be left in the groove. Light-duty and medium-duty cutting and gouging typically require approximately 8 cubic feet per minute (cfm) [227 liters per minute (lpm)] to 35 cfm (991 lpm) of airflow at 40 psig (2.76 bars) to 80 psig (5.52 bars).

Table 1 Welding Current Types for Different Metals

BASE METAL	ELECTRODE TYPE	CURRENT TYPE AND POLARITY	ADDITIONAL RECOMMENDATIONS
Carbon steel and low alloyed steel	DC	DCEP*	Use DC electrodes with DCEP.
	AC	AC or DCEN**	AC electrodes with a (C) transformer can be used, but AC will only be half as efficient as DC.
Stainless steel	AC	DCEP	Use DC electrodes with DCEP.
	DC	AC or DCEN	AC electrodes with a (C) transformer can be used, but AC will only be half as efficient as DC.
Iron (cast, ductile, or malleable)	AC	AC or DCEN	Best practice to use 1/2" (12.7 mm) or if possible, larger electrodes. These must be used at highest rated amperage.
	DC	DCEP (Max Amps Rated)	
Copper alloys (content 60% and above)	DC	DCEN (Max Amps Rated)	Use at the electrode's highest rated amperage.
	AC	AC	Use AC electrodes with AC.
Copper alloys (content 60% or below)	DC	DCEN (Max Amps Rated)	Use at the electrode's highest rated amperage.
Aluminum bronze and aluminum nickel bronze	DC	DCEN	For special navy propeller alloy.
Nickel alloys (content is over 80% of mass)	AC	AC	
Nickel alloys (content is less than 80% of mass)	DC	DCEP	
Aluminum	DC	DCEP	Use stainless steel wire brush on base metal before welding. Note: Electrode length between the torch and base metal must not exceed 3" (76.2 mm).
	AC	AC or DCEN	

*Direct Current Electrode Positive
**Direct Current Electrode Negative

29104-15_T01.EPS

Compressor Sizes

A 1 horsepower (hp) air compressor can typically supply sufficient air pressure and volume for smaller electrodes. Larger electrodes may require an air compressor of 10hp or greater to supply sufficient air pressure and volume. The torch manufacturer's recommendations for air pressure and flow rates should be followed.

Heavy-duty cutting can require up to 50 cfm (1,416 lpm) of airflow at 80 psig (5.52 bars) to 100 psig (6.89 bars). The regulator should be set for the lowest pressure that will reliably do the job.

1.3.0 Electrodes

A-CAC electrodes used for cutting and gouging are composed of a blend of carbon and graphite with a binder material, which is baked to produce a uniform structure. They may be plain or coated with copper, also referred to as copper-clad. Electrodes designed for A-CAC use are copper-coated and contain a rare-earth, arc-stabilizing metal. Note that rare-earth metals are not necessarily rare, but they are usually widely dispersed rather than concentrated, and are challenging to separate from each other. Coated and uncoated electrodes are available in various types, styles, and sizes. *Figure 6* shows the two types of coated electrodes that are most commonly used.

WARNING!

Do not use wet carbon arc electrodes, as they can explode. The electrodes must be kept dry.

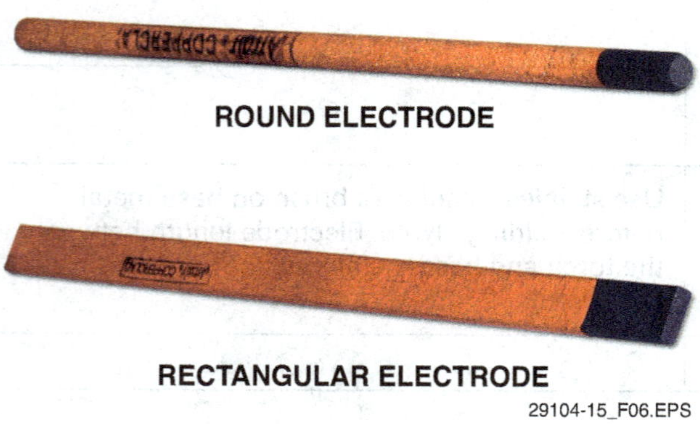

Figure 6 Copper-coated electrodes.

1.3.1 Electrode Types

A-CAC electrodes are of the following three basic types:

- *Plain for direct current (DC)* – Plain carbon electrodes for DC use do not have a copper coating. This type of electrode runs at hotter surface temperatures, **oxidizes** along its length, and loses diameter (reduces in thickness) between the torch and the arc tip. This electrode carries less current and burns more easily.
- *Copper-coated for direct current (DC)* – Copper-coated electrodes for DC use are the most common. These electrodes are superior for most applications because they can carry more current, can operate at cooler temperatures, and do not lose diameter through oxidation. However, they cannot be used where copper contamination could be a problem.
- *Copper-coated for alternating current (AC)* – Copper-coated AC electrodes differ from copper-coated DC electrodes because they contain rare-earth materials that help stabilize the arc; copper-coated DC electrodes do not.

1.3.2 Electrode Styles

Electrodes are manufactured in several styles, each with a range of sizes:

- *Round electrodes* – Round electrodes are the most common type used for manual cutting or gouging. They are manufactured with an uncoated tip that tapers as they begin to be consumed.
- *Round-jointed electrodes* – Round-jointed electrodes are designed for use with continuous-feed automatic and semi-automatic torches. Each electrode has a slightly conical tenon machined on one end and a matching socket machined into the opposite end. These ends allow the electrodes to be joined end-to-end to allow continuous cutting or gouging without interruptions for electrode changes. This system also eliminates stub waste.
- *Special shapes* – Special-shape electrodes are designed to make special-shaped grooves or cuts. Special shapes include flat (rectangular) and half-round. The flat electrodes are designed for close tolerance metal removal. Their uses include removing weld buildups, shaping dies, removing welded lugs, and beveling edges. Half-round electrodes are designed for cutting wide grooves with torches designed for smaller electrodes. *Figure 7* shows the cross-section shapes of special electrodes.

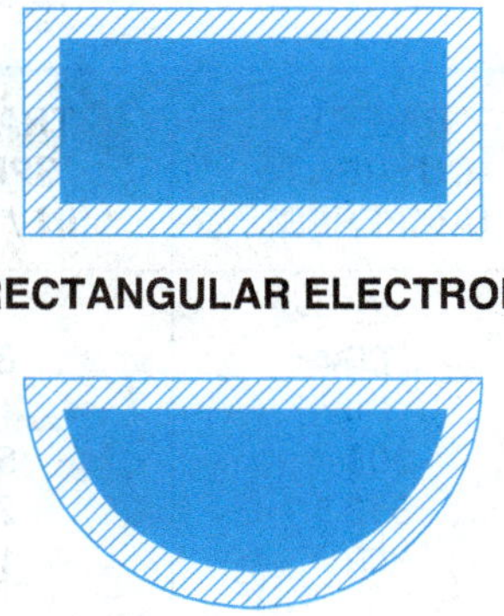

Figure 7 Cross-section shapes of special electrodes.

Commonly Available Carbon Electrodes

The carbon electrodes that are usually readily available from general welding suppliers are the copper-coated round and rectangular electrodes used with direct current electrode positive (DCEP) current. Other types and styles, including jointed, plain, semi-round, or copper-coated AC versions, are not commonly used and must be special-ordered from suppliers.

1.3.3 Electrode Sizes and Amperages

Round electrode sizes are specified by diameter; flat electrodes are specified by the widths of their widest side.

Plain and copper-coated AC electrodes range in size from ⁵⁄₃₂" (4 mm) to ⅝" (16 mm). They are 12" (300 mm) long.

DC copper-coated electrodes range in size from ⅛" (3.2 mm) to 1" (25 mm). The smaller-sized round and flat electrodes are 12" (300 mm) long. Some of the larger-sized jointed electrodes are 17" (432 mm) long. AC copper-coated electrodes range in size from ³⁄₁₆" (4.8 mm) to ½" (12.7 mm).

Electrode current ratings increase with size. Plain electrodes and AC copper-coated electrodes operate at lower amperages than the same size of DC copper-coated electrodes.

Table 2 shows some typical current ratings for plain DC and copper-covered AC electrodes. These ratings will vary with different manufacturers. For DC copper-covered electrodes, add 10 percent to the maximum amperage values shown in the table.

1.3.4 Electrode Selection

When performing A-CAC cutting or gouging, the minimum groove width is controlled by the size of the electrode. Groove depth and contour are controlled by electrode angle, travel speed, electrode side-to-side movement, current setting, and air pressure.

With copper-coated electrodes up to ½" (13 mm) in size on carbon steel, the minimum groove width will run about ⅛" (≈3 mm) wider than the electrode size.

Table 3 lists copper-coated electrode sizes and travel speeds for various groove sizes in carbon steel.

Table 2 Electrode Current Ratings

ELECTRODE TYPE	ELECTRODE SIZE		MINIMUM CURRENT	MAXIMUM CURRENT*
	IN	MM		
PLAIN DC OR COPPER-CLAD AC	⅛"	3.2	30 AMP	60 AMP
	⁵⁄₃₂"	4.0	90 AMP	150 AMP
	³⁄₁₆"	4.8	150 AMP	200 AMP
	¼"	6.4	200 AMP	400 AMP
	⁵⁄₁₆"	8.0	250 AMP	450 AMP
	⅜"	9.5	350 AMP	600 AMP
	½"	13	600 AMP	1000 AMP
	⅝"	15	800 AMP	1200 AMP
	¾"	19	1200 AMP	1600 AMP
	1"	25	1800 AMP	2200 AMP

*(For copper-clad DC, add 10% to maximum current value)

29104-15_T02.EPS

Table 3 Groove Size, Electrode Size, Amperes, and Travel Speed

GROOVE WIDTH (IN, MM)	GROOVE DEPTH (IN, MM)	ELECTRODE SIZE (DIA.) (IN, MM)	DIRECT CURRENT (AMPERES)	TRAVEL SPEED (IN / MIN)
¼, 6.4	¹⁄₁₆, 1.6	³⁄₁₆, 4.8	200	82
⁵⁄₁₆, 7.9	⅛, 3.2	¼, 6.4	300	51
⁵⁄₁₆, 7.9	¼, 6.4	¼, 6.4	320	29
⁵⁄₁₆, 7.9	⅜, 9.5	¼, 6.4	300	15
⅜, 9.5	⅛, 3.2	⁵⁄₁₆, 7.9	320	65
⅜, 9.5	¼, 6.4	⁵⁄₁₆, 7.9	420	31
⅜, 9.5	½, 12.7	⁵⁄₁₆, 7.9	540	27
⁷⁄₁₆, 11.1	⅛, 3.2	⅜, 9.5	560	82
⁷⁄₁₆, 11.1	¼, 6.4	⅜, 9.5	560	75
⁷⁄₁₆, 11.1	½, 12.7	⅜, 9.5	560	15
⁷⁄₁₆, 11.1	¹¹⁄₁₆, 17.5	⅜, 9.5	560	12
⁹⁄₁₆, 14.3	¼, 6.4	½, 12.7	1200	22
⁹⁄₁₆, 14.3	⅜, 9.5	½, 12.7	1200	21
⁹⁄₁₆, 14.3	½, 12.7	½, 12.7	1200	18
⁹⁄₁₆, 14.3	¾, 19.1	½, 12.7	1200	12
¹³⁄₁₆, 20.6	¼, 6.4	⅝, 15.9	1300	29
¹³⁄₁₆, 20.6	½, 12.7	⅝, 15.9	1300	14
¹³⁄₁₆, 20.6	¾, 19.1	⅝, 15.9	1300	11

29104-15_T03.EPS

To select an electrode, follow these steps:

Step 1 Identify the base metal type or alloy.

WARNING!

Never cut or burn cadmium material without proper ventilation and respiratory protection. The process produces poisonous gases that can cause severe personal injury.

CAUTION

Most manufacturers do not recommend using inverters or electronically controlled welding machines for carbon arc cutting. Peak currents are produced that are not usually filtered by the machine, which can damage the internal circuitry.

Step 2 Select a welding machine with the recommended current type (AC or DC), adequate ampere rating, and duty cycle for the base metal type and cut or gouge size.

Step 3 Select an electrode type compatible with the weld current type.

Step 4 Select the electrode size to produce the desired groove width.

1.3.5 Electrode Installation

An A-CAC electrode is usually gripped about 6"(≈150 mm) from the arc tip, as shown in *Figure 8*. For aluminum, a 3" (≈76 mm) stick-out is recommended. As the electrode is consumed, it is periodically advanced in the torch handle to maintain the proper projection.

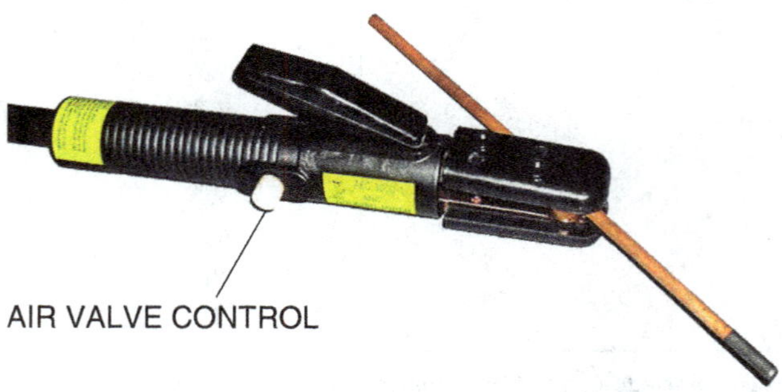

Figure 8 Properly gripped electrode.

1.4.0 Safety Practices

The following is a summary of safety procedures and practices that must be observed while cutting or welding. Keep in mind that this is just a summary. Complete safety coverage is provided in the *Welding Safety* module. If you have not completed that module, do so before continuing. Above all, be sure to wear appropriate protective clothing and equipment when welding or cutting.

1.4.1 Protective Clothing and Equipment

Welding activities can cause injuries unless you wear all of the protective clothing and equipment that is designed specifically for the welding industry. The following information includes important safety guidelines about protective clothing and equipment:

- Always use safety glasses with a full face shield or a helmet. The glasses, face shield, or helmet lens must have the proper light-reducing tint for the type of welding or cutting being performed. Use corrective glasses if needed; note that some companies will not allow contact lenses. Never directly or indirectly view an electric arc without using a properly tinted lens.
- Wear proper protective leather and/or flame-retardant clothing along with welding gloves that will protect you from flying sparks and molten metal, as well as heat.
- Wear high-top safety shoes or boots. Make sure that the tongue and lace area of the footwear will be covered by a pant leg. If the tongue and lace area is exposed, or if the footwear must be protected from burn marks, wear leather spats under your pants or chaps and over the top front area of the footwear.
- Wear a 100-percent cotton cap with no mesh material included in its construction. The bill of the cap points to the rear. If a hard hat is required for the environment, use one that allows the attachment of rear deflector material and a face shield. A hard hat with a rear deflector is generally preferred when working overhead, and may be required by some employers and job sites.

WARNING!

Do not wear a cap with a button in the middle. The conductive metal button beneath the fabric represents a safety hazard.

- Wear a face shield over snug-fitting cutting goggles or safety glasses for gas welding or cutting. Either the face shield or the lenses of the welding goggles must be an approved shade for the application. A welding hood equipped with a properly tinted lens is also acceptable. See *Table 4* for a list of lens shades based on arc current.

Table 4 Guide for Shade Numbers

From *AWS F2.2 Lens Shade Selector*. Shade numbers are given as a guide only and may be varied to suit individual needs.

PROCESS	ELECTRODE SIZE IN (MM)	ARC CURRENT (AMPERES)	MINIMUM PROTECTIVE SHADE	SUGGESTED* SHADE NO. (COMFORT)
Shielded Metal Arc Welding (SMAW)	Less than 3⁄32 (2.4)	Less than 60	7	–
	3⁄32–5⁄32 (2.4–4.0)	60–160	8	10
	5⁄32–1⁄4 (4.0–6.4)	160–250	10	12
	More than 1⁄4 (6.4)	250–550	11	14
Gas Metal Arc Welding (GMAW) and Flux Cored Arc Welding (FCAW)		Less than 60	7	–
		60–160	10	11
		160–250	10	12
		250–500	10	14
Gas Tungsten Arc Welding (GTAW)		Less than 50	8	10
		50–150	8	12
		150–500	10	14
Air Carbon Arc Cutting (CAC-A)	(Light)	Less than 500	10	12
	(Heavy)	500–1000	11	14

*As a rule of thumb, start with a shade that is too dark to see the weld zone. Then go to a lighter shade which gives sufficient view of the weld zone without going below the minimum. In oxyfuel gas welding, cutting, or brazing where the torch and/or the flux produces a high yellow light, it is desirable to use a filter lens that absorbs the yellow or sodium line of the visible light spectrum.

29104-15_T04.EPS

WARNING!

Because of the high temperature and the intensity of the arc, choose a shade of helmet lens that is darker than the normal shade you would use for welding on the same thickness of metal. A number 10 to 14 lens shade is recommended for air-carbon arc cutting and gouging, depending on the current level.

- Wear earplugs to protect ear canals from sparks. Wear hearing protection to protect against the consistent sound of the torch.

WARNING!

Ear protection is essential to protect ears from the noise of the torch. Other personal protective equipment (PPE) must be worn to protect the operator from hot metal and slag.

1.4.2 Fire/Explosion Prevention

Welding activities usually involve the use of fire or extreme heat to melt metal. Whenever fire is used, it must be controlled and contained. Welding activities are often performed on vessels that may once have contained flammable or explosive materials. Residues from those materials can catch fire or explode when a welder begins work on such a vessel. The following are fire and explosion prevention guidelines associated with welding:

- Never carry matches or gas-filled lighters in your pockets. Sparks can cause the matches to ignite or the lighter to explode, causing serious injury.
- Always comply with all site and/or employer requirements for a hot-work permit and a fire watch.
- Never use oxygen to blow dirt or dust off clothing. The oxygen can remain trapped in the fabric for a time. If a spark hits the oxygen in the fabric, the clothing can burn rapidly and violently.
- Make sure that any flammable material in the work area is moved or shielded by a fire-resistant covering.
- Approved fire extinguishers must be available before attempting any heating, welding, or cutting operations. Make sure the extinguisher is charged, the inspection tag is valid, and any individual that may be required to operate it knows how to do so.
- Never release a large amount of oxygen or use oxygen in place of compressed air. The presence of oxygen around flammable materials or sparks can cause rapid and uncontrolled combustion. Keep oxygen away from oil, grease, and other petroleum products.
- Never release a large amount of fuel gas, especially acetylene. Methane and propane are heavier than air and tend to migrate to and concentrate in low areas. As a result, they can ignite at a considerable distance from the release point. Acetylene is lighter than air but is even more dangerous than methane; when mixed with air or oxygen, it will explode at much lower concentrations than any other common fuel gas.
- To prevent fires, maintain a neat and clean work area, and make sure that any metal scrap or slag is cold before disposal.

DCEP Electrode Consumption

When making a gouge that equals the electrode diameter, you will get approximately 8" (≈200 mm) of groove for every inch of electrode consumed. However, this will vary widely depending on electrode angle, speed, current, and the material being cut.

Before cutting or welding containers such as tanks or barrels, find out if they contained any explosive, hazardous, or flammable materials, including petroleum products, citrus products, or chemicals that decompose into toxic fumes when heated. Proper procedures for cutting or welding hazardous containers are described in the *American Welding Society (AWS) F4.1, Safe Practices for the Preparation of Containers and Piping for Welding and Cutting,* and *ANSI Z49.1*. As a standard practice, always clean and then fill any tanks or barrels with water, or purge them with a flow of inert gas such as nitrogen to displace any oxygen.

> **WARNING!**
>
> Welding or cutting must never be performed on drums, barrels, tanks, vessels, or other containers until they have been emptied and cleaned thoroughly, eliminating all flammable materials and all substances (such as detergents, solvents, greases, tars, or acids) that might produce flammable, toxic, or explosive vapors when heated. Do not assume that a container that has held combustibles is clean and safe until proven so by proper tests. Do not weld in places where dust or other combustible particles are suspended in air or where explosive vapors are present.

Containers must be cleaned by steam cleaning, flushing with water, or washing with detergent until all traces of the material have been removed.

> **WARNING!**
>
> Clean containers only in well-ventilated areas. Vapors can accumulate during cleaning, causing explosions or injury.

After cleaning the container, fill it with water or a purging gas, such as carbon dioxide, argon, or nitrogen to displace the explosive fumes. Air, which contains oxygen, is displaced from inside the container by the water or inert gas. Without oxygen, combustion cannot take place.

A water-filled vessel is the best alternative. When using water, position the container to minimize the air space. When using an inert gas, provide a vent hole so the inert gas can push the air and other vapors out to the atmosphere. Keep in mind, though, that even these precautions do not guarantee the absence of flammable materials inside. For that reason, these types of activities should not be done without proper supervision and the use of proper testing methods.

1.4.3 Work Area Ventilation

Vapors and fumes tend to rise in the air from their sources. Welders often have to work above the welding area where the fumes are being created. Welding fumes can cause personal injuries. Good work area ventilation helps to remove the vapors and to protect the welder. The following is a list of work area ventilation guidelines to consider before and during welding activities:

- Make sure confined space procedures are followed before conducting any welding or cutting in the confined space.
- Always perform cutting or welding operations in a well-ventilated area. Cutting or welding operations involving materials such as coatings containing cadmium, lead, zinc, and chromium will result in toxic fumes. For cutting or welding of such materials, always wear an approved respirator as directed by your employer with a filter that may be used on a standard respirator.
- Make sure confined spaces are ventilated properly for cutting or welding purposes.
- Never use pure oxygen from cylinders for the purpose of ventilation or breathing air.

Because the molten metal cut during most A-CAC cutting or gouging activities tends to splatter, make sure that fire-resistant shields or screens are set up to protect any equipment or personnel near the work area. Also, make sure that a fire watch is posted.

Additional Resources

ANSI C5.3, Recommended Practices for Air Carbon Arc Gouging and Cutting. Latest Edition. Miami, FL: American Welding Society.

Air Carbon-Arc Guide, Form Number: 89-250-008. Denton, Texas: Victor Technologies, Inc.

2014 Technical Training Guide. The Lincoln Electric Company, Cleveland, OH. USA. **www.lincolnelectric.com**.

1.0.0 Section Review

1. One of the primary uses for air-carbon arc cutting (A-CAC) is _____.
   a. to gouge weld grooves, to back-gouge welds, and to gouge out defective welds
   b. together with compressed air, to control base metal temperature during cutting
   c. to introduce additional metal alloys into the base metal
   d. to reduce the amperage used in the cutting and gouging process

2. When the electrode is gripped in the torch handle, the electrode tip should be no farther from the heads than _____.
   a. 3" (7.6 cm)
   b. 6" (15 cm)
   c. 8" (20 cm)
   d. 12" (31 cm)

3. Electrodes designed for A-CAC use are copper-coated and contain an arc-stabilizing ingredient referred to as _____.
   a. graphite
   b. copper
   c. carbon
   d. a rare-earth metal

4. One of the procedures to prevent fires or explosions during A-CAC work is to make sure _____.
   a. that a certified welder is a member of the fire watch posted
   b. that the gouge equals the electrode diameter
   c. that the carbon arc electrodes to be used are completely dry
   d. to not use inverters or electronically controlled welding machines

Section Two

2.0.0 A-CAC Setup, Operation, and Care

Objective

Describe how to set up, safely operate, and care for air-carbon arc cutting equipment.

a. Describe how to prepare the equipment and work area for air-carbon arc cutting.
b. Describe how to wash and gouge metals.
c. Describe how to care for air-carbon arc cutting equipment.

Performance Tasks

2. Prepare the work area and air-carbon arc cutting equipment for safe operation.
3. Use air-carbon arc cutting equipment for washing.
4. Use air-carbon arc cutting equipment for gouging.
5. Perform storage and housekeeping activities for air-carbon arc cutting equipment.

Trade Terms

Concentric cable system: A-CAC configuration in which a unique combination fitting is used to connect the torch cable to welding power in order to enable compressed air passage through the power conductor.

Polarity: The condition of a system in which it has opposing physical properties at different points, such as an electric charge.

Now that basic A-CAC equipment has been described, this section will focus on setting up the equipment and the techniques to successfully gouge and wash.

2.1.0 Preparing to Cut

A-CAC equipment includes the torch and its attached cable. The torch cable must be connected to a welding machine and a compressed air source. Some torch cables contain a separate air hose and electrical cable within a common jacket. Others use a special concentric cable system that routes the compressed air through the center of a hollow electrical cable. *Figure 9* shows an inside view of both a concentric cable and a cable with separate air hose and electrical cable within a common jacket.

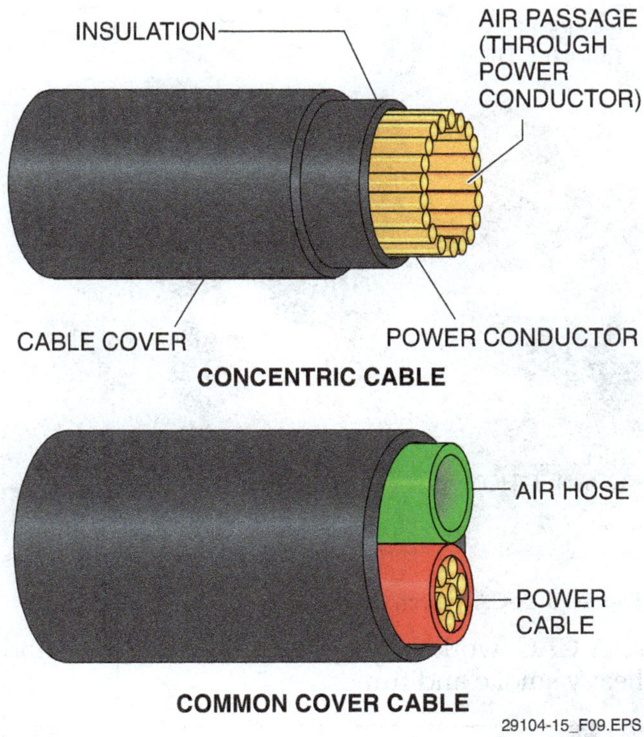

Figure 9 Inside view of concentric cable and common cover cable.

With the concentric system, a unique combination fitting is used to connect the torch cable to welding power and compressed air, as shown in *Figure 10*. It consists of a special female connector assembly and a hose nut with a pierced tab extending from one side. Heavy-duty, high-current foundry torches may have two tabs on the nut. The special fitting is electrically connected to the hollow electrical conductor in the torch cable. A standard threaded air hose fitting attaches to the connector, while the welding machine's electrode cable bolts to the lug on the side of the fitting. A protective rubber boot on the torch cable covers the connections.

2.1.1 Preparing the Work Area for A-CAC

The A-CAC equipment should be set up at a convenient distance from the welding machine. The area should be well ventilated and clear of all combustible material. The work to be cut or gouged should be placed at a comfortable working position.

A-CAC work will blow molten metal considerable distances. It is important that everything combustible be removed from the range of the blown molten metal. Flame-resistant shields or curtains must be erected to protect any workers or equipment within the range of the molten metal. It is also important to station a fire watch in the area. Make sure a fully charged fire extinguisher is at hand.

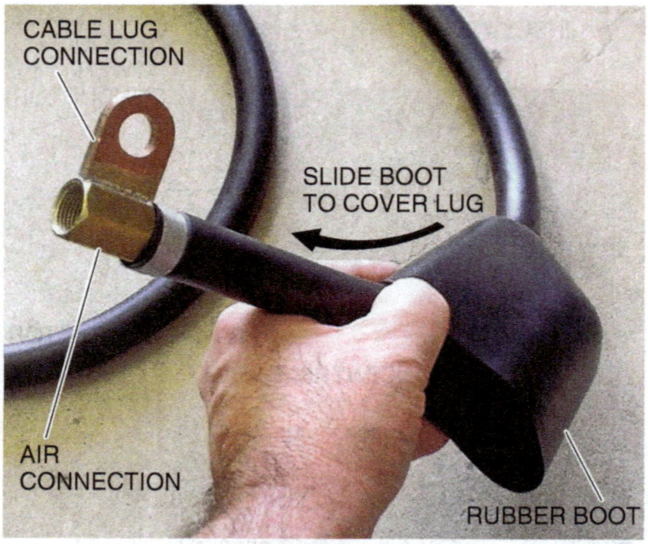

Figure 10 A-CAC torch combination power and air fitting

A-CAC work also generates loud noise and heavy smoke and fumes.

> **WARNING!**
> Ear protection is essential to protect ears from the noise. Other personal protective equipment must be worn to protect the operator from hot metal and slag.

If the site ventilation cannot keep the smoke and fumes away from the operator, an approved respirator must be worn.

2.1.2 Test Operating A-CAC Equipment

A-CAC equipment is simple to operate. To operate the equipment, follow these steps:

Step 1 Verify that the work area is properly shielded.

Step 2 Set the compressed air regulator to the desired pressure and open all valves supplying the A-CAC torch. Operate the air valve to ensure proper flow.

Step 3 Adjust the welding machine for the desired polarity and amperage.

Step 4 Put on a welding shield.

> **WARNING!**
> Always use proper hearing protection and appropriate PPE.

Step 5 Put on heat-insulating gauntlet-type welding gloves.

Step 6 Attach the welding machine workpiece clamp to the workpiece.

> **CAUTION**
> Be sure to position the workpiece clamp where the welding current will not pass through seals, bearings, or other contacting surfaces that could be damaged from arcing.

Step 7 Open the air jet spool valve. Position the torch at a 55-degree push angle with the air jets below the electrode and between the workpiece and the electrode.

Step 8 Strike an arc by touching the electrode to the workpiece. Do not draw the electrode back after the arc is established. Maintain a short arc and begin washing or gouging as described in the next section.

> **WARNING!**
> Do not use wet carbon-arc electrodes, as they can explode. The electrodes must be kept dry.

2.2.0 Washing and Gouging with A-CAC Equipment

Before using A-CAC equipment for washing or gouging activities, always evaluate the work to be done and plan the best way to achieve the desired results. Always be sure you understand how to position the A-CAC electrode to be most effective. It is also important to inspect the surfaces that need to be cut or gouged. Planning the work saves time, effort, and materials.

2.2.1 A-CAC Washing

Washing is the process of cutting objects or projections, such as bolt heads, old welds, and other unwanted materials from the surface of a base metal to leave a flush surface. Typically, the electrode is used at a push angle of about 55 degrees to the surface that contains the projection. The air orifices must blow air between the electrode and the base metal. The washing is performed by weaving the electrode from side to side across the projection. On some torches with available special washing heads, the air jet orifices are located on one side of the electrode. This arrangement reduces the tendency to gouge, but limits the torch to cutting from only one direction. *Figure 11* shows the results that can be achieved with A-CAC washing.

Figure 11 The results of A-CAC washing.

To perform washing with an A-CAC torch, follow these steps:

> **WARNING!**
>
> Wear eye and ear protection and protective clothing during this procedure.

Step 1 Select the proper electrode and set the correct polarity and amperage at the welding machine.

Step 2 Turn on the welding machine and start the air flowing through the jet air orifices.

Step 3 Holding the electrode at about a 55-degree push angle, strike an arc at the edge of the projection to be removed. As the arc melts the metal, swing it gradually across the projection. Continue the process until the projection is removed flush with the surface.

2.2.2 A-CAC Gouging

Gouging is a process of cutting a groove into a surface, usually to prepare it for welding. With A-CAC gouging, the electrode size determines the minimum groove width. Special-shaped electrodes can be used to cut special-shaped grooves. Groove width can be increased by weaving the electrode tip slightly from side to side. *Figure 12* shows the result of an A-CAC gouging process.

Groove depth and contour are controlled by electrode travel angle, tip side movement, and travel speed. For a shallow gouge, the electrode is advanced at a rapid rate. To increase the groove depth, decrease the advance speed or increase the current setting. To increase the width of the groove, move the electrode tip from side to side or use a larger electrode. Practicing A-CAC skills, like all welding skills, will help you to see the effect of very small and different hand motions.

Figure 12 A-CAC gouging.

> **Vertical Washing and Gouging**
>
> Positioning the workpiece in a vertical position simplifies metal ejection by using gravity in addition to the air jets. Make sure that adequate shoes and shoe top coverings are worn if vertical washing and gouging positions are used.

To gouge with an A-CAC torch, follow these steps:

> **WARNING!**
>
> Wear eye and ear protection and protective clothing during this procedure.

Step 1 Select the proper size and shape electrode and set the correct polarity and amperage at the welding machine.

Step 2 Turn on the welding machine and start the air flowing through the air jet orifices.

Step 3 Holding the electrode at about a 55-degree push angle, strike an arc where the groove is to begin. As the arc melts the metal, advance it steadily along the base metal. Keep the electrode at a 0-degree work angle with respect to the groove direction.

2.2.3 Inspecting A-CAC Cut Surfaces

Inspect the A-CAC cut surface for the following conditions:

- Smooth edges that are relatively free of notches
- No dross
- No carbon deposits

Notches left in a base metal have a tendency to trap slag during welding. Dross or carbon deposits left on a surface to be welded will combine with the base metal to form hard, brittle zones and porosity in the weld zone. The topic of weld quality will be covered in depth in another module. Check your site or employer's quality standards for the base metal being cut. Cleaning is always necessary after A-CAC operations to ensure that all carbon and/or dross left by the electrode is removed prior to welding. A-CAC cutting is often used to quickly remove material, shortening the time required for grinding and other preparation tasks.

2.3.0 A-CAC Equipment Care

Proper equipment storage and housekeeping are essential for work efficiency and safety. Good housekeeping requires the following steps:

Step 1 Disconnect the A-CAC torch from the welding machine cable and air hose and store the torch in its assigned location.

Step 2 Neatly store the air hose, welding cables, and any related tools used in their assigned locations.

Step 3 Clean off the welding area and sweep up the slag and debris that were blown around the area.

Step 4 Make sure any remaining electrodes are stored in the proper conditions.

2.3.1 Repair of A-CAC Equipment

Following the proper equipment storage and housekeeping procedures usually results in trouble-free cutting and minimal repairs. However, cables and air hoses may become damaged and require replacement, especially if they have been run over by forklifts or other mobile equipment, been dragged across hot metal, or burned by molten-metal cutting sparks.

Before replacing a cable, always make sure the electrical disconnect and any other sources of energy such as an air supply is in the Off position. Air pressure should be vented from the hose. Remove the cable from the machine using the correct size wrench. If an electrode holder becomes cracked or worn, it may be possible to replace only the damaged portion.

> **Automatic Gouging System**
>
> Manual A-CAC is not usually thought of as a precise way to groove or cut. However, when the process is automated, it produces faster gouging speeds and the accuracy is both surprising and exceptional. A video that demonstrates precision gouging using equipment from Arcair can be viewed at **www.victortechnologies.com/arcair**. From the Home page, look for Media and Downloads under the Featured Products tab.

Additional Resources

ANSI C5.3, Recommended Practices for Air Carbon Arc Gouging and Cutting. Latest Edition. Miami, FL: American Welding Society.

2014 Technical Training Guide. The Lincoln Electric Company, Cleveland, OH. USA. **www.lincolnelectric.com**.

Air Carbon-Arc Guide, Form Number: 89-250-008. Denton, Texas: Victor Technologies, Inc.

2.0.0 Section Review

1. An A-CAC concentric system has compressed air passing through _____.
   a. an air hose that runs next to the power conductor
   b. a special male connector assembly and a hose nut with a pierced fitting
   c. the power conductor
   d. a separate filter to ensure clean compressed air

2. Washing is a process used to prepare the surface of the base metal _____.
   a. and introduce additional metal alloys such as copper to the surface
   b. by creating a narrow groove
   c. by heating it up and then allowing even cooling of the surface
   d. by removing objects and projections from the surface

3. If an electrode holder becomes cracked or worn, _____.
   a. the cable and torch assembly must be replaced
   b. it may be possible to replace only the damaged portion
   c. it can be exchanged with one designed for SMAW welding
   d. it can be exchanged with one designed for GTAW welding

Summary

The A-CAC process is very useful for washing surfaces flush or for gouging out cracks to prepare them for welding repairs. However, because the torch air jets are continuously blasting molten metal from the arc, good site fire-prevention practices and use of appropriate PPE are essential elements. Ear protection is especially important to prevent noise damage from the air blast and to prevent sparks or slag from entering the ears.

Review Questions

1. The A-CAC process uses a high-current electric arc between the base metal and the end of a(n) _____.
   a. carbon-graphite electrode to melt the base metal
   b. high-frequency pilot arc to melt the base metal
   c. plasma jet to melt the base metal
   d. idle arc to melt the base metal

2. An A-CAC torch requires compressed air to blow the _____.
   a. carbon dioxide away from the cut
   b. molten metal away from the cut
   c. nitrogen away from the cut
   d. carbon arc away from the cut

3. When cutting and gouging with A-CAC equipment, the jet orifices of the torch head(s) should be located _____.
   a. beside the electrode
   b. above the electrode
   c. ahead of the electrode
   d. between the electrode and the base metal surface

4. To prevent overheating, the electrode tip to the V-groove head should be no farther than _____
   a. 1" (2 cm)
   b. 2" (5 cm)
   c. 4" (10 cm)
   d. 6" (15 cm)

5. Light-duty gouging or cutting is typically done with a _____.
   a. 10A welding machine
   b. 200A welding machine
   c. 600A welding machine
   d. 1,200A welding machine

6. Heavy-duty cutting can require a torch air-flow rate up to _____.
   a. 20 cfm (566 lpm)
   b. 35 cfm (991 lpm)
   c. 40 cfm (1,133 lpm)
   d. 50 cfm (1,416 lpm)

7. Compressed air is supplied to the torch through a hose attached to a fitting on the _____.
   a. welding machine
   b. end of the torch cable
   c. air jet orifices
   d. workpiece

8. Electrodes designed for A-CAC contain a rare-earth, _____.
   a. magnet metal
   b. undercurrent ingredient
   c. amplifier ingredient
   d. arc-stabilizing metal

9. The most common style of A-CAC electrode used for manual cutting or gouging is the _____.
   a. flat electrode
   b. round electrode
   c. half-round electrode
   d. round-jointed electrode

10. When performing A-CAC cutting or gouging, the minimum groove width is controlled by the _____.
    a. size of the torch
    b. size of the electrode
    c. speed of travel
    d. angle of the electrode holder

11. Because A-CAC activities tend to blow molten metal considerable distances, it is important to station a(n) _____.
    a. observer in the A-CAC work area
    b. slag trap in the A-CAC work area
    c. oxygen monitor in the A-CAC work area
    d. fire watch in the A-CAC work area

12. To wash a projection from a metal surface using A-CAC equipment, the electrode is typically used at a push angle of about _____.
    a. 30 degrees to the base metal surface
    b. 45 degrees to the base metal surface
    c. 55 degrees to the base metal surface
    d. 90 degrees to the base metal surface

13. Torches with special washing heads in which the air jet orifices are located on only one side of the electrode are limited to cutting from _____.

    a. one direction
    b. two directions
    c. three directions
    d. four directions

14. In A-CAC gouging, groove depth is influenced by the electrode _____.

    a. travel angle and speed
    b. size and length
    c. weave and angle
    d. length and weight

15. Dross or carbon deposits left on a surface to be welded will combine with the base metal to form _____.

    a. hard, brittle zones and porosity
    b. oxidation and/or corrosion
    c. soft areas and poor adhesion
    d. an excellent, strong weld

Trade Terms Quiz

Fill in the blank with the correct term that you learned from your study of this module.

1. To combine with oxygen, such as in burning or rusting (corrosion), is to _______.
2. A mixture of soft amorphous carbon and hard graphite carbon, possibly coated in copper, are the ingredients for a(n) _______.
3. A unique combination fitting used to connect the torch cable to welding power that also enables compressed air passage through the power conductor is part of a(n) _______.
4. The condition of opposing physical properties at different points, such as an electric charge, is referred to as _______.

Trade Terms

Carbon-graphite electrode
Concentric cable system
Oxidize
Polarity

Appendix

Performance Accreditation Tasks

The American Welding Society (AWS) School Excelling through National Skills Standards Education (SENSE) program is a comprehensive set of minimum Standards and Guidelines for Welding Education programs. The following performance accreditation tasks are aligned with and designed around the SENSE program.

The Performance Accreditation Tasks (PATs) correspond to and support the learning objectives in *AWS EG2.0, Guide for the Training and Qualification of Welding Personnel: Entry-Level Welder.*

Note that in order to satisfy all learning objectives in *AWS EG2.0*, the instructor must also use the PATs contained in the second level of the NCCER Welding curriculum.

PAT 1 and 2 correspond to *AWS EG2.0, Module 8 – Thermal Cutting Processes, Unit 4 – Air-Carbon Arc Cutting*, Key Indicators 4, and 5.

PATs provide specific acceptable criteria for performance and help to ensure a true competency-based welding program for students.

The following tasks are designed to develop your competency in preparing base metal using A-CAC processes. Practice each task until you are thoroughly familiar with the procedure.

As you complete each task, take it to your instructor for evaluation. Do not proceed to the next task until instructed to do so by your instructor.

A-CAC WASHING AND GOUGING

Perform A-CAC Washing

Using any of the materials identified below, perform A-CAC washing to remove the portion identified by the instructor. Materials that can be used for this task include:

- Steel backing strip on a butt weld
- Excess buildup on the face of a weld
- Rivets or bolts in a plate
- Blocks, angles, clips, eyes, D-rings, or items welded to a plate

Criteria for Acceptance

- Material removed flush with the base metal surface __________
- No notching in the surface of the base metal __________

Perform A-CAC gouging

Using mild steel plate ½" (12.7 mm) thick or thicker, gouge a U-groove at least 8" (20 cm) long, as shown in the figure, in the 1G and 2G positions.

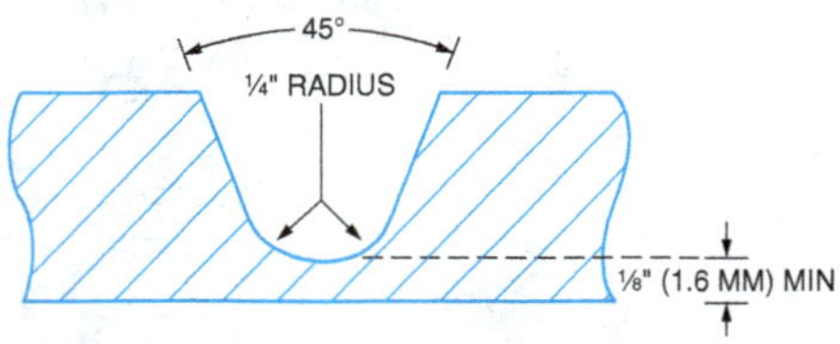

NOTE: GOUGE A WELDING PASS AT LEAST 8" LONG, AS SHOWN IN THE FIGURE BELOW, IN THE 1F AND 2F POSTIONS.

NOTE: BASE METAL = CARBON STEEL PLATE AT LEAST ¼" (6.4 MM) THICK

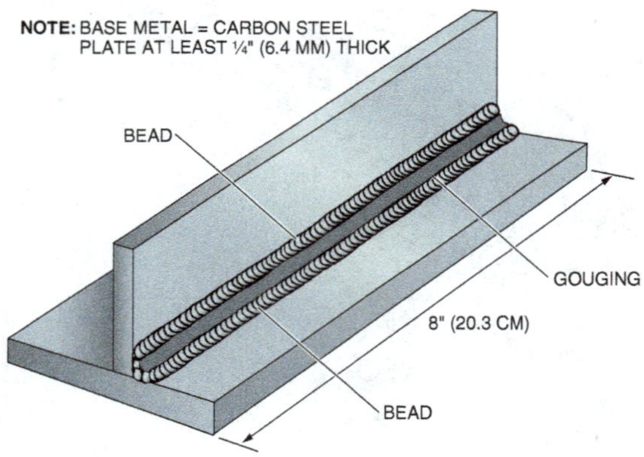

Criteria for Acceptance

- Groove width and depth are uniform __________
- Groove walls are smooth and uniform __________
- No dross within the groove __________

Trade Terms Introduced in This Module

Carbon-graphite electrode: An electrode composed of a mixture of soft amorphous carbon and hard graphite carbon that may be coated with copper.

Concentric cable system: A-CAC configuration in which a unique combination fitting is used to connect the torch cable to welding power in order to enable compressed air passage through the power conductor.

Oxidize: To combine with oxygen, such as in burning (rapid oxidation) or rusting (slow oxidation).

Polarity: The condition of a system in which it has opposing physical properties at different points, such as an electric charge.

Additional Resources

This module presents thorough resources for task training. The following resource material is suggested for further study.

ANSI C5.3, Recommended Practices for Air Carbon Arc Gouging and Cutting. Latest Edition. Miami, FL: American Welding Society.

2014 Technical Training Guide. The Lincoln Electric Company, Cleveland, OH. USA. **www.lincolnelectric.com**.

Air Carbon-Arc Guide, Form Number: 89-250-008. Denton, Texas: Victor Technologies, Inc.

Figure Credits

The Lincoln Electric Company, Cleveland, OH, USA, Module Opener, Figures 3 (photo), 11, 12

Topaz Publications, Inc., Figures 2, 6, 8, 10

Victor Technologies, SA01

AWS F2.2:2001, Lens Shade Selector, reproduced with permission from the American Welding Society (AWS), Miami, FL, USA, Table 4

Section Review Answer Key

Answer	Section Reference	Objective
Section One		
1. a	1.0.0	1a
2. b	1.2.1	1b
3. d	1.3.0	1c
4. c	1.3.0	1d
Section Two		
1. c	2.1.0	2a
2. d	2.2.1	2b
3. b	2.3.1	2c

This page is intentionally left blank.

NCCER CURRICULA — USER UPDATE

NCCER makes every effort to keep its textbooks up-to-date and free of technical errors. We appreciate your help in this process. If you find an error, a typographical mistake, or an inaccuracy in NCCER's curricula, please fill out this form (or a photocopy), or complete the online form at **www.nccer.org/olf**. Be sure to include the exact module ID number, page number, a detailed description, and your recommended correction. Your input will be brought to the attention of the Authoring Team. Thank you for your assistance.

Instructors – If you have an idea for improving this textbook, or have found that additional materials were necessary to teach this module effectively, please let us know so that we may present your suggestions to the Authoring Team.

NCCER Product Development and Revision
13614 Progress Blvd., Alachua, FL 32615
Email: curriculum@nccer.org
Online: www.nccer.org/olf

❑ Trainee Guide ❑ Lesson Plans ❑ Exam ❑ PowerPoints Other ______________

Craft / Level: Copyright Date:

Module ID Number / Title:

Section Number(s):

Description:

____

____

____

____

Recommended Correction:

____

____

____

____

Your Name:

Address:

____

Email: Phone:

This page is intentionally left blank.

Base Metal Preparation

Overview

This module introduces information related to the preparation of metals for welding, including the identification of common base metals, the physical characteristics and mechanical properties of different metals, and cleaning techniques. Welding joint preparation, codes, and specifications related to metal preparation are also described.

Module 29105

Trainees with successful module completions may be eligible for credentialing through the NCCER Registry. To learn more, go to **www.nccer.org** or contact us at **1.888.622.3720**. Our website has information on the latest product releases and training, as well as online versions of our *Cornerstone* magazine and Pearson's product catalog.

Your feedback is welcome. You may email your comments to **curriculum@nccer.org**, send general comments and inquiries to **info@nccer.org**, or fill in the User Update form at the back of this module.

This information is general in nature and intended for training purposes only. Actual performance of activities described in this manual requires compliance with all applicable operating, service, maintenance, and safety procedures under the direction of qualified personnel. References in this manual to patented or proprietary devices do not constitute a recommendation of their use.

29105 V5

29105
Base Metal Preparation

Objectives

When you have completed this module, you will be able to do the following:

1. Identify safety practices related to preparing base metals and describe basic cleaning procedures.
   a. Identify safety practices related to preparing base metals.
   b. Describe the basic properties and types of carbon and stainless steel.
   c. Describe basic metal cleaning procedures and concerns.
2. Identify and describe basic weld joint design and types of welds.
   a. Identify and describe the loads that are routinely placed on weld joints.
   b. Identify and describe the various types of weld joints.
   c. Describe a welding procedure specification (WPS) and the information it provides.
3. Describe how to prepare joints for welding.
   a. Describe how to mechanically prepare joints for welding.
   b. Describe how to thermally prepare joints for welding.

Performance Tasks

Under the supervision of your instructor, you should be able to do the following:

1. Mechanically or hand grind a bevel on the edge of a ¼"- to ¾"-thick mild steel plate (6 to 20 mm metric plate) at 22½ degrees.
2. Thermally bevel the edge of a ¼"- to ¾"-thick mild steel plate (6 to 20 mm metric plate) at 22½ degrees.

Trade Terms

Annealed
Austenitic
Backing
Base metal
Castings
Distortion
Ductile
Ferritic
Ferrous
Load
Malleable
Martensitic
Melt-through
Nonferrous metal
Oxide
Piping porosity
Porosity
Quench
Root face
Root opening
Surfacing
Tempering
Tensile strength
Weathering steel
Welding procedure specification (WPS)
Weldment
Wrought

Industry Recognized Credentials

If you are training through an NCCER-accredited sponsor, you may be eligible for credentials from NCCER's Registry. The ID number for this module is 29105. Note that this module may have been used in other NCCER curricula and may apply to other level completions. Contact NCCER's Registry at 888.622.3720 or go to **www.nccer.org** for more information.

Contents

Topics to be presented in this module include:

Figures and Tables

SECTION ONE

1.0.0 BASIC WELDING SAFETY AND CLEANING

Objective

Identify safety practices related to preparing base metals and describe basic cleaning procedures.

a. Identify safety practices related to preparing base metals.
b. Describe the basic properties and types of carbon and stainless steel.
c. Describe basic metal cleaning procedures and concerns.

Trade Terms

Annealed: To free from internal stress by heating and gradually cooling.

Austenitic: Containing austenite (a solid solution of carbon, or of carbon and other elements in a ferrous alloy) added through heating.

Base metal: As used in this module, metal to be welded, cut, or brazed.

Castings: Something cast; any article that has been cast in a mold.

Distortion: The expansion and contraction of welded parts caused by the heating and subsequent cooling of the weld joint.

Ductile: Able to undergo change of form without breaking.

Ferritic: Steel containing less than 0.10 percent carbon and is magnetic. This steel can't be hardened via heat treatment.

Ferrous: Containing iron.

Malleable: Capable of being extended or shaped by hammering or by pressure from rollers.

Martensitic: Steel that shares some characteristics with ferritic, but has a higher level of carbon, up to a full 1 percent. It can be tempered and hardened and is used where strength is more important than a resistance to oxidation.

Nonferrous metal: A metal, such as aluminum, copper, or brass, lacking sufficient quantities of iron to have any effect on its properties.

Oxide: The scale that forms on metal surfaces when they are exposed to oxygen or air containing oxygen.

Piping porosity: A form of porosity having a length greater than its width and that is approximately perpendicular to the weld face.

Porosity: Gas pockets, or voids in the weld metal.

Quench: To cool suddenly by plunging into a liquid.

Tempering: To impart strength or toughness to (steel or cast iron) by heating and cooling.

Weathering steel: Steel alloy that, under specific conditions, is designed to form a very dense oxide layer on its outer surfaces, which retards further oxidation.

Weldment: An assembly that is fastened together by welded joints.

Wrought: Produced or shaped by beating with a hammer, as with iron.

To ensure that the safest and highest quality welds are produced, and to comply with welding codes, **base metals** must be properly prepared prior to welding. Safety must be priority one, and proper safety precautions add to the welder's organizational skills during the preparation process. The type of preparation required depends on the governing code requirements, the base metal type, the condition of the base metal, the welding process to be used, and the equipment available.

1.1.0 Basic Welding Safety

Complete safety coverage is provided in the *Welding Safety* module. If you have not completed that module, do so before continuing.

Welders routinely use manual and powered tools to clean and shape metals. *Figures 1* and *2* are examples of metal being prepared for welding. Everyone entering the welding trade starts by doing basic tasks, which includes preparing the metals for welding.

When using tools to clean metals prior to welding, wires from rotating wire brushes and particles from abrasive discs can become flying objects. Metal particles ground from the workpieces are usually hot and can cause burns, severe eye injuries, or fires. Welding screens or similar protective devices are used to keep the hot particles from igniting combustible materials and harming other workers nearby. Most companies require a fire extinguisher and a fire watch posted near the welding and cutting activities. To prevent injury from the flying particles and debris, always wear the proper personal protective equipment (PPE).

Before working on welding equipment and power tools, ensure that they are disconnected from the power supply. When the welding machine is energized, the open circuit between the electrode holder and the workpiece clamp represents a significant electrical hazard.

While much of the preparation work is performed in the shop where parts can be secured in clamping devices, some of it must be performed in the field on structures or equipment. Welders must sometimes perform at heights where fall prevention and fall arrest equipment is required. Always make sure that fall protection devices are undamaged and the inspections are current before using them.

The following is a summary of safety procedures and practices that must be observed while cutting, cleaning, shaping, and welding. Keep in mind that this is a summary. Complete safety coverage is provided in the *Welding Safety* module. If you have not completed that module, do so before continuing.

1.1.1 Protective Clothing and Equipment

Welding activities can cause injuries unless you wear all of the protective clothing and equipment that is designed specifically for the welding industry. The following information includes important safety guidelines about protective clothing and equipment:

29105-15_F01.EPS

Figure 1 Cutting metal with an oxyfuel torch.

29105-15_F02.EPS

Figure 2 Grinding metal.

- Always use safety glasses with a full face shield or a helmet. The glasses, face shield, or helmet lens must have the proper light-reducing tint for the type of welding or cutting being performed. Use corrective lenses if needed; some companies will not allow contact lenses to be worn during cutting or welding activities. Never view an electric arc directly or indirectly without using a properly tinted lens.
- Wear proper protective leather and or flame-retardant clothing along with welding gloves that will protect you from flying sparks and molten metal (*Figure 3*).

29105-15_F03.EPS

Figure 3 Appropriate personal protective equipment (PPE).

- Wear high-top safety shoes or boots. Make sure that the tongue and lace area of the footwear will be covered by a pant leg or other protective material. Boots with no laces at all are generally preferred for both welding and cutting activities. If the tongue and lace area is exposed, or if the footwear must be protected from burn marks, wear leather spats under your pants or chaps and over the top front area of the footwear.
- Wear a 100-percent cotton cap with no ventilation mesh material included in its construction. The bill of the cap points to the rear. If a hard hat is required for the environment, use one that allows the attachment of rear deflector and a face shield. A hard hat with a rear deflector is generally preferred when working overhead, and may be required by some employers and job sites.

WARNING!

Do not wear a cap with a button in the middle. The conductive metal button beneath the fabric represents a safety hazard.

- Wear a face shield over snug-fitting cutting goggles or safety glasses for gas welding or cutting. Either the face shield or the lenses of the welding goggles must be an approved shade for the application. A welding hood with a properly tinted lens is also acceptable. See *Table 1* for a list of lens shades based on arc current.
- Wear earplugs to protect ear canals from sparks. Wear hearing protection to protect against the consistent sound of the torch.

WARNING!

Ear protection is essential to protect ears from the noise of the torch. Other personal protective equipment (PPE) must be worn to protect the operator from hot metal and slag.

1.1.2 Fire and Explosion Prevention

Welding activities usually involve the use of fire or extreme heat to melt metal. Although most work site fires are caused by cutting torches and the hot material they produce, welding sparks and spatter can also start fires easily. Whenever fire is used, it must be controlled and contained. Welding or cutting activities are often performed on vessels that may once have contained flammable or explosive materials. Residues from those materials can catch fire or explode when a welder begins work on such a vessel. The following are fire and explosion prevention guidelines associated with welding:

- Never carry matches or gas-filled lighters in your pockets. Sparks can cause the matches to ignite or the lighter to explode, causing serious injury.
- Always comply with any site requirement for a hot-work permit and/or a fire watch.
- Never use oxygen to blow dirt or dust off clothing. The oxygen can remain trapped in the fabric for a time. If a spark hits the oxygen in the fabric, the clothing can burn rapidly and violently.
- Never release a large amount of oxygen or use oxygen in place of compressed air. The presence of oxygen around flammable materials or sparks can cause rapid and uncontrolled combustion. Keep oxygen away from oil, grease, and other petroleum products.
- Make sure that any flammable material in the work area is moved or shielded by a fire-resistant covering.
- Approved fire extinguishers must be available before attempting any heating, welding, or cutting operations. Make sure the extinguisher is charged, the inspection tag is valid, and any individual that may be required to operate it knows how to do so.
- Never release a large amount of fuel gas, especially acetylene. Propylene and propane are heavier than air and tend to migrate to and concentrate in low areas. As a result, they can ignite at a considerable distance from the release point. Acetylene is lighter than air but is even more dangerous than methane; when mixed with air or oxygen, it will explode at much lower concentrations than any other common fuel gas.
- To prevent fires, maintain a neat and clean work area, and make sure that any metal scrap or slag is cold before disposal.

Before cutting or welding containers such as tanks or barrels, find out if they contained any explosive, hazardous, or flammable materials, including petroleum products, citrus products, or chemicals that decompose into toxic fumes when heated. Proper procedures for cutting or

Table 1 Guide for Shade Numbers

From *AWS F2.2 Lens Shade Selector.* Shade numbers are given as a guide only and may be varied to suit individual needs.

Process	Electrode Size in (mm)	Arc Current (Amperes)	Minimum Protective Shade	Suggested* Shade No. (Comfort)
Shielded Metal Arc Welding (SMAW)	Less than 3⁄32 (2.4)	Less than 60	7	–
	3⁄32–5⁄32 (2.4–4.0)	60–160	8	10
	5⁄32–1⁄4 (4.0–6.4)	160–250	10	12
	More than 1⁄4 (6.4)	250–550	11	14
Gas Metal Arc Welding (GMAW) and Flux Cored Arc Welding (FCAW)		Less than 60	7	–
		60–160	10	11
		160–250	10	12
		250–500	10	14
Gas Tungsten Arc Welding (GTAW)		Less than 50	8	10
		50–150	8	12
		150–500	10	14
Air Carbon Arc Cutting (CAC-A)	(Light)	Less than 500	10	12
	(Heavy)	500–1000	11	14
Plasma Arc Welding (PAW)		Less than 20	6	6 to 8
		20–100	8	10
		100–400	10	12
		400–800	11	14
Plasma Arc Cutting (PAC)		Less than 20	4	4
		20–40	5	5
		40–60	6	6
		60–80	8	8
		80–300	8	9
		300–400	9	12
		400–800	10	14
Torch Brazing (TB)		–	–	3 or 4
Torch Soldering (TS)		–	–	2
Carbon Arc Welding (CAW)		–	–	14

Process	Plate Thickness in	Plate Thickness mm	Suggested* Shade No. (Comfort)
Oxyfuel Gas Welding (OFW)			
Light	Under 1⁄8	Under 3	4 or 5
Medium	1⁄8 to 1⁄2	3 to 13	5 or 6
Heavy	Over 1⁄2	Over 13	6 or 8
Oxygen Cutting (OC)			
Light	Under 1	Under 25	3 or 4
Medium	1 to 6	25 to 150	4 or 5
Heavy	Over 6	Over 150	5 or 6

*As a rule of thumb, start with a shade that is too dark to see the weld zone. Then go to a lighter shade which gives sufficient view of the weld zone without going below the minimum. In oxyfuel gas welding, cutting, or brazing where the torch and/or the flux produces a high yellow light, it is desirable to use a filter lens that absorbs the yellow or sodium line of the visible light spectrum.

29105-15_T01.EPS

welding hazardous containers are described in the *American Welding Society (AWS) F4.1, Safe Practices for the Preparation of Containers and Piping for Welding and Cutting,* and *ANSI Z49.1.* As a standard practice, always clean and then fill any tanks or barrels with water, or purge them with a flow of inert gas such as nitrogen to displace any oxygen.

> **WARNING!**
>
> Welding or cutting must never be performed on drums, barrels, tanks, vessels, or other containers until they have been emptied and cleaned thoroughly, eliminating all flammable materials and all substances (such as detergents, solvents, greases, tars, or acids) that might produce flammable, toxic, or explosive vapors when heated. Do not assume that a container that has held combustibles is clean and safe until proven so by proper tests. Do not weld in places where dust or other combustible particles are suspended in air or where explosive vapors are present.

Containers must be cleaned by steam cleaning, flushing with water, or washing with detergent until all traces of the material have been removed.

> **WARNING!**
>
> Clean containers only in well-ventilated areas. Vapors can accumulate during cleaning, causing explosions or injury.

After cleaning the container, fill it with water or a purging gas, such as carbon dioxide, argon, or nitrogen to displace the explosive fumes. Air, which contains oxygen, is displaced from inside the container by the water or inert gas. Without oxygen, combustion cannot take place.

A water-filled vessel is the best alternative. When using water, position the container to minimize the air space. When using an inert gas, provide a vent hole so the inert gas can push the air and other vapors out to the atmosphere. Keep in mind, though, that even these precautions do not guarantee the absence of flammable materials inside. For that reason, these types of activities should not be done without proper supervision and the use of proper testing methods.

> **GOING GREEN**
>
> **Welding Water**
>
> To better protect the environment, send any water used in a welding/cutting process or used to clean tanks to a waste treatment facility instead of simply allowing it to run into a storm drain or onto the ground.

1.1.3 Work Area Ventilation

Vapors and fumes tend to rise in the air from their sources. Welders often have to work above the welding area where the fumes are being created. Welding fumes can cause personal injuries. Good work area ventilation helps to protect the welder by removing the vapors. The following is a list of work area ventilation guidelines to consider before and during preparation activities:

- Make sure confined space procedures are followed before conducting any welding or cutting in a confined space.
- Always perform cutting or welding operations in a well-ventilated area. Cutting or welding operations involving materials such as coatings that contain cadmium, lead, zinc, or chromium will result in toxic fumes. For cutting or welding of such materials, always wear an approved respirator with an appropriate filter as directed by your employer.
- Grinding wheels containing aluminum oxide, silicon carbide and other materials have a composition that requires the use of respiratory protection. Consult the SDS/MSDS for warnings before starting any grinding activities.
- Make sure confined spaces are ventilated properly for cutting or welding purposes.
- Never use oxygen in confined spaces for purposes of ventilation.

> **GOING GREEN**
>
> **Welding Fumes**
>
> To better protect the environment, perform any welding activities under a vapor extraction system that can filter the welding fumes before they reach the atmosphere.

1.2.0 Properties of Steel

Metals are a collection of chemical elements bound together by natural forces or manmade production processes. The chemical elements in some metals make them stronger and give them the ability to withstand stress or force better than other metals. The composition of common welding base metals varies widely, from metals made of essentially one metallic element, to alloys (mixtures) of metallic and nonmetallic components.

Metals are classified into two basic groups: **ferrous** metals, which are composed mainly of iron, and **nonferrous metals**, which contain very little or no iron. Ferrous metals include all steel, cast iron, **wrought** iron, **malleable** iron, and **ductile** (nodular) iron. Nonferrous metals and their alloys include the light metals (aluminum, magnesium, titanium); the heavy metals (copper, nickel, lead, tin, zinc); and the precious metals (platinum, gold, silver). Ferrous metals contain mostly iron and some carbon. Special steels may contain as much as 2 percent carbon. Carbon steel, the largest group of the ferrous-based metals, generally contains less than 1.7 percent carbon.

At low levels of carbon content, steel has some of the properties of cast iron, but is more ductile and more easily formed. Above 1.4 percent carbon, steel becomes stronger and more wear-resistant, although it becomes less ductile (difficult to form) and more difficult to weld. The higher carbon content lowers the melting temperature. The carbon content of a particular carbon steel product is the primary factor in determining its use.

Steel **castings** usually contain a higher percentage of carbon than rolled plate and other rolled shapes. Casting is done instead of rolling or shaping due to the loss of ductility in high-carbon steels. The casting foundry can usually provide the alloy formula, including the carbon content.

Steels, including carbon steels, low-alloy carbon steels, alloy steels, and stainless steels, are classified by various systems, such as by specification

Carbon and Alloy Steels

Steel is an alloy of iron and carbon typically containing less than 1 percent carbon. All steels also contain varying amounts of other elements such as manganese, silicon, phosphorus, sulfur, and oxygen. In addition, some standard alloy steels can also contain elements such as nickel, chromium, and molybdenum. There are currently about 3,500 different grades of steel with many different properties. Seventy percent of these steels have been developed in the past 25 years. Most of the steels produced today are carbon and alloy steels. Ninety-five percent of the construction and fabrication metals used worldwide consist of these materials.

29105-15_SA01.EPS

number and grade, or by manufacturer's trade name and number. Steel classification systems include those developed by the American Iron and Steel Institute (AISI), the American Society for Testing Materials International (ASTM), and the Unified Numbering System (UNS), as well as manufacturer's trade names and identification numbers. Stainless steels have their own classification systems. They are sometimes referred to by the percentages of their chromium and nickel content, such as 18/8, 25/20, and 18/10, but this system has largely been replaced by the AISI stainless steel classification system. Common classification systems for steels are explained in the following sections.

1.2.1 Carbon Steel Classification

The principal classification system for carbon steels is the AISI numerical designation of Standard Carbon and Alloy Steels. It was originally a four-digit system developed by the Society of Automotive Engineers (SAE) for carbon steels commonly used in structural shapes, plate, strip, sheet, and welded tubing. Five digits are now used for some alloys. The classification system may be referred to as the AISI, SAE, or the AISI and SAE system.

The AISI and SAE designations are essentially the same, except that the AISI system sometimes uses a letter prefix to indicate the manufacturing process that produced the steel. The following are examples of letter designations:

- A – Open-hearth steel
- B – Acid-Bessemer carbon steel
- C – Basic open-hearth carbon steel
- D – Acid open-hearth carbon steel
- E – Electric furnace steel

The absence of a letter prefix indicates basic open-hearth or acid-Bessemer carbon steel. The prefix letter (if any) designates the manufacturing process. The first two numerical digits represent the series (type and class) of steel. The third and fourth (and sometimes fifth) numerical digits specify the approximate percentage of carbon content.

The following examples explain the AISI and SAE numbers for carbon steels:

- *AISI number C1020*:
  C = Indicates basic open-hearth carbon steel
  10 = Carbon steel, nonresulfurized
  20 = Contains approximately 0.20 percent carbon
- *AISI number E2512*:
  E = Indicates electric furnace steel
  25 = Designates steel alloyed with approximately 5 percent nickel
  12 = Designates steel containing approximately 0.12 percent carbon
- *AISI number E52100*:
  E = Indicates electric furnace steel
  5 = Contains either approximately 0.50 percent, 1.00 percent, or 1.45 percent chromium, designated by the next digit: 0, 1, or 2, respectively
  2 = Designates approximately 1.45 percent chromium
  100 = Designates approximately 1 percent carbon

The common group classification of carbon steels is primarily based on carbon content. Plain carbon steels (AISI series 10XX through 11XX) are basically iron-carbon alloys. High-strength low alloy (HSLA) carbon steels (AISI series 13XX through 98XX) have small amounts of alloying elements to improve strength, hardness, and toughness or to increase resistance to oxidation, heat, and environmental damage. The plain carbon steels are commonly grouped as follows:

- *Low-carbon* – 0.10 percent to 0.15 percent carbon, 0.25 percent to 1.50 percent manganese
- *Mild-carbon* – 0.15 percent to 0.30 percent carbon, 0.60 percent to 0.70 percent manganese
- *Medium-carbon* – 0.30 percent to 0.50 percent carbon, 0.60 percent to 1.65 percent manganese
- *High-carbon* – 0.50 percent to 1.00 percent carbon, 0.30 percent to 1.00 percent manganese

The HSLA carbon steels are alloyed with one or more of the elements manganese, nickel, chromium, or molybdenum to provide higher strength, better toughness, weldability, and, in some cases, greater resistance to oxidation. As the percentage of these elements increases, changes in the weldability of the low-alloy steels make electrode selection and welding procedures more critical. HSLA steels include the following:

- Low-nickel steels
- Low-nickel chrome steels
- Low-manganese steels
- Low-alloy chromium steels
- **Weathering steels**

Another class of steel in the AISI series is sulfur steel. It is not an alloy steel but is included in the series.

1.2.2 Common Grade Stainless Steel Classifications

Stainless steels are typically classified as being a common grade or a specialty grade. Common-grade stainless steels (oxidation-resistant steels) are iron-based alloys that normally contain at least 11 percent chromium. Other alloying elements, including nickel, carbon, manganese, and silicon, may be present in varying quantities, depending on the specific type of stainless steel, to enhance its physical and mechanical properties. The main physical characteristics of all stainless steels are their resistance to oxidation and heat. Some have good low- and high-temperature mechanical properties. When compared with mild steels, stainless steels have the following characteristics:

- Lower coefficients of thermal conductivity that increase the chances of **distortion**
- Higher coefficients of thermal expansion that increase the chances of distortion
- Higher electrical resistances that increase the tendency to build up heat from welding current

Common-grade stainless steels are classified by their grain structures. The type of grain structure is determined by the specific alloy content of the stainless steel and its heat treatment during manufacture. Based on their microcrystalline structures, the classification of common-grade stainless steels is divided into these three groups:

- **Austenitic**
- **Ferritic**
- **Martensitic**

Austenitic stainless steels are non-magnetic in the **annealed** condition and not able to be hardened by heat treatment. However, they can be hardened significantly by cold working. Austenitic stainless steels combine excellent oxidation and heat resistance with good mechanical properties over a broad temperature range. Austenitic stainless steels make up the largest of the three common-grade stainless steel groups. For this reason, austenitic stainless steel is encountered most often by welders. They include the AISI 200 and 300 series stainless steels that all contain significant amounts of both chromium and nickel.

Austenitic steels are sometimes further subdivided in two classifications based on their compositions: chromium-nickel-magnesium (AISI 200 series) and chromium-nickel (AISI 300 series). The range of applications for austenitic stainless steels includes housewares, containers, industrial piping and vessels, and architectural facades. Austenitic stainless steels are also available in low-carbon (L-grade) and high-carbon (H-grade) types. The letter L after a stainless steel type, such as 304L, indicates a low-carbon content of 0.03 percent or under. Similarly, the letter H after a stainless steel type indicates a high-carbon content, ranging between 0.04 and 0.10 percent. L-grades are typically used when annealing after welding is impractical, such as in the field where pipe and fittings are welded. H-grades are used when the steel will be subjected to extreme temperatures; the higher carbon content helps it retain strength.

Ferritic stainless steels are always magnetic. They are hardened to some extent by cold working, not by heat treatment. Ferritic stainless steels combine oxidation and heat resistance with fair mechanical properties over a narrower temperature range than austenitic stainless steels. Ferritic stainless steels are straight chromium stainless steels containing 11.5 to 27 percent chromium, about 1 percent manganese, and little or no nickel. The carbon content is 0.20 percent or less. Examples of ferritic stainless steels include AISI types 405, 409, 430, 442, 444, and 446. They typically are used for decorative trim, sinks, and automotive applications, particularly exhaust systems.

Martensitic stainless steels are also magnetic and can be hardened by **quenching** and **tempering**. They are excellent for use in mild environments such as the atmosphere, freshwater, steam, and with weak acids. However, they are not resistant to severely corrosive solutions. Martensitic stainless steels comprise two groups: martensitic and chromium-martensitic stainless steels.

Martensitic stainless steels contain only 4 to 6 percent chromium and no nickel. Chromium-martensitic stainless steels contain from 11.5 to 18 percent chromium, about 1 percent manganese, and in some cases 0 to 2.5 percent nickel. Because of the low chromium and no nickel, martensitic stainless steels are not considered true stainless steels, although their oxidation resistance is much greater than mild carbon steels, even at elevated temperatures.

Examples of chromium-martensitic stainless steels are AISI types 403, 410, 414, 420, 422, 431, and 440. Examples of martensitic stainless steels are AISI types 501 and 502.

1.3.0 Base Metal Cleaning

All base metals must be cleaned before welding. Even new materials that may look clean often are not. They pick up contaminants and surface oxidation during shipping and handling. When performing maintenance welding, a welder often comes across components that have been exposed to surface contamination such as oxidation, paint, oil, and grease. The heat generated by the welding process could cause paints and coatings to give off toxic fumes, endangering the worker performing the welding.

Do not assume a metal's surface is oxidation-free just because it is shiny. Oxidation on stainless steel and aluminum is difficult to see; it may have a white color or it may be transparent. To ensure quality welds and to conform to code requirements, surface contaminants and **oxides** must be removed prior to welding. In general, the codes state that the surface to be thermally cut or welded must be clean and free from paint, oil, rust, scale, and other material that would be harmful either to the weld or to the base metal when heat is applied. Cleaning is typically performed by mechanical and or chemical means.

Most metal has some surface oxidation. Oxidation occurs when metal is exposed to oxygen in the air or from other sources. Some oxidation exists as a thin, hard film that merely stains the metal and, in fact, acts as a retardant to further oxidation. Examples of this type of oxidation are found on stainless steel, copper, and aluminum. The thin, hard layer of oxidation bonds tightly to the surface and protects the metal.

A more recognizable type of oxidation found on mild steel is called rust (*Figure 4*). Rust is a very coarse type of oxidation that tends to flake easily, exposing more of the base metal to oxidation. Alloys such as chromium and copper are often added to mild steel to protect it from this coarse type of oxidation. These types of steels are called alloy steels. The alloys cause the rust to be very fine, bonding to the surface and protecting the base metal beneath. Weathering steel is an example of a copper-alloyed steel that protects itself from oxidation by forming a hard, tough layer of brown oxidation (rust) on its surface. Weathering steels are used outdoors and require no painting.

29105-15_F04.EPS

Figure 4 Examples of oxidation.

Regardless of the type of contamination, it must be removed prior to welding. Any type of contamination can cause weld defects. The most common defect caused by surface contamination is **porosity**. Examples of porosity are shown in *Figure 5*. Porosity occurs when gas pockets or voids appear in the weld metal. When the porosity has a length greater than its width and is approximately perpendicular to the weld face, it is called **piping porosity**. Piping porosity is formed as the gas pocket or bubble floats toward the surface of the weld, leaving a void. The gas pocket is trapped as the weld metal solidifies, but as the next layer of weld is deposited, the bubble often continues to float up. Porosity may not always be visible on the surface of the weld.

Surface contamination can easily result from any of several causes, including the following:

- Use of gloves contaminated with oil or other substances
- Allowing alloys to come into contact with carbon steel objects, including chain falls, jack stands, and forklifts
- The use of chloride-based markers to write on the base metal

Because of the risk of contamination, metals such as carbon steel, stainless steel, and aluminum must be stored separately. Aluminum must be stored indoors.

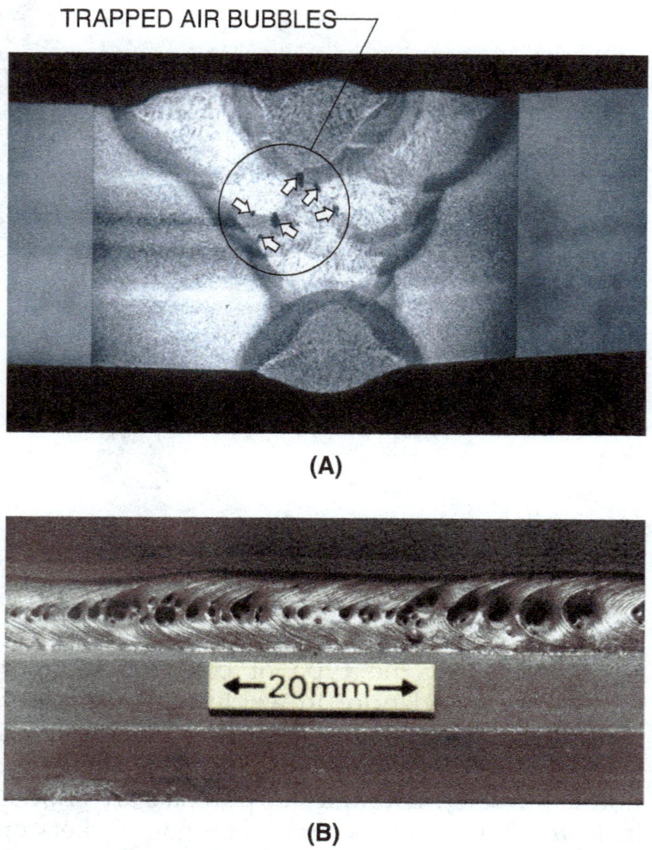

Reproduced by permission TWI Ltd.

29105-15_F05.EPS

Figure 5 Porosity.

1.3.1 Mechanical Cleaning

Mechanical cleaning is the most common method of removing surface contamination. Tools used for mechanical cleaning include hand tools, power tools, and special sandblasting equipment. When performing mechanical cleaning, be sure to wear safety glasses and a face shield for protection from the flying particles produced during the cleaning operation. In addition, special clothing is required for sandblasting.

Flexible scrapers are used to remove dirt and grease from weldments or weldment components. Rigid scrapers can be used to remove hardened paint or dirt.

Wire brushes are used to remove paint and light-to-medium surface oxidation. Wire brushes are available with bristles made from a variety of metals, including carbon steel, stainless steel, and brass. Use only stainless steel wire brushes on stainless steel and aluminum, and make sure that the stainless steel wire brush used on these metals has never been used with another metal. Once brushes, grinding wheels, and files have been used on carbon steel and other metals, they should never be used on stainless steel or aluminum. The remaining particles of other metals on the tools will contaminate stainless steel and aluminum, as will a wire brush made of carbon steel. The contamination leads to weld defects.

Files are another way to remove surface oxidation. When filing aluminum or stainless steel, do not use files that have been used on other metals. Before using a file, always be sure it has a handle installed on the file tang. Use a fine file and keep it clean to prevent scratching the base metal surface. Take care not to damage or scratch the base metal outside the weld zone. Clean files by using stainless steel brushes or file cards that either are new or have only been used to brush the same type of metal (*Figure 6*). Even particles from a file card can contaminate a file, which in turn can contaminate metals on which the file is used.

Tight oxidation must be removed by filing, grinding, or sandblasting. For large jobs or jobs where speed is important, power tools are most efficient. Power tools may be electrically or pneumatically powered.

> **NOTE**
> Sandblasting is not used on aluminum.

29105-15_F06.EPS

Figure 6 File and cleaning tools.

> **WARNING!**
>
> To contain sparks when cleaning metal with power tools, use a welding screen. Always wear the appropriate PPE when using power tools.

Angle grinders (*Figure 7*) are very effective in removing large areas of surface contamination. Die grinders and small angle grinders work very well for small areas such as weld grooves and bevel edges. Grinders have various attachments that can be used for special applications. These attachments include grinding disks and wheels, wire brushes, rotary files, and flapper wheels (*Figure 8*).

> **WARNING!**
>
> If a grinding wheel or disk shatters while in use, the pieces of the wheel or disk become deadly projectiles as the weights of the pieces are multiplied by the centrifugal force and the speed of the grinding motor. Never grind without full face protection. Avoid standing in the orbital path of the wheel or disk.

Grinding disks and wheels are made for specific types of metals. Be sure you have the correct disk for the type of metal being ground. Always use aluminum oxide disks for grinding aluminum or stainless steel. For stainless steel or aluminum, do not use wheels that have been used on other metals. They will contaminate the surface of aluminum and stainless steel.

When using wire brush attachments on stainless steel or aluminum, brushes with stainless steel bristles must be used. Only use stainless steel brushes that have not been used on other metals. Remember that carbon steel bristles will contaminate stainless steel and aluminum. Over time, the contamination will reveal itself by iron oxide (rust) developing at the contamination sites of the joint.

When grinding, care must be taken to prevent grinding the base metal below the minimum allowable base metal thickness. If this should happen, the base metal will have to be discarded or, if allowable, built up with weld metal. Both of these alternatives are expensive.

Cutoff wheels are used to cut through metals and other materials. They are made for specific types of metals, or for other materials such as stone. Be sure you have the correct cutoff wheel for the type of material being cut.

Weld flux chippers and needle scalers are usually pneumatic rather than electric. These tools are used by welders to clean surfaces and remove dross from cuts and welds. Weld flux chippers, and needle scalers are also excellent for removing paint, heavy scale, or hardened dirt, but they are not very effective for removing surface oxidation. Weld flux chippers have a single chisel; needle scalers have about 18 to 20 blunt steel needles about 10 inches (≈25 cm) long. Many weld flux chippers can be converted to needle scalers with an attachment. A weld flux chipper and a needle scaler are shown in *Figure 9*.

(A) ELECTRIC ANGLE GRINDER

(B) ELECTRIC DIE GRINDER

(C) PNEUMATIC ANGLE GRINDER

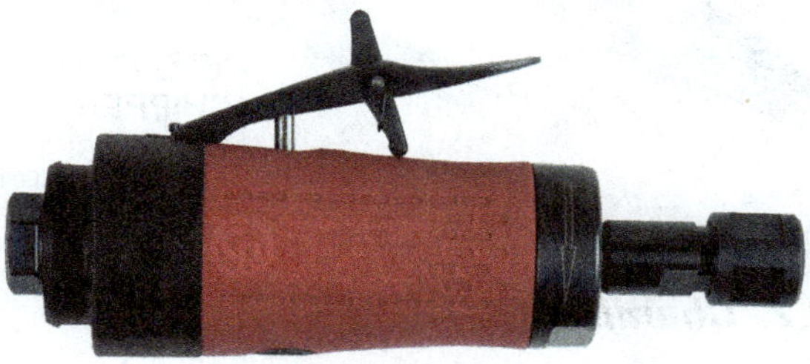

(D) PNEUMATIC DIE GRINDER

29105-15_F07.EPS

Figure 7 Handheld grinders.

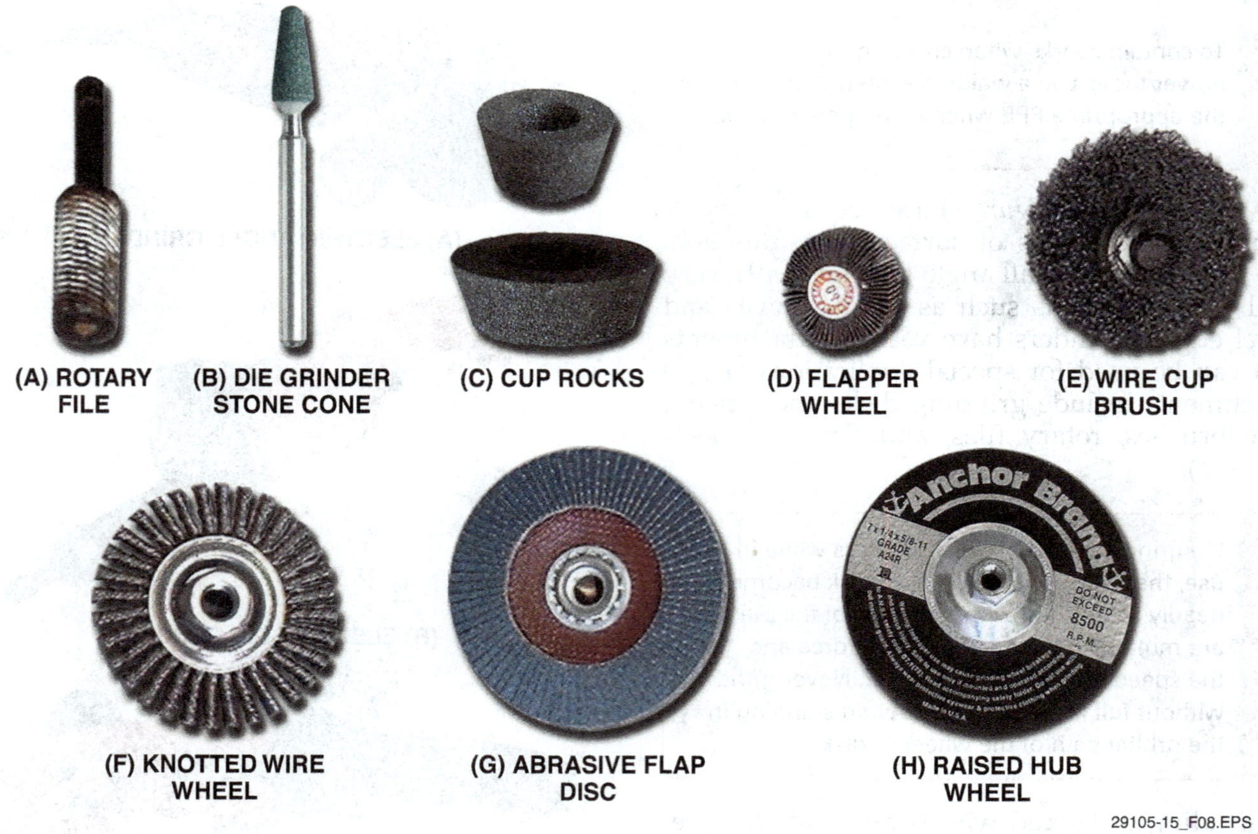

Figure 8 Grinder attachments.

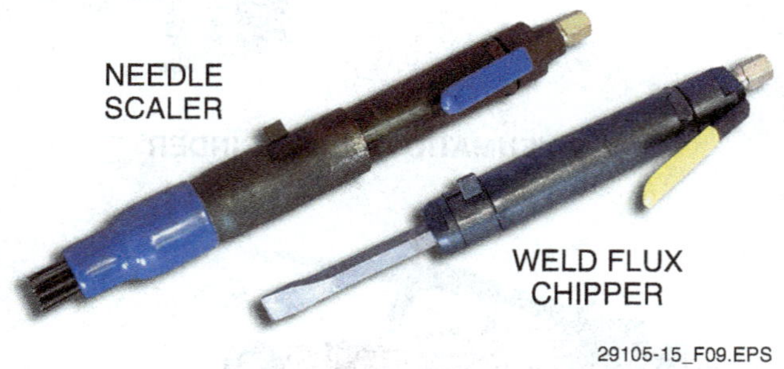

Figure 9 Weld flux chipper and needle scaler.

1.3.2 Chemical Cleaning

There will be times when mechanical cleaning is not enough. Residues left on a metal can contaminate the weld. Even the oil from the skin of the welder handling the metal can contaminate a weld. Welders may have to use some form of chemical cleaner to ensure that the metals being prepared are very clean.

It is essential that only approved cleaners are used. Some chemicals can create hazardous fumes during the welding process. Always follow the prescribed specifications of the approved cleaners and the related SDS/MSDS. The two chemical cleaners most often used by welders are acetone and butanone. Butanone is better known as methyl ethyl ketone, or MEK.

WARNING!

Do not use brake part cleaning fluid to clean surfaces prior to welding. Some brake part cleaning fluids contain chlorinated hydrocarbons. Phosgene gas is formed by the decomposition of chlorinated hydrocarbon solvents by ultraviolet radiation. Phosgene gas reacts with moisture in the lungs to produce hydrogen chloride, which in turn destroys lung tissue. For this reason, any use of chlorinated solvents for any purpose should be kept away from welding operations.

Acetone is a flammable, colorless liquid. Although a manufactured product, it can also be found in some natural environments. It is a primary ingredient in nail polish remover. Anyone who has worked in an auto repair shop has probably used acetone to remove over-sprayed paint from car windows. It will dissolve resins, epoxies, and many plastics. It is a great degreaser that is often used to prepare metals before painting.

From a safety standpoint, acetone is an irritant. Because it can cause permanent eye damage (corneal clouding), workers handling it or working near it must wear safety goggles. Acetone vapors should be avoided, and a suitable respirator is recommended. Prolonged exposure may eventually cause liver damage. Check the product SDS or MSDS closely to ensure that you understand how to safely work with acetone.

MEK is a manufactured organic chemical. It is a colorless liquid that dissolves substances such as gums, resins, vinyl film coatings, and other similar materials that may still be on the metals. MEK is highly flammable, but it presents much less of a health risk than acetone. However, anyone working with MEK must review the SDS or MSDS to make sure all safety precautions are followed.

WARNING!

Acetone and MEK are both extremely flammable. Keep these products away from all potential sources of ignition.

After the metal parts have been mechanically cleaned, the chemical cleaner is applied, usually with a brush or spray. The chemical is allowed to work for a short time before being removed. A clean cloth is used to remove the cleaning agent along with any contamination loosened from the metal. When cleaning metals with chemicals, be aware that the contaminants being removed may give off vapors that could be dangerous. Be sure to use the proper PPE, including respiratory protection in some conditions.

Additional Resources

Steel Metallurgy for the Non-Metallurgist. John D. Verhoeven; Prepared under the direction of the ASM International Technical Book Committee. ASM International® Materials Park, OH 44073-0002, USA. **www.asminternational.org**

American Iron and Steel Institute, 25 Massachusetts Avenue, NW Suite 800, Washington, DC 20001, USA. **www.steel.org**

Association for Iron & Steel Technology, 186 Thorn Hill Road Warrendale, PA, USA. **www.aist.org**

1.0.0 Section Review

1. Grinding wheels containing aluminum oxide, silicon carbide and other materials have a composition that require _____.
   a. a simple dust mask that is comfortable for the wearer
   b. a mask made of 100-percent cotton, such as a handkerchief, or similar type of cover
   c. consultation of the SDS/MSDS for recommended respiratory protection
   d. a safety shield only; a respiratory mask is not necessary

2. Which letter prefix, when used in the AISI numbering system, indicates acid-Bessemer carbon steel?
   a. The letter A
   b. The letter B
   c. The letter C
   d. The letter D

3. When brushing aluminum, use stainless steel brushes that have only been used on _____.
   a. stainless steel
   b. aluminum
   c. bronze
   d. carbon steel

Section Two

2.0.0 Joint Design and Weld Types

Objective

Identify and describe basic weld joint design and types of welds.

a. Identify and describe the loads that are routinely placed on weld joints.
b. Identify and describe the various types of weld joints.
c. Describe a welding procedure specification (WPS) and the information it provides.

Trade Terms

Backing: A weldable or non-weldable material used behind a root opening to allow defect-free welding at the open root of a joint.

Load: The amount of force applied to a material or a structure.

Melt-through: Complete joint penetration.

Root face: A small flattened area on the end of a bevel for a groove weld.

Root opening: The space between the base metal pieces at the bottom or root of the joint.

Surfacing: The application by welding, brazing, or thermal spraying of a layer of material to a surface to obtain desired properties or dimensions.

Tensile strength: The measure of the ability of a material to withstand a longitudinal stress without breaking.

Welding procedure specification (WPS): The document containing all the detailed methods and practices required to produce a sound weld.

Welded joints are selected primarily for the safety and strength required for the conditions that will be encountered. When selecting the joint for a particular application, many factors must be taken into account, including load considerations, the environment in which it will be used, materials, processes, and cost.

2.1.0 Load Considerations

The loads in a welded steel structure or component are transferred through the welded joint. The welded joint must be designed to withstand the stresses caused by the loads.

Any or all of these forces may act on welded parts. When two parts are welded together, in whatever manner, they are intended to become one part. If the surface areas of the two parts being welded are not properly prepared, the welds may fail when one or more of these forces is applied. The following are brief explanations of load terms as they relate to welds. Refer to *Figure 10* for a visual representation of these forces:

- *Tensile* – When metals are pulled in opposite directions by forces (F_1 vs. F_2), they will eventually break. Their strength to remain one is referred to as their tensile strength. When metals are heated, as when they are welded, the molecules in the metals change. When the metals are cooled, their molecules change again. These heating and cooling cycles can weaken the tensile qualities of metals. When two pieces of metal are welded end-to-end, they become one piece. When they are pulled at each end, the tensile strength of the two pieces, and the weld, are tested. If the weld is a good weld, one of the base metal pieces will usually break before the weld breaks.
- *Compression* – The reduction of volume or mass. Metals can be compressed. If you strike a steel nail or bolt with a hammer, you can eventually compress the nail or bolt into a different shape. When two pieces of metal are welded end-to-end and forces (F_1 and F_2) are applied inwardly on each end, the pieces and their weld are compressed.
- *Bending* – To bend is to make curved. For example, you may be able to manually bend a section of straight lightweight flat bar around a 6-inch (DN150) pipe. The pipe in this case is a force (F_2) pushing against the center of the bar as you pull or apply force (F_1) in the opposite direction at each end of the bar. The result of such action is that the flat bar becomes curved. When two pieces of metal are welded end-to-end, they may be tested in what is known as a bend test.
- *Torsion* – Twisting. Torsion is the amount of stress the welded bars can withstand as a counterclockwise force (F_1) is applied at one end and a clockwise force (F_2) is applied at the other end. For example, the driveshaft of a truck is twisted as it transmits the rotational movements of the drive motor and transmission to the drive axle.

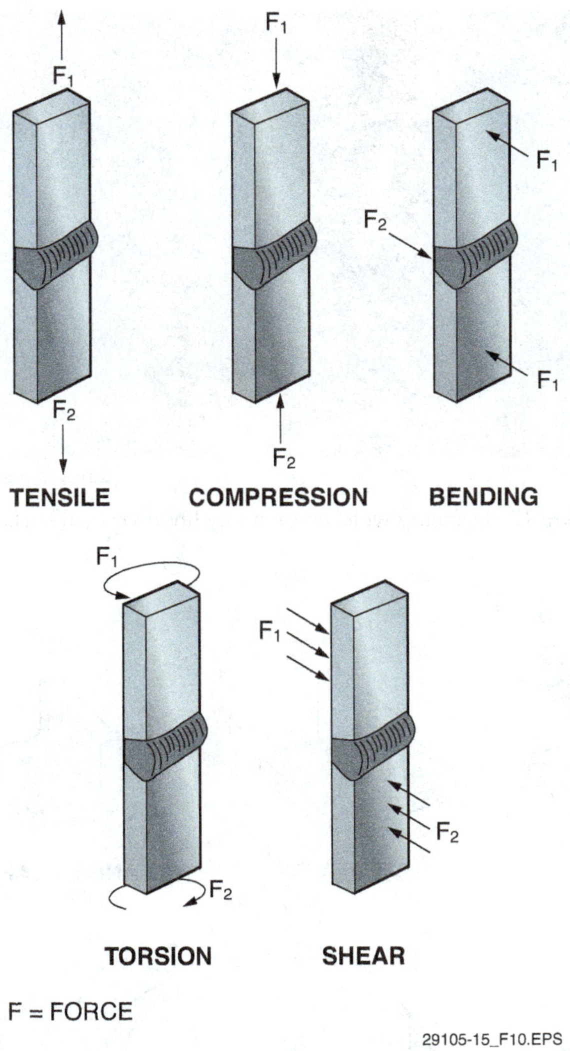

Figure 10 Forces that act on welded joints.

- *Shear* – To cut off or trim. When a piece of sheet metal is placed into a press brake machine, one side of the sheet metal is resting against a stationary surface of the machine. That stationary surface is applying some force to that side of the sheet metal. That stationary force is shown as F_1. When the brake is activated, a moveable part of the machine applies force to the opposite side of the sheet metal. F_2 represents the forced created by that moveable part. When the moveable force (F_2) is increased, it eventually shears or cuts the metal because the stationary surface (F_1) is applying force to the opposite side of the sheet metal.

2.2.0 Types of Welds

Most types of welds require some degree of base metal edge preparation depending on the type of joint being welded. Common types of welds include the following: **surfacing**, plug or slot, fillet, square groove, bevel groove, V-groove, J-groove, and U-groove. If a particular type of weld requires a specified **root opening**, it is typically shown on the welding drawing or **welding procedure specification (WPS)**. Base metals may require that some type of bevel be cut or ground on one or both edges, which form a V-groove when they are mated. Edge preparations will be discussed for each type of weld as it is covered in this section.

There are five basic types of joints (*Figure 11*). They are the butt joint, lap joint, T-joint, edge joint, and corner joint (both inside and outside corners). Each of these joints may be accomplished using various types of welds.

2.2.1 Surfacing Welds

Surfacing is the only type of weld that is applied to just one base metal surface, since it is not used to join two or more components. It is used either to build up a base surface that has become worn below the desired thickness or dimension, or to add material to a base to increase its strength and stability. An example of surfacing is shown in *Figure 12*. When you begin making your first welds, you will build what is referred to as a pad—a section of base metal with weld beads run parallel to each other along its length. This is an example of surface welding.

Did You Know?

Surfacing

Surfacing, as defined by the American Welding Society, is the application by welding, brazing, or thermal spraying of a layer of material to a surface to obtain desired properties or dimensions, as opposed to creating a joint between two or more pieces.

Before applying the first layer of multi-layered welds, the base metal surface must be cleaned to remove contaminants, such as oxidation and dirt. When applying the surfacing weld using an oxyfuel process, the surface should be evenly preheated before surfacing to eliminate warping of the base material.

For proper penetration and layering to take place, each new layer of weld surfacing must be properly chipped, scraped, and thoroughly cleaned before applying additional layers of surfacing.

Surfacing is often applied in the tool and die industry to rebuild relatively expensive tools and dies that have worn down. It is also used in the repair of heavy equipment to build up areas that have experienced excessive wear.

29105-15_F12.EPS

Figure 12 Surfacing welding on an inclined vertical surface.

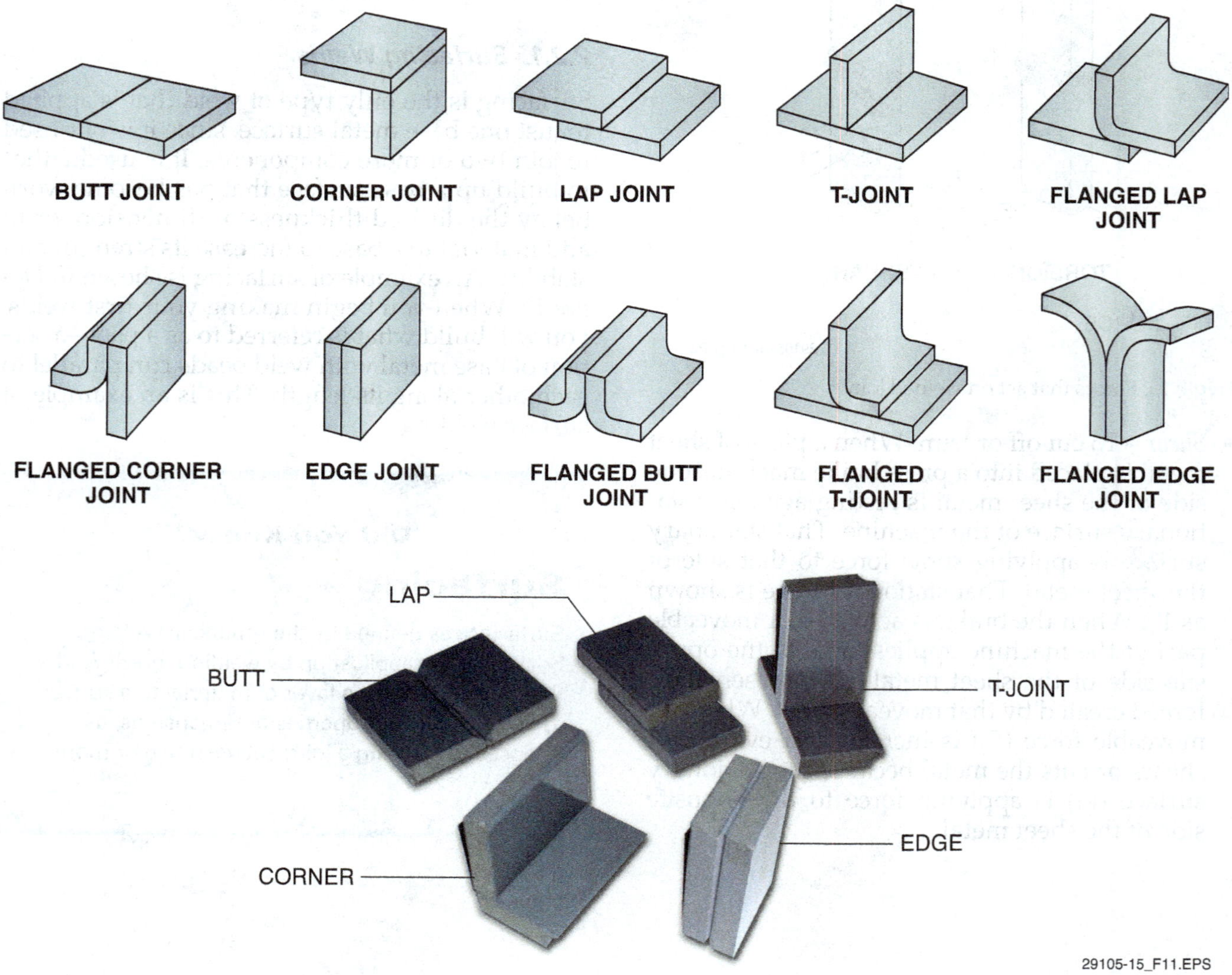

29105-15_F11.EPS

Figure 11 Five basic joint types with variations.

2.2.2 Plug and Slot Welds

Plug and slot welds are used to join metal pieces when the edges cannot be welded. The hole or slot can either be completely or partially filled with weld material when joining it to the other base metal. If the hole is round, the weld is referred to as a plug weld. If the weld is slotted or elongated, it is called a slot weld. *Figure 13* illustrates how a plug or slot weld may be used on a lap joint. A typical application is to overfill the hole or slot, then grind the excess material to a level, finished surface.

Preparation of the base metals for this type of weld requires proper cleaning and removal of any oxides or dirt from both pieces. The hole or slot is then drilled to receive the weld. The walls of the hole or slot must also be cleaned of any oxides, dirt, and oils prior to welding.

Plug and slot welds are commonly used where a finished surface is required, or where overall dimensions are not permitted to exceed those of the plate thickness.

2.2.3 Fillet Welds

Fillet welds may be applied to lap joints, T-joints, or corner joints. The fillet weld does not require any base metal edge preparation other than removing contaminants by cleaning with a wire brush or other appropriate cleaning tools. The two base metals are positioned together with or without a root opening, and one or more passes (welding beads) are applied at the intersection of the two base metal edges (*Figure 14*). Whenever more than one pass is made on a fillet weld—or on any weld—the previous pass must be cleaned of all contaminants such as dross, slag, and oxides before the next pass is applied.

When using a fillet weld on outside corner welds two types of fillet welds may be used. The half-lap joint (*Figure 15*) is easier to assemble, requires less welding material, and is less likely to burn through the corner. The half-lap does require a second weld on the inside of the corner. If a half-lap fit is used, allowances in the plate dimensions must be made for the lap.

The other fillet weld is sometimes called the corner-to-corner joint. The corner-to-corner joint is difficult to assemble because neither plate can support the other, and care must be used when welding to prevent burning through the corner.

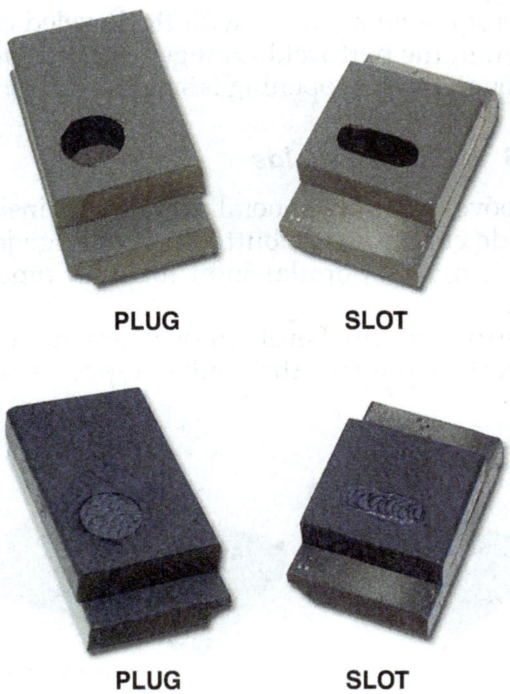

Figure 13 Plug and slot welds.

Surface Welding

Certain shafts located within gearboxes used in the marine industry turn at relatively high rpms (revolutions per minute) and will corrode and wear rapidly if subjected to salt water. It is not always cost effective to discard and replace each shaft with a new one. Surface welding may be used to bring the shafts back to the desired dimensions by building them up and re-machining them. They are then classified as rebuilt shafts. All rotating parts must be balanced before being installed. Otherwise they will vibrate, which in turn will destroy bearings and may cause the part itself to break.

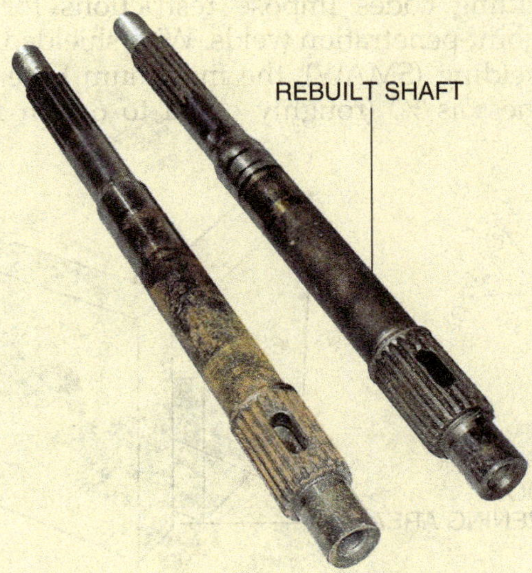

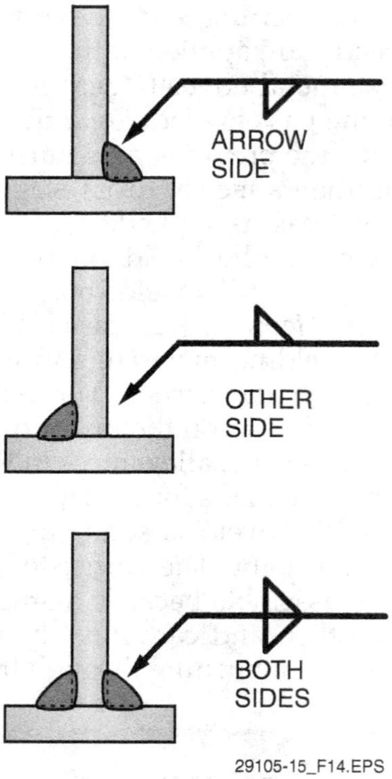

Figure 14 Fillet welds and drawing symbols.

2.2.4 Square-Groove Welds

Square-groove welds can be used with butt joints, corner joints, T-joints, and edge joints. The difference between the fillet and the square groove is that a square groove requires a root opening to be set or prepared between the two base pieces prior to welding, as shown in *Figure 16*. This allows penetration into a greater portion of the surface area of the two base metals. A partial joint-penetration weld will have much less strength than a complete joint-penetration weld.

Welding codes impose restrictions for complete joint-penetration welds. With shielded metal arc welding (SMAW), the maximum base metal thickness is ¼" (roughly equal to 6 mm metric

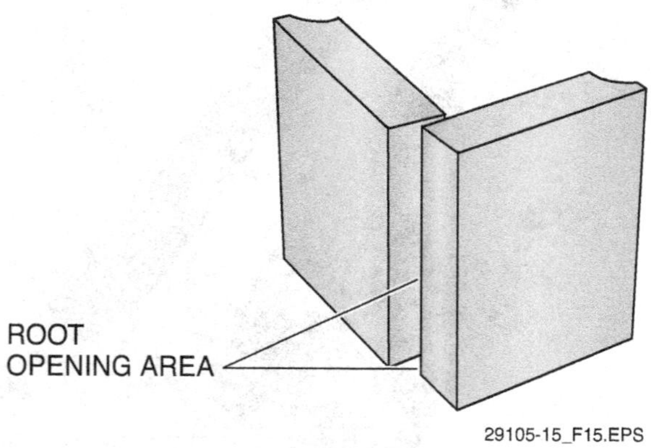

Figure 15 Fit-up of half-lap fillet welds.

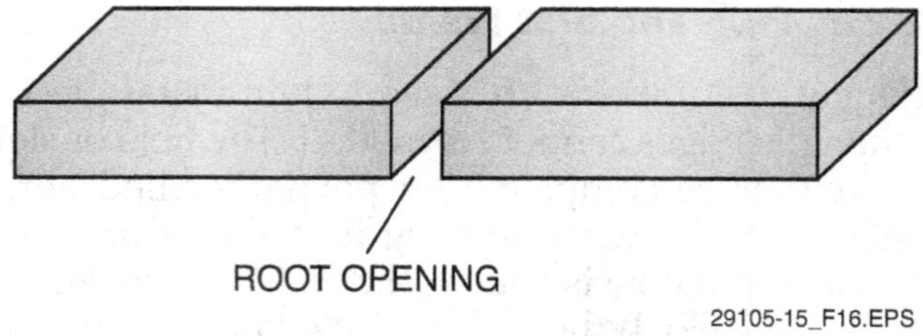

Figure 16 Preparation for a square groove on a butt weld.

steel plate), and welding from both sides is required. In addition, a root opening of half the thickness of the base metal is required, and the root of the first weld must be gouged before the second weld is made. For gas metal arc welding (GMAW) and flux-cored arc welding (FCAW), the maximum base metal thickness is ⅜" (≈10 mm metric plate), with the same requirements for welding from both sides and for back gouging the first pass.

Welding codes and specifications will be discussed in greater detail later in this module.

2.2.5 Single-Bevel Groove Welds

Single-bevel groove welds are commonplace in the welding industry because the base metal preparation is relatively simple and provides greater surface area penetration than the square groove. The single-bevel groove weld can be applied to all five of the basic weld joints. Typically, a single bevel is cut on the edge of one of the base metal pieces. The preparation may include a specified root opening along with the beveled cut, as shown in the butt weld arrangement in *Figure 17*. Although the root opening is small, it can be seen.

2.2.6 V-Groove Welds

V-groove welds are generally applied to inside or outside corner joints, butt joints, or edge joints. This is a very popular joint for both pipe and plate.

V-grooves and bevel grooves are more economical to prepare than other types of welds.

Figure 17 Workpieces for corner and butt bevel-groove welds.

Figure 18 shows a prepared V-groove on a butt joint to be welded. Typically, a 22½- or 30-degree bevel angle must be cut on both workpiece edges to prepare the base metals to receive the weld. In preparing inside or outside corner joints or butt joints for V-groove welds, the angle may not extend from the top of the bevel edge to the bottom of the bevel edge on each piece, but may be cut short of the bottom corner if a **root face** is required. In this edge preparation, the flat surface from the point where the angle stops to the bottom of the piece is referred to as the root face.

2.2.7 Single Versus Double V-Groove Welds

When possible, a double V-groove weld should be used in place of a single V-groove (*Figure 19*).

Although it may seem unusual, the double V-groove requires half the weld metal compared to the single V-groove. Welding from both sides also reduces distortion, because the forces of distortion work against each other.

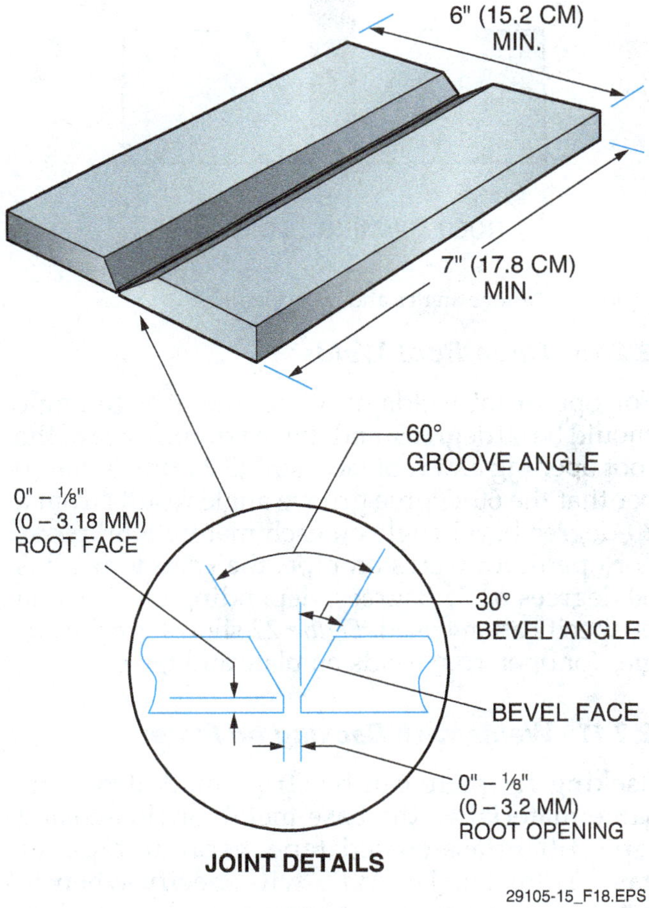

Figure 18 Open root V-groove weld.

2.2.8 J- and U-Groove Welds

J- and U-groove welds (*Figure 20*) are similar. The main difference is that a J-groove weld requires only one base metal to have its edge grooved in the form of a J. A U-groove is formed by preparing two matching J-grooves, one on each base metal edge. Even though both types require much less weld material than bevel or V-groove welds, they require much more preparation time because of the shape of the grooves. A J-groove weld can be applied to all five types of weld joints. The U-groove can only be used on butt, inside and outside corner, and edge joints.

2.2.9 Groove Angles and Root Openings

The purpose of the groove angle is to allow access to the root of the weld. The root preparation is sized to control **melt-through**.

Increasing or decreasing the root preparation (root opening and root face) will result in excess melt-through or insufficient root penetration. As a general rule, as the groove angle decreases, the root opening increases to compensate (*Figure 21*). Root faces are used with open-root joints, but not when metal **backing** strips are used.

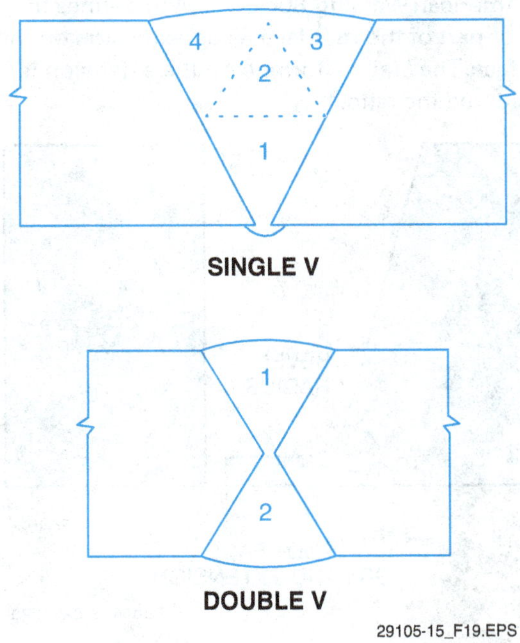

Figure 19 Comparison of double and single V-grooves.

Figure 20 J- and U-groove welds.

Root Face or Land

The root face is also widely known as the land. The American Welding Society (AWS) defines the land or part of the root face as an extension on the root face. The "table" formed on the extension is considered the land.

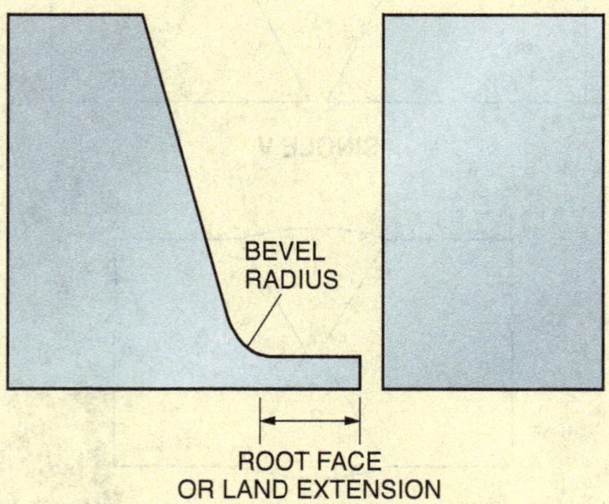

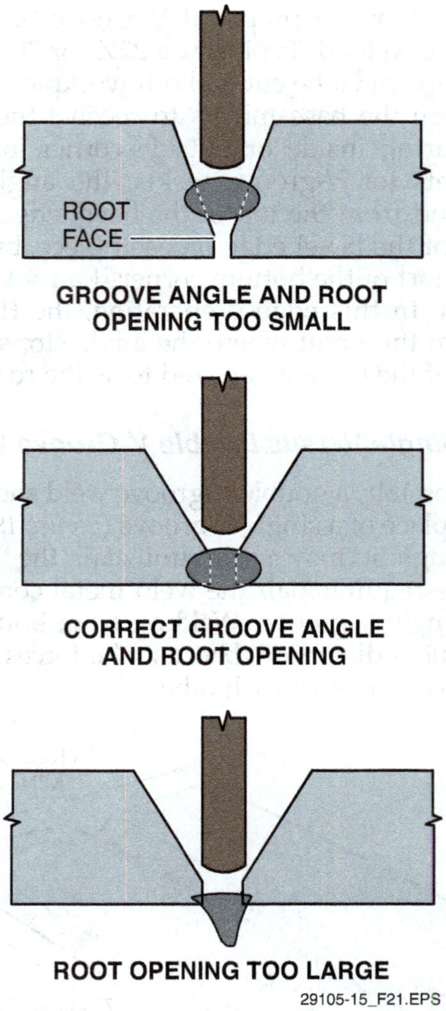

Figure 21 Groove angles and root openings.

2.2.10 Open-Root Welds

For open-root welds on plate, the groove angle should be 60 degrees, and the maximum size of the root opening and root face is ⅛" (3.2 mm). Remember that the 60-degree groove angle results from a 30-degree bevel angle on each mating workpiece. For open-root welds on pipe, the groove angle is 60 degrees or 75 degrees, depending on the code or specifications used. *Figure 22* shows groove angles for open root welds on plate and pipe.

2.2.11 Welds with Backing on Plate

Backing for plate can be strips made from the same material as the base metal, or flux-coated tape, fiberglass-coated tape, ceramic tape, or gas (*Figure 23*). The WPS will specify when to use backing and what type is to be applied.

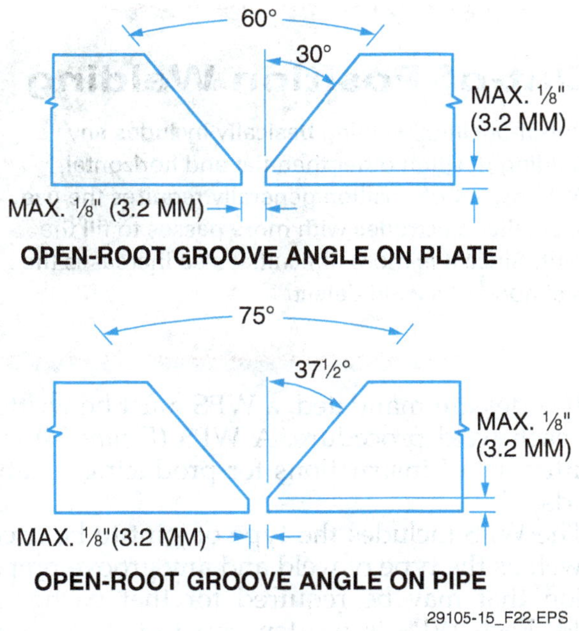

Figure 22 Open root joint preparation.

When gas is used as a backing, the same groove angles and root preparation used for open-root welds are used. When flux-coated tape, fiberglass-coated tape, or ceramic tape is used as a backing, the groove angle should be 60 degrees, the maximum root opening should be 3/16" (4.8 mm), and the maximum root face should also be 3/16" (4.8 mm). When applying tape, be sure to clean the surface to ensure that the tape adheres tightly, and make sure that the tape is centered on the weld joint.

When strips made from the same material as the base metal are used, a groove angle of 45 degrees, a root opening of 1/4" (6.4 mm), and a feather edge (no root face) are used. The backing strip must be thick enough and wide enough to prevent burn-through at the root and to absorb and dissipate the heat of the root pass.

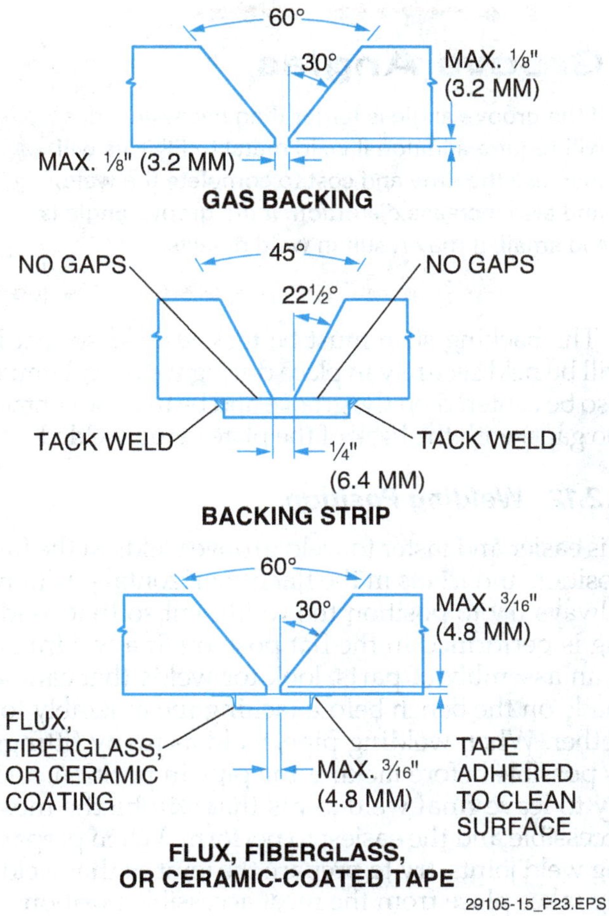

Figure 23 Plate joint preparation for backing.

> **NOTE**
>
> Check the WPS or site procedures for the backing size requirements for your site.

Backing Gas Hazards

Nitrogen and argon are used as backing and purge gases for some welding applications. These gases will displace all the oxygen in the space where they are used. At a refinery in Corpus Christi, Texas, while the welder was at lunch, a helper entered a large section of pipe in which argon had been used as a backing gas. When the welder returned, he found the helper lifeless in the pipe.

Nitrogen is lighter than air and, if allowed, will float upward. Argon, on the other hand, is heavier than air and will settle in low-lying areas. When removing nitrogen from a confined space, open the top to allow it to float out. When removing argon from a confined space, provide some way for it to flow out of the bottom. This is often impractical, however; forced ventilation is by far the most effective method.

Groove Angles

If the groove angle is larger than necessary, it will require additional weld metal to fill. This will increase the time and cost to complete the weld, and also increase distortion. If the groove angle is too small, it may result in weld defects.

The backing strip must be tack welded so that it will be held securely in place during welding. It must also be centered on the groove and be in close contact (no gaps) with the back of the plate being welded.

2.2.12 Welding Position

It is easier and faster to weld groove welds in the flat position and fillets in the flat or horizontal position. Always try to position the weldment so that welding is performed in the flat position. If a weldment is an assembly of parts, look for welds that can be made on the bench before tacking the assembly together. When welding pipe, weld as many fittings as possible before tacking the pipe in position, and try to leave final weld joints that will be the most accessible and the easiest to perform. When preparing weld joints, try to prepare the joint so that welding takes place from the most accessible position.

2.3.0 Codes and Welding Procedure Specifications

A welding code is a detailed listing of the rules and principles that apply to specific welded products. Codes ensure that safe and reliable welded products will be produced and that persons associated with the welding operation will be safe. Clients specify which welding codes should be followed when they place their orders or award project contracts.

Out-of-Position Welding

Out-of-position welding basically includes any welding position other than flat and horizontal. Welding out-of-position generally requires the use of smaller electrodes with more passes to fill the joint. Since it is more difficult, it also increases the likelihood of a weld defect.

If codes are mandated, a WPS must be written for each weld procedure. A WPS (*Figure 24*) is a written set of instructions for producing reliable welds.

The WPS includes the type of joint to be used, as well as the type of weld and any groove preparation that may be required for that particular weld. Each WPS is written and tested in accordance with a particular welding code or specification. All welding requires that acceptable industry standards be followed, but not all welds are governed by codes. The requirement for use of a WPS is often listed on blueprints as a note or in the tail of the welding symbol. If you are unsure whether or not the welding being performed requires a WPS, you should not proceed until you verify this information.

Note that the WPS shown in *Figure 24* uses metric dimensions. Electrodes and filler metals are basically the same size internationally, but they are called out using the metric equivalent of the size in most areas outside of the United States. For example, the 1.6 mm filler metal—wire in this case, for TIG welding—shown on the WPS is actually $\frac{1}{16}$" wire; the 1.2 mm filler is $\frac{3}{64}$" filler metal. These are the rounded metric equivalents of their actual manufactured Imperial size.

Weld Procedure Number	30 P1 TIG 01 Issue A
Qualifying Welding Procedure (WPAR)	WP T17/A

Manufacturer:	National Fabs Ltd 25 Lane End Birkenshaw Leeds	Method Of Preparation and Cleaning:	Machine and Degrease
Location:	Workshop	Parent Metal Specification:	Grade 304L Stainless Steel
Welding Process:	Manual TIG	Parent Metal Thickness	3 to 8mm Wall
Joint Type:	Single Sided Butt Weld	Pipe Outside Diameter	25 to 100mm
		Welding Position:	All Positions
		Welding Progression:	Upwards

Joint Design	Welding Sequences
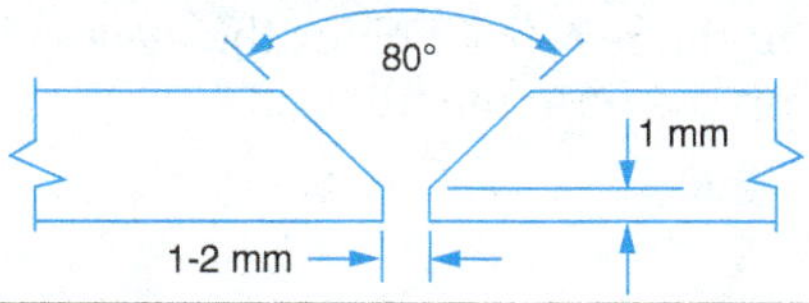	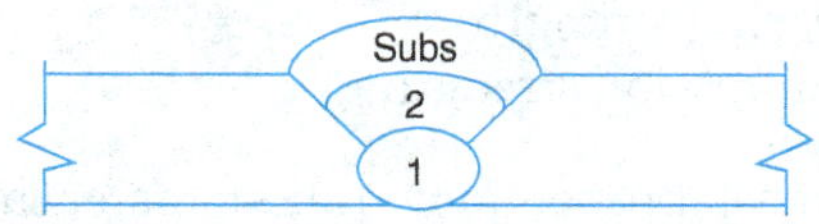

Run	Process	Size of Filler Metal	Current A	Voltage V	Type of Current/Polarity	Wire Feed Speed	Travel Speed	Heat Input
1 2 And Subs	TIG TIG	1.2 mm 1.6 mm	70 - 90 80 - 140	N/A	DC– DC–	N/A	N/A	N/A

Welding Consumables:- Type, Designation Trade Name: Any Special Baking or Drying: Gas Flux: Gas Flow Rate - Shield: - Backing: Tungsten Electrode Type/ Size: Details of Back Gouging/Backing: Preheat Temperature: Interpass temperature: Post Weld Heat Treatment Time, temperature, method: Heating and Cooling Rates*:	 BS 2901 Part 2 : 308S92 No Argon 99.99% Purity 8 - 12 LPM 5 LPM 2% Thoriated 2.4mm Dia Gas Backing 5°C Min 200°C Max Not Required	**Production Sequence** 1. Clean weld and 25mm borders to bright metal using approved solvent. 2. Position items to be welded ensuring good fit up and apply purge. 3. Tack weld parts together using TIG, tacks to at least 5mm min length. 4. Deposit root run using 1.2mm dia. wire. 5. Inspect root run internally. 6. Complete weld using 1.6mm dia wire using stringer beads as required. 7. 100% Visual inspection of completed weld

Revision History

Date	Issue	Changes	Authorization
26/11/2000	A	First changes	Jack Straw *Jack Straw*

29105-15_F24.EPS

Figure 24 Example of a WPS.

Additional Resources

2014 Technical Training Guide. Current Edition. The Lincoln Electric Company, Cleveland, OH. USA. **www.lincolnelectric.com**

Welding Handbook. Current Edition. Miami, FL: The American Welding Society.

2.0.0 Section Review

1. When a metal piece is pulled in opposite directions to determine its breaking point, it is a test of its ____ .
   a. tensile strength
   b. torsion strength
   c. shearing strength
   d. compression strength

2. To achieve proper penetration for each layer during multi-layer surface welding, ____ .
   a. the previous weld must be cleaned before each layer application
   b. no cleaning is required since the heat of welding will remove any impurities
   c. post-heating the base metal is needed before layer application
   d. every layer must first be gouged

3. The requirement for use of a WPS is often listed ____ .
   a. within the company's hand book
   b. on blueprints, notes, or on the tail of the welding symbol
   c. in the SDS/MSDS for the electrode
   d. on the base metal

SECTION THREE

3.0.0 WELDING JOINT PREPARATION

Objective

Describe how to prepare joints for welding.

a. Describe how to mechanically prepare joints for welding.
b. Describe how to thermally prepare joints for welding.

Performance Tasks

1. Mechanically or hand grind a bevel on the edge of a ¼"- to ¾"-thick mild steel plate (6 to 20 mm metric plate) at 22½ degrees.
2. Thermally bevel the edge of a ¼"- to ¾"-thick mild steel plate (6 to 20 mm metric plate) at 22½ degrees.

There are many different ways to prepare a joint for welding. The joint can be prepared mechanically using nibblers, grinders, or cutters. It can also be prepared thermally using an oxyfuel, plasma arc, or carbon arc cutting torch. The method used to prepare the joint will generally be determined by the base metal type, ease of use, and code or procedure specifications.

Before starting a joint preparation, determine if the joint is covered by a WPS or site quality standard. If this is the case, the joint type and welding parameters will be specified. These specifications must be followed; the guidance provided by a WPS is mandatory. If you are unsure whether or not a joint is covered by a WPS or site quality standard, check with your supervisor before proceeding.

3.1.0 Mechanical Joint Preparation

Mechanical joint preparation is most often used on alloy steel, stainless steel, and nonferrous metal piping or plate. It is slower than thermal methods such as oxyfuel, carbon arc, and plasma arc, but has the advantages of high precision, low heat input, and the absence of oxides commonly left by thermal methods.

3.1.1 Grinders

Handheld electric or air-operated grinders are used in welding shops, and even more often in the field, to cut and bevel pipe and plate to prepare them for welding. *Figure 25* shows a handheld angle grinder being used to prepare a stainless steel elbow for welding.

WARNING!

Cutoff and grinding wheels must be selected based on the material to be cut or beveled. Using an improper wheel may damage the wheel or workpiece, and can create hazardous conditions for the operator should the wheel shatter during operation. Refer to the manufacturer's recommendations and warnings when selecting grinding wheels. Ensure that the maximum revolutions per minute (rpm) rating of the grinding or cutoff wheel is equal to or greater than the speed of the grinder in use.

29105-15_F25.EPS

Figure 25 Preparing a weld joint.

3.1.2 Pipe Beveling Machines

Welded piping is found extensively in all industries. Nearly every piece of pipe that is welded to a fitting, to another piece of pipe, or to any other connection requires that the end be cut square and beveled according to specifications before welding. Welded pipe fittings are typically beveled from the factory and simply need to be cleaned of paint and other contaminants before welding. If a mechanical cutting and beveling process is used (*Figure 26*), the procedure may be done using an electric, hydraulic, or pneumatic beveling machine. Many of these are portable machines that operate much like a lathe, and they are often referred to as pipe-end prep lathes. They have mandrels that hold various cutting tools that cut metal away, as well as numerous adjusting mechanisms that make it easy to set the bevel angle and depth. Various models are available to cut and bevel both large and small pipe. Boiler tube ends are typically prepared using these smaller machines. When specifications call for boiler tubing to be welded, the machine can face, square, and chamfer the ends of the tubes to prepare them.

29105-15_F26.EPS

Figure 26 Pipe beveling machine.

WARNING!

Always follow all manufacturers' safety procedures when using grinders and cutoff machines. Failure to follow the manufacturer's safety recommendations could result in serious personal injury.

Special mechanical cutoff machines (*Figure 27*) are also made to be mounted on the outside of the pipe. They can be mounted with a ring or a special chain with rollers. A cutoff blade or tool is used to make the cut.

3.1.3 Nibblers

Nibblers prepare the edge of a plate or pipe with a reciprocal punch that cuts or bites off a chip with each stroke. The bevel angle is set by adjusting the nibbler. A nibbler must have access to an edge on which it will be used (*Figure 28*).

WARNING!

Always follow all manufacturers' safety procedures when using nibblers and cutters. Failure to follow the manufacturer's safety recommendations could result in serious personal injury.

3.2.0 Thermal Joint Preparation

Thermal joint preparation involves preparing a joint with the oxyfuel or plasma arc processes. It can be done using the air-carbon arc process, but the result is typically poor. Because of carbon deposits and the inaccuracy of the result, the air-carbon arc cutting process is best used for gouging seams or cracks only.

29105-15_F27.EPS

Figure 27 Pipe cutoff machine.

Thermal Joint Preparation Precautions

When preparing a joint with the oxyfuel, plasma arc, or air-carbon arc cutting processes, all dross must be removed prior to welding. Any dross remaining on the joint during welding will cause porosity in the weld. Joints prepared with a carbon arc cutting torch must be carefully inspected for carbon deposits. It is common for small carbon deposits to be left behind as the carbon electrode is consumed during the cutting or gouging operation. These carbon deposits will cause defects such as hard spots, loss of ductility, and cracking in the weld. Before welding, use a grinder to clean surfaces prepared with most thermal processes.

The torch used for oxyfuel or plasma arc cutting can be handheld or mounted on a motorized or manual carriage. These are often referred to as track burners. The motorized carriage for plate cutting runs on flat tracks positioned on the surface of the plate to be cut (*Figure 29*). The carriage has an On or Off switch, forward or reverse switches, and a speed adjustment. The torch position can be adjusted as necessary for the proper bevel angle. The carriage can carry a special oxyfuel or plasma arc cutting torch designed to fit into the torch holder.

Other equipment designs are used for cutting pipe (*Figure 30*). A steel ring or special chain with rollers is attached to the outside of the pipe.

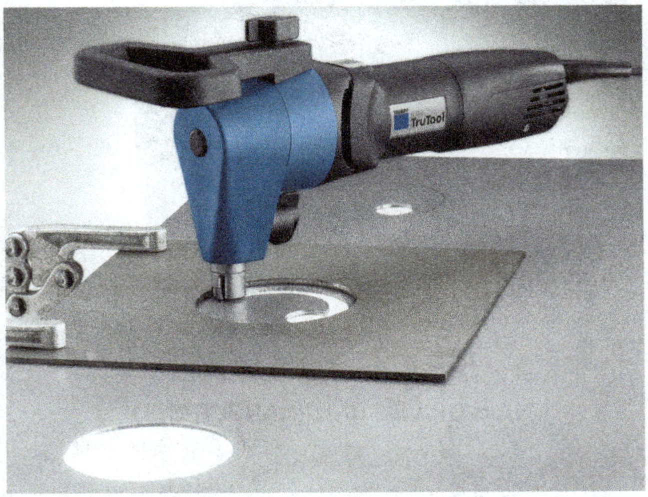

29105-15_F28.EPS

Figure 28 Nibbler.

29105-15_F29.EPS

Figure 29 Track burner thermally cutting and beveling plate.

The torch is carried around the pipe on a torch holder that is powered by electricity, air, or a hand crank. The bevel angle is set by pivoting the torch holder, and the torch is adjusted vertically with a handwheel. An out-of-round attachment can be used to compensate for pipe that is out-of-round.

For very large-diameter pipe, special internal cutting and beveling equipment is available (*Figure 31*). The motorized unit shown here uses rare-earth magnets in the wheels, developing 300 pounds (136 kg) of force. No rings, bands, or chains are required. It can also be used for smaller pipe by traveling on the outside surface, and for long cuts on plate.

Automatic Pipe Bevelers

Automatic versions of pipe bevelers and cutters are an improvement over pattern cutters and other similar equipment. This is because there is no need to reset the preheat flame before each cut. Once an automatic pipe beveler has been initially set up, all subsequent cuts can be made with the same settings. These systems also save time and cutting gases, because when the system is switched off, all gases are immediately shut off as well.

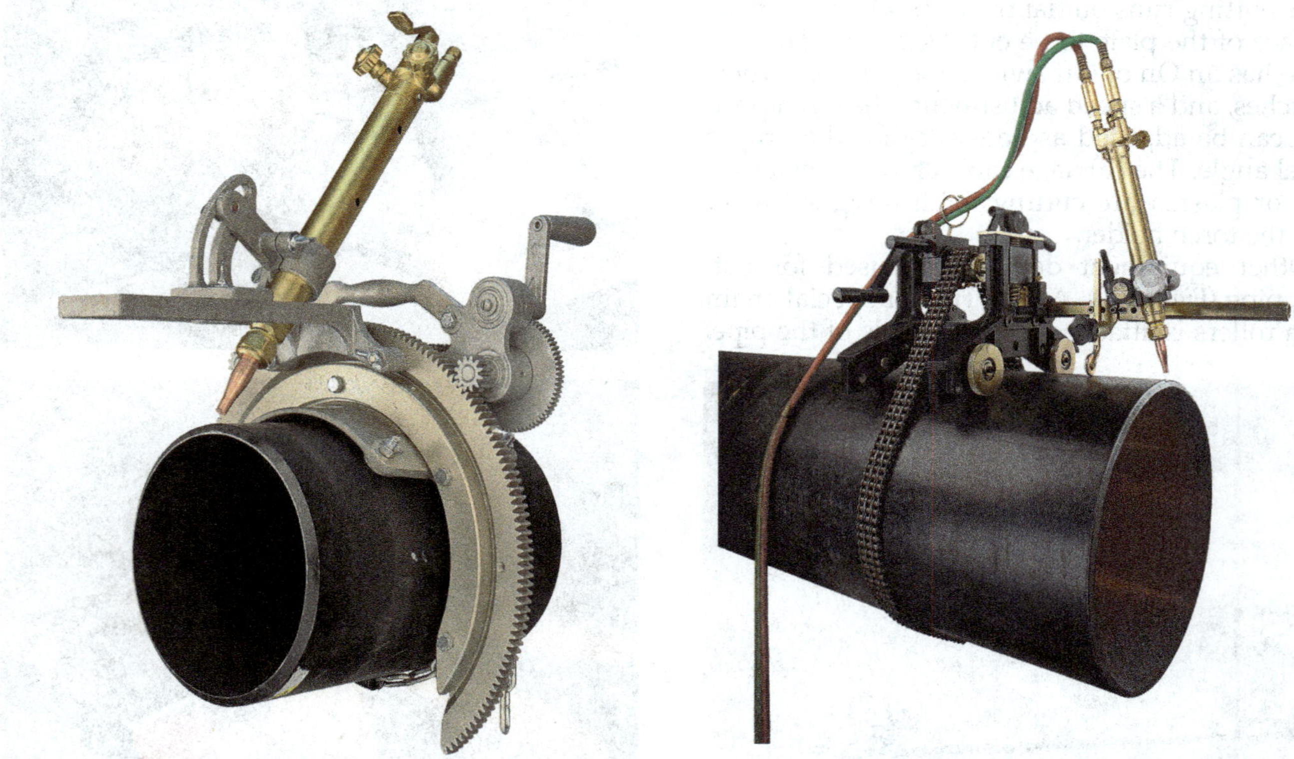

GEAR-DRIVEN UNIT (HANDCRANK)

CHAIN-DRIVEN UNIT (HANDCRANK)

29105-15_F30.EPS

Figure 30 Pipe cutting and beveling equipment.

29105-15_F31.EPS

Figure 31 Thermal cutting from inside the pipe.

Additional Resources

2014 Technical Training Guide. Current Edition. The Lincoln Electric Company, Cleveland, OH 44117-1199 USA. **www.lincolnelectric.com**

Welding Handbook. Current Edition. Miami, FL: The American Welding Society.

3.0.0 Section Review

1. Which of the following is an advantage of mechanical joint preparation over the thermal method?
   a. It is more precise with less heat input.
   b. It is faster and less labor intensive.
   c. Mechanically prepared joints require less weld metal.
   d. Mechanically prepared joints require fewer weld passes.

2. Which of the following is a thermal process seldom used due to the deposits left behind which can lead to weld defects?
   a. Air-carbon arc cutting
   b. Oxyfuel cutting
   c. Plasma arc cutting
   d. Pipe-end prep lathe

Summary

The importance of safety cannot be overemphasized within the welding environment. It should be the highest priority at all times.

Performing chemical or mechanical cleaning removes base metal surface contamination. To create reliable weld joints that meet specifications, proper cleaning techniques must always be used.

A great deal of load stress may be placed on a weld joint. Understanding the different types of stress helps engineers to determine what type of welding joint and technique is best for the application. The specifications for a given weld joint are communicated to the welder through the welding procedure specification (WPS).

Weld joints can be prepared mechanically and thermally. Mechanical methods do have advantages, but they can also be labor-intensive. Thermal methods include the use of oxyfuel cutting processes, with the torch mounted on a carriage that allows it to rotate smoothly around the pipe to make a consistent cut and/or bevel.

Review Questions

1. Never directly or indirectly view an electric arc without _____.
   a. having your eyes examined
   b. wearing sunglasses
   c. being at least 10 feet away
   d. using a properly tinted lens
2. Medium-carbon steel contains _____.
   a. 0.10 percent to 0.15 percent carbon
   b. 0.15 percent to 0.30 percent carbon
   c. 0.30 percent to 0.50 percent carbon
   d. 0.50 percent to 1.00 percent carbon
3. A type of copper-alloyed steel that protects itself from damaging oxidation by creating a hard, tough layer of oxidation on its surface is called _____.
   a. spring steel
   b. weathering steel
   c. mild steel
   d. stainless steel
4. A pneumatic tool that uses a number of blunt steel needles to remove surface oxidation is called a _____.
   a. needle scaler
   b. weld flux chipper
   c. needle nose chipper
   d. contamination tattoo scaler
5. Torsion applied to a weld joint refers to a force _____.
   a. pulling in opposite directions
   b. twisting the metal in opposite directions
   c. pushing against a stationary surface
   d. attempting to create a bend
6. The five basic types of joints are butt, lap, corner, T-joint, and _____.
   a. chamfer
   b. square
   c. bevel
   d. edge
7. Lap joints, T-joints, or corner joints can be welded using _____.
   a. surface welds
   b. butt welds
   c. fillet welds
   d. spot welds
8. U-groove welds are formed by preparing two matching J-grooves that require _____.
   a. less preparation time but use more weld material than a V-groove weld
   b. more preparation time but use less weld material than a V-groove weld
   c. more preparation time and use more weld material than a V-groove weld
   d. no preparation time and use the same weld material as a V-groove weld
9. Which of the following materials is *not* a common joint backing material?
   a. The same base metal type
   b. A gas
   c. Ceramic tape
   d. Cast iron
10. When a backing strip of the same material as the base metal is used, the root face should be _____.
    a. a feather edge
    b. 1⁄16" (1.6 mm)
    c. 1⁄8" (3.2 mm)
    d. 1⁄4" (6.4 mm)
11. Grinding wheels selected for use must be rated equal to or greater than the grinder's _____.
    a. current
    b. horsepower
    c. rotational speed
    d. air pressure
12. Before a pipe can be butt-welded to a fitting or other connection, its edge must be _____.
    a. burnished
    b. hardened
    c. ground to fit
    d. cut square and beveled
13. Beveling equipment made for pipe is sometimes called _____.
    a. J-prep cutters
    b. ID-prep cutter
    c. pipe-end prep lathes
    d. compound-bevel prep lathes

14. Which of the following processes produces the least desirable result, and is therefore *not* a popular choice for general metal cutting?
    a. Plasma arc cutting
    b. Band saw
    c. Oxyfuel cutting
    d. Air-carbon arc cutting

15. The best approach to gouging seams or cracks is _____.
    a. using a nibbler
    b. carbide-chisel cutting
    c. grinder cutting
    d. air-carbon arc cutting

Trade Terms Quiz

Fill in the blank with the correct term that you learned from your study of this module.

1. An assembly that is fastened together by weld joints is referred to as a(n) ________.
2. A weldable or non-weldable material used behind a root opening to allow defect-free welding at the open root of a joint is a(n) ________.
3. The document containing all the detailed methods and practices required to produce a sound weld is a(n) ________.
4. The primary metal to be welded, cut, or brazed is referred to as the ________.
5. The steel alloy that, under specific conditions, is designed to form a very dense oxide layer on its outer surfaces to retard further oxidation is called ________.
6. The expansion and contraction of welded parts caused by the heating and subsequent cooling of the weld joint is ________.
7. The space between the base metal pieces at the bottom or root of a joint is called a(n) ________.
8. A steel that contains less than 0.10 percent carbon and is magnetic is called ________.
9. The amount of force applied to a material or structure is called a(n) ________.
10. A small flattened area on the end of a bevel for a groove weld is called a(n) ________.
11. Complete joint penetration results in ________.
12. Gas pockets, or voids in the weld metal are called ________.
13. A metal, such as aluminum or brass, lacking sufficient quantities of iron to have any effect on its properties is called a(n) ________.
14. A form of porosity having a length greater than its width and is approximately perpendicular to the weld face is known as a(n) ________.
15. The scale that forms on metal surfaces when they are exposed to the oxygen or air containing oxygen is called a(n) ________.
16. A steel that shares some characteristics with ferritic, but boasts higher levels of carbon, up to a full 1 percent is called ________.
17. To impart strength or toughness to steel or cast iron by heating and cooling is referred to as ________.
18. When a metal is heated and cooled to free it from internal stress, it is said to be ________.
19. Any article that has been removed from a mold is a(n) ____ .
20. A metal that contains austenite (a solid solution of carbon, or of carbon and other elements, in a ferrous alloy), added through heating is called ________.
21. When a metal is ________, it is able to undergo a change of form without breaking.
22. To ________ a piece of metal, you will need a bucket of water.
23. A metal that has been shaped by beating it with a hammer, as is common with iron, is referred to as being ________.
24. Only ________ metals contain any substantial iron.
25. The measure of the ability of a material to withstand a longitudinal stress, without breaking is considered the material's ________.

26. The application by welding, brazing, or thermal spraying of a layer of material to a surface to obtain desired properties or dimensions is _______.

27. Capable of being extended or shaped by hammering or by pressure from rollers, this material is considered to be _______.

Trade Terms

Annealed
Austenitic
Backing
Base metal
Castings
Distortion
Ductile
Ferritic
Ferrous
Load
Malleable
Martensitic
Melt-through
Nonferrous metal
Oxide
Piping porosity
Porosity
Quench
Root face
Root opening
Surfacing
Tempering
Tensile strength
Weathering steel
Welding procedure specification (WPS)
Weldment
Wrought

Appendix

Performance Accreditation Tasks

The American Welding Society (AWS) School Excelling through National Skills Standards Education (SENSE) program is a comprehensive set of minimum Standards and Guidelines for Welding Education programs. The following performance accreditation is aligned with and designed around the SENSE program.

The Performance Accreditation Tasks (PATs) correspond to and support the learning objectives in *AWS EG2.0, Guide for the Training and Qualification of Welding Personnel: Entry-Level Welder*.

Note that in order to satisfy all learning objectives in *AWS EG2.0*, the instructor must also use the PATs contained in the second level of the NCCER Welding curriculum.

PAT 1 corresponds to no *AWS EG2.0* reference.

PAT 2 corresponds to *AWS EG2.0, Module 8 – Thermal Cutting Processes, Unit 2 – Manual OFC (e.g., Track Burner)*, Key Indicators 4, 5, and 6.

PATs provide specific acceptable criteria for performance and help to ensure a true competency-based welding program for students.

The following tasks are designed to develop your competency in preparing base metal. Practice each task until you are thoroughly familiar with the procedure.

As you complete each task, take it to your instructor for evaluation. Do not proceed to the next task until instructed to do so by your instructor.

PREPARE PLATE JOINTS MECHANICALLY

Using a nibbler, cutter, or grinder, mechanically or manually prepare the edge of a ¼"- to ¾"-thick carbon steel plate (≈6 to 20 mm metric plate) with a bevel angle of 22½ or 30 degrees, at the discretion of the instructor.

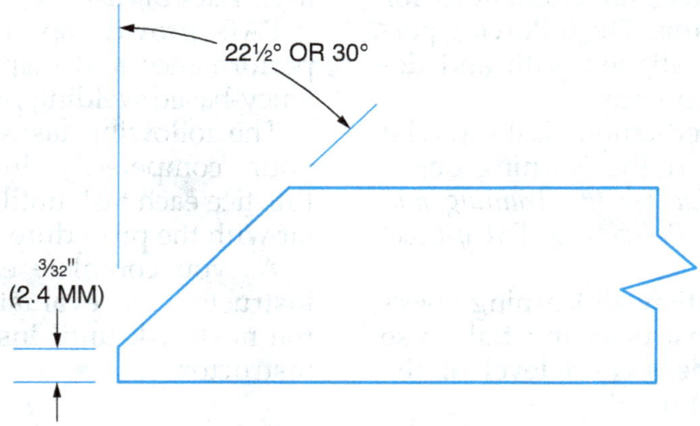

NOTE: BASE METAL = CARBON STEEL PLATE

29105-15_A01.EPS

Criteria for Acceptance:

- Bevel angle ±2½" ________
- Bevel face smooth and uniform to 1⁄16" (1.6 mm) ________
- Root face ±1⁄32" (±0.8 mm) ________

PREPARE PLATE JOINTS THERMALLY

Using oxyfuel or plasma arc cutting equipment, thermally prepare the edge of a ¼"- to ¾"-thick carbon steel plate (≈6 to 20 mm metric plate) with a bevel angle of 22½ or 30 degrees, at the discretion of the instructor.

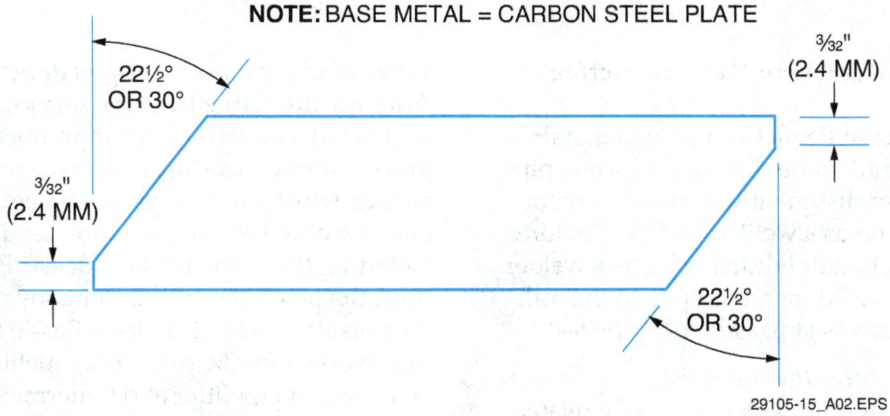

Criteria for Acceptance:

- Bevel angle ±2½" ______
- No dross ______
- Minimal notching not exceeding 1⁄16" (1.6 mm) deep on the kerf face ______
- Root face ±1⁄32" (±0.8 mm) ______

Cornerstone of Craftsmanship

Frank Johnson

PBF Energy Toledo Refinery Inc.
Welding Trainer and CWE Weld Procedure Writer

How did you choose a career in the construction industry?
I took welding courses at Penta County Vocational High School. Upon graduation, I passed the state pipe welding test, an accomplishment which was rare for a man my age. I worked as a welder for a year before joining the military. Though I didn't work as a welder during this time, I received increased pay in the military because of my success at passing the pipe test.

Who inspired you to enter the industry?
The reason for choosing this career is directly related to who inspired me to enter the industry. My grandfather owned a weld shop and I began welding when I was a young boy of just 13. My grandfather pushed me to read and learn as much about welding as possible so that I could solve problems as they arose. I was also inspired by my hobby of steam tractors because I had to weld them whenever they were in need of repair.

How important are education and training in construction?
I think training is absolutely essential in order to be a good welder. It is very important to know the codes, variables, and demands of the industry and to be able to perform any type of welding in any type of situation. I feel that the NCCER curriculum is important because I use it to train welders. The NCCER standardized curriculum, developed by expert welders, ensures that employees are trained properly.

What kinds of work have you done in your career?
After my discharge from the service, I started and completed a boilermaking apprenticeship. Six months into my apprenticeship, I was recognized as a skilled welder, which enabled me to be paid as a journeyman. I worked at various refineries and power plants, including the Davis Bessie Nuclear Plant. At that time, the pipefitters at the company needed welders. As a result, I welded at Davis Bessie for about a year, then worked for 2½ years as a pipefitter. I earned journeyman pipefitter and boilermaker journey level cards. Next, I worked at Enrico Fermi Nuclear Power Plant in Monroe, Michigan. While there, I was asked by Detroit Edison to work with their welding engineer because he qualified 95 percent of their welding procedures. I am now a certified welder, journey-level boilermaker, and pipefitter. I have created and qualified 17 welding procedures.

Tell us about your present job.
In my capacity as a trainer, I test and qualify welders, write welding procedures, qualify welding procedures, and serve as a certified welding instructor. The work is challenging and I really enjoy it.

What factors have contributed most to your success?
Factors that have contributed the most to my success are my natural curiosity, as well as my experience working in a nuclear plant where I admired the excellent craftsmanship of seasoned welders. Their professionalism gave me the desire to achieve their level of expertise. In that environment, quality standards are extremely high, so only top-notch welders will succeed there.

Would you suggest construction as a career to others? Why?
I would suggest construction as a career because if you are a hard worker you can attain a well-paying, secure job. I believe that a career in construction is full of challenges that craftworkers can overcome. They can also experience the satisfaction of seeing a project through, which yields them the satisfaction of putting a project into operation and seeing it work as it should. When recommending a career in construction, I've said that it is "Good work, challenging work, and satisfying work that will support you with a good wage."

What advice would you give to those new to the field?
My advice to those new to the field is: "Don't be afraid to try. Even if you don't succeed the first time, you will make it eventually, and you will be successful."

Tell us an interesting career-related fact or accomplishment.
I serve as an advisor to the Historical Boiler Board for the State of Ohio and on the National Board for Historical Boilers, an honor bestowed upon me due to my successful welding career. I was asked to serve as the subject matter expert on the committee that shapes the Welding curriculum for NCCER. Looking back, I never thought I'd be in a position like this. Welding has taken me a long way.

How do you define craftsmanship?
A craftsman is one who takes pride in his or her work.

Trade Terms Introduced in This Module

Annealed: To free from internal stress by heating and gradually cooling.

Austenitic: Containing austenite (a solid solution of carbon or of carbon and other elements in a ferrous alloy), added through heating.

Backing: A weldable or nonweldable material used behind a root opening to allow defect-free welding at the open root of a joint

Base metal: Metal to be welded, cut, or brazed.

Castings: Something cast; any article that has been cast in a mold.

Distortion: The expansion and contraction of welded parts caused by the heating and subsequent cooling of the weld joint.

Ductile: Able to undergo change of form without breaking.

Ferritic: Steel containing less than 0.10 percent carbon and is magnetic. This steel can't be hardened via heat treatment.

Ferrous: Containing iron.

Load : The amount of force applied to a material or a structure.

Malleable: Capable of being extended or shaped by hammering or by pressure from rollers.

Martensitic: Steel that shares some characteristics with ferritic, but has a higher levels of carbon, up to a full 1 percent. It can be tempered and hardened and is used where strength is more important than a resistance to oxidation.

Melt-through: Complete joint penetration.

Nonferrous metal: A metal, such as aluminum, copper, or brass, lacking sufficient quantities of iron to have any effect on its properties.

Oxide: The scale that forms on metal surfaces when they are exposed to oxygen or air containing oxygen.

Piping porosity: A form of porosity having a length greater than its width and that is approximately perpendicular to the weld face.

Porosity: Gas pockets, or voids in the weld metal.

Quench: To cool suddenly by plunging into a liquid.

Root face: A small flattened area on the end of a bevel for a groove weld.

Root opening: The space between the base metal pieces at the bottom or root of the joint.

Surfacing: The application by welding, brazing, or thermal spraying of a layer of material to a surface to obtain desired properties or dimensions.

Tempering: To impart strength or toughness to (steel or cast iron) by heating and cooling.

Tensile strength: The measure of the ability of a material to withstand a longitudinal stress without breaking.

Weathering steel: Steel alloy that, under specific conditions, is designed to form a very dense oxide layer on its outer surfaces, which retards further oxidation.

Welding procedure specification (WPS): The document containing all the detailed methods and practices required to produce a sound weld.

Weldment: An assembly that is fastened together by welded joints.

Wrought: Produced or shaped by beating with a hammer, as with iron.

Additional Resources

This module presents thorough resources for task training. The following resource material is suggested for further study.

2014 Technical Training Guide. Current Edition. The Lincoln Electric Company, Cleveland, OH. USA. **www.lincolnelectric.com**

American Iron and Steel Institute, 25 Massachusetts Avenue, NW Suite 800, Washington, DC 20001, USA. **www.steel.org**

Association for Iron & Steel Technology, 186 Thorn Hill Road Warrendale, PA, USA. **www.aist.org**

Steel Metallurgy for the Non-Metallurgist. John D. Verhoeven; Prepared under the direction of the ASM International Technical Book Committee. ASM International® Materials Park, OH 44073-0002. **www.asminternational.org**

Welding Handbook. Current Edition. Miami, FL: The American Welding Society.

Figure Credits

Topaz Publications, Inc., Module Opener, Figures 4, 6, 8A, 8C–8H, 9, 11 (photos), 12, 13, 17, 20, 25, 29, SA02

The Lincoln Electric Company, Cleveland, OH, USA, Figures 1–3

Reproduced by permission TWI Ltd., Figure 5

Courtesy of Milwaukee Electric Tool Corporation, Figures 7A, 7B

Courtesy of Chicago Pneumatic Tool Co., Figures 7C, 7D

Courtesy of Dremel, Figure 8B

Courtesy of Tri Tool Inc., Figures 26, 27

TRUMPF Inc., Figure 28

Courtesy of Mathey Dearman, Figures 30, 31

AWS F2.2:2001, Lens Shade Selector, reproduced with permission from the American Welding Society (AWS), Miami, FL, USA, Table 1

Photo Courtesy of U.S. Army, SA01

Section Review Answer Key

Answer	Section Reference	Objective
Section One		
1. c	1.1.3	1a
2. b	1.2.1	1b
3. b	1.3.1	1c
Section Two		
1. a	2.1.0	2a
2. a	2.2.1	2b
3. b	2.3.0	2c
Section Three		
1. a	3.1.0	3a
2. a	3.2.0	3b

NCCER CURRICULA — USER UPDATE

NCCER makes every effort to keep its textbooks up-to-date and free of technical errors. We appreciate your help in this process. If you find an error, a typographical mistake, or an inaccuracy in NCCER's curricula, please fill out this form (or a photocopy), or complete the online form at **www.nccer.org/olf**. Be sure to include the exact module ID number, page number, a detailed description, and your recommended correction. Your input will be brought to the attention of the Authoring Team. Thank you for your assistance.

Instructors – If you have an idea for improving this textbook, or have found that additional materials were necessary to teach this module effectively, please let us know so that we may present your suggestions to the Authoring Team.

NCCER Product Development and Revision

13614 Progress Blvd., Alachua, FL 32615

Email: curriculum@nccer.org

Online: www.nccer.org/olf

❑ Trainee Guide ❑ Lesson Plans ❑ Exam ❑ PowerPoints Other ____________

Craft / Level: Copyright Date:

Module ID Number / Title:

Section Number(s):

Description:

Recommended Correction:

Your Name:

Address:

Email: Phone:

Weld Quality

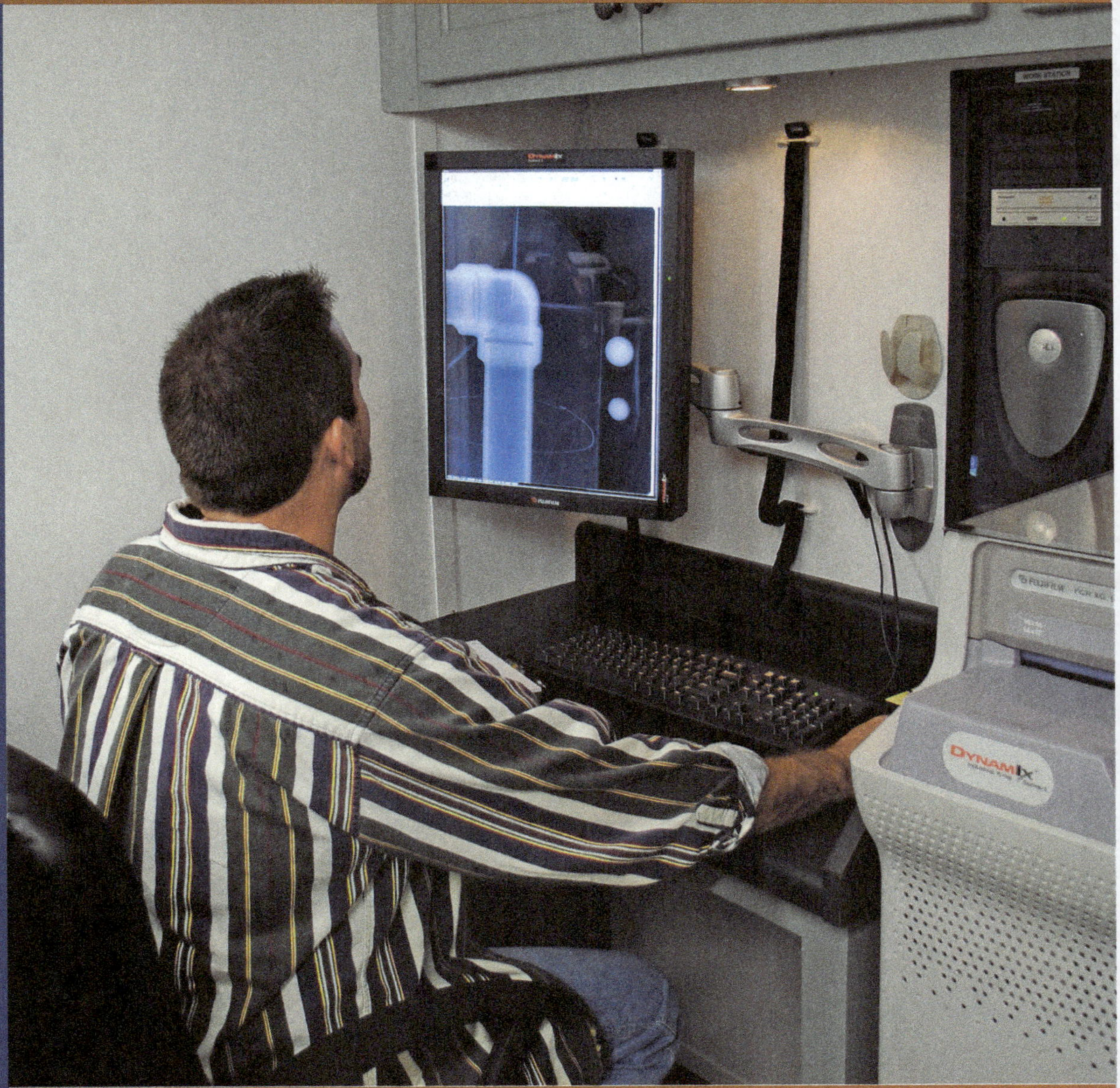

Overview

The ability to make high-quality welds that will consistently pass inspection is a necessary skill for all welders. An important part of achieving this level of skill is the ability to recognize weld imperfections, as well as their causes.

Module 29106

Trainees with successful module completions may be eligible for credentialing through the NCCER Registry. To learn more, go to **www.nccer.org** or contact us at **1.888.622.3720**. Our website has information on the latest product releases and training, as well as online versions of our *Cornerstone* magazine and Pearson's product catalog.

Your feedback is welcome. You may email your comments to **curriculum@nccer.org**, send general comments and inquiries to **info@nccer.org**, or fill in the User Update form at the back of this module.

This information is general in nature and intended for training purposes only. Actual performance of activities described in this manual requires compliance with all applicable operating, service, maintenance, and safety procedures under the direction of qualified personnel. References in this manual to patented or proprietary devices do not constitute a recommendation of their use.

29106 V5

29106
Weld Quality

Objectives

When you have completed this module, you will be able to do the following:

1. Identify and describe the various code organizations that apply to welding and their basic elements.
   a. Identify the various welding code organizations and their sponsoring organizations.
   b. Identify and describe the basic provisions of welding codes.
2. Identify and describe weld discontinuities and their causes.
   a. Identify and describe discontinuities related to porosity and inclusions.
   b. Identify and describe discontinuities that result in cracking.
   c. Identify and describe discontinuities related to joint penetration, fusion, and undercutting.
   d. Identify and describe acceptable and unacceptable weld profiles.
3. Describe various nondestructive and destructive weld examination practices.
   a. Describe basic visual inspection methods including measuring devices and liquid penetrants.
   b. Describe magnetic particle and electromagnetic inspection processes.
   c. Describe the radiographic and ultrasonic inspection processes.
   d. Describe destructive testing processes.
4. Describe the welder performance testing process.
   a. Describe the qualification of welders by position.
   b. Describe welder qualification testing to meet American Welding Society (AWS) and American Society of Mechanical Engineers (ASME) requirements.
   c. Describe the process for completing a weld test.

Performance Task

Under the supervision of your instructor, you should be able to do the following:

1. Perform a visual inspection (VT) on a fillet and/or groove weld and complete an inspection report.

Trade Terms

Arc blow
Brazing
Code
Defect
Discontinuity
Embrittled
Essential variables
Hardenable materials
Homogeneity
Inclusion
Laminations
Non-essential variables
Notch toughness
Procedure qualification record (PQR)
Radiographic
Specification
Standard
Supplemental essential variables
Transducer
Underbead cracking
Welding procedure qualification

Industry Recognized Credentials

If you are training through an NCCER-accredited sponsor, you may be eligible for credentials from NCCER's Registry. The ID number for this module is 29106. Note that this module may have been used in other NCCER curricula and may apply to other level completions. Contact NCCER's Registry at 888.622.3720 or go to **www.nccer.org** for more information.

Contents

Topics to be presented in this module include:

Figures and Tables

Figures and Tables (Continued)

SECTION ONE

1.0.0 WELDING CODES AND PROVISIONS

Objective

Identify and describe the various code organizations that apply to welding and their basic elements.

a. Identify the various welding code organizations and their sponsoring organizations.
b. Identify and describe the basic provisions of welding codes.

Trade Terms

Brazing: A method of joining metal using heat and a filler metal with a melting point above 842°F (450°C). Unlike welding, the base metal is not melted during the brazing process.

Code: A document that establishes the minimum requirements for a product or process. Codes can be, and often are, adopted as laws.

Essential variable: Items in a welding procedure specification (WPS) that cannot be changed without requalifying the WPS.

Non-essential variable: Items in a welding procedure specification (WPS) that can be changed without requalifying the WPS.

Notch toughness: The ability of a material to absorb energy in the presence of a flaw such as a notch.

Procedure qualification record (PQR): The document containing the results of the nondestructive and destructive testing required to qualify a welding procedure specification (WPS).

Specification: A document that defines in detail the work to be performed or the materials to be used in a product or process.

Standard: A document that defines how a code is to be implemented.

Supplemental essential variable:: The variables that must be considered when notch toughness requirements are invoked.

Welding procedure qualification:: A demonstration through testing that welds made following a specific process can meet prescribed standards.

Welding criteria have been established in codes and standards produced by a number of organizations. In addition, each country may have its own welding standards. Welding codes govern welding activities, qualification requirements, and tests that can be performed on weldments to identify imperfections in welds. Standards define how the code requirements are to be achieved. Welders must be familiar with the codes and standards that apply to their work.

1.1.0 Codes and Standards Governing Welding

Welding work is governed by codes, standards, and specifications. A welding code is a set of requirements covering permissible materials, service limitations, fabrication, inspection, testing procedures, and qualification of welders. Welding codes ensure that safe and reliable welded products are produced and that persons associated with the welding operation are safe. Codes are often adopted into law. A standard is a document that defines how a code is to be implemented. Standards are developed by organizations that bring together professionals from all areas of an industry to create standards for that industry. Specifications are detailed instructions for producing a product or performing a particular task. One important difference to note is that standards represent the cooperative effort of people from all parts of an industry, and are published and maintained by organizations established for that purpose. Specifications, on the other hand, can be produced by an individual company or organization without industry consensus and can apply to a single project.

Clients specify in the contract the codes, standards, and specifications that will be used on the project. Since there are a number of codes and each is updated periodically, welders must know which welding codes and code year apply to the project on which they are working. In addition to codes and standards, there are generally specifications that apply to each job. A specification is a document that defines in detail how the work is to be performed, but it is not intended to replace codes and standards.

Codes and standards that apply to welding safety and quality have been developed and are published by a number of nationally recognized agencies. To eliminate the necessity of writing a code for each new job, sections of these existing codes are referenced by the project contract. Agencies and societies that have established widely used welding codes and standards in the United States include the following:

- American Society of Mechanical Engineers (ASME)
- American Welding Society (AWS)
- American Petroleum Institute (API)
- American National Standards Institute (ANSI)

It is beyond the scope of this module to present all the details of the codes and standards organizations mentioned here. However, some of the most important will be discussed along with their impact on the industry. There are also a number of other organizations that have developed codes and standards used internationally. Some of those organizations will also be introduced in this section.

1.1.1 American Society of Mechanical Engineers

The American Society of Mechanical Engineers publishes two codes that welders must be aware of: the *ASME Boiler and Pressure Vessel Code and ASME B31, Code for Pressure Piping.* Both of these codes are endorsed by the American National Standards Institute.

The *ASME Boiler and Pressure Vessel Code (BPVC)* contains eleven sections. The sections most frequently referenced by welders are as follows:

- *Section II, Material Specifications* – This section contains the specifications for acceptable ferrous (Part A) and nonferrous (Part B) base metals and for acceptable welding and brazing filler metals and fluxes (Part C). Many of these specifications are identical to and have the same number designation as AWS specifications for welding consumables. This section is used to match base metals and filler metals.
- *Section V, Nondestructive Examination* – This section covers the methods and standards for nondestructive examination of boilers and pressure vessels.
- *Section IX, Welding and Brazing Qualifications* – This section covers the qualification of welders, welding operators, brazers, and **brazing** operators. It also covers the welding and brazing procedures that must be used for welding or brazing boilers or pressure vessels. This section of the code is often cited in other codes and standards as the welding qualification standard.

ASME B31, Code for Pressure Piping, consists of eight sections. Each section gives the minimum requirements for the design, materials, fabrication, erection, testing, and inspection of a particular type of piping system. In particular, *B31.1, Power Piping,* covers power and auxiliary service systems for electric generation stations. *B31.3, Process Piping,* covers chemical plant and petroleum refining piping. Other sections include the following:

- *B31.2, Fuel Gas Piping*
- *B31.4, Pipeline Transportation Systems for Liquid Hydrocarbons and Other Liquids*
- *B31.5 Refrigeration Piping and Heat Transfer Components*
- *B31.8 Gas Transmission and Distribution Piping Systems*
- *B31.9 Building Services Piping*
- *B31.11 Slurry Transportation Piping Systems*

All sections of *ASME B31, Code for Pressure Piping,* require qualification of the welding procedures and testing of welders and welding operators. Some sections require these qualifications to be performed in accordance with Section IX of the *ASME Boiler and Pressure Vessel Code,* while in others it is optional.

The Importance of Welding Inspection

It is important that you learn to recognize the various types of weld discontinuities and identify their causes so that you can be the first line of defense against problem welds.

Welding is used to join metals in many critical applications, including buildings, bridges, pipelines, motor vehicles, and heavy machinery. Failure of a weld in any of these applications could mean loss of life or major property damage. For this reason, welding inspectors will frequently check your work to make sure your welds meet the standards established for the project. The greater the potential hazard or cost risk, the more frequent and intense the inspections will be.

Inspection Standards

Among the many documents and standards published by the American Welding Society are documents covering weld inspection. These include the following:

- *AWS B1.10, Guide for the Nondestructive Inspection of Welds*
- *AWS B1.11, Guide for the Visual Inspection of Welds*
- *AWS B5.1, Specification for the Qualification of Welding Inspectors*
- *AWS Welding Inspection Technology*

1.1.2 American Welding Society

The American Welding Society publishes numerous welding-related documents. These documents include codes, standards, specifications, recommended practices, and guides. *AWS D1.1, Structural Welding Code – Steel,* is the code most frequently referenced. It covers welding and qualification requirements for welded structures of carbon and low-alloy steels. It is not intended to apply to pressure vessels, pressure piping, or base metals less than ⅛" (<3.2 mm metric plate) thick. Other commonly used AWS welding codes include the following:

- *AWS D1.2 Structural Welding Code – Aluminum*
- *AWS D1.3 Structural Welding Code – Sheet Steel*
- *AWS D1.6 Structural Welding Code – Stainless Steel*
- *AWS D1.5 Bridge Welding Code*

1.1.3 American Petroleum Institute

The American Petroleum Institute (API) publishes numerous documents in all areas related to petroleum production. *API 1104, Standard for Welding of Pipelines and Related Facilities,* applies to arc and oxyfuel gas welding of piping, pumping, transmission, and distribution systems for petroleum. It presents methods for making acceptable welds by qualified welders using approved welding procedures, materials, and equipment. It also presents methods for ensuring proper analysis of weld quality.

1.1.4 American National Standards Institute

The American National Standards Institute (ANSI) is a private organization that does not actually prepare standards. Instead, it adopts standards that it feels are of value to the public interest. ANSI standards deal with dimensions; ratings; terminology and symbols; test methods; and performance, as well as safety specifications for materials, equipment, components, and products in many fields, including construction. Many codes used today have been adopted as ANSI standards.

1.1.5 Other Standards Organizations

A number of national and international organizations publish standards and specifications governing welding. These organizations include the following:

- *International Standards Organization (ISO)* – This is an independent, non-governmental standards organization with 163 member countries. ISO identifies itself as the world's largest developer of voluntary standards, including nearly 300 welding standards.
- *ASTM International* – Formerly known as the American Society for Testing and Materials, ASTM International develops standards aimed at improving product quality and safety and facilitating market access and trade.
- *American Society for Nondestructive Testing (ASNT)* – ASNT is an organization for professionals involved in nondestructive testing. ASNT provides a forum for the exchange of NDT-related technical information; provides educational materials and programs; and publishes standards related to NDT. The key ASNT standards are as follows:
  - *ANSI/ASNT CP-189, ASNT Standard for Qualification and Certification of Nondestructive Testing Personnel*
  - *ANSI/ASNT CP-105, ASNT Standard Topical Outlines for Qualification of Nondestructive Testing Personnel*

Many countries produce their own standards. Examples include the GOST in Russia; the Canadian Welding Bureau; the European Union (CEN), and many others.

Did You Know?

PHMSA

The US Department of Transportation (DOT) established the Pipeline and Hazardous Materials Safety Administration (PHMSA) in 2004 to develop and enforce regulations for the safe, reliable, and environmentally sound operation of the 2.6 million miles of pipeline operated by 3,000 companies in the United States. Among other PHMSA missions is the inspection of pipelines to ensure compliance with safety regulations.

1.1.6 Maritime Welding Guides and Specifications

The AWS website lists the guides and specifications associated with marine welds and provides the following summary statements for each code:

- *AWS D3.5-93R, Guide for Steel Hull Welding* – This guide provides information to users in the marine construction industry about the best and most practical methods to weld steel hulls for ships, barges, mobile offshore drilling units, and other marine vessels. It provides information on steel plates, shapes, castings, and forging, as well as their selection and weldability. It discusses welding processes and proper design for welding. Hull construction is presented in terms of preparation of materials, erection and fitting, and control of distortion. Qualification of procedures and personnel are outlined, and inspection methods are discussed. A common shipyard problem, stray current protection, is discussed as is the health and safety of the workforce. Supplementary non-mandatory appendices are provided for informational purposes.
- *AWS D3.6M, Specification for Underwater Welding* – This specification covers the requirements for welding structures or components under the surface of the water. It includes welding in both dry and wet environments. Sections 1 through 6 constitute the general requirements for underwater welding, while Sections 7 through 10 contain the special requirements applicable to four individual classes of weld: Class A – Comparable to above-water welding; Class B – For less critical applications; Class C – Where load bearing is not a primary consideration; and Class O – To meet the requirements of another designated code or specification.
- *AWS D3.7, Guide for Aluminum Hull Welding* – This guide provides information on the welding of seagoing aluminum hulls and other structures in marine construction. Included are sections on hull materials, construction preparation, welding equipment and processes, qualification requirements, welding techniques, and safety precautions.

Anyone involved in marine welding must be knowledgeable of these welding codes. When in doubt, find a current copy of the applicable code and review it. In addition, welders working on US Navy shipbuilding projects must recognize that the Navy has their own welding specifications, which can differ significantly from those used in commercial work. The main point to remember is to always ensure that the work is being done to the correct codes, standards, and specifications.

1.1.7 Quality Workmanship

The codes and standards discussed in this module were written to ensure that welders consistently make quality welds. Many weldments will be examined and tested. However, due to time and cost, it is not feasible to examine every one, so inspection is done on a sampling basis. Regardless, quality workmanship is expected in every weld a welder makes.

The welder should be able to work with site representatives to ensure that quality work is achieved. To do this, the site organizational structure needs to be understood. If quality problems arise, the welder needs to follow the appropriate chain of command to eliminate any problems. However, there may be instances when the welder should bypass the chain of command. Examples of such instances include the following:

- You have been directed to perform an unsafe act. If you cannot resolve the matter with your immediate supervisor, it is your responsibility to go to the general foreman, superintendent, project manager, or safety engineer.
- You have been directed to perform a weld that requires a specific certification, and you are not certified to perform it. If you cannot resolve the matter with your immediate supervisor, it is your responsibility to go to the general foreman, superintendent, project manager, or quality engineer.

1.2.0 Basic Provisions of Welding Codes

A major function of welding codes is to establish qualification for operators to perform a particular welding operation. All welding codes provide detailed information about qualification in the following general areas:

- Welding procedure qualification
- Welder performance qualification
- Welding operator qualification

Machine welding is covered in some codes but is not common to all codes. Each type of qualification mentioned is different and is subject to different requirements.

> **NOTE**
> The information in this module is provided as a general guideline only. Check with your supervisor if you are unsure of the codes and specification requirements for your project.

1.2.1 Welding Procedure Qualification

A welding procedure is a written document that contains materials, methods, processes, electrode types, techniques, and all other necessary and relevant information about the weldment. Welding procedures must be qualified before they can be used. Welding procedure qualification has nothing to do with the skills of the individual welder, but deals only with the process itself.

Welding procedure qualifications are limiting instructions written to explain how a welding operation will be done. These limiting instructions are listed in a document known as a welding procedure specification (WPS). *Figure 1* shows a sample WPS. The purpose of the WPS is to define and document in detail the variables related to project-specific welds. The WPS lists the following information in detail:

- Base metals to be joined by welding
- Filler metal to be used
- Range of preheat and postheat treatment
- Thickness and other variables for each welding process

WPS variables are identified either as essential or nonessential variables. Essential variables are items in the welding procedure specification that cannot be changed without requalifying the welding procedure. Supplemental essential variables are those variables that must be considered when notch toughness requirements are invoked.

Essential variables vary by code or welding process. Refer to the specific code or welding process for the relevant essential variables. The following are some of the essential variables in a welding procedure.

- Filler metal classification
- Material thickness
- Joint design
- Type of base metal
- Welding process
- Current type
- Pre- and post-heat treatment

Nonessential variables are items in the WPS that may be changed within a range identified by the code, but that do not affect the qualification status.

Examples of nonessential variables that may be changed without having to requalify the welding procedure include the following:

- Amperage
- Travel speed
- Shielding gas flow (if applicable)
- Electrode and filler wire size
- Rod travel angle

> **CAUTION**
> Do not change any essential or supplemental essential variable without discussing it with your supervisor.

The WPS is qualified for use by welding test coupons as the WPS instructs, and then by testing the coupons in accordance with the applicable code. A test weld is made and test coupons are cut from it. The test coupons are used to make tensile tests, root bends, and face bends as required by the code. The test results are then recorded on a document known as a procedure qualification record (PQR). This documents the tested specimen, testing method, and the results. If the weldment produced by the WPS-guided procedure meets the code requirements, the procedure becomes qualified. Under most codes, each WPS must have a matching PQR to document the quality of the weld produced. *Figure 2* shows a sample PQR.

The methods used to qualify procedures are more detailed and thorough than those used to qualify either welders or welding machine operators. This is because procedures must qualify physical and metallurgical properties.

ANNEX B—SAMPLE FORMS

WELDING PROCEDURE SPECIFICATION (WPS) Yes ☐
PREQUALIFIED ________ QUALIFIED BY TESTING X
or PROCEDURE QUALIFICATION RECORDS (PQR) Yes ☐

Identification # PQR 231
Revision 1 Date 12-1-07 By W. Lye
Authorized by J. Jones Date 1-18-08
Type—Manual ☐ Semi-Automatic ☐
Machine ☐ Automatic ☐

Company Name Red Inc.
Welding Process(es) FCAW
Supporting PQR No.(s) –

JOINT DESIGN USED
Type: Butt
Single ☒ Double Weld ☐
Backing: Yes ☒ No ☐
Backing Material: ASTM A131A
Root Opening 1/4" Root Face Dimension –
Groove Angle: 35° Radius (J–U) –
Back Gouging: Yes ☐ No ☒ Method –

BASE METALS
Material Spec. ASTM A131
Type or (Grade) A
Thickness: Groove 1" Fillet –
Diameter (Pipe) –

FILLER METALS
AWS Specification A5.20
AWS Classification E71T-1

SHIELDING
Flux – Gas CO_2
Composition
Electrode-Flux (Class) ____ Flow Rate 30-40 cfh
Gas Cup Size 5/8" or 3/4"

PREHEAT
Preheat Temp., Min 75° Ambient
Interpass Temp., Min 75°F Max 350° F

POSITION
Position of Groove: 4G Fillet: –
Vertical Progression: Up ☐ Down ☐

ELECTRICAL CHARACTERISTICS
Transfer Mode (GMAW) Short-Circuiting ☐
Globular ☒ Spray ☐
Current: AC ☐ DCEP ☒ DCEN ☐ Pulsed ☐
Other
Tungsten Electrode (GTAW)
Size:
Type:

TECHNIQUE
Stringer or Weave Bead: Stringer
Multi-pass or Single Pass (per side) Multipass
Number of Electrodes 1
Electrode Spacing Longitudinal –
Lateral –
Angle –

Contact Tube to Work Distance 3/4-1"
Peening None
Interpass Cleaning: Wire Brush

POSTWELD HEAT TREATMENT
Temp. N.A.
Time

WELDING PROCEDURE

Pass or Weld Layer(s)	Process	Filler Metals: Class	Filler Metals: Diam.	Current: Type & Polarity	Current: Amps or Wire Feed Speed	Volts	Travel Speed	Joint Details
1	FCAW	E71T-1	.045"	DC+	180	26	8	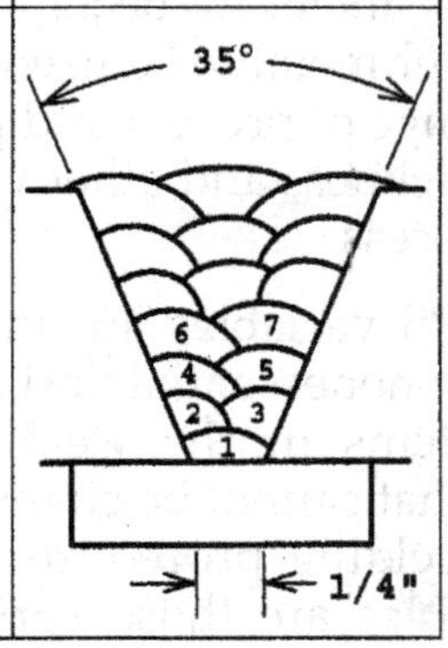
2-8	"	"	"	"	200	27	10	
9-11	"	"	"	"	200	27	11	
12-15	"	"	"	"	200	27	9	
16	"	"	"	"	200	27	11	

Form E-1 (Front)

29106-15_F01.EPS

Figure 1 Example of a welding procedure specification (WPS).

Procedure Qualification Record (PQR) # 231
Test Results

TENSILE TEST

Specimen No.	Width	Thickness	Area	Ultimate Tensile Load, lb	Ultimate Unit Stress, psi	Character of Failure and Location
231-1	.75"	1.00"	.75"	52 500	70 000	Ductile
231-3	.75"	1.00"	.75"	52 275	69 700	Ductile

GUIDED BEND TEST

Specimen No.	Type of Bend	Result	Remarks
231-2	Side	Pass	
231-4	Side	Pass	Small (< 1/16") opening acceptable
231-6	Side	Pass	
231-5	Side	Pass	

VISUAL INSPECTION
Appearance acceptable
Undercut acceptable
Piping porosity none
Convexity none
Test date 12-3-2002
Witnessed by D. Davis

Radiographic-ultrasonic examination
RT report no.: D231 Result passed
UT report no.: ______ Result ______

FILLET WELD TEST RESULTS
Minimum size multiple pass Macroetch
1. ______ 3. ______
2. ______
Maximum size single pass Macroetch
1. ______ 3. ______
2. ______

Other Tests

All-weld-metal tension test
Tensile strength, psi 83,100
Yield point/strength, psi 72,600
Elongation in 2 in., % 28
Laboratory test no. PW 231

Welder's name W. T. Williams Clock no. 261 Stamp no. ______
Tests conducted by RED Inc. & ABC Testing Laboratory
Test number PQR 231
Per D. Miller

We, the undersigned, certify that the statements in this record are correct and that the test welds were prepared, welded, and tested in conformance with the requirements of Section 4 of AWS D1.1/D1.1M, (2002 (year)) Structural Welding Code—Steel.

Signed RED Inc.
Manufacturer or Contractor
By R. M. Boncrack
Title Q.C. Mgr.
Date 12-15-2002

29106-15_F02.EPS

Figure 2 Example of a procedure qualification record (PQR).

> **GOING GREEN**
>
> **Test Coupons**
>
> When welding coupons are no longer needed or usable, they should be properly recycled, to save the company money and to keep salvageable items out of the landfills.

1.2.2 Welder Performance Qualification

Once a procedure has been qualified, the welder using it must be qualified to use that procedure by passing a welding performance qualification (WPQ) test. Because no single performance test can qualify welders for all the different types of welding that must be done, a welder may be required to pass a number of performance qualification tests. Performance tests used to qualify welders are covered later in this module.

1.2.3 Welding Machine Operator Qualification

When automatic welding equipment is used, the operators of the equipment must demonstrate their ability to set up and monitor the equipment so that it will produce acceptable welds. The codes also contain qualification tests for these operators.

Additional Resources

2014 Technical Training Guide. Current Edition. Cleveland, OH, USA: The Lincoln Electric Company. **www.lincolnelectric.com**

1.0.0 Section Review

1. Which of the following documents is often adopted into law?
   a. Specifications
   b. Codes
   c. Standards
   d. WPSs

2. A WPQ test is used to qualify a welder to perform a specific procedure.
   a. True
   b. False

SECTION TWO

2.0.0 WELD DISCONTINUITIES AND THEIR CAUSES

Objective

Identify and describe weld discontinuities and their causes.

a. Identify and describe discontinuities related to porosity and inclusions.
b. Identify and describe discontinuities that result in cracking.
c. Identify and describe discontinuities related to joint penetration, fusion, and undercutting.
d. Identify and describe acceptable and unacceptable weld profiles.

Trade Terms

Arc blow: The deflection of an arc from its intended path by magnetic forces.

Defect: A discontinuity or imperfection that renders a part of the product or the entire product unable to meet minimum acceptable standards or specifications.

Discontinuity: A change or break in the shape or structure of a part that may or may not be considered a defect, depending on the code.

Embrittled: Metal that has been made brittle and that will tend to crack with little bending.

Hardenable materials: Metals that have the ability to be made harder by heating and then cooling.

Homogeneity: The quality or state of having a uniform structure or composition throughout.

Inclusion: Foreign matter introduced into and remaining in a weld.

Underbead cracking: Cracking in the base metal near the weld, but under the surface.

Codes and standards define the quality requirements necessary to achieve the integrity and reliability of a weldment. These quality requirements help ensure that welded joints are capable of serving their intended function for the expected life of the weldment. Weld discontinuities can prevent a weld from meeting the minimum quality requirements.

AWS defines a **discontinuity** as an interruption of the typical structure of a weldment, such as a lack of **homogeneity** in the mechanical, metallurgical, or physical characteristics of the material or weldment. A discontinuity is not necessarily a **defect**. A defect found during inspection will require the weld to be rejected. A single excessive discontinuity or a combination of discontinuities can make the weldment defective (unable to meet minimum quality requirements). However, a weld can have one or more discontinuities and still be acceptable.

The welder should be able to identify discontinuities and understand the effect they have on weld integrity. Some can be determined from visual inspection. Those that are internal to the weldment can only be detected through other testing methods. The most common weld discontinuities are the following:

- Porosity
- **Inclusions**
- Cracks
- Incomplete joint penetration
- Incomplete fusion
- Undercuts
- Arc strikes
- Spatter
- Unacceptable weld profiles

Ideally, a weld should not have any discontinuities, but most will have one or more. When evaluating a weld, it is important to note the type, size, and location of the discontinuity. Any one of these factors, or all three, can change a discontinuity into a defect, requiring the weld to be rejected during the inspection process. For example, discontinuities located at stress points tend to expand and thus have higher risk than those in other locations. Surface or near-surface discontinuities may be more harmful than similarly shaped internal discontinuities.

2.1.0 Porosity and Inclusions

Porosity and inclusions are weld defects that generally result from improper welding technique. Welders must recognize the causes of these defects and the techniques required to avoid them.

2.1.1 Porosity

Porosity is the presence of voids or empty spots in the weld metal. It is the result of gas pockets being trapped in the weld as it is being made. As the molten metal hardens, the gas pockets form voids.

Unless the gas pockets work up to the surface of the weld and burst open before the metal hardens, porosity cannot be detected through visual inspection.

Porosity can be grouped into the following major types:

- *Linear porosity* – May be aligned along a weld interface, the root of a weld, or a boundary between weld beads (*Figure 3[A]*).
- *Uniformly scattered porosity* – May be located throughout single-pass welds or throughout several passes in multiple-pass welds (*Figure 3[B]*).
- *Clustered porosity* – A localized grouping of pores that result from improperly starting or stopping the welding.
- *Piping porosity* – Normally extends from the root of the weld toward the face. These elongated gas pores are also called wormholes. They do not often extend to the surface, and the porosity cannot be visually detected (*Figure 3[C]*).

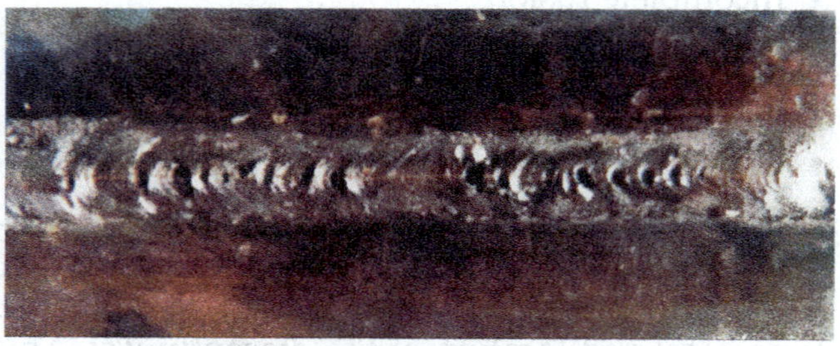

(A) LINEAR POROSITY

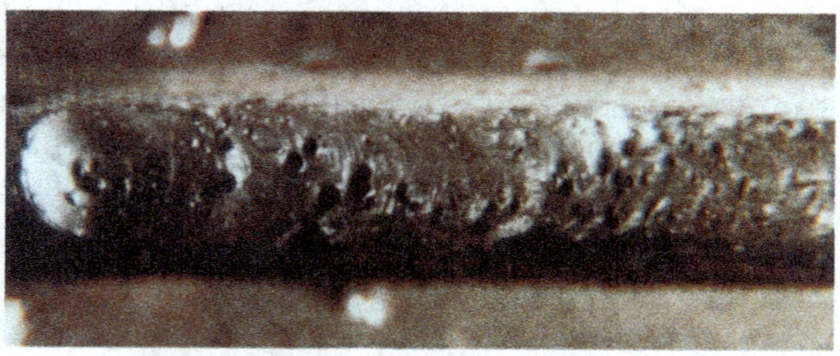

(B) SCATTERED SURFACE POROSITY

(C) PIPING POROSITY

29106-15_F03.EPS

Figure 3 Examples of porosity.

Most porosity is caused by improper welding technique or contamination. Improper welding techniques may cause an inadequate amount of shielding gas to be formed. As a result, parts of the weld site are left unprotected. Oxygen in the air at the weld site, or moisture in the flux or on the base metal that dissolves in the weld pool, can become trapped and produce porosity.

The intense heat of the weld can decompose paint, dirt, oil, or other contaminants, producing hydrogen. This gas can become trapped in the solidifying weld pool and produce porosity.

Excessive porosity has a serious effect on the mechanical properties of the joint. Although some codes permit a certain amount of porosity in welds, it is best to have as little as possible. This can be accomplished by properly cleaning the base metal, avoiding excessive moisture in the electrode covering, and using proper welding techniques. Any porosity that occurs must be ground out until it is removed.

2.1.2 Inclusions

Inclusions are foreign matter trapped in the weld metal (*Figure 4*), between weld beads, or between the weld metal and the base metal. Inclusions are sometimes jagged and irregularly shaped. Sometimes they form in a continuous line. This concentrates stresses in one area and reduces the structural integrity (strength) of the weld.

Inclusions generally result from faulty welding techniques, improper access to the joint for welding, or both. A typical example of an inclusion is slag, which normally forms over a deposited weld. If the electrode is not manipulated correctly, the force of the arc will cause some of the slag particles to be blown into the molten pool. If the pool solidifies before the inclusions can float to the top, they become lodged in the metal, producing a discontinuity. Sharp notches in joint boundaries or between weld passes also can result in slag entrapment.

Inclusions are more likely to occur in out-of-position welding because the tendency is to keep the molten pool small and allow it to solidify rapidly to prevent it from sagging. Tungsten inclusions may occur in GTAW.

With proper welding technique, along with use of the correct electrode and the proper setting, inclusions can be avoided or kept to a minimum. Other preventive measures include the following:

- Positioning the work to maintain slag control
- Changing the electrode to improve control of molten slag

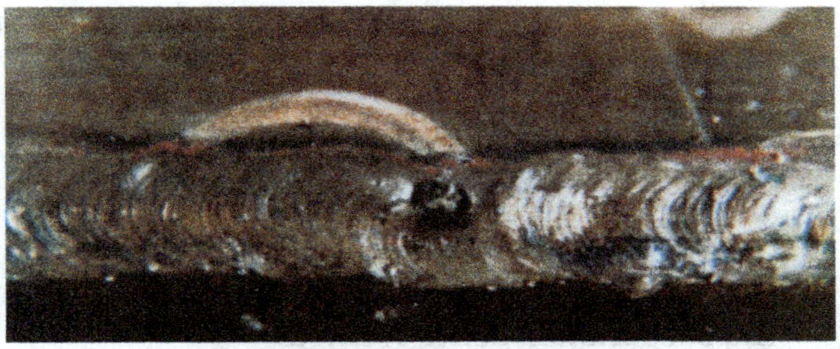

SURFACE SLAG INCLUSIONS

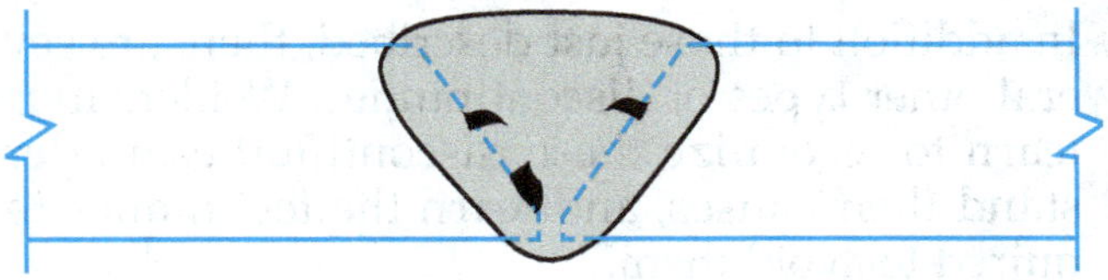

29106-15_F04.EPS

Figure 4 Examples of nonmetallic inclusions.

- Thoroughly removing slag between weld passes
- Grinding or sanding the weld surface if it is rough and likely to entrap slag
- Removing heavy mill scale or rust on weld preparations
- Avoiding the use of electrodes with damaged coverings

2.2.0 Cracks

Cracks are narrow breaks that occur in the weld metal, in the base metal, or in the crater formed at the end of a weld bead (*Figure 5*). Cracks occur when localized stresses exceed the ultimate strength of the metal. Cracks are generally located near other weld or base metal discontinuities.

2.2.1 Weld Metal Cracks

Three basic types of cracks can occur in weld metal: transverse, longitudinal, and crater. As seen in *Figure 5*, weld metal cracks are named to correspond with their location and direction.

Transverse cracks run across the face of the weld and may extend into the base metal. They are more common in joints that have a high degree of restraint.

Longitudinal cracks are usually located in the center of the weld deposit. They may be the continuation of crater cracks or cracks in the first layer of welding. Cracking of the first pass is likely to occur if the bead is thin. If this cracking is not eliminated before the other layers are deposited, the crack will progress through the entire weld deposit.

Case History

Mexico City Refinery

In Mexico City in 1984, a weld on a liquid propane gas (LPG) tank cracked. The escaping gas cloud drifted toward a nearby residential area until it encountered an ignition source. When the gas ignited, the fireball quickly burned its way back to the leak source. The resulting explosion destroyed the refinery and many residential neighborhoods, leaving 503 people dead and more than 4,000 injured.

The Bottom Line: Welds must be made correctly and inspected properly to prevent injury, loss of property, and sometimes death.

Crater cracks have a tendency to form in the crater whenever the welding operation is interrupted. These cracks usually proceed to the edge of the crater and may be the starting point for longitudinal weld cracks. Crater cracks can be minimized or prevented by filling craters to a slightly convex shape prior to breaking the welding arc.

Figure 6 shows various kinds of weld metal cracks. Weld metal cracking can usually be reduced by taking one or more of the following actions:

- Improve the contour or composition of the weld deposit by changing the electrode manipulation or electrical conditions
- Increase the thickness of the deposit and provide more weld metal to resist the stresses by decreasing the travel speed
- Reduce thermal stress by preheating
- Use low-hydrogen electrodes
- Balance shrinkage stress by sequencing welds
- Avoid rapid cooling conditions

2.2.2 Base Metal Cracks

Base metal cracking usually occurs within the heat-affected zone of the metal being welded. The possibility of cracking increases when working with **hardenable materials**. These cracks usually occur along the edges of the weld and through the heat-affected zone into the base metal. Types of base metal cracking include **underbead cracking** and toe cracking.

Underbead cracks are limited mainly to steel. They are usually found at regular intervals under the weld metal and usually do not extend to the surface. Because of this, they cannot be detected by visual inspection.

Hot and Cold Cracks

Hot cracks occur while the weld is solidifying. They can be caused by insufficient ductility at high temperature. Cold cracks occur after the weld has solidified. They are often caused by improper welding techniques.

Toe cracks are generally the result of strains caused by thermal shrinkage acting on a heat-affected zone that has been **embrittled**. They sometimes occur when the base metal cannot handle the shrinkage strains that are imposed by welding.

Base metal cracking can usually be reduced or eliminated by one of the following methods:

- Controlling the cooling rate by preheating
- Controlling heat input
- Using the correct electrode
- Controlling welding materials
- Properly matching the electrode filler metals to the base metals being welded

2.3.0 Other Discontinuities

In addition to those just described, there are several other types of discontinuities. Welders must learn to recognize these discontinuities, understand their causes, and learn the techniques required to avoid them.

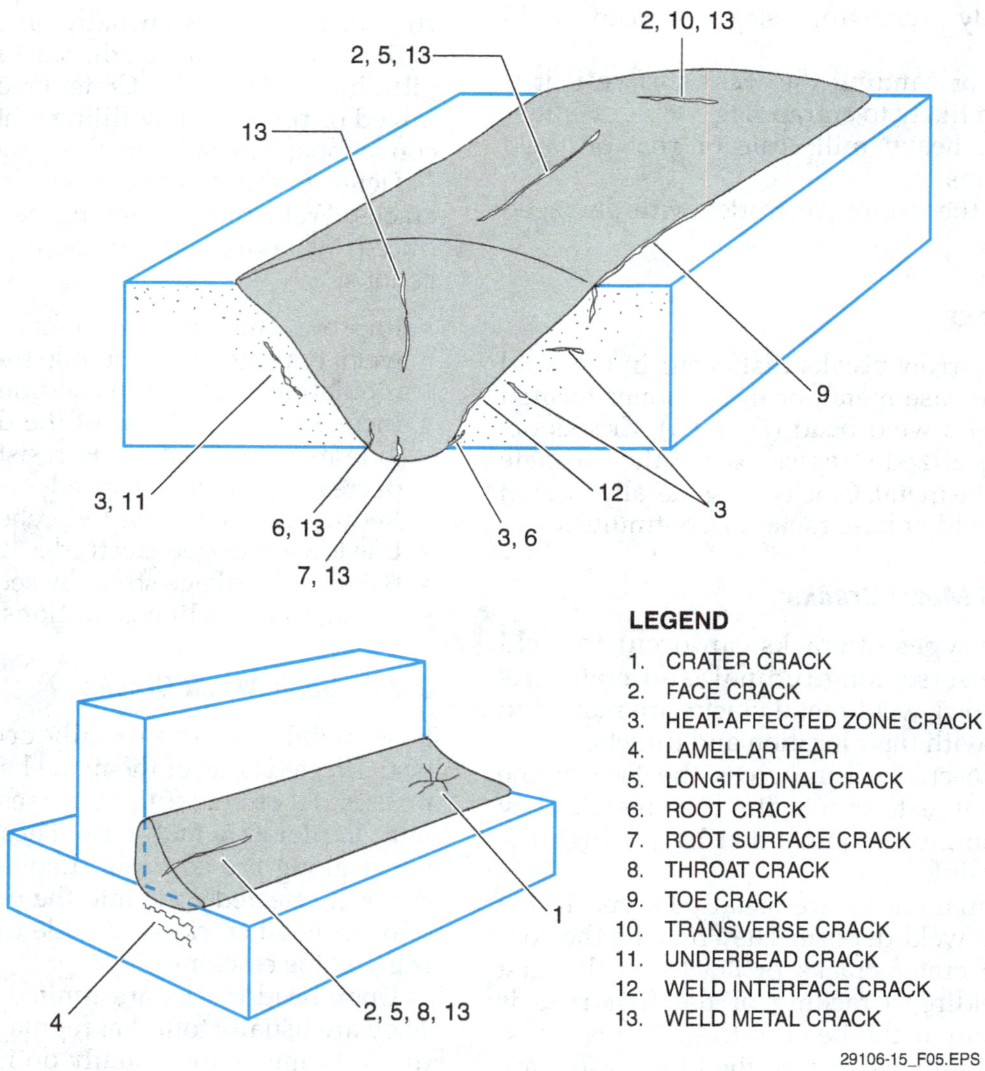

Figure 5 Types of weld metal and base metal cracks.

Figure 6 Examples of weld metal cracks.

2.3.1 Incomplete Joint Penetration

Incomplete joint penetration (*Figure 7*) occurs when the filler metal fails to penetrate and fuse with an area of the weld joint. Incomplete penetration will cause weld failure if the weld is subjected to tension or bending stresses.

Insufficient heat at the root of the joint is a frequent cause of incomplete joint penetration. If the metal being joined first reaches the melting point at the surfaces above the root of the joint, molten metal may bridge the gap between these surfaces and screen off the heat source before the metal at the root melts.

Improper joint design is another leading cause of incomplete joint penetration. If the joint is not prepared or fitted accurately, an excessively thick root face or an insufficient root gap may cause incomplete penetration. Incomplete joint penetration is likely to occur under the following conditions:

- If the root face dimension is too big, even though the root opening is adequate
- If the root opening is too small
- If the included angle of a V-groove is too small

Figure 8 is a labeled diagram showing correct and incorrect joint designs.

Even if the welding heat is correct and the joint design is adequate, incomplete penetration can result from poor control of the welding arc. Examples of poor control include the following:

- Using an electrode that is too large
- Excessive travel speed
- Using a welding current that is too low

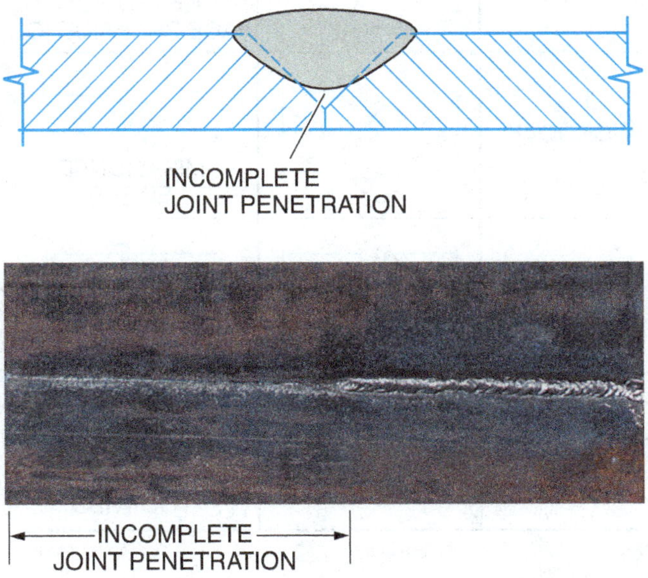

Figure 7 Incomplete joint penetration.

Incomplete penetration is always undesirable in welds, especially in single-groove welds where the root of the weld is subject either to tension or bending stresses. It can lead directly to weld failure or can cause a crack to start at the unfused area.

2.3.2 Incomplete Fusion

Many welders confuse incomplete joint penetration with incomplete fusion. It is possible to have good penetration without complete root fusion. Incomplete fusion is the failure of a welding process to fuse, or join together, layers of weld metal or weld metal and base metal.

Incomplete fusion may occur at any point in a groove or fillet weld, including the root of the weld. Often, the weld metal simply rolls over onto the plate surface. This is generally referred to as overlap. In many cases, the weld has good fusion at the root and at the plate surface, but because of poor technique and insufficient heat, the toe of the weld does not fuse. *Figure 9* shows incomplete fusion and overlap. Causes for incomplete fusion include the following:

- Insufficient heat as a result of low welding current, high travel speeds, or an arc gap that is too close
- Wrong size or type of electrode
- Failure to remove oxides or slag from groove faces or previously deposited beads
- Improper joint design
- Inadequate gas shielding
- Improper electrode angle
- **Arc blow**

Incomplete fusion discontinuities affect weld joint integrity in much the same way as porosity and slag inclusion.

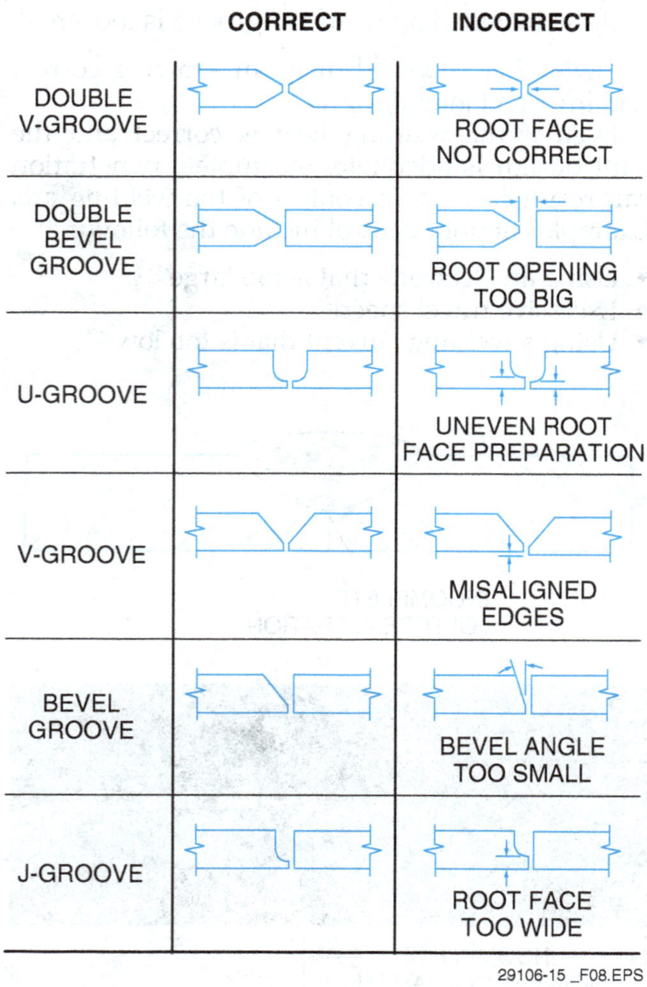

Figure 8 Correct and incorrect joint designs.

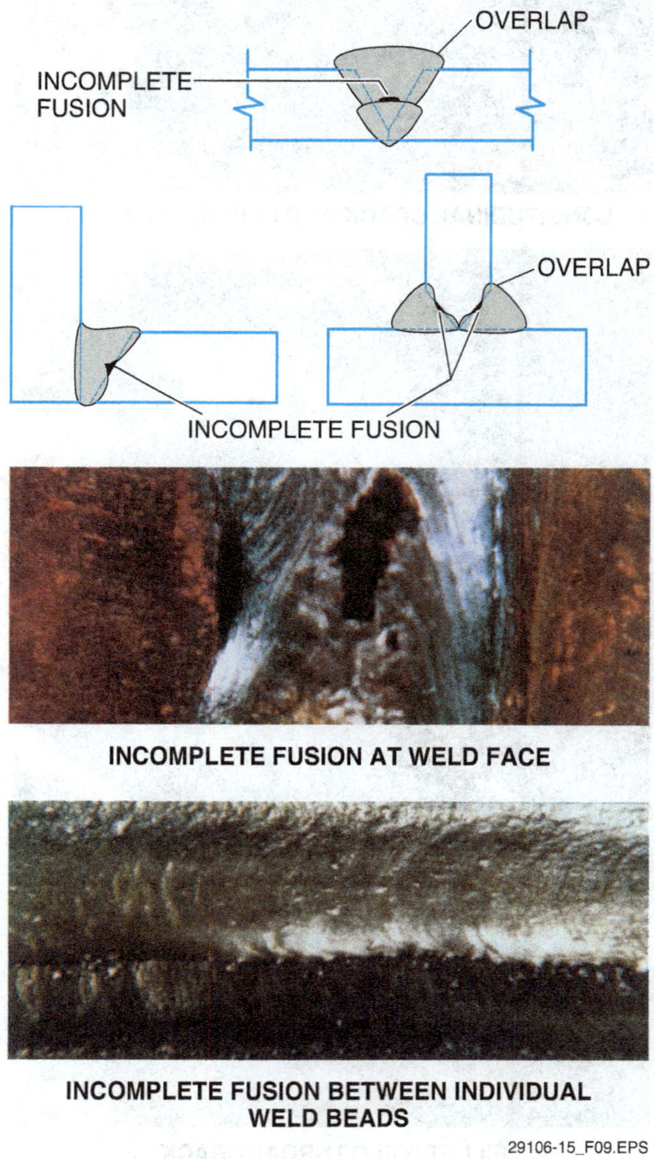

Figure 9 Incomplete fusion and overlap.

2.3.3 Undercut

Undercut is a groove melted into the base metal beside the weld. It is the result of the arc removing more metal from the joint face than is replaced by weld metal. On multilayer welds, it may also occur at the point where a layer meets the wall of a groove (*Figure 10*).

Undercutting is usually caused by improper electrode manipulation. Other causes of undercutting include the following:

- Using a current adjustment that is too high
- Having an arc gap that is too long
- Failing to fill up the crater completely with weld metal
- Improper electrode angle
- Incorrect rod travel speed

Most welds have some undercut that can be found upon careful examination. When it is controlled within the limits of the specifications and does not create a sharp or deep notch, undercut is usually not considered a weld defect. However, when it exceeds the limits, undercutting can be a serious defect because it reduces the strength of the joint.

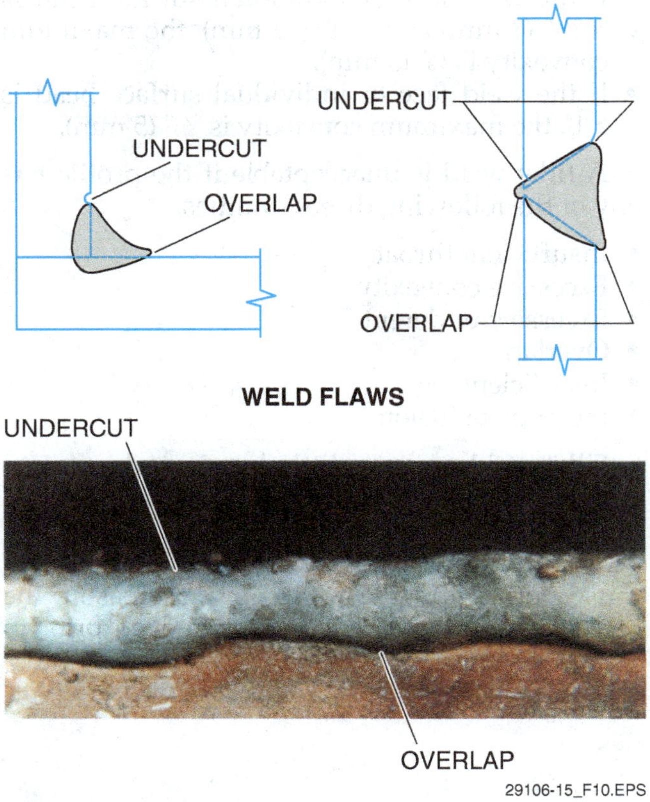

Figure 10 Undercut and overlap.

Causes of Incomplete Joint Penetration

Incomplete joint penetration is generally associated with groove welds. It may result from insufficient welding heat, improper joint design (too much metal for the welding arc to penetrate), or poor control of the welding arc.

2.3.4 Arc Strikes

Arc strikes (*Figure 11*) are small, localized points where surface melting occurs away from the joint. These spots may be caused by accidentally striking the arc in the weld zone or by faulty ground connections. The weld zone is the area where the weld bead should be, and/or where the weld metal is to be deposited.

Striking an arc on base metal that will not be fused into the weld metal should be avoided. Arc strikes can cause hardness zones in the base metal and can become the starting point for cracking. Arc strikes can cause a weld to be rejected.

2.3.5 Spatter

Spatter is made up of very fine particles of metal on the plate surface adjoining the weld area. It is usually caused by high current; a long arc; an irregular and unstable arc; or improper shielding. Spatter makes a poor appearance on the weld and base metal and can make it difficult to inspect the weld. Spatter can also cause coating failure.

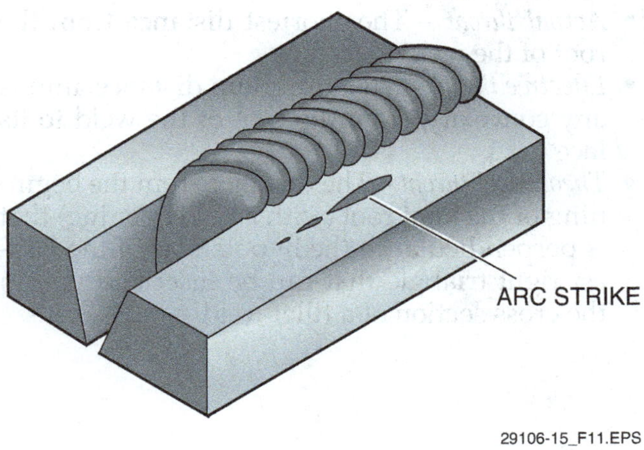

Figure 11 Arc strike.

2.4.0 Acceptable and Unacceptable Weld Profiles

The profile of a finished weld can affect the performance of the joint under load as much as other discontinuities affect it. This applies to the profile of a single-pass weld and to a layer of a multiple-pass weld. An unacceptable profile for a single-pass or multiple-pass weld could lead to the formation of discontinuities such as incomplete fusion or slag inclusions as the other layers are deposited. *Figure 12* shows acceptable and unacceptable weld profiles for both fillet and groove welds.

2.4.1 Fillet Welds

A fillet weld is a weld that is approximately triangular in cross section and is used with T-, lap, and corner joints. The sizes and locations of fillet welds are given as welding symbols. The two types of fillet welds are convex and concave (*Figure 13*). A convex fillet weld has its surface bowed out like the outside surface of a ball. A concave fillet weld has its surface bowed in like the inside surface of a bowl.

The following terms are used to describe a fillet weld:

- *Weld face* – The exposed surface of the weld
- *Leg* – The distance from the root of the joint to the toe of a fillet weld
- *Weld toe* – The junction between the face of a weld and the base metal
- *Weld root* – The point shown in cross section at which the weld metal intersects with the base metal and extends farthest into the weld joint
- *Size* – The leg lengths of the largest right triangle that can be drawn within the cross section of a fillet weld
- *Actual throat* – The shortest distance from the root of the weld to its face
- *Effective throat* – The minimum distance, minus any convexity, from the root of the weld to its face
- *Theoretical throat* – The distance from the beginning of the joint root (with a zero opening) that is perpendicular to the hypotenuse of the largest right triangle that can be inscribed within the cross section of a fillet weld.

Preferred Fillet Weld Contours

In single-pass, weave-bead fillet welds where two workpieces are being joined at an angle (not lap joints), flat or slightly convex faces are usually preferred because weld stresses are more uniformly distributed through the fillet and workpieces.

Fillet welds may be either equal leg or unequal leg (*Figure 14*). The face may be slightly convex, flat, or slightly concave. Welding codes require that fillet welds have a uniform concave or convex face, although a slightly non-uniform face is acceptable. The maximum convexity of a fillet weld or individual surface bead is dependent on the width of the weld face or individual surface bead, as follows:

- If the weld face or individual surface bead is ≤ 5⁄16" (8 mm), the maximum convexity is 1⁄16" (1.6 mm).
- If the weld face or individual surface bead is > 5⁄16" (8 mm) and < 1" (25 mm), the maximum convexity is 1⁄8" (3 mm).
- If the weld face or individual surface bead is ≥ 1", the maximum convexity is 3⁄16" (5 mm).

A fillet weld is unacceptable if the profile has any of the following discontinuities:

- Insufficient throat
- Excessive convexity
- Excessive undercut
- Overlap
- Insufficient leg
- Incomplete fusion

Fillet welds require little base metal preparation except for cleaning the weld area and removing all dross from cut surfaces. Any dross from oxyfuel, plasma arc, or carbon arc cutting will cause porosity in the weld. For this reason, the codes require that all dross be removed prior to welding.

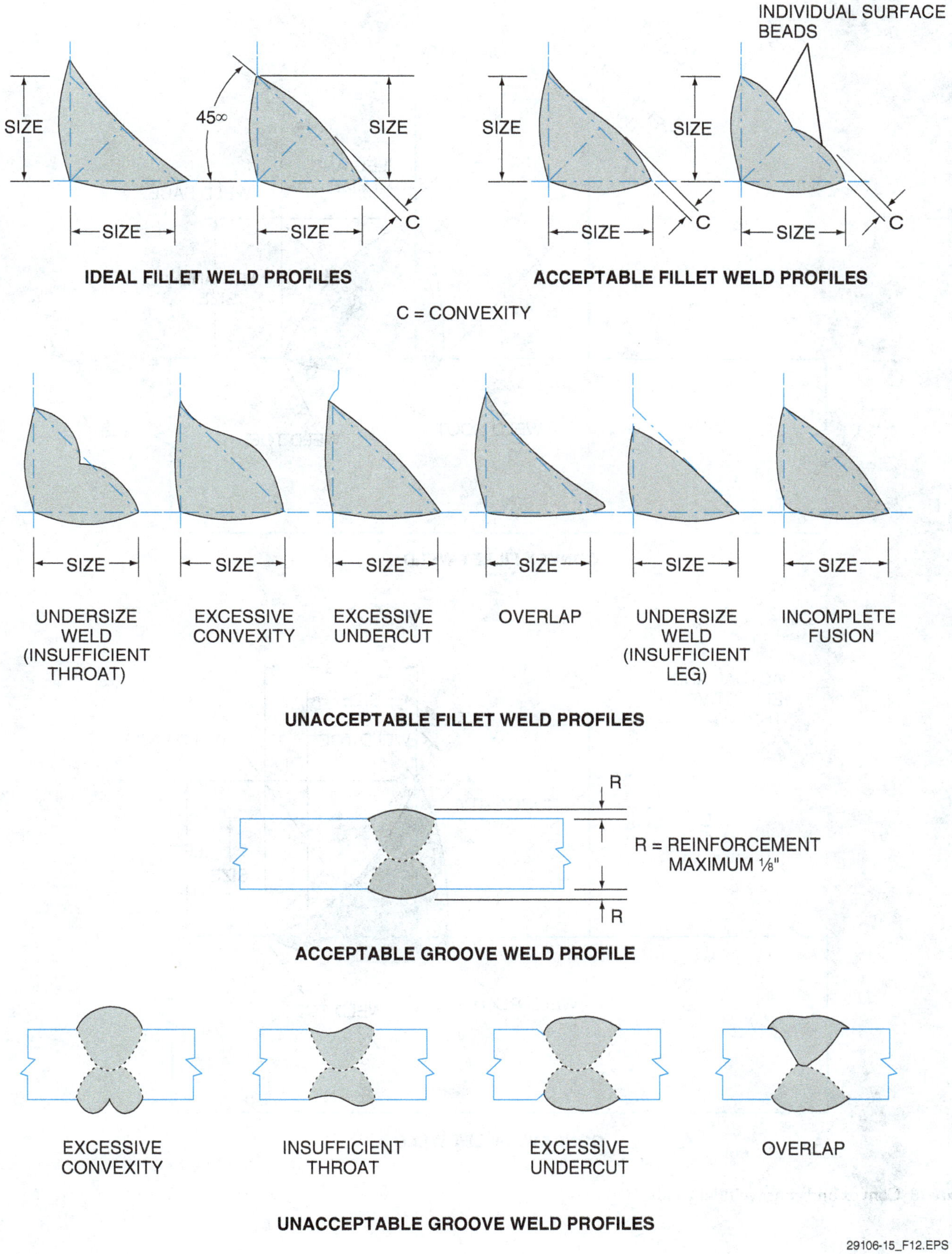

Figure 12 Acceptable and unacceptable weld profiles.

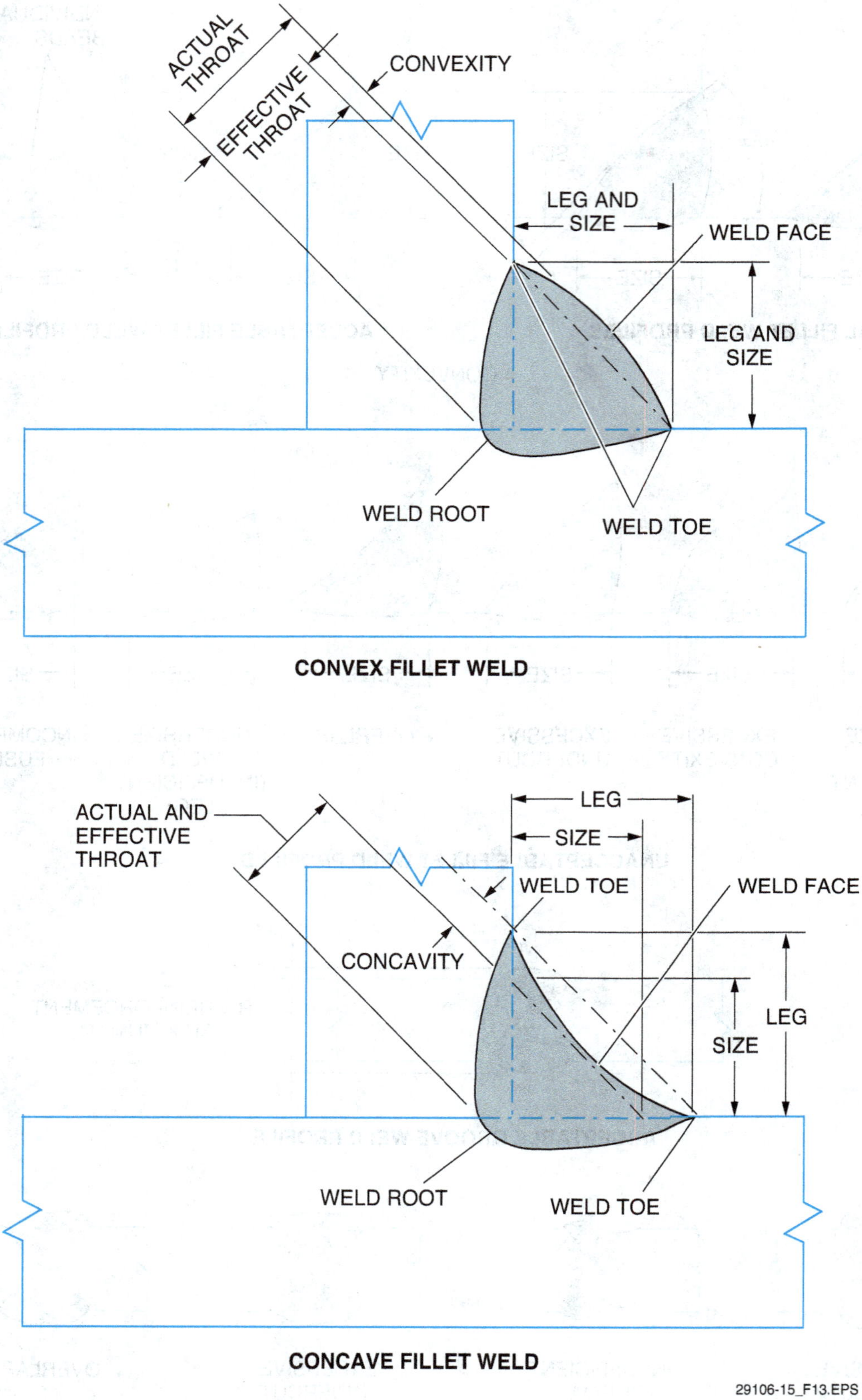

Figure 13 Convex and concave fillet welds.

2.4.2 Groove Welds

Groove welds should be made with reinforcement not exceeding ⅛" (3.2 mm) and a gradual transition to the base metal at each toe. Groove welds should not have excess reinforcement, insufficient throat, excessive undercut, or overlap. If a groove weld has any of these defects, it is considered unacceptable. The bead width should not exceed the groove width by more than ⅛" (3.2 mm).

> **NOTE**
> Refer to your site's WPS for specific requirements on fillet and butt welds. The information provided here is only a general guideline. The site WPS or quality specifications must be followed for all welds. Check with your supervisor if you are unsure of the specifications for your application.

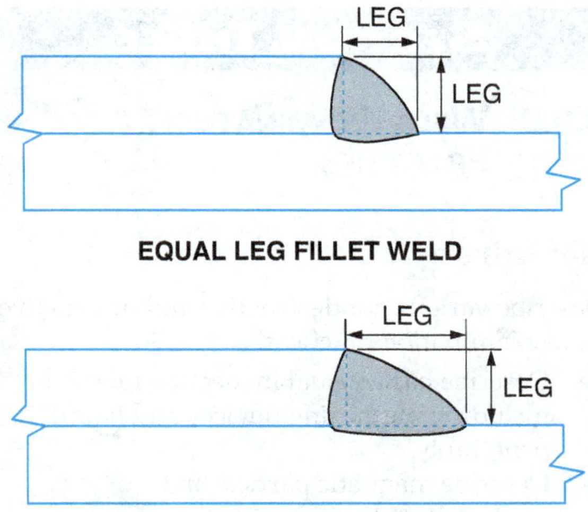

Figure 14 Equal leg and unequal leg fillet welds.

Additional Resources

Visual Inspection Workshop Reference Manual, AWS VIW-M-2008. Miami, FL: American Welding Society.

Welding Inspection Technology. Miami, FL: American Welding Society.

2.0.0 Section Review

1. The entrapment of gas pockets in a weld will cause _____.
   a. inclusions
   b. cracks
   c. porosity
   d. undercut

2. Which of these discontinuities *cannot* be detected by visual inspection?
   a. Toe cracks
   b. Underbead cracks
   c. Transverse cracks
   d. Face cracks

3. Undercut is usually caused by improper electrode manipulation.
   a. True
   b. False

4. Which of the following describes the leg of a weld?
   a. The shortest distance from the root of the weld to its face
   b. The junction between the face of the weld and the base metal
   c. The minimum distance, minus any convexity, from the root of the weld to its face
   d. The distance from the root of the joint to the toe of a fillet weld

SECTION THREE

3.0.0 WELD EXAMINATION PRACTICES

Objective

Describe various nondestructive and destructive weld examination practices.

a. Describe basic visual inspection methods including measuring devices and liquid penetrants.
b. Describe magnetic particle and electromagnetic inspection processes.
c. Describe the radiographic and ultrasonic inspection processes.
d. Describe destructive testing processes.

Performance Task

1. Perform a visual inspection (VT) on a fillet and/or groove weld and complete an inspection report.

Trade Terms

Laminations: Cracks in the base metal formed when layers separate.

Radiographic: Describes images made by passing X-rays or gamma rays through an object and recording the variations in density on photographic film.

Transducer: A device that converts one form of energy into another.

Nondestructive examination (NDE), sometimes referred to as nondestructive testing (NDT) or nondestructive inspection, is a term used for those inspection methods that allow materials to be examined without changing or destroying them. NDE methods can usually detect the discontinuities and defects previously described.

Nondestructive examination is usually performed by the site quality group as part of the site quality program. Inspectors trained in the proper test methods conduct the examinations.

The welder should be familiar with the following basic nondestructive examination practices:

- Visual inspection (VT)
- Liquid penetrant inspection (PT)
- Magnetic particle inspection (MT)
- Radiographic inspection (RT)
- Ultrasonic inspection (UT)
- Electromagnetic (eddy current) inspection (ET)
- Leak testing (LT)

There is considerable overlap in the application of nondestructive and destructive tests. Destructive tests, which destroy the weld, are frequently used on several sample weldments to supplement, confirm, or establish the limits of nondestructive tests. Destructive testing is also used to provide supporting information. Once this information has been established, nondestructive examinations can be made on similar welds to locate all discontinuities above the critical defect size that was determined by the destructive tests.

3.1.0 Visual Inspection

In visual inspection, the surface of the weld and the base metal are observed for visual imperfections. Certain tools and gauges may be used during the inspection. Visual inspection is the examination method most commonly used by welders and inspectors. It is the fastest and most inexpensive method for examining a weld. However, it is limited to what can be detected by the naked eye or through a magnifying glass.

Properly done before, during, and after welding, visual inspection can detect more than 75 percent of discontinuities before they are found by more expensive and time-consuming nondestructive examination methods.

Prior to welding, the base metal should be examined for conditions that may cause weld defects. The required dimensions of the material, including edge preparations, should also be confirmed by measurements. If problems or potential problems are found, corrections should be made before proceeding any further.

After the parts are assembled for welding, the weld joint should be visually checked for a proper root opening and any other aspects that might affect the quality of the weld. Visually check for the following conditions:

- Proper cleaning
- Joint preparation and dimensions
- Clearance dimensions for backing strips, rings, or consumable inserts
- Alignment and fit-up of the pieces being welded
- Welding procedures and machine settings
- Specified preheat temperature (if applicable)
- Tack-weld quality

Fixed Weld Fillet Gauge

This fixed weld fillet gauge can be used to measure 11 fillet weld sizes in fractions or decimals. Metric versions are also available.

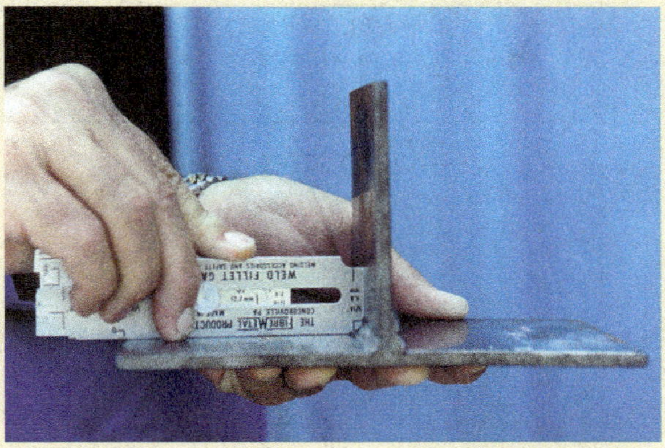

29106-15_SA01.EPS

During the welding process, visual inspection is the primary method for controlling quality. Some of the aspects that should be visually examined include the following:

- Quality of the root pass and the succeeding weld layers
- Sequence of weld passes
- Interpass cleaning
- Root preparation prior to welding a second side
- Conformance to the applicable procedure

After the weld has been completed, the weld surface should be thoroughly cleaned. A complete visual examination may disclose weld surface defects such as cracks, shrinkage cavities, undercuts, incomplete penetration, incomplete fusion, overlap, and crater deficiencies before they are discovered using other nondestructive inspection methods.

An important aspect of visual examination is checking the dimensional accuracy of the weld after it has been completed. Dimensional accuracy is determined by the use of measuring gauges. The purpose of using the gauges is to determine if the weld is within allowable limits as defined by the applicable codes and specifications.

Some of the more common welding gauges are the following:

- Undercut gauge
- Butt weld reinforcement gauge
- Fillet weld blade gauge set

3.1.1 Undercut Gauge

An undercut gauge is used to measure the amount of undercut on the base metal. Typically, codes allow for undercut to be between 0.01" and 0.031" deep. These gauges have a pointed end that is pushed into the undercut. The reverse side of the gauge indicates the measurement in either inches or millimeters. Two types of undercut gauges currently used are the bridge cam gauge and the V-WAC gauge (*Figure 15*). These gauges can be used for measuring undercut and for many other measurements.

3.1.2 Butt Weld Reinforcement Gauge

The butt weld reinforcement gauge has a sliding pointer calibrated to several different scales that are used to measure the size of a fillet weld or the reinforcement of a butt weld. To use the gauge for a fillet weld, position it as shown in *Figure 16*, and then slide the pointer to contact the base metal or weld metal. Be sure to read the correct scale for the measurement being taken. The other end of the gauge is used for butt welds.

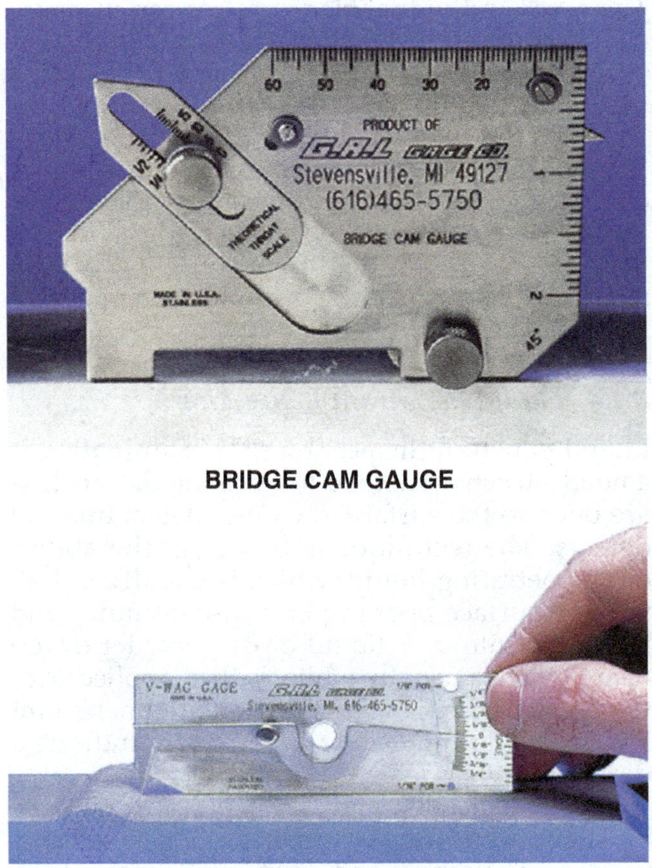

29106-15_F15.EPS

Figure 15 Undercut gauges.

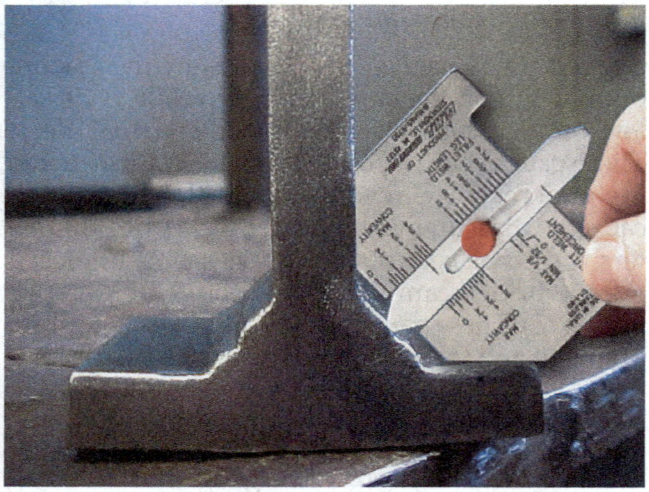

29106-15_F16.EPS

Figure 16 Automatic weld size gauge (AWS).

3.1.3 Fillet Weld Blade Gauge Set

The fillet weld blade gauge set has seven individual blade gauges for measuring convex and concave fillet welds. The individual gauges are held together by a screw secured with a knurled nut. The seven individual blade gauges can measure eleven concave and convex fillet weld sizes: ⅛", $^{3}/_{16}$", ¼", $^{5}/_{16}$", ⅜", $^{7}/_{16}$", ½", ⅝", ¾", ⅞", and 1" as well as their metric equivalents.

To use the fillet weld blade gauge set, identify the type of fillet weld to be measured (concave or convex) and the size. Select the appropriate blade and position it. Be sure the gauge blade is flush to the base metal with the tip touching the vertical member. *Figure 17* shows an application of a fillet weld blade gauge.

3.1.4 Liquid Penetrant Inspection

Liquid penetrant inspection (PT) examination is a nondestructive method for locating defects that are open to the surface. It cannot detect internal defects. The technique is based on the ability of a penetrating liquid, which is usually red, to wet the surface opening of a discontinuity and be drawn into it. A liquid or dry powder developer, which is usually white, is then applied over the metal. If the flaw is significant, red penetrant bleeds through the white developer to indicate a discontinuity or defect.

The dye, cleaner, and developer are available in spray cans for convenience. Some solvents used in the cleaners and developers contain high amounts of chlorine, a known health hazard, to make the liquids nonflammable.

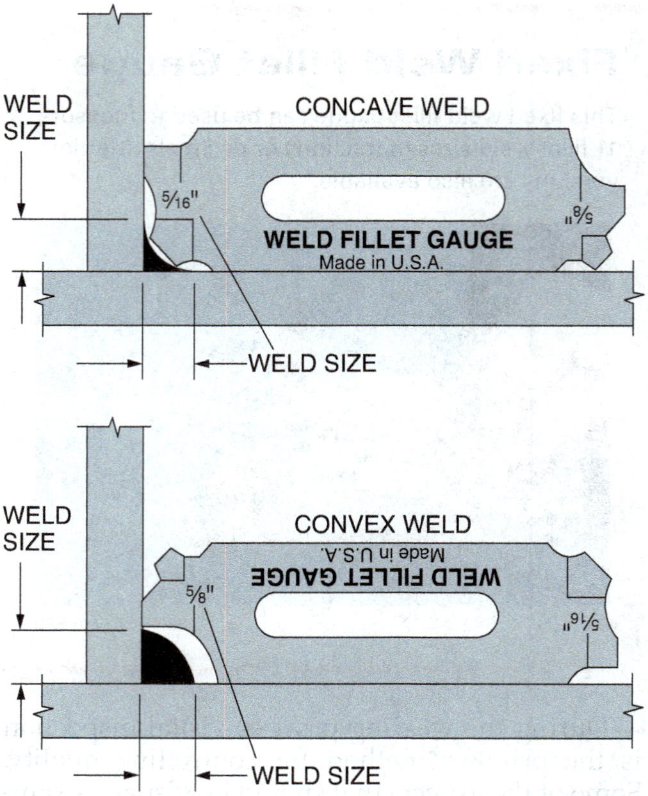

29106-15_F17.EPS

Figure 17 Fillet weld blade gauge.

WARNING!

Refer to the safety data sheet (SDS/MSDS) for hazards associated with the liquid penetrant solvent.

The most common defects found using this process are surface cracks. Most cracks exhibit an irregular shape. The width of the bleed-out (the red dye bleeding through the white developer) is a relative measure of the depth of a crack.

Surface porosity, metallic oxides, and slag will also hold penetrant and cause bleed-out. These indications are usually more circular and have less width than a crack. *Figure 18* shows liquid penetrant materials and an example of the results of liquid penetrant inspection.

The advantages of liquid penetrant inspection are that it can find small defects not visible to the naked eye; it can be used on most types of metals; it is inexpensive; and it is fairly easy to use and interpret. PT is most useful in examining welds that are susceptible to surface cracks. Except for visual inspection, it is perhaps the most commonly used nondestructive examination method for surface inspection.

29106-15_F18.EPS

Figure 18 Liquid penetrant materials and inspection example.

The disadvantages of liquid penetrant inspection are that it takes more time to use than visual inspection, and it can only find surface defects. The presence of weld bead ripples and other irregularities can also hinder the interpretation of indicators. Because chemicals are used, care must be taken when performing the inspection. When testing rough, irregular surfaces, the presence of irrelevant indicators may also make interpretation difficult.

3.1.5 Leak Testing

Leak testing (*Figure 19*) is used to determine the ability of a pipe or vessel to contain a gas or liquid under pressure. Testing methods vary depending on the application of the weldment. In some cases, the vessel is pressurized and tested by immersing it in water or by applying a soap bubble solution to the weld. An open tank can be tested using water that contains fluorescein, which can be detected by ultraviolet light.

A method called the vacuum box test is used to test a vessel when only one side of the weld is accessible. The base of a storage tank is an example. The vacuum box is a transparent box with a soft rubber seal. A vacuum pump is used to extract all the air from the box. A leak is indicated by the presence of bubbles.

In the helium spectrometer leak test, helium is used as a tracer gas. Because of the small size of helium atoms, they can pass through an opening so small that it might not be detectable by other test methods. Sensitive instruments are used to detect the presence of helium.

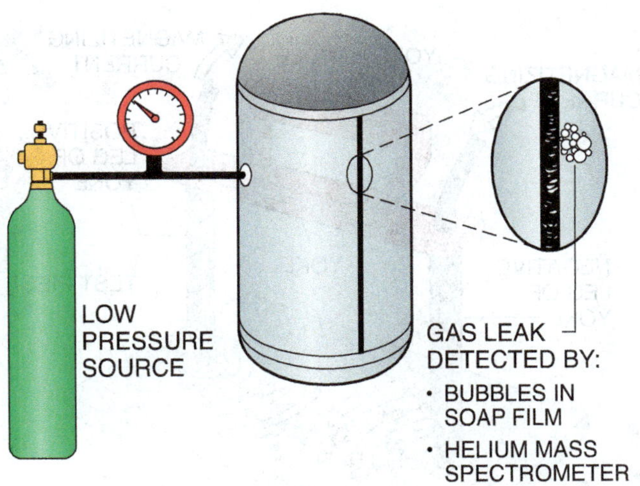

29106-15_F19.EPS

Figure 19 Example of a leak test.

3.2.0 Magnetic Particle and Electromagnetic Inspection

These non-destructive methods make use of electromagnetism to perform weld examination. Magnetic particle inspection methods are only effective when used with ferrous metals. Electromagnetic inspection can be used with non-ferrous metals as well as ferrous metals.

3.2.1 Magnetic Particle Inspection

Magnetic particle inspection (MT) is a nondestructive examination method that uses electricity to magnetize the weld that will be examined (*Figure 20*). After the metal has been magnetized, metal particles are sprinkled onto the weld surface. If there are defects in the surface or just below it, the metal particles will be grouped into a pattern around the defect. The defect can be identified by the shape, width, and height of the particle pattern.

Magnetic particle inspection is used to test welds for such defects as surface cracks, incomplete fusion, porosity, and slag inclusion. It can also be used to inspect plate edges for surface imperfections prior to welding. Defects can be detected only at or near the surface of the weld. Defects much deeper than this are not likely to be found. Certain discontinuities exhibit characteristic powder patterns that can be identified by a skilled inspector.

For magnetic particle examination, the part to be inspected must be ferromagnetic (made of steel or a steel alloy), smooth, clean, dry, and

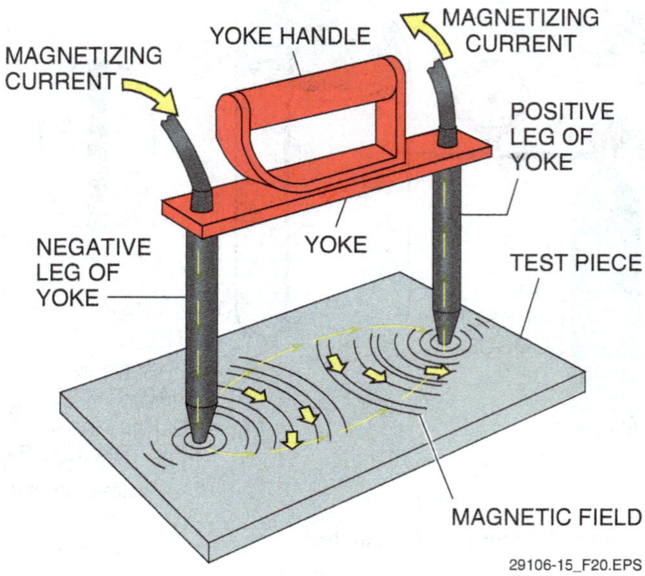

Figure 20 Magnetic particle examination with electromagnetic yokes.

free from oil, water, and excess slag. The part is magnetized by using an electric current to set up a magnetic field within the material. The magnetized surface is covered with a thin layer of magnetic powder. If there is a defect, the powder is held to the surface at the defect because of the powerful magnetic field.

When this examination method is used, there is normally a code or standard that governs both the method and the acceptance/rejection criteria of indications.

The advantages of magnetic particle inspection are that it can find small defects not visible to the naked eye, and it is faster than liquid penetrant inspection.

Disadvantages of magnetic particle inspection are that the materials must be capable of being magnetized; the inspector must be skilled in interpreting indications; rough surfaces can interfere with the results; the method requires an electrical power source; and it cannot identify internal discontinuities located deep in the weld.

3.2.2 Electromagnetic (Eddy Current) Inspection

Like magnetic particle testing, electromagnetic, or eddy current inspection (ET) uses electromagnetic energy to detect defects in the joint. A coil, which produces a magnetic field, is placed on or around the part being tested. After being calibrated, the coil is moved over the part to be inspected. The coil produces a current in the metal through induction. The induced current is called an eddy current. If a discontinuity is present in the test part, it will interrupt the flow of the eddy currents. This change can be observed on the oscilloscope display.

Eddy currents only detect discontinuities near the surface of the part. This method is suitable for both ferrous and nonferrous materials, and it is used in testing welded tubing and pipe. It can determine the physical characteristics of a material and the wall thickness in tubing. It can also check for porosity, pinholes, slag inclusions, internal and external cracks, and incomplete fusion.

The advantage of eddy current inspection is that it can detect surface and near-surface weld defects. It is particularly useful in inspecting circular parts like pipes and tubing.

The disadvantage of eddy current inspection is that eddy currents decrease with depth, so defects farther from the surface may go undetected. The accuracy of the examination depends in large part on the calibration of the instrument and the ability of the inspector. *Figure 21* shows an eddy current test being conducted.

29106-15_F21.EPS

Figure 21 Eddy current test.

3.3.0 Radiographic and Ultrasonic Inspection

These non-destructive inspection methods rely on technology that is commonly used in medical diagnostics. The ability to apply these methods and to correctly interpret the results requires extensive training and experience.

3.3.1 Radiographic Inspection

Radiography (RT) is a nondestructive examination method that uses radiation (X-rays or gamma rays) to penetrate the weld and produce an image. Like medical X-rays, this weld inspection technology originally worked by exposing treated film to the X-rays. In traditional RT work, X-ray film is placed against the weld and exposed (*Figure 22*). When a joint is radiographed, the radiation source is placed on one side of the weld and the film on the other. The joint is then exposed to the radiation source. The radiation penetrates the metal and produces an image on the film. The film is called a radiograph and provides a permanent record of the weld quality.

Modern RT offers two additional options: digital detector array (DDA) and computed radiography (CR). DDA radiography, also referred to as digital radiography, uses a semiconductor array instead of silver-based film to capture the image. Like film radiography, digital radiography is considered a form of direct radiography. Once the DDA is exposed, the image it captures can be transferred directly to a computer and enhanced for viewing. DDA radiography uses the same type of X-ray source as film radiography.

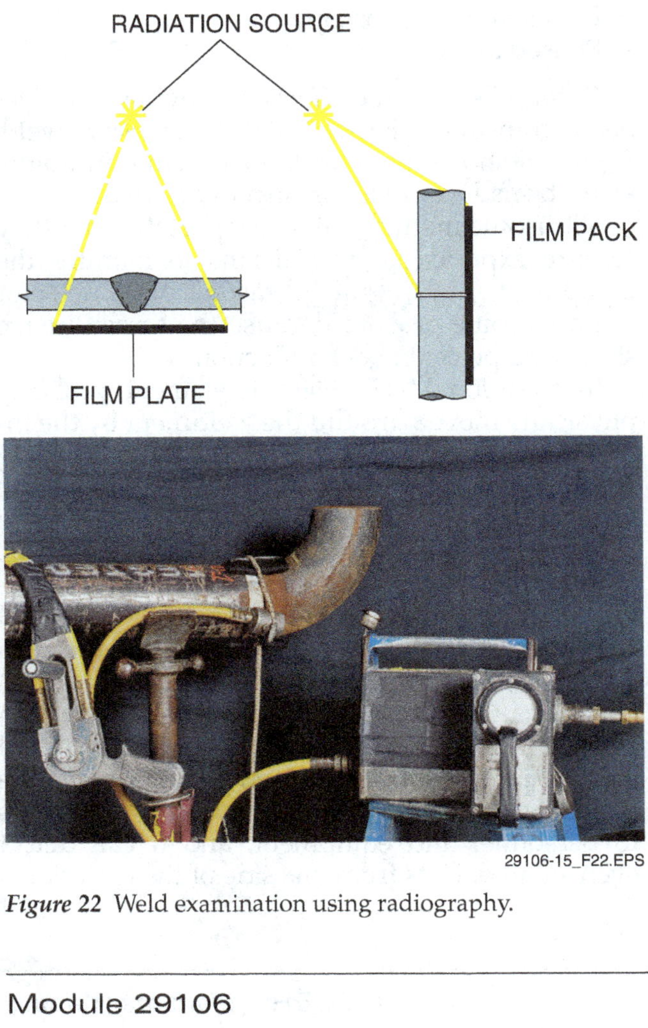

29106-15_F22.EPS

Figure 22 Weld examination using radiography.

CR generally uses a cassette that contains a reusable imaging plate. Once the plate is exposed, it is processed through a special laser scanner that captures the image and allows it to be transferred to a computer for enhancement and interpretation. *Figure 23* shows CR radiography being performed.

Radiographic inspection can produce a visible image of weld discontinuities, both surface and subsurface, when they are different in density from the base metal and different in thickness parallel to the radiation. Surface discontinuities are better identified by visual, penetrant, or magnetic particle examination. Radiographic weld inspection requires specialized skill. It should therefore be done and interpreted only by trained, qualified personnel.

The advantages of radiographic inspection are that it provides a permanent record of the weld quality, stored either on film or as a computer image. Through radiography, the entire thickness can be examined, and it can be used on all types of metals.

The disadvantages of radiographic inspection are that it is a slow and expensive method for inspecting welds; some joints are inaccessible to radiography; and excessive exposure to radiation of any type is very hazardous to humans. Cracks can frequently be missed if they are very small or are not aligned with the radiation beam.

3.3.2 Ultrasonic Inspection

Ultrasonic inspection (UT) is a relatively low-cost, nondestructive examination method that uses sound wave vibrations to find surface and subsurface defects in the weld material. Ultrasonic waves are passed through the material being tested and are reflected back by any density change caused by a defect. The reflected signal is shown on the screen display of the instrument (*Figure 24*).

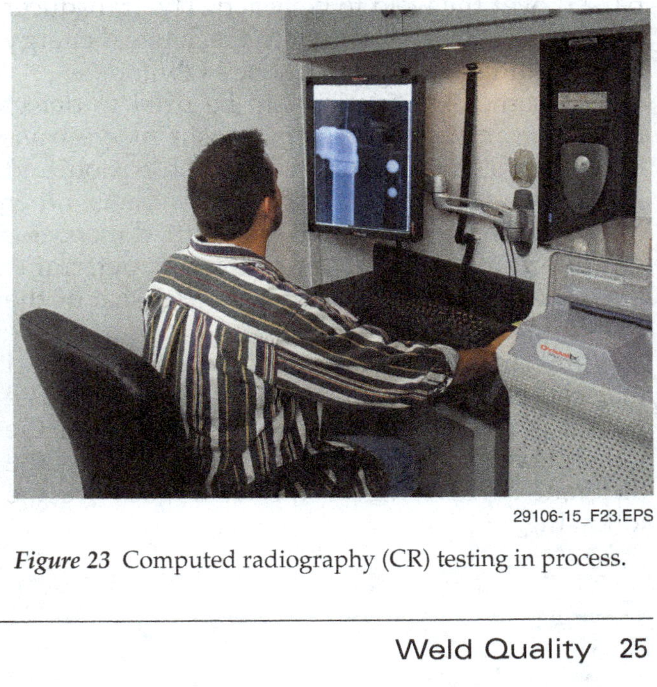

29106-15_F23.EPS

Figure 23 Computed radiography (CR) testing in process.

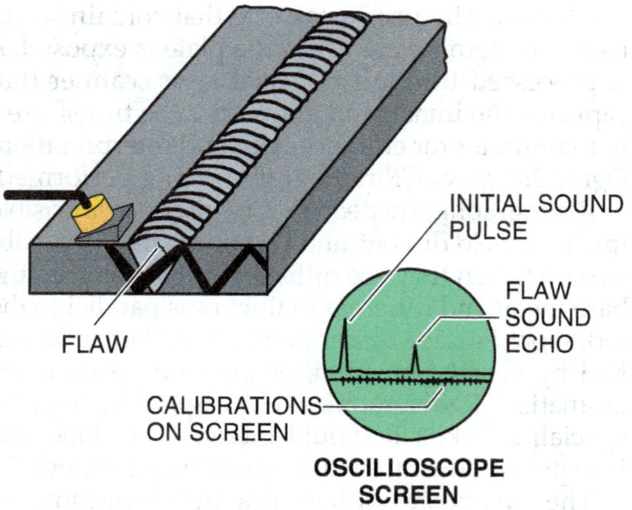

29106-15_F24.EPS

Figure 24 Portable ultrasonic device.

The term *ultrasonic* indicates frequencies above those heard by the human ear. Ultrasonic devices operate very much like depth sounders or fish finders. The key element is a **transducer** that is passed over the weld to be tested. The transducer converts electrical energy into mechanical energy that is emitted as high-frequency vibrations.

Ultrasonic examination can be used to detect and locate cracks, **laminations**, shrinkage cavities, pores, slag inclusions, incomplete fusion, and incomplete joint penetration, as well as other discontinuities in the weld. A qualified inspector can interpret the signal on a screen to determine the approximate position, depth, and size of the discontinuity.

Ultrasonic beams can deeply penetrate weld material with high sensitivity and accuracy. The output is readily digitized, and operation of the monitor is entirely nonhazardous. Some of the different types of ultrasonic techniques are the following:

Self-Propelled NDT

This ultrasonic test instrument is often used to inspect pipe and tanks. Its magnetic wheels allow it to cling to the surface as it moves along, controlled by a joystick mechanism.

29106-15_SA02.EPS

- Pulse-echo
- Pulse-transmission
- Ultrasonic attenuation
- Continuous wave resonance
- Ultrasonic spectroscopy
- Phased array

Figure 25 shows the effects of moving an angle-beam transducer back and forth across a weld. *Figure 26* shows the effects of moving an angle-beam transducer in an arc around a weld.

Ultrasonic monitors are very portable, but they require experienced technicians to monitor the equipment. Reading an ultrasonic flaw detector requires some practice because the screen image shows the peaks of back reflection.

In traditional UT, a single-element probe is physically moved around the weldment by the inspector to create a cross-section of the joint being tested. In contrast, the newer phased array (PA) ultrasonic testing method (*Figure 27*) uses a probe containing an array of ultrasonic transducers. The beam created by these transducers is electronically steered. The data from the individual elements is combined electronically to create a cross-section of the weld.

Among the advantages of ultrasonic inspection are that it can find defects throughout the material being examined; it can be used to check materials that cannot be radiographed; it is nonhazardous to personnel and equipment; and it can detect even small defects from one side of the material.

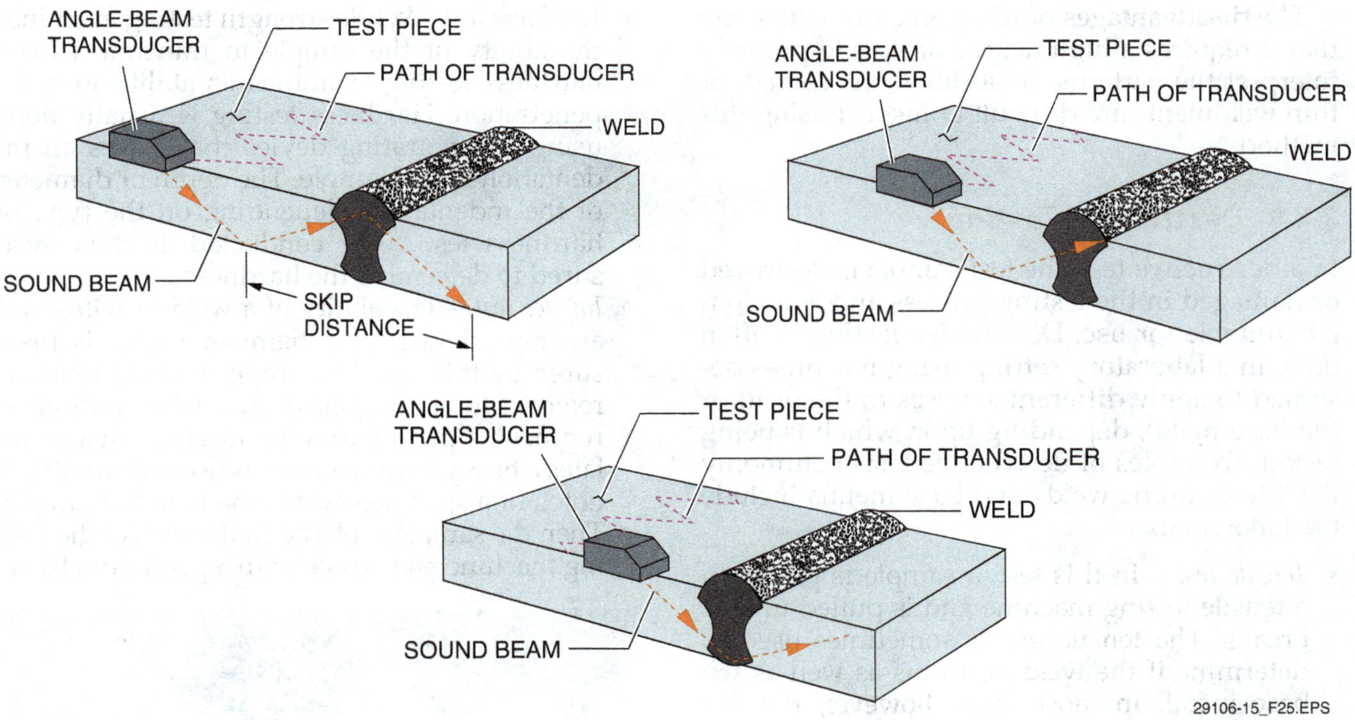

Figure 25 Moving a transducer back and forth.

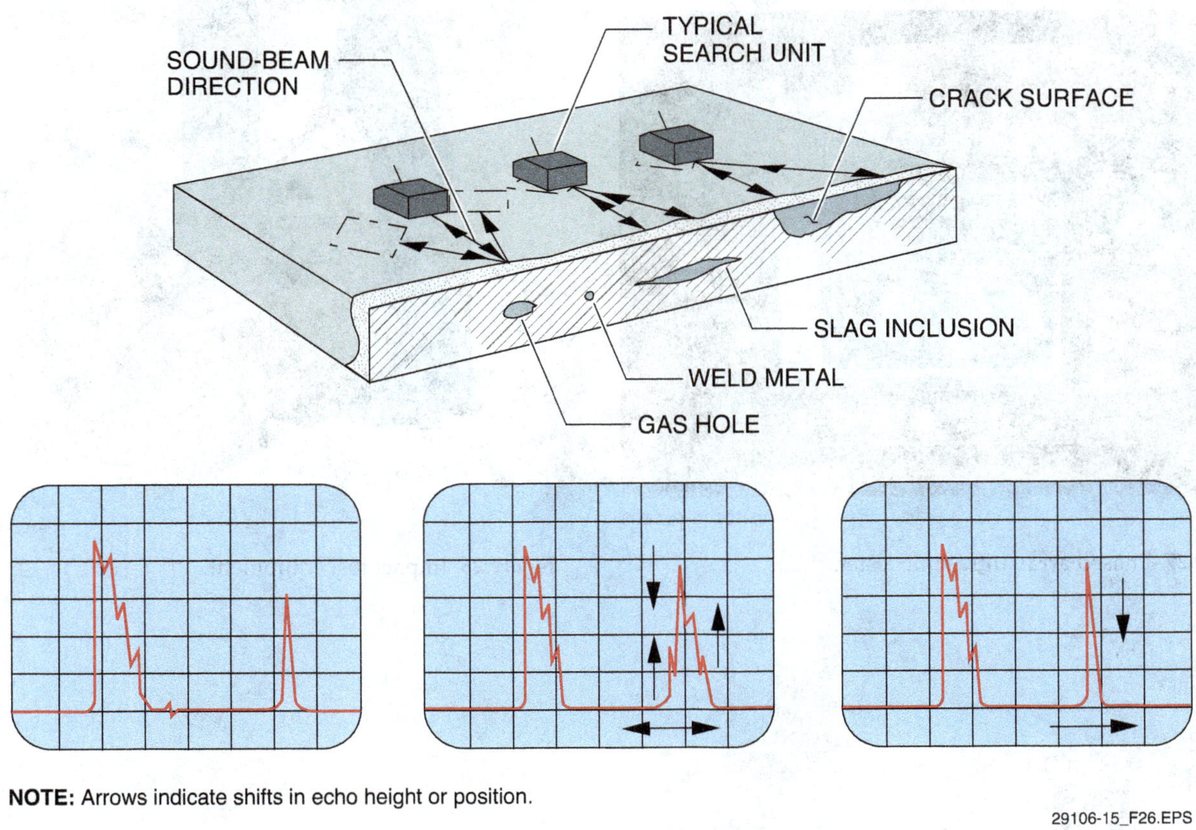

Figure 26 Moving a transducer in an arc.

The disadvantages of ultrasonic inspection are that it requires a high degree of skill to properly interpret the patterns. In addition, very small or thin weldments are difficult to inspect using this method.

3.4.0 Destructive Testing

In a destructive test, the test sample is destroyed or damaged in the testing process and is no longer suitable for use. Destructive testing is often done in a laboratory setting using machines designed to apply different stresses to the weld or the base metal, depending upon which is being tested. Examples of destructive tests commonly used to examine welds and base metals include the following:

- *Tensile test* – In this test, a sample is placed in a tensile testing machine and is pulled until it breaks. The tensile test is sometimes used to determine if the weld performs as well as the base metal. In most cases, however, the test is performed to determine specific properties of the weldment, such as strength and ductility (the amount the sample will stretch before breaking).
- *Hardness test* – While strength testing examines the ability of the sample to transmit a load, hardness testing examines its ability to resist penetration. Hardness testing is usually done using a penetrating device that leaves an indentation in the sample. The depth or diameter of the indentation, depending on the type of hardness test being conducted, is then measured to determine the hardness.
- *Impact test* – The ability of a weld to withstand an impact, such as a hammer strike, is measured by this test. The property being tested is referred to as toughness. In a laboratory environment, a pendulum-type machine that simulates a heavy hammer blow is used (*Figure 28*). A notch of a specified size is made in the sample. Then the sample is placed in the jaws of the testing machine and struck with a pendulum blow.

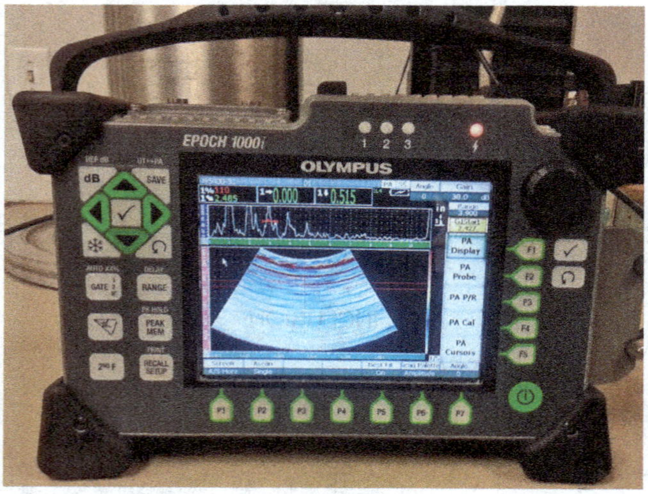

29106-15_F27.EPS

Figure 27 Phased array ultrasonic tester.

29106-15_F28.EPS

Figure 28 Impact test equipment.

- *Soundness test* – The three types of soundness tests include bend, nick-break, and fillet weld break tests.
  - Bend testing is the test most commonly used in determining the qualifications of a welder or welding procedures. Bend tests are performed by placing the sample in a special fixture and applying stress in a way that causes the sample to bend 180 degrees. The bend is then inspected for weld defects. Guided bend tests (*Figure 29*) are used to evaluate groove test welds on both plates and pipes. In this method, specimens are bent into a U-shape with a device called a jig. The bending action places stress on the weld metal and reveals any discontinuities in the weld. The jig has a plunger and die that are dimensioned for the thickness of the specimen being bent (*Figure 30*). Refer to the welding code being used for testing for the required dimensions of the bending jig's plunger and die.

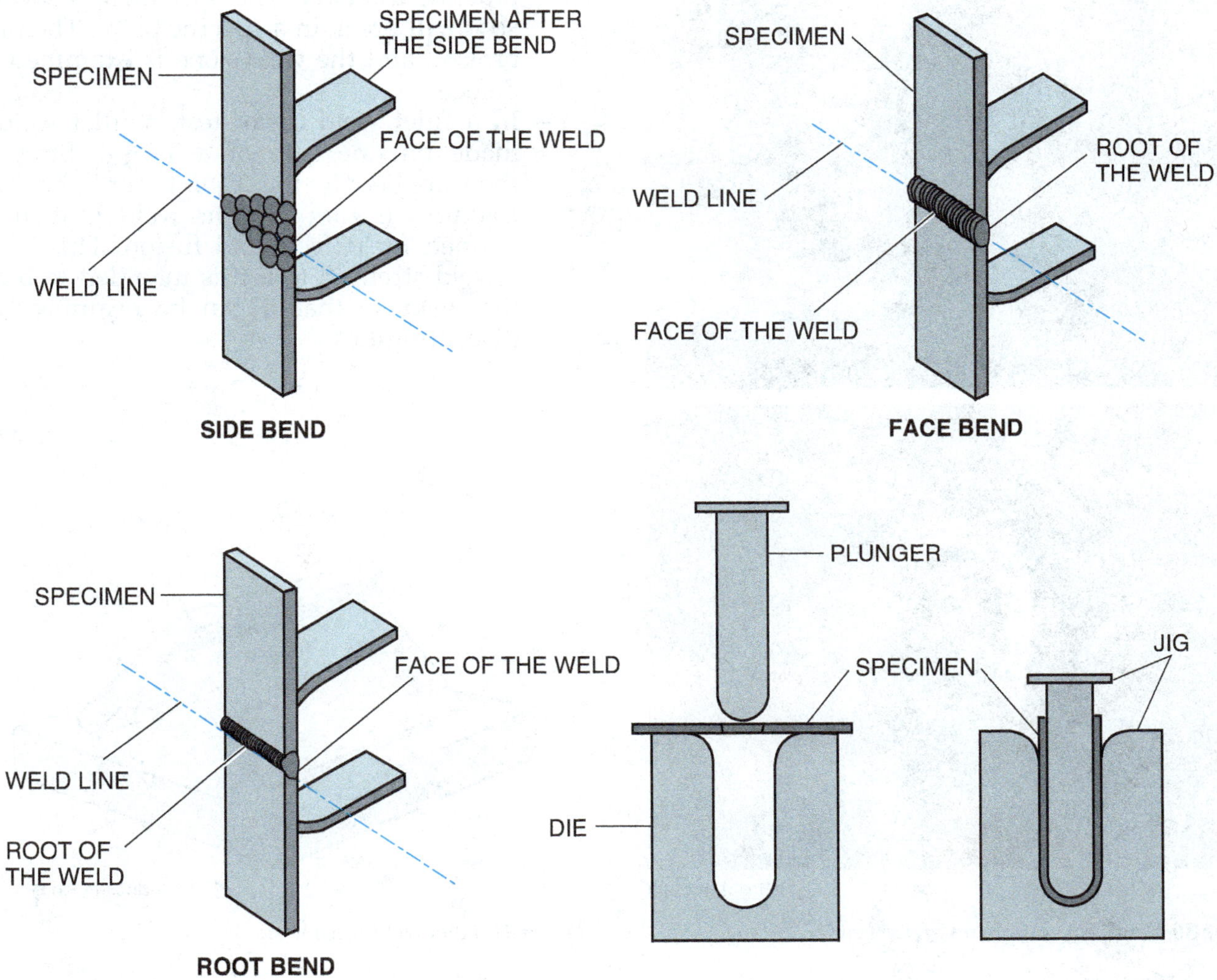

Figure 29 Guided bend test method and samples.

- The three types of tests performed on the jig are root bends, face bends, and side bends. These bends test for penetration and fusion throughout the root joint. They also test for porosity, inclusions, and other defects, as well as the quality of the fusion to the side walls and the face of the weld joint. Face and root bend tests are used for materials up to ⅜" (≈10 mm metric plate) thick. For materials thicker than this, side bend testing may be used.
- Nick-break testing is used primarily in the pipeline industry. The specimen is saw-cut so it will break in a specific place. Then it is broken, and the weld zone is examined for defects.
- In a fillet weld break test, a fillet weld is made on one side of a T-joint. Stress is then applied to the T-joint until the weld fractures (*Figure 31*). The weld is then examined for defects and fusion. This is not a weld strength test; it is intended to break the weld so that it can be examined for discontinuities.

29106-15_F30.EPS

Figure 30 Bend test equipment and specimen.

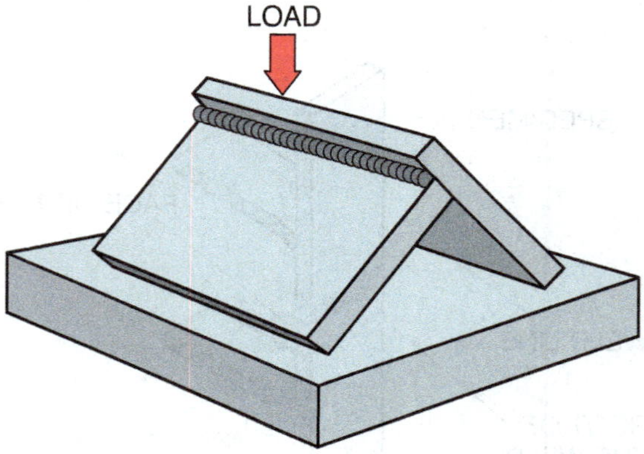

29106-15_F31.EPS

Figure 31 Fillet weld break test.

Additional Resources

Visual Inspection Workshop Reference Manual, AWS VIW-M-2008. Miami, FL: American Welding Society.

Certification Manual for Welding Inspectors. Miami, FL: American Welding Society.

Welding Inspection Technology. Miami, FL: American Welding Society.

B1.10M/B1.10:2009 Guide for the Nondestructive Examination of Welds. Miami, FL: American Welding Society.

3.0.0 Section Review

1. A bridge cam gauge is used to measure _____.
   a. undercut
   b. fillet weld depth
   c. butt weld reinforcement
   d. overlap

2. Electromagnetic inspection can be used to test welds in non-ferrous metals.
   a. True
   b. False

3. Which of these test methods uses an imaging plate placed in a cassette?
   a. Ultrasonic
   b. Phased array UT
   c. Computed radiography
   d. Digital radiography

4. A tensile test is one in which the test coupon is _____.
   a. bent
   b. pulled
   c. struck
   d. compressed

SECTION FOUR

4.0.0 WELDER PERFORMANCE QUALIFICATION TESTS

Objective

Describe the welder performance testing process.

a. Describe the qualification of welders by position.
b. Describe welder qualification testing to meet AWS and ASME requirements.
c. Describe the process for completing a weld test.

The purpose of welder performance qualification is to measure the proficiency of individual welders. As previously discussed, codes require that welders take a test to qualify for performing a welding procedure.

> **NOTE**
> Various codes and specifications often require similar methods for qualifying welders. The applicable code or specification should be consulted for specific details and requirements. Ask your supervisor if you are unsure about which codes or specifications apply to your project.

4.1.0 Welding Position Qualification

The welder or welding machine operator is qualified by welding position. Welders may be qualified to perform a welding procedure in only one position, or possibly all positions by passing a welding qualification test. The qualification tests are designed to measure the welder's ability to make groove and fillet welds in different positions on plate, pipe, or both in accordance with the applicable code. Each welding position is designated by a number and a letter, such as 1G. These designations are standard for all codes.

The letter G designates a groove weld. The letter F designates a fillet weld. For plate welding, the positions are designated by the following numbers:

- 1 – Flat position welding
- 2 – Horizontal position welding
- 3 – Vertical position welding
- 4 – Overhead position welding

Figure 32 shows the plate welding positions for both fillet welds and groove welds.

For pipe welding, there are these additional positions: 5F, 5G, 6G, and 6GR. The number indicates that multiple-position welds are required. Also, in the 1G (flat groove) and 1F (flat fillet) positions, the pipe is rotated during welding. *Figure 33* shows the pipe welding positions for both fillet welds and groove welds.

A welder who qualifies in one position does not automatically qualify to weld in all positions. However, in most cases qualification for groove welds qualify the welder for fillet welds; qualification for pipe qualify the welder for plate. Qualification in one code may or may not qualify the welder in other codes. The qualification requirements between codes may not match. Refer to your site requirements/code for qualifying requirements.

4.2.0 Code-Required Testing

AWS, ASME, and API codes require that welders be qualified through testing to perform specific types of welds based on base metal, welding position, joint design, type of electrode, and other variables.

4.2.1 AWS Structural Steel Code

The AWS structural steel code provides information concerning the qualification of welding procedures, welders, and welding operators for the types of welding done by contractors and fabricators in building and bridge construction. Qualification for plate welding also qualifies the welder for rectangular tubing.

The mild steel electrodes used with shielded metal arc welding (SMAW) are classified by F numbers: F1, F2, F3, and F4. Qualification with an electrode in a particular F-number classification will qualify the welder with all electrodes identified in that classification and in lower F-number classifications. *Table 1* shows AWS F-number electrode classifications.

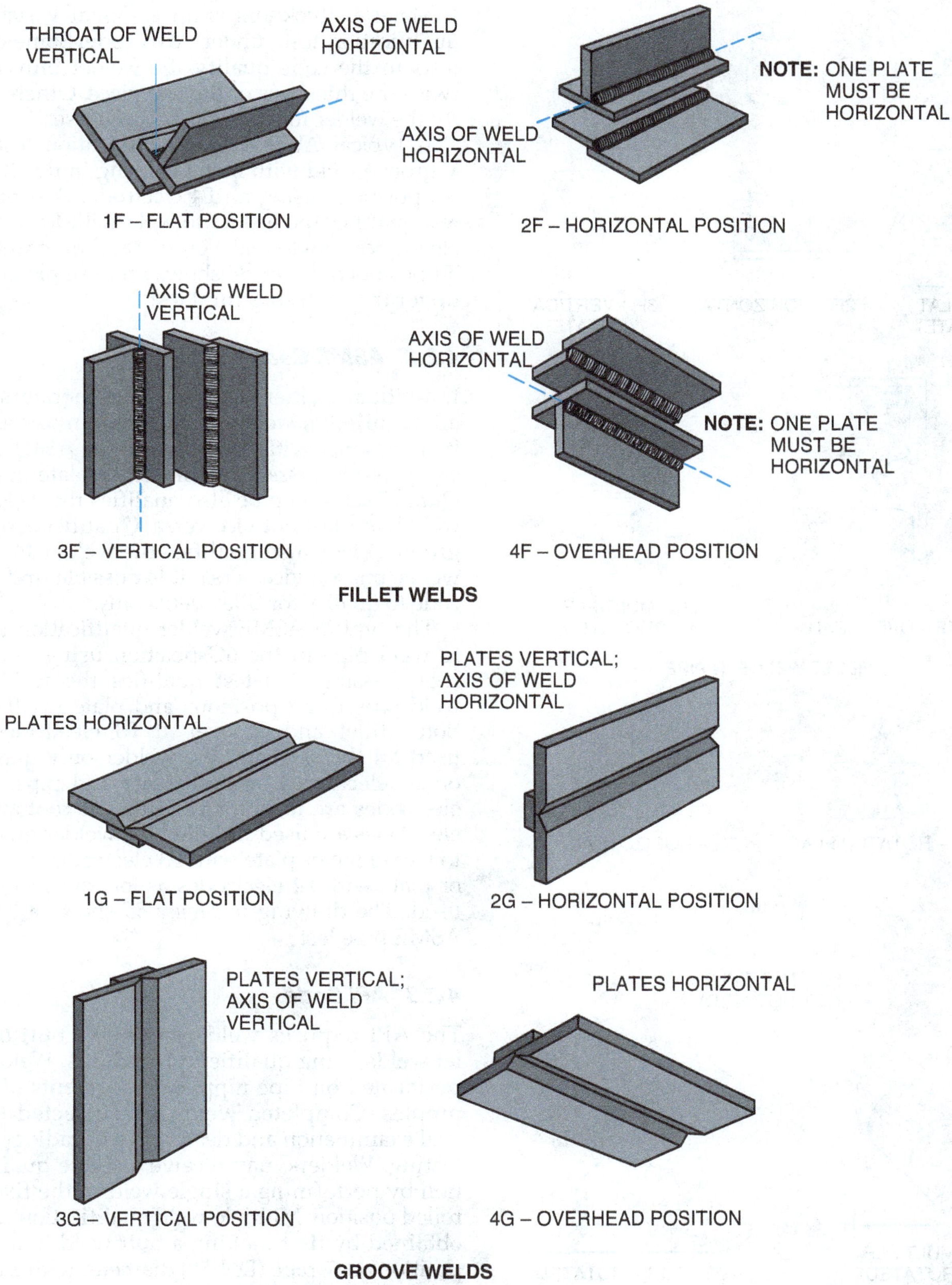

Figure 32 Welding positions for plate.

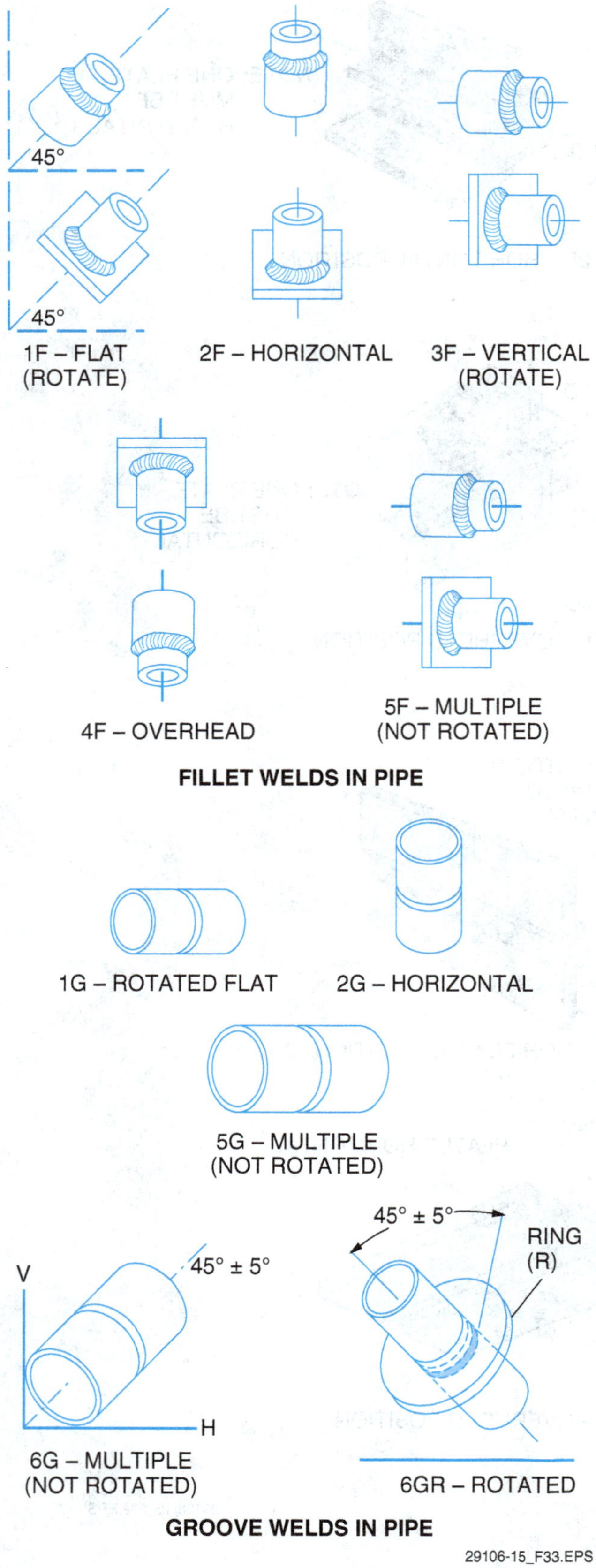

Figure 33 Welding positions for pipe.

Material thickness is an essential variable in qualification tests under AWS code. Some of the tests in the code qualify the welder only up to twice the thickness of the test piece. Others qualify the welder for unlimited thicknesses.

A typical AWS welder qualification test is a V-groove weld with metal backing in the 3G and 4G positions using an F4 electrode. Passing this test qualifies the welder to weld with F4 or lower electrodes, and to make groove and fillet welds in all positions. *Figure 34* shows an example of a fit-up for the AWS structural test.

4.2.2 ASME Code

Individual welders and welding operators who are required to weld to ASME code must qualify in accordance with Section IX of the *ASME Boiler and Pressure Vessel Code* on either plate or pipe. Qualification on pipe also qualifies the welder to weld plate, but not vice versa. Qualification with groove welds also qualifies the welder for fillet welds, but not vice versa. It is possible under the code to qualify for fillet welds only.

The typical ASME welder qualification test is to weld pipe in the 6G position using an open root. Passing this test qualifies the welder to weld pipe in all positions and plate in all positions (fillet and groove). If F3 electrodes are used all the way out, the welder only qualifies on F3 electrodes on both plate and pipe. If F3 electrodes are used for welding the root and F4 electrodes are used for filler, the welder qualifies to weld pipe or plate with F3 electrodes and pipe or plate with F4 electrodes as long as backing is used. The drawing in *Figure 35* shows a typical ASME pipe test.

4.2.3 API Code

The API requires welders to make butt or fillet welds using qualified procedures. Welds are performed on pipe nipples or segments of pipe nipples. Completed welds are subjected to visual examination and destructive or radiographic testing. Welders may receive a single qualification by performing a single weld in the fixed or rolled position. Multiple-weld qualification can be obtained by first making a butt weld in a fixed position on 6" pipe (DN150) diameter with a minimum thickness of 0.250" (6.4 mm). A second test requires the welder to cut, fit, and weld a pipe of the same size without a backing strip.

Table 1 AWS F-Number Electrode Classification

Group	AWS Electrode Classification				
F4	EXX15	EXX16	EXX18		Low-Hydrogen
F3	EXX10	EXX11			Fast-Freeze
F2	EXX12	EXX13	EXX14		Fill-Freeze
F1	EXX20	EXX24	EXX27	EXX28	Fast-Fill

Figure 34 Typical AWS plate test coupon.

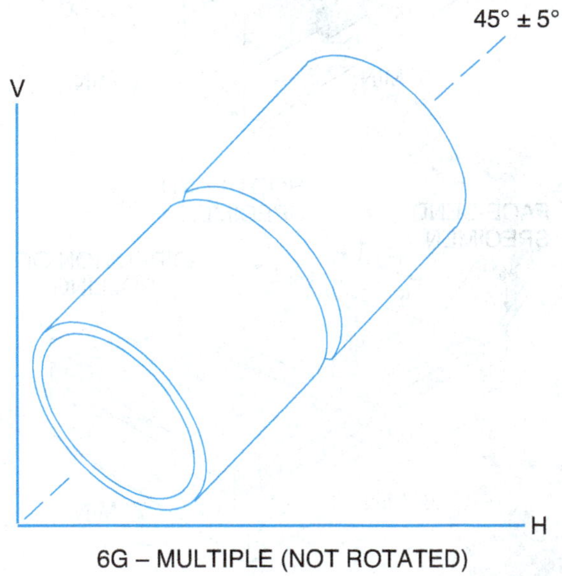

Figure 35 Typical ASME pipe test.

4.3.0 Welder Qualification Tests

The welder becomes qualified by successfully completing a weld made in accordance with the WPS. It is general practice for code welding to qualify welders on groove weld tests. Passing these tests also permits the welder to perform fillet welds.

4.3.1 Making the Test Weld

Although qualification tests are designed to determine the capability of welders, some welders fail for reasons not related to their welding ability. This is due principally to carelessness in the application of the weld and in the preparation of the test specimen. It is important to note prior to welding where the test strips will be cut from the weld coupon. By doing this, you can avoid potential problems such as restarts in the area of the test strips. The following sections explain how to prepare a test specimen.

4.3.2 Removing Test Specimens

After making the qualification test weld, the test specimens are cut from the test pipe or plate by any suitable means. There are specific locations where the test specimen is cut from the pipe or plate. Refer to the applicable code for details related to specimen location and quantity.

Typical specimen locations for plate welds are shown in *Figure 36*. For material ⅜" (10 mm metric plate) thick, a face bend and a root bend are required. For material thicker than ⅜" (10 mm metric plate), two side bends are required.

Note that tests are usually given on plate thicknesses of ⅜" (10 mm metric plate) for limited thickness and 1" (25 mm metric plate) for unlimited thickness qualifications.

4.3.3 Preparing the Specimens for Testing

After the specimen has been cut from the test piece, it must be properly prepared for testing (*Figure 37*). Poor specimen preparation can cause a sound weld metal to fail. For example, a slight nick may open up under the severe bending stress of the test, causing the specimen to fail. To properly prepare the test specimen, do the following:

- Grind or machine the surface to a smooth finish. All grinding and machining marks must be lengthwise on the sample. Otherwise, they produce a notch effect, which may cause failure.

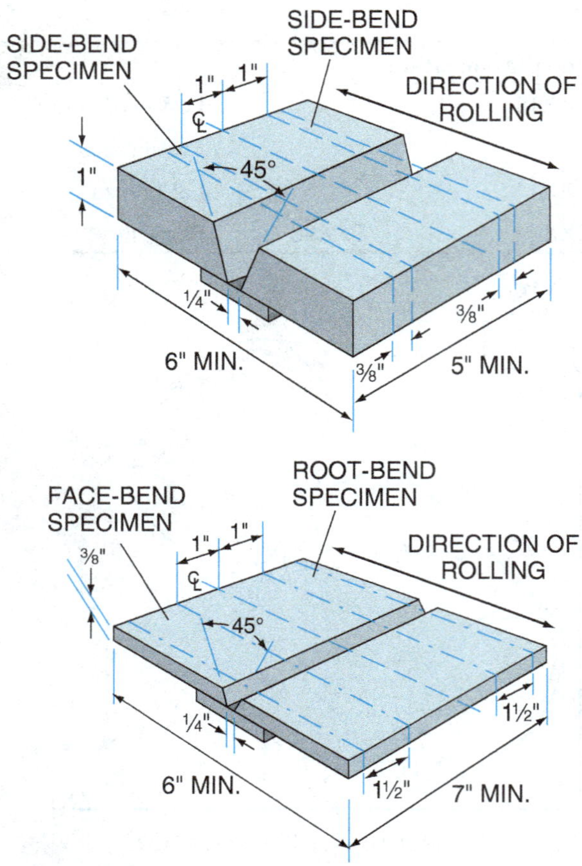

METRIC EQUIVALENTS	
1/4"	6.4 MM
3/8"	10 MM
1"	25 MM
1 1/2"	38 MM
6"	15.2 CM
7"	19.8 CM

29106-15_F36.EPS

Figure 36 Specimen locations for plate welds.

- Remove any face or root reinforcement from the weldment. This is part of the test requirement and, more important to the welder, failure to do so can cause the failure of a good weld. Grinding of the weld reinforcement may not exceed 1⁄32" (0.8 mm) or 5 percent of the thickness of the base material, according to AWS standards.
- Round the edges to a smooth 1⁄16" (1.6 mm) radius. This can be done with a file. Rounded edges help prevent failure caused by cracks starting at a sharp corner.
- Refrain from quenching specimens you are grinding when they are hot. Quenching may create small surface cracks that become larger during the bend test.

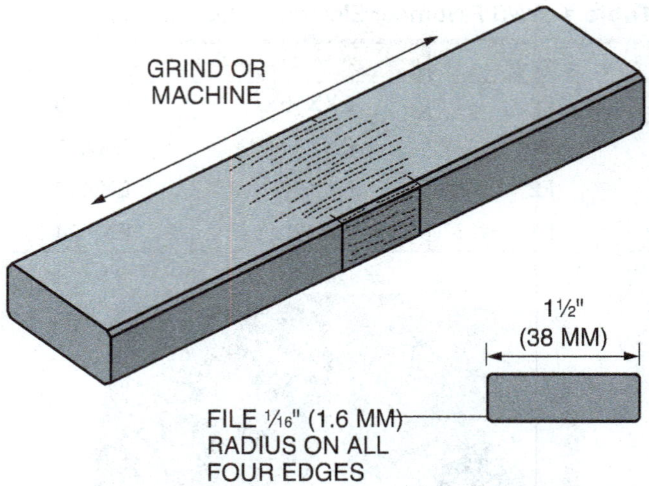

29106-15_F37.EPS

Figure 37 Example of a prepared test specimen.

After bending, the specimen is evaluated by measuring the discontinuities that are exposed. The criteria for acceptance can vary by code or site quality standards. AWS standards require that the surface shall contain no discontinuities exceeding the following dimensions:

- *1⁄8" (3.2 mm)* – Measured in any direction on the surface.
- *3⁄8" (10 mm)* – Sum of the greatest dimensions of all discontinuities exceeding 1⁄32" (0.8 mm) but less than or equal to 1⁄8" (3.2 mm).
- *1⁄4" (6.4 mm)* – Maximum corner crack, except when the corner crack results from visible slag inclusion or other fusion-type discontinuities, then a 1⁄8" (3.2 mm) maximum shall apply. A specimen with corner cracks exceeding 1⁄4" (6.4 mm), with no evidence of slag inclusions or other fusion-type discontinuities may be discarded, and a replacement test specimen from the original weldment shall be tested.

NOTE

In some cases, a radiographic inspection will be used instead of the guided bend test. This allows the entire weld to be examined and can detect small discontinuities at any location within the weld.

When the welder passes the qualification tests, the test results and the WPSs that the welder may weld are listed on a record that is kept by the company. This record becomes part of the quality documentation, and the welder becomes qualified to weld to that procedure.

AWS vs. ASME Pipe Welds

AWS and ASME codes have different requirements. The AWS version is shown on the left, while the ASME version is shown on the right.

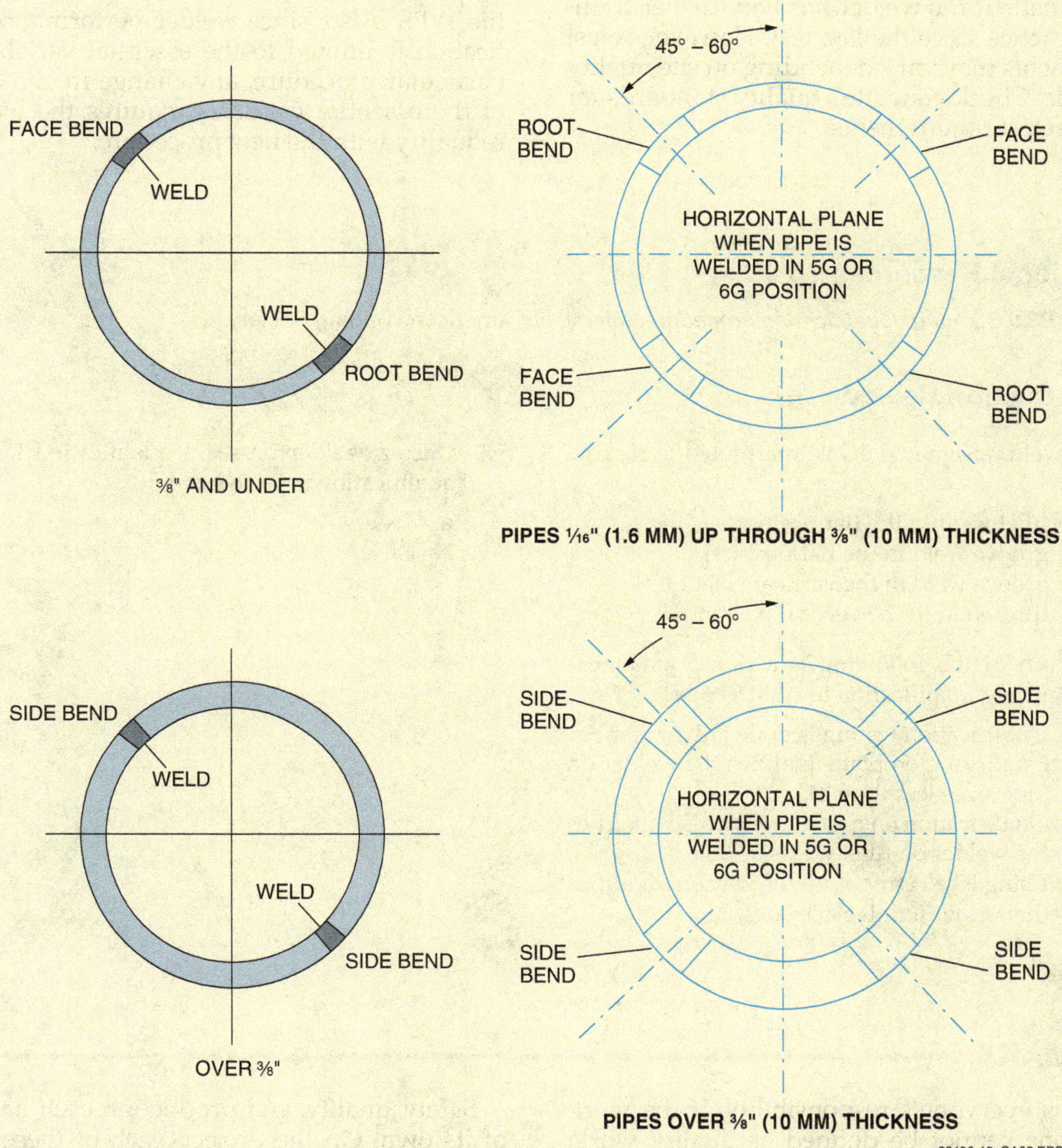

29106-15_SA03.EPS

4.3.4 Welder Qualification Limits

Welders may retest if they initially fail the test. An immediate retest consists of two test welds of each type of test that was failed. All the test specimens must pass this retest. A complete retest may be made if the welder has had further training or practice since the last test. However, retest requirements may vary depending on site quality standards. Check your site's quality standards for specific retest requirements.

After welders have qualified, they may have to requalify if they have not used the specific process for a certain time period. This time period varies depending on the codes. Welders may also be required to requalify if there is a reason to question their ability to make welds that meet the WPS. Also, since welder performance qualification is limited to the essential variables of a particular procedure, any change in one or more of the essential variables requires the welder to requalify with the new procedure.

Additional Resources

AWS B3.0 Standard Qualification Procedure. Miami, FL: American Welding Society.

4.0.0 Section Review

1. A weld designated 3G is interpreted to mean a _____.
   a. fillet weld in the flat position
   b. groove weld in the flat position
   c. groove weld in the vertical position
   d. fillet weld in the vertical position

2. Which of the following is a *correct* statement regarding qualification to ASME code?
   a. Qualification testing is done only on pipe.
   b. Qualifying on plate qualifies the welder on pipe sizes less than 24".
   c. Qualification on groove welds also qualifies the welder on fillet welds.
   d. Using F3 electrodes all the way out qualifies the welder for all electrodes.

3. Once a welder passes a qualification test, the qualification is permanent.
   a. True
   b. False

Summary

Quality is everyone's responsibility. If the work being done cannot be defined as quality work, it reflects on all those involved in the process. One essential trait of a craftsperson is a sense of quality workmanship. The craftsperson is generally closest to the work and will therefore have a major impact on product quality. Keeping quality in mind as you perform each step of your job will help you identify and correct small problems before they become major ones. This will make everyone's job easier and instill a sense of pride in what has been accomplished.

Safety, quality, and production each has a cost of its own. On the project, each of these factors should have proper guidelines. At the completion of the project, when all records for safety, cost, planning, scheduling, and effectiveness have been evaluated, the papers are usually filed away, but quality remains indefinitely for all eyes to see. Quality is, perhaps, the major reason for repeat business. How well the craftsperson performed will be noticed long after the project has been completed.

Review Questions

1. Which section of the *ASME Boiler and Pressure Vessel Code* contains the specifications for acceptable ferrous (Part A) and nonferrous (Part B) base metals?
   a. I
   b. II
   c. V
   d. IX

2. The minimum requirements for the design, materials, fabrication, erection, testing, and inspection of various types of piping systems are covered by _____.
   a. *ASME B31*
   b. *AWS D1.1*
   c. *AWS B1.11*
   d. *API Std. 1104*

3. The code that covers the requirements for welding structures or components under the surface of water is _____.
   a. *ASME B31*
   b. *AWS D1.1*
   c. *AWS D3.6M*
   d. *API 1104*

4. If you are directed to perform a weld for which you are not qualified, you should _____.
   a. bypass the chain of command
   b. report directly to the safety engineer as soon as possible
   c. try to resolve the matter with your immediate supervisor
   d. perform the weld to the best of your ability

5. Which of the following is considered a nonessential variable for a welding performance qualification?
   a. Material thickness
   b. Post-heat treatment
   c. Travel speed
   d. Welding process

6. The presence of voids or empty spots in the weld metal is called _____.
   a. porosity
   b. inclusion
   c. defect
   d. discontinuity

7. The type of porosity in which the weld exhibits a localized grouping of pores is known as _____.
   a. uniformly scattered porosity
   b. clustered porosity
   c. linear porosity
   d. piping porosity

8. The type of weld discontinuity referred to as wormholes is _____.
   a. clustered porosity
   b. piping porosity
   c. cold cracks
   d. incomplete fusion

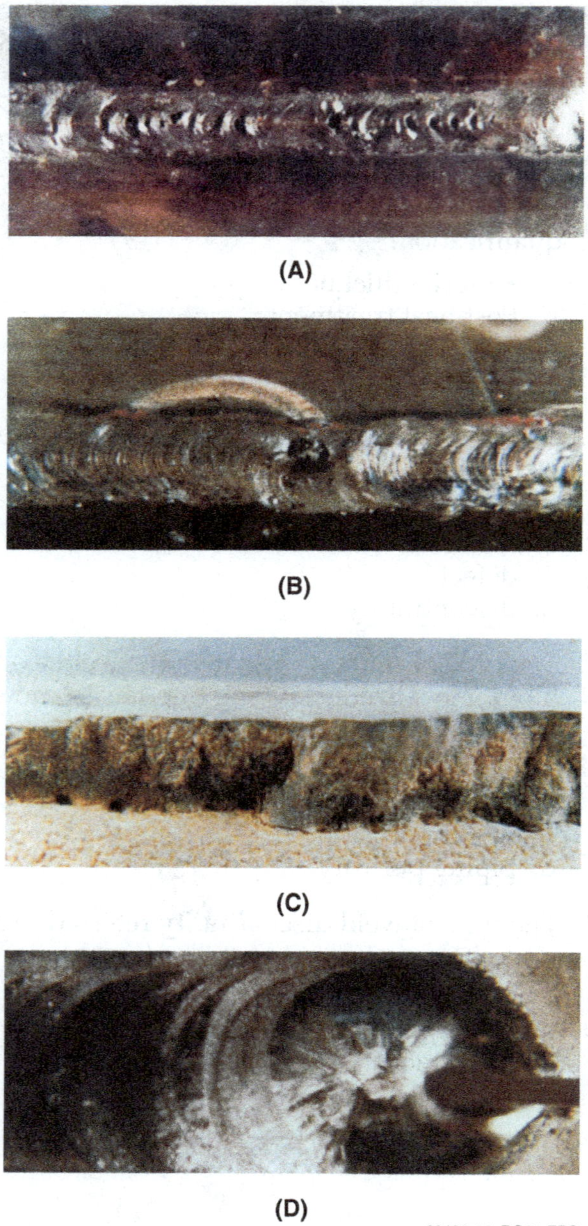

29106-15_RQ01.EPS

Figure 1

For Questions 9 through 11, match a picture in Review Question Figure 1 to each discontinuity listed by putting the correct letter in the blank.

9. Linear porosity ________

10. Surface slag inclusion ________

11. Toe crack ________

12. One method of reducing weld metal cracking is to use _____.
    a. fast-fill electrodes
    b. fill-freeze electrodes
    c. low-hydrogen electrodes
    d. fast-freeze electrodes

13. Dross from flame cutting, if not removed, is likely to cause _____.
    a. porosity
    b. spatter
    c. undercut
    d. incomplete penetration

14. Dimensional accuracy of a weld is checked by _____.
    a. visual inspection
    b. using a tape measure
    c. radiography
    d. using welding gauges

15. A vacuum box is used to check for _____.
    a. inclusions
    b. porosity
    c. leaks
    d. incomplete fusion

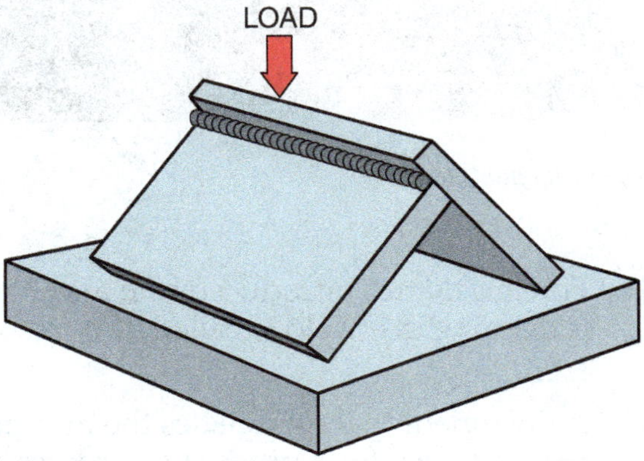

Figure 2

16. The test method shown in *Review Question Figure 2* is the _____.
    a. nick-break test
    b. fillet weld break test
    c. impact test
    d. hardness test

17. The nick-break test is a type of _____.
    a. tensile test
    b. hardness test
    c. soundness test
    d. impact test

18. A typical AWS welder qualification test is a V-groove weld with metal backing in the 3G and 4G positions using an _____.
    a. F4 electrode
    b. F3 electrode
    c. F2 electrode
    d. F1 electrode

19. An ASME welder qualification that qualifies a welder to weld pipe in all positions and plate in all positions is _____.
    a. 1F position test
    b. 6G position test
    c. 4F position test
    d. 4G position test

20. To properly prepare a test specimen, it is important to _____.
    a. quench the specimen while it is hot
    b. grind it width-wise to produce a notch effect
    c. remove any face or root reinforcement from the weldment
    d. make sure the edges are sharp

Trade Terms Quiz

Fill in the blank with the correct term that you learned from your study of this module.

1. _______ is the demonstration that welds made following a specific procedure can meet prescribed standards.
2. A(n) _______ is a discontinuity or imperfection that renders a part of the product or the entire product unable to meet minimum acceptable standards or specifications.
3. The type of discontinuity that is identified by cracking in the base metal near the weld, but under the surface is _______.
4. Metals that have the ability to be made harder by heating and then cooling are called _______.
5. The document containing the results of the nondestructive and destructive testing required to qualify a welding procedure specification (WPS) is called a(n) _______.
6. _______ is a change or break in the shape or structure of a part that may or may not be considered a defect, depending on the code.
7. The quality or state of having a uniform structure or composition throughout is known as _______.
8. The testing method is which images are made by passing X-rays or gamma rays through an object and recording the variations in density on photographic film is known as _______.
9. Cracks in the base metal formed when layers separate are called _______.
10. _______ metal has been made brittle and will tend to crack with little bending.
11. Foreign matter introduced into and remaining in a weld is called a(n) _______.
12. A document that establishes the minimum requirements for a product or process and can be adopted into law is a(n) _______.
13. A welding parameter that cannot be changed without requalifying a procedure is a(n) _______.
14. The ability of a material to absorb energy in the presence of a flaw is known as _______.
15. A welding parameter that can be changed without requalifying a procedure is a(n) _______.
16. The variables that must be considered when notch toughness requirements are invoked are called _______.
17. A document that defines how a code is to be implemented is a(n) _______.
18. A(n) _______ is a document that defines in detail the work to be performed or the materials to be used in a product or process.
19. A device that converts one form of energy into another is called a(n) _______.
20. _______ is a method of joining metal using heat and a filler metal with a melting point above 842°F (450°C).
21. The deflection of an arc from its intended path by magnetic forces is known as _______.

Trade Terms

Arc blow
Brazing
Code
Defect
Discontinuity
Embrittled
Essential variables
Hardenable materials
Homogeneity
Inclusion
Laminations
Non-essential variables
Notch toughness
Procedure qualification record (PQR)
Radiographic
Specification
Standard
Supplemental essential variables
Transducer
Underbead cracking
Welding procedure qualification

Appendix A

Procedure Qualification Record

FORM QW-483 SUGGESTED FORMAT FOR PROCEDURE QUALIFICATION RECORDS (PQR)
(See QW-200.2, Section IX, ASME Boiler and Pressure Vessel Code)
Record Actual Variables Used to Weld Test Coupon

Organization Name ______
Procedure Qualification Record No. ______ Date ______
WPS No. ______
Welding Process(es) ______
Types (Manual, Automatic, Semi-Automatic) ______

JOINTS (QW-402)

Groove Design of Test Coupon
(For combination qualifications, the deposited weld metal thickness shall be recorded for each filler metal and process used.)

BASE METALS (QW-403)
Material Spec. ______
Type/Grade, or UNS Number ______
P-No. ______ Group No. ______ to P-No. ______ Group No. ______
Thickness of Test Coupon ______
Diameter of Test Coupon ______
Maximum Pass Thickness ______
Other ______

POSTWELD HEAT TREATMENT (QW-407)
Temperature ______
Time ______
Other ______

GAS (QW-408)

	Gas(es)	Percent Composition (Mixture)	Flow Rate
Shielding			
Trailing			
Backing			
Other			

FILLER METALS (QW-404)

	1	2
SFA Specification		
AWS Classification		
Filler Metal F-No.		
Weld Metal Analysis A-No.		
Size of Filler Metal		
Filler Metal Product Form		
Supplemental Filler Metal		
Electrode Flux Classification		
Flux Type		
Flux Trade Name		
Weld Metal Thickness		
Other		

ELECTRICAL CHARACTERISTICS (QW-409)
Current ______
Polarity ______
Amps. ______ Volts ______
Tungsten Electrode Size ______
Mode of Metal Transfer for GMAW (FCAW) ______
Heat Input ______
Other ______

POSITION (QW-405)
Position of Groove ______
Weld Progression (Uphill, Downhill) ______
Other ______

TECHNIQUE (QW-410)
Travel Speed ______
String or Weave Bead ______
Oscillation ______
Multipass or Single Pass (Per Side) ______
Single or Multiple Electrodes ______
Other ______

PREHEAT (QW-406)
Preheat Temperature ______
Interpass Temperature ______
Other ______

(07/13)

29106-15_A01A.EPS

FORM QW-483 (Back)

Tensile Test (QW-150)

PQR No. ______

Specimen No.	Width	Thickness	Area	Ultimate Total Load	Ultimate Unit Stress, (psi or MPa)	Type of Failure and Location

Guided-Bend Tests (QW-160)

Type and Figure No.	Result

Toughness Tests (QW-170)

Specimen No.	Notch Location	Specimen Size	Test Temperature	Impact Values			Drop Weight Break (Y/N)
				ft-lb or J	% Shear	Mils (in.) or mm	

Comments ______

Fillet-Weld Test (QW-180)

Result — Satisfactory: Yes ______ No ______ Penetration into Parent Metal: Yes ______ No ______

Macro — Results ______

Other Tests

Type of Test ______

Deposit Analysis ______

Other ______

....

Welder's Name ______ Clock No. ______ Stamp No. ______

Tests Conducted by ______ Laboratory Test No. ______

We certify that the statements in this record are correct and that the test welds were prepared, welded, and tested in accordance with the requirements of Section IX of the ASME Boiler and Pressure Vessel Code.

Organization ______

Date ______ Certified by ______

(Detail of record of tests are illustrative only and may be modified to conform to the type and number of tests required by the Code.)

(07/13)

29106-15_A01B.EPS

Appendix B

Performance Accreditation Task – Visual Weld Test Inspection

The American Welding Society (AWS) School Excelling through National Skills Standards Education (SENSE) program is a comprehensive set of minimum Standards and Guidelines for Welding Education programs. The following performance accreditation is aligned with and designed around the SENSE program.

The Performance Accreditation Tasks (PATs) correspond to and support the learning objectives in *AWS EG2.0, Guide for the Training and Qualification of Welding Personnel: Entry-Level Welder*.

Note that in order to satisfy all learning objectives in *AWS EG2.0*, the instructor must also use the PATs contained in the second level of the NCCER Welding curriculum.

PAT 1 corresponds to *AWS EG2.0, Module 9 – Welding Inspection and Testing Principles, Unit 4 Welding Inspection and Testing*, Key Indicators 1 and 2.

PATs provide specific acceptable criteria for performance and help to ensure a true competency-based welding program for students.

The following task is designed to develop your competency in performing a visual weld test inspection. Practice this task until you are thoroughly familiar with the procedure.

As you complete the task, take it to your instructor for evaluation. Do not proceed to the next task until instructed to do so by your instructor.

VISUAL WELD TEST INSPECTION

Obtain a completed fillet weld on ¼" plate (3.2 mm metric plate) a minimum of 6" (15.2 cm) long, and a complete joint-penetration groove weld minimum 6" (15.2 cm) long, and then perform a visual test inspection (VT) on each weld using the following guide and acceptance criteria.

Visual Test Inspection Report

Trainee Inspector ____________________ **Instructor** ____________________

Discontinuity Category	Acceptance Criteria	¼" Fillet Weld	CJP Groove Weld
Crack Prohibition	Any crack shall be unacceptable regardless of size or location	Acceptable ________ or Reason for rejection ________	Acceptable ________ or Reason for rejection ________
Weld/Base Metal Fusion	Complete fusion shall exist between adjacent layers of weld metal and base metal	Acceptable ________ or Reason for rejection ________	Acceptable ________ or Reason for rejection ________
Crater Cross Section	All craters shall be filled to provide the specified weld size	Acceptable ________ or Reason for rejection ________	Acceptable ________ or Reason for rejection ________
Weld Profiles	Weld profiles shall be in accordance with page 6.15 figure 14	Acceptable ________ or Reason for rejection ________	Acceptable ________ or Reason for rejection ________
Fillet Weld Size	The specified nominal fillet weld size tolerance is ±¹⁄₁₆ inch	Acceptable ________ or Reason for rejection ________	N/A
Undercut	Undercut shall not exceed ¹⁄₃₂" for any accumulated length up to 2 inches	Acceptable ________ or Reason for rejection ________	Acceptable ________ or Reason for rejection ________
Porosity	The sum of visible porosity ¹⁄₃₂" or greater shall not exceed ⅜" in any linear inch and shall not exceed ¾" in any linear 12 inches of weld	Acceptable ________ or Reason for rejection ________	Acceptable ________ or Reason for rejection ________
Complete Weld	Circle accept or reject for fillet and groove final inspection	Acceptable ________ or Reason for rejection ________	Accept or Reject

Trainee Inspector's Signature ____________________ **Date** ____________

Instructor's Verification: PASS/FAIL Signature ____________________

Trade Terms Introduced in This Module

Arc blow: The deflection of an arc from its intended path by magnetic forces.

Brazing: A method of joining metal using heat and a filler metal with a melting point above 842°F (450°C). Unlike welding, the base metal is not melted during the brazing process.

Code: A document that establishes the minimum requirements for a product or process. Codes can be, and often are, adopted as laws.

Defect: A discontinuity or imperfection that renders a part of the product or the entire product unable to meet minimum acceptable standards or specifications.

Discontinuity: A change or break in the shape or structure of a part that may or may not be considered a defect, depending on the code.

Embrittled: Metal that has been made brittle and that will tend to crack with little bending.

Essential variable: Items in a welding procedure specification (WPS) that cannot be changed without requalifying the WPS.

Hardenable materials: Metals that have the ability to be made harder by heating and then cooling.

Homogeneity: The quality or state of having a uniform structure or composition throughout.

Inclusion: Foreign matter introduced into and remaining in a weld.

Laminations: Cracks in the base metal formed when layers separate.

Non-essential variable: Items in a welding procedure specification (WPS) that can be changed without requalifying the WPS.

Notch toughness: The ability of a material to absorb energy in the presence of a flaw such as a notch.

Procedure qualification record (PQR): The document containing the results of the nondestructive and destructive testing required to qualify a WPS.

Radiographic: Describes images made by passing X-rays or gamma rays through an object and recording the variations in density on photographic film.

Specification: A document that defines in detail the work to be performed or the materials to be used in a product or process.

Standard: A document that defines how a code is to be implemented.

Supplemental essential variable: The variables that must be considered when notch toughness requirements are invoked.

Transducer: A device that converts one form of energy into another.

Underbead cracking: Cracking in the base metal near the weld, but under the surface.

Welding procedure qualification: The demonstration that welds made following a specific process can meet prescribed standards.

Additional Resources

This module presents thorough resources for task training. The following resource material is suggested for further study.

2014 Technical Training Guide. Current Edition. Cleveland, OH, USA: The Lincoln Electric Company. **www.lincolnelectric.com**

AWS B3.0 Standard Qualification Procedure. Miami, FL: American Welding Society.

B1.10M/B1.10:2009 Guide for the Nondestructive Examination of Welds. Miami, FL: American Welding Society.

Certification Manual for Welding Inspectors. Miami, FL: American Welding Society.

Visual Inspection Workshop Reference Manual, AWS VIW-M-2008. Miami, FL: American Welding Society.

Welding Inspection Technology. Miami, FL: American Welding Society.

Figure Credits

Applied Technical Services, Module Opener, Figures 21, 23, 24 (photo), 28, SA02

AWS D1.1/D1.1M: 2008, Welding Procedure Specification, Reproduced with permission from the American Welding Society (AWS), Miami, FL, USA, Figure 1

AWS D1.1/D1.1M: 2008, Procedure Qualification Record #231, Reproduced with permission from the American Welding Society (AWS), Miami, FL, USA, Figure 2

AWS B1.11-2000, Figures 1, 2, and 3, Reproduced with permission from the American Welding Society (AWS), Miami, FL, USA, Figure 3

AWS B1.11-2000, Figure 30, Reproduced with permission from the American Welding Society (AWS), Miami, FL, USA, Figure 4

AWS B1.11-2000, Figure 20, Reproduced with permission from the American Welding Society (AWS), Miami, FL, USA, Figure 5

AWS B1.11-2000, Figures 3, 24, 25, 26, and 28, Reproduced with permission from the American Welding Society (AWS), Miami, FL, USA, Figure 6

Topaz Publications, Inc., Figures 7 (photo), 18, 22 (photo), 30, SA01

AWS B1.11-2000, Figures 7 and 8, Reproduced with permission from the American Welding Society (AWS), Miami, FL, USA, Figure 9

AWS B1.11-2000, Figure 18, Reproduced with permission from the American Welding Society (AWS), Miami, FL, USA, Figure 10

G.A.L. Gage Co, Figures 15, 16

Certification Manual for Welding Inspectors, Figure 12.11, Reproduced with permission from the American Welding Society, Miami, FL, Figure 19

Olympus NDT Inc., Figure 27

Zachry Industrial, Inc., Figure 34

Reprinted from ASME 2013 BPVC, Section IX, by permission of The American Society of Mechanical Engineers. All rights reserved.

Section Review Answer Key

Answer	Section Reference	Objective
Section One		
1. b	1.1.0	1a
2. b	1.2.1	1b
Section Two		
1. c	2.1.1	2a
2. b	2.2.2	2b
3. a	2.3.3	2c
4. d	2.4.1	2d
Section Three		
1. a	3.1.1	3a
2. a	3.2.0	3b
3. c	3.3.1	3c
4. b	3.4.0	3d
Section Four		
1. c	4.1.0	4a
2. c	4.2.2	4b
3. b	4.3.4	4c

NCCER CURRICULA — USER UPDATE

NCCER makes every effort to keep its textbooks up-to-date and free of technical errors. We appreciate your help in this process. If you find an error, a typographical mistake, or an inaccuracy in NCCER's curricula, please fill out this form (or a photocopy), or complete the online form at **www.nccer.org/olf**. Be sure to include the exact module ID number, page number, a detailed description, and your recommended correction. Your input will be brought to the attention of the Authoring Team. Thank you for your assistance.

Instructors – If you have an idea for improving this textbook, or have found that additional materials were necessary to teach this module effectively, please let us know so that we may present your suggestions to the Authoring Team.

NCCER Product Development and Revision

13614 Progress Blvd., Alachua, FL 32615

Email: curriculum@nccer.org

Online: www.nccer.org/olf

❑ Trainee Guide ❑ Lesson Plans ❑ Exam ❑ PowerPoints Other ____________

Craft / Level: Copyright Date:

Module ID Number / Title:

Section Number(s):

Description:

Recommended Correction:

Your Name:

Address:

Email: Phone:

SMAW – Equipment and Setup

Overview

Good welding results from a welder's skill, knowledge, and the proper use of technology. Practice develops skill and study increases knowledge. This module presents the tools and equipment used for shielded metal arc welding (SMAW) and how to prepare them for use.

Module 29107

Trainees with successful module completions may be eligible for credentialing through the NCCER Registry. To learn more, go to **www.nccer.org** or contact us at **1.888.622.3720**. Our website has information on the latest product releases and training, as well as online versions of our *Cornerstone* magazine and Pearson's product catalog.

Your feedback is welcome. You may email your comments to **curriculum@nccer.org**, send general comments and inquiries to **info@nccer.org**, or fill in the User Update form at the back of this module.

This information is general in nature and intended for training purposes only. Actual performance of activities described in this manual requires compliance with all applicable operating, service, maintenance, and safety procedures under the direction of qualified personnel. References in this manual to patented or proprietary devices do not constitute a recommendation of their use.

29106 V5

29107

SMAW – Equipment and Setup

Objectives

When you have completed this module, you will be able to do the following:

1. Identify SMAW-related safety practices and explain how electrical characteristics apply to SMAW.
   a. Define SMAW and identify related safety practices.
   b. Explain how various electrical characteristics apply to SMAW.
2. Identify and describe SMAW equipment.
   a. Identify and describe various types of SMAW machines.
   b. Identify and describe SMAW welding cable and connectors.
   c. Identify common tools used to clean a weld.
3. Explain how to set up and start SMAW equipment.
   a. Explain how to set up SMAW equipment.
   b. Explain how to start, stop, and maintain SMAW equipment.

Performance Task

Under the supervision of your instructor, you should be able to do the following:

1. Set up a machine for SMAW.

Trade Terms

Conductor
Duty cycle
Governor
Motor-generator
Polarity
Rectifier
Step-down transformer

Industry Recognized Credentials

If you are training through an NCCER-accredited sponsor, you may be eligible for credentials from NCCER's Registry. The ID number for this module is 29107. Note that this module may have been used in other NCCER curricula and may apply to other level completions. Contact NCCER's Registry at 888.622.3720 or go to **www.nccer.org** for more information.

Contents

Topics to be presented in this module include:

Figures and Tables

SECTION ONE

1.0.0 SMAW

Objective

Identify SMAW-related safety practices and explain how electrical characteristics apply to SMAW.

a. Define SMAW and identify related safety practices.
b. Explain how various current characteristics apply to SMAW.

Trade Terms

Conductor: A material that will support the flow of an electrical current. Copper wire is the most common conductor.

Polarity: Refers to the direction of electrical current flow in a DC welding circuit.

Rectifier: An electronic device that converts AC voltage to DC voltage.

Step-down transformer: An electrical device that uses two wire coils of different sizes, referred to as the primary and secondary windings, to convert a higher voltage to a lower voltage through induction. Primary windings are connected to the primary power source; the secondary windings provide the reduced voltage to the process.

Shielded metal arc welding (SMAW) is often referred to as stick welding. This welding process can use either alternating current (AC) or direct current (DC) power, with adjustments made on the welding machine to regulate the level of current. The electrodes used, also referred to as welding rods, are typically coated with a clay-like mixture of fluorides, carbonates, oxides, metal alloys, and binding materials to provide a gas shield for the weld puddle. This coating hardens and covers the weld during its cooling cycle. The coating protects the weld from the atmosphere, which can cause weld degradation; then it is removed to expose the weld beneath.

1.1.0 SMAW Safety

Welding can be a dangerous profession, not only because of the welding itself, but also due to the environments in which the work is often performed. Any welder must consider a number of safety issues associated with welding. Complete safety coverage is provided in *Welding Safety.* If you have not completed that module, do so before continuing. Primary safety concerns involve the following:

- Moving welding equipment
- Electrical hazards
- Lifting hazards
- Working at heights
- Welding hazards

1.1.1 Moving Welding Equipment

Welding in the shop usually means that you are using a welding machine that is either set up on a stationary platform in or near a welding booth, or is on a portable cart that can be rolled around the shop floor. When welding is performed in the field, the welding machines may be mounted on the back of a truck, on a skid, or on a mobile wagon or trailer. *Figure 1* shows a welder using a mobile SMAW welding machine outdoors.

There will be times when the welding machines must be lifted to some elevated location. Whenever a welding machine is moved, it becomes a potential hazard. Machines can fall off their carts or be dropped. Even a 40-pound welding machine can seriously injure a foot or hand when dropped from its cart. Consider the damage a mobile welding machine could do if it detached from its tow vehicle, if the brakes or chocks allowed it to roll down an incline, or if it fell due to poor rigging. Be cautious and observant as welding equipment is being moved by any means.

29107-15_F01.EPS

Figure 1 A mobile SMAW machine being used outdoors.

1.1.2 Electrical Hazards

Electric welding machines typically require 110VAC, 120VAC, 208/240VAC, or 480VAC power sources. Regardless of the voltage level, be cautious of the power supply. Electrical shock or electrocution can result from contact with the electrical power supplied to a welding machine. Always know where the main circuit breaker is located for the outlet supplying power to your machine. Whenever possible, turn that breaker off before connecting or disconnecting the welding machine's corded plug. Always make sure the machine itself is switched off before connecting and disconnecting power cords.

Remember that an arc welder operates on the principle of creating an electrical arc. The potential for the arc occurs between the two welding machine leads. Perhaps more importantly, that potential exists for any other **conductor** of electricity that bridges the gap between the leads. Respect the welding leads whenever the machine is on; and always be aware of where they are and what is nearby when laying them down.

Another concern with electricity is moisture. Electricity, water, and the human body are a hazardous mix. Make sure that all the welding cables are in good condition, without nicks in the insulation that could allow water in.

1.1.3 Lifting Hazards

Welders sometimes forget the basics of lifting as they transport their welding machines, cables, and even the parts to be welded. Use rigging or lifting devices whenever needed. Get help if you must lift a heavy object by hand. Always remember to use tag lines when hoisting heavy objects that tend to spin or swing.

1.1.4 Working at Heights

Welders may have to be suspended from the side of a structure to weld something. Remember to wear approved fall arrest and fall protection devices when performing such work. Make sure to inspect your fall protection equipment items each time you use them, and be certain that they are current and in the best possible condition.

Figure 2 Appropriate personal protective equipment (PPE).

1.1.5 Other Welding Hazards

Although this module addresses the preparation of SMAW equipment, there may be a time when you must test-operate a welding machine before using it. When that situation occurs, make sure that you are wearing all the required personal protective equipment (PPE) and that the work area meets all safety requirements. Do not compromise safety, even for a test run.

Many welding machines are coupled with gas- or diesel-fired combustion engines. The engines turn a generator to create the electrical power needed to operate the machine. Remember that combustion engines are extremely hot and have their own set of safety concerns. For example, operating a diesel-fired welding machine in an enclosed or stagnant area could lead to breathing trauma. Flammable fuels must also be handled carefully. Be familiar with all of the relevant safety precautions for the specific equipment being used.

The best way to prevent an accident is to avoid putting yourself in a compromising position. PPE (*Figure 2*) is essential; it will help keep you from being injured, or worse. While welding, always use safety glasses with a full face shield or a helmet. Use corrective lenses if needed; some companies will not allow contact lenses to be worn during welding activities. Never view an electric arc directly or indirectly without using a properly tinted lens. Wear high-top safety shoes or boots. Make sure that the tongue and lace area of the footwear will be covered by a pant leg or spats. Wear a 100-percent cotton cap with no synthetic mesh material included in its construction. The bill of the cap should point to the rear to cover the back of the neck.

1.2.0 Welding Current

In electrical terms, current is the flow of electrons or electrical charge from one point to another. A source of stored electrical energy, such as the battery in your car, would be where the flow of electrons starts. Electrical current will not flow unless a complete circuit is made. In the case of a car, for example, the battery has positive and negative connections. This enables the electric charge to complete a path or circuit from pole to pole, while passing through devices on your car, such as loads and control switches.

The unit of measurement for current is the ampere (abbreviated amp or A). The number of amps produced by the welding machine determines the amount of heat available to melt the workpiece and the electrode. The current is increased or decreased according to the size of electrode being used and the position in which the welding is being performed.

For arc welding to work, an arc must first be created. It is the intense heat of this arc that melts the base metal. This arc is created by bringing the positive and negative electrical connections of the power source (the welding machine) close enough together that the circuit can be completed. To arc weld, an electrical current flow between the two welding machine leads must be established. The base metal being welded and the electrodes used as conductors complete the circuit.

The arc that is produced during SMAW is the result of an electrical current that arcs across a gap between the tip of the electrode and the surface of the workpiece, as long as the workpiece clamp from the SMAW lead is attached to the workpiece (*Figure 3*). The current originates in the transformer or generator of the welding machine, is routed through one welding lead, then through the workpiece, and finally returns to the transformer or generator through the other welding lead. If the electrode tip is touched and held against the workpiece instead of allowing for a gap between the two, the electrode will pass the electricity directly to the workpiece without establishing an arc. As long as the electrode is in physical contact with the base metal, it has the same effect as connecting two wires together. The maximum available current from the welding machine will be passed through this completed circuit. However, no welding will take place in this condition as no continuous arc has been established. The intense heat of an arc crossing the gap is needed to heat the workpiece and the electrode to a molten state. Therefore, the electrode must remain close to the metal, but without making direct contact.

29107-15_F03.EPS

Figure 3 Striking an arc.

> **WARNING!**
> Never remove your welding helmet if the electrode is stuck or frozen to the workpiece. You must free the electrode with your helmet in place because the electrode will flash as it breaks free. First remove the holder from the electrode, then remove the electrode with pliers.

1.2.1 Types of Welding Current

SMAW power supplies are designed to produce an output that has a nearly constant current. The current may be either AC or DC. The type of current used depends on the type of welding to be done and the welding equipment available. Voltage is the measure of the electromotive force or pressure that causes current to flow in a circuit.

Alternating current (AC) in welding machines is derived from either a transformer-type welding machine or an alternator-type machine. The transformer-type machine changes high-voltage, low-current commercial AC power to the low voltage, high-current power required for welding. An alternator-type machine uses an electric or fuel-driven motor to turn a generator to produce the current needed for welding.

An AC voltage is developed by rotating conductors in a magnetic field, as shown in *Figure 4(A),* which symbolizes a generator in its simplest form and the alternating current waveform it produces. AC power alternates constantly from positive to negative, as shown. The waveform is known as a sine wave. The top half of the sine wave is considered positive. The current follows the voltage. In the bottom half of the cycle, the voltage is considered negative, with the current continuing to follow the voltage. In one complete cycle, the waveform crosses the zero point twice. The peaks or extreme high or low points on the wave indicate the maximum level of the voltage, either positive (on the top) or negative (on the bottom). The number of cycles completed in one second is called the frequency. Frequency varies depending on the number of poles (magnetic field coils) in the generator and on the rotational speed of the rotor. Frequency, or cycles per second, is stated in Hertz (Hz). The standard frequency in the United States is 60 Hz. In many other regions of the world, the standard power frequency is 50 Hz.

In a power plant, huge turbine generators, like the one shown in *Figure 4(B)* create the AC voltage that is distributed to homes and businesses.

Direct current (DC) is electrical current that has no frequency. It does not travel from positive to negative and negative to positive within the same operation. Rather, the waveform for DC power resembles a straight line. Direct current in welding is derived either from a transformer-**rectifier** type of system or from a generator. In a transformer-rectifier type of system, AC voltage first is reduced (transformed) to a lower voltage level by a transformer, then it is converted (rectified) into DC voltage. Most engine-powered SMAW welding machines produce DC current through a generator. SMAW units that plug into 60 Hz AC power generally produce DC current using a transformer and a rectifier.

> ## Striking an Arc
> Two general methods of striking the initial arc are scratching the electrode on the workpiece and tapping the electrode on the workpiece. The scratching method is generally used by most beginners and even by many experienced welders. However, it is usually not acceptable for code or finished-product welding if arc strikes are within the area covered by the weld. Some of the newer inverter power sources do not activate the open-circuit voltage until after an electrode touches a surface. With these machines, a scratching motion must be accomplished in order to activate the required open-circuit voltage and start the arc.

1.2.2 Polarity

Polarity only applies to DC current. Polarity in welding is determined by the way the welding leads are connected either to the welding machine or to the workpiece. DC current travels from the negative pole to the positive pole. The minus-marked terminal on the welding machine indicates the negative pole; the plus-marked terminal indicates the positive pole. In welding or in any other DC circuit, one cannot change that process. You can, however, change the path the current takes in going from the negative pole to the positive pole. In welding, this can be done simply by changing the welding lead connections at the machine or, in newer machines, by changing a switch position on the machine.

Most welding leads are configured with one end of one lead equipped with an electrode holder and the other lead equipped with a clamping device to attach to the workpiece. Both of the other ends of these two cables typically are equipped with ring-type or plug-in connectors for attaching the leads to the welding machine.

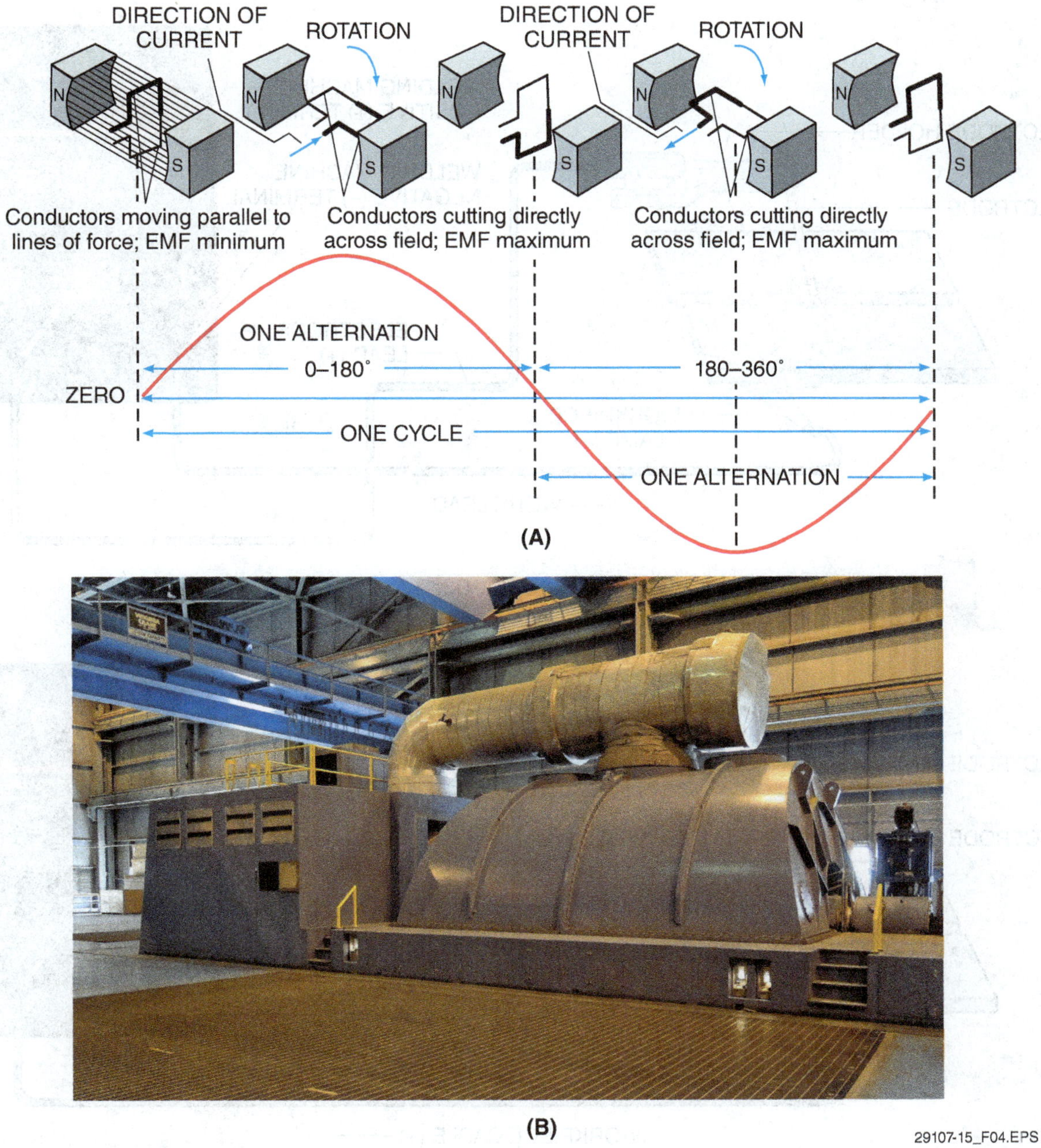

Figure 4 AC development of an AC voltage waveform.

When the electrode-holder lead is connected to the plus (positive) terminal on the machine and the lead with the clamping device is connected to the minus (negative) terminal on the welding machine, direct current electrode positive (DCEP) polarity is established. If these leads are manually switched at the welding machine terminals or through the operation of the manual polarity switch located on the machine, the electrode-holder lead becomes the negative lead and the workpiece clamping device lead becomes the positive lead. This is referred to as direct current electrode negative (DCEN) polarity. *Figure 5* shows labeled diagrams of DCEP and DCEN hookups.

CAUTION

On machines that have polarity switches, do not switch the polarity during operation. This may damage the machine.

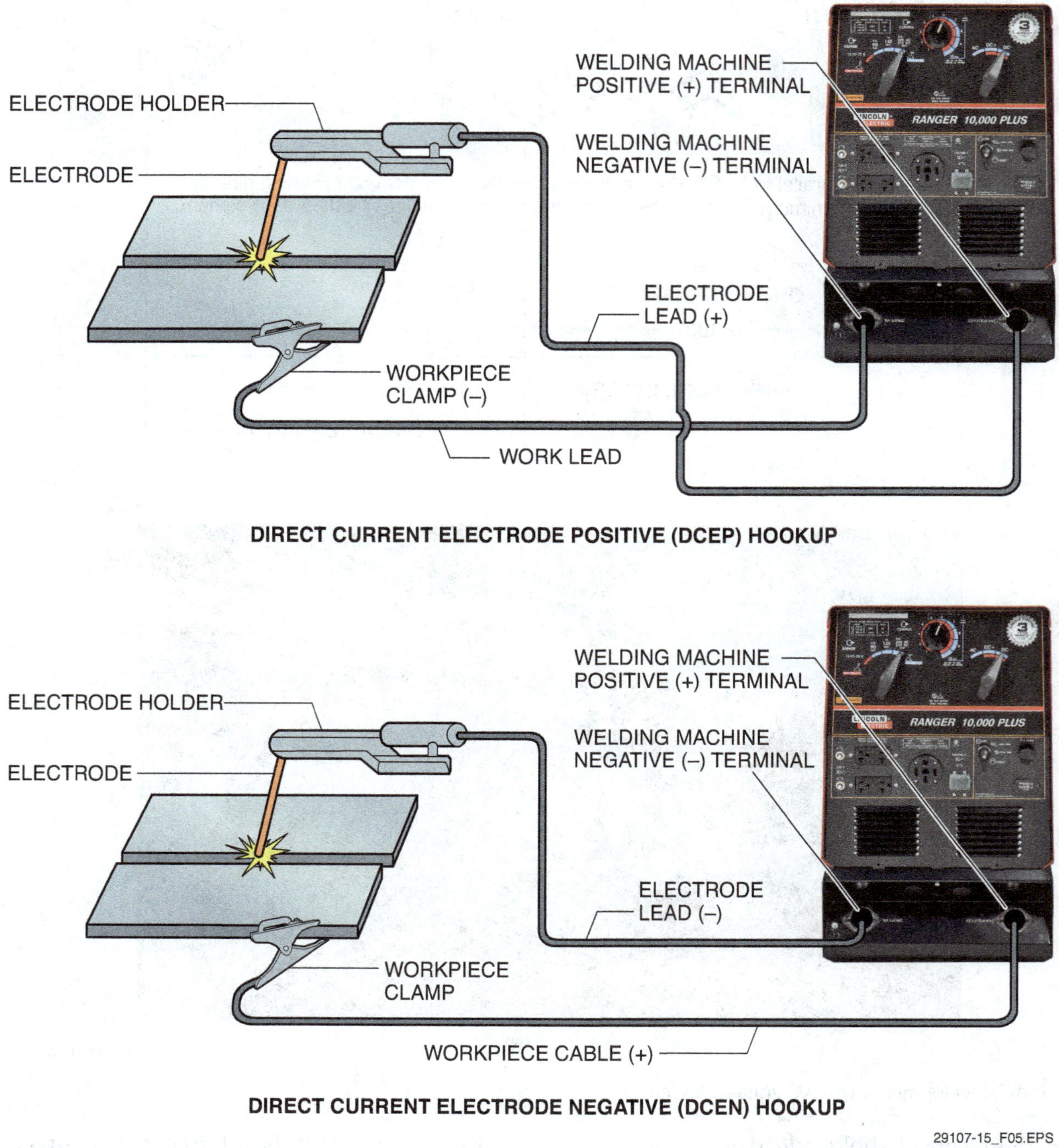

Figure 5 DCEP and DCEN hookups.

1.2.3 Characteristics of Welding Current

The voltage supplied to a welding machine is reduced by a **step-down transformer**. The step-down transformer reduces the voltage level and raises the current level. The higher current allows the arc to occur at a lower voltage potential that reduces the risk of electrocution. However, electrical shock in welding remains a potential hazard that can be reduced or eliminated only by using proper PPE and applying safe welding practices.

Figure 6(A) shows a representation of a step-down transformer. *Figure 6(B)* shows the physical appearance of a transformer. It consists of coils of wire wound around an iron coil.

Voltage is the measure of the electromotive force or pressure that causes current to flow in a circuit. Two types of voltage are associated with welding current: open-circuit voltage and operating voltage. Open-circuit voltage is the voltage present when the welding machine is on but no arc is present. For SMAW, the open-circuit voltage potential between the two leads is usually between 50 and 80 volts.

Operating voltage, or arc voltage, is the voltage measured after the arc has been struck. This voltage is lower than open-circuit voltage and usually measures between 18 and 45 volts.

A higher, open-circuit voltage is required to establish the arc because the air gap between the electrode and the work has high resistance to current flow. Once the arc has been established, less voltage is needed, and the welding machine automatically compensates by lowering the voltage. This characteristic of an SMAW machine is called variable-voltage, constant-current power and is shown as a volt-ampere output curve in *Figure 7*. Operating voltage is dependent on the current range selected. However, SMAW voltage can be altered somewhat by a change in the arc gap controlled by the welder.

Amperage is a measurement of the electric current flowing in a circuit. The unit of measurement for current is the ampere (amp or A). The number of amps produced by the welding machine determines the amount of heat available to melt the workpiece and the electrode. The current is increased or decreased according to the size of electrode being used and the position in which the welding is being performed.

Properties of an Arc

An arc produces a temperature in excess of 6,000°F (3,316°C) at the tip of an electrode. This extreme heat is what melts the base metal and electrode, producing a pool of molten metal often referred to as a weld puddle. As the weld puddle solidifies behind the weld, the bond is formed.

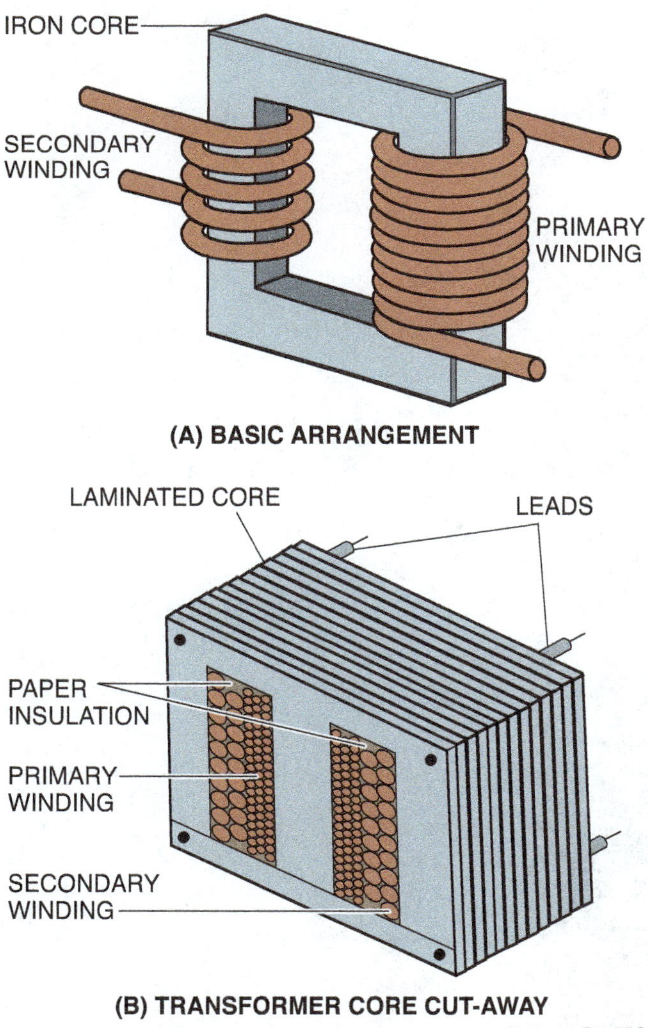

Figure 6 A step-down transformer.

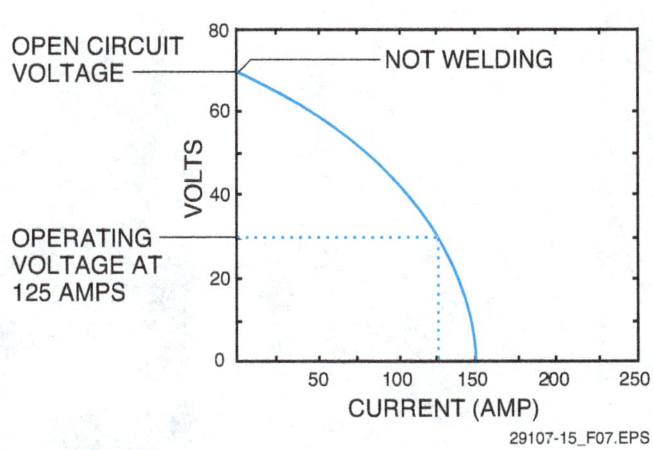

Figure 7 Volt-ampere output curve.

Additional Resources

2014 Technical Training Guide. Current Edition. Cleveland, OH, USA: The Lincoln Electric Company. **www.lincolnelectric.com**

Stick Electrode Product Catalog. Current Edition. Cleveland, OH, USA: The Lincoln Electric Company. **www.lincolnelectric.com**

1.0.0 Section Review

1. When hoisting heavy objects that tend to spin or swing, always use a(n) _____.
   a. pickup line
   b. chain bridle
   c. tag line
   d. synthetic web sling

2. When viewing an alternating current (AC) sine wave, the top half of the cycle of the voltage path is considered _____.
   a. a variable-voltage
   b. the positive half
   c. a constant-current
   d. the negative half

SECTION TWO

2.0.0 SMAW EQUIPMENT

Objective

Identify and describe SMAW equipment.

a. Identify and describe various types of SMAW machines.
b. Identify and describe SMAW welding cable and connectors.
c. Identify common tools used to clean a weld.

Trade Terms

Duty cycle: The percentage of a ten-minute period that a welding machine can continuously produce its rated amperage without overheating.

Governor: A device used to limit engine speed.

Motor-generator: A combination device in which the motor turns the shaft of a generator, which in turn produces an AC voltage.

Welding machines, like any other tools, have specific classifications, special types, and rated capacities. This section will explain the use of SMAW welding machines within these specified areas.

2.1.0 SMAW Machines

SMAW welding machines are classified by the type of welding current they produce: AC, DC, or AC/DC.

An SMAW welding machine classified as an AC machine will produce only AC welding voltage and current, and a DC machine will produce only DC welding voltage and current. If a welding machine produces both AC and DC welding voltage and current, it is classified as an AC/DC welding machine.

Welding machines that produce only DC welding voltage and current can be further classified by the characteristics of the welding current they produce. If the welding current can be varied between the higher open-circuit voltage and the lower operating voltage, it is classified as a variable-voltage constant-current power source. This type of welding machine is used for SMAW. If open-circuit and operating voltage ratings are nearly the same, the machine is classified as a constant-voltage or constant-potential DC welding machine. This type of machine is used for GMAW and FCAW. *Figure 8* shows the difference in outputs between constant-current and constant-voltage welding machines.

Several different types of basic welding machines are available to produce the current necessary for SMAW. They include transformers, transformer rectifiers, inverter power sources, and engine-driven generators.

Transformers, transformer-rectifiers, and electric **motor-generators** all require electrical power from commercial power lines to operate. This high-voltage, relatively low-amperage current from commercial power lines coming into the welding machine is called the primary current. This is related to the transformer in the machine. The windings connected to the incoming power source are called the primary windings.

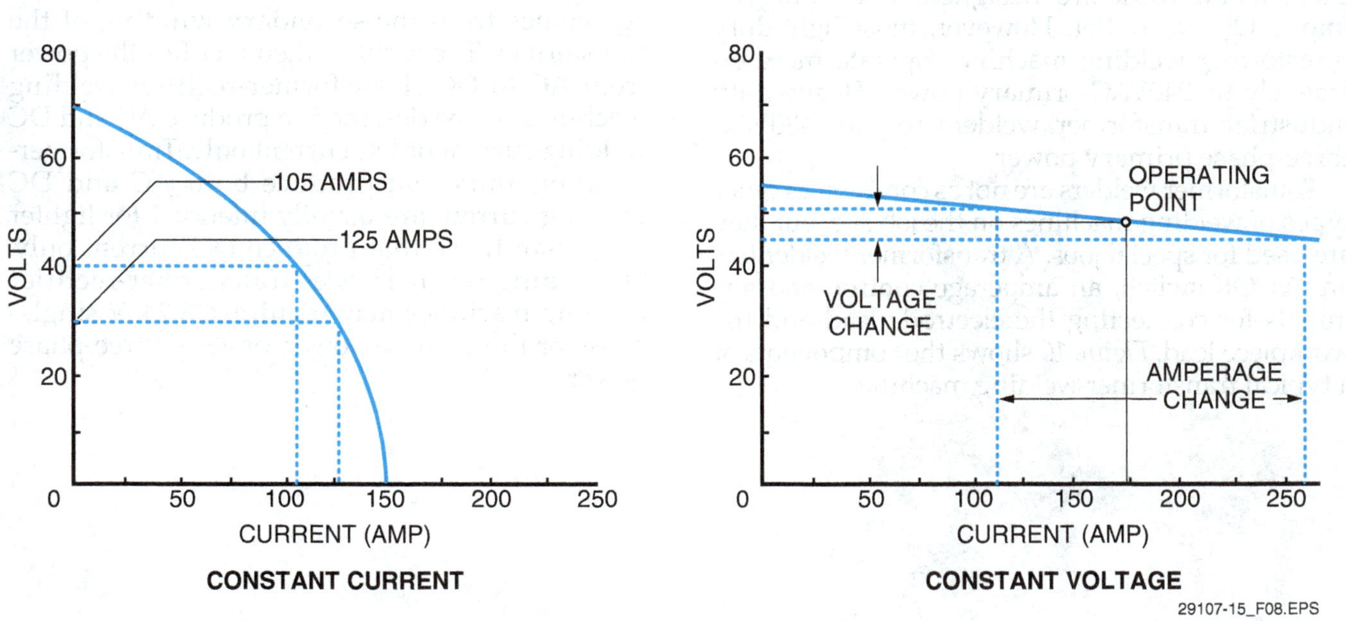

Figure 8 Constant-current and constant-voltage output curves.

The primary current first flows from the power lines through an electrical circuit breaker or fused disconnect switch. From the disconnect switch, the current flows to a receptacle. Alternatively, the welding machine is permanently and directly wired into the power disconnect. The disconnect switch can be used to shut off the flow of power. If necessary, the welding machine power cord is plugged into the receptacle. The primary current required for welding machines can be 120VAC single-phase, 208–240VAC single-phase or three-phase, or 480VAC three-phase. *Figure 9* shows a typical welding machine circuit.

WARNING!

Coming into contact with the primary power source of a welding machine can cause electrocution. Ensure that welding machines are properly grounded.

2.1.1 Transformer Welding Machines

Transformer welding machines without rectifiers produce AC welding current only. They use a step-down transformer, which converts high voltage (line voltage) from commercial power lines to a lower voltage. The construction of the transformer also allows the output current to be significantly higher than the current coming into the machine. This lower-voltage, high-current power supply is needed to arc weld.

The primary power required for a transformer welder may be 240VAC single-phase or three-phase, or 480VAC three-phase. Special light-duty transformer-based welding machines used for sheet metal work are designed to be plugged into a 120VAC outlet. However, most light-duty transformer welding machines operate more effectively on 240VAC primary power. Heavy-duty industrial transformer welders require 480VAC three-phase primary power.

Transformer welders are not as common as other types of welding machines on the job site, but they are used for special jobs. A transformer welder has an On/Off switch, an amperage control, and terminals for connecting the electrode lead and the workpiece lead. *Figure 10* shows the components of a typical transformer welding machine.

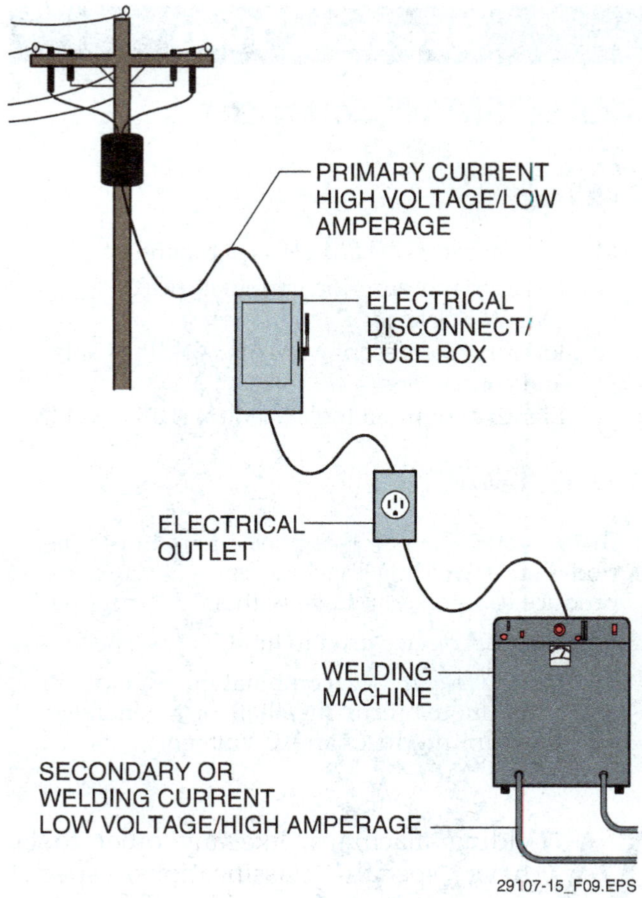

Figure 9 Welding machine primary circuit.

2.1.2 Transformer-Rectifier Welding Machines

A rectifier is a device that converts AC current to DC current. A transformer-rectifier welding machine uses a transformer to reduce the primary voltage to the lower welding voltage. This voltage comes from the secondary winding of the transformer. The rectifier then rectifies the power from AC to DC. Transformer-rectifier welding machines can be designed to produce AC and DC welding current or DC current only. Transformer-rectifier units that produce both AC and DC welding current are usually intended for lighter duty than those that produce DC current only. Depending on their size, transformer-rectifier welding machines may require 208-240V single-phase or three-phase power, or 480V three-phase power.

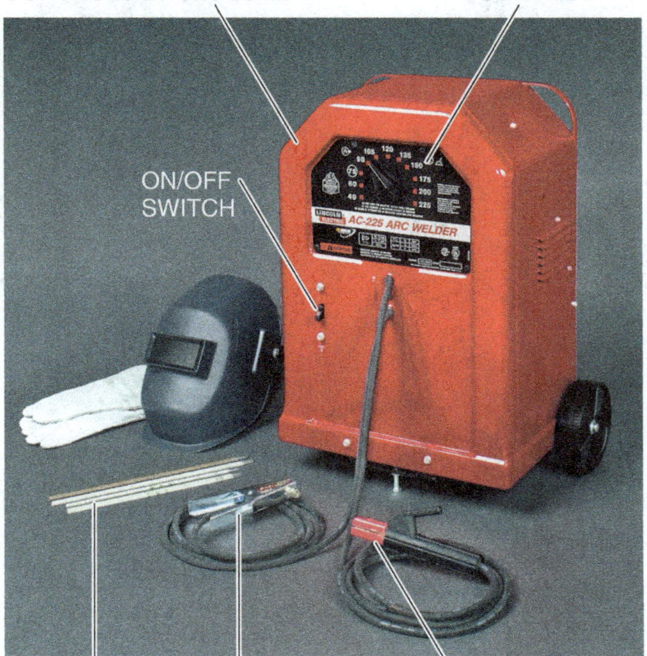

Figure 10 A basic transformer welding machine.

> **WARNING!**
>
> Coming into contact with the primary power serving a welding machine can cause electrocution. Ensure that you take proper safety precautions to avoid contact.

Transformer-rectifier units can also be designed to produce variable-voltage constant-current or constant-voltage DC welding current. Some are even designed to provide both variable and constant voltage; those are referred to as multi-process power sources since they can be used for many welding processes. Refer to the manufacturer's documentation to determine how the machine is configured.

Transformer-rectifiers have an On/Off switch and an amperage control. The welding cables (electrode lead and workpiece lead) are connected to terminals. They often have selector switches to select DCEP or DCEN, or to select AC welding current if it is available. If there is no selector switch, the cables must be manually changed on the machine terminals to select DCEP or DCEN operation. *Figure 11* shows the various parts of a typical transformer-rectifier welding machine.

Multiple transformer-rectifiers are available grouped into a single cabinet, called a pack.

Figure 11 A transformer-rectifier welding machine.

When the pack contains six welding machines, it is called a six-pack; when it contains eight welding machines, it is called an eight-pack (*Figure 12*).

The pack in *Figure 12* has eight individual welding machines. They receive their electrical power from a power distribution panel at the end of the rack. The panel has one switch that allows all the machines to be turned off at once. Each individual welding machine has its own amperage control and welding cables. In addition, each welding machine has a receptacle for a remote current control. When the remote current control is plugged in, it can be used to adjust the amperage from a remote location without going back to the pack. The pack unit also has 120V receptacles for power tools and lights.

Packs are mounted on skids and have attachments for lifting so they can be moved easily. They require only one primary power connection, which is normally 480VAC three-phase, since a significant power supply is needed to power all the units at once. Packs are common on construction sites and existing facilities during rebuilds when many welding machines are required.

AC/DC Machines

Transformer-rectifier welding machines that can provide either AC or DC current bypass the rectifier circuit when welding with AC current. In this situation, the welding machine functions like a transformer welding machine.

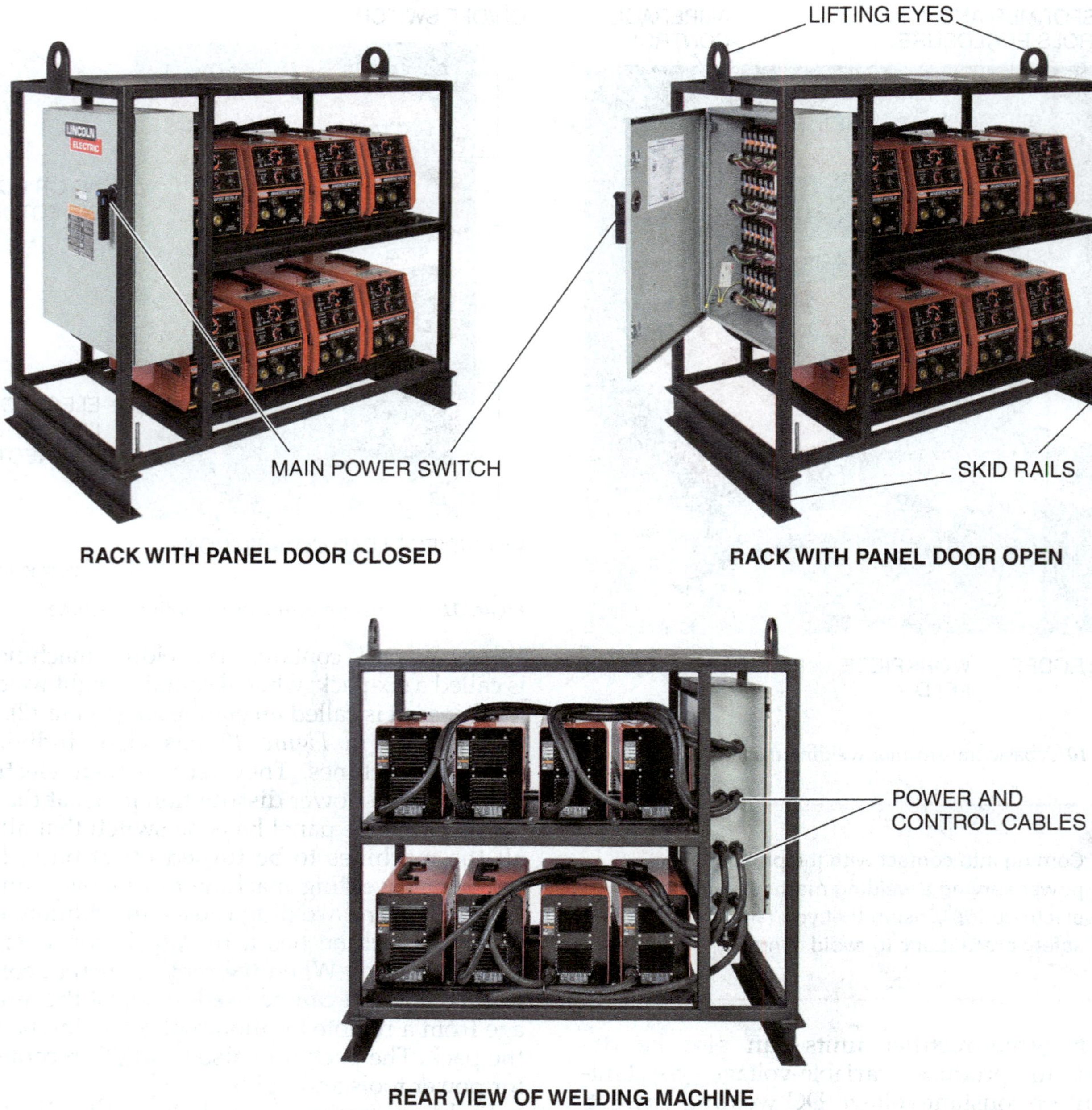

Figure 12 Eight-pack welding machine.

2.1.3 Inverter Power Sources

Inverter-type welding machines provide a smaller, lighter power source with a faster response time and much more waveform control. Inverter power sources are used where limited space and portability are important factors. In contrast to the transformer-types welding machine, an inverter machine first converts the AC input voltage to DC in a rectifier. The resulting DC is then switched on and off at a high rate, resulting in a high-frequency DC voltage, which is then fed into a transformer. The output of the transformer is then converted to DC in a rectifier. A significant advantage of inverter power supplies is that they produce a very stable DC output, without the ripple effect that results at 60 Hz. This results in a very smooth and steady DC welding arc. The controls on these machines vary according to size and application. *Figure 13* shows a close-up view of an inverter power source.

Another advantage lies in size reduction; inverter machines use much smaller transformers than transformer machines, because much of the work is done by lightweight electronic components.

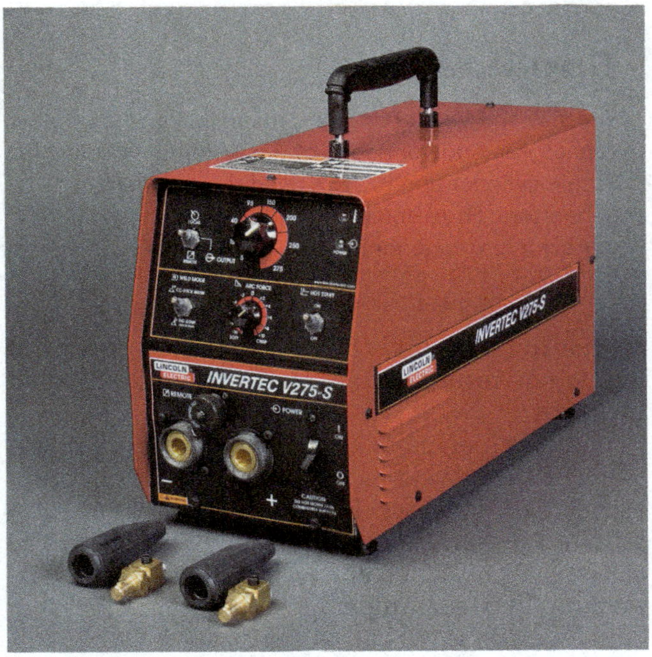

29107-15_F13.EPS

Figure 13 An inverter power source.

The inverter power source shown is designed for outdoor use. It is used either as a single, standalone unit or can be mounted in a rack with others. It supports both gas tungsten arc welding (GTAW) and SMAW processes, and can be used for air-carbon arc gouging. With up to 275 amps of output power, it allows the welder to choose from a greater variety of stick electrodes, especially the E6010 and E7018 electrodes.

Another advantage of the inverter power supply is cost of operation. Inverter technology is more energy-efficient than a transformer-based system. An inverter unit may draw roughly 29 amps when operating at a 200 amp output on 230VAC single-phase power. The older transformer-based designs typically draw 50 to 60 amps at the same conditions. Over time, the electrical cost savings is quite significant, helping to return the investment in cost.

2.1.4 Engine-Driven Generator Machines

Welding machines can also be powered by gasoline or diesel engines. The engine is connected to a generator. Engine-driven generators produce DC welding current.

To produce welding current, the generator must turn at a required number of revolutions per minute. Engines that turn generators have governors to control the engine speed. Most governors have a welding speed switch. The switch can be set to idle the engine when no welding is taking place. When the electrode is touched to the base metal, the governor will automatically increase the speed of the engine to the required rpm for welding. After about 15 seconds of a no-welding condition, the engine will automatically return to idle. This saves fuel and reduces wear on the generator. The switch can also be set to enable the engine to run continuously at welding speed. *Figure 14* shows an example of an engine-driven generator welding machine.

Engine-driven generators often have an auxiliary power unit to produce 120VAC current for lighting, power tools, and other electrical equipment. When 120VAC power is required to operate tools, the engine-driven generator must run continuously at its welding speed.

Engine-driven generators have both engine controls and welding machine controls. The engine controls vary with the type and size but normally include the following items:

- Starter
- Voltage gauge
- Temperature gauge
- Fuel gauge
- Hour meter

Engine-driven generators have an amperage control. To change polarity, there may be a polarity switch, or you may have to manually exchange the welding cables at the terminals.

Many engine-driven generators are mounted on trailers. This makes them portable so they can be used in the field where electricity is not available for other types of welding machines.

29107-15_F14.EPS

Figure 14 Engine-driven welding machine.

The disadvantage of engine-driven generators is that they are costly to purchase, operate, and maintain because of the added cost of the drive unit. However, for field operations where electrical power sources are often very limited, they are essential.

2.1.5 SMAW Machine Ratings

The capacity of a welding machine is determined by the amperage output of the machine at a given duty cycle. The duty cycle of a welding machine is based on a ten-minute period of time. It is the percentage of a ten-minute period that the machine can continuously produce its rated amperage without overheating. For example, a machine with a rated output of 300A at a 60-percent duty cycle can deliver 300A of welding current for six minutes out of every ten (60 percent of ten minutes) without overheating.

The duty cycle of a welding machine will generally be 10 percent, 20 percent, 30 percent, 40 percent, 60 percent, or 100 percent. A welding machine having a duty cycle of from 10 to 40 percent is considered a light-duty to medium-duty machine. Most industrial, heavy-duty machines for manual welding have at least a 60-percent duty cycle. Machines designed for automatic welding operations must be capable of a 100-percent duty cycle.

With the exception of 100-percent duty-cycle machines, the maximum amperage that a welding machine will produce is always higher than its rated capacity. A welding machine rated 300A with a 60-percent duty cycle will generally produce a maximum of 375A to 400A. However, since the duty cycle is a function of its rated capacity, the duty cycle will decrease as the amperage is raised over 300A. Welding at 375A with a welding machine rated at 300A will lower the duty cycle from 60 percent to about 30 percent.

GOING GREEN

Welding Machine Maintenance

When performing maintenance on fuel-powered welding machine engines, make sure to prevent fuel and oil spills. Such spills, even small ones, can harm the environment. Always clean up and properly dispose of any spilled materials.

Engines

The size and type of engine chosen is based on the capacity and power needs of the welding machine. Single-cylinder engines are used to power small generators. Four- and six-cylinder engines are used to power larger generators.

If welding continues for more than three out of every ten minutes, the machine will probably overheat.

If the amperage is set below the rated amperage, the duty cycle increases. Setting the amperage at 200A for a welding machine rated at 300A with a 60-percent duty cycle will increase the duty cycle to 100 percent. *Figure 15* shows the relationship between amperage and duty cycle.

2.2.0 Welding Cable and Connectors

This section covers the cables and connectors that place welding power into the operator's hands. In addition, this section describes some other common tools used in the welding process.

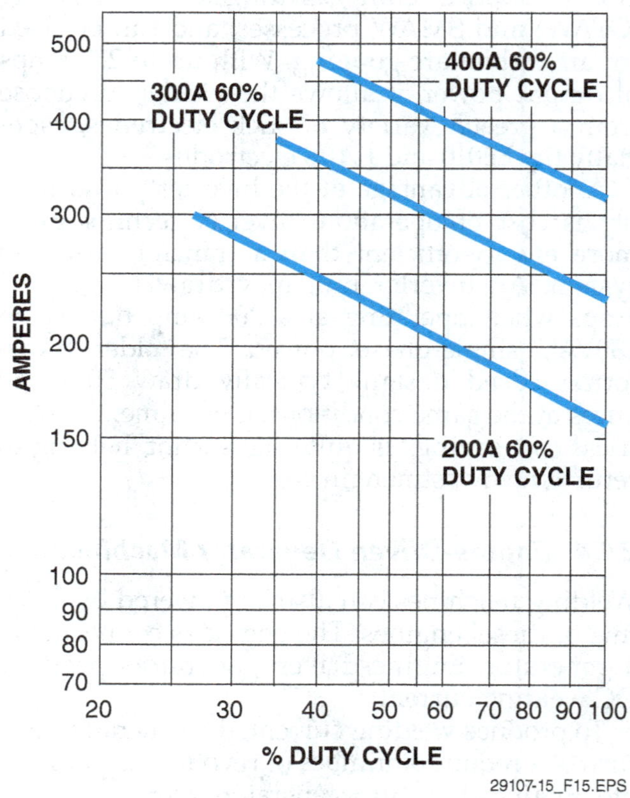

Figure 15 Amperage and duty cycle chart.

Welding Machine Circuit Breakers

Most welding machines have a heat-activated circuit breaker that will shut off the machine automatically when it overheats. The machine cannot be turned on again until it has cooled off and resets.

2.2.1 Welding Cable

Cables that carry welding current (*Figure 16*) are designed for maximum strength and flexibility. The conductors inside the cable are made of fine strands of copper wire. The copper strands are covered with layers of rubber reinforced with nylon or Dacron® cord.

The size of a welding cable is based on the number of copper strands it contains. Large-diameter cable has more copper strands and can carry more welding current. Typically the smallest welding cable size is No. 8 while the largest is No. 4/0 (spoken as *four-aught*).

When selecting welding cable size, the amperage load and the distance the current will travel must be taken into account. The longer the distance the current has to travel, the larger the cable must be to reduce voltage drop and heating caused by the extra electrical resistance of a long cable. This is similar to any other drop cord used on the job site. Longer cords must be a larger wire size to accommodate the load of the connected tools on the other end.

Figure 16 Welding cable.

When selecting welding cable, use the rated capacity of the welding machine for the cable amperage requirement. To determine the distance, measure both the electrode and ground leads and add the two lengths together. To identify the welding cable size required, refer to a recommended welding cable size table furnished by most welding cable manufacturers. *Table 1* shows typical welding cable sizes for various welding currents and cable distances.

Welding cables must be equipped with the proper end connections. End connections used on welding cables include lugs, quick disconnects, ground clamps, and electrode holders.

CAUTION

If the end connection is not tightly secured to the cable, the connection will overheat. An overheated connection will cause variations in the welding current and permanent damage to the connector and/or cable. Repair any connectors that overheat.

2.2.2 Lugs and Quick Disconnects

Lugs are used at the end of the welding cable to connect the cable to the welding machine current terminals. Lugs come in various sizes to match the welding cable size and are mechanically crimped onto the welding cable. Lugs, as shown in *Figure 17*, are normally ringed terminals to prevent the welding lead from pulling loose from the stud connection on the machine. Due to the current that passes through these connections, a sound electrical connection must be made.

Quick disconnects are also mechanically connected to the cable ends. They are insulated and serve to splice two lengths of cable together. Quick disconnects are connected or disconnected with a half twist. When using quick disconnects, make sure that they are tightly connected to prevent overheating or arcing in the connector. The DINSE®-style connector offers increased efficiency with a short, strong turn for a positive stop-lock action and maximum conductivity. Like all connectors, they must be kept clean of dirt, sand, mud, and other contaminants to ensure a good connection.

Table 1 Welding Cable Sizes

Recommended Cable Sizes for Manual Welding						
Machine Size in Amperes	Duty Cycle (%)	Copper Cable Sizes for Combined Lengths of Electrodes Plus Ground Cable				
		Up to 50 Feet	50–100 Feet	150 Feet	150–200 Feet	200–250 Feet
100	20	#8	#4	#3	#2	#1
180	20	#5	#4	#3	#2	#1
180	30	#4	#4	#3	#2	#1
200	50	#3	#3	#2	#1	#1/0
200	60	#2	#2	#2	#1	#1/0
225	20	#4	#3	#2	#1	#1/0
250	30	#3	#3	#2	#1	#1/0
300	60	#1/0	#1/0	#1/0	#2/0	#3/0
400	60	#2/0	#2/0	#2/0	#3/0	#4/0
500	60	#2/0	#2/0	#3/0	#3/0	#4/0
600	60	#3/0	#3/0	#3/0	#4/0	* * *
650	60	#3/0	#3/0	#4/0	* *	* * *

* * Use Double Strand of #2/0
* * * Use Double Strand of #3/0

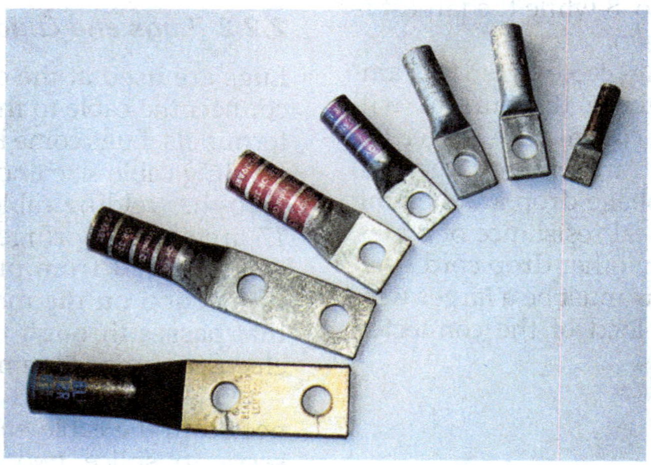

(A) LUGS

MALE

FEMALE

(B) QUICK DISCONNECTS

29107-15_F17.EPS

Figure 17 Lugs and quick disconnects.

2.2.3 Workpiece Clamps

Workpiece clamps (*Figure 18*) establish the connection between the end of the workpiece and the workpiece lead. Workpiece clamps are mechanically connected to the welding cable and come in a variety of shapes and sizes. The capacity or rating of a workpiece clamp is based on the amperage it can carry consistently without overheating.

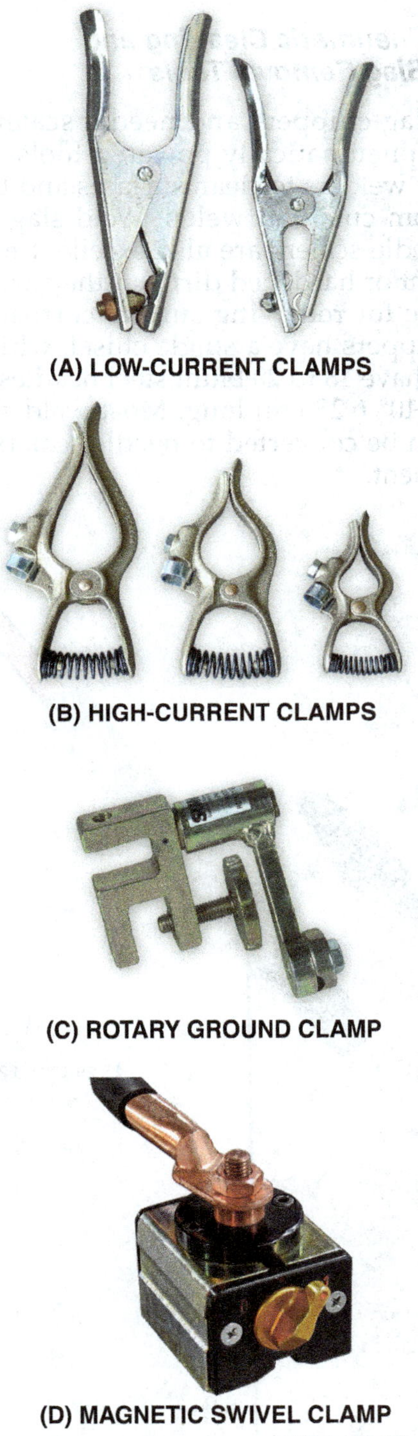

Figure 18 Workpiece clamps.

When selecting a workpiece clamp, be sure that it is rated at least as high as the rated capacity of the welding machine on which it will be used.

2.2.4 Electrode Holders

Electrode holders grasp the electrode and provide the electrical contact between the electrode and the end of the welding cable. Electrode holders are also mechanically connected to the welding cable and come in a variety of styles and sizes. Like the workpiece clamp, the rating of an electrode holder is the rated amperage it will consistently carry without overheating. The physical size of an electrode holder is generally proportional to its current capacity. A smaller 200 amp electrode holder will be light and easy to handle, but will not handle higher currents. When selecting an electrode holder, select the size for the welding current and electrodes being used. *Figure 19* shows two styles of electrode holders. The collet-style typically allows the use of more electrode, as it grips an electrode close to its tip.

2.3.0 Tools for Cleaning Welds

Even though some of the hand tools used to clean welds were presented in another module, it is worthwhile to mention some of them again and explain their functions. Hand tools such as files, chipping hammers, and wire brushes, as well as small power tools such as slag chippers and scalers, are common tools used by welders to prepare surfaces before welding and to clean welds after they are completed.

WARNING!

Always wear eye and hand protection when preparing and cleaning weldment surfaces. Remember that finished welds remain extremely hot for extended periods and can cause serious burns. Chipping slag and other coatings from finished welds is very hazardous and requires proper eye protection, including face shields.

2.3.1 Hand Tools

It is extremely important that the base metals are thoroughly cleaned and prepared before welding. Dirty or corroded surfaces are very difficult, if not impossible, to weld properly. Also, before any additional weld beads can be applied over previous beads, the previous pass must be cleaned of any slag that was deposited with first bead.

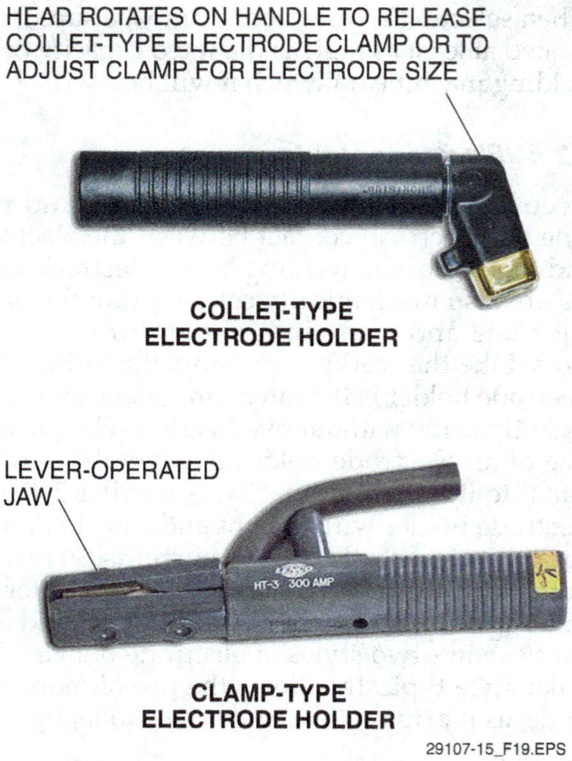

Figure 19 Electrode holders.

Many mechanical methods have been developed to assist in these cleaning and slag-removing processes, including sandblasting and ultrasonic cleaning. However, many welders prefer simple manual procedures, using tools such as files, wire brushes, and chipping hammers. *Figure 20* shows some of the popular hand tools used by welders to prepare base metals and remove slag after a bead has been laid.

2.3.2 Pneumatic Cleaning and Slag Removal Tools

Weld slag chippers and needle scalers (*Figure 21*) are pneumatically powered tools. They are used by welders to clean surfaces and to remove slag from cuts and welds. Weld slag chippers and needle scalers are also excellent for removing paint or hardened dirt, but they are not very effective for removing surface corrosion. Weld slag chippers have a single chisel, while needle scalers have 18 to 20 blunt steel needles approximately 10" (≈25 cm) long. Most weld slag chippers can be converted to needle scalers with an attachment.

Figure 20 Typical hand tools used in welding.

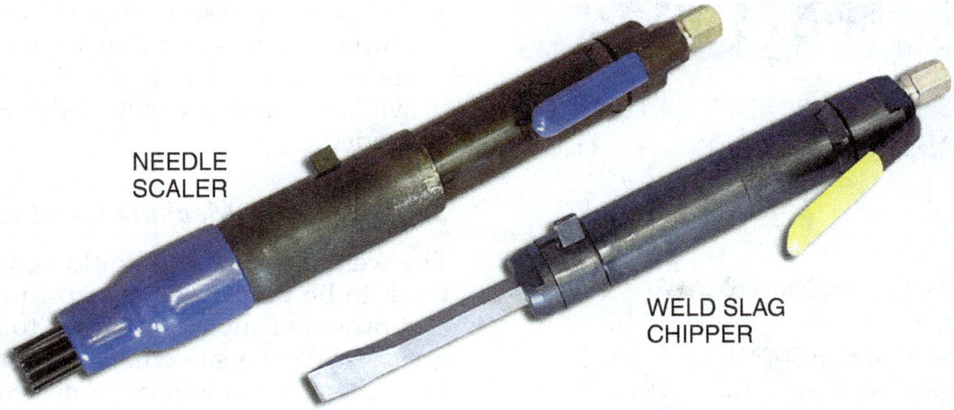

Figure 21 Pneumatic weld slag chipper and needle scaler.

Additional Resources

2014 Technical Training Guide. Current Edition. Cleveland, OH, USA: The Lincoln Electric Company. **www.lincolnelectric.com**

2.0.0 Section Review

1. Heavy-duty industrial transformer welders are most likely to require _____.
   a. 120VAC single-phase power supply
   b. 208VAC single-phase power supply
   c. 240VAC single-phase power supply
   d. 480VAC three-phase power supply

2. The longer the welding leads, the larger the cable should be to _____.
   a. increase conductor resistance
   b. decrease conductor insulation
   c. increase current drop
   d. reduce voltage drop

3. Which of the following tools are more likely to be used for cleaning slag from a weld?
   a. Electric hammer drill, sander, or propane torch
   b. Files, wire brushes, and chipping hammers
   c. A metal pick, steel pry-bar, and drag chain
   d. Cleaning solutions, including alcohol, water, and detergent

SECTION THREE

3.0.0 SMAW EQUIPMENT SETUP AND STARTUP

Objective

Explain how to set up and start SMAW equipment.

a. Explain how to set up SMAW equipment.
b. Explain how to start, stop, and maintain SMAW equipment.

Performance Task

1. Set up a machine for SMAW.

Because of the nature of electricity in general and specifically as it relates to SMAW, using proper procedure is imperative. SMAW welding equipment is not complicated but there are necessary preparations. This section covers some of the typical setup requirements and the operation of SMAW equipment.

3.1.0 SMAW Equipment Setup

In order to safely and efficiently weld using the SMAW process, the equipment must be properly selected and set up. Accomplishing proper SMAW equipment setup involves the following procedures:

- Selecting the proper SMAW equipment
- Choosing the machine's location
- Moving a welding machine into place
- Stringing welding cable
- Locating the workpiece clamp

3.1.1 Selecting the Proper SMAW Equipment

To select the proper SMAW machine, the following factors must be considered:

- *The welding process* – SMAW requires a variable-voltage power source with a constant current. Other processes may require constant-voltage power sources.
- *The type of welding current (AC or DC)* – Most SMAW processes use DC current.
- *The maximum amperage required* – Consider the composition of the material, its thickness, and other material properties to determine the amperage requirements of the machine.
- *The primary power requirements* – If an AC power source with the proper current rating and receptacle for the welding machine is not available, then an engine-driven unit will be needed.

3.1.2 Welding Machine Location

The welding machine should be located near the work to be performed to minimize lead length and provide convenient access to it as work progresses. Select a site where the machine will not be in the way but will be protected from welding or cutting sparks. There should be good air circulation to keep the machine cool, and the environment should be free from explosive or corrosive fumes. It should also be as free as possible from airborne dust and dirt. Welding machines have internal cooling fans that will pull these materials into the machine if they are present. The site should be free of standing water or water leaks. If an engine-driven generator is used, locate it so it can be easily refueled and serviced.

If the machine is to be plugged into an outlet, be sure the outlet has been properly installed by a licensed electrician. Also, be certain to identify the location of the electrical disconnect or circuit breaker for the outlet before plugging in the welding machine.

WARNING!

Never run an engine-driven generator indoors, due to the production of hazardous carbon monoxide (CO) and other pollutants.

3.1.3 Moving a Welding Machine

Large, engine-driven generators are mounted on a trailer frame and can be easily moved by a pickup truck or tractor using a trailer hitch. Other types of welding machines may have a skid base or may be mounted on steel or rubber wheels for manual positioning. Be careful when moving welding machines that are mounted on wheels. Some machines are top-heavy and may fall over in a tight turn or if the floor or ground is uneven or soft.

WARNING!

If a welding machine starts to fall over, do not attempt to hold it. Welding machines are very heavy and can cause severe injury if they roll over.

Many welding machines have lifting eyes. Lifting eyes are usually provided to move machines mounted on skids. Before lifting any welding machine, refer to the equipment specifications for the weight. Be sure the lifting device and tackle are rated at more than the weight of the machine. When lifting a welding machine, always use a proper shackle to connect rigging to the lifting eyes. Never attempt to lift a machine by placing the lifting hook directly in the machine eye. Before lifting or moving a welding machine, make sure the welding cables and other accessories are coiled and secure. *Figure 22* shows how to lift a welding machine. Lifts should be prepared and executed by qualified riggers.

3.1.4 Stringing Welding Cable

Before stringing welding cable, inspect the cable for damage. Cuts or breaks in the insulation must be repaired according to designated procedures. Breaks in the welding cable insulation could arc to equipment or to any metal surface the welding contacts. This will damage the surface and cause additional damage to the welding cable. If the welding cable cannot be repaired, it should be recycled.

Welding cables should be long enough to reach the work area but not so long that they must always be coiled. Welding cables must be strung to prevent tripping hazards or damage from traffic. Keep welding cables out of walkways and aisles. If welding cables must cross a walkway or aisle, string the cables overhead. If the cables cannot be strung overhead, use boards or ramps to prevent workers from tripping and to protect the cable from foot and vehicular traffic. *Figure 23* shows how to protect welding cables with a ramp.

When stringing welding cables, take care not to damage site equipment. Do not string welding cable over instrumentation wires and tubing. These can be easily damaged by the weight or movement of the heavy cables. If there are moving parts, keep the welding cables well away from pinch points and moving parts. If the equipment suddenly starts, the welding cables could be pulled into the equipment. When there is new construction or rebuild work taking place, equipment such as cable trays, conduit, piping, and pipe hangers may not be permanently attached. The weight of welding cables could cause these items to collapse. Check such equipment before stringing welding cables.

When stringing welding cable overhead or between floors, use rope to tie off the cables. The weight of the welding cables could damage connections or pull an individual holding onto the electrode holder off balance. Rope is relatively nonconductive (when dry) and non-abrasive, and will support the weight of the welding cable without damaging it. No significant weight should be placed on the connectors.

Moving Welding Machines on Wheels

It is not uncommon on industrial construction projects to place many welding machines using mobile or fixed cranes, even those on trailers or machines with wheels, because of the limited access for vehicular traffic on these sites.

Figure 22 Lifting a welding machine.

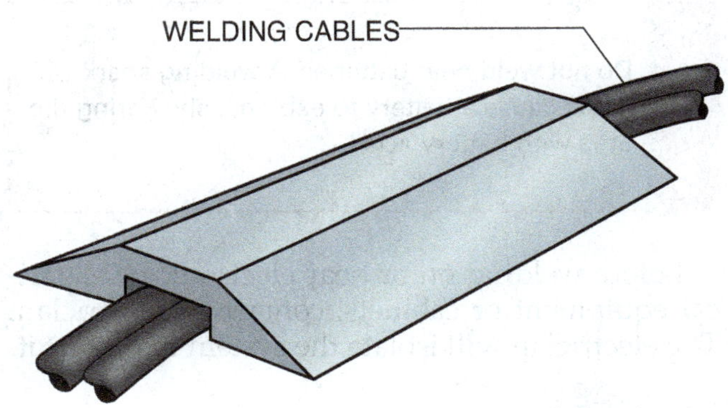

Figure 23 Protecting welding cables.

GOING GREEN

Recycling

Welding cables contain a tremendous amount of copper. The value of copper has been very high for some time now, even encouraging its consistent theft. They should never be disposed of. Instead, the cable should be turned over to a facility so that the copper can be recovered and recycled.

3.1.5 *Locating the Workpiece Clamp*

The workpiece clamp must be properly located to prevent damage to surrounding equipment. If the welding current travels through any type of bearing, seal, valve, or contacting surface, it could cause severe damage from arcing at these points, which would require replacement of the item. Carefully check the area to be welded and position the workpiece clamp so the welding current will not pass through any materials or components unnecessarily. If in doubt, ask your supervisor for assistance before proceeding.

CAUTION

Welding current can severely damage bearings, seals, valves, and contacting surfaces. Position the workpiece clamp to prevent welding current from passing through them.

Welding current passing through electrical or electronic equipment will also cause severe damage. Disconnect any other power sources or batteries to protect the electrical system before welding on any type of equipment. If welding will be performed near a battery, remove the battery completely from the equipment and/or work area. Batteries produce hydrogen gas, which is extremely explosive. A welding spark could cause the battery to explode.

WARNING!

Do not weld near batteries. A welding spark could cause a battery to explode, showering the area with battery acid.

Before welding on or near electronic or electrical equipment or cabinets, contact an electrician. The electrician will isolate the system to protect it.

CAUTION

The slightest spark of welding current can destroy sensitive electronic or electrical equipment. Have an electrician check the equipment and, if necessary, isolate the system before welding.

Workpiece clamps must never be connected to pipes carrying flammable or corrosive materials. The welding current could cause overheating or sparks, resulting in an explosion or fire.

The workpiece clamp must make good electrical contact when it is connected. Dirt and paint will inhibit the connection and cause arcing, resulting in overheating of the workpiece clamp. Dirt and paint also affect the welding current and can cause defects in the weld. Clean the surface before connecting the workpiece clamp. If the workpiece clamp is damaged and does not close securely onto the surface, repair or replace it. Many workpiece clamps have replaceable contact assemblies.

3.2.0 Starting SMAW Welding Machines

Starting an electrically powered welder is a simple matter of flipping a switch and making adjustments. The source of energy is constantly available. Starting a motor-driven welding machine is a bit more complicated than starting an electrically powered machine.

Electrically powered welding machines are plugged into an electrical outlet or hard-wired to a disconnect switch. The electrical requirements vary and will be on the equipment data plate displayed prominently on the machine. Machines requiring single-phase power will have a three-prong plug. Machines requiring three-phase power will have a four-prong plug. *Figure 24* shows examples of some grounded electrical plugs for welding machines. Each plug pattern relates to a particular current capacity for the power supply. This prevents connecting a welding machine to an insufficient power source.

WARNING!

Never use a welding machine until you identify the electrical disconnect location. In the event of an emergency, you must be able to quickly turn off the power to the welding machine at the disconnect to prevent injury or electrocution.

(A) SINGLE PHASE AND THREE PHASE PLUGS

(B) VARIATION OF A THREE PHASE PLUG

29107-15_F24.EPS

Figure 24 Grounded electrical plugs.

If a welding machine does not have a power plug, an electrician must connect the machine. The electrician will add a plug or will hardwire the machine directly into a disconnect switch.

Before welding can take place with an engine-driven welding machine, the engine must be checked and then started. As with a car engine, the engine powering the welding machine must also have routine maintenance performed.

3.2.1 Prestart Checks

Many facilities have prestart checklists that must be completed and signed prior to starting and operating an engine-driven welding machine. If your site has such a checklist, you must complete and sign it. If your site does not have a prestart checklist, perform the following basic checks before starting the engine:

- *Engine oil* – Check the oil when the engine has been shut off for at least 10 minutes. This allows the engine oil to drain back into the crankcase. Check the oil level using the engine oil dipstick. If the oil is low, add the proper amount of appropriate grade oil. Do not overfill.
- *Engine coolant* – Do not remove the radiator cap on a hot engine. Check the coolant level in the radiator if the engine is liquid-cooled. The best time to check the engine coolant is before the engine has been used, such as in the morning when the engine is cold. If the coolant level is low, add coolant. Make sure the radiator cap is seated properly.

Case History

Trailer Mud Flaps

A welder was given the job of reattaching a mud flap to the back of a refrigerated trailer. The loaded trailer was parked right next to the welding machine, and it looked to be a very simple job. Because it was so close, the employee reached over and connected the workpiece clamp to the front of the trailer and then pulled the cable with the electrode holder to the back end. After cleaning and positioning the bracket, the mud flap was welded to the trailer. The entire job took less than 15 minutes to complete.

When the customer returned to pick up the trailer, he noticed that the refrigeration unit was off and there was no power at the display panel. A service technician determined that the microprocessor on the refrigeration unit had failed and needed to be replaced. After further investigation, it was determined that the pressure and temperature sensors on the refrigeration unit had been in the path of the welding current. The current was too high for the microprocessor and burned out some of its components. As a result, the welding shop was required to pay more than $1,000 in parts and labor to replace the microprocessor.

The Bottom Line: If the welder had isolated the electronic components from the welding circuit, or at least had located the workpiece clamp close to the weld point, the damage would not have occurred.

CAUTION

Do not add plain water to radiators that contain antifreeze. Antifreeze not only protects radiators from freezing in cold weather, it also has rust inhibitors and additives to aid in cooling. If the antifreeze is diluted, it will not provide sufficient protection. If the weather turns cold, the system may freeze, causing damage to the radiator, engine block, and water pump.

- *Engine fuel* – Check the fuel level. The unit likely has a fuel gauge or sight glass. If the fuel is low, add the correct fuel (diesel or gasoline) to the fuel tank. The type of fuel required should be marked on the fuel tank. If not, contact your supervisor to verify the fuel required, and have the unit marked.

CAUTION

Adding gasoline to a diesel engine or diesel to a gasoline engine will cause severe engine problems and can cause a fire hazard. Always be sure to add the correct type of fuel to the fuel tank.

- *Engine battery* – If the battery is lead-acid, check the battery water level unless the battery is sealed. Add water if the battery water level is low.

WARNING!

Battery cell vapors can be explosive. Use extreme care when working with or around a lead-acid battery. Avoid exposing the battery to sparks or arcs of any type. Many serious accidents are reported annually due to batteries exploding due to accidental arcing.

- *Electrode holder* – Look at the electrode holder and make sure it is not touching the workpiece. If the electrode holder is touching the workpiece, it will arc and overheat the welding system when the welding machine is started.
- *Fuel shutoff valve* – Open the fuel shutoff valve if the equipment has one. If there is a fuel shutoff valve, it will be located in the fuel line between the fuel tank and the carburetor or fuel injector(s).
- *Hour meter* – Record the hours from the hour meter if the equipment has one. An hour meter records the total number of hours the engine runs. This information is used to determine when the engine needs to be serviced. The hours will be displayed on a gauge similar to an odometer.
- *Engine appearance* – Clean the engine and equipment after each use. Potential problems can be discovered before they can become larger issues if this is part of the daily maintenance routine. Refer to the manufacturer's instructions for specific areas of attention.

3.2.2 Starting the Engine

Most engines will have an On/Off ignition switch and a starter. They may be combined into a key switch similar to the ignition switch on a car. To start the engine, turn on the ignition switch and press the starter. Release the starter when the engine starts. The engine speed will be controlled by the governor. If the governor switch is set for idle, the engine will slow to an idle after a few seconds. If the governor is set to welding speed, the engine will continue to run at welding speed.

Small engine-driven welding machines may have an On/Off switch and a pull rope. These machines are started by turning on the ignition switch and pulling the cord, which is similar to starting a lawn mower. These engines do not have a battery.

Engine-driven welding machines should be started about five to ten minutes before they are needed for welding. This will allow the engine to warm up before a welding load is placed on it.

WARNING!

When you start the welding machine, always make sure the welding electrode and/or electrode holder are secure and not touching any nearby worker or any conductive surface. Engine cowlings, or covers, should be closed and latched prior to starting the machine to protect workers from accidental contact with moving parts.

3.2.3 Stopping the Engine

If no welding is required for 30 minutes or longer, stop the engine by turning off the ignition switch. If you are finished with the welding machine for the day, also close the fuel valve if there is one. Allow the machine to cool before storing welding leads inside the engine compartment.

3.2.4 Preventive Maintenance

Engine-driven welding machines require regular preventive maintenance to keep the equipment operating properly. Most sites will have a preventive maintenance schedule based on the hours that the engine operates. In severe conditions, such as a very dusty environment or cold weather, maintenance may have to be performed more frequently. The responsibility for performing preventive maintenance will vary from site to site. Check with your supervisor to determine who is responsible for performing preventive maintenance.

When performing preventive maintenance, follow the manufacturer's guidelines. Typical tasks to be performed as part of a preventive maintenance schedule include the following:

- Changing the oil, gas filter, and air filter
- Checking/changing the antifreeze
- Greasing the undercarriage
- Repacking the wheel bearings
- Replacing worn drive belts
- Maintaining/replacing the spark plugs
- Maintaining the battery
- Cleaning surfaces

CAUTION

To prevent equipment damage, perform preventive maintenance as recommended by the site procedures or the manufacturer's maintenance schedule in the equipment manual.

GOING GREEN

Fuel-Powered Machine Maintenance

Proper maintenance of fuel-powered welding machine engines helps to protect the environment. If a fuel-powered engine is smoking or leaking fluids, have it turned in for proper maintenance or repair.

Additional Resources

Covering the Basics – Engine Driven Welder Maintenance. Al Nystrom, Technical Service Representative. Cleveland, OH, USA: The Lincoln Electric Company. **www.lincolnelectric.com**

3.0.0 Section Review

1. When lifting a welding machine, connect the lifting eye to the rigging using a _____.
   a. cable lug
   b. shackle
   c. quick disconnect
   d. swivel clamp

2. To prevent a battery from exploding when welding, you must avoid _____.
   a. exposing the battery to extremely cold temperatures
   b. locating the workpiece clamp close to the joint being welded
   c. disconnecting it from the circuit
   d. exposing the battery to sparks or arcs of any type

SUMMARY

Shielded metal arc welding is often referred to as stick welding. SMAW remains one of the most widely used forms of welding in the industry because of its simplicity and relatively low cost. However, it is not as foolproof as some of the newer technologies now used in the welding industry. It requires practice to be a successful SMAW welder. Whatever type of welding is being used, proper selection and setup of equipment and accessories are necessary to achieve a reliable weld while maintaining a safe working environment.

Included in this module are the welding currents associated with SMAW, the various types of SMAW machines, and some of the accessories needed to connect SMAW equipment. The steps in selecting and setting up SMAW equipment are also covered. SMAW is the basis upon which all other welding processes have been developed. As a widely used welding technique, it is important for you to become familiar with the equipment and develop the skills needed to become a successful SMAW welder.

Review Questions

1. SMAW is often referred to as _____.
   a. AC welding
   b. stick welding
   c. gap welding
   d. resistive welding

2. Make sure that your fall protection equipment _____.
   a. is tagged before each use
   b. is inspected prior to each use
   c. is made of nylon
   d. is rated to 600 lbs

3. In the SMAW process, an electrical path must be completed from the tip of the electrode to the surface of the workpiece through the creation of a(n) _____.
   a. gas
   b. arc
   c. electrode holder
   d. transformer

4. In AC current, the number of cycles completed in one second is called the _____.
   a. polarity
   b. travel
   c. frequency
   d. current value

5. The type of power that has no frequency since its waveform is actually a straight line is _____.
   a. direct current (DC)
   b. alternating current (AC)
   c. single-phase power
   d. three-phase power

6. A device that converts AC power to DC power is known as a(n) _____.
   a. conductor
   b. transformer-rectifier
   c. alternator
   d. generator

7. Which of the following terms only applies to DC current?
   a. Frequency
   b. Polarity
   c. Amperage
   d. Voltage

8. The primary voltage in a welding machine is reduced by a(n) _____.
   a. governor
   b. generator
   c. alternator
   d. step-down transformer

9. The voltage that is present at the leads when the welding machine is on but no arc is present is _____ .
   a. open-circuit voltage
   b. operating voltage
   c. zero voltage
   d. minimum voltage

10. If a SMAW welding machine is equipped with a transformer-rectifier, it can be designed to produce _____.
    a. only DC welding current
    b. only AC welding current
    c. both AC and DC welding current
    d. high-frequency pulses for pulse welding

11. Governors on engine-driven generator machines directly control the _____.
    a. engine temperature
    b. engine speed
    c. rectifier
    d. primary power

12. The rated duty cycle of a welding machine is based on _____.
    a. 10 minutes
    b. 15 minutes
    c. 20 minutes
    d. 30 minutes

13. To prevent the weld lead from pulling loose from the welding machine stud connection, terminal welding cables generally use _____.

    a. slotted terminals
    b. soldered-in-place terminals
    c. ringed terminals
    d. spade terminals

14. Engine-driven welding machines should not be operated in poorly ventilated areas because they can produce hazardous _____.

    a. carbon monoxide (CO)
    b. carbon dioxide (CO_2)
    c. hydrogen peroxide (H_2O_2)
    d. methyl ethyl ketone (MEK)

15. Machines requiring single-phase power will have a _____.

    a. two-prong plug
    b. three-prong plug
    c. four-prong plug
    d. five-prong plug

Trade Terms Quiz

Fill in the blank with the correct term that you learned from your study of this module.

1. The direction of flow of electrical current in a DC welding circuit is referred to as the ________.
2. An electrical device that uses two wire coils of different sizes to convert a higher-voltage power source to a lower voltage is a(n) ________.
3. A combination device in which the motor turns the shaft of a generator, which in turn produces an AC voltage is a(n) ________.
4. A material that will support the flow of an electrical current is a(n) ________.
5. A device used to limit engine speed is known as is a(n) ________.
6. An electronic device that converts AC voltage to DC voltage is a(n) ________.
7. The percentage of a ten-minute period that a welding machine can continuously produce its rated amperage without overheating is its ________.

Trade Terms

Conductor
Duty cycle
Governor
Motor-generator
Polarity
Rectifier
Step-down transformer

Cornerstone of Craftsmanship

Jerry Trainor

Anchorage, Alaska

Welding Instructor/College Administrator

Currently semi-retired and working as a part-time welding instructor, Jerry started working as a welder in 1959. Since then, he has developed an impressive list of credentials as a vocational trainer, teacher, and college administrator. Jerry was born in Pablo, Montana, and grew up in Washington's Yakima Valley. He learned on the job to become a journeyman welder/fabricator. He even spent some time as a fourth-grade teacher, where he learned he had a natural talent for teaching. He eventually was appointed assistant professor for welding technology at Lewis-Clark State College in Idaho and has since held several positions as an academic administrator.

How did you choose a career in the construction industry?

I got interested in welding as a result of taking a wood and metal class during my senior year in high school. Then I got my first welding job in Davis, California helping to build a food processing plant for Hunt's Foods. From that time on until I started teaching in college, I welded on many construction and production shop projects.

What kinds of work have you done in your career?

I started as a fabricator/welder working at lumber mills, power plants, wood and pulp plants, and a variety of other facilities over a 24-year period. In 1983, I began my career as an educator, first teaching welding at technical schools and colleges in Idaho and the Pacific Northwest. From 1993 to 2012, I served in administrative roles. My last administrative position was as head of the Construction Trades Technology Academic Program at the University of Alaska-Fairbanks.

What advice would you give to those new to the field?

My advice would be to become as proficient as possible in all phases of welding, from using a cutting torch to welding in SMAW, GMAW, and GTAW. Take an interest in the metallurgy and production of different metal materials, understand complex blueprints, and be open to taking advanced theory courses through a community college or other provider. Always be ready to go the extra mile and put in a full day's work for a full day's pay. Be proud of your skills and abilities and take pride in always doing the best job possible. Be able to walk away from a job knowing that your part of the construction project was done with your best expertise and efforts.

Tell us an interesting career-related fact or accomplishment.

In my position as head of construction trades technology, I analyzed construction projects planned for the Interior and Aleutian regions of Alaska. In these remote areas, contractors usually bring workers with them into the villages whenever a school, home, health clinic, water treatment plant, or power plant needs to be built. The purpose of the Construction Trades Technology (CTT) program is to train local people and to decrease the need for outside workers.

The Aleutians are a chain of 14 large islands and 55 smaller ones that extend about 1,200 miles west from the southwestern tip of Alaska. The program has been very successful, mainly because the training program is delivered in the villages, so the students do not have to leave their environment. Instructors spend from 8 to 10 weeks living in a village and taking part in the local community.

What do you think it takes to be a success in the welding trade?

To be successful in welding, a person has to take pride in doing the best job possible, understand the field of metallurgy, and be able to fabricate simple and complex projects. I enjoy fabricating and designing different implements, working on complex drawings of steel structures, and working with other professionals.

Trade Terms Introduced in This Module

Conductor: A material that will support the flow of an electrical current. Copper wire is the most common conductor.

Duty cycle: The percentage of a ten-minute period that a welding machine can continuously produce its rated amperage without overheating.

Governor: A device that limits engine speed.

Motor-generator: A combination device in which the motor turns the shaft of a generator, which in turn produces an AC voltage.

Polarity: Refers to the direction of electrical current flow in a DC welding circuit.

Rectifier: An electronic device that converts AC voltage to DC voltage.

Step-down transformer: An electrical device that uses two wire coils of different sizes, referred to as the primary and secondary windings, to convert a higher voltage to a lower voltage through induction. Primary windings are connected to the primary power source; the secondary windings provide the reduced voltage to the process.

Additional Resources

This module presents thorough resources for task training. The following resource material is suggested for further study.

2014 Technical Training Guide. Current Edition. Cleveland, OH, USA: The Lincoln Electric Company. **www.lincolnelectric.com**

Covering the Basics – Engine Driven Welder Maintenance. Al Nystrom, Technical Service Representative. Cleveland, OH, USA: The Lincoln Electric Company. **www.lincolnelectric.com**

Stick Electrode Product Catalog. Current Edition. Cleveland, OH, USA: The Lincoln Electric Company. **www.lincolnelectric.com**

Figure Credits

The Lincoln Electric Company, Cleveland, OH, USA, Module Opener, Figures 1–3, 5 (photos), 10–14, 17B, 22 (photo), 24

Topaz Publications, Inc., Figures 4 (photo), 17A, 18A, 18B, 19–21

Courtesy of Sumner Manufacturing Company, Inc., Figure 18C

PowerBase™ Grounding Magnet, Strong Hand Tools, **www.StrongHandTools.com**, Figure 18D

Section Review Answer Key

Answer	Section Reference	Objective
Section One		
1 c	1.1.3	1a
2. b	1.2.1	1b
Section Two		
1. d	2.1.1	2a
2. d	2.2.1	2b
3. b	2.3.1	2c
Section Three		
1. b	3.1.3	3a
2. d	3.2.1	3b

NCCER CURRICULA — USER UPDATE

NCCER makes every effort to keep its textbooks up-to-date and free of technical errors. We appreciate your help in this process. If you find an error, a typographical mistake, or an inaccuracy in NCCER's curricula, please fill out this form (or a photocopy), or complete the online form at **www.nccer.org/olf**. Be sure to include the exact module ID number, page number, a detailed description, and your recommended correction. Your input will be brought to the attention of the Authoring Team. Thank you for your assistance.

Instructors – If you have an idea for improving this textbook, or have found that additional materials were necessary to teach this module effectively, please let us know so that we may present your suggestions to the Authoring Team.

NCCER Product Development and Revision

13614 Progress Blvd., Alachua, FL 32615

Email: curriculum@nccer.org
Online: www.nccer.org/olf

❑ Trainee Guide ❑ Lesson Plans ❑ Exam ❑ PowerPoints Other ____________

Craft / Level: Copyright Date:

Module ID Number / Title:

Section Number(s):

Description:

Recommended Correction:

Your Name:

Address:

Email: Phone:

SMAW Electrodes

Overview

Shielded metal arc welding (SMAW) processes depend on the use of stick-like electrodes that provide the filler metal for the weld. However, the welding electrode is designed to contribute more than just filler metal to the welding process. There are many different types of electrodes, each of which has its advantages and disadvantages when coupled with the various base metals, welding positions, and weld types. This module provides information on the classification and selection of SMAW electrodes, and how to identify the various types.

Module 29108

Trainees with successful module completions may be eligible for credentialing through the NCCER Registry. To learn more, go to **www.nccer.org** or contact us at **1.888.622.3720**. Our website has information on the latest product releases and training, as well as online versions of our *Cornerstone* magazine and Pearson's product catalog.

Your feedback is welcome. You may email your comments to **curriculum@nccer.org**, send general comments and inquiries to **info@nccer.org**, or fill in the User Update form at the back of this module.

This information is general in nature and intended for training purposes only. Actual performance of activities described in this manual requires compliance with all applicable operating, service, maintenance, and safety procedures under the direction of qualified personnel. References in this manual to patented or proprietary devices do not constitute a recommendation of their use.

29108 V5

29108
SMAW ELECTRODES

Objectives

When you have completed this module, you will be able to do the following:

1. Describe the SMAW electrode classification system and how to select the proper electrode for the task.
   a. Describe the AWS filler metal specification system and various electrode characteristics.
   b. Describe the characteristics of the four main electrode groups.
2. Explain how to select electrodes and describe their proper care and handling.
   a. Identify various considerations in the selection of the proper electrode.
   b. Describe the proper handling and storage of electrodes.

Performance Tasks

This is a knowledge-based module; there are no Performance Tasks.

Trade Terms

Alloy
Condensation
Ductility
Flux
Heat-affected zone
Hermetically sealed
Low-hydrogen electrode
Notch toughness
Traceability
Vertical welding

Industry Recognized Credentials

If you are training through an NCCER-accredited sponsor, you may be eligible for credentials from NCCER's Registry. The ID number for this module is 29108. Note that this module may have been used in other NCCER curricula and may apply to other level completions. Contact NCCER's Registry at 888.622.3720 or go to **www.nccer.org** for more information.

Contents

Topics to be presented in this module include:

Figures and Tables

SECTION ONE

1.0.0 SMAW ELECTRODES

Objective

Describe the SMAW electrode classification system and how to select the proper electrode for the task.

a. Describe the AWS filler metal specification system and various electrode characteristics.
b. Describe the characteristics of the four main electrode groups.

Trade Terms

Alloy: A metal that has had other elements added to it that substantially change its mechanical properties.

Ductility: Refers to the mechanical property of a material that allows it to be bent or shaped without breaking.

Flux: Material used to prevent, dissolve, or facilitate the removal of oxides and other undesirable substances on a weld or base metal.

Heat-affected zone: The part of the base metal that has been altered by heating, but not melted by the heat.

Hermetically sealed: Having an airtight seal.

Low-hydrogen electrode: An electrode specially manufactured to contain little or no moisture.

Notch toughness: The ability of a material to resist breaking at points where stress is concentrated.

Traceability: The ability to verify that a procedure has been followed by reviewing the documentation step-by-step.

Vertical welding: Welding with an upward or downward progression.

Shielded metal arc welding (SMAW) cannot take place without electrodes. SMAW has often been referred to as stick welding, and the electrodes are referred to as stick electrodes or as filler metal. However, there are many different types of electrodes, including wire rolls used in other welding processes. Stick electrodes are the type used in SMAW and the focus of this module.

The welder must be able to distinguish different types of electrodes and select the correct electrode for the job. In addition, the welder must know how to handle and store electrodes, some of which require special considerations. As simple as they appear, their storage and condition at the time of use is essential to weld quality.

This module explains the electrode classification system, electrodes, and the considerations for selecting the correct electrode for the job. It also explains how to properly store electrodes in compliance with the appropriate welding codes.

1.1.0 Electrode Basics and the AWS Specification System

A SMAW electrode has a wire core coated with flux. The wire core transfers the welding current from the electrode holder to the workpiece. The arc at the end of the electrode melts the metal core, the flux coating, and the base metal simultaneously. This occurs at a temperature exceeding 6,000°F (3,316°C). The wire core mixes with the melted base metal, forming and building the weld. The flux coating also has important functions. As the flux coating is melted, it does the following:

- Produces a gaseous shield of carbon dioxide that protects the molten weld by pushing away harmful oxygen and hydrogen present in the atmosphere.
- Acts as all fluxing agents do to clean and deoxidize the molten metal.
- Stabilizes the arc and reduces spatter.
- Forms a slag covering over the weld to further protect it from the atmosphere and slow the cooling rate.

Figure 1 shows an electrode in use.

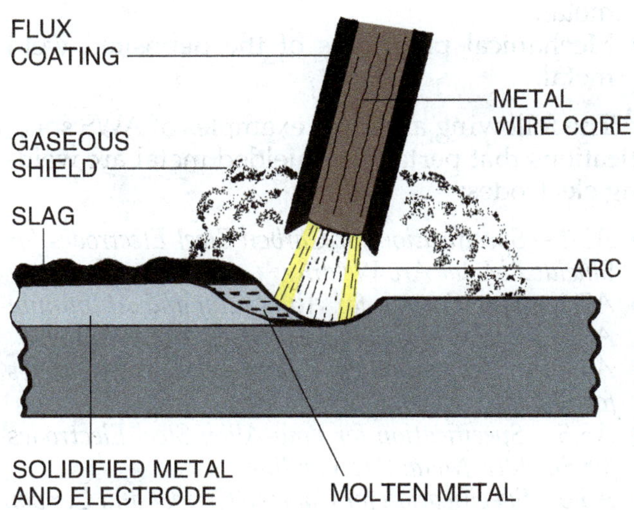

Figure 1 Electrode in use.

Some flux coatings also have powdered metal and alloying elements in them. The powdered metal adds filler metal and helps improve the appearance of the weld. The alloying elements change the chemical composition and strength of the deposited weld metal. **Alloys** are normally added to electrodes so the deposited weld metal will match the base metal chemical composition and strength.

1.1.1 The AWS Filler Metal Specification System

The American Welding Society (AWS) writes specifications for welding consumables, such as electrodes, filler rods, and fluxes, used with all the various welding processes. There are a series of different specifications identified as *AWS A5.xx.* These specifications are used in all major welding codes and accepted in all industries except where additional approvals are specified, such as the American Bureau of Shipping (ABS), the US Coast Guard, the Federal Highway Administration, Military Specifications (MIL), and Lloyds of London. The American Society of Mechanical Engineers (ASME) Boiler Code and the European Norm (EN) electrode specifications closely follow the AWS specifications.

The purpose of the specifications is to set standards that all manufacturers must follow when manufacturing welding consumables of all types. This ensures consistency for the user regardless of which company manufactured the product. The specifications set standards for the following:

- Classification system, identification, and marking.
- Chemical composition of the deposited weld metal.
- Mechanical properties of the deposited weld metal.

The following are some examples of AWS specifications that pertain to shielded metal arc welding electrodes:

- *A5.1 – Specification for Carbon Steel Electrodes for Shielded Metal Arc Welding*
- *A5.3 – Specification for Aluminum and Aluminum Alloy Electrodes for Shielded Metal Arc Welding*
- *A5.4 – Specification for Stainless Steel Electrodes for Shielded Metal Arc Welding*
- *A5.5 – Specification for Low-Alloy Steel Electrodes for Shielded Metal Arc Welding*
- *A5.6 – Specification for Covered Copper and Copper Alloy Arc Welding Electrodes*
- *A5.11 – Specification for Nickel and Nickel-Alloy Welding Electrodes for Shielded Metal Arc Welding*
- *A5.15 – Specification for Welding Electrodes and Rods for Cast Iron*
- *A5.16 – Specification for Titanium and Titanium Alloy Welding Electrodes and Rods*
- *A5.21 – Specification for Bare Electrodes and Rods for Surfacing*
- *A5.24 – Specification for Zirconium and Zirconium Alloy Welding Electrodes and Rods*

Obviously an in-depth study of each classification would require a significant amount of time. As your training and experience progress, a deeper understanding of the less common electrode types will likely be required. The most common electrodes are discussed in the *A5.1* and *A5.5* specifications. These are the electrodes that are most important at this point.

The electrodes described in the *A5.1* specification are used to weld carbon (mild) steel. The electrodes in the *A5.5* specification are used to weld low-alloy, high-strength steels. These two specifications will be explored further in this module.

The specification number is generally followed by the year of publication, such as *A5.1:2004* or *A5.5:2006.* The year of publication is the year that the specification was last revised. These documents are normally on a five-year revision cycle. Always refer to the latest version when information is needed.

1.1.2 A5.1 and A5.5 Classification Systems

The classification system that identifies *AWS Specification A5.1* carbon steel electrodes and *A5.5* low-alloy electrodes uses the prefix E followed by four or five numbers, such as E6010 or E10018. This is printed on each electrode near the end where the electrode holder grasps the electrode. These markings are shown in *Figure 2*.

Refer to *Figure 3* to review the meaning of the markings:

- *E* stands for electrode. Welding current flows through an electrode.
- The first two digits, or three if 100 or over, multiplied by 1,000 designate the minimum tensile strength in pounds per square inch of the deposited weld metal. For example:
  - E6010 has a minimum tensile strength of 60,000 psi.
  - E7018 has a minimum tensile strength of 70,000 psi.
  - E12018 has a minimum tensile strength of 120,000 psi.

29108-15_F02.EPS

Figure 2 Electrode markings.

Sizing and Packaging of SMAW Electrodes

SMAW electrodes are sized according to the wire core diameter, and are commonly packed in 10- and 50-pound cans, as shown in the photo.

29108-15_SA01.EPS

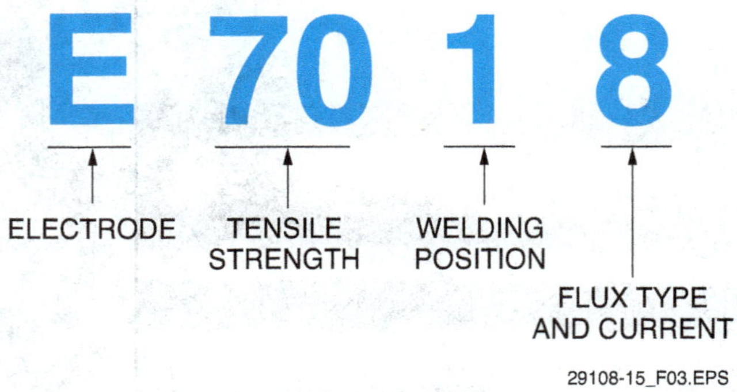

Figure 3 Electrode marking system.

- The third or fourth digit indicates the possible welding positions:
  - *1* can be used to weld in all positions.
  - *2* can be used to weld in flat and horizontal fillet-weld positions only.
  - *4* can be used to weld in flat, horizontal, overhead, and vertical-up positions.
- The fourth or fifth digit indicates special characteristics of the flux coating and the type of welding current required, such as alternating current, direct current electrode negative (DCEN), or direct current electrode positive (DCEP).

Table 1 shows the characteristics of mild steel electrodes covered by the *A5.1* specification. Note that the Xs shown in the AWS Class column refer to the first two digits of the number, which may be 60, 70, or other digits. For example, E6018, E7018, and E12018 electrodes share the same flux characteristics shown in the bottom row of the table.

The low-alloy steel electrodes that are covered by AWS Specification *A5.5* have a suffix in addition to the standard classification numbers. The suffix has a letter and a number. The letter indicates the chemical composition of the deposited weld metal, and the number indicates the composition of the chemical classification. *Figure 4* shows an electrode classification marking from the *A5.5* specification.

The suffix letters indicate the following:

- *A* – carbon-molybdenum alloy steel
- *B* – chromium-molybdenum alloy steel
- *C* – nickel steel alloy
- *D* – manganese-molybdenum alloy steel
- *G* – other alloys with minimal elements

The meaning of the numbers varies by alloy type, but always indicates the percentages of the following alloying elements:

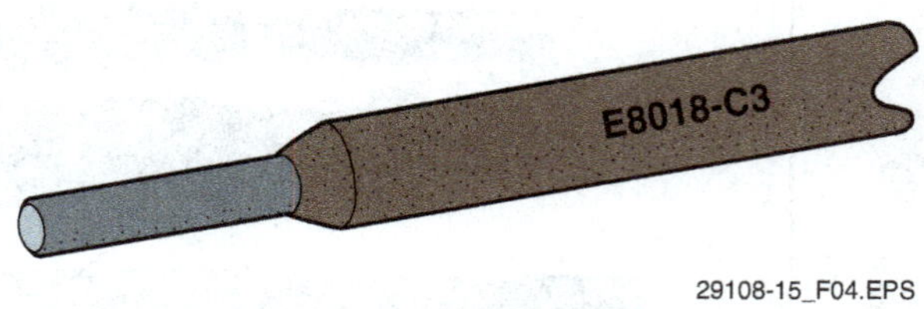

Figure 4 Markings on a low-alloy, high-strength electrode.

Table 1 Mild Steel Electrode Characteristics

AWS Class	Position	Flux Coating	Current Requirements	Characteristics
EXX 10 EXX 20	ALL FLAT, HOR. FILLET	CELLULOSE SODIUM	DCEP	DEEP PENETRATION, FLAT BEADS
EXX 11	ALL	CELLULOSE POTASSIUM	AC, DCEP	DEEP PENETRATION, FLAT BEADS
EXX 12	ALL	TITANIA SODIUM	AC, DCEN	MEDIUM PENETRATION
EXX 13	ALL	TITANIA POTASSIUM	AC, DCEP, DCEN	SHALLOW PENETRATION
EXX 14 EXX 24	ALL FLAT, HOR. FILLET	TITANIA IRON POWDER	AC, DCEP, DCEN	MEDIUM PENETRATION, FAST DEPOSIT
EXX 15	All	LOW-HYDROGEN SODIUM	DCEP	MODERATE PENETRATION
EXX 16	All	LOW-HYDROGEN POTASSIUM	AC, DCEP	MODERATE PENETRATION
EXX 27	FLAT, HOR. FILLET	IRON POWDER, IRON OXIDE	AC, DCEP, DCEN	MEDIUM PENETRATION
EXX 18 EXX 28	ALL FLAT, HOR. FILLET	IRON POWDER LOW-HYDROGEN	AC, DCEP	SHALLOW TO MEDIUM PENETRATION

- Carbon (C)
- Manganese (Mn)
- Phosphorus (P)
- Sulphur (S)
- Silicon (Si)
- Nickel (Ni)
- Chromium (Cr)
- Molybdenum (Mo)
- Vanadium (V)

Figure 5 shows the electrode marking system for low-alloy, high-strength electrodes.

> **CAUTION**
>
> Be very careful when selecting electrodes. For example, a 7018-B2 cannot be used in place of a 7018 electrode because the additional alloy will cause a significant metallurgical change in the weld. Never assume that electrodes with similar numbers are interchangeable.

Figure 6 shows *A5.5* electrodes listed by their alloy content. Remember that these are subject to change; as mentioned previously, the specifications are periodically updated.

1.1.3 Manufacturers' Classification

All electrodes used for code-governed work must have the AWS classification number printed on them as well as on the container in which they are shipped. In addition, electrode manufacturers generally print their own unique classification name or number on the electrode and on the container in which the electrodes are shipped. Manufacturers may also make more than one electrode within the same AWS classification. For example, Lincoln Electric manufactures two electrodes with the AWS classification E6010. One is called Fleetweld 5P, and the other is called Fleetweld 5P+. Words such as Fleetweld and Pipemaster are trade names and are not related to the AWS specification.

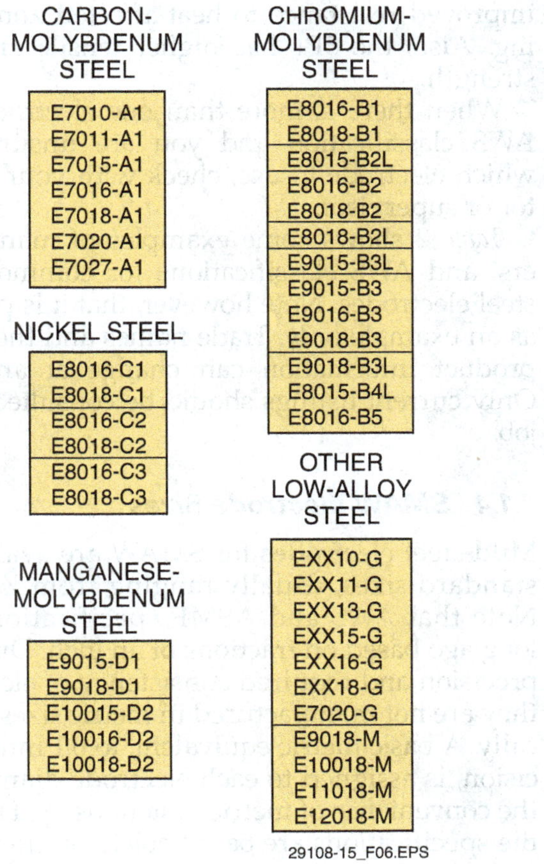

Figure 6 Electrodes listed by alloy.

Lincoln Electric recommends Fleetweld 5P for general fabrication and maintenance welding and states that it is good for pipe. However, Lincoln Electric states that Fleetweld 5P+ is the best choice for pipe. The major difference between the two electrodes is the mechanical properties of the deposited weld metal. Even though both fall into the E6010 classification, the 5P+ electrode has improved characteristics to decrease cracking possibilities in **heat-affected zones**. It is also better suited for open root joints. When 5P+ is used for the root pass, its greater elongation results in

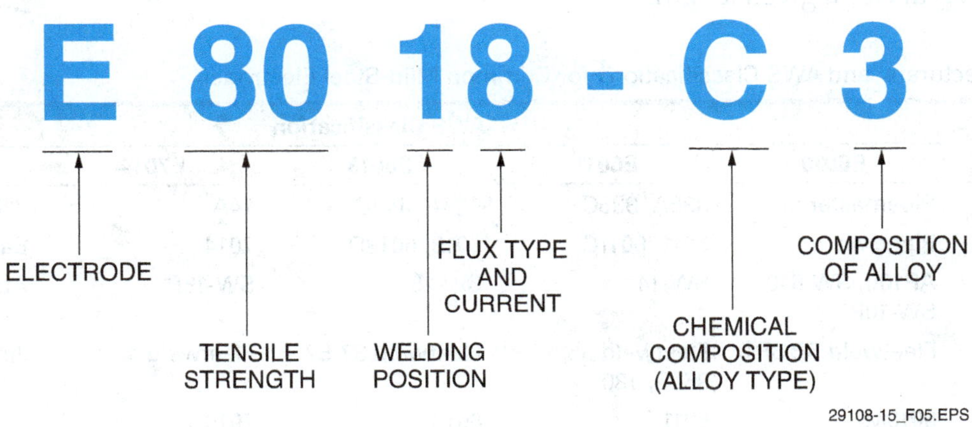

Figure 5 Marking system for low-alloy, high strength electrodes.

improved resistance to heat-affected zone cracking. Also, the 5P+ has higher tensile and yield strength.

When there is more than one electrode in an AWS classification and you are unsure about which electrode to use, check with your instructor or supervisor.

Table 2 shows some examples of manufacturers' and AWS classifications for common mild-steel electrodes. Note however, that it is provided as an example only. Trade names and the related product information can change at any time. Only current listings should be consulted on the job.

1.1.4 SMAW Electrode Sizes

Mild steel electrodes for SMAW are available in standard sizes, usually ranging from 3⁄32" to 1⁄4". Note that AWS and ASME specifications were long ago based on fractions of an inch. Due to the precision and required consistency of electrodes, they are not manufactured in metric sizes specifically. A basic metric equivalent, to 0.1 mm of precision, is assigned to each electrode diameter for the convenience of metric system users. However, the specifications are based solely on fractions of an inch.

SMAW electrodes are typically packaged in 10- and 50-pound cans, but also come in smaller containers, as shown in *Figure 7*. The amperage application for which the electrodes are used is proportional to the diameter size of the electrode. The number of electrodes in a 50-pound can varies with the diameter and length of the electrodes. Note that the length of the electrode may vary from manufacturer to manufacturer. *Table 3* is an example of a manufacturer's table listing some mild steel electrodes and the number of electrodes in one pound. Note that there are lengths of 14" (≈36cm) and 18" (≈46cm) listed. Electrodes from other manufacturers will be of very similar weights for a given length.

1.1.5 Filler Metal Traceability Requirements

Welding codes and the Welding Procedure Specification (WPS) on a project or job site normally specify the type of electrode to be used. Once a weld is made, there must be a means to trace and document the type of electrode that was used to make the weld or welds in order to verify that the correct welding codes and WPS were followed.

Traceability requirements vary according to client requirements and job specifications. On many job sites, these traceability requirements may be found in the quality control documentation related to the project. One of the more common requirements followed is found in the *ASME Boiler and Pressure Vessel Code, Section II, Material Specifications*. This section of the ASME code lists acceptable base metal and filler metal combinations.

CAUTION

The traceability requirements in site quality control manuals must be followed. The consequences for violation of the requirements can be severe. Understand traceability requirements before proceeding with a welding job. Verify this information with your supervisor, project engineer, or welding inspector.

1.2.0 Electrode Groups

Because so many different electrodes are available, the welder must be able to distinguish between them and select the correct electrode for the job. The following sections will explain how to select electrodes.

Electrodes are often classified into one of four groups based on their general characteristics:

- Fast-freeze
- Fast-fill
- Fill-freeze
- Low-hydrogen

Table 2 Manufacturers' and AWS Classifications for Common Mild-Steel Electrodes

Manufacturer	AWS Classification				
	E6010	E6011	E6013	E7014	E7018
Hobart	Pipemaster 60	335A, 335C	447A, 447C	14A	718, 718C
Airco	Pipecraft	6011, 6011C	6013, 6013D	7014	Easy Arc 7018MR
ESAB	AP100, SW-610, SW-10P	SW-14	SW-15	SW-15P	Atom Arc 7018
Lincoln	Fleetweld 5P, 5P1	Fleetweld 35, 35LS, 180	Fleetweld 37, 57	Fleetweld 47	Jetweld LH-70
McKay	6010M	6011	6013	7014	—

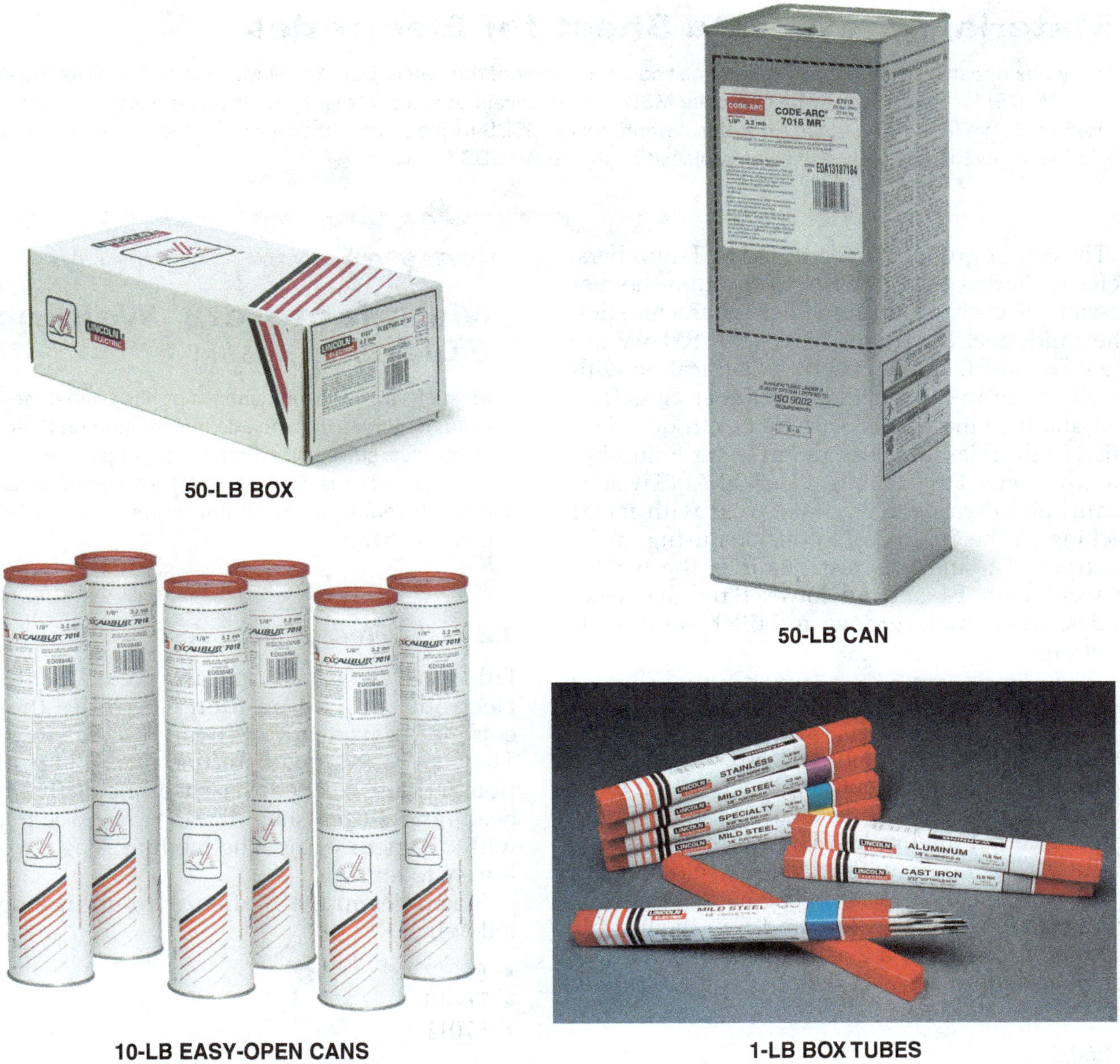

Figure 7 Electrode containers.

Table 3 Number of Electrodes per Pound

Electrode Type	Diameter:	3/32" (2.4mm)	1/8" (3.2mm)	5/32" (4.0mm)	3/16" (4.8mm)	7/32" (5.6mm)	1/4" (6.4mm)
	Length:	14"	14"	14"	14"	18"	18"
6010 PM		30	17	12	8	—	—
6011, Soft-Arc 6011		25	15	11	7	—	—
6013		25	15	10	7	—	—
7014		24	13	9	6	—	—
7024		—	10	7	—	4	2
7018 XLM, Soft-Arc 7018-1		32	15	10	7	—	3
7018 AC, Millennia 7018		32	15	10	7	4	3

Material Safety Data Sheet for Electrodes

Every box or can of welding electrodes must contain a copy of the Safety Data Sheet/Material Safety Data Sheet (SDS/MSDS) for that product. The complete MSDS for an Excalibur electrode with an AWS classification of E7018-1H4R is shown in the *Appendix*. As of this writing, many MSDS forms from numerous manufacturers remain valid and in use. Eventually, they will all be replaced by the newer SDS format.

These four groups are identified by F-numbers. Refer to *Table 4* as you progress through the discussion of each group and their characteristics. The mild steel electrodes used with SMAW are classified as F1, F2, F3, and F4. Qualification with an electrode in a particular F-number classification qualifies the welder with all electrodes identified in that classification and in lower F-number classifications. For example, a typical AWS welder qualification test is a V-groove weld with metal backing in the 3G and 4G positions using an F4 electrode. Passing this test qualifies the welder to weld with F4 and all lower F-number electrodes, and to make groove and fillet welds in all positions.

1.2.1 Fast-Fill Electrodes

Fast-fill electrodes (F1) have powdered iron in the flux, which gives them high deposition rates but requires that they only be used for flat welds and horizontal fillet welds. These electrodes have shallow penetration with excellent weld appearance and almost no spatter. The heavy slag covering is easy to remove. Electrodes in the fast-fill group will also have a 2 as the next-to-last number, as shown in *Table 4*. They include electrodes such as the following:

- E6027
- E7024
- E7028 (also in the low-hydrogen F4 group)

Typical applications for fast-fill electrodes are production welds on plate more than ¼" thick and flat and horizontal fillets and lap welds.

Manufacturers' Welding Guides

Most electrode manufacturers publish paperback welding guides that include information on their electrodes, such as recommended amperage settings and uses. These booklets can be obtained through your local distributor for free or at a nominal charge.

1.2.2 Fill-Freeze Electrodes

Fill-freeze electrodes (F2), also called fast-follow electrodes, are all-position electrodes that are commonly used for vertical and flat welding. They have medium deposition rates and penetration and are excellent for sheet metal. The weld bead ranges from smooth and ripple-free to even with distinct ripples. The medium slag covering is easy to remove.

Electrodes in the fill-freeze group include the following:

- E6012
- E6013
- E7014

Typical applications for fill-freeze electrodes are the following:

- Vertical fillet or lap welds
- Sheet metal lap and fillet welds
- Joints having poor fit-up in the flat position
- General purpose welding

Table 4 AWS F-Number Electrode Classification

Group	AWS Electrode Classification				
F4	EXX15	EXX16	EXX18		Low-Hydrogen
F3	EXX10	EXX11			Fast-Freeze
F2	EXX12	EXX13	EXX14		Fill-Freeze
F1	EXX20	EXX24	EXX27	EXX28	Fast-Fill

1.2.3 Fast-Freeze Electrodes

Fast-freeze electrodes (F3) are all-position electrodes that provide deep penetration. However, the arc creates a lot of fine weld spatter. The weld bead is flat with distinct ripples and a light slag coating that can be difficult to remove. The arc is easy to control, making fast-freeze electrodes good for overhead and vertical welding. These more challenging positions often make it harder for the welder to maintain a precise arc gap. Electrodes in the fast-freeze group include the following:

- E6010
- E7010
- E6011

Typical applications for fast-freeze electrodes are as follows:

- Vertical and overhead plate welding
- Pipe welding (cross-country and in-plant)
- Welds made on galvanized, plated, or painted surfaces
- Joints requiring deep penetration, such as square butts
- Sheet metal welds

> **WARNING!**
>
> Many surface coatings, such as galvanized coatings, give off very toxic fumes when heated, cut, or welded. Fumes from certain types of electrodes may also be hazardous. Breathing these fumes can cause damage to the lungs and possibly other organs. Identify any hazard for the coatings with which you are working by referring to the SDS/MSDS for that particular coating. Also, read the SDS/MSDS contained in the electrode package. Use proper ventilation and take the necessary precautions to protect yourself and other workers in the immediate area.

1.2.4 Low-Hydrogen Electrodes

Low-hydrogen electrodes (F4) are designed for welding high-sulphur, phosphorus, and medium- to high-carbon steels, which have a tendency to develop porosity and cracks under the weld bead. These problems are caused by hydrogen that is absorbed during welding. Low-hydrogen electrodes are also available with fast-fill and fill-freeze characteristics. They produce dense, X-ray quality welds with excellent notch toughness and ductility. Electrodes in the low-hydrogen group include the following:

- E7015
- E7016
- E7018
- E7028
- E7048

Typical applications for low-hydrogen electrodes are the following:

- X-ray quality welds
- Welds requiring high mechanical properties
- Crack-resistant welds in medium- to high-carbon steels
- Welds that resist hot-short cracking in phosphorus steels
- Welds that minimize porosity in sulphur steels
- Welds in thick or highly restrained mild or alloy steels to minimize the danger of weld cracking
- Welds in alloy steels requiring a strength of 70,000 psi (≈4,826 bars) or more
- Multiple-pass welds in all positions

If hydrogen is present in the weld zone, it will create defects. The greatest source of hydrogen is airborne water vapor, which is composed of oxygen and hydrogen. Low-hydrogen electrodes are carefully manufactured to eliminate moisture in the flux coating. Low hydrogen is defined as being less than 16 milliliters (ml) per 100 grams of weld metal. The amount of hydrogen in these electrodes varies from 4 to 16 ml per 100 grams of metal weld.

Low-hydrogen electrodes have two unique suffixes. The M suffix is for compliance. Most military codes require an M suffix. When the suffix H and a number are added, the numbers relate to the amount of hydrogen in the electrodes. As a result, the electrodes will have suffixes such as H4, H8, and H16. These three suffixes indicate the dryness of the electrodes, identifying the milliliters of hydrogen per 100 grams of metal weld they contain. An additional suffix may also be added to these dryness suffixes. A suffix such as H4R would be one example. The R indicates that the electrodes are moisture resistant. Without an R suffix, the electrodes may only be exposed to the atmosphere for a maximum of 4 hours. Those with an R suffix have an allowable exposure of up to 8 hours. *Figure 8* shows a 7018 electrode with the H4R suffix.

If low-hydrogen electrodes are exposed to the atmosphere, they will begin to absorb moisture and will no longer be low-hydrogen as a result. This is the reason for their short atmospheric exposure

Figure 8 Electrode with suffix.

limits. For this reason, all low-hydrogen electrodes are shipped in **hermetically sealed** metal containers. Once opened, the electrodes must be stored in a heated oven. *Figure 9* shows low-hydrogen electrodes properly stored. The temperature range of this oven is 100°F to 550°F (38°C to 288°C). However, it is important to note that placing the oven in areas below freezing may not allow it to maintain its maximum temperature, if that setting is necessary. Additional guidelines for the storage and handling of low-hydrogen electrodes are covered later in this module.

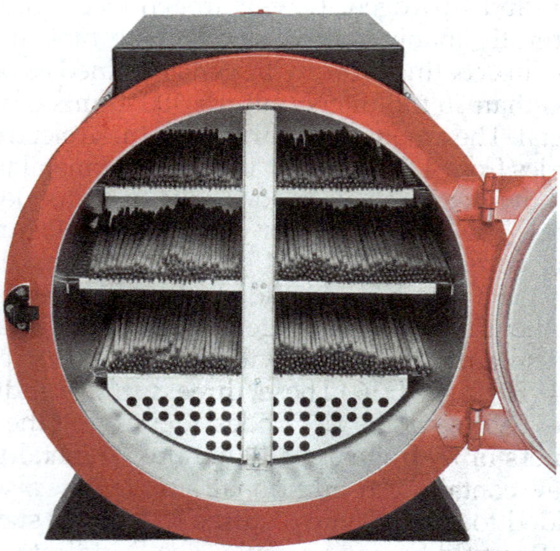

Figure 9 Low-hydrogen electrode storage.

CAUTION

The welding codes establish strict guidelines for the storage, handling, and re-drying of low-hydrogen electrodes. These procedures will be referenced in your site quality standards and the WPS. Severe penalties will result from violation of these standards. Improperly storing these electrodes will lead to poor-quality welds and the possibility of having numerous welds rejected. Be sure you read and understand your site standards before working with low-hydrogen electrodes.

Portable Electrode Storage

Electrodes can also be kept in a portable rod oven for convenience on the job. This ensures that the lowest possible level of hydrogen content is maintained. The model shown has a spring-loaded rod elevator that pushes the rods up for easy grasping when the lid is opened. This is a valuable feature when welding gloves are being worn. It maintains a fixed temperature of 300°F (149°C) when powered. Non-powered dry containers, without electric heating elements, are also available.

Additional Resources

The Welding Handbook for Maritime Welders. Current Edition. Lysaker, Norway: Wilhelmsen Ships Service. **www.wilhelmsen.com**

Welding Consumables Catalog. Current Edition. Cleveland, OH, USA: The Lincoln Electric Company. **www.lincolnelectric.com**

1.0.0 Section Review

1. The slag over a fresh weld primarily forms due to ____ .
   a. impurities in the base metal
   b. impurities in the electrode metal
   c. the flux coating of the electrode
   d. hydrogen and nitrogen in the atmosphere

2. An H8 suffix on the end of an electrode classification number means ____ .
   a. there are 8 grams of hydrogen in each stick
   b. there are 8 ml of hydrogen per 100 grams of weld metal
   c. there is 8 times as much hydrogen in the electrode than normal
   d. the electrode can be exposed to the atmosphere up to 8 minutes

SECTION TWO

2.0.0 ELECTRODE SELECTION AND CARE

Objective

Explain how to select electrodes and describe their proper care and handling.

a. Identify various considerations in the selection of the proper electrode.
b. Describe the proper handling and storage of electrodes.

Trade Term

Condensation: The process in which atmospheric water vapor condenses (returns to its liquid state) on a cool surface.

With some basic information concerning electrodes used in SMAW, it will be easier to understand the selection process. The importance of properly caring for these simple consumable items cannot be overstressed. This section provides additional information that will help ensure that electrode quality standards are maintained from the point of manufacture to the job site.

2.1.0 Electrode Selection Considerations

Many factors enter into the proper selection of electrodes for the job. The available welding equipment and its current ranges limit the choice of electrodes, and base metal properties play an important role in the selection as well. Also, the position in which the weld must be made and design factors affect the choice of electrodes for a given application. *Table 5* indicates position, polarity, and common usages for some steel electrodes, listed by their AWS classification.

2.1.1 Welding Procedure Specification

If there is a WPS for the weld to be made, the filler metal will be specified. It will be given as an AWS classification and/or manufacturer's standard and will sometimes even include a manufacturer's name. The filler metal specified in the WPS must be used.

2.1.2 Base Metal Type

The filler metal should match the base metal's chemical composition and mechanical properties. The chemical composition and mechanical properties of the base metal can be obtained by referring to the base metal mill specifications. The mill specifications can then be compared to various tables that list base metal and filler metal compatibility. These tables are available in a number of codes, such as the *AWS Structural Welding Code – Steel D1.1, Table 3.1 Prequalified Base Metal – Filler Metal Combinations for Matching Strength,* and the *ASME Boiler and Pressure Vessel Code, Section II, Parts A, B, and C, Material Specifications.* Tables are also readily available from the electrode and/or base metal manufacturers. Manufacturers acknowledge that an informed user has a great deal to do with the success of their product. An improperly applied filler metal that results in poor weld quality can also reflect on their product, regardless of the circumstances. If the filler metal and base metal are not compatible, the weld produced will be defective.

A weld made on mild steel with a high-tensile electrode such as E12018 may actually create more stress on the joint. The high-tensile weld will be so strong that when stress is placed on the weld, it will not give, transferring all of the stress to the heat-affected zone along the weld. This will cause cracking. A weld made with E7018 will give along with the base metal, distributing the stress across the weld, making a stronger joint. The opposite occurs when mild steel electrodes are used to weld high-strength steel. All the stress occurs in the weld because the surrounding base metal is so much stronger, causing the weld to fail prematurely.

Table 5 Common Steel Electrode Selection Factors

AWS Class	Position	Polarity	Usage
E6011	All	AC, DCEP	Good for dirty, rusty steel
E6013	All	AC, DCEN, DCEP	Good for thin steel and poor fit-up
E7014	All	AC, DCEN, DCEP	Good for thin steel
E7018	All	AC, DCEP	Good for high deposition; good for medium- or high-carbon steels

2.1.3 Base Metal Thickness

Normally, an electrode that is smaller in diameter than the thickness of the metal to be welded is used. For sheet metal, a $\frac{3}{32}$" E6013 electrode may be recommended because it has shallow penetration. For very thick sections, always use a low-hydrogen electrode for its ductile qualities. To minimize joint preparation, a deep penetration E6010 or E7010 electrode can be used for the root pass.

2.1.4 Base Metal Surface Condition

The base metal surface should always be cleaned before welding. Surface corrosion, plating, coatings, and paint should be removed. If this is not possible for some reason, use a deep penetrating electrode, such as an E6010.

> **WARNING!**
>
> Many surface coatings, such as galvanized coatings, give off toxic fumes when heated, cut, or welded. Breathing these fumes can cause damage to the lungs and other organs. Identify any hazard for the coatings with which you are working by referring to the MSDS for that particular coating. Use proper ventilation and take the necessary precautions to protect yourself.

2.1.5 Welding Position

The position of welding is important because some electrodes, such as E7024 and E7028, can only be used in the flat and horizontal fillet weld positions. In addition, welding in vertical and overhead positions should never be attempted with electrodes larger than $\frac{3}{16}$", except for low-hydrogen electrodes. The limitation for low-hydrogen electrodes and EXX14 electrodes is $\frac{5}{32}$". As a general rule, the joint should be positioned for flat welding so you can use the largest electrode possible.

2.1.6 Joint Design

The type of joint to be welded is an important consideration when selecting an electrode. When T- and lap joints can be positioned flat, they are usually welded with E7024 or E7028 electrodes. Butt joints that need deep penetration should be welded with E6010 electrodes. If the joint has poor fit-up, use a shallow penetrating electrode such as an E6012 or E6013. If low-hydrogen electrodes are used for open root joints, back-gouging may be required. To eliminate the need for back-gouging and to make the root pass easier to run, use metal backing or put the root in with an E6010 electrode and then fill the weld out with a low-hydrogen electrode.

2.1.7 Welding Current

The most common type of welding current is DC. Because all electrodes can be run on DC, there is no problem when DC current is available. Welding machines that can produce only AC current cannot use E6010 or E7010 electrodes. Although many other electrodes will run on AC, they will run better using DC current.

The amount of current produced by the welding machine is also a consideration. Running large-diameter electrodes that require high amperages near the maximum output of the welding machine can cause the welding machine to overheat and cut out. Check the welding machine current type, rated output, and duty cycle before selecting the electrode size.

2.2.0 Filler Metal Storage and Control

Electrodes used for SMAW must be dry. The dry mineral flux coatings on electrodes start to absorb moisture as soon as they are exposed to the atmosphere. In as little as 30 minutes, some electrodes can pick up enough moisture from the atmosphere to cause weld defects. Note too, that the local conditions have a lot to do with the rate at which moisture is absorbed from the air. For example, electrodes will absorb moisture faster and in greater volume on a rainy day in Florida than they will on a clear summer day in Arizona. Some of the defects caused, such as under-bead cracking, may not be visible to the naked eye.

Low-hydrogen electrodes pick up moisture faster than any other type of electrode. Also, because of the nature of the base metals welded with low-hydrogen electrodes, a smaller amount of moisture will cause defects. For these reasons, special care must be used when handling and storing low-hydrogen electrodes.

2.2.1 Code Requirements

All welding codes require electrodes that have a low-hydrogen coating to receive special handling. To begin, they must be shipped in hermetically sealed containers. Immediately after opening the sealed containers, low-hydrogen electrodes must be stored in special ovens at a minimum temperature of 250°F (120°C). When low-hydrogen electrodes are removed from the oven, they can be exposed to the atmosphere only for a designated period of time. This time period varies with the classification of the electrode, but generally ranges from one-half hour to ten hours. *AWS D.1.1, Table 5.1 Allowable Atmospheric Exposure of Low-Hydrogen Electrodes* gives allowable exposure times (*Table 6*). The electrodes must then be returned to the oven and generally require a drying process at temperatures above 250°F (>120°C). The exact procedures for handling filler metal at your site are typically written out in a quality standard. These procedures must be followed. If you are unsure of the requirements at your site, check with your supervisor.

CAUTION

The exposure times shown are general guidelines only. Excessive exposure times are known to cause weld defects. Check your site quality standards for the actual exposure times for your location. To the degree possible, do not allow electrodes to remain exposed any longer than absolutely necessary.

CAUTION

The consequences of not following your site's standards on electrode handling and storage may be severe. Be sure you understand your site's standards before handling electrodes. Improper handling of electrodes can cause weld defects. The information provided in this module is of a general nature. Your site's particular quality standards must be followed.

2.2.2 Receiving Electrodes

Inspect electrodes as they are received. All low-hydrogen electrodes (EXX15, 16, or 18), stainless steel electrodes, and nickel alloy coated electrodes must be supplied in hermetically sealed containers. If the seal has been broken, reject the electrodes. Mark them "Not to Be Used" and notify your supervisor.

All electrodes other than low-hydrogen electrodes must be shipped in moisture-resistant containers. Inspect these containers for damage. The electrodes can be used if their coatings are not damaged and they have not come into direct contact with water or oil. Mark electrodes that cannot be used and notify your supervisor.

Each container must be marked with at least the following information:

- ASME, AWS, or other electrode specification number and electrode classification designation
- Electrode size
- Weight

Table 6 Allowable Atmospheric Exposure of Low-Hydrogen Electrodes.

Electrode	Column A (hours)	Column B (hours)
A5.1		
E70XX	4 max	
E70XXR	9 max	Over 4 to 10 max
E70XXHXR	9 max	
E7018M	9 max	
A5.5		
E70XX-X	4 max	Over 4 to 10 max
E80XX-X	2 max	Over 2 tp 10 max
E90XX-X	1 max	Over 1 to 5 max
E100XX-X	½ max	Over ½ to 4 max
E110XX-X	½ max	Over ½ to 4 max

Notes:
1. Column A: Electrodes exposed to atmosphere for longer periods than shown shall be redried before use.
2. Column B: Electrodes exposed to atmosphere for longer periods than those established by testing shall be redried before use.
3. Electrodes shall be issued and held in quivers, or other small open containers. Heated containers are not mandatory.
4. The optional supplemental designator, R, designates a low-hydrogen electrode which has been tested for covering moisture content after exposure to a moist environment for 9 hours and has met the maximum level allowed in *AWS A5.1, Specification for Carbon Steel Electrodes for Shielded Metal Arc Welding*.

29108-15_T06.EPS

If the containers are not properly marked, mark them "Not to Be Used" and notify your supervisor.

2.2.3 Storing Filler Metal

All filler metal must be stored in a warm (40°F [4.5°C] minimum), dry storage area. Filler metals are not to be stored directly on the floor. Electrodes that are stored on the floor are more likely to form condensation. Use pallets or shelves to store electrodes off the floor. They must also be stored in such a manner that they will not be damaged. Once low-hydrogen, stainless steel, or nickel-alloy electrode containers have been opened, the electrodes must immediately be stored in an oven.

Electrode Mix-Up

Do not mix electrode types in an oven if possible. If electrodes of various types must be stored in the same oven, keep them carefully organized by placing them on separate shelves. Labels on the shelves are another way that mistakes can be avoided.

WARNING!

Storage ovens are extremely hot. Interior surfaces, as well as the stored contents, can cause severe burns. Be careful when filling or removing stock from an oven. Wear the appropriate PPE, including gloves capable of withstanding the heat of electrodes that may exceed 500°F (260°C). Remember than even portable ovens and their contents become extremely hot inside.

CAUTION

Do not put cellulose electrodes into ovens. Doing so will destroy the operating characteristics of the electrode coating.

2.2.4 Storage Ovens

Electrode storage ovens are typically electrical and are controlled by a thermostat. Storage ovens for large shops can hold large quantities of electrodes and can possibly reach temperatures exceeding 800°F (427°C). This maximum temperature is too high for most electrodes. Always follow the electrode manufacturer's guidelines for storing electrodes in ovens. Smaller portable ovens and dry containers that are designed to be used on the job may hold only a few pounds of electrodes. Depending on the WPS or job quality standards, portable electrode ovens may be required at the point of use. The electrodes remaining in the portable ovens after performing the task must be returned to the holding oven. *Figure 10* shows several examples of electrode storage ovens.

Storage ovens should typically be set between 250°F and 300°F (121°C to 149°C). The oven must remain on at all times. Never unplug an oven or plug it into an outlet that could accidentally be shut off.

2.2.5 Exposure and Drying

The time that a low-hydrogen, stainless steel, or nickel alloy electrode can be exposed to the atmosphere is limited and depends on the electrode. It is the welder's responsibility to take from the oven only the number of electrodes that can be used before the maximum exposure time is reached. As a best practice, do not try to push the limits, as the minimal time savings is not worth the risk. This is especially true in wet, humid weather. The exposure time is defined as the period of time that the electrode is not in a heated holding or portable oven.

Samples of electrodes being used at a site are often collected by the quality control group for moisture testing. Depending on the results of this testing, the maximum exposure times may be adjusted higher. However, if high moisture content is found in the electrodes, the welds made with the electrodes may have to be replaced. This can be an extremely costly situation, as well as frustrating to the welder and other affected workers.

Depending on the site quality standards, electrodes that exceed the maximum exposure time must either be destroyed or properly dried. Check your site quality standards for the proper actions. Low-hydrogen electrodes that are exposed to water must be destroyed; drying them to an accepted standard is not practical.

Electrodes that exceed the maximum exposure time can be dried once. They are dried by placing them in a holding oven at an elevated temperature for a period of time. The amount of time and temperature depend on the nature of the moisture pickup and the electrode.

CAUTION

Do not attempt to dry electrodes before determining whether your site quality standards allow drying. Improperly drying electrodes can cause weld defects. Check with your supervisor.

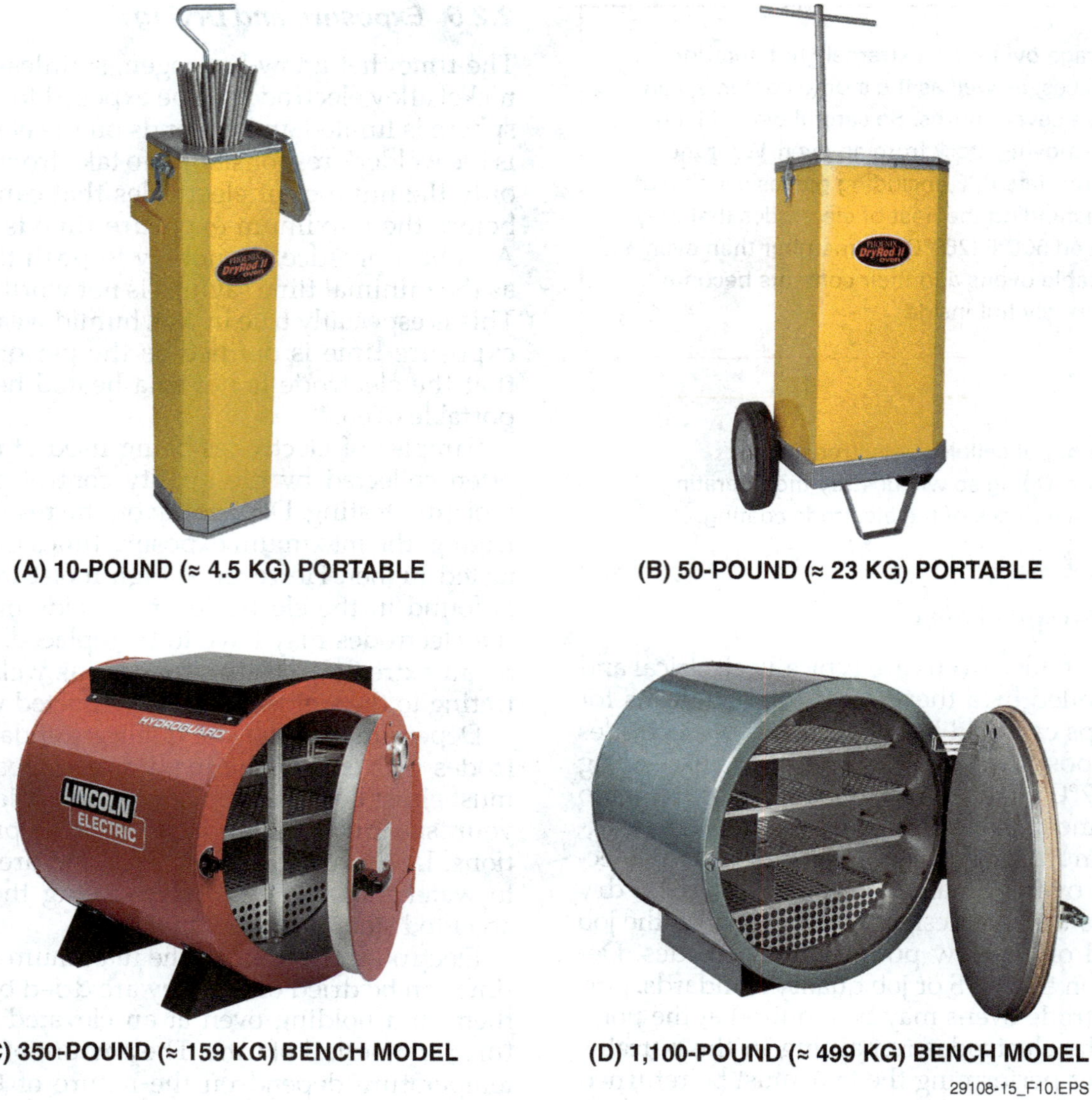

Figure 10 Electrode storage ovens.

Additional Resources

The Welding Handbook for Maritime Welders. Current Edition. Lysaker, Norway: Wilhelmsen Ships Service. **www.wilhelmsen.com**

Welding Consumables Catalog. Current Edition. Cleveland, OH, USA: The Lincoln Electric Company. **www.lincolnelectric.com**

2.0.0 Section Review

1. Which of the following electrodes would be the best choice for medium- or high-carbon steels?
   a. E6011
   b. E6013
   c. E7014
   d. E7018

2. How many times can electrodes that have exceeded their maximum period of exposure be dried?
   a. Once
   b. Twice
   c. Three times
   d. Four times

SUMMARY

Many different electrodes are produced by many different manufacturers. To ensure quality and uniformity, the American Welding Society has established a classification system for electrodes. Understanding this classification system and why electrodes are placed in a particular classification is essential for the welder to correctly identify and select electrodes for the weld to be made.

Welders must be able to select proper types of filler metals and electrodes to make welds on specific materials. It is also important for welders to understand how to properly store electrodes to prevent damage or deterioration. Properly stored, good quality SMAW electrodes do not guarantee an excellent weld; a great deal of the result is a reflection of the welder's skill. However, electrodes that are poorly maintained and do not meet standards will result in poor, unacceptable welds regardless of the welder's skill level.

Review Questions

1. Welding current is transferred from the electrode holder and into the workpiece through the ____.
   a. base metal
   b. flux coating
   c. gas shield
   d. wire core of the electrode

2. The harmful oxygen and hydrogen present in the atmosphere surrounding a molten SMAW weld is pushed away by the ____.
   a. gaseous shield of carbon dioxide
   b. gaseous shield of nitrogen
   c. constant welding spatter
   d. intense heat of the arc

3. The purpose of the specifications written by the American Welding Society (AWS) is to set standards that all manufacturers must follow when manufacturing ____.
   a. welding consumables
   b. welding machines
   c. chemicals
   d. metals

4. The number placed at the end of a complete AWS specification number indicates the ____.
   a. type of electrodes discussed in the specification
   b. year the specification was last revised
   c. number of chapters in the document
   d. year the specification was written

5. The traceability requirements of filler metals vary according to ____.
   a. the electrode trade name
   b. the welder's qualifications
   c. the client requirements and job specifications
   d. state and local laws

6. All-position electrodes that provide deep penetration belong to which electrode group?
   a. F1 fast-fill
   b. F2 fill-freeze
   c. F3 fast-freeze
   d. F4 low-hydrogen

7. Electrodes designed for welding high-sulphur, phosphorus, and medium- to high-carbon steels belong to which electrode group?
   a. F1 fast-fill
   b. F2 fill-freeze
   c. F3 fast-freeze
   d. F4 low-hydrogen

8. Of all electrodes, the ones that collect moisture faster than any other type are the ____.
   a. F1 fast-fill electrodes
   b. F2 fill-freeze electrodes
   c. F3 fast-freeze electrodes
   d. F4 low-hydrogen electrodes

9. The exposure time for low-hydrogen electrodes, per AWS standards, typically ranges from ____.
   a. ½ hour to 4 hours
   b. ½ hour to 10 hours
   c. 2 to 5 hours
   d. 3 to 6 hours

10. All filler metal must be stored in a dry storage area at a minimum temperature of ____.
    a. 30°F (–1°C)
    b. 40°F (4.5°C)
    c. 50°F (10°C)
    d. 60°F (15.5°C)

Trade Terms Quiz

Fill in the blank with the correct term that you learned from your study of this module.

1. Welding upward or downward is called _______ .
2. A(n) _______ is a metal that has had other elements added to it that substantially change its mechanical properties.
3. The part of a base metal that has been altered but not melted by the heat is called the _______ .
4. _______ is the ability of a material to resist breaking at points where stress is concentrated.
5. The process in which water vapor from the atmosphere returns to its liquid state is called _______ .
6. A container that is closed airtight is said to be _______ .
7. _______ is the ability to verify that a procedure has been followed by reviewing the documentation step-by-step.
8. _______ refers to the mechanical property of a material that allows the material to be bent or shaped without breaking.
9. An electrode that is carefully manufactured to contain little or no moisture is called a(n) _______ .
10. A material used to prevent, dissolve, or facilitate the removal of oxides and other undesirable substances from the area of a weld is called _______ .

Trade Terms

Alloy
Condensation
Ductility
Flux
Heat-affected zone
Hermetically sealed
Low-hydrogen electrode
Notch toughness
Traceability
Vertical welding

Sample Material Safety Data Sheet

Date:	**1/21/2013**	**MSDS No.:**	**US-M292**
Trade Name:	**Excalibur 7018-1 MR**		
Sizes:	**All**		
Supersedes:	**8/29/2011**		

MATERIAL SAFETY DATA SHEET

For Welding Consumables and Related Products

Conforms to Hazard Communication Standard 29CFR 1910.1200 Rev. October 1988

SECTION I - IDENTIFICATION

Manufacturer/ Supplier: The Lincoln Electric Company 22801 St. Clair Avenue Cleveland, OH 44117-1199 (216) 481-8100	**Product Type:** Covered Electrode **Classification:** AWS E7018-1H4R

SECTION II - HAZARDOUS MATERIAL (1)

IMPORTANT!

This section covers the materials from which this product is manufactured. The fumes and gases produced during welding with the normal use of this product are covered by Section V; see it for industrial hygiene information.

CAS Number shown is representative for the ingredients listed. All ingredients listed may not be present in all sizes.

(1) The term "hazardous" in "Hazardous Materials" should be interpreted as a term required and defined in the Hazard's Communication Standard and does not necessarily imply the existence of any hazard. All materials are listed on the TSCA inventory.

Ingredients:	CAS No.	Wt.%	TLV mg/m^3	PEL mg/m^3
Iron	7439-89-6	15	10*	15*
Limestone and/or calcium carbonate	1317-65-3	10	10*	15
Titanium dioxides	13463-67-7	< 5	10	15
Fluorides (as F)	7789-75-5	< 5	2.5	2.5
Silicates and other binders	1344-09-8	< 5	10*	15*
Manganese and/or manganese alloys and compounds (as Mn)*****	7439-96-5	< 5	0.2	5 (c)
Mineral silicates	1332-58-7	< 5	5**	5**
Silicon and/or silicon alloys and compounds (as Si)	7440-21-3	1	10*	15*
Zirconium alloys and compounds (as Zr)	12004-83-0	1	5	5
Cellulose and other carbohydrates	65996-61-4	0.5	10*	15*
Quartz	14808-60-7	< 0.5	#0.025**	#0.1**
Molybdenum alloys (as Mo)	7439-98-7	< 0.5	10	10
Lithium compounds (as Li)	554-13-2	< 0.5	10*	15*
Carbon steel core wire	7439-89-6	55	10*	15*

Supplemental Information:

(*) Not listed. The OSHA PEL for nuisance particles is 15 milligrams per cubic meter. The ACGIH guideline for total particulate is 10 milligrams per cubic meter. PEL value for iron oxide is 10 milligrams per cubic meter. TLV value for iron oxides is 5 milligrams per cubic meter.

(**) As respirable dust.

(*****) Subject to the reporting requirements of Sections 311, 312, and 313 of the Emergency Planning and Community Right-to-Know Act of 1986 and of 40CFR 370 and 372.

(c) Value is for manganese fume. Present PEL is 5 milligrams per cubic meter (ceiling value). Values proposed by OSHA in 1989 were 1.0 milligrams per cubic meter TWA and 3.0 milligrams per cubic meter STEL (Short Term Exposure Limit).

(#) Crystalline silica (quartz) is on the IARC (International Agency for Research on Cancer) and NTP (National Toxicology Program) lists as posing a carcinogenic risk to humans.

SECTION III - HAZARD DATA

Non Flammable; Welding arc and sparks can ignite combustibles and flammable products. See Z49.1 referenced in Section VI. Product is inert, no special handling or spill procedures required. Not regulated by DOT.

Rev 9/07

(CONTINUED ON SIDE TWO)

29108-15_A01A.EPS

Product:	Excalibur 7018-1 MR
Date:	1/21/2013

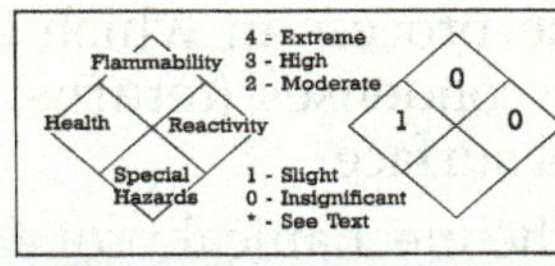

SECTION IV - HEALTH HAZARD DATA

Threshold Limit Value: The ACGIH recommended general limit for Welding Fume NOS- (Not Otherwise Specified) is 5 mg/m³. ACGIH-1999 preface states that the TLV-TWA should be used as guides in the control of health hazards and should not be used as fine lines between safe and dangerous concentrations. See Section V for specific fume constituents which may modify this TLV. Threshold Limit Val ues are figures published by the American Conference of Government Industrial Hygienists. Units are milligrams per cubic meter of air.

Effects of Overexposure: Electric arc welding may create one or more of the following health hazards:
Fumes and Gases can be dangerous to your health. Common entry is by inhalation. Other possible routes are skin contact and ingestion.

Short-term (acute) overexposure to welding fumes may result in discomfort such as metal fume fever, dizziness, nausea, or dryness o r irritation of nose, throat, or eyes. May aggravate pre-existing respiratory problems (e.g. asthma, emphysema).

Long-term (chronic) overexposure to welding fumes can lead to siderosis (iron deposits in lung) and may affect pulmonary function. Manganese overexposure can affect the central nervous system, resulting in impaired speech and movement. Bronchitis and some lung fibrosis have been reported. Repeated exposure to fluorides may cause excessive calcification of the bone and calcificatio n of ligaments of the ribs, pelvis and spinal column. May cause skin rash. Titanium dioxide is listed by the IARC (International Agency for Research on Cancer) as a Group 2B carcinogen (possibly carcinogenic to humans based on animal studies). Respiratory exposure to the crystalline silica present in this welding electrode is not anticipated during normal use. Respiratory overexposure to airborne crystalline silica is known to cause silicosis, a form o f disabling pulmonary fibrosis which can be progress ive and may lead to death. Crystalline silica is on the IARC (International Agency for Research on Cancer) and NTP (National Toxicology Program) lists as posing a cancer risk to humans. WARNING: This product, when used for welding or cutting, produces fumes or gases which contain chemicals known to the State of California to cause birth defects and, in some cases, cancer. (California Health & Safety Code Section 25249.5 et seq.)

Arc Rays can injure eyes and burn skin. *Skin cancer has been reported.*
Electric Shock can kill. If welding must be performed in damp locations or with wet clothing, on metal structures or when in cramped positions such as sitting, kneeling or lying, or if there is a high risk of unavoidable or accidental contact with workpiece, use the following equipment: Semiautomatic DC Welder, DC Manual (Stick) Welder, or AC Welder with Reduced Voltage Control.

Emergency and First Aid Procedures: Call for medical aid. Employ first aid techniques recommended by the American Red Cross
IF BREATHING IS DIFFICULT give oxygen. IF NOT BREATHING employ CPR (Cardiopulmonary Resuscitation) techniques.
IN CASE OF ELECTRICAL SHOCK, turn off power and follow recommended treatment. In all cases call a physician.

SECTION V - REACTIVITY DATA

Hazardous Decomposition Products: Welding fumes and gases cannot be classified simply. The composition and quantity of both are dependent upon the metal being welded, the process, procedure and electrodes used.

Other conditions which also influence the composition and quantity of the fumes and gases to which workers may be exposed include: coatings on the metal being welded (such as paint, plating, or galvanizing), the number of welders and the volume of the worker area, the quality and amount of ventil ation, the position of the welder's head with respect to the fume plume, as well as the presence of contaminants in the atmosphere (such as chlorinated hydrocarbon vapors from cleaning and degreasing activities.)

When the electrode is consumed, the fume and gas decomposition products generated are different in percent and form from the ingredients listed in Section II. Decomposition products of normal operation include those originating from the volatilizatbn, reaction, or oxidation of the materials shown in Section II, plus those from the base metal and coating, etc., as noted above.

Reasonably expected fume constituents of this product would include: Primarily iron oxide and fluorides; secondarily compbx oxides of manganese, potassium, silicon, sodium, and titanium.

Maximum fume exposure guideline for this product (based on manganese content) is 4.0 milligrams per cubic meter.

Gaseous reaction products may include carbon monoxide and carbon dioxide. Ozone and nitrogen oxides may be formed b the radiation from the arc.

Determine the composition and quantity of fumes and gases to which workers are exposed by taking an air sample from inside the welder's helmet if worn or in the worker's breathing zone. Improve ventilation if exposures are not below limits. See ANSI/AWS F1.1, F1.2, F1.3 and F1.5, available from the American Welding Society, 550 N.W. LeJeune Road, Miami, FL 33126.

SECTION VI AND VII
CONTROL MEASURES AND PRECAUTIONS FOR SAFE HANDLING AND USE

Read and understand the manufacturer's instruction and the precautionary label on the product. Request Lincoln Safety Publication E205. See American National Standard Z49.1, "Safety In Welding, Cutting and Allied Processes" published by the American Welding Society, 550 N.W. LeJeune Road, Miami, FL, 33126 (both available for free download at http://www.lincolnelectric.com/community/safety/) and OSHA Publication 2206 (29CFR1910), U.S. Government Printing Office, Superintendent of Documents, P.O. Box 371954, Pittsburgh, PA 15250-7954 for more details on many of the following:

Ventilation: Use enough ventilation, local exhaust at the arc, or both to keep the fumes and gases from the worker's breathing zone and the general area. Train the welder to keep his head out of the fumes. *Keep exposure as low as possible.*

Respiratory Protection: Use respirable fume respirator or air supplied respirator when welding in confined space or generalwork area when local exhaust or ventilation does not keep exposure below TLV.

Eye Protection: Wear helmet or use face shield with filter lens shade number 12 or darker. Shield others by providing screens and flash goggles.

Protective Clothing: Wear hand, head, and body protection which help to prevent injury from radiation, sparks and electrical shock. See Z49.1. At a minimum this includes welder's gloves and a protective face shield, and may include arm protectors, aprons, hats, shoulder protection, as well as dark substantial clothing. Train the welder not to permit electrically live partsor electrodes to contact skin . . . or clothing or gloves if they are wet. Insulate from work and ground.

Disposal Information: Discard any product, residue, disposable container, or liner as ordinary waste in an environmentally acceptable manner according to Federal, State and Local Regulations unless otherwise noted. No applicable ecological information available.

29108-15_A01B.EPS

Trade Terms Introduced in This Module

Alloy: A metal that has had other elements added to it that substantially change its mechanical properties.

Condensation: The process in which atmospheric water vapor condenses (returns to its liquid state) on a cool surface.

Ductility: Refers to the mechanical property of a material that allows it to be bent or shaped without breaking.

Flux: Material used to prevent, dissolve, or facilitate the removal of oxides and other undesirable substances on a weld or base metal.

Heat-affected zone: The part of the base metal that has been altered by heating, but not melted by the heat.

Hermetically sealed: Having an airtight seal.

Low-hydrogen electrode: An electrode specially manufactured to contain little or no moisture.

Notch toughness: The ability of a material to resist breaking at points where stress is concentrated.

Traceability: The ability to verify that a procedure has been followed by reviewing the documentation step by step.

Vertical welding: Welding with an upward or downward progression.

Additional Resources

This module presents thorough resources for task training. The following resource material is suggested for further study.

The Welding Handbook for Maritime Welders. Current Edition. Lysaker, Norway: Wilhelmsen Ships Service. **www.wilhelmsen.com**

Welding Consumables Catalog. Current Edition. Cleveland, OH, USA: The Lincoln Electric Company. **www.lincolnelectric.com**

Figure Credits

The Lincoln Electric Company, Cleveland, OH, USA, Module Opener, Figures 2, 7–9, 10C, SA01, SA02, Appendix

Phoenix International Inc., Figures 10A, 10B, 10D

AWS D1.1/D1.1M: 2008, Table 5.1, Allowable atmospheric exposure of low-hydrogen electrodes, Reproduced with permission from American Welding Society (AWS), Miami, FL, USA., Table 6

Section Review Answer Key

Answer	Section Reference	Objective
Section One		
1. c	1.1.0	1a
2. b	1.2.4	1b
Section Two		
1. d	2.1.0	2a
2. a	2.2.5	2b

NCCER CURRICULA — USER UPDATE

NCCER makes every effort to keep its textbooks up-to-date and free of technical errors. We appreciate your help in this process. If you find an error, a typographical mistake, or an inaccuracy in NCCER's curricula, please fill out this form (or a photocopy), or complete the online form at **www.nccer.org/olf**. Be sure to include the exact module ID number, page number, a detailed description, and your recommended correction. Your input will be brought to the attention of the Authoring Team. Thank you for your assistance.

Instructors – If you have an idea for improving this textbook, or have found that additional materials were necessary to teach this module effectively, please let us know so that we may present your suggestions to the Authoring Team.

NCCER Product Development and Revision

13614 Progress Blvd., Alachua, FL 32615

Email: curriculum@nccer.org

Online: www.nccer.org/olf

❑ Trainee Guide ❑ Lesson Plans ❑ Exam ❑ PowerPoints Other ____________

Craft / Level: Copyright Date:

Module ID Number / Title:

Section Number(s):

Description:

Recommended Correction:

Your Name:

Address:

Email: Phone:

SMAW – Beads and Fillet Welds

Overview

This module provides trainees with their first opportunities to experiment with SMAW processes in hands-on practice. Common preparations for welding are reviewed, along with a brief review of the important safety considerations. The methods of striking an arc are presented, along with the importance of this simple action in weld quality. Both stringer and weave beads are explained, and trainees create many of their own weld beads through a significant amount of practice in the welding booth.

Module 29109

29109 V5

29109

SMAW – Beads and Fillet Welds

Objectives

When you have completed this module, you will be able to do the following:

1. Explain how to prepare for SMAW welding and how to strike an arc.
   a. Identify safety practices related to SMAW.
   b. Explain how to prepare the area and equipment for welding.
   c. Explain how to strike an arc and respond to arc blow.
2. Explain how to successfully complete various types of beads and welds.
   a. Explain how to properly restart and terminate a weld pass.
   b. Describe the technique required to produce stringer beads.
   c. Describe the technique required to produce weave and overlapping beads.
   d. Describe the techniques required to produce fillet welds in various positions.

Performance Tasks

Under the supervision of your instructor, you should be able to do the following:

1. Set up welding equipment.
2. Strike an arc.
3. Make stringer, weave, and overlapping beads using E6010 and E7018 electrodes.
4. Make corner welds on an angle iron section end welded to a plate coupon.
5. Make fillet welds using E6010 and E7018 electrodes in the specified positions:
   - Flat (1F)
   - Horizontal (2F)
   - Vertical (3F)
   - Overhead (4F)

Trade Terms

Arc blow
Arc strike
Drag angle
Oscillation
Push angle
Quench
Restart
Stringer bead
Undercut
Weave bead
Weld axis
Weld coupon
Work angle

Industry Recognized Credentials

If you are training through an NCCER-accredited sponsor, you may be eligible for credentials from NCCER's Registry. The ID number for this module is 29109. Note that this module may have been used in other NCCER curricula and may apply to other level completions. Contact NCCER's Registry at 888.622.3720 or go to **www.nccer.org** for more information.

Contents

Topics to be presented in this module include:

Figures and Tables

SECTION ONE

1.0.0 SMAW PREPARATIONS

Objective

Explain how to prepare for SMAW welding and how to strike an arc.

a. Identify safety practices related to SMAW.
b. Explain how to prepare the area and equipment for welding.
c. Explain how to strike an arc and respond to arc blow.

Performance Tasks

1. Set up welding equipment.
2. Strike an arc.

Trade Terms

Arc blow: The deflection of the arc from its intended path due to magnetic forces that develop from welding current flow.

Arc strike: A discontinuity consisting of localized melting of the base metal or finished weld caused by the initiation of an arc. It remains visible as a result of striking the arc outside the area to be welded and can lead to rejection of an otherwise good weld.

Weld coupon: Metal pieces to be welded together as a test or practice.

One of the most basic welds is the fillet weld. Before trainees can make a fillet weld, they must be able to set up arc welding equipment, strike an arc, and maintain that arc to run a welding bead. This section reviews the necessary preparations for SMAW, through the establishment of the all-important arc. SMAW is often referred to by the nonstandard name stick welding.

1.1.0 Safety Summary

The following is a summary of safety procedures and practices that must be observed when cutting or welding. Keep in mind that this is a summary only. Complete safety coverage is provided in the *Welding Safety* module. If you have not completed that module, do so before continuing. Above all, be sure to wear the appropriate personal protective equipment (PPE) when welding or cutting.

1.1.1 Protective Clothing and Equipment

To maintain safety and prevent injury, it is essential that you wear the appropriate PPE when cutting or welding metals (*Figure 1*). Be sure to follow these guidelines during all phases of cutting or welding:

- Always use safety glasses with a full face shield or a welding helmet. The glasses, face shield, or helmet lens must have the proper light-reducing tint for the type of welding or cutting being performed. Never view an electric arc, directly or indirectly, without using a properly tinted lens.
- Wear proper protective leather and/or flame retardant clothing along with welding gloves that will protect you from flying sparks and molten metal as well as heat.
- Wear high-top safety shoes or boots. Make sure that the tongue and lace area of the footwear will be covered by a pant leg. If the tongue and lace area is exposed, or if the footwear must be protected from burn marks, wear leather spats under your pants or chaps and over the front of the footwear.
- Wear a 100-percent cotton cap with no mesh material included in its construction. The bill of the cap points to the rear. If a hard hat is required for the environment, use one that allows the attachment of rear deflector material and a face shield. A hard hat with a rear deflector is generally preferred when working overhead, and may be required by some employers and job sites.

29109-15_F01.EPS

Figure 1 Welder working in full protective equipment.

> **WARNING!**
> Do not wear a cap with a button in the middle. The conductive metal button beneath the fabric represents a safety hazard.

- Wear a face shield over snug-fitting cutting goggles or safety glasses for gas welding or cutting. Either the face shield or the lenses of the welding goggles must be an approved shade for the application. A welding hood equipped with a properly tinted lens is also acceptable.
- Wear earplugs to protect ear canals from sparks. Wear hearing protection to protect against the consistent sound of the torch.

> **WARNING!**
> Ear protection is essential to protect ears from the noise of the torch. Other personal protective equipment (PPE) must be worn to protect the operator from hot metal and slag.

1.1.2 Fire and Explosion Prevention

Welding activities involve the use of fire or extreme heat to melt metal. Whenever fire is used, it must be controlled and contained. Welding activities are often performed on vessels that may once have contained flammable or explosive materials. Residues from those materials can catch fire or explode when a welder begins work on such a vessel. The following are basic fire and explosion prevention guidelines associated with welding:

- Never carry matches or gas-filled lighters in your pockets. Sparks can cause the matches to ignite or the lighter to explode, causing serious injury.
- Always comply with all site and/or employer requirements for a hot-work permit and a fire watch.
- Never use oxygen to blow dust or dirt off clothing. The oxygen can remain trapped in the fabric for a time. If a spark hits the oxygen in the fabric, the clothing can burn rapidly and violently.
- Make sure that any flammable material in the work area is moved or shielded by a fire-resistant covering.
- Approved fire extinguishers must be available before attempting any heating, welding, or cutting operations. Make sure the extinguisher is charged, the inspection tag is valid, and any individual that may be required to operate it knows how to do so.
- Never release a large amount of oxygen or use oxygen in place of compressed air. The presence of oxygen around flammable materials or sparks can cause rapid and uncontrolled combustion. Keep oxygen away from oil, grease, and other petroleum products.
- Never release a large amount of fuel gas, especially acetylene. Methane and propane are heavier than air and tend to migrate to and concentrate in low areas. As a result, they can ignite at a considerable distance from the release point. Acetylene is lighter than air but is even more dangerous than methane; when mixed with air or oxygen, it will explode at much lower concentrations than any other common fuel gas.
- To prevent fires, maintain a neat and clean work area, and make sure that any metal scrap or slag is cold before disposal.

Before cutting containers such as tanks or barrels, check to see if they have contained any explosive, hazardous, or flammable materials, including petroleum products, citrus products, or chemicals that decompose into toxic fumes when heated. Proper procedures for cutting or welding hazardous containers are described in the *American Welding Society (AWS) F4.1, Safe Practices for the Preparation of Containers and Piping for Welding and Cutting*, and *ANSI Z49.1*. As a standard practice, always clean and then fill any tanks or barrels with water, or purge them with a flow of inert gas such as nitrogen to displace any oxygen.

> **WARNING!**
> Welding or cutting must never be performed on drums, barrels, tanks, vessels, or other containers until they have been emptied and cleaned thoroughly, eliminating all flammable materials and all substances (such as detergents, solvents, greases, tars, or acids) that might produce flammable, toxic, or explosive vapors when heated. Do not assume that a container that has held combustibles is clean and safe until proven so by proper tests. Do not weld in places where dust or other combustible particles are suspended in air or where explosive vapors are present.

Containers must be cleaned by steam cleaning, flushing with water, or washing with detergent until all traces of the material have been removed.

> **WARNING!**
> Clean containers only in well-ventilated areas. Vapors can accumulate during cleaning, causing explosions or injury.

After cleaning the container, fill it with water or a purging gas, such as carbon dioxide, argon, or nitrogen to displace the explosive fumes. Air, which contains oxygen, is displaced from inside the container by the water or inert gas. Without oxygen, combustion cannot take place.

A water-filled vessel is the best alternative. When using water, position the container to minimize the air space. When using an inert gas, provide a vent hole so the inert gas can push the air and other vapors out to the atmosphere. Keep in mind, though, that even these precautions do not guarantee the absence of flammable materials inside. For that reason, these types of activities should not be done without proper supervision and the use of proper testing methods.

1.1.3 Work Area Ventilation

Vapors and fumes tend to rise in the air from their sources, and welders often have to work above welding areas where fumes are being created. Welding fumes can be harmful. Good work area ventilation helps to remove the vapors and protect the welder. The following is a list of work area ventilation guidelines to consider before and during welding activities:

- Make sure confined space procedures are followed before conducting any welding or cutting in the confined space.
- Make sure confined spaces are ventilated properly for cutting or welding purposes.
- Never use oxygen from cylinders for the purpose of ventilation or breathing air.
- Always perform cutting or welding operations in a well-ventilated area. Cutting or welding operations involving materials, coatings, or electrodes that contain cadmium, manganese, mercury, lead, zinc, chromium, and beryllium produce toxic fumes. For cutting or welding of such materials, always use proper area ventilation and wear an approved full-face, supplied-air respirator (SAR) that provides breathing air from outside of the work area. For occasional, very short-term exposure to fumes from zinc- or copper-coated materials, a high-efficiency, particulate arresting (HEPA) or metal-fume filters may be used on a standard respirator.

1.2.0 SMAW Equipment Setup

Before welding can take place, the area has to be made ready, the welding equipment must be set up, and the metal to be welded must be prepared. The following sections explain how to set up equipment for welding.

1.2.1 Preparing the Welding Area

To practice welding, a welding table, bench, or stand is needed (*Figure 2*). The welding surface must be steel, and provisions must be made for placing **weld coupons** out of position.

To set up the area for welding, follow these steps:

Step 1 Ensure that the area is properly ventilated. Make use of doors, windows, and fans.

Step 2 Check the area for fire hazards. Remove any flammable materials before proceeding.

Step 3 Determine the location of the nearest fire extinguisher. Do not proceed unless the extinguisher is charged and you know how to use it.

Step 4 Set up flash shields around the welding area.

1.2.2 Preparing the Weld Coupons

The weld coupons should be carbon steel, ¼" to ¾" (≈6 to 19 mm) thick. Use a wire brush or grinder to remove heavy mill scale or corrosion. Prepare weld coupons to practice the welds indicated as follows:

- *Striking an arc* – The coupons can be any size or shape that is easily handled.

29109-15_F02.EPS

Figure 2 Welding station.

- *Running beads* – The coupons can be any size or shape that is easily handled.
- *Overlapping beads* – The coupons can be any size or shape that is easily handled.
- *Fillet welds* – Cut the metal into 4" × 6" rectangles for the base and 3" × 6" rectangles for the web. To make best use of available materials for welding practice, the instructor may specify different sizes.

Figure 3 shows the weld coupons for fillet welding.

Steel for practice welding is expensive and difficult to obtain. Every effort should be made to conserve and not waste the material that is available. Reuse weld coupons until all surfaces have been welded upon. Weld on both sides of the joint, and then cut the weld coupon apart and reuse the pieces. Use material that cannot be cut into weld coupons to practice striking an arc and running beads.

1.2.3 Electrodes

Obtain a small quantity of the electrodes to be used. Electrodes are sometimes referred to by the nonstandard term welding rods or simply rods. For the welding exercises in this module, 3⁄32" (2.4 mm) to 5⁄32" (4.0 mm) E6010 and E7018 electrodes will be used. Obtain only the electrodes to be used for a particular welding exercise at one time. The E7018 electrodes should remain in an electrode oven or portable caddy until they are used. Have some type of pouch or rod holder in which to store the electrodes, to prevent them from becoming damaged. Never store electrodes to roll around loose on a table. They may fall on the floor and become damaged or create a tripping hazard. Some type of metal container, such as a pail, must also be available to discard hot electrode stubs.

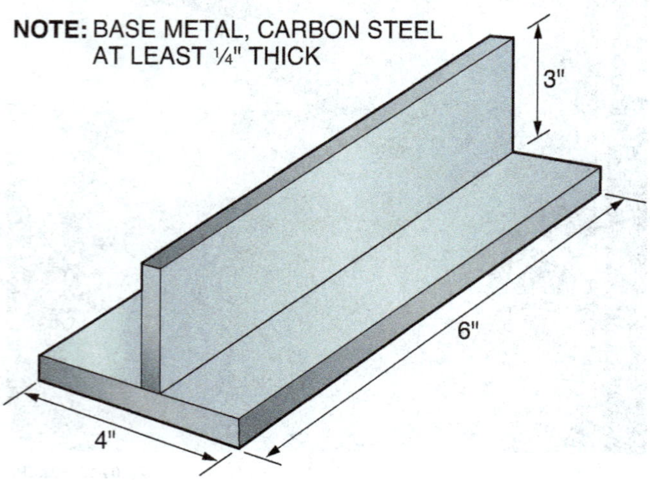

Figure 3 Fillet weld coupons.

Return unused electrodes to the proper storage location.

> **NOTE**
> The electrode sizes indicated are recommendations. Other electrode sizes may be substituted, depending on site conditions. Confirm the electrode size to use with your instructor.

> **WARNING!**
> Do not throw electrode stubs on the floor. They easily roll, and could cause someone to slip and fall.

1.2.4 Preparing the Welding Machine

Welders can expect to find different types and makes of welding machines in welding shops and on jobs in the field. Most machines will have no operator manuals with them. As a trainee, and even as a trained welder, you must be able to recognize the different types of welding machines and figure out how to safely set up and operate them. Select a welding machine to use (*Figure 4*) and then follow these steps to set it up for welding:

Step 1 Verify that the welding machine can be used for DC welding.

> **NOTE**
> An AC welding machine can be used if a DC machine is not available. If an AC welding machine is used, use E6011 and E7018 electrodes.

Step 2 Verify the location of the primary power disconnect to the machine.

Step 3 Check the area for proper ventilation.

Step 4 Set the polarity to direct current electrode positive (DCEP).

Step 5 Connect the clamp of the workpiece lead to the workpiece.

> **WARNING!**
> Even though workpiece leads and clamps are sometimes called ground leads or clamps, they may not actually be grounded. In this case, the full open-circuit voltage of the welding machine may exist between the workpiece lead or clamp and any grounded object.

Work Area Ventilation

There are many different breathing hazards that can be associated with welding processes. Vaporizing a number of base metal and electrode components allows these materials to be inhaled. Chromium, for example, is considered a carcinogenic risk by international organizations. Manganese affects the central nervous system, leading to problems in coordination and walking. The effects are irreversible.

Although employers and organizations generally recognize the potential hazards and provide the correct PPE and ventilation systems to protect welders, you must also take personal responsibility for your safety and well-being. Know the components of base metals and electrodes you are working with, and consider proper ventilation in the work area to be as essential to the task as the welding machine. This welder is using both source capture and a personal air-filtration system to ensure a safe work area. Proper ventilation is also proven to help keep the entire work area cleaner and present a more professional image.

29109-15_SA01.EPS

Step 6 Set the amperage for the electrode type and size to be used. Typical settings are shown in *Table 1*.

NOTE

Amperage recommendations vary by manufacturer, position, current type, and electrode brand. For specific recommendations, refer to the manufacturer's literature for the electrode being used.

Step 7 Ensure that the electrode holder is not touching the workpiece lead clamp, the workpiece, or a grounded object.

Step 8 Turn on the welding machine.

1.3.0 Striking an Arc

The first step in starting a weld is striking the arc. To strike an arc, touch the end of the electrode to the base metal and then quickly raise it to the correct arc length. The general rule is that the arc length should be the diameter of the electrode being used. Thus, if a ⅛" (3.2 mm) electrode is being used, the arc length should be ⅛" (3.2 mm). The arc length is measured from the end of the electrode core to the base metal. If the electrode has a heavier flux coating, such as E7018 or E6013, the end of the electrode core will be slightly recessed in the flux coating. For these electrodes, the visible arc length (from the end of the coating to the base metal) should be slightly less to compensate for the part of the arc that is recessed. This will ensure that the actual arc length is correct. *Figure 5* shows arc lengths for a ⅛" (3.2 mm) electrode.

There are two ways to strike an arc: the scratching method and the tapping method.

GOING GREEN

Recycling Metal

Salvage any coupons or scrap metal that is no longer usable for welding practice and place it in a recycling bin. Almost all metals are recyclable and valuable. One technical school took in donated steel, used it for cutting and welding practice, and then sold all their recyclable metals to fund their school's annual picnic. The bottom line is that metal that is no longer usable for cutting or welding practice can be recycled and kept out of the landfills.

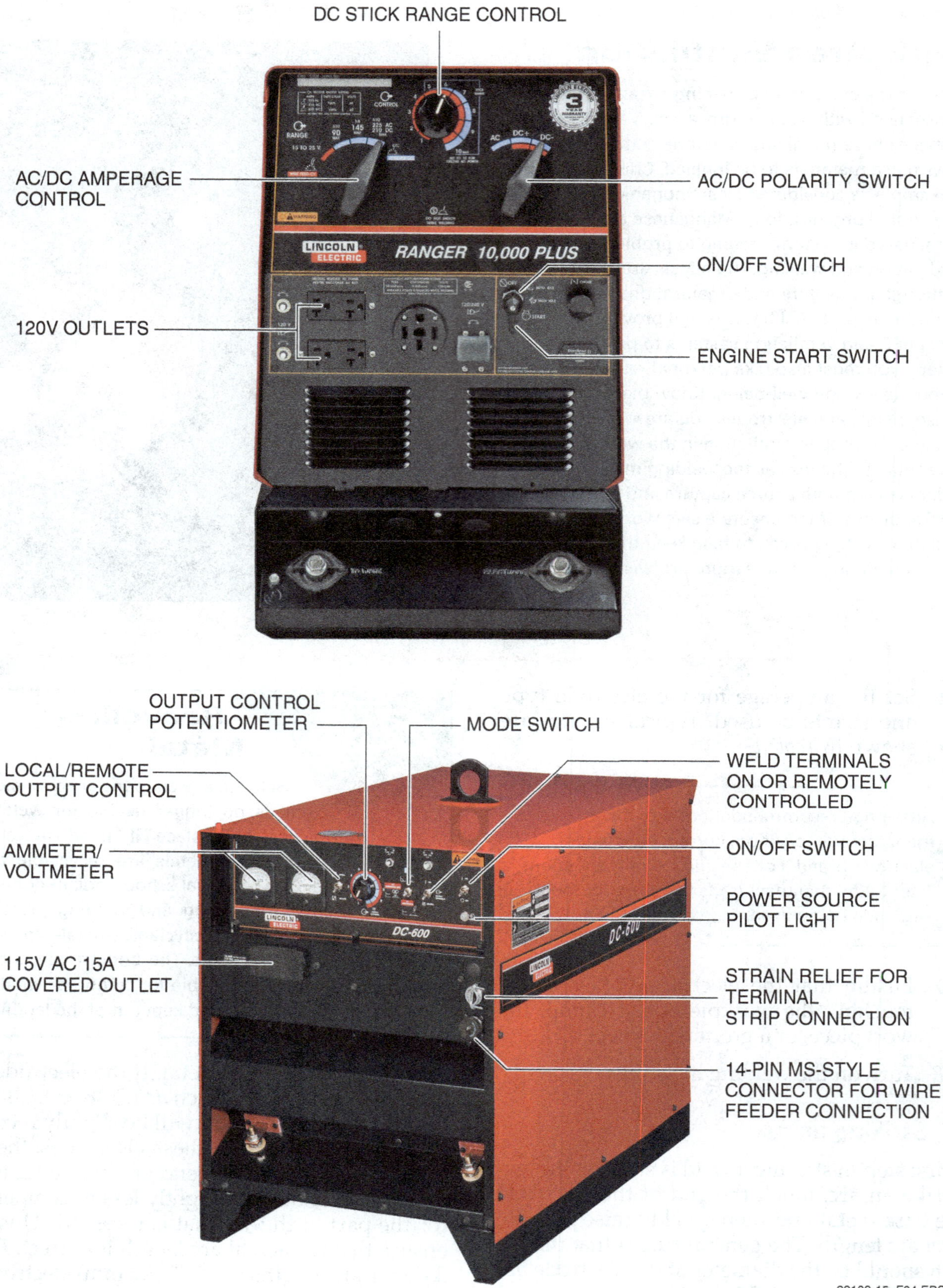

Figure 4 Typical AC/DC and DC welding machines.

Table 1 Amperage Settings for Electrodes

Electrode	Size	Amperage
E6013	⅛"	75A to 130A
E6013	5⁄32"	90A to 175A
E7018	⅛"	75A to 120A
E7018	5⁄32"	90A to 160A
E6011	⅛"	90A to 150A
E6011	5⁄32"	120A to 190A
E6010	⅛"	110A to 150A
E6010	5⁄32"	150A to 200A

1.3.1 Scratching Method

The scratching method (*Figure 6*) is the easiest way to strike an arc, and it is used by trainees who are learning to weld. It is similar to striking a match. The end of the electrode is simply scratched along the base metal to establish an arc. When the arc has been established, the electrode is raised to establish the correct arc length.

1.3.2 Tapping Method

The tapping method (*Figure 7*) is the best method of establishing an arc when using a transformer DC welding machine. It is more difficult and takes more practice to perfect, but as welders develop their skills, it is the method that should be used. The scratching method leaves arc strikes on the base metal that are not allowed by the welding codes unless they occur within the welded area.

Rod Holders

Several different types of rod holders can be purchased. One type is a leather pouch as shown here. Another is a sealable holder that can be used to prevent the electrodes from getting wet. These non-heated rod holders help protect the electrodes from physical damage and at least slow the absorption of moisture.

29109-15_SA02.EPS

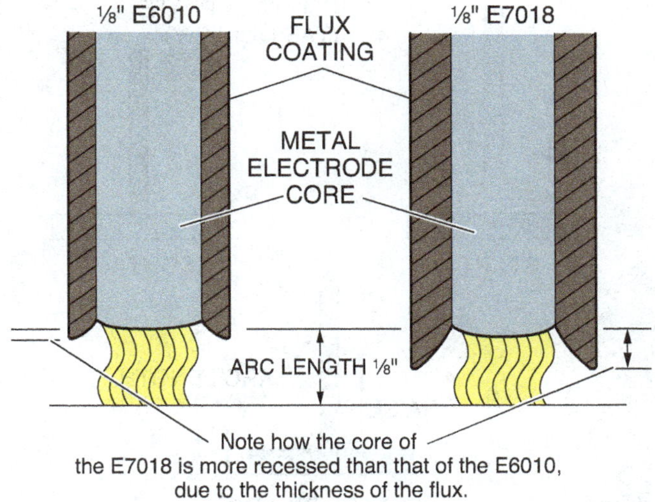

Figure 5 Arc lengths.

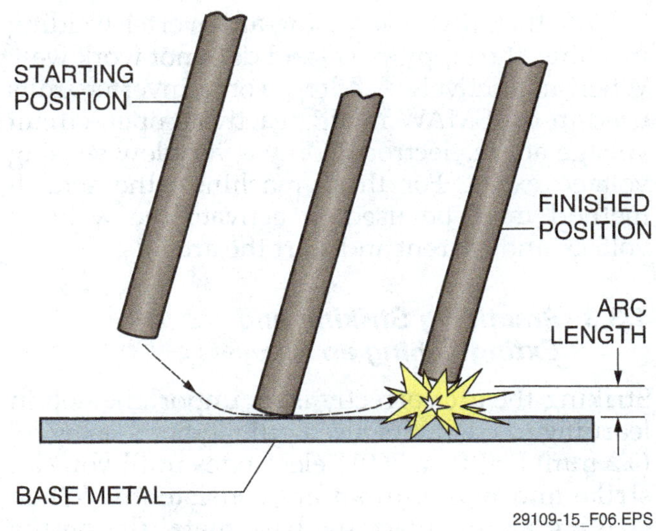

Figure 6 Scratching method of striking an arc.

With the tapping method, the arc is usually established just after the point where the welding should begin (¼" to ½", or ≈6.3 to 12.6 mm). The arc is then moved back to the correct starting point as it stabilizes. To strike an arc with the tapping method, move the electrode quickly to the base metal, lightly tap it, and then raise the electrode to establish the correct arc length.

Economizing with Electrodes

GOING GREEN

Electrodes are expensive, so they should be used economically. One way to conserve electrodes is to save the stubs from a welding project for use in making tack welds.

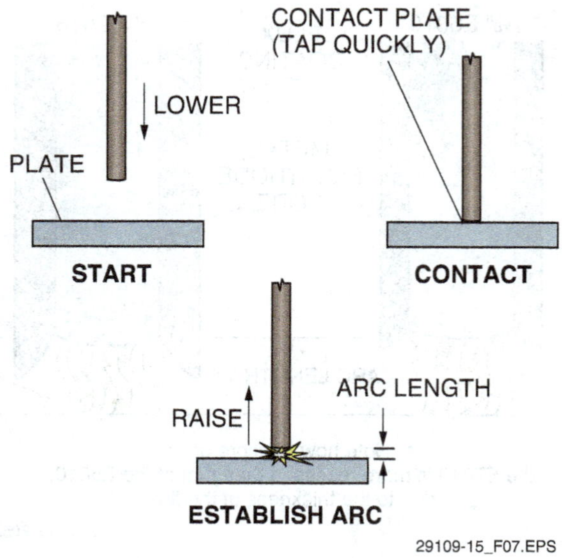

Figure 7 Tapping method of striking an arc.

Note that, if you are using an inverter welding machine, the tapping method does not work well. When not actively welding, some inverter units used in the SMAW mode deactivate open-circuit voltage at the electrode. Only a very low sensing voltage exists. For these machines, the scratch method must be used to activate the welding voltage and current and start the arc.

1.3.3 Practicing Striking and Extinguishing an Arc

Striking the arc correctly is an important step in learning to weld. Practice both methods using ⅛" (3.2 mm) E6010 or E6011 electrodes until you can strike and maintain an arc consistently. When striking an arc, place the base metal flat on the welding table. To extinguish the arc at the end of a weld, quickly lift the electrode away from the work.

Striking an arc with E7018 (F4 low-hydrogen) electrodes is more difficult than with most other electrodes because they tend to stick to the base metal more easily. Practice striking an arc with E7018 after you become more skilled striking and running beads with E6010 or E6011.

When practicing striking an arc, you will generally experience these two problems:

- The end of the electrode will instantly weld itself to the base metal. This is caused by keeping the electrode in contact with the base metal too long or by trying to maintain too short an arc length. If this condition occurs, free the electrode by quickly moving the electrode holder from side to side. If this fails, release the electrode from the holder and then remove the electrode from the base metal with pliers. Note that when the electrode is released from the holder, arcing generally occurs between the holder jaws and the electrode. Repeated electrode releases can damage the electrode holder jaws. Inspect the holder jaws for any heavy pitting or gouging damage, and replace them if necessary.
- The arc will extinguish. This is caused by raising the end of the electrode too far above the base metal.
- These problems will be eliminated as you gain experience in striking and controlling the arc gap.

1.3.4 Arc Blow

When current flows, especially DC current, strong magnetic fields may be created. The magnetic fields tend to concentrate in corners, in deep grooves, or in the ends of the base metal. When the arc approaches these concentrated magnetic fields, it is deflected. This phenomenon is known as **arc blow**. In some cases, ferrous metal welding fixtures or jigs that are part of a consistent DC welding current path can become magnetized and contribute to arc blow. In AC welding, arc blow is rarely a problem because the magnetic field is constantly reversing at twice the frequency of the primary power source. This effectively cancels any strong magnetic field effects. DC arc blow can cause defects such as excessive weld spatter and porosity.

Constant-Current Machines

The AC or AC/DC machines desired for SMAW or gas tungsten arc welding (GTAW) are basically constant-current types of machines, some of which are referred to as droop-current machines or droopers. In a true constant-current machine, the output current varies little over a relatively wide range of circuit voltage. Because of this, raising and lowering the electrode from the workpiece (changing the gap) has little effect on the weld quality. In addition, striking an arc is easier with a true constant-current machine because of the high open-circuit voltage. On the other hand, droopers can be used to control the molten weld-pool temperature more easily. This is because the current and resultant weld heat changes much more with a smaller change in voltage that is caused by raising or lowering the electrode in relation to the workpiece.

Restriking a Cold Electrode

When attempting to strike an arc using a cold, partially consumed electrode, the cup formed by the hardened flux sheath extending beyond the metal core may not allow the core to contact the workpiece to establish an arc. Scratch striking or hard tapping can be used to break off the sheath. Unfortunately, this sometimes causes large pieces of the flux coating to break off the sides, exposing part of the core. Electrodes with missing flux are difficult to strike. If an arc is struck, it will be unstable and will blow or wander until the core is consumed up to where the entire flux sheath cup is reestablished. During this time, poor welds will occur. One method of removing the flux sheath without chipping flux away from the sides of the electrode is to draw the end of it across the face of a medium-coarse file. If code requirement welding is being performed, electrodes with chipped, missing, or cracked flux coatings must be discarded.

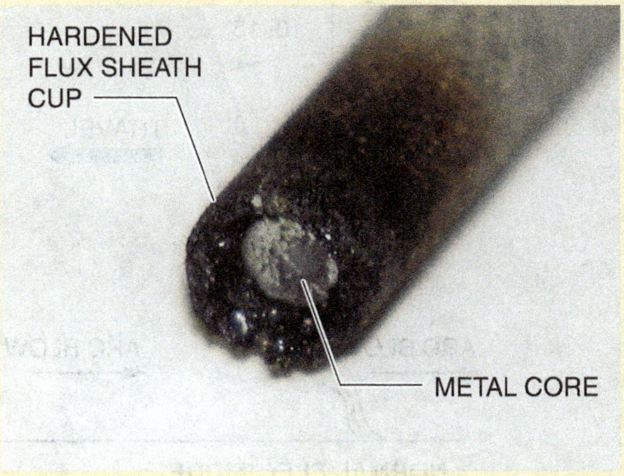

29109-15_SA03.EPS

If arc blow occurs, try one or more of these methods to control it:

- If possible, tack-weld the workpieces together at the ends and in the middle.
- Shorten the arc length. With a shorter arc length, the magnetic field will have less effect on the arc.
- Change the angle of the electrode (*Figure 8*). The normal electrode angle is 10 degrees to 15 degrees in the direction of travel. Raising the electrode angle toward 90 degrees or, in extreme cases, up to 20 degrees in the opposite direction of travel will compensate for the arc blow.
- Change the position of the workpiece lead clamp. This will change the flow of welding current, affecting the way the magnetic fields are created.
- Reducing the weld current when practical may also reduce arc blow.

Accurate Tap-Striking of an Electrode

An accurate method of tap-striking an electrode at an exact spot is to rest the electrode on the finger of your gloved free hand, similar to a pool cue. Angle the rod in the direction of travel. Then move the electrode quickly down and up to strike the arc. This method is best if your helmet can be lowered by nodding your head, or if you are using an auto-darkening lens helmet.

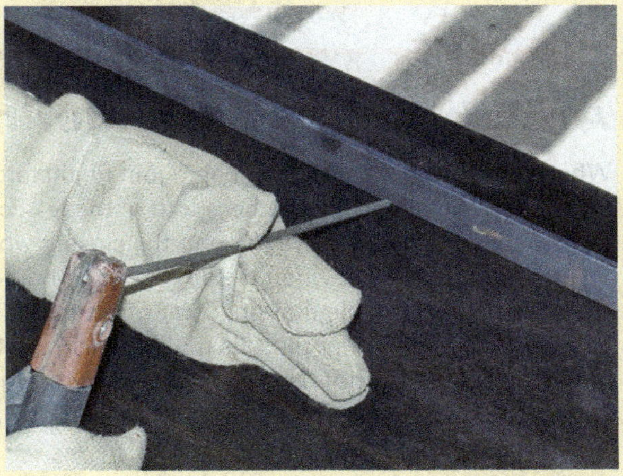

29109-15_SA04.EPS

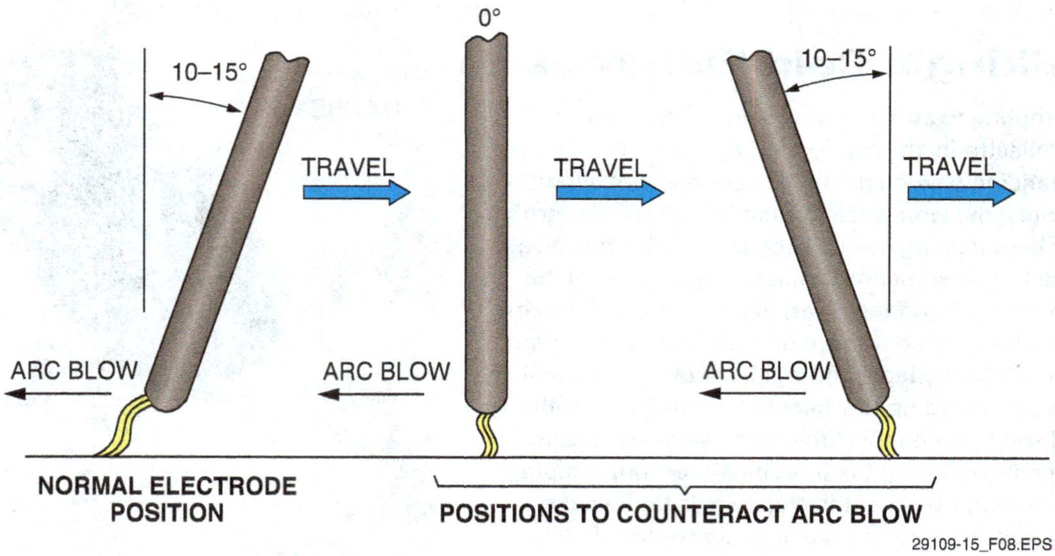

Figure 8 Controlling arc blow.

Reusing a Stuck Electrode

Carefully inspect the end of any electrode that has been stuck to the workpiece and freed by bending. If any of the flux coating is missing, cracked, or badly burned, discard the electrode.

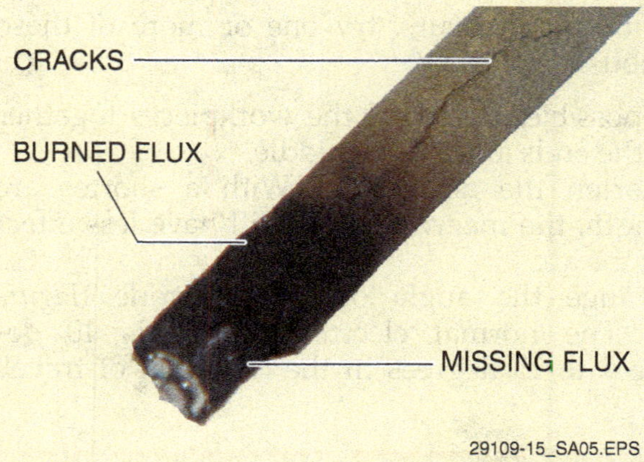

Arc Shifting or Wandering

When two workpieces are first being joined, the arc may deposit material on one piece only or wander back and forth from one piece to the other. Both of these occurrences are usually the result of too long an arc. If this is the case, the arc will always jump to the closest workpiece. This is especially noticeable with electrodes that are ⅛" (3.2 mm) or smaller when welding gapped-butt or T-joints. If the arc initially deposits material on only one workpiece, it is possible that only one of the workpieces is electrically connected to the workpiece lead from the welder and the other piece is not in the circuit. In this case, weaving the electrode back and forth across the workpiece at the start of the weld will usually bridge the gap between the workpieces and tack the pieces together so that they are both in the circuit. Other solutions are to clamp a piece of metal across both pieces, or to perform the welding on a metal tabletop.

Additional Resources

2014 Technical Training Guide. Current Edition. Cleveland, OH. USA: The Lincoln Electric Company. **www.lincolnelectric.com**

1.0.0 Section Review

1. For electric arc welding, wear safety glasses and a welding hood with a lens tinted to shade _____.
   a. 1 to 4
   b. 4 to 8
   c. 9 to 14
   d. 12 to 16

2. Electrode stubs should be _____.
   a. discarded to the floor for later collection
   b. welded together to form additional electrodes
   c. disposed of in a suitable metal bucket or container
   d. placed in water

3. With the tapping method of striking an arc, the arc is struck _____.
   a. ¼" to ½", or ≈6.3 to 12.6 mm, past where the weld is to begin
   b. ¼" to ½", or ≈6.3 to 12.6 mm, ahead of where the weld is to begin
   c. ½" to 1", or ≈12.6 to 25.3 mm, past where the weld is to begin
   d. ½" to 1", or ≈12.6 to 25.3 mm, ahead of where the weld is to begin

Section Two

2.0.0 SMAW Techniques

Objective

Explain how to successfully complete various types of beads and welds.

a. Explain how to properly restart and terminate a weld pass.
b. Describe the technique required to produce stringer beads.
c. Describe the technique required to produce weave and overlapping beads.
d. Describe the techniques required to produce fillet welds in various positions.

Performance Tasks

3. Make stringer, weave, and overlapping beads using E6010 and E7018 electrodes.
4. Make corner welds on an angle iron section end welded to a plate coupon.
5. Make fillet welds using E6010 and E7018 electrodes in the specified positions:
   - Flat (1F)
   - Horizontal (2F)
   - Vertical (3F)
   - Overhead (4F)

Trade Terms

Drag angle: Describes the travel angle when the electrode is pointing in a direction opposite to the welding bead's progression.

Oscillation: A repetitive and consistent side-to-side motion.

Push angle: Describes the travel angle when the electrode is pointing in the same direction as the welding bead's progression.

Quench: To rapidly cool a hot component such as a freshly welded coupon. Note that only coupons should be quenched; real welds with a function should never be quenched.

Restart: The action of restarting a weld bead that is already in progress that requires the crater left from the termination of the previous arc.

Stringer bead: A type of weld bead made without any significant weaving motion. With SMAW, stringer beads are not more than three times the diameter of the electrode.

Undercut: A defect that occurs when the cross-sectional thickness of the base metal is reduced at the edges of the weld. Excessive weld current can cause this as it causes the base metal at the edges to melt and drain into the weld puddle. A downward sloped edge toward the weld bead is the visual evidence.

Weave bead: A type of weld bead made by transverse oscillation of the electrode in a particular pattern.

Weld axis: A straight line drawn through the center of a weld along its length.

Work angle: An angle less than 90 degrees between a line perpendicular to the major workpiece surface and a plane determined by the electrode axis and the weld axis. 0-degree work angle is common. In a T-joint or corner joint, the line is perpendicular to the non-butting member. The definition of work angle for a pipe weld is covered in a later module, as it differs from that of plate welding.

Now that the work area and equipment have been properly prepared and you understand how to strike an arc and control arc blow, the next step is to begin creating weld beads.

2.1.0 Terminations and Restarts

How a weld bead is terminated is important to the quality of the weld. Since electrodes burn away rather quickly and the welder must stop and replace the electrode repeatedly, the bead must also be restarted with care.

2.1.1 Terminating a Weld

Terminations are made at the end of a weld. A termination leaves a crater. When making a termination, the welding codes require that the crater be filled to the full cross section of the weld. This can be difficult because most terminations are at the edge of a plate where welding heat tends to build up, making filling the crater more difficult.

The technique for making a termination is basically the same for all SMAW. Refer to *Figure 9* as the following steps are reviewed.

Step 1 As you approach the end of the weld, begin to stand the electrode up toward 0 degrees, and slow the forward travel.

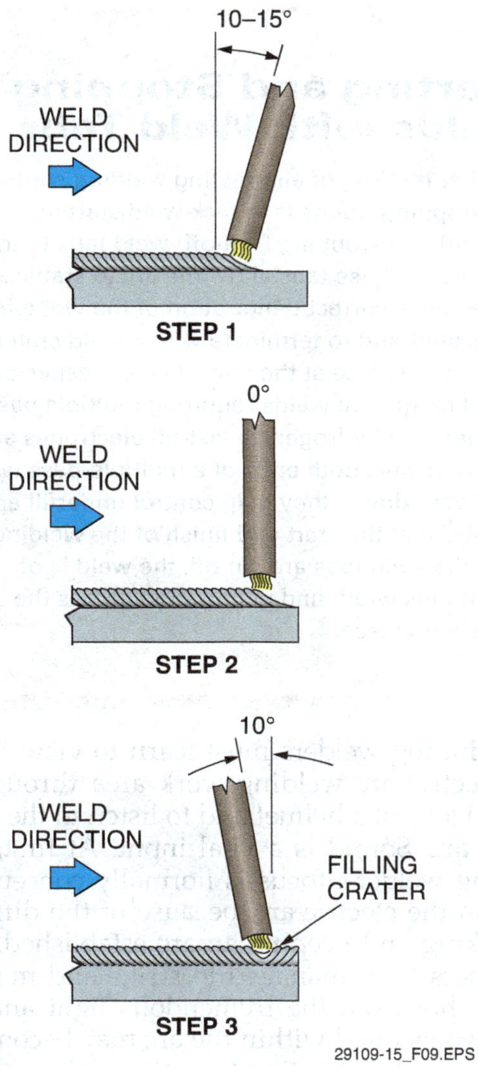

Figure 9 Weld termination.

Step 2 Stop forward movement about ⅛" (≈3.2 mm) from the end of the plate and slowly angle the electrode to about 10 degrees toward the start of the weld.

Step 3 Move about ⅛" (≈3.2 mm) toward the weld, and break the arc when the crater has been filled.

Step 4 Inspect the termination. The crater should be filled to the full cross section of the weld.

Step 5 Remove the electrode from the electrode holder and discard as soon as you stop welding to prevent arcing to nearby surfaces, and place the stub in a proper container.

WARNING!

Do not throw electrode stubs on the floor. They roll easily and could cause someone to slip and fall.

2.1.2 Restarting a Weld

A **restart**, sometimes referred to as a tie-in, is the point where one weld bead stops and another starts (*Figure 10*). Restarts are important because most SMAW welds cannot be made without at least one restart, and an improperly made restart will create a weld defect. A restart must be made so that it blends smoothly into the rest of the weld and does not stand out. The technique for making a restart is the same for both **stringer beads** and **weave beads**. Follow these steps to make a restart:

Step 1 Just before the previous electrode is used up, quickly increase the weld speed to taper the weld for ¼" to ⅜" (≈6.3 to 9.5 mm), and then break the arc.

Step 2 Chip any slag from the tapered section and crater.

Step 3 Remove the electrode from the electrode holder and discard as soon as you stop welding to prevent arcing to nearby surfaces, and place the stub in a proper container.

Step 4 With a new electrode, restrike the arc in front of the crater and in line with the weld. Remember that welding codes do not allow arc strikes outside of the area to be welded.

Step 5 Move the electrode back onto the tapered weld section and develop the weld puddle with a slight circular motion, maintaining the correct arc length and electrode angles.

Step 6 Start the forward motion as the bead achieves the proper width, and continue to weld.

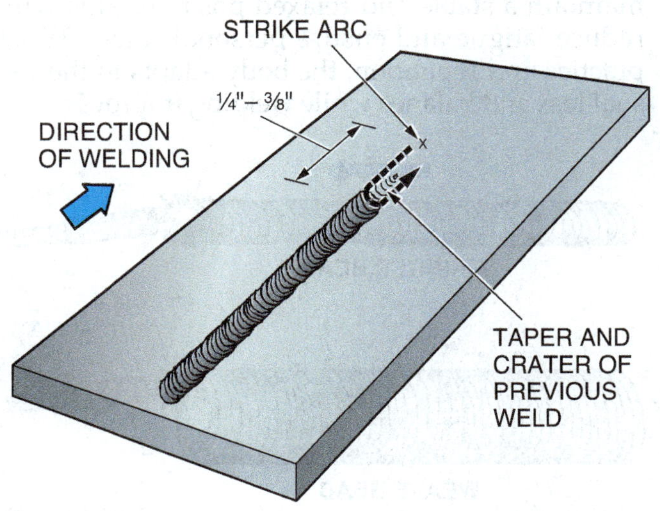

Figure 10 Making a restart.

Step 7 Discard the electrode stub from the holder and inspect the restart. A properly made restart will blend into the previous bead, making it hard to detect.

If the restart has **undercut**, not enough time was spent on the taper or the crater to fill it. If the undercut is on one side or the other, use more of a side-to-side motion as you move into the taper. If the restart has a hump, it was overfilled; too much time was spent before resuming the forward motion.

Continue to practice restarts until they are correct.

2.2.0 Stringer Beads

A stringer bead is a weld bead that is made with very little or no side-to-side motion of the electrode. The width of a stringer bead will vary with the electrode type. A weave bead is a weld bead that is made with a side-to-side motion of the electrode. Outside of welding joints, stringer and weave beads (*Figure 11*) are used for resurfacing or hard-surfacing. This section will focus on the stringer bead.

The width of weld beads is usually specified in the drawing welding symbols, welding code, or welding procedure specification (WPS) being used for the work. Do not exceed the widths specified for the work.

2.2.1 Beginning to Weld

A welder needs to be in a relaxed, comfortable position whenever possible. Because of the limited view of the welding area through a dark helmet lens, a beginning welder may sway a little. This is due to the loss of a visual component in the body's ability to maintain balance. To counteract this, a welder should sit or lean against something to achieve and maintain a stable and relaxed position. This will reduce fatigue and ensure personal safety. With practice and repetition, the body adapts to the visual loss and balance while welding improves.

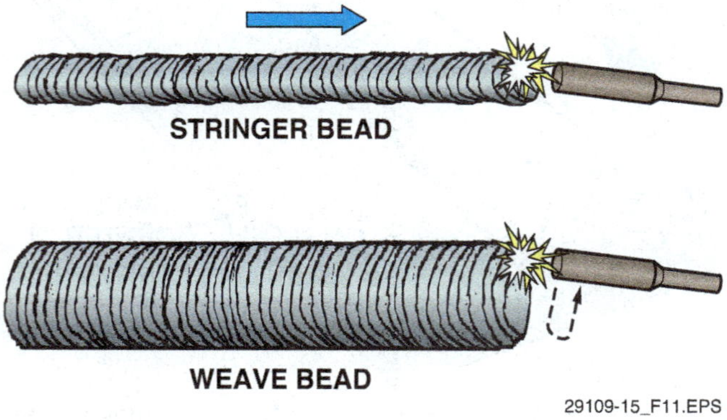

Figure 11 Stringer and weave beads.

Starting and Stopping Welds with Weld Tabs

Another method of eliminating welding starting and stopping points is to tack-weld starting (run-on) and stopping (run-off) weld tabs to the workpiece. These tabs allow the arc to stabilize and achieve correct penetration of the workpiece at the start and to terminate with a weld crater off the workpiece at the end. They are especially useful on groove welds requiring multiple passes or when low-hydrogen or fast-fill electrodes are used. Because both ends of a multiple-pass weld may taper down, they help control underfill and back-burn at the start and finish of the welding. After the weld tabs are cut off, the weld is of continuous width and penetration across the entire workpiece.

Beginning welders must learn to view the entire electric arc welding work area through the shaded lens of a helmet and to listen to the sound of the arc. Sound is a vital input. At first, a beginning welder's focus is normally concentrated only on the electric arc, because of the difficulty of striking and keeping an arc established. Once beginners have managed to strike and maintain an arc, however, the tremendous light and heat energy generated within the arc may become fascinating and distracting for a time. With practice, the beginning welder will gradually be able to shift focus and see all sides of a molten weld pool, the weld buildup at the trailing edge of the pool, the cooling slag form over the weld buildup, and even the adjacent welding area on the workpiece. The welder becomes attuned to the correct sound of the arc for the rod and work being welded, and the arc itself may not even be noticed. Once the welder learns to view the entire weld area and listen to the arc, running straight and correct welds is relatively easy.

Before running the first bead, a word of caution is in order. On occasion in this text, you may be instructed to **quench** your work. This means to dunk the workpiece in water briefly so it can be more easily and safely handled. However, it is very important to understand that this practice relates only to welding coupons used for practice. Real-world welds should never be quenched due to the potential for weld failure due to the rapid temperature change.

2.2.2 Practicing Stringer Beads with E6010

Practice running stringer beads in the flat position using ⅛" (3.2 mm) E6010 or E6011 electrodes. After striking the arc, the electrode axis angle should be a 10-degree to 15-degree **drag angle** in the direction of travel for the weld and at a 0-degree **work angle**. Note that a 0-degree work angle means the electrode is perpendicular (90 degrees) to the base metal. *Figure 12* shows the electrode axis angles.

Travel angles used for SMAW and other processes are categorized as either drag or **push angles**. An electrode drag angle is a travel angle in which the electrode axis points at the weld bead during the running of a weld bead. A push angle is the opposite condition; it occurs when the electrode axis points away from the weld bead and toward the direction of travel when running a weld bead. Except for vertical welds, SMAW is usually accomplished with an electrode drag angle.

The electrode work angle is the side-to-side tilt of a plane containing the electrode axis from a perpendicular line to the major work surface. The plane is formed by the intersection of the **weld axis** and the centerline of the electrode. Work angles typically range from 0 to 25 degrees. In

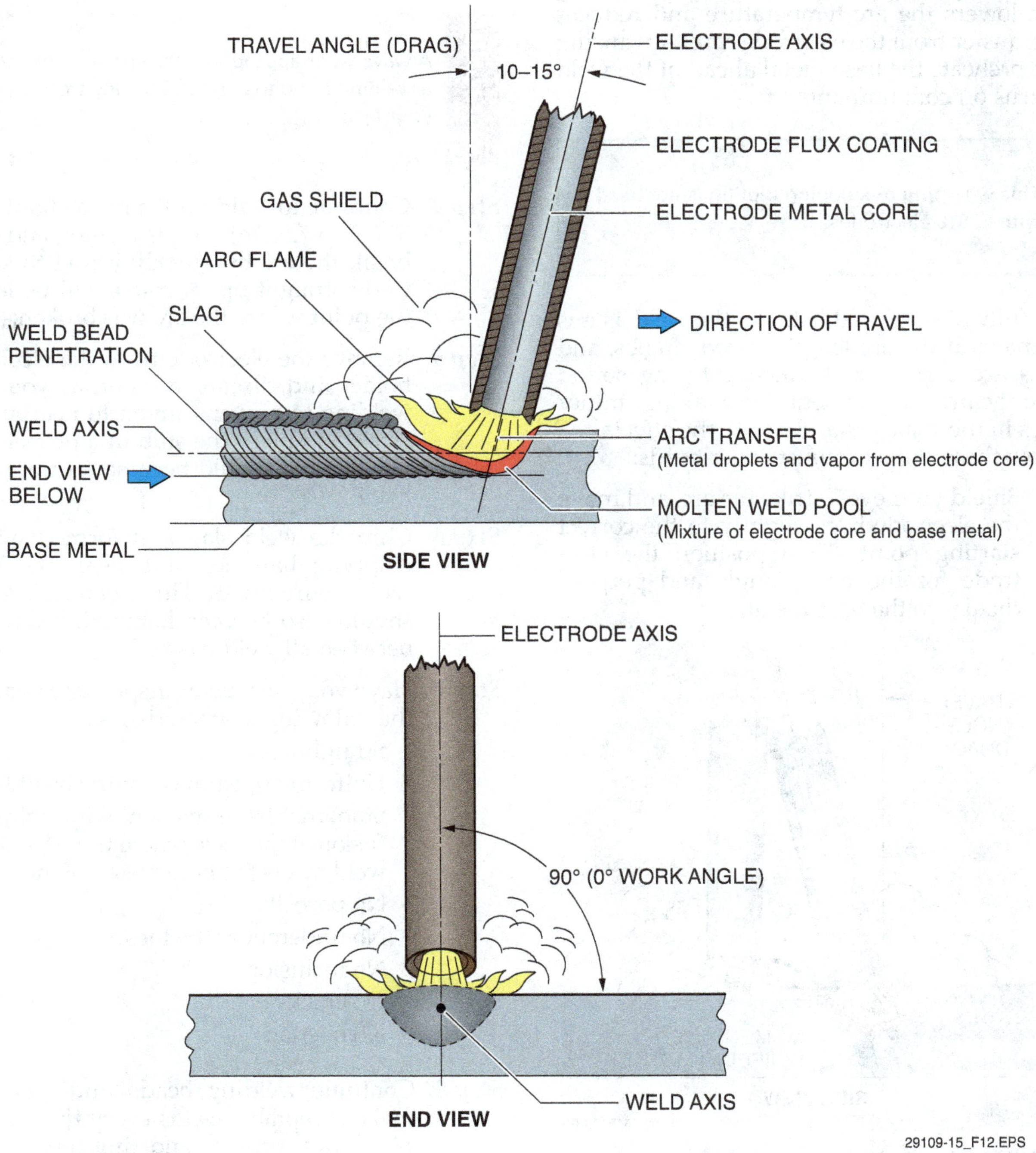

Figure 12 SMAW characteristics with proper travel and work angle.

the case of a single bead being run on a flat plate, the work angle is zero. The work angle for a fillet weld in a T- or corner joint is always taken from a perpendicular line to the non-butting member.

The whipping motion (*Figure 13*), also called a stepping motion, can be used when depositing the stringer bead to control the weld puddle. It is performed by moving the electrode up and forward about ¼" (≈6.3 mm), then down and backward about 3⁄16" (≈4.8 mm). A short pause at the end of the backward travel deposits the weld metal. The exact length of travel can vary. By momentarily lengthening and advancing the arc, the molten weld pool is allowed to cool. Lengthening the arc lowers the arc temperature and reduces metal transfer from the electrode core. Advancing the arc preheats the base metal ahead of the weld and burns off contaminants.

NOTE

This whipping or stepping motion is not used with E7018 electrodes.

Carefully observe and listen to the weld as it is being made. If the arc length, speed, angles, and motions are correct, a distinctive frying sound will be heard. Experiment by making minor changes in these factors and notice the effects.

Follow these steps to run stringer beads:

Step 1 Shield your eyes, strike the arc, and move the electrode back slightly to the correct starting point. Then position the electrode for the travel angle and perpendicular to the base metal.

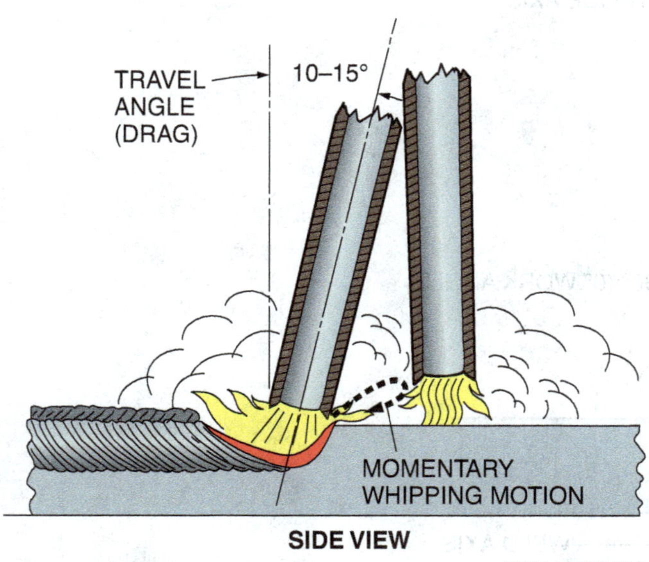

Figure 13 Whipping motion.

Step 2 Hold the arc in place until the weld puddle widens to about two times the diameter of the electrode's metal core.

Step 3 Slowly move the arc forward, keeping a constant arc length. Since the electrode is constantly adding filler metal to the weld, it becomes shorter. The welder, therefore, must be constantly closing in on the workpiece to keep the arc gap (the length of the arc) consistent. As necessary, use the whipping motion and adjust the forward speed to control buildup and preheat the weld zone.

WARNING!

Always wear appropriate face protection and/or a welding hood to prevent hot slag from striking your face.

Step 4 Continue to weld until a bead about 2" to 3" (≈5 to 7.5 cm) long is formed and then break the arc by quickly lifting the electrode straight up. A crater will be left at the point where the arc was broken.

Step 5 Remove the electrode from the electrode holder and discard as soon as you stop welding to prevent arcing to nearby surfaces, and place the stub in a proper container. This should become an unbroken habit.

Step 6 Chip the weld slag that formed with a chipping hammer and clean the bead with a wire brush. This cleaning process should also become habitual, conducted between all weld passes.

Step 7 Have your instructor inspect the bead for the following characteristics:

- Straightness
- Uniform appearance on the bead face
- Smooth, flat transition with complete fusion at the toes (where the edge of the weld meets the base metal) of the weld
- No porosity
- No undercut at the toes
- No inclusions
- No cracks
- No overlap

Step 8 Continue welding beads until you can make acceptable welds every time. *Figure 14* shows proper and improper weld beads.

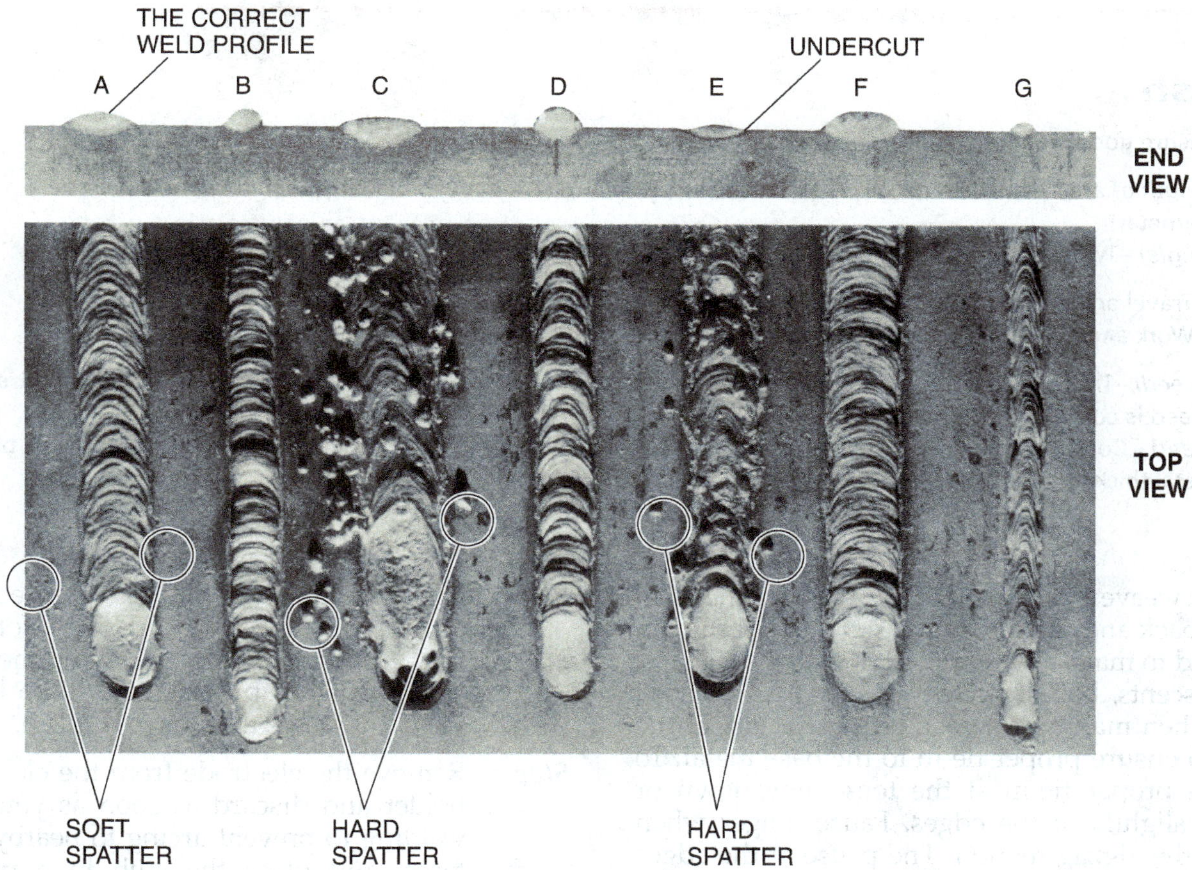

A = Correct current, arc length, and travel speed. Note the easily removed spatter (soft spatter).

B = Current set too low.

C = Current set too high. Note the hard spatter (spatter firmly bonded to base material that must be ground or chiseled off). Note the pointed ends of the bead indicating the weld pool was too hot and cooled too slowly. Impurities are usually trapped in the weld due to the slow cooling.

D = Arc length too short (narrow, high bead caused by arc pressure).

E = Arc length too long. Note the hard spatter and bead undercut of the edges.

F = Travel speed too slow (wide high bead).

G = Travel speed too fast. Note the pointed ends of bead.

29109-15_F14.EPS

Figure 14 Effect of current, arc length, and travel speed on SMAW beads.

2.2.3 Practicing Stringer Beads with E7018

Continue running stringer beads, this time using ⅛" (3.2 mm) E7018 electrodes. Use the same electrode angles, but remember not to whip the E7018 electrode. When running all low-hydrogen (F4) electrodes, the arc should never leave the weld puddle, and the visible arc should be shorter than when using the E6010 electrode. This is due to the thicker, longer sheath of the flux coating. Weld defects, such as porosity or hydrogen embrittlement, can occur if the arc leaves the weld puddle or if the arc is too long. The arc can be moved within the weld puddle to control the bead shape.

2.3.0 Weave Beads and Overlaps

Running weave beads is significantly different from stringer beads. There are a number of different weave patterns that can be used.

2.3.1 Practicing Weave Beads with E6010

Practice running weave beads in the flat position using ⅛" (3.2 mm) E6010 or E6011 electrodes. After striking the arc, the electrode angle should be 10 to 15 degrees in the direction of travel and at a 0-degree work angle.

Lash

To ensure good welding results, remember LASH.

L *(length of arc)* –The distance between the electrode and the base metal (usually one times the electrode diameter).

A *(angle)* –Two angles are critical:

- Travel angle –The longitudinal angle of the electrode in relation to the axis of the weld joint
- Work angle –The traverse angle of the electrode in relation to the axis of the weld joint

S *(speed)* –Travel speed is measured in inches per minute (IPM). The width of the weld will determine if the travel speed is correct.

H *(heat)* – Controlled by the amperage setting and dependent upon the electrode diameter, base metal type, base metal thickness, and the welding position.

The weave bead is made by moving the electrode back and forth. Many different patterns can be used to make a weave bead, including zigzags, Js, crescents, boxes, circles, and figure 8s (*Figure 15*). When making a weave bead, use care at the toes to ensure proper tie-in to the base metal. To ensure proper tie-in at the toes, slow down or pause slightly at the edges. Pause longer when using the zigzag motion. The pause at the edges will also flatten out the weld, giving it the proper profile.

Carefully observe and listen to the weld as it is being made. If the arc length, speed, angles, and motions are correct, a distinctive frying sound will be heard. Experiment by making minor changes in these factors and notice the effects.

Follow these steps to run weave beads (*Figure 16*):

Step 1 Strike the arc and position the electrode for a drag angle. Make sure the electrode is perpendicular to the base metal (0-degree work angle).

Step 2 Hold the arc in place until the weld puddle widens to about two times the diameter of the electrode's metal core.

Step 3 Slowly move the arc forward in a weaving motion, keeping a constant arc length.

WARNING!

Always wear the proper PPE to prevent hot slag from hitting your face.

Step 4 Continue to weld until a weave bead about 2" to 3" (≈5 to 7.5 cm) long is formed, and then break the arc by lifting the electrode straight up. A crater will be left at the point where the arc was broken.

Step 5 Remove the electrode from the electrode holder and discard as soon as you stop welding to prevent arcing to nearby surfaces, and place the stub in a proper container.

Step 6 Chip, clean, and wire brush the crater and bead.

Step 7 Make a restart, and continue welding to the end of the plate.

Step 8 Have the instructor inspect the bead, which should have the following characteristics:

- Weld bead straight to within ⅛" (≈3.2 mm)
- Uniform appearance on the bead face
- Smooth, flat transition with complete fusion at the toes of the weld
- Crater and restarts filled to the full cross section of the weld
- No porosity
- No excessive undercut at the toes
- No inclusions
- No cracks
- No overlap

Continue welding weave beads until you can make acceptable welds every time.

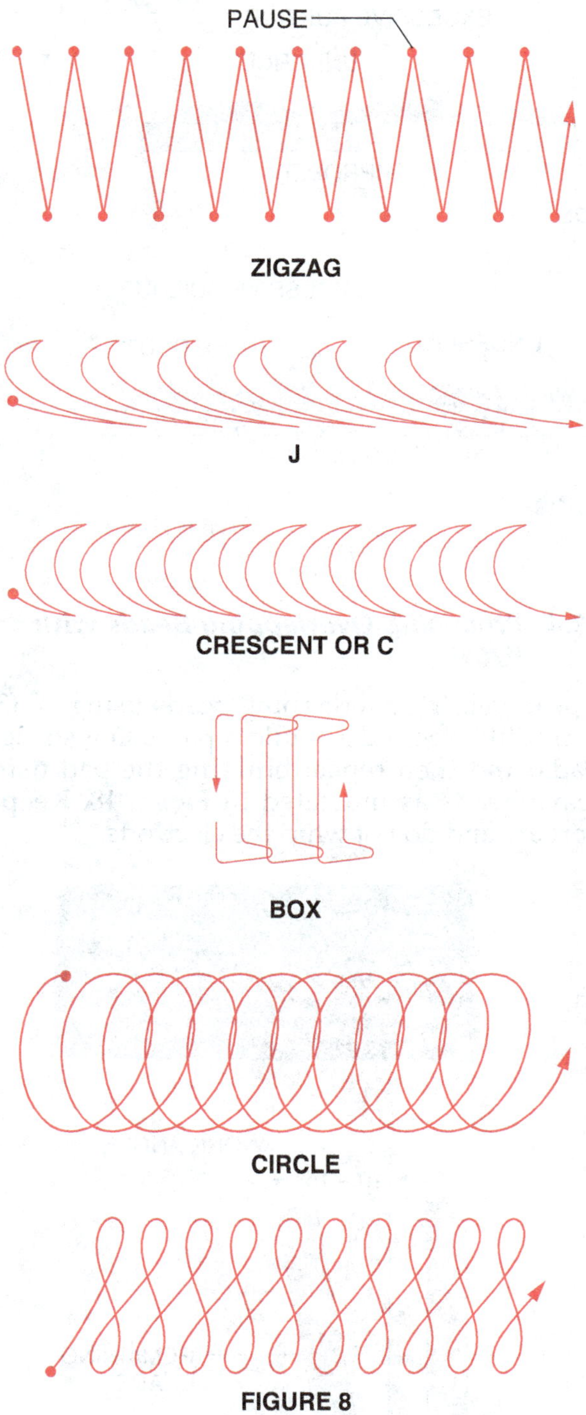

Figure 15 Weave motions.

Weave Bead Width

Although the usual width of a straight bead is two to three times the diameter of the electrode's metal core, the bead width for a weave bead can usually be up to, but should not exceed, eight times the electrode diameter.

Figure 16 Weave bead.

2.3.2 Practicing Weave Beads with E7018

Repeat running weave beads using ⅛" (3.2 mm) E7018 electrodes. Use the same electrode angles, but do not whip the E7018 electrode. When running low-hydrogen electrodes, the arc should never leave the weld puddle, and the visible arc should be shorter than when using E6010. Weld defects, such as porosity or hydrogen embrittlement, can occur if the arc is too long or if it leaves the weld puddle. The arc can be moved within the weld puddle to control the bead shape.

It is important to note that there are several weave techniques that should not be used with the E7018 electrode. They are the circle, box, and figure 8 weave techniques.

2.3.3 Practicing Overlapping Beads with E6010

Overlapping beads are made by depositing connective weld beads parallel to one another. The parallel beads overlap, forming a flat surface (*Figure 17*). This technique is also called padding. Overlapping beads are used to build up a surface and to make multi-pass welds. Both stringer and weave beads can be overlapped. Properly overlapped beads, when viewed from the end, will form a relatively flat surface.

Overlapping stringer beads must be cleaned carefully to prevent slag inclusions. The thickness of the overlap must be fairly consistent, so that it will not be necessary to machine the surface. Follow these steps to weld overlapping stringer beads using ⅛" (3.2 mm) E6010 electrodes:

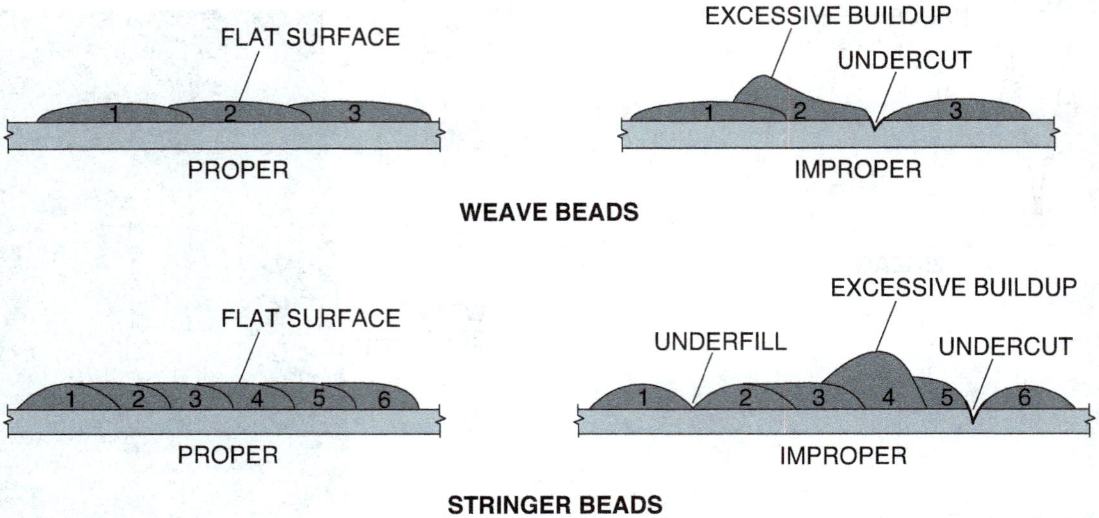

Figure 17 Proper and improper overlapping beads.

Step 1 Mark out a 4" (102 mm) square on a piece of steel.

Step 2 Weld a stringer bead along one edge.

Step 3 Properly dispose of the electrode stub immediately and chip/clean the weld bead.

Step 4 After striking the arc for the next stringer bead, and with the proper travel angle, position the electrode at a work angle of 10 to 15 degrees to the side of the previous bead to obtain proper tie-in (*Figure 18*).

Step 5 Continue running stringer beads until the marked square is covered.

WARNING!

The base metal will get very hot as it is built up. Using pliers, cool it by plunging it into water. However, be careful of the resulting plume of hot steam.

CAUTION

Only welding coupons for practice should be quenched. Never quench actual welding work.

Step 6 Continue building layers of stringer beads, one on top of the other, until the technique is perfected.

Continue welding overlapping beads using the weaving technique. Remember to angle the electrode toward the previous bead to obtain a good tie-in.

2.3.4 Practicing Overlapping Beads with E7018

Repeat welding overlapping beads using ⅛" (3.2 mm) E7018 electrodes. Build a pad using stringer beads, and then repeat building the pad using weave beads as indicated in *Figure* 18. Keep a short arc and do not whip the electrode.

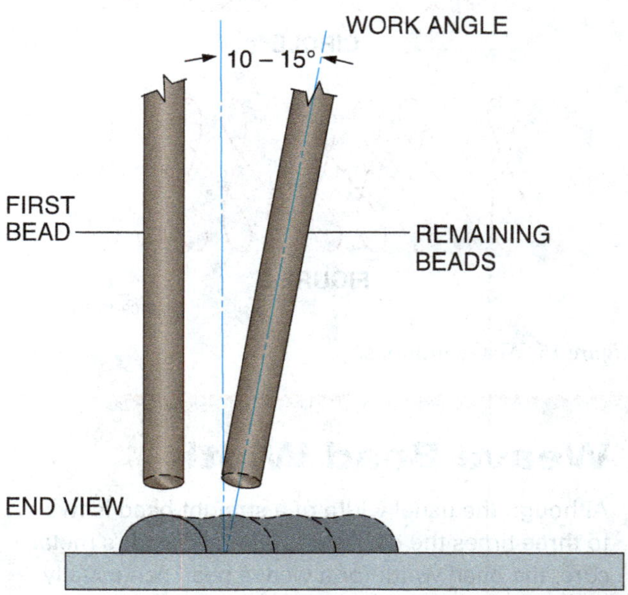

Figure 18 Electrode work angle for overlapping beads.

2.4.0 Fillet Welds

A fillet weld is a weld that is approximately triangular in cross section and is used with T-, lap, and corner joints. The sizes and locations of fillet welds are given as welding symbols. The two types of fillet welds are convex and concave (*Figure 19*). A convex fillet weld has its surface bowed out like the outside surface of a ball. A concave fillet weld has its surface bowed in like the inside surface of a bowl. The following terms are used to describe various aspects of fillet welds:

- *Weld face* – The exposed surface of the weld
- *Leg* – The distance from the root of the joint to the toe of a fillet weld
- *Weld toe* – The junction between the face of a weld and the base metal
- *Weld root* – The point shown in cross section in which the weld metal intersects with the base metal and extends farthest into the weld joint
- *Size* – The leg lengths of the largest right triangle that can be drawn within the cross section of a fillet weld
- *Actual throat* – The shortest distance from the root of the weld to its face
- *Effective throat* – The minimum distance, minus any convexity, from the root of the weld to its face
- *Theoretical throat* – The distance from the beginning of the joint root (with a zero opening) that is perpendicular to the hypotenuse of the largest right triangle that can be inscribed within the cross section of a fillet weld

As shown in *Figure 20*, fillet welds may be either equal leg or unequal leg. The face may be slightly convex, flat, or slightly concave. Welding codes require that fillet welds have a uniform concave or convex face, although a slightly nonuniform face is acceptable. The convexity of a fillet weld or individual surface bead should not exceed 0.07 times the actual face width or individual surface bead plus 0.06" (≈0.15 mm).

CAUTION

Refer to your site's WPS for specific requirements on fillet welds. The information in this module is provided as a general guideline only. The site WPS or quality specifications must be followed for all welds. Check with your supervisor if you are unsure of the specifications for your application.

A fillet weld is unacceptable if the profile has insufficient throat, excessive convexity, excessive undercut, overlap, insufficient leg, or incomplete fusion, as shown in *Figure 21*.

Fillet welds require little base metal preparation except for cleaning the weld area and removing all dross from cut surfaces. Any dross from oxyfuel, plasma arc, or carbon arc cutting will cause porosity in the weld. For this reason, the codes require that all dross be removed completely prior to welding.

2.4.1 Fillet Weld Positions

The most common fillet welds are made in lap and T-joints. The weld position for plate is determined by the weld axis and the orientation of the workpiece. The positions for fillet welding on plate are flat, or 1F (the letter F stands for fillet); horizontal, or 2F; vertical, or 3F; and overhead, or 4F (*Figure 22*). In the 1F and 2F positions, the weld axis can be inclined up to 15 degrees. Any weld axis inclination for the other positions varies with the rotational position of the weld face as specified in AWS standards.

2.4.2 Practicing Horizontal Fillet Welds with E6010 (2F Position)

Practice horizontal fillet (2F) welding by placing multiple-pass fillet welds in a T-joint using ⅛" (3.2 mm) E6010 electrodes. When making horizontal fillet welds, pay close attention to the electrode angles and travel speed. For the first bead, the electrode work angle is 45 degrees. The work angle is adjusted for all other welds. Increase or decrease the travel speed to control the amount of weld-metal buildup.

Follow these steps to make a horizontal fillet weld:

Step 1 Tack two plates together to form a T-joint for the fillet weld coupon (*Figure 23*). Clean the tack welds.

Step 2 Position the coupon on the welding table.

Preferred Fillet Weld Contours

In single-pass, weave-bead fillet welds where two workpieces are being joined at an angle (not lap joints), flat or slightly convex faces are usually preferred because weld stresses are more uniformly distributed through the fillet and workpieces.

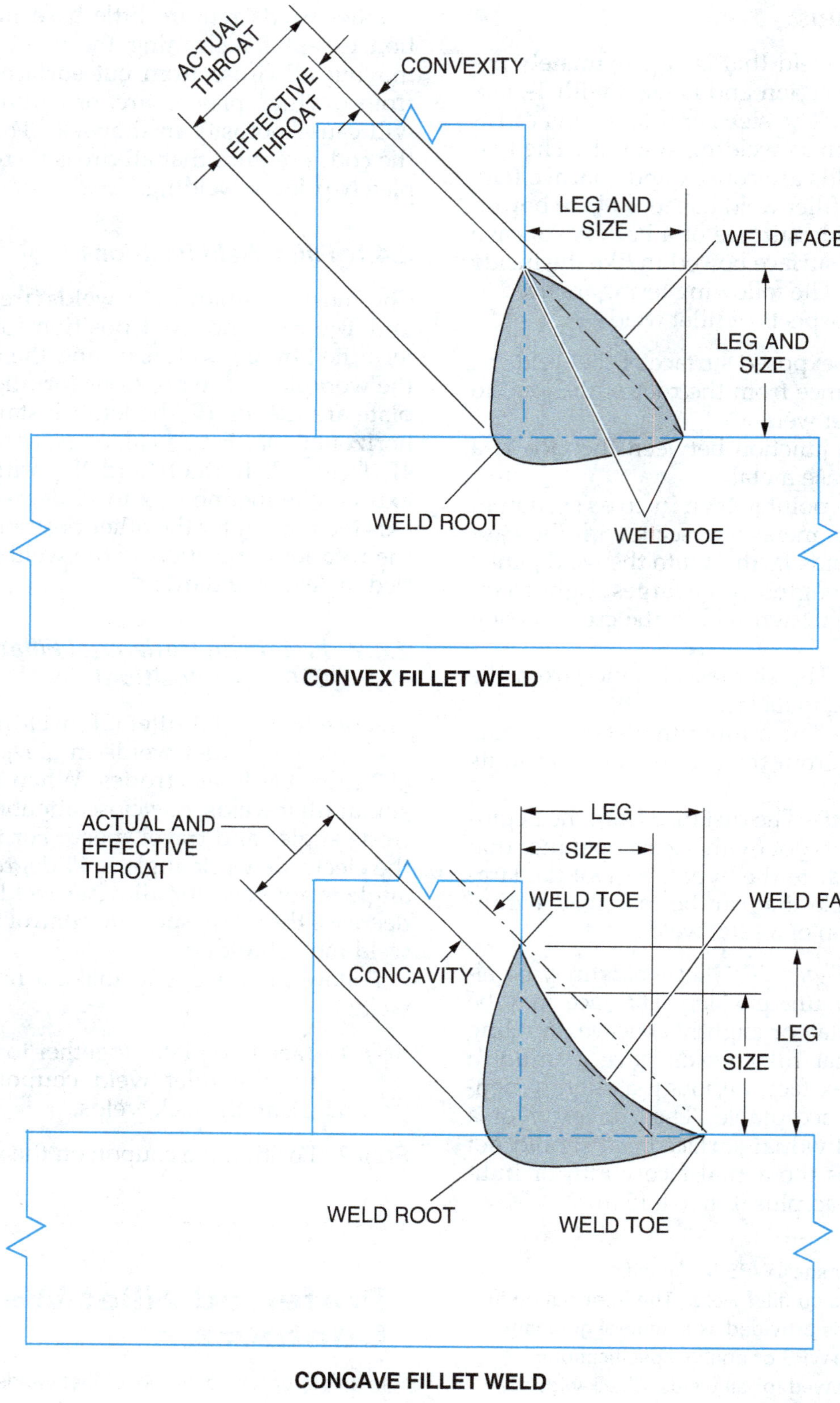

Figure 19 Convex and concave fillet welds.

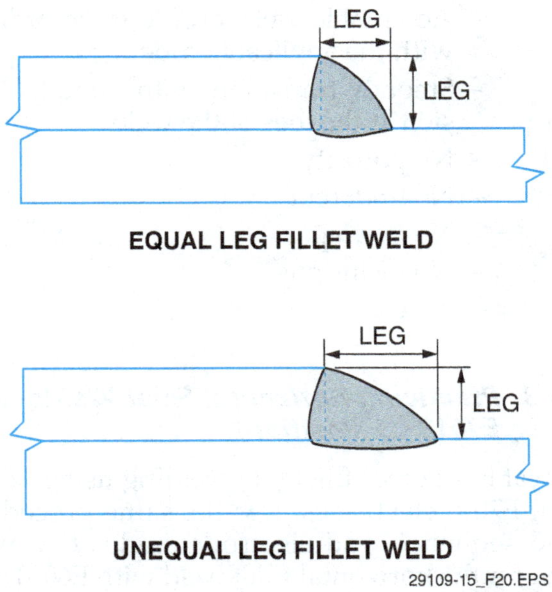

Figure 20 Equal leg and unequal leg fillet welds.

Step 3 Run the first bead along the root of the joint using an electrode work angle of approximately 45 degrees with a 10- to 15-degree drag angle. Use a C- or J-weave or stepping motion and push the arc into the root. Keep the root of the joint fusing together or else a notch will appear on the leading edge of the bead.

Step 4 Properly dispose of the electrode stub immediately and chip/clean the weld bead.

Step 5 Using a slight oscillation or stepping motion, run the second bead along the bottom toe of the first weld, overlapping about 75 percent of the first bead. Use the electrode work angle shown in *Figure 24*.

Step 6 Properly dispose of the electrode stub immediately and chip/clean the weld bead.

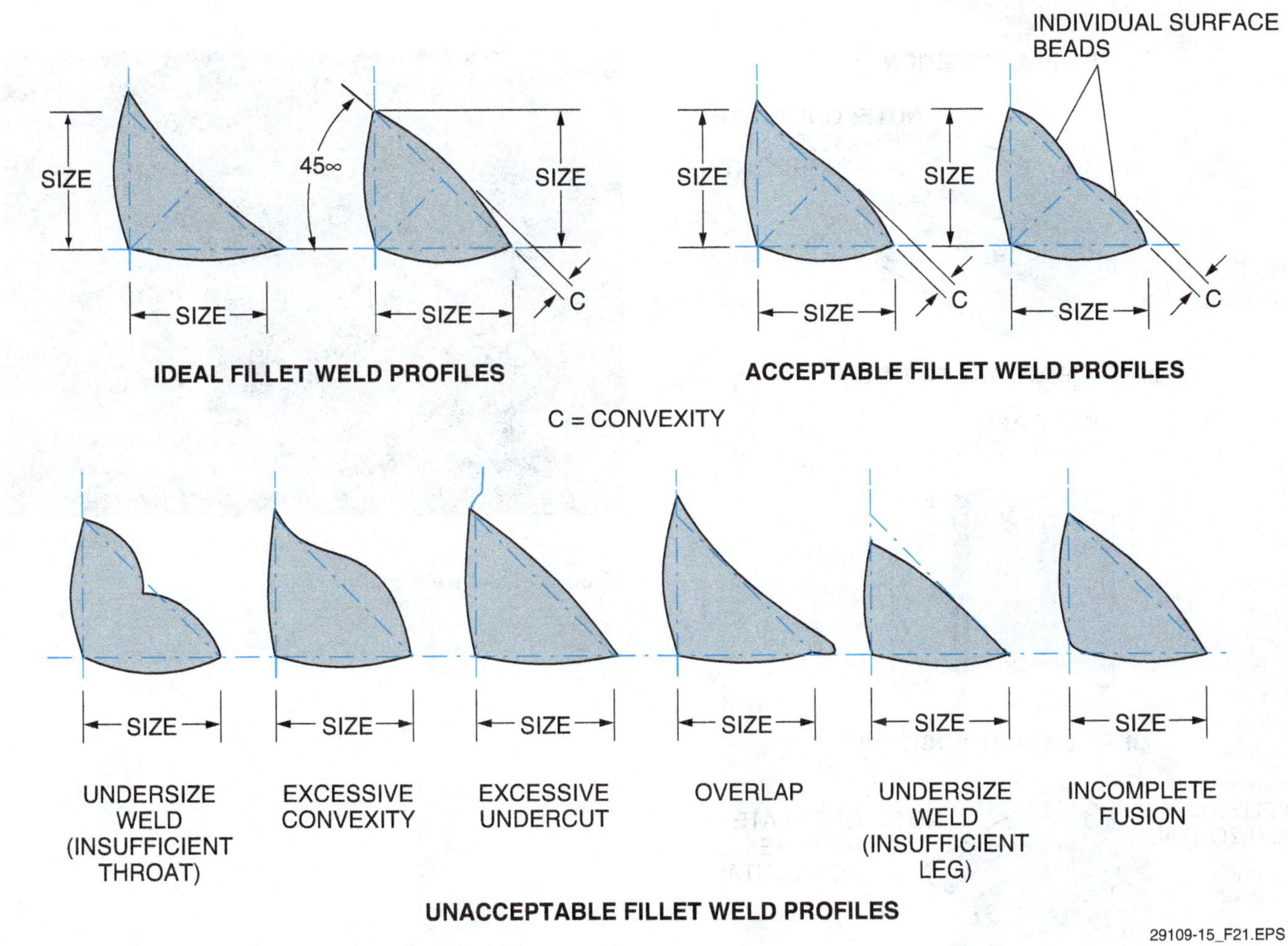

Figure 21 Acceptable and unacceptable fillet weld profiles.

Step 7 Repeat Steps 5 and 6 for each of the remaining bead passes. Run the beads along the toes of the underlying beads and overlap them about 50 percent.

Step 8 Have the instructor inspect the weld. The weld is acceptable if it has the following characteristics:

- Uniform appearance on the bead face
- Craters and restarts filled to the full cross section of the weld
- Uniform weld size ±1/16" (±1.6 mm)
- Acceptable weld profile in accordance with the applicable code
- Smooth transition with complete fusion at the toes of the weld
- No porosity
- No undercut
- No overlap
- No inclusions
- No cracks

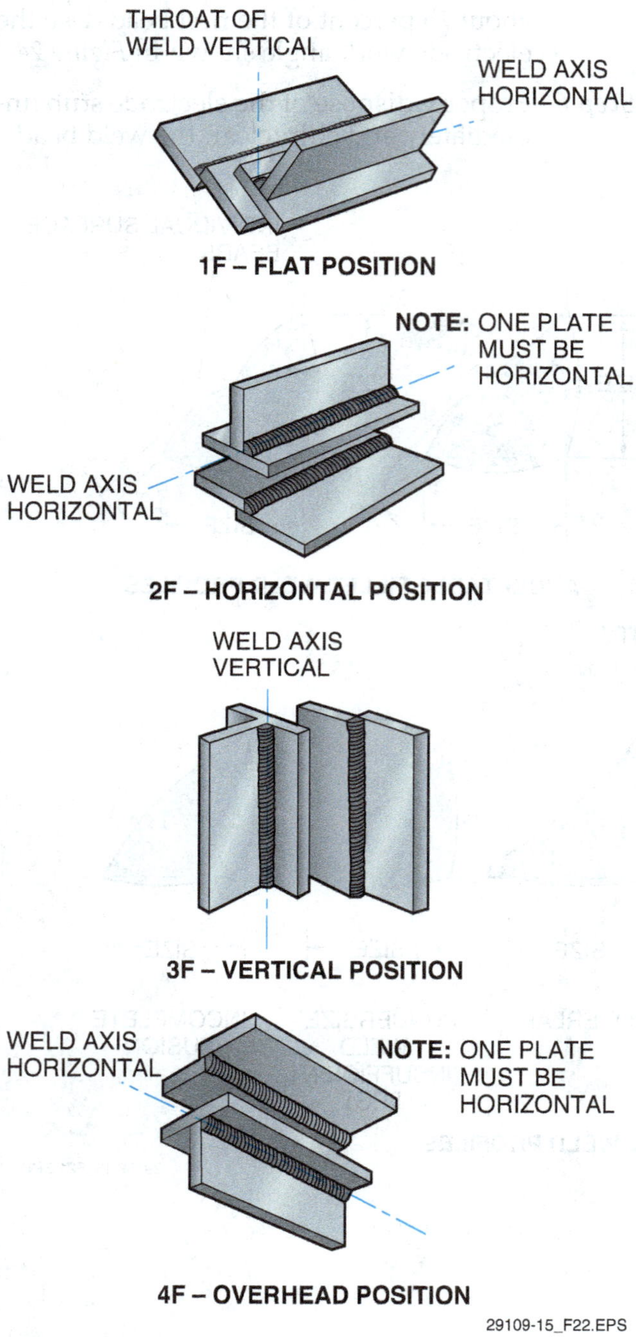

Figure 22 Fillet welding positions for plate.

2.4.3 Practicing Horizontal Fillet Welds with E7018 (2F Position)

Repeat horizontal fillet (2F) welding using 1/8" (3.2 mm) E7018 electrodes. Use the same procedure, bead sequence, and electrode angles that were used for the horizontal fillet weld with E6010 electrodes. Use a short arc and remember not to whip the E7018 electrode.

29109-15_F23.EPS

Figure 23 Fillet weld coupon.

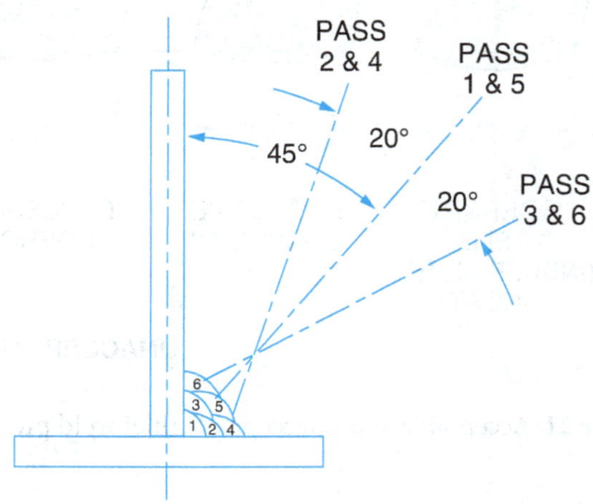

Figure 24 Multiple-pass 2F weld sequences and work angles.

2.4.4 *Practicing Vertical Fillet Welds with E6010 (3F Position)*

Practice vertical fillet (3F) welding by placing multiple-pass fillet welds in a T-joint using ⅛" (3.2 mm) E6010 electrodes. Normally, vertical welds are accomplished by welding uphill from the bottom to the top using an electrode push angle (up-angle). Because of the uphill welding and push angle, this type of weld is sometimes called vertical-up fillet welding. When vertical welding, either stringer or weave beads can be used. On the job, the site WPS or quality standard will specify which technique to use. Typically, weave beads are used with E6010 electrodes and, for welding carbon steel, with E7018 electrodes. Stringer beads are generally called for when welding alloy steels with low-hydrogen (F4) electrodes.

> **NOTE**
> Check with your instructor to see if you should run stringer beads, run weave beads, or practice both techniques.

Follow these steps to make a vertical-up fillet weld:

Step 1 Tack two plates together to form a T-joint for the fillet weld coupon.

Step 2 Tack-weld the coupon in the vertical position.

Step 3 Run the first bead along the root of the joint (starting at the bottom) using an electrode- (work) angle of approximately 45 degrees with a 0- to 10-degree push angle. Use a whipping motion by quickly raising the electrode about ¼" (≈6 mm) and then dropping it back into the weld puddle. Pause in the weld puddle to fill the crater. For a stringer bead, use the same technique. See *Figure 25* for bead placement.

Step 4 Properly dispose of the electrode stub immediately and chip/clean the weld bead.

Step 5 Run the second bead using a weave technique, such as a C-pattern. Use a slow motion across the face of the weld, pausing at each toe for penetration and to fill the crater. A slight whip can be used to control the puddle when you reach the toe. Adjust the travel speed across the face of the weld to control the buildup. *Figure 26* shows the bead sequence and electrode angles for weave beads (all degrees shown are approximate). In practice, all bead passes run the entire length of the weld.

Step 6 Properly dispose of the electrode stub immediately and chip/clean the weld bead.

Step 7 Continue to run weld beads as shown in *Figure 26*.

Step 8 Properly dispose of the electrode stub immediately and chip/clean the weld bead.

Step 9 Have the instructor inspect the weld. The weld is acceptable if it has the following characteristics:

- Uniform appearance on the bead face
- Craters and restarts filled to the full cross section of the weld
- Uniform weld size ±1⁄16" (±1.6 mm)
- Acceptable weld profile in accordance with the applicable code
- Smooth transition with complete fusion at the toes of the weld
- No porosity
- No undercut
- No overlap
- No inclusions
- No cracks

Tacking and Aligning Workpieces

When tacking workpieces together, both sides of the workpieces are usually tacked with welds that are about ½" long. The tacks are used to position the workpieces and minimize distortion as the final welds are made. After the first tack weld, use a hammer or other tool to align the workpieces side-to-side and end-to-end; then tack the opposite side. Tack the far ends of the workpieces in the same manner. Intermediate tack welds can be made every 5" to 6" as necessary, to minimize lengthwise distortion.

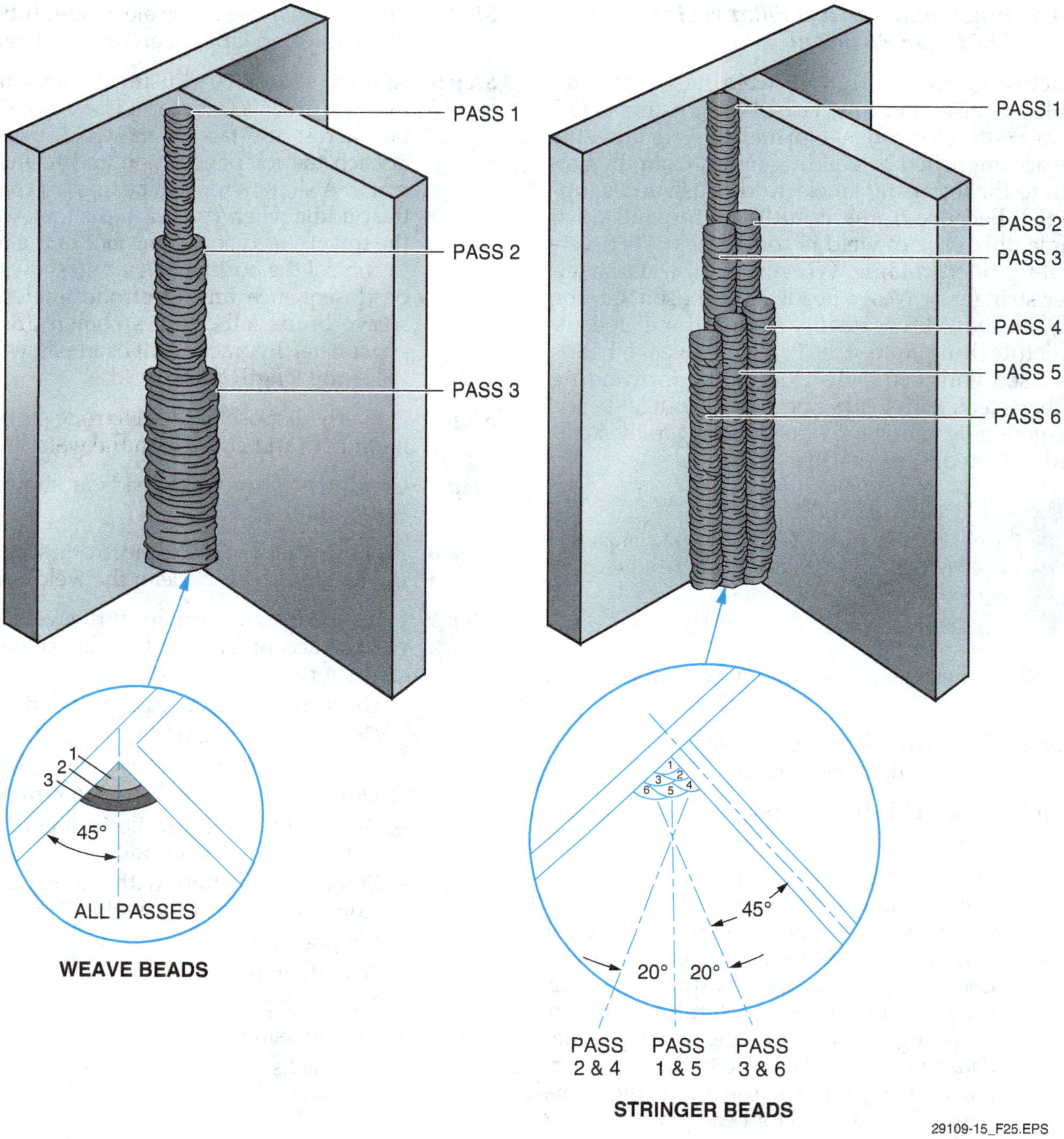

Figure 25 Bead placement for the 3F position.

Test Joint Heat Dissipation

In T-joints, the welding heat dissipates more rapidly in the thicker or the non-butting member. On various bead passes, the arc may have to be concentrated slightly more on the thicker or the non-butting member to compensate for the heat loss.

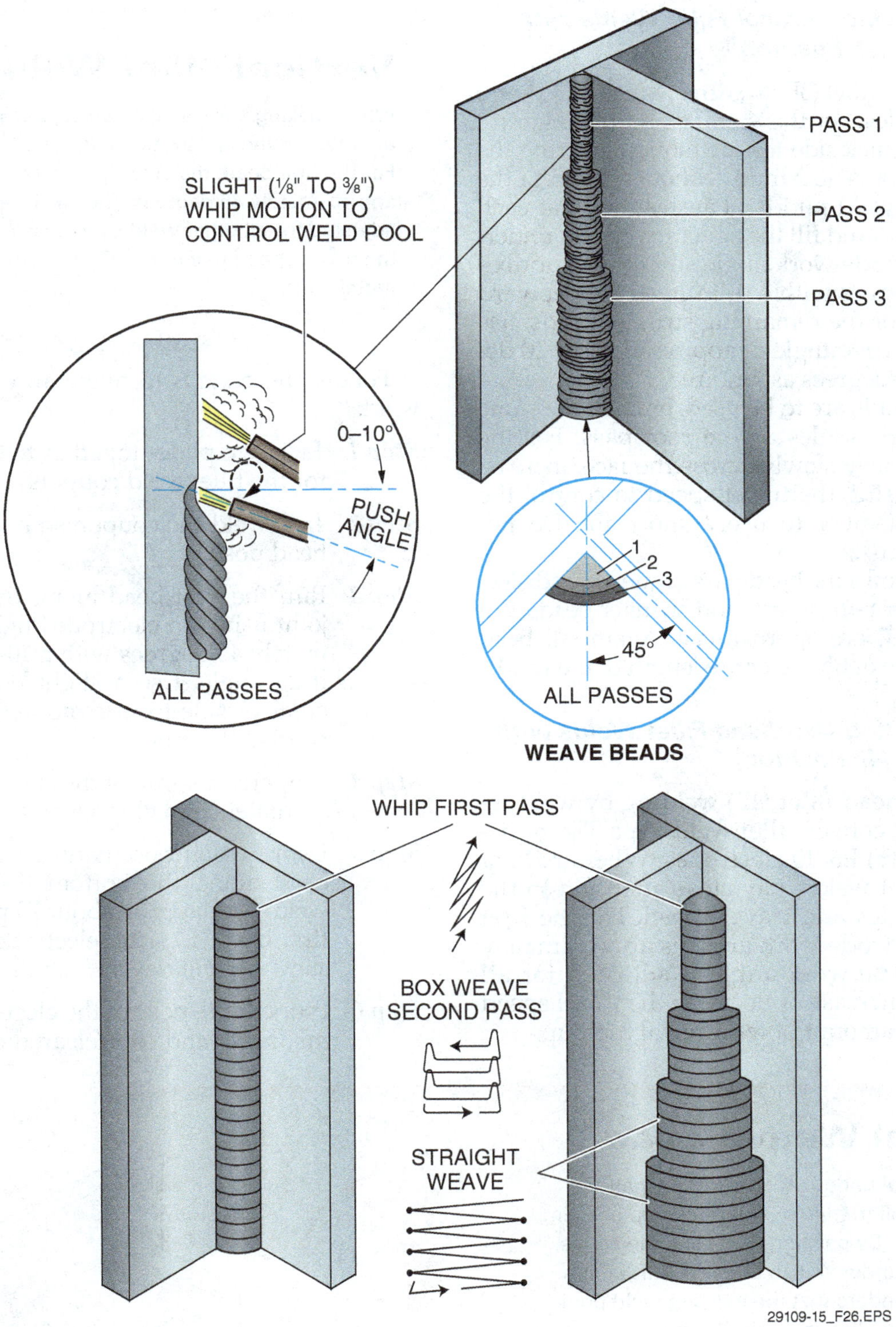

Figure 26 Bead sequence and work angles for weave beads.

T-Joint Vertical Plate Undercut

The most common defect for T-joints is undercut on the vertical plate of the joint. Use of a J-weave usually eliminates the problem. However, if the problem persists, angling the arc slightly toward the vertical plate and the bead at the top of the weave will force more metal into the bead at the top edge of the weld

2.4.5 Practicing Vertical Fillet Welds with E7018 (3F Position)

Repeat vertical fillet (3F) welding using ⅛" (3.2 mm) E7018 electrodes. For the root pass and stringer fill beads, use a quick side-to-side motion, moving the electrode about ⅛" (3.2 mm) without removing the arc from the weld puddle. Pause slightly at each toe to penetrate and fill the crater to prevent undercut. The electrode work angle should be approximately 45 degrees with a 0- to 10-degree upward push angle. For the remaining stringer beads, use an electrode work angle of approximately ±20 degrees from 45 degrees as required.

If weave beads are to be used, employ the same electrode work angles as the root pass. For the filler beads, move slowly across the face, increasing or decreasing the travel speed to control the buildup. Remember to use a short arc. Do not whip the electrode.

Figure 25 shows the bead pass sequence and electrode angle for both weave and stringer beads. All degrees shown are approximate. Again, all bead passes would extend the entire length of the weld.

2.4.6 Practicing Overhead Fillet Welds with E6010 (4F Position)

Practice overhead fillet (4F) welding by welding multiple-pass convex fillet welds in a T-joint using ⅛" (3.2 mm) E6010 electrodes. When making overhead fillet welds, pay close attention to the electrode angles and travel speed. For the first bead, the electrode work angle is approximately 45 degrees. The work angle is adjusted for all other welds. Increase or decrease the travel speed to control the amount of weld metal buildup.

Vertical Fillet Welds

When making vertical fillet welds, pay close attention to the electrode angles and travel speed. For the first bead, the electrode work angle is approximately 45 degrees. The work angle is adjusted for all other welds. Increase or decrease the travel speed to control the amount of weld metal buildup.

Follow these steps to make an overhead fillet weld:

Step 1 Tack two plates together to form a T-joint for the fillet weld coupon.

Step 2 Tack-weld the coupon so it is in the overhead position.

Step 3 Run the first bead along the root of the joint using an electrode angle of approximately 45 degrees with a 10- to 15-degree drag angle. Use a slight oscillation (circular or side-to-side motion) to tie-in the weld at the toes.

Step 4 Properly dispose of the electrode stub immediately and chip/clean the weld bead.

Step 5 Using a slight oscillation, run the second bead along the bottom toe of the first weld, overlapping about 75 percent of the first bead. Use the electrode work angle shown in *Figure 27*.

Step 6 Properly dispose of the electrode stub immediately and chip/clean the weld bead.

Vertical Weave Bead

To help control undercut, an alternate pattern such as a small triangular weave can be used in vertical welds. By pausing the rod at the edges, the previous undercut can be filled. This action also creates undercut at the existing weld pool, but that will be filled in by the next weave.

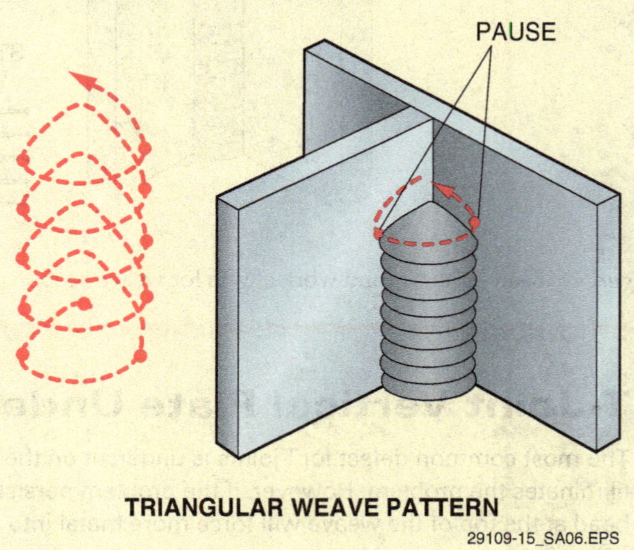

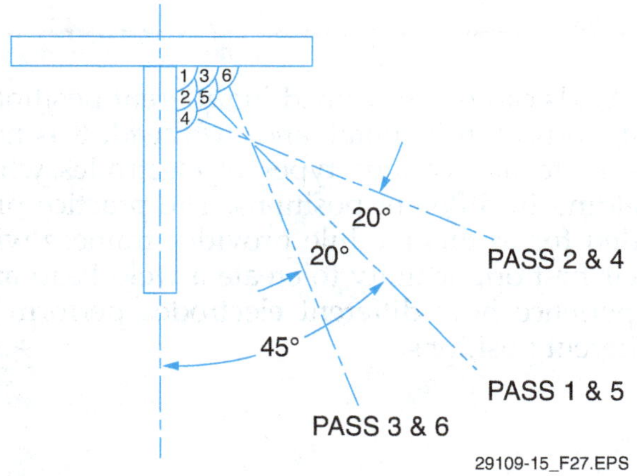

Figure 27 Multiple-pass 4F weld sequences and work angles.

Step 7 Repeat Steps 5 and 6 for each of the remaining bead passes. Run the beads along the weld toes of the underlying beads and overlap them about 75 percent.

Step 8 Have the instructor inspect the weld. The weld is acceptable if it has the following characteristics:

- Uniform appearance on the bead face
- Craters and restarts filled to the full cross section of the weld
- Uniform weld size ±1⁄16" (±1.6 mm)
- Acceptable weld profile in accordance with the applicable code
- Smooth transition with complete fusion at the toes of the weld
- No porosity
- No undercut
- No overlap
- No inclusions
- No cracks

2.4.7 Practicing Overhead Fillet Welds with E7018 (4F Position)

Repeat overhead fillet (4F) welding using 1⁄8" (3.2 mm) E7018 electrodes. Use the same procedure, bead sequence, and electrode angles that were used for the overhead fillet weld with E6010 electrodes. Use a short arc and do not whip the electrode.

Additional Resources

2014 Technical Training Guide. Latest Edition. Cleveland, OH. USA: The Lincoln Electric Company. **www.lincolnelectric.com**

2.0.0 Section Review

1. On a restart, restrike the arc _____.
   a. in front of the crater and in line with the weld
   b. behind the crater and in line with the weld
   c. in front of the crater, just outside the existing bead
   d. behind the crater, just outside the existing bead

2. The whipping motion in SMAW welding helps control the weld puddle by _____.
   a. dramatically increasing the weld current
   b. taking advantage of the E7018 electrode characteristics
   c. depositing additional hydrogen into the puddle
   d. momentarily lengthening and advancing the arc

3. You are working on overlapping stringer bead practice with E6010. After striking the arc for the next stringer bead, and with the proper travel angle, position the electrode at a work angle of _____.
   a. 10 to 15 degrees to the side of the previous bead
   b. 15 to 20 degrees to the side of the previous bead
   c. 25 to 30 degrees to the side of the previous bead
   d. 40 to 45 degrees to the side of the previous bead

4. The position designation for a vertical fillet weld is _____.
   a. 1F
   b. 2F
   c. 3F
   d. 4F

SUMMARY

Striking an arc, running stringer and weave beads, and making fillet welds are basic and essential skills a welder must develop. Even the method used to strike an arc can result in a rejected weld. Developing and honing these skills allows welders to progress to more difficult procedures. It is important to practice these welds until acceptable welds are produced.

Welds can be performed in different positions: flat, vertical, horizontal, and overhead. It is necessary to use various types of electrodes when welding in different positions. The practice provided for in this module provides trainees with their first opportunity to create a weld bead and experience how different electrodes perform in different positions.

Review Questions

1. Which of the following is a gas that would be used to purge a container before welding or cutting?
   a. Oxygen
   b. Methane
   c. Nitrogen
   d. Propane

2. Which of the following compounds or elements are most likely to be encountered in welding and can create toxic fumes?
   a. Cadmium
   b. Brass
   c. Helium
   d. Argon

3. Which of the following electrodes must remain in an oven until they are needed?
   a. E6010
   b. E6011
   c. E7018
   d. E6013

4. The diameter of a SMAW electrode's metal core is also representative of _____.
   a. one-half the proper arc length
   b. twice the proper arc length
   c. four times the arc length
   d. the proper arc length

5. The easiest method of striking an arc is the _____.
   a. scratching method
   b. grounding method
   c. weaving method
   d. tapping method

6. E7018 electrodes are more difficult to strike an arc with because _____.
   a. the flux coating will not break away from the tip
   b. they stick to the base metal more easily
   c. the welding current has to be very low
   d. the welding current has to be very high

7. Arc blow can cause defects such as excessive weld spatter and _____.
   a. porosity
   b. burn through
   c. increased magnetic fields
   d. decreased magnetic fields

8. Which of the following is *not* a common method of controlling or eliminating arc blow?
   a. Tacking the workpieces together.
   b. Changing from an AC welding process to a DC process.
   c. Shortening the length of the arc.
   d. Reducing the weld current.

9. When preparing to terminate a weld bead, _____.
   a. increase the forward travel speed and change from a push angle to a drag angle
   b. begin to lay the electrode down to a more acute work angle
   c. decrease the current setting on the welding machine
   d. slow the forward travel speed and begin to stand the electrode up

10. The point where one weld bead stops and another begins to continue its progression is called a _____.
    a. retrace
    b. restart
    c. weave
    d. blend

11. A weld bead with very little or no side-to-side motion of the electrode required is called a(n) _____.
    a. arc bead
    b. weave bead
    c. stringer bead
    d. whipping bead

12. The width of stringer and weave beads is usually specified in the drawing weld symbols or the _____.
    a. SDS/MSDS for the electrode in use
    b. electrode manufacturer's documentation
    c. welding procedure specification (WPS)
    d. welding machine documentation

13. The whipping motion used to control the weld puddle should not be used with _____.
    a. E6010 electrodes
    b. E6011 electrodes
    c. E7018 electrodes
    d. E6013 electrodes

14. Overlapping beads are made by depositing connective weld beads parallel to one another to form a(n) _____.
    a. flat surface
    b. undercut surface
    c. concave surface
    d. convex surface

15. The part of a fillet weld that can be convex or concave is called the _____.
    a. leg
    b. throat
    c. face
    d. root

Trade Terms Quiz

Fill in the blank with the correct term that you learned from your study of this module.

1. When magnetic forces deflect the arc from its intended path, the action is called _______.
2. A straight line drawn through the center of a weld and along its length is called the _______.
3. _______ is a discontinuity consisting of localized melting of the base metal or finished weld caused by the initiation of an arc.
4. A type of weld bead made by transverse oscillation of the electrode is called a(n) _______.
5. _______ is an angle that is less than 90 degrees between a line perpendicular to the surface of the workpiece and a plane determined by the electrode axis and the weld axis.
6. _______ is the travel angle at which the electrode is pointing in the same direction as the welding bead progression.
7. A consistent and repetitive side-to-side motion is called _______.
8. _______ is the travel angle in which the electrode is pointing in a direction opposite to the welding bead progression.
9. A piece of metal to be welded upon or welded together with another piece as a test or practice is called a(n) _______.
10. _______ is a type of weld bead made without any significant weaving motion.
11. When the cross-sectional thickness of the base metal has been reduced along the edge of a weld bead, this defect is referred to as _______.
12. A _______ is where the welder begins to add to an already existing weld bead.
13. A welder should never _______ a weld that has a function, but it is acceptable for practice coupons.

Trade Terms

Arc blow
Arc strike
Drag angle
Oscillation
Push angle
Quench
Restart
Stringer bead
Undercut
Weave bead
Weld axis
Weld coupon
Work angle

Appendix

Performance Accreditation Tasks

The American Welding Society (AWS) School Excelling through National Skills Standards Education (SENSE) program is a comprehensive set of minimum Standards and Guidelines for Welding Education programs. The following performance accreditation tasks are aligned with and designed around the SENSE program.

The Performance Accreditation Tasks (PATs) correspond to and support the learning objectives in *AWS EG2.0, Guide for the Training and Qualification of Welding Personnel: Entry-Level Welder*.

Note that in order to satisfy all learning objectives in *AWS EG2.0*, the instructor must also use the PATs contained in the second level of the NCCER Welding curriculum.

PATs 1 and 2 correspond to *AWS EG2.0, Module 4—Shielded Metal Arc Welding*, Key Indicators: 3 and 4.

PATs 3 through 8 correspond to *AWS EG2.0, Module 4—Shielded Metal Arc Welding*, Key Indicators: 3, 4, and 5.

PATs provide specific acceptable criteria for performance and help to ensure a true competency-based welding program for students.

The following tasks are designed to evaluate your ability to run beads and fillet welds with SMAW equipment and techniques. Perform each task when you are instructed to do so by your instructor. As you complete each task, bring it to your instructor for evaluation. Do not proceed to the next task until instructed to do so.

BUILD A PAD WITH E6010 OR E6011 ELECTRODES IN THE FLAT POSITION

Using ³⁄₃₂" to ⁵⁄₃₂" (2.4 to 4.0 mm) E6010 or E6011 electrodes, build up a pad of weld metal on carbon steel plate as indicated.

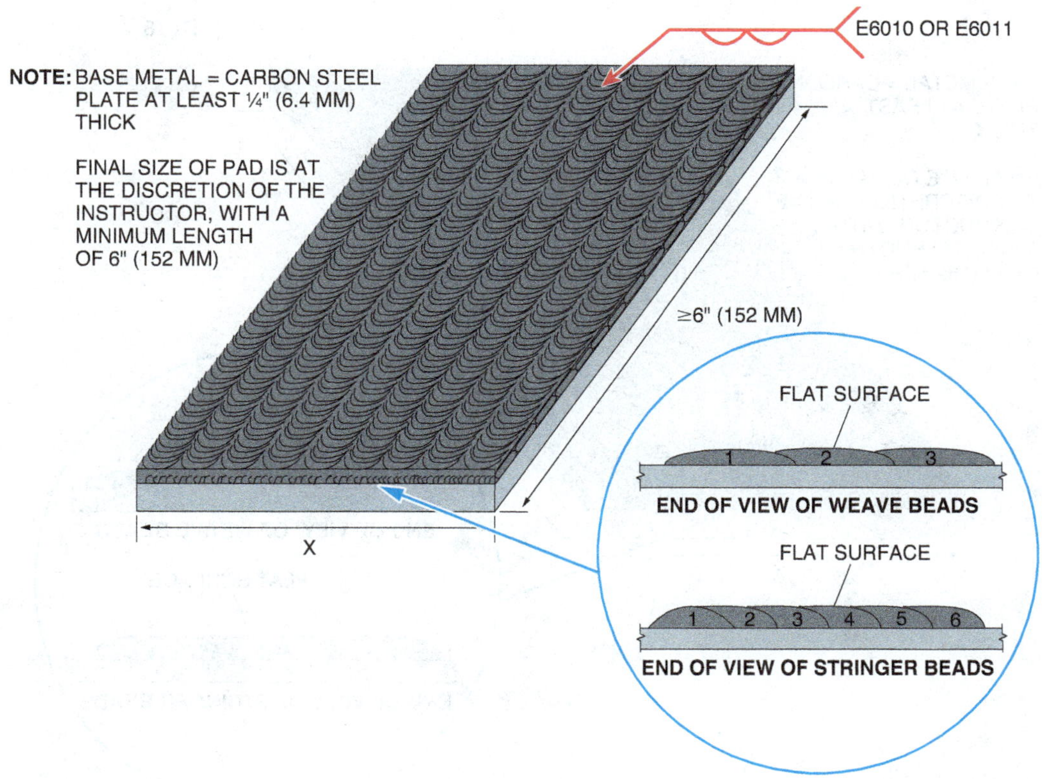

29109-15_A01.EPS

Criteria for Acceptance:

- Weld beads straight to within ⅛" (3.2 mm) ________
- Uniform rippled appearance on the bead face ________
- Craters and restarts filled to the full cross section of the weld ________
- Face of the pad flat to within ⅛" (3.2 mm) ________
- Smooth flat transition with complete fusion at the toes of one bead into the face of the previous bead ________
- No porosity ________
- No overlap at weld toes ________
- No excessive undercut ________
- No inclusions ________
- No cracks ________

BUILD A PAD WITH E7018 ELECTRODES IN THE FLAT POSITION

Using 3⁄32" to 5⁄32" (2.4 to 4.0 mm) E7018 electrodes, build up a pad of weld metal on carbon steel plate as indicated.

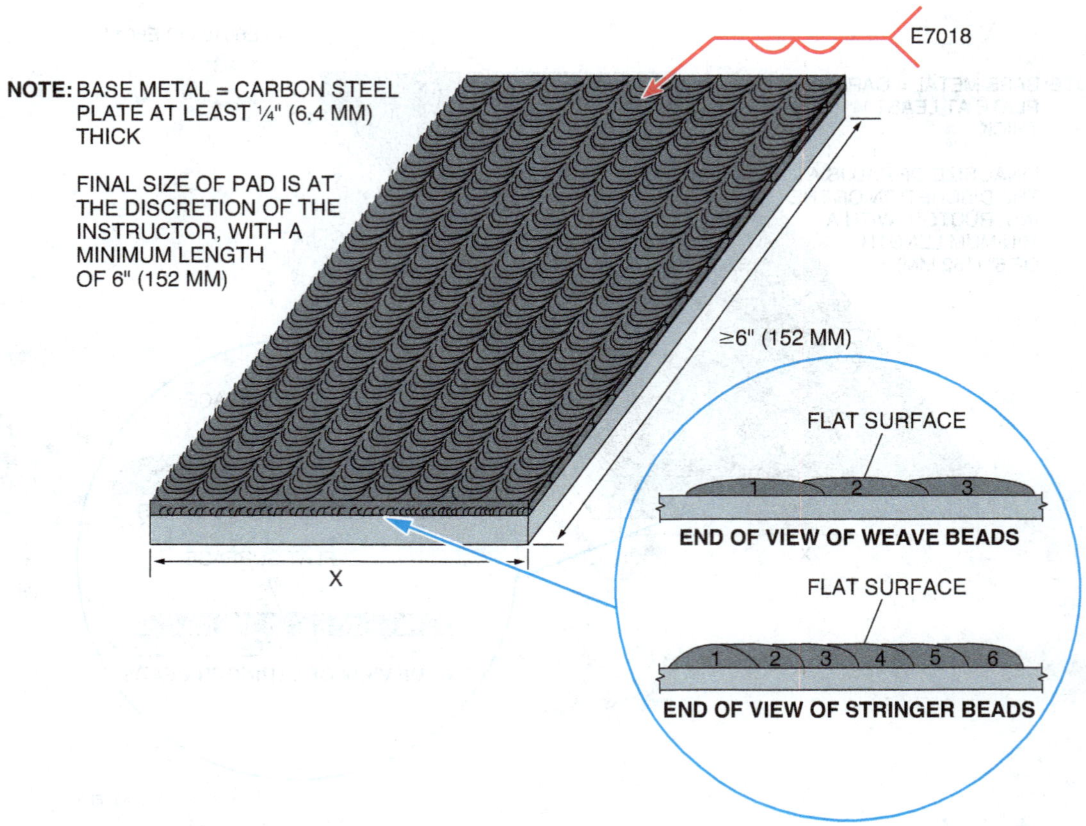

29109-15_A02.EPS

Criteria for Acceptance:

- Weld beads straight to within 1⁄8" (3.2 mm) ________
- Uniform rippled appearance on the bead face ________
- Craters and restarts filled to the full cross section of the weld ________
- Face of the pad flat to within 1⁄8" (3.2 mm) ________
- Smooth flat transition with complete fusion at the toes of one bead into the face of the previous bead ________
- No porosity ________
- No overlap at weld toes ________
- No excessive undercut ________
- No inclusions ________
- No cracks ________

HORIZONTAL (2F) FILLET WELD WITH E6010 OR E6011 ELECTRODES

Using 3⁄32" to 5⁄32" (2.4 to 4.0 mm) E6010 or E6011 electrodes, make a horizontal fillet weld as indicated.

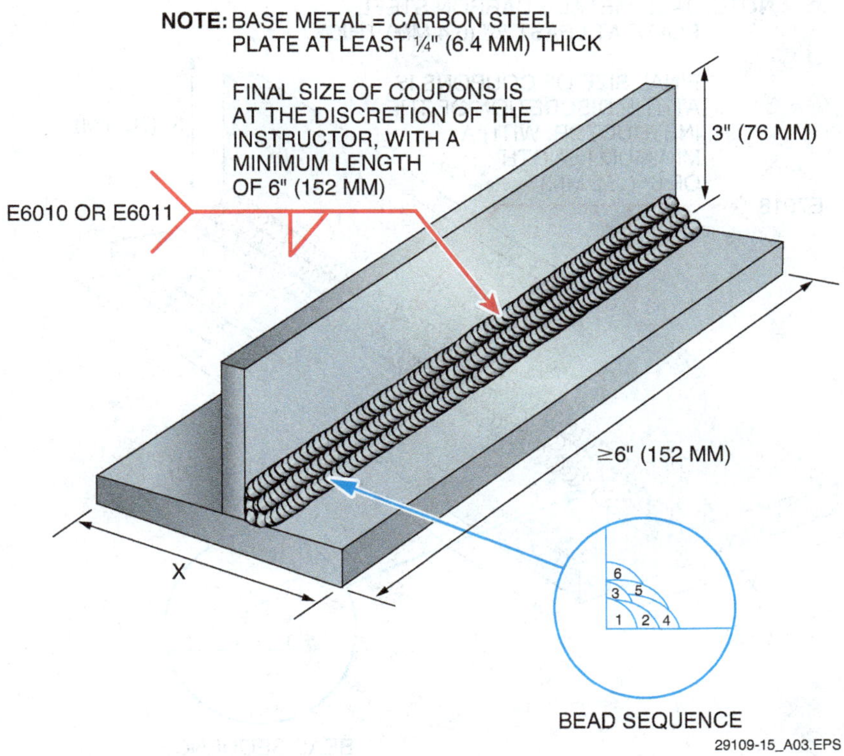

Criteria for Acceptance:

- Uniform rippled appearance on the bead face ______
- Craters and restarts filled to the full cross section of the weld ______
- Uniform weld size, ±1⁄16" (1.6 mm) ______
- Acceptable weld profile in accordance with *AWS D1.1* ______
- Smooth transition with complete fusion at the toes of the weld ______
- No porosity ______
- No excessive undercut ______
- No inclusions ______
- No cracks ______
- No overlap ______

HORIZONTAL (2F) FILLET WELD WITH E7018 ELECTRODES

Using ³⁄₃₂" to ⁵⁄₃₂" (2.4 to 4.0 mm) E7018 electrodes, make a horizontal fillet weld as indicated.

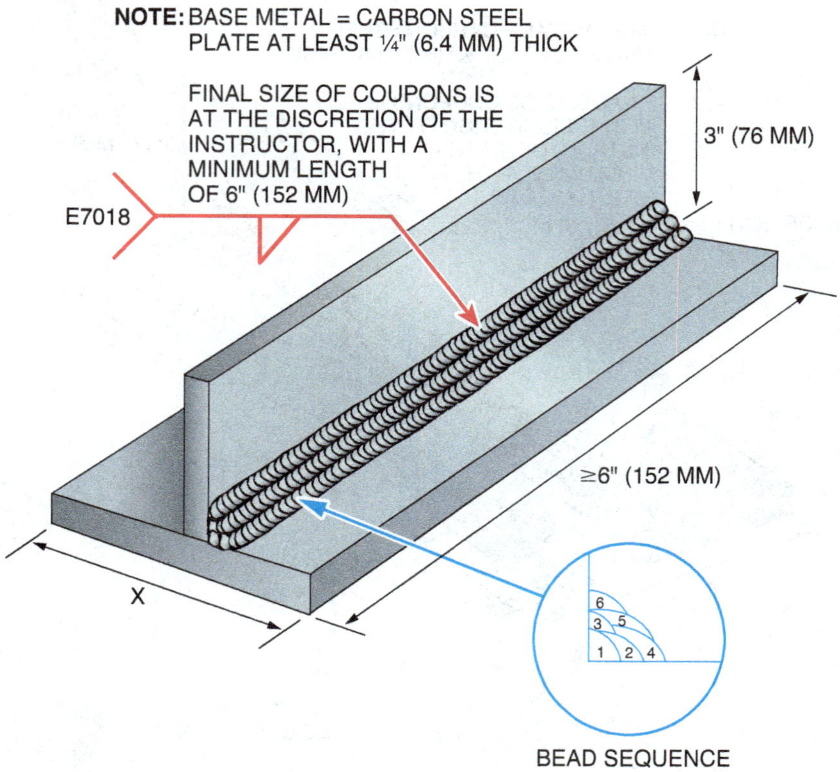

29109-15 A04.EPS

Criteria for Acceptance:

- Uniform rippled appearance on the bead face ________
- Craters and restarts filled to the full cross section of the weld ________
- Uniform weld size, ±¹⁄₁₆" (1.6 mm) ________
- Acceptable weld profile in accordance with *AWS D1.1* ________
- Smooth transition with complete fusion at the toes of the weld ________
- No porosity ________
- No excessive undercut ________
- No inclusions ________
- No cracks ________
- No overlap ________

VERTICAL (3F) FILLET WELD WITH E6010 OR E6011 ELECTRODES

Using 3⁄32" to 5⁄32" (2.4 to 4.0 mm) E6010 or E6011 electrodes, make a vertical fillet weld as indicated.

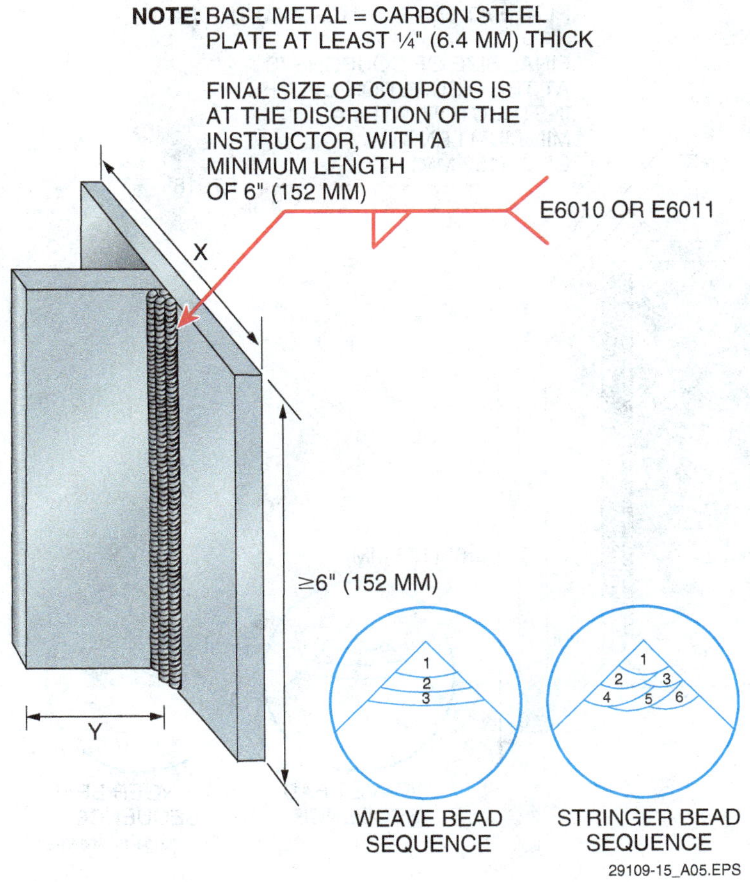

Criteria for Acceptance:

- Uniform rippled appearance on the bead face ________
- Craters and restarts filled to the full cross section of the weld ________
- Uniform weld size, ±1⁄16" (1.6 mm) ________
- Acceptable weld profile in accordance with *AWS D1.1* ________
- Smooth transition with complete fusion at the toes of the weld ________
- No porosity ________
- No excessive undercut ________
- No inclusions ________
- No cracks ________
- No overlap at weld toes ________

VERTICAL (3F) FILLET WELD WITH E7018 ELECTRODES

Using 3⁄32" to 5⁄32" (2.4 to 4.0 mm) E7018 electrodes, make a vertical fillet weld as indicated.

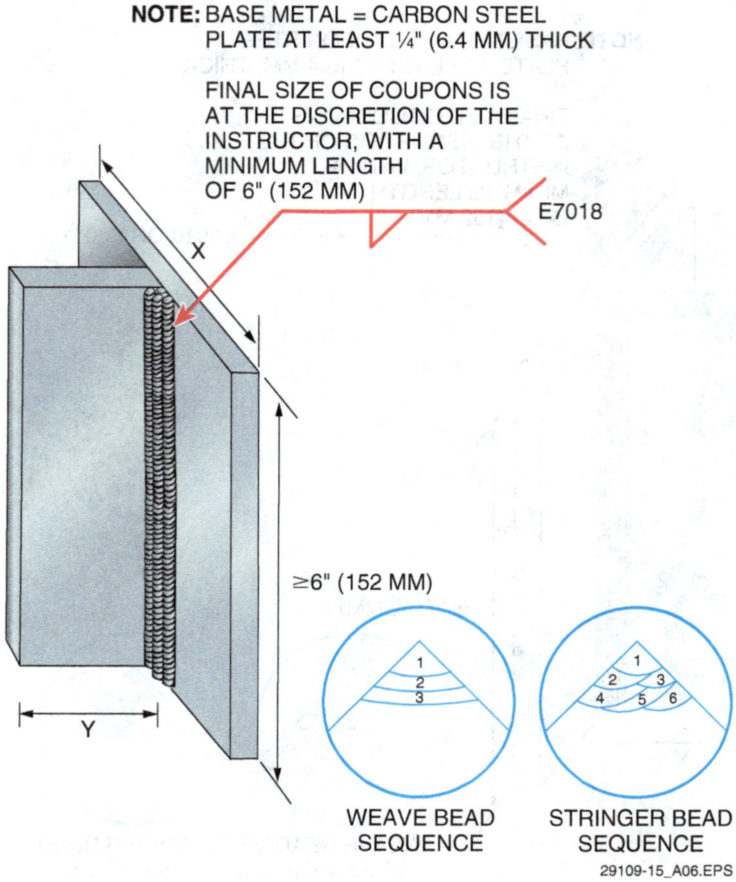

Criteria for Acceptance:

- Uniform rippled appearance on the bead face ____________
- Craters and restarts filled to the full cross section of the weld ____________
- Uniform weld size, ±1⁄16" (1.6 mm) ____________
- Acceptable weld profile in accordance with *AWS D1.1* ____________
- Smooth transition with complete fusion at the toes of the weld ____________
- No porosity ____________
- No excessive undercut ____________
- No inclusions ____________
- No cracks ____________
- No overlap at weld toes ____________

OVERHEAD (4F) FILLET WELD WITH E6010 OR E6011 ELECTRODES

Using 3⁄32" to 5⁄32" (2.4 to 4.0 mm) E6010 or E6011 electrodes, make an overhead fillet weld as indicated.

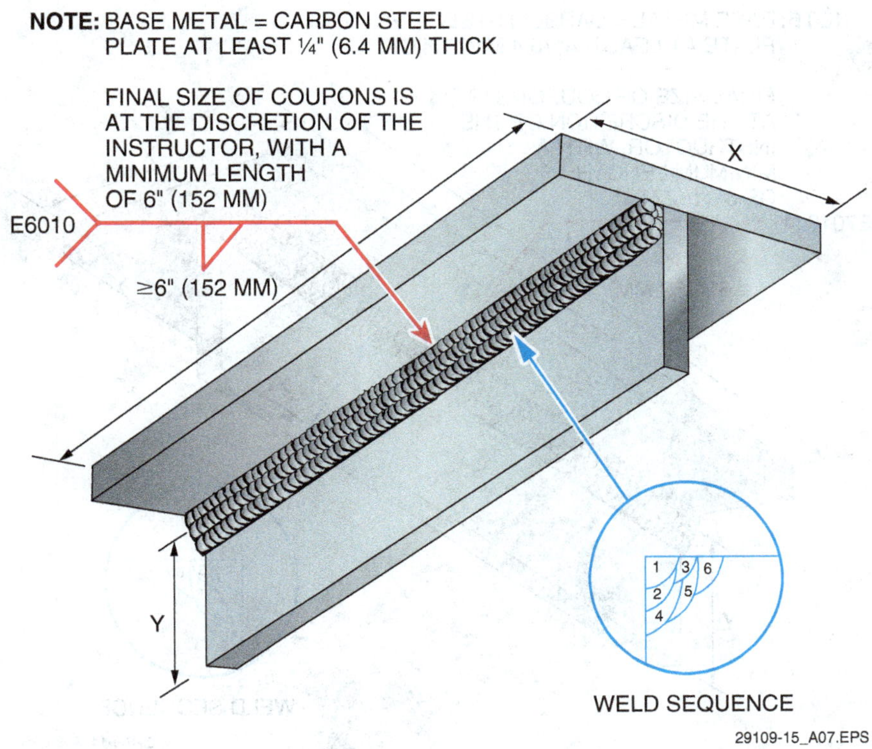

29109-15_A07.EPS

Criteria for Acceptance:

- Uniform rippled appearance on the bead face ______
- Craters and restarts filled to the full cross section of the weld ______
- Uniform weld size, ±1⁄16" (1.6 mm) ______
- Acceptable weld profile in accordance with *AWS D1.1* ______
- Smooth transition with complete fusion at the toes of the weld ______
- No porosity ______
- No excessive undercut ______
- No inclusions ______
- No cracks ______
- No overlap ______

OVERHEAD (4F) FILLET WELD WITH E7018 ELECTRODES

Using ³⁄₃₂" to ⁵⁄₃₂" (2.4 to 4.0 mm) E7018 electrodes, make an overhead fillet weld as indicated.

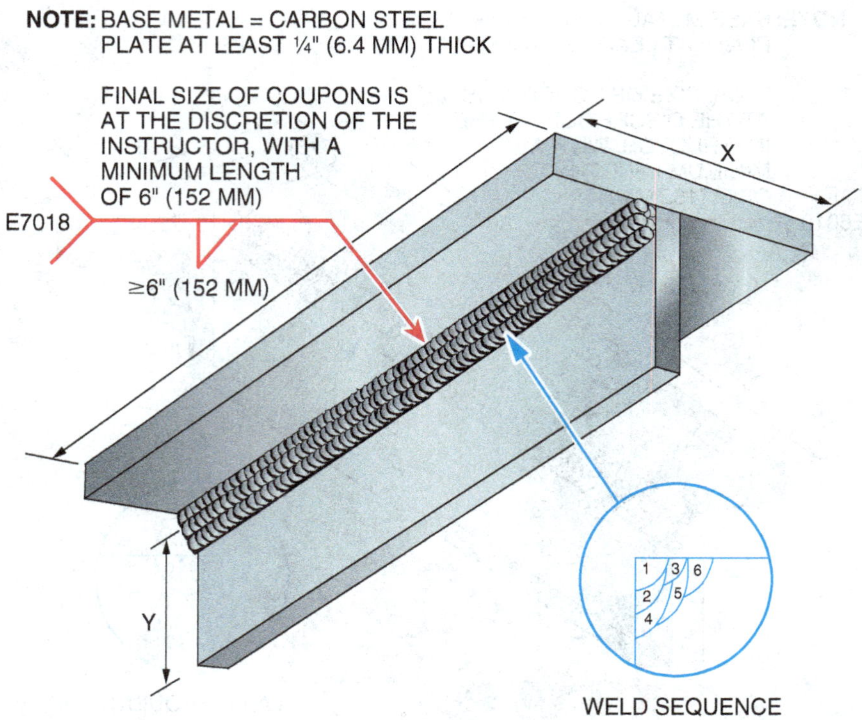

29109-14_A08.EPS

Criteria for Acceptance:

- Uniform rippled appearance on the bead face ______
- Craters and restarts filled to the full cross section of the weld ______
- Uniform weld size, ±¹⁄₁₆" (1.6 mm) ______
- Acceptable weld profile in accordance with AWS D1.1 ______
- Smooth transition with complete fusion at the toes of the weld ______
- No porosity ______
- No excessive undercut ______
- No inclusions ______
- No cracks ______
- No overlap ______

FLAT (1F) FILLET WELD WITH E6010 OR E6011 ELECTRODES

Using ³⁄₃₂" to ⁵⁄₃₂" (2.4 to 4.0 mm) E6010 or E6011 electrodes, make a flat fillet weld as indicated.

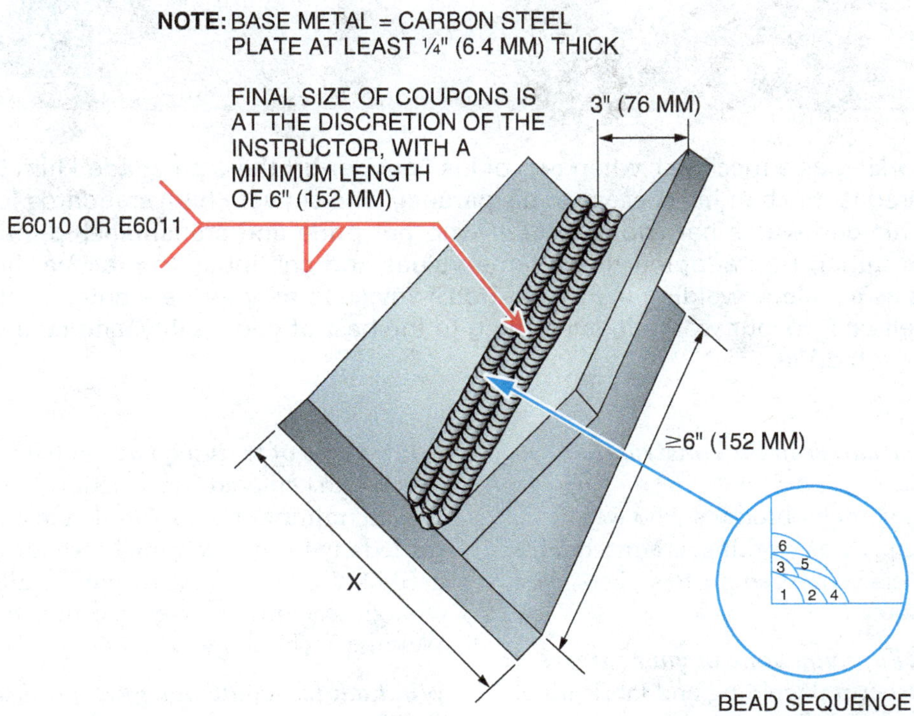

29109-15_A09.EPS

Criteria for Acceptance:

- Uniform rippled appearance on the bead face ____________
- Craters and restarts filled to the full cross section of the weld ____________
- Uniform weld size, ±¹⁄₁₆" (1.6 mm) ____________
- Acceptable weld profile in accordance with *AWS D1.1* ____________
- Smooth transition with complete fusion at the toes of the weld ____________
- No porosity ____________
- No excessive undercut ____________
- No inclusions ____________
- No cracks ____________
- No overlap ____________

Cornerstone of Craftsmanship

Rod Hellyer

Lee College, Baytown, TX
Department Chair – Industrial Studies

Rod was already working as a mechanic when two of his younger brothers persuaded him to try a career in construction. Rod credits much of his success to his parents, who set very high standards for their children. According to Rod, his dad was a hardworking man who put pride and craftsmanship into everything he did. Inspired by his father, Rod adopted these same values and continues to practice them today as he celebrates 31 years as a college welding instructor. Rod's advice to new welders entering the field is this—take pride in yourself and in your work, do everything to the best of your ability, and be a worker who can be trusted to get the job done.

How did you choose a career in the construction industry?
I owe it all to my two younger brothers who were welders and pipefitters. I was working as a mechanic at the time. My brothers convinced me to switch gears and go into construction.

What kinds of work have you done in your career?
I started out doing structural welding and fabricating. I worked on anything and everything—ship loaders, conveyors, oil storage tanks, towers, pipelines, boilers—you name it. I've performed work in accordance with a wide variety of welding codes.

Tell us about your present job.
I've been a welding instructor at Lee College for more than 30 years. This includes 15 years working at a local high school teaching students interested in welding. I have a deep admiration for welding and cutting, and I try to communicate this to my students and to the instructors I supervise as chair of the industrial studies department. We teach classes in SMAW, GTAW, and GMAW. Since 2008 I've served as department chair of industrial studies, in which I'm responsible for the welding, pipefitting, millwright, and machine shop programs.

What factors have contributed most to your success?
I attribute most of my success to my parents, who had very high moral standards. My dad worked hard all his life. He put a lot of pride and craftsmanship into everything he did, and he instilled the same values in us. Dad believed that anything worth doing was worth doing well. He required that even the least important jobs be done to the best of our ability. Today, I continue to practice the values instilled in me by my father.

What types of training have you been through?
When I first entered construction, I received some on-the-job training. Then I enrolled in a local college and started my formal training in welding. I took classes in SMAW, GTAW, GMAW, and metallurgy, along with enough academic courses to complete my A.A.S. in Welding Technology.

What advice would you give to those new to the field?
I would stress the importance of taking pride in yourself and in your work. Do everything to the best of your ability. You will be judged on your work by your peers and your employers. Don't worry about the work of others. Just focus on the job you are doing. Be one of those workers who will be called upon to tackle the jobs that no one else can get done.

Tell us an interesting career-related fact or accomplishment.
Some of the most interesting work I've done was on pipelines inside the plants making tie-ins on live units. I also enjoyed the manufacturing of products. And what an experience it is to see a new plant built and then started up for operation—knowing that you were a part of building something and transforming it from a bare spot of ground to a large industrial unit.

Trade Terms Introduced in This Module

Arc blow: The deflection of the arc from its intended path due to magnetic forces that develop from welding current flow.

Arc strike: A discontinuity consisting of localized melting of the base metal or finished weld caused by the initiation of an arc. It remains visible as a result of striking the arc outside the area to be welded and can lead to rejection of an otherwise good weld.

Drag angle: Describes the travel angle when the electrode is pointing in a direction opposite to the welding bead's progression.

Oscillation: A repetitive and consistent side-to-side motion.

Push angle: Describes the travel angle when the electrode is pointing in the same direction as the welding bead's progression.

Quench: To rapidly cool a hot component such as a freshly welded coupon. Note that only coupons should be quenched; real welds with a function should never be quenched.

Restart: The action of restarting a weld bead that is already in progress that requires the crater left from the termination of the previous arc.

Stringer bead: A type of weld bead made without any significant weaving motion. With SMAW, stringer beads are not more than three times the diameter of the electrode.

Undercut: A defect that occurs when the cross-sectional thickness of the base metal is reduced at the edges of the weld. Excessive weld current can cause this as it causes the base metal at the edges to melt and drain into the weld puddle. A downward sloped edge toward the weld bead is the visual evidence.

Weave bead: A type of weld bead made by transverse oscillation of the electrode in a particular pattern.

Weld axis: A straight line drawn through the center of a weld along its length.

Weld coupon: Metal pieces to be welded together as a test or practice.

Work angle: An angle less than 90 degrees between a line perpendicular to the major workpiece surface and a plane determined by the electrode axis and the weld axis. 0-degree work angle is common. In a T-joint or corner joint, the line is perpendicular to the non-butting member. The definition of work angle for a pipe weld is covered in a later module, as it differs from that of plate welding.

Additional Resources

This module is intended as a thorough resource for task training. The following reference works are suggested for further study.

2014 Technical Training Guide. Latest Edition. Cleveland, OH. USA: The Lincoln Electric Company. **www.lincolnelectric.com**

Figure Credits

The Lincoln Electric Company, Cleveland, OH, USA, Module Opener, Figures 1, 4, 18 (photo), SA01

Topaz Publications, Inc., Figures 2, 16, SA02–SA05

©American Welding Society (AWS) *Welding Handbook* 1991, Welding Processes Volume No. 2, Edition No. 8, Miami: American Welding Society., Figure 14

Holley Thomas, SME, Figure 23

Section Review Answer Key

Answer	Section Reference	Objective
Section One		
1 c	1.1.1	1a
2. c	1.2.3	1b
3. a	1.3.2	1c
Section Two		
1. a	2.1.2	2a
2. d	2.2.2	2b
3. a	2.3.3	2c
4. c	2.4.1	2d

NCCER CURRICULA — USER UPDATE

NCCER makes every effort to keep its textbooks up-to-date and free of technical errors. We appreciate your help in this process. If you find an error, a typographical mistake, or an inaccuracy in NCCER's curricula, please fill out this form (or a photocopy), or complete the online form at **www.nccer.org/olf**. Be sure to include the exact module ID number, page number, a detailed description, and your recommended correction. Your input will be brought to the attention of the Authoring Team. Thank you for your assistance.

Instructors – If you have an idea for improving this textbook, or have found that additional materials were necessary to teach this module effectively, please let us know so that we may present your suggestions to the Authoring Team.

NCCER Product Development and Revision

13614 Progress Blvd., Alachua, FL 32615

Email: curriculum@nccer.org
Online: www.nccer.org/olf

❑ Trainee Guide ❑ Lesson Plans ❑ Exam ❑ PowerPoints Other ______________________

Craft / Level: Copyright Date:

Module ID Number / Title:

Section Number(s):

Description:

Recommended Correction:

Your Name:

Address:

Email: Phone:

Joint Fit-Up and Alignment

Overview

Regardless of a welder's talent and experience, a joint that is poorly fit or aligned is subject to failure and a liability. All welded joints must be properly fit according to codes, specifications, and industry standards. Due to the intense heat of the welding process, care must also be taken to ensure that a properly fitted joint is not distorted and pulled out of alignment due to expansion and contraction. This module introduces welders to the basic tools and practices related to the proper fit of both plate and pipe joints.

Module 29110

Trainees with successful module completions may be eligible for credentialing through the NCCER Registry. To learn more, go to **www.nccer.org** or contact us at **1.888.622.3720**. Our website has information on the latest product releases and training, as well as online versions of our *Cornerstone* magazine and Pearson's product catalog.

Your feedback is welcome. You may email your comments to **curriculum@nccer.org**, send general comments and inquiries to **info@nccer.org**, or fill in the User Update form at the back of this module.

This information is general in nature and intended for training purposes only. Actual performance of activities described in this manual requires compliance with all applicable operating, service, maintenance, and safety procedures under the direction of qualified personnel. References in this manual to patented or proprietary devices do not constitute a recommendation of their use.

29110 V5

29110
Joint Fit-up and Alignment

Objectives

When you have completed this module, you will be able to do the following:

1. Identify and describe various types of fit-up and alignment tools.
   a. Identify and describe various fit-up gauges and measuring devices.
   b. Identify and describe common weldment positioning equipment.
   c. Identify and describe various plate alignment tools.
   d. Identify and describe various pipe and flange alignment tools.
2. Describe techniques to avoid weldment distortion and describe the role of codes and specifications.
   a. Describe the causes of weldment distortion.
   b. Describe the techniques and tools used to control weldment distortion.
   c. Describe the role of codes and specifications in welding procedures and techniques.

Performance Tasks

Under the supervision of your instructor, you should be able to do the following:

1. Fit up joints using plate and pipe fit-up tools.
2. Check the joint for proper fit-up and alignment using gauges and measuring devices.

Trade Terms

Addendum
Ambient temperature
Coefficient of thermal expansion (linear)
Consumable insert
High-low
Level
Peening
Plumb
Residual stress
Specific heat per unit volume
Tack weld

Industry Recognized Credentials

If you are training through an NCCER-accredited sponsor, you may be eligible for credentials from NCCER's Registry. The ID number for this module is 29110. Note that this module may have been used in other NCCER curricula and may apply to other level completions. Contact NCCER's Registry at 888.622.3720 or go to **www.nccer.org** for more information.

Contents

Topics to be presented in this module include:

Figures and Tables

SECTION ONE

1.0.0 JOINT FIT-UP AND ALIGNMENT TOOLS

Objective

Identify and describe various types of fit-up and alignment tools.

a. Identify and describe various fit-up gauges and measuring devices.
b. Identify and describe common weldment positioning equipment.
c. Identify and describe various plate alignment tools.
d. Identify and describe various pipe and flange alignment tools.

Performance Tasks

1. Fit up joints using plate and pipe fit-up tools.
2. Check the joint for proper fit-up and alignment using gauges and measuring devices.

Trade Terms

Consumable insert: Preplaced filler metal that is completely fused into the root of the joint during welding, becoming part of the weld.

High-low: The discrepancy in the internal alignment of two sections of pipe.

Level: A line on a horizontal axis parallel with the earth's surface (horizon). Also refers to the tool used to check if an object is level.

Plumb: A line on a vertical axis perpendicular to the earth's surface (horizon).

Tack weld: A weld made to hold parts of a weldment in proper alignment until the final weld is made.

Joint design and setup affect the safety and quality of a completed weldment. Because joint design and setup are so important, they are covered by written codes and specifications that must be followed. Special tools to measure and support weldment fit-up are also available. This module explains the codes and introduces special tools and measuring devices that aid setup. In addition, joint inspection techniques and procedures to ensure that proper joint setup is maintained during welding will also be explained. This first section of the module focuses on the tools used in fit-up.

1.1.0 Fit-Up Gauges and Measuring Devices

Before making a weld, the joint must be fit up and checked to ensure it conforms to the Welding Procedure Specification (WPS) or site quality standards. The most common tools used to lay out and check joint fit-ups are straightedges, squares, levels, and Hi-Lo gauges.

Before proceeding, it is important to note that fit-up tools and instruments are made from a variety of metals. Of course, weldments can consist of a wide variety of metals as well. The tools and instruments used in the fit-up and preparation of a joint must be compatible with the base metal to avoid contamination. Most combinations are not a large problem, but tools and instruments made from carbon steel must not be used on stainless steel. Stainless steel devices can be used on carbon steel base metals without concern. However, it is important to note that some stainless steel tools, such as wire brushes, should not be used on stainless steel once they have been used with carbon steel. The concern is the contamination of stainless steel by carbon steel, but not the reverse.

1.1.1 Straightedges

Straightedges are used to scribe straight lines and check joint alignment. Many have calibrations along their length for measuring. Straightedges, particularly longer ones, are typically fabricated on the job from small channel or angle iron. When using a straightedge, be careful not to apply heat to it by placing it on hot metal or near the flame of a cutting torch or welding arc. Heat can cause the straightedge to become permanently distorted. Before using a straightedge, check it for straightness by visually sighting down the edge. If it is distorted and cannot be straightened, destroy it.

1.1.2 Squares

Two types of squares are commonly used for layout: pipefitter's squares and combination squares (*Figure 1*). Pipefitter's squares, which are used to measure angles and check squareness, are available in a variety of sizes. Like a carpenter's framing square, the pipefitter's square has a great deal of helpful information stamped into the metal.

Combination squares are smaller than pipefitter's squares, with blades typically 12" or 18" long (≈30 cm or 45 cm). They have replaceable attachments that slide and lock along a groove on the blade. Attachments include a combination 90-degree/45-degree level head, a centering head, and a protractor head. The combination attachment is used to check and lay out 90-degree and

45-degree angles, to check level, and to measure depth. The V-shaped centering head is used to measure round stock and to locate the center of shafts or other round objects. The protractor head is used to lay out and check angles.

When using squares, avoid overheating them as it could cause distortion. Also, protect squares from welding spatter that could stick to measuring surfaces, causing a false reading. To preserve their accuracy, squares should be stored in a dry, protected area when not in use. Carbon steel squares should be kept free of corrosion.

1.1.3 Levels

Levels come in a variety of sizes and shapes (*Figure 2*). Most levels designed for plate and piping work are made from magnesium or aluminum. Some have magnetized bases. Levels are used to check that layouts are level (horizontal) and **plumb** (vertical). Levels use a bubble in a glass vial to check level and plumb. Centering the bubble between the lines marked on the vial indicates level when the level is in the horizontal position and plumb when the level is in the vertical position.

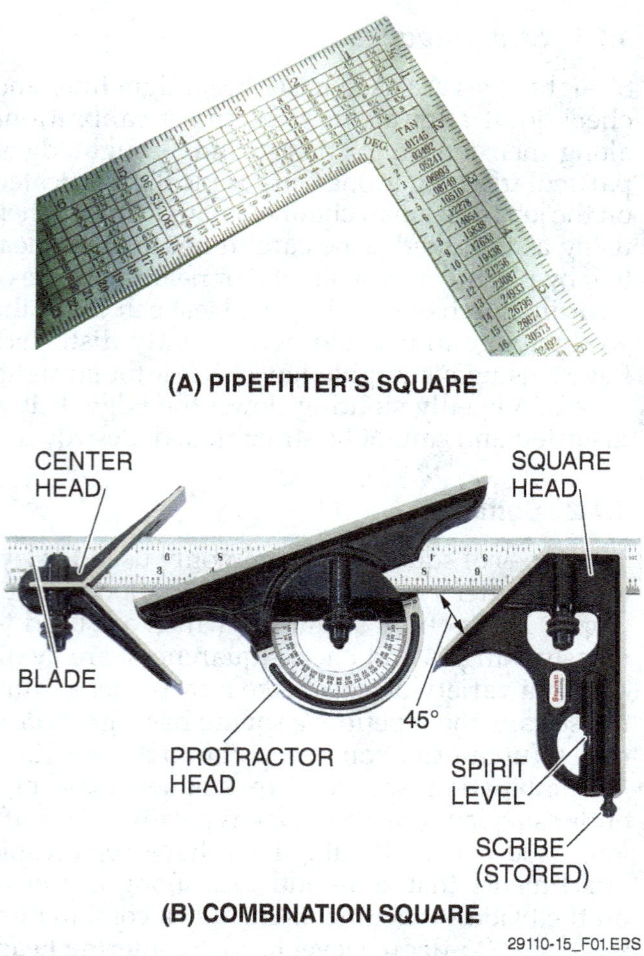

Figure 1 Pipefitter's square and combination square.

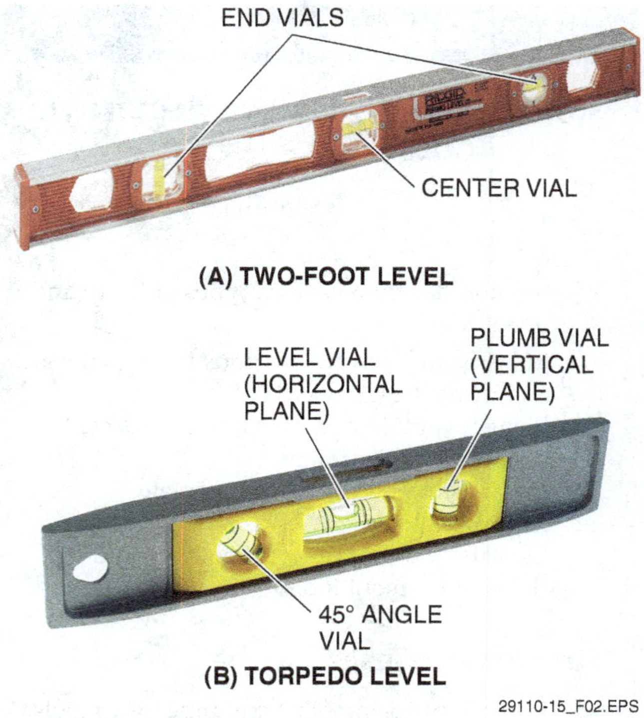

Figure 2 Common levels.

Some levels also have a vial set at 45 degrees to check 45-degree angles. There are also special levels that have an adjustable protractor scale that can be set to check any angle.

The torpedo level is about 9" (≈23 cm) long and tapered on each end. The torpedo level has three vials: one to check level, one to check plumb, and one to check 45-degree layouts.

Protect levels from heat and weld spatter, which can break the glass tube. Also, be aware that weld spatter can stick to the bottom of the level, giving a false reading. The edge of the level that is designed to be placed against the surface in question must be kept clean and free of significant damage that might alter the reading.

1.1.4 Hi-Lo Gauges

The primary purpose of a Hi-Lo gauge is to check for pipe joint misalignment, although plate joint misalignment can also be checked. The name of the gauge comes from the relationship between the alignment of one pipe or plate to the other, which is called **high-low**. When either type of weldment is misaligned, one half is considered low and the other half is high.

To check for internal misalignment of pipe, Hi-Lo gauges have two prongs, or alignment stops, that are pulled tightly against the inside diameter of the joint so that one stop is flush with each side of the joint. The variation between the two stops is read on a scale marked on the gauge.

To measure pipe misalignment with a Hi-Lo gauge, insert the prongs of the gauge into the root opening. Pull up on the gauge until the prongs are snug against both inside surfaces, and read the misalignment indicated on the scale. *Figure 3* shows checking internal misalignment with a Hi-Lo gauge.

Many Hi-Lo gauges also have the capability to check the following:

- Root opening width
- Material thickness
- Weld reinforcement
- Scribe lines for socket welds

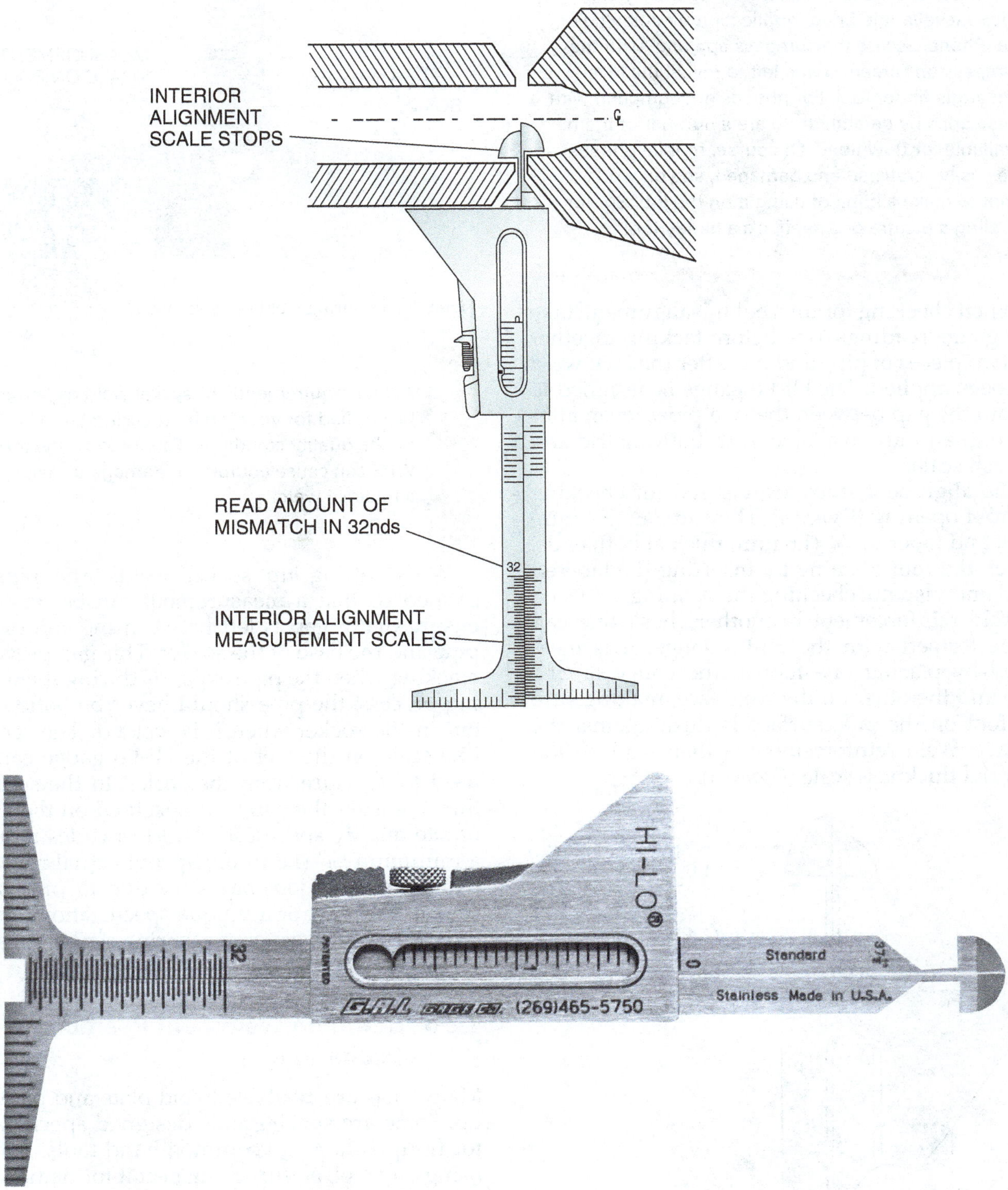

29110-15_F03.EPS

Figure 3 Checking internal misalignment.

Handy Level and Angle Finder

There might be an angle finder and level closer to you than you might think. The next time you need one, reach instead for your mobile phone. A number of smartphones, including the iPhone, offer a level/angle finder application by default. For the iPhone, access the compass app and, with the compass on screen, swipe left to reveal the level and angle finder tool. For phones not equipped with these apps by default, there are a number of them available for download. Of course, mobile phones are easily scratched and damaged, so you won't want to make a habit of using it on the job. But for leveling a picture or a shelf, it's a handy tool.

When checking for internal misalignment, take two gauge readings: one before tacking together the two pieces of pipe and one after the **tack weld** has been applied. The Hi-Lo gauge is designed to fit into the gap between the two pipes, even after tacking, and are available with both metric and English scales.

The alignment stops are tapered for checking the root opening (*Figure 4*). They are 3⁄32" (2.4 mm) thick and taper to 1⁄16" (1.6 mm) thick at both ends. Check the root opening by inserting the tapered head and visually checking the opening.

Weld reinforcement is another check that can be performed with the Hi-Lo gauge. It is measured by placing one foot of the gauge on the pipe and the other on the weld face, making sure the foot on the pipe surface is flush against the surface. Weld reinforcement is then read on the material thickness scale (*Figure 5*).

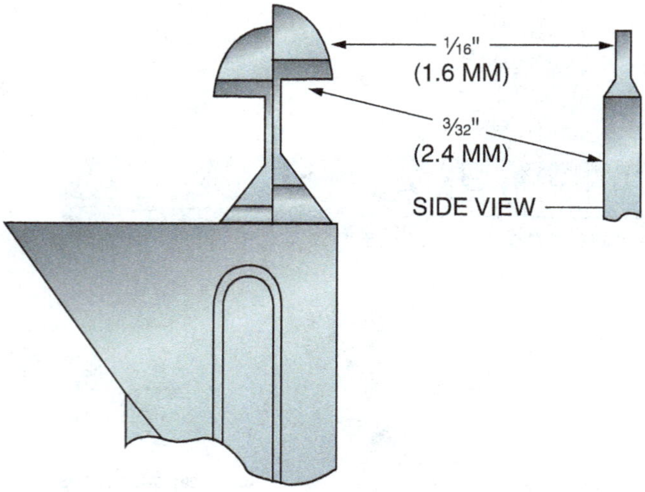

Figure 4 Tapered alignment stops.

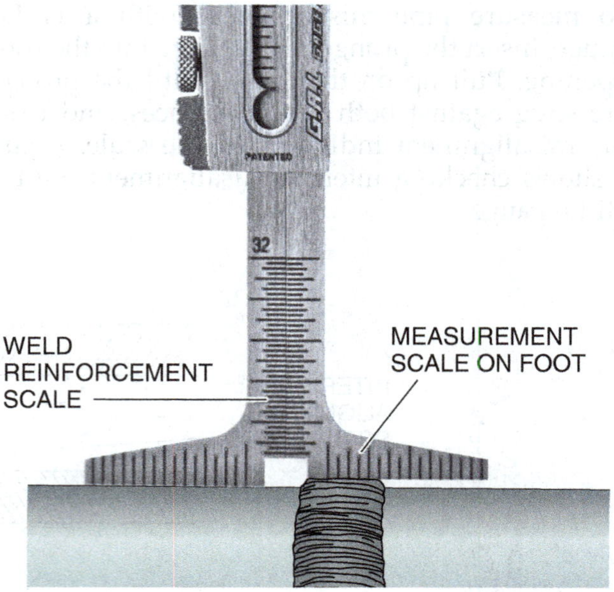

Figure 5 Measuring weld reinforcement.

CAUTION

Actual requirements for socket weld gaps must be verified for your job by checking the WPS or site quality standards. Failure to follow the WPS can cause equipment damage or result in a rejected weld.

When fitting up socket welds, the pipe is scribed so that a measurement can be made to ensure that a gap is left between the end of the pipe and the base of the socket. This gap prevents cracking when the pipe expands during the welding process; the pipe should never be bottomed-out in the socket when it is welded. The scribe-line scale on the feet of the Hi-Lo gauge can be used to measure from the socket to the scribed line to ensure the gap is as specified on the WPS or site quality specifications. Most codes require a minimum 1⁄16" (1.6 mm) gap, but actual requirements for your job may vary due to pipe size, base metal type, or a unique specification. *Figure 6* shows a Hi-Lo gauge measuring scribe lines on a socket weld.

1.2.0 Common Weldment Positioning Equipment

Many tools are available to aid plate and pipe fit-up. Some are special tools designed specifically for fit-up; others are common hand tools. Before using any tool, be sure to inspect it for damage or missing parts. Never use a tool for a job for which it was not intended. Be sure to clean and return all tools when you have finished using them.

Also remember to ensure the tool material is compatible with the base metal being fitted. Carbon steel tools, as well as some stainless steel tools that have been used with carbon steel, should not be used on stainless steel.

Hydraulic jacks, chain falls, and come-alongs are sometimes used to position parts of a weldment. These types of positioning tools are described in the following sections.

1.2.1 Hydraulic Jacks

When using hydraulic jacks, never weld directly on the jack base or ram. To secure the jack, a ring can be tack welded to act as a socket for the base and/or ram. When welding near hydraulic jacks, protect the ram from weld spatter. If the ram is off the floor, be sure to secure it with rope or chain. Also monitor the jack for oil leaks. Oil will create a fire hazard, and oil in the area to be welded will create weld defects from porosity. Oil leaks must be cleaned up before welding. *Figure 7* shows using a hydraulic jack to aid joint fit-up.

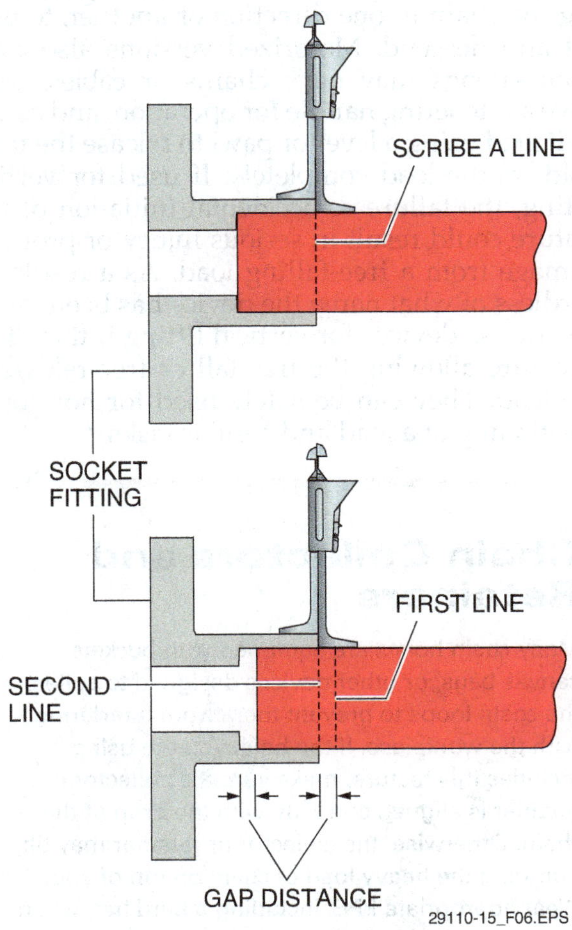

Figure 6 Measuring the distance between scribe lines on a socket weld.

WARNING!

Make sure the current path does not pass through the hydraulic jack supporting the load you are welding. Extreme heat generated by the current flow can cause the hydraulic jack to fail, releasing the load support. This can cause serious injury or even death.

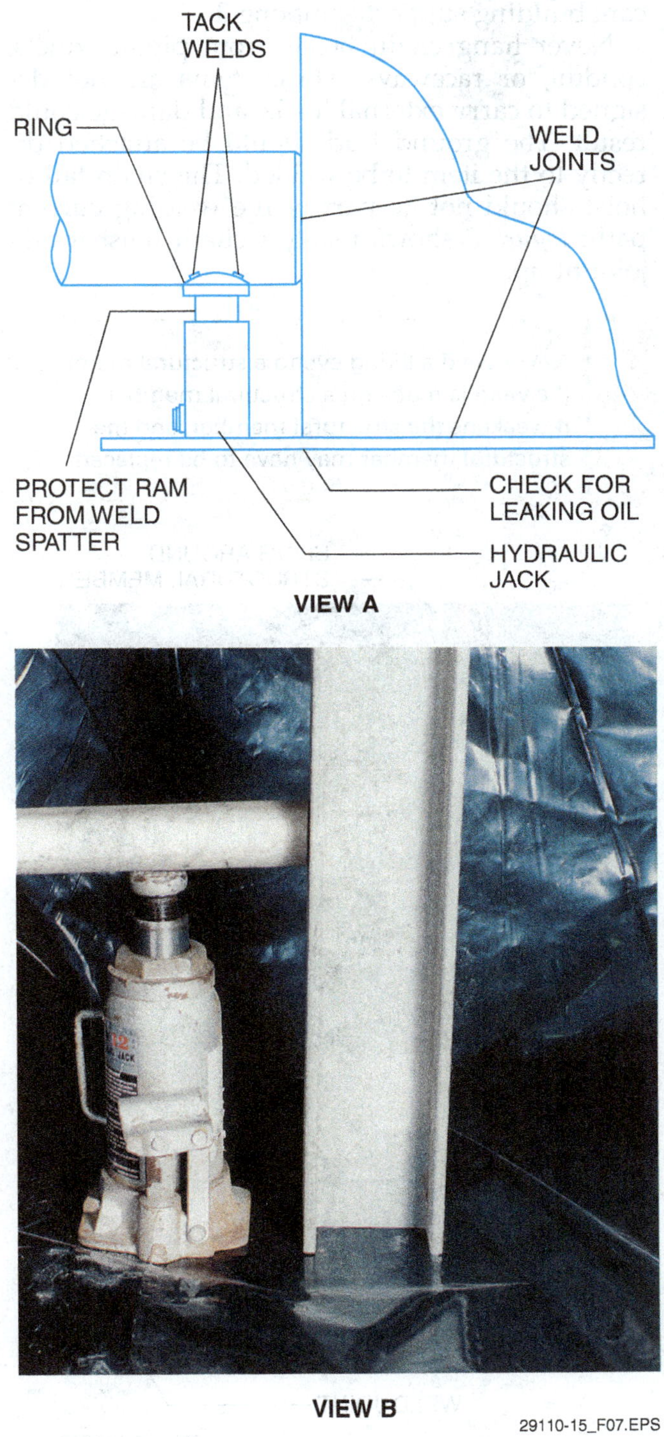

Figure 7 Proper use of a hydraulic jack for joint fit-up.

1.2.2 Chain Hoists, Chain Falls, and Come-Alongs

Chain hoists, also called chain falls, are used to lift or lower weldment parts. Always position the chain hoist directly over the center of gravity of the object being lifted. Secure the chain hoist over the weldment with an approved sling wrapped around a structural member. A suitable structural member may be a large I-beam or other significant building support component.

Never hang chain hoists from piping, ducts, conduit, or raceways. These items are not designed to carry external loads, and damage could result. The ground lead should be attached directly to the item to be welded. The chain fall or hoist should not be part of the welding current path. *Figure 8* shows using a chain hoist to aid joint fit-up.

CAUTION

Never weld a lifting eye to a structural member. If a weld is made on a structural member, it weakens the structural member, and the structural member may have to be replaced.

Figure 8 Chain hoist lifting a pipe into position.

WARNING!

Make sure the current path does not pass through the chain hoist supporting the load you are welding. Extreme heat generated by the current flow can cause the chain hoist to fail, releasing the load support. This can cause serious injury or even death.

Often, more than one come-along or chain fall may be used to precisely position a weldment. Before welding, check the ground lead connection location to be sure the welding current will not pass through the come-along. The ground lead should be attached directly to the item to be welded.

The difference in chain falls, chain hoists, and come-alongs can be a confusing topic. Some manufacturers label their products in ways that adds to the confusion. Some may call an item a chain fall while another item with the same features is referred to as a come-along. As a general rule, chain hoists and chain falls do not have a setting or feature that allows for the free-fall of the load. The device must be operated manually, by pulling the chain in one direction or another, to both lift and descend. Motorized versions also exist. Come-alongs may have chains or cables. Most have a ratcheting handle for operation, and can be switched using a lever or pawl to release the tool's hold on the load completely. If used for vertical lifting, the failure or accidental initiation of this feature could result in serious injury or property damage from a free-falling load. As a result, regardless of what name the device has been given, do not use devices for vertical lifting if they have a feature allowing the free-fall or free-release of the load. They can be safely used for horizontal positioning of a load and similar tasks.

Chain Collectors and Retainers

Many chain hoists are equipped with buckets, canvas bags, or other devices designed to collect the chain loops to prevent them from interfering with the workpiece. If the hoist you are using includes this feature, make sure the collector or retainer is aligned correctly with the drop of the chain. Otherwise, the collector or retainer may tilt, dumping the heavy load of chain on top of you. Wear appropriate PPE, including a hard hat, when using chain hoists or falls.

> **WARNING!**
>
> Before lifting any weldment part, add the weight of the part to the weight of the rigging (lifting) devices to determine the overall weight that will be placed on the structural member. Make sure that the structural member and all associated rigging devices are strong enough to handle the overall weight being lifted. Always inspect all rigging devices for damage and make sure that they are approved for the intended use.

1.3.0 Plate Fit-Up Tools

The most common method of holding a joint in place after it has been fitted up is to tack weld it in place. This works very well for small weldments and joints that are straight. However, large weldments or long or thick joints often require some type of mechanical means in addition to tack welding to set up and hold them in place for welding. The most common tools for plate fit-up are strongbacks, clips, yokes, and wedges. In addition, special plate alignment tools are available.

1.3.1 Strongbacks

Strongbacks (*Figure 9*) are typically made on the job site from heavy bar stock. They are notched at the weld joint to allow access to the joint so that welds can be made without interference. The strongback can be placed on the face or root side of the weldment. The plates to be joined are clamped or tack welded flush against the strongback, which holds the joint in alignment. If possible, when tack welding strongbacks, place the tack welds on only one side of the strongback. This will make the strongback easier to remove.

A good way to remove a strongback that has been tack welded is to use a grinder to remove or greatly reduce the tack welds. After the tack welds are minimized with a grinder, the tap of a hammer should break the strongback free of the base metal. After the strongback is removed, the grinder can be used to grind the tacks down flush, but care must be taken to not gouge the base metal during the grinding process.

1.3.2 Clips, Yokes, and Wedges

Clips, yokes, and wedges are also made on the job site. They can be used to align joints and then hold them in place during welding. Clips are welded to the edge of one plate and then wedges are positioned on the other plate and driven under the clips to force the joint into alignment. Yokes work in a similar manner. A yoke is welded to one plate. A slotted plate is then placed over the yoke, and a wedge is driven under the yoke to force the joint into alignment. *Figure 10* shows yokes and wedges being used to align and hold a joint.

1.3.3 Combining Alignment Tools

If the joint needs to be forced into alignment, yokes and wedges or bolts can be used with a strongback (*Figure 11*). When yokes and wedges are used, the strongback is positioned on the weldment straddling the joint. Yokes are positioned over the strongback and welded to the plates. Wedges are then driven under the yokes to force the joint into alignment. When bolts are used, the strongback must be made of channel stock or reinforced angle iron. Holes are cut in the channel or angle iron so that the bolts can pass through. The strongback is then positioned on the weldment straddling the joint. The position of the bolts is marked, and the strongback is removed. Bolts long enough to protrude through the strongback or threaded stock are then welded onto the plates.

The strongback is then positioned with the bolts protruding. Nuts are placed on the bolts and tightened with a ratchet or impact wrench, pulling the joint into position.

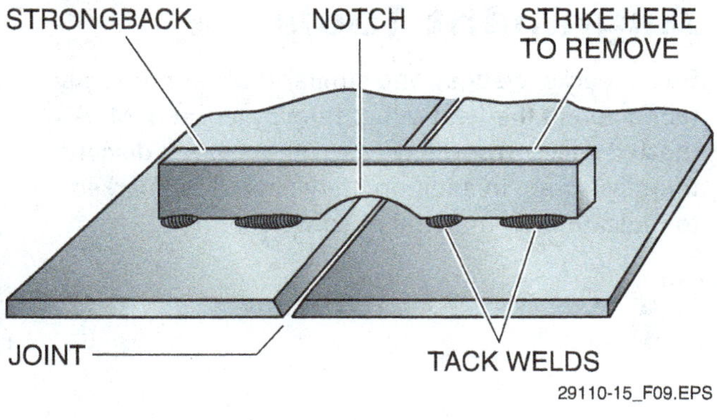

Figure 9 Strongback.

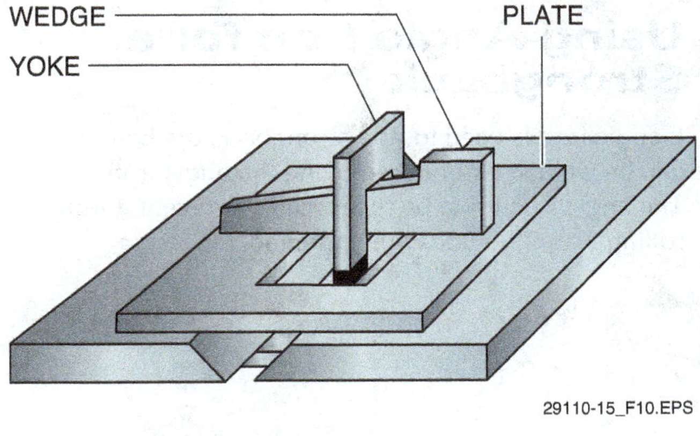

Figure 10 Yokes and wedges.

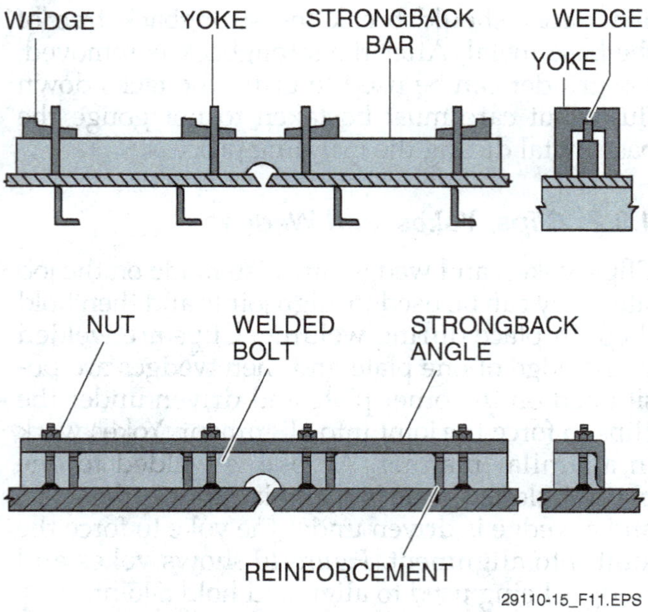

Figure 11 Aligning a joint using several tools.

1.3.4 Plate Alignment Tools

Special tools are manufactured for aligning plate that can make the task easier. A typical configuration of an alignment tool consists of a yoke, threaded adjusting rod, gap plate, and root bar (*Figure 12*). The alignment tool is used by straddling the joint opening with the yoke. The gap plate can be changed to match the specified root opening. The threaded adjusting rod is lowered until the opening in the gap plate is below the joint and the root bar can be inserted. The threaded adjusting rod is then tightened, bringing the joint into alignment. If necessary, the root bar can be adjusted to compensate for uneven joint thicknesses. Remember to properly match the tool materials with the base metal to eliminate contamination. Alternatively, isolating the two surfaces from each other so they don't touch while in use is also an option.

> **Using Angle Iron for a Strongback**
>
> If angle iron is used for the strongback, the bolts can be welded without removing the angle iron. The angle iron must be reinforced to prevent it from rolling when the bolts are tightened.

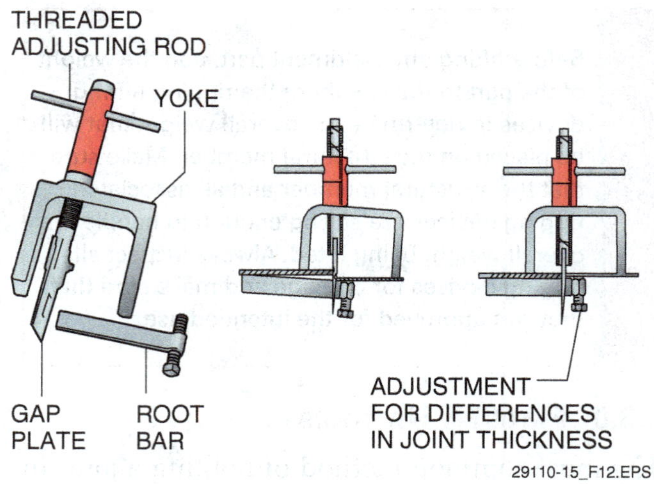

Figure 12 Plate alignment tool.

1.4.0 Pipe and Flange Fit-Up Tools

There are many different manufacturers of pipe fit-up tools. The tools they produce are similar, but before using a tool, be sure to review the guidelines provided by its manufacturer.

1.4.1 Pipe Jacks and Rollers

Pipe jack stands and rollers (*Figure 13*) are used to support pipe for fit-up and welding. Pipe jack stands typically have either a V-head or roller head and a height adjustment. Rollers, which can be floor-stand or table models, can be adjusted horizontally for various pipe diameters.

Adjustable jack stands and rollers are necessary tools for pipe fabrication, especially in the shop environment. They allow the welder to position various sizes of pipe for alignment and to manually roll the pipe as needed so the weld can be performed from the top, which is the preferred position. When fabricating piping systems in place, however, jack stands and rollers cannot

> **Marking Fabricated Alignment Tools**
>
> Strongbacks, wedges and similar tools are typically fabricated in the field rather than purchased. Mark these devices in some way so others won't discard them as scrap. In addition, they should be marked to indicate their material of construction.

Figure 13 Pipe jack stands and rollers.

be used because of positioning limitations. One way to deal with this is to weld straight sections of piping systems using jack stands and rollers. The straight sections can then be lifted and welded in place.

WARNING!

Jack stands and roller assemblies should not be field-fabricated. Load capacities and other safety considerations are built into these devices by their manufacturers, so only commercial jack stands and rollers should be used. Field-fabricated jack stands and rollers can fail and cause injury or even death. For pipe over 10" (DN250), OSHA requires that four-legged jack stands and roller assemblies be used.

CAUTION

When using pipe rollers to tack or weld pipe, be sure to connect the ground lead directly to the pipe to prevent arcing damage to the rollers and roller bearings.

1.4.2 Chain Clamps

Chain clamps are used to align and hold pipe for fit-up and tacking. Some manufacturers use link chain, others use chain similar to bicycle chain. Regardless of the type of chain used, the basic procedure for using chain clamps is the same. The chain, which is anchored to one side of the clamp, is passed around the pipe and secured. The slack in the chain is then taken up using a screw jack to pull the pipe tightly against the clamp.

Several types of chain clamps can be used to provide precise fit and alignment:

- Single-chain, single-jackscrew clamps
- Single-chain, double-jackscrew clamps
- Double-chain, double-jackscrew clamps
- Rim clamps

The single-chain, single-jackscrew clamp shown in *Figure 14* will align and even re-form out-of-round pipe up to Schedule 40 wall thickness. It can also be used on pipe heavier than Schedule 40 as long as no re-forming is required. Single-chain, double-jackscrew clamps have even more power to reform pipe. These clamps have a one or two jackscrews on each jack bar with optional spacing screws that allow the gap to be set quickly and precisely. A spacer may be used to maintain the required gap in the joint. Both the single- and double-jackscrew clamps are shown in *Figure 14*.

In addition to fitting up straight pipe, attachments for the precision fit-up clamps allow them to be used to fit up elbows, Ts, and flanges. The attachments connect by chain to the main block and to the elbow, T, or flange with a chain or clamp, as shown in *Figure 14*. The attachment has a fine adjustment crank to remove slack from the chain to support the elbow, T, or flange.

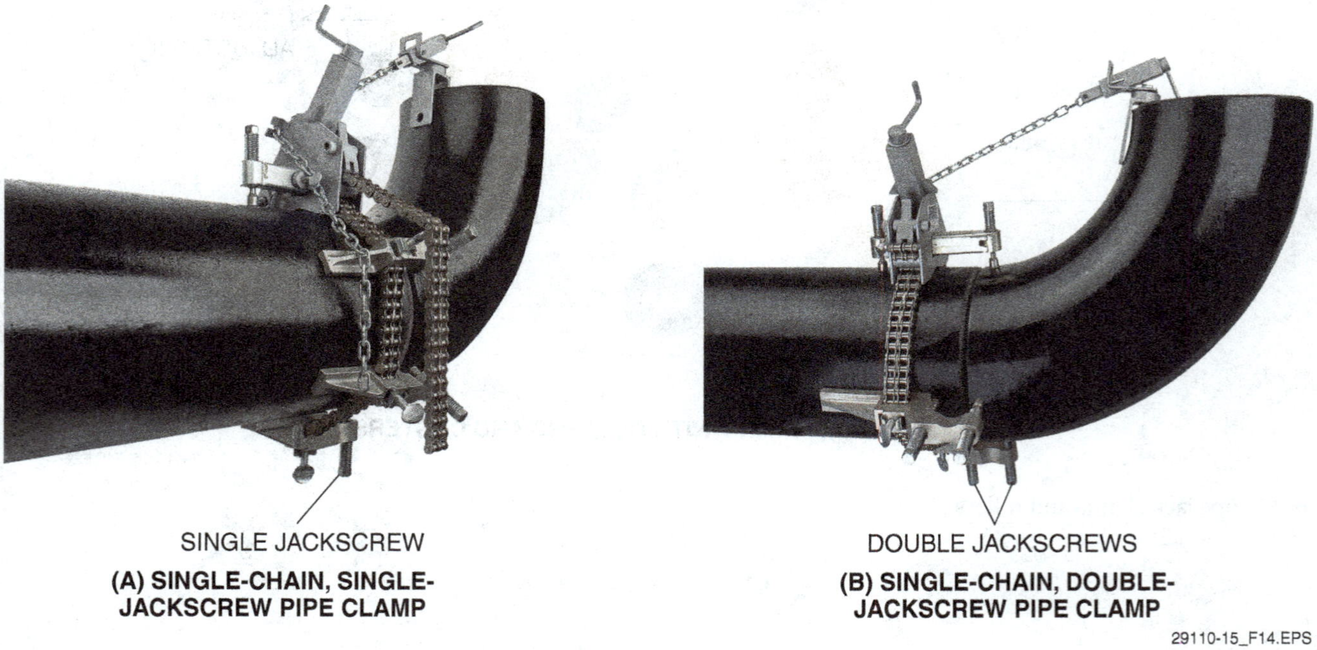

Figure 14 Chain clamps.

1.4.3 Other Pipe Alignment and Clamping Tools

For one section of pipe to be welded end-to-end to another section of pipe, their ends must be cut or ground so that they mate properly. As you have seen in previous sections, jack stands and rollers are used to support sections of pipe to be welded. A chain clamp is one way to secure the two sections of pipe together. Cage clamps are another style of clamp that can also be used for the same purpose. Cage clamps come in a variety of styles and sizes. All must be installed manually, but the ways they are secured to the pipe sections vary. The latches on some of the old-style cage clamps had to be hammered on or off, but hammering tends to damage the clamps. Damaging the clamp can create problems joining the pipe sections.

Newer cage clamps are now available that allow the cages to be custom fitted to the pipes. *Figure 15* shows a cage clamp installed on two pipe sections, a clamp that is manually closed and opened, and a cage clamp that is tightened by a locking screw turned by a ratchet handle.

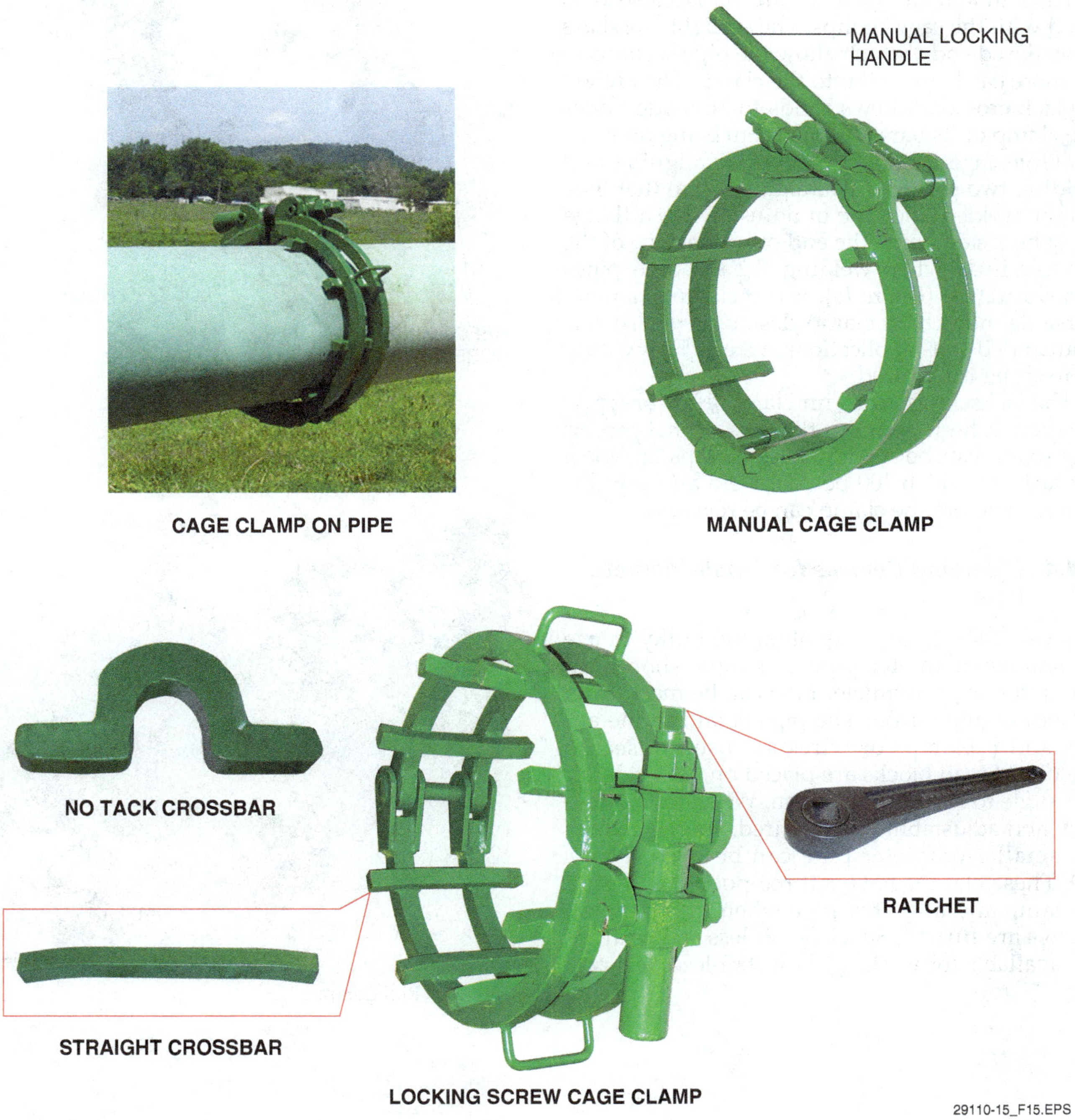

Figure 15 Cage clamps.

Nonferrous Alignment Tools

In some industries, fabricated alignment tools such as wedges, strongbacks, and yokes must be constructed of nonferrous materials to prevent sparking. Brass is a popular nonferrous metal for this purpose. This is a common requirement in some segments of the petrochemical industry.

Also shown in *Figure 15* are replaceable bars used with the cage clamps. The straight crossbars have flared ends, which allow the pipe sections to be more easily moved into the clamp. The arched no-tack crossbar allows welds to be made under the clamp or its bars without them being moved.

While cage clamps are good for aligning and holding two sections of straight pipe so that they can be welded, they are of no use when a flange must be welded onto the end of a pipe. One of the devices used when welding flanges onto pipes is a rim clamp (*Figure 16*). A rim clamp is a more versatile, non-chain clamp design. It is also recommended for applications where heavy-duty re-forming is required.

The jackscrews on a rim clamp exert pressure on specific high points of the pipe so that precise alignment can be made. These clamps are ideal for tasks in which 100 percent weld and grind is required before the clamp can be released.

1.4.4 Clamping Devices for Small-Diameter Pipe

Typical chain clamps are often too bulky to use on smaller-diameter pipe. A simple shop-built clamp for small-diameter pipe can be made from a piece of angle iron. The pipe is laid in the angle, and C-clamps or wires are used to secure the pipe. Small blocks are placed on both sides of the angle to hold it in position. When more accuracy and adjustability is required, special clamps for smaller-diameter pipe can be used (*Figure 17*). These clamps have a three-point jaw design to clamp and align the pipe when the clamping screws are turned. Special stainless steel clamps are available for working with stainless steel piping.

1.4.5 Pipe Pullers

Pipe pullers are clamps, often chain-type, that are used to pull together two pipe workpieces so that they can be joined by welding. These types of pullers are often applied in field fabrication or addition projects.

Figure 16 Rim clamp.

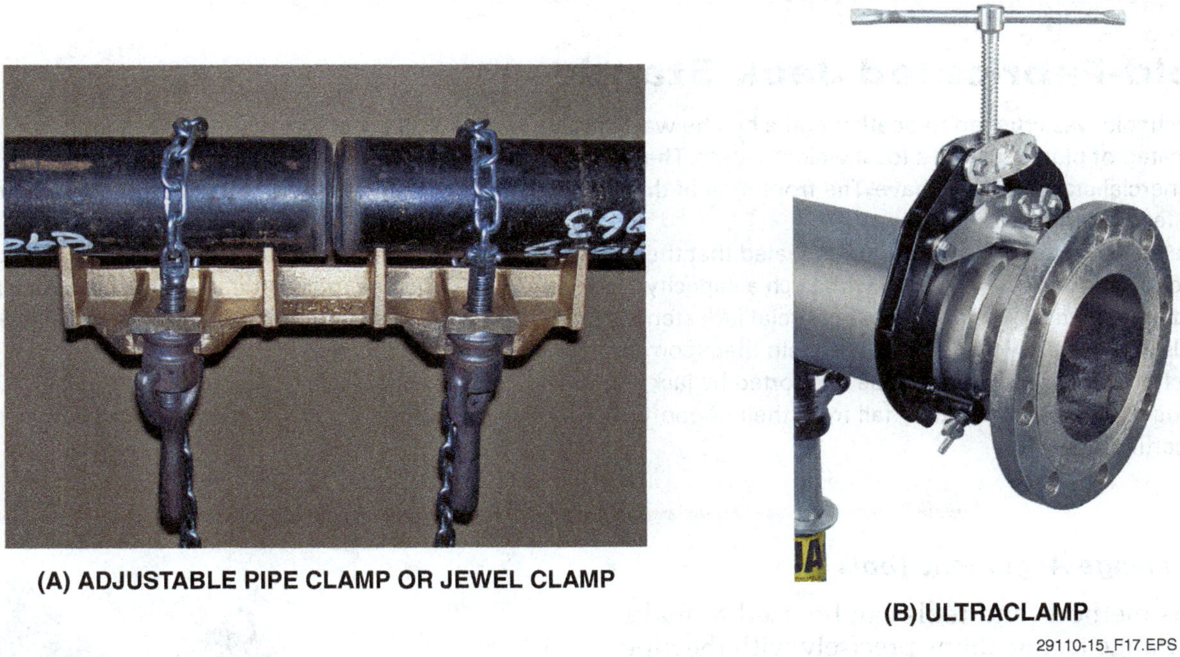

Figure 17 Pipe clamps for smaller pipe.

In the petrochemical industry, for example, a new unit may be added to an existing facility. The piping must be joined through the use of manifolds and other piping designs. In order to make the final tie-ins from existing piping to the new piping, pipe pullers are often used to align the joining parts precisely.

CAUTION

When new piping will be coupled to machinery such as valves or pumps, it is extremely important that the flange of the pipe be perfectly aligned to the flange of the valve or pump. Using a come-along, jack, or other such device can put excessive stress on both the piping and the machine. Piping stress on pumps in particular will make them misalign, which in turn will destroy their bearings.

Do not force joints into alignment and then hold them in place with welds. Stress will be exerted at the welded joint in such an alignment, and cracking could result. Precise alignment and fitting without significant stress should be accomplished before finalizing the installation with the weld.

Pipe pullers (*Figure 18*) are frequently used to pull together sections of large-diameter pipe for welding. Fillers or **consumable inserts** may be installed between the faces of the joints to achieve precisely the necessary root opening between the parts. These inserts or fillers may become a permanent part of the finished weld.

CAUTION

Pipe pullers may not be permitted in some applications or projects.

Figure 18 Pipe puller.

Field-Fabricated Jack Stands

A mechanic was crushed to death when a bus he was working on fell off a set of jack stands. The stands were fabricated of plate steel by a local welding shop. The top plate of each was completely flat; they had no lips, which commercial stands always have. The front tires of the bus were not chocked, and there was nothing to prevent it from falling off the stands.

The investigation of the accident revealed that the jack stands had not been tested or certified for their rated capacity, nor were they marked with such a capacity. They were not fabricated in accordance with commercial jack stand construction. The lips on commercial jack stands cradle the area being supported. Also, commercial jack stands normally have three legs that help them compensate for irregular surfaces.

Sections of large-diameter pipe supported by jack stands may not be as large as buses, but they can cause serious or fatal injuries if they fall from their supports. Always use commercial jack stands that have been tested and certified.

1.4.6 Flange Alignment Tools

Various methods and tools can be used to make sure a flange facing aligns precisely with the pipe before it is permanently welded in place. Flange pins can be installed in the flange bolt holes to provide a reference point from which leveling can be measured for tack welds or complete welding (*Figure 19*). It is important to use a minimum of two flange pins installed in two of the bolt holes to provide a precise reference point from the center of one hole to another. From this flat, level benchmark, other tools can be used to adjust the flange to the pipe for true and precise alignment. The tool shown in *Figure 20* is also used to precisely align a flange for welding.

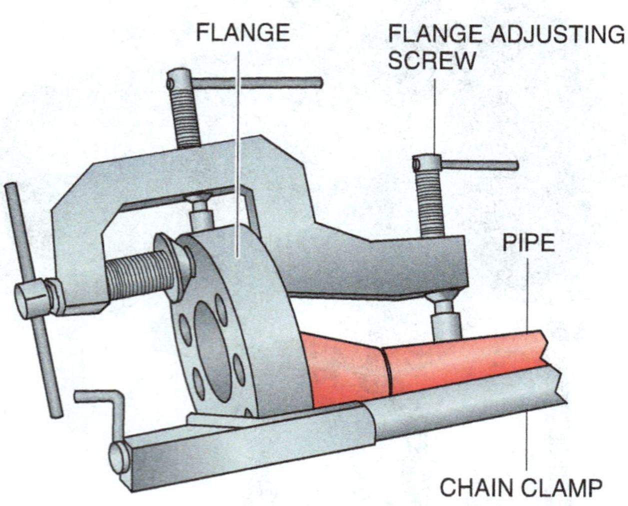

Figure 20 Flange alignment tool.

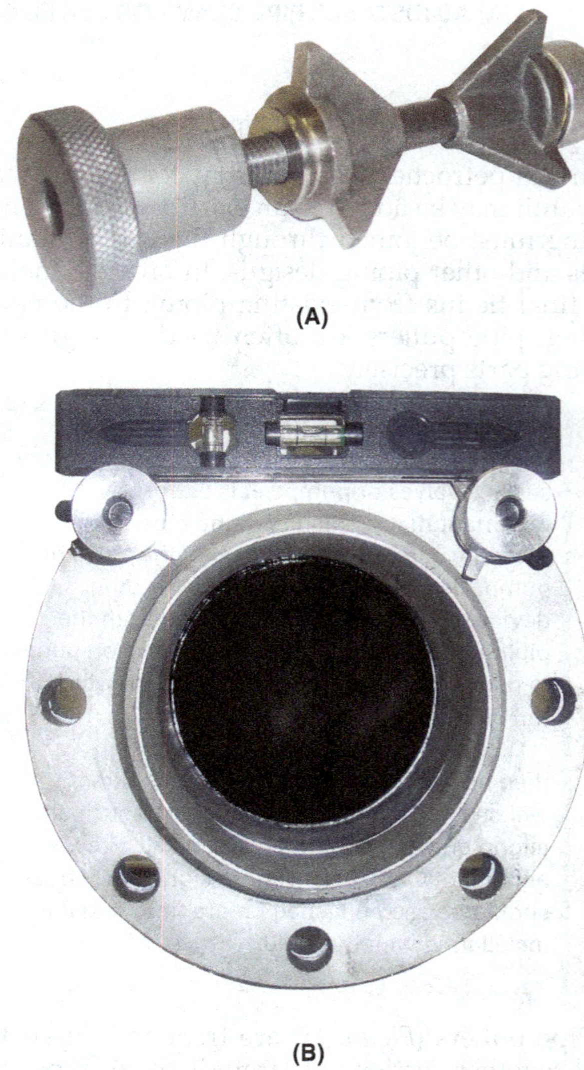

Figure 19 Using flange pins.

Types of Welded Flanges

Some of the more common flanges generally installed using welded connections are the welding neck flange, the lap joint flange, the socket weld flange, and the slip-on flange. Threaded flanges may also be back-welded, as determined by the process. Welding neck flanges are preferred for use in severe service applications, such as those involving high pressure, subzero temperatures, or elevated temperatures.

Think About It

Root Openings in Aligning and Fitting

Why is it necessary to maintain a root opening between two pieces, especially pipe pieces, when aligning and fitting prior to welding?

Additional Resources

2014 Technical Training Guide. Current Edition. Cleveland, OH. USA: The Lincoln Electric Company. **www.lincolnelectric.com**

Welding Handbook. Current Edition. Miami, FL: The American Welding Society.

Mathey Dearman Product Manuals. The Mathey Dearman Company. **www.matheydearman.com**

1.0.0 Section Review

1. Most levels designed for plate and piping work are made from _____.
   a. plastic
   b. wood
   c. aluminum or magnesium
   d. fiberglass or carbon steel

2. When lifting components into place to prepare for welding, vertical lifts should *not* be conducted using _____.
   a. a device with a free-fall mode or setting
   b. a device without a free-fall mode or setting
   c. any type of chain fall
   d. any type of synthetic lifting strap

3. Clips, yokes, and wedges are generally _____.
   a. purchased by the bag
   b. made on site
   c. shipped with the base metal
   d. made of aluminum

4. A simple shop-built clamp can be made from angle iron to help align _____.
   a. heavy plate
   b. large-diameter pipe
   c. small-diameter pipe
   d. sheet metal

SECTION TWO

2.0.0 WELDMENT DISTORTION

Objective

Describe techniques to avoid weldment distortion and describe the roles of codes and specifications.

a. Describe the causes of weldment distortion.
b. Describe the techniques and tools used to control weldment distortion.
c. Describe the role of codes and specifications in welding procedures and techniques.

Trade Terms

Addendum: Supplementary information typically used to describe corrections or revisions to documents.

Ambient temperature: The temperature of the atmosphere surrounding an object or individual.

Coefficient of thermal expansion (linear): The change in length per unit length of material for a 1°C change in temperature.

Peening: The mechanical working of metal using repeated hammer blows, typically using a ball-shaped hammer head such as that found on a ball peen hammer. Peening may be done for both metallurgical and appearance reasons.

Residual stress: Stress remaining in a weldment as a result of heat.

Specific heat per unit volume: The quantity of heat required to raise one unit mass of a substance by one unit degree.

Besides the obvious cosmetic problems resulting from poor alignment and fit, the proper application of fit-up methods must be applied to deal with distortion and other problems. Distortion is the expansion and contraction of metal as it responds to changes in temperature. Tools such as chain clamps, rim clamps, C-clamps, strongbacks, and other devices can be used to redirect this expansion or contraction. This is referred to as controlling the distortion rather than preventing it. Distortion cannot be prevented as long as the metal experiences changes in temperatures created by the welding process. It can, however, be controlled, resulting in a weldment that is relatively free of damaging stress.

2.1.0 Causes of Distortion

Distortion is caused by the non-uniform expansion and contraction of the weld metal and adjacent base metal during the heating and cooling cycles of welding. When a metal block is heated, it expands uniformly in all directions. As the metal block cools, it contracts uniformly in all directions. If the metal block is restricted on two sides as it is heated, expansion cannot take place in the direction of restriction. Because the metal block must expand the same amount, all expansion occurs in the unrestricted directions. As the metal block cools, it contracts uniformly in all directions, leaving it narrower in the restricted direction and longer in the unrestricted direction. *Figure 21* illustrates the expansion, contraction, and distortion control of a block of steel.

During welding, these same forces act on the weld metal and base metal. As the weld metal is deposited and fuses with the base metal, it is at its maximum expanded state. Upon cooling, it attempts to contract but is restricted by the surrounding base metal. Stresses develop within the weld until the stress reaches the yield point of the base metal, causing the base metal and weld metal to stretch. These same stresses also cause the base metal to move. When the base metal is heated during the welding process, it

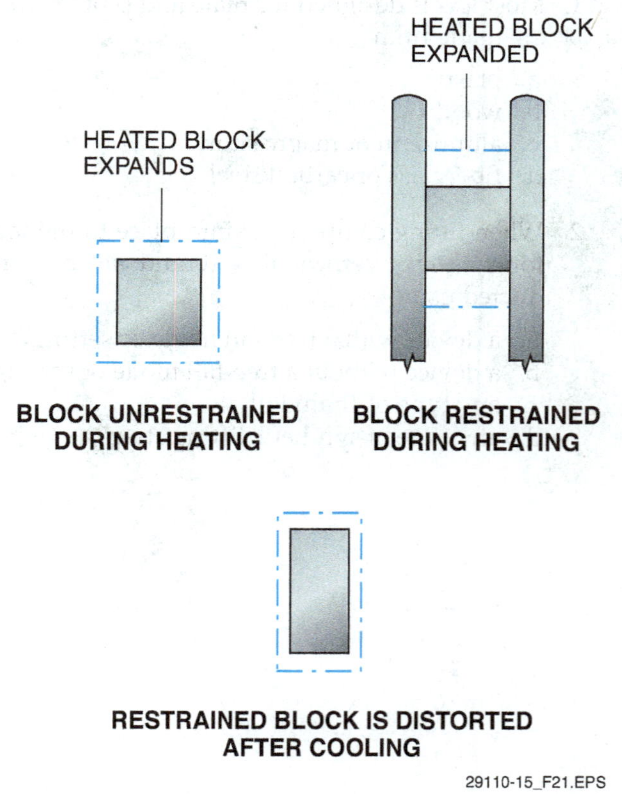

Figure 21 Controlling distortion.

will attempt to expand but will be restricted by the surrounding cooler base metal. Some movement will occur, but as the weld cools, the base metal that was welded will contract more due to the stress caused by the restriction of the surrounding base metal. This causes distortion. Even when the weld is at room temperature, stress equal to the strength of the base metal will be locked in the weldment. This is called **residual stress**. *Figure 22* shows how stress and distortion affect a weld.

The degree of distortion is directly related to the stresses generated during welding. Two of the metal properties that figure prominently in these stresses are the **coefficient of thermal expansion (linear)** and the **specific heat per unit volume**. Metals with higher values for these two properties will experience a higher degree of distortion than those with lower values. Stainless steel, for example, has a much higher coefficient of thermal expansion and specific heat per unit volume than carbon steel. It is therefore more likely to be affected by distortion. *Table 1* compares the coefficients of thermal expansion and the specific heats per unit volume for some common metals.

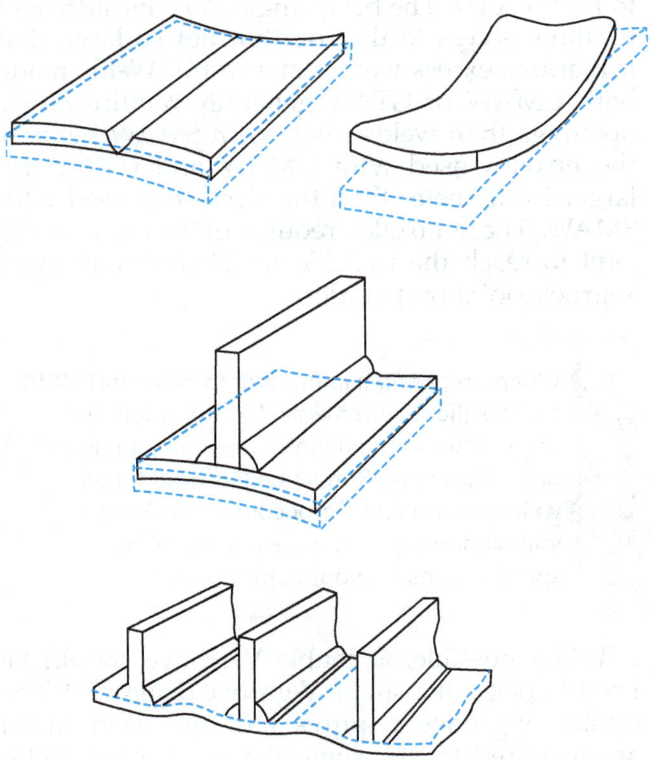

Figure 22 Stress and distortion in welds.

2.2.0 Controlling Distortion

Distortion can be controlled using a variety of techniques and tools. The following sections explain how to control distortion to minimize its effect on weldments.

2.2.1 Clamping and Bracing

Chain clamps, rim clamps, strongbacks, and other field-fabricated alignment devices are often used to hold weldments firmly in place during the welding process in order to prevent warping caused by distortion. These devices must be designed, constructed, and installed such that their overall support can withstand the stresses of welding. They must also be left in place long enough to allow the weldment to cool to **ambient temperature**. Some distortion may still take place once the devices are removed. This is caused by residual stress present in the weldment.

2.2.2 Tack Welding

A tack weld holds parts of a weldment in proper alignment until the finish welds are made. In order for them to be effective, pay attention to the number of tack welds, their length, and the distance between them. If too few tacks are made, the chance of the joint closing up as the welding proceeds is much greater. Clamps and other fit-up and aligning devices must be used to maintain proper alignment and the proper gap during the process.

Latent Distortion

If a weldment is clamped firmly during welding to prevent any movement, it will stay straight as long as the clamps are in place. Even after cooling, when the clamps are removed, residual stress can cause some distortion. When weldments that have residual stress are machined, they distort as the machining process removes metal, allowing the residual stress to overcome the strength of the base metal. This is referred to as latent distortion since it occurs long after the stress was created.

Table 1 Thermal Expansion and Specific Heat per Unit Volume for Common Metals

Metal	Coefficient of Thermal Expansion at 20°C	Specific Heat Per Unit Volume J/g°C
Aluminum	$22–24 \times 10^{-6}/°C$	0.900
Stainless steel (316)	$14–17 \times 10^{-6}/°C$	0.500
Steel	$12–13 \times 10^{-6}/°C$	0.448
Copper	$16–17 \times 10^{-6}/°C$	0.390

If tack welds are to be included with the main weld, they must be applied by qualified welders. They must also follow the qualifying WPS for the project, if one applies. Likewise, the filler metals must meet the same filler metal requirements as the filler metal for the finished welds. For those tack welds that require removal prior to the finish weld, use extreme care to avoid causing defects in the material.

2.2.3 Amount of Weld Material

The more weld metal placed in a joint, the greater the forces of shrinkage. Properly sizing a weld not only minimizes distortion, it also saves weld metal and time. Excess reinforcement on the face of a weld increases the forces of distortion and adds nothing to the strength of the weld. Excess face reinforcement actually reduces the strength of a weld and is therefore prohibited by welding codes. The face of fillet welds may be slightly convex, flat, or slightly concave. Groove welds should have slight reinforcement of no more than ⅛" (3.2 mm) for butt or corner welds. *Figure 23* shows acceptable weld reinforcement profiles.

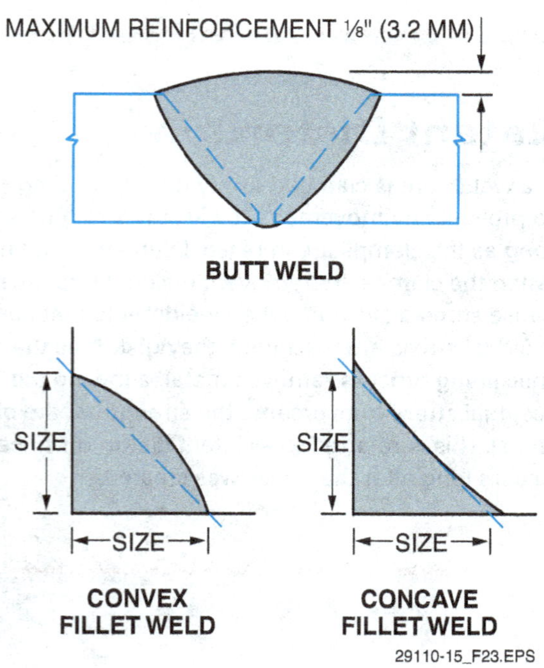

Figure 23 Weld reinforcement.

NOTE

Refer to your site's WPS for specific requirements on weld reinforcement. The information in this module is provided as a general guideline only. The site WPS must be followed for all welds covered by a procedure. Check with your supervisor if you are unsure of the specifications for your application.

Proper fit-up and edge preparation also reduce the amount of weld required. Open-root joints should have a root opening from 1⁄16" to ⅛" (1.6 to 3.2 mm). To control melt-through, a root face of 1⁄16" to ⅛" (1.6 to 3.2 mm) is used. Each side of the open-root joint is beveled from 30 to 37½ degrees (included angle of 60 to 75 degrees). The greater the angle of bevel, the more weld will be required to fill the joint. The bevel angle must be sufficient to allow access to the root but not so large that it requires excess weld metal to fill. Welds made with GMAW or GTAW generally require larger openings than welds made with SMAW because the nozzles used with GMAW and GTAW are larger in diameter than the electrodes used with SMAW. These nozzles require more room in the joint to reach the root. *Figure 24* shows an open V-groove joint preparation.

CAUTION

When preparing a joint, refer to your site's WPS for specific requirements. The information in this manual is provided as a general guideline only. The site WPS must be followed for all welds covered by a procedure. Check with your supervisor if you are unsure of the specifications for your application.

When possible, a double V-groove should be used in place of a single V-groove (*Figure 25*). The double V-groove requires half the weld metal as compared to the single V-groove. Also, welding from both sides reduces distortion since the forces of distortion will be working against each other. On thick joints, U- and J-grooves also require less weld metal than V-grooves or bevel grooves, although U- and J-grooves require more preparation time.

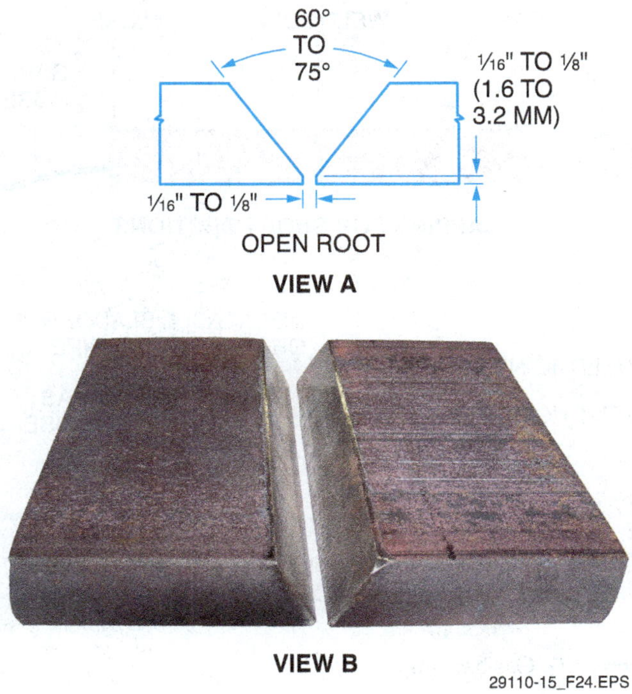

Figure 24 Open V-groove preparation.

2.2.4 Backing Strips on Groove Welds

Various forms of backing materials or strips can be applied to the back side of plate weldments. Temporary strips typically have a groove machined directly into them that controls the shape and size of the penetration material once it cools. Backing strips can be purchased in rolls from which the desired length can be cut and then taped directly to the back of the prepared joint. If the backing strip is designed to become part of the permanent weldment, it must be made of material similar to the alloy or metal being welded. If the backing strip is to be removed, it normally is made of dissimilar metal to prevent it from being welded into the joint. The backing strip can then be peeled from the back side of the finished weldment. Backing strips of copper, for example, are often used on weldments of stainless steel.

> **WPS**
>
> The qualifying WPS that specifies the tack-welding procedures, as well as the consumables to be used, is typically designed as part of the engineering plans. It is not normally a function or responsibility of the welder. However, the welder must be qualified to perform the tasks and follow the procedures specified in the WPS.

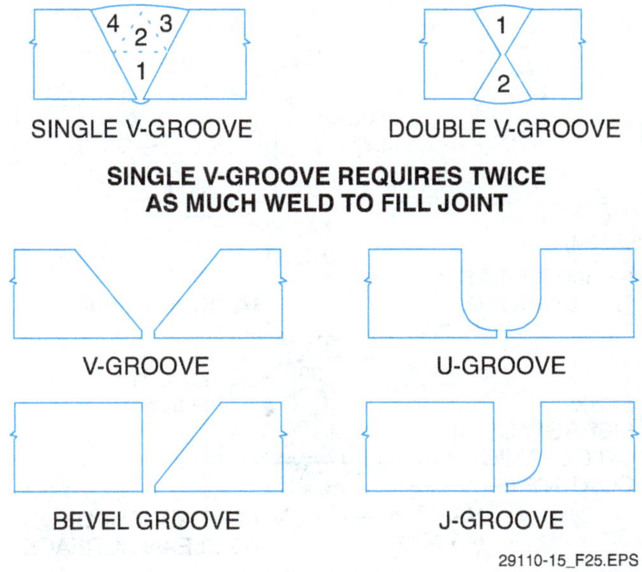

Figure 25 Types of groove styles.

When thick metal backing strips are to be used on groove welds, the root opening normally opens to ¼" (6.4 mm). The bevel is reduced to 22½ degrees (a total angle of 45 degrees) as shown in *Figure 26*. A root face is not required because the backing will prevent melt-through. Distortion can be reduced considerably through this control of the size and shape of the penetration.

2.2.5 Open Root Pipe Welds

It is important to note here that, in pipe welding, the alignment of the inside diameter is the most important, as opposed to the alignment of the outside diameter. In some cases, counter-boring of the pipe can be used to achieve alignment. However, there are limitations to counter-boring, since a minimum wall thickness must be maintained.

Molten metal that extends beyond the back or opposite side of the groove is called root reinforcement. Controlling the amount and shape of this root reinforcement is important to the function of the weldment. It is also important because controlling the root reinforcement helps reduce distortion by controlling the weld material uniformly. When welding pipe, the amount of root reinforcement on the inside of the pipe must be controlled. Excessive root reinforcement in the interior of the pipe interferes with the flow within the pipe, causing turbulence, pressure drop, and possibly other problems. Controlling this melt-through is also necessary to reduce stress on joints. However, it is difficult, if not impossible, to gain access to the interior of the pipe. Various methods are used to provide backing and to control penetration in pipe welding.

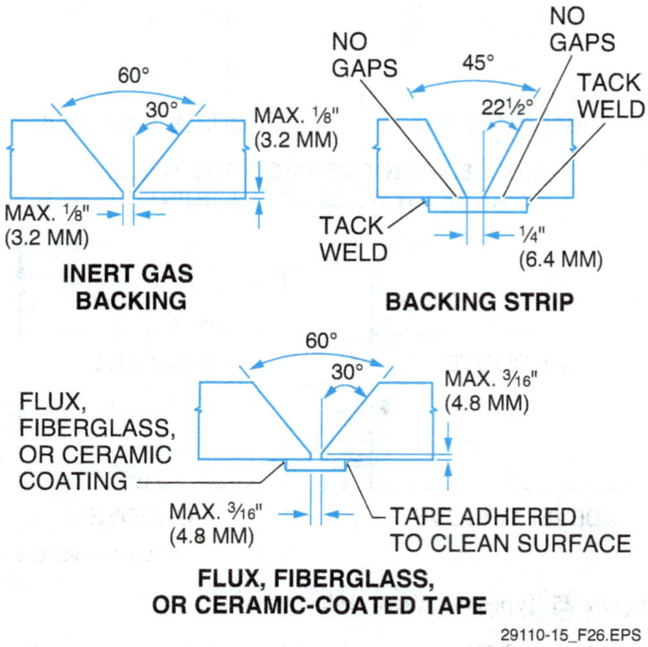

Figure 26 Groove welds with backing strip.

Gas backing using gases such as nitrogen or argon is one method used when welding pipe joints. When gas backing is used, some method of containing the gas inside the pipe must be used (*Figure 27*). For a short section of pipe, the ends of the pipe can be capped by taping or clamping a metal disk over the ends. For larger or longer pipe sections, however, this method may not be feasible due to the amount of gas that would be required to fill the pipe. For these cases, inflatable bladders, water-soluble plugs, or soft plastic bags can be inserted into the pipe near the weld joint. The inflatable bladders can be fished out of the pipe after the weld cools. The water-soluble plugs or soft plastic bags can either be blown or washed out of the pipe after the weld cools.

CAUTION

The pipe end opposite the end into which gas enters must have a vent hole to purge the gas and prevent pressure from building up. What is desired is an inert gas atmosphere in the pipe, rather than gas under pressure.

The type of gas used will depend on the base metal of the pipe being welded. Generally, nitrogen or carbon dioxide is used for carbon steel piping; argon is used for stainless steel, aluminum, and alloy steel piping.

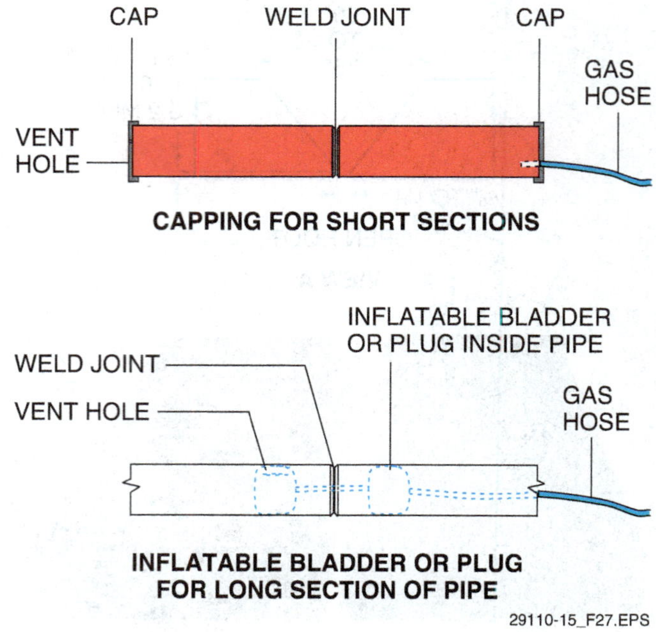

Figure 27 Gas backing.

2.2.6 Backing Rings

Backing rings (*Figure 28*) are flat metal strips that have been rolled to fit inside a pipe. They can be ordered in a variety of base metal types to match the base metal being welded. Some of these rings are split, making them easier to insert and adjust to piping that may be slightly out of round. Backing rings are available that have three or more nubs around the outside of the ring. The diameter of the nub is the root opening required for the pipe diameter being welded. Nubs can be pegs, buttons, or indents punched into the ring. Pegs must be removed after the joint is tack welded. They are removed by striking them with a chipping hammer; they snap off rather easily. Buttons can be removed or left in position to be melted during the root pass.

The joint preparation for backing rings (*Figure 29*) is similar to the preparation for open-root joints. The joint angle should be 60 or 75 degrees with a root opening and root face up to ⅛" (3.2 mm), or as specified on the WPS. If backing rings with nubs are used, the root opening will be set automatically by making sure the beveled ends of the pipe are positioned firmly against the nubs.

2.2.7 Consumable Inserts on Pipe Welds

Consumable inserts are similar to backing rings. As the name implies, they are completely consumed during welding and become part of the weld metal.

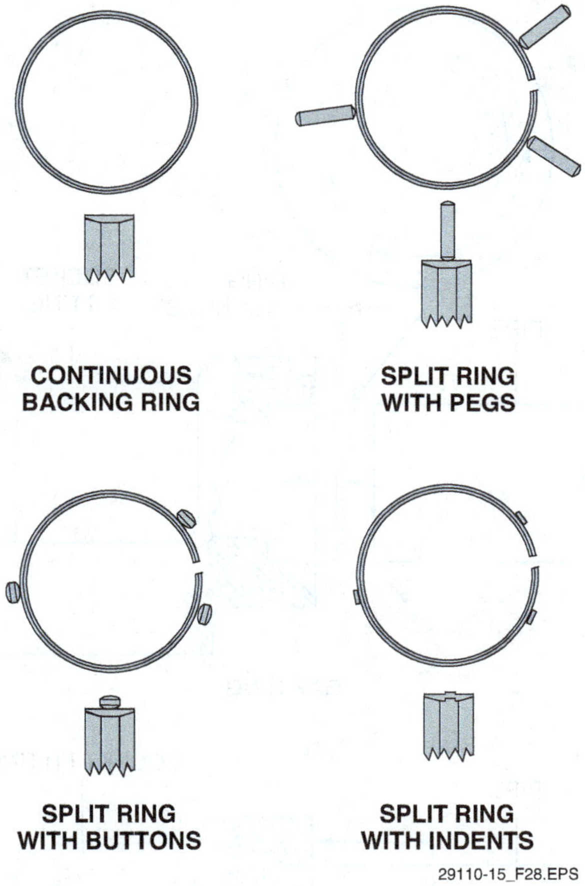

Figure 28 Examples of backing rings.

Because they are part of the finished weld, they must match the filler metal requirements for the weld being made. Gas backing often is required when using consumable inserts. Consumable inserts come in a variety of shapes and are identified by class numbers, as shown in *Figure 30.*

Consumable inserts can be purchased in coils and in preformed split rings. The rectangular Class 3 insert is also available as a solid ring.

Joint preparation for consumable inserts (*Figure 31*) can consist of a V-groove or a J-groove, with the V-groove having a joint angle of 75 degrees and a root face of 1⁄16" (1.6 mm). The J-groove joint angle is 20 degrees, also with a 1⁄16" (1.6 mm) root face. The consumable insert usually is tacked against the prepared ends of the pipe. The WPS will specify the tacking requirements.

2.2.8 Inserts on Socket Joints

Socket joints are generally used on pipe that is 5" (DN125) or smaller in diameter. A socket joint uses a prefabricated fitting containing sockets on the ends; the pipe slips into the sockets. The fitting and pipe are joined using a fillet weld.

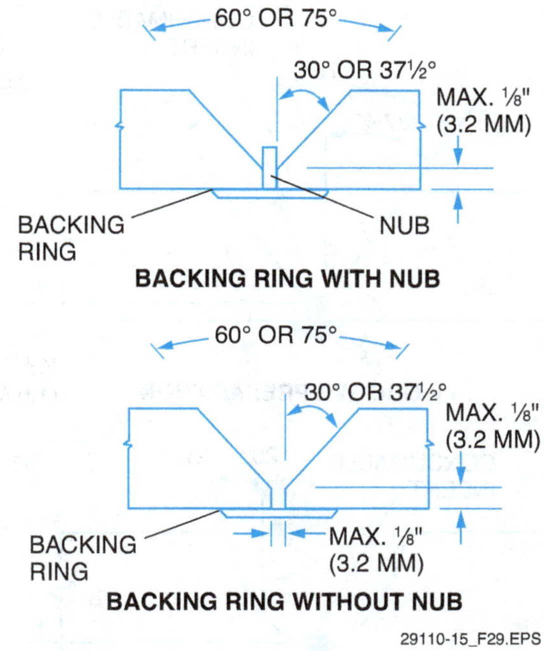

Figure 29 Pipe joint with backing ring.

All the common fittings, such as elbows, flanges, couplings, reducers, and even valves, are available as socket fittings. Using socket joints is a quick and easy method for joining pipe evenly.

In order to eliminate stress, distortion, and possible cracking by expansion, the end of the pipe must not touch the bottom of the socket. Depending on the code, socket welds require a 1⁄16" to ⅛" (1.6 to 3.2 mm) gap between the end of the pipe and the bottom of the socket fitting.

When preparing the joint, there are two ways to ensure that the gap between the end of the pipe and the socket fitting is maintained.

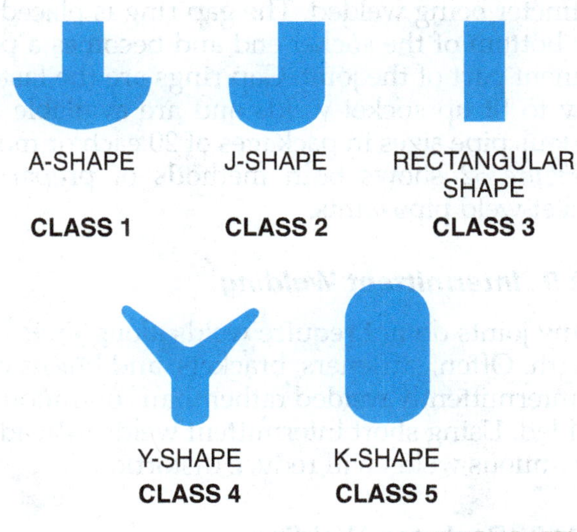

Figure 30 Consumable insert profiles.

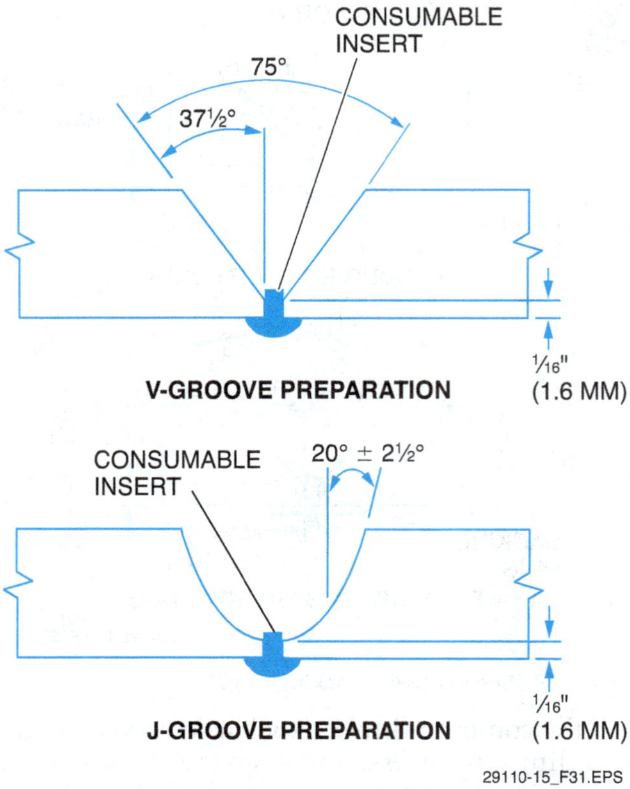

Figure 31 Joint preparation for consumable inserts.

One way, described earlier, is to scribe the pipe a preset distance from the end. The gap can be checked by measuring from the scribed line to the socket and then adding the socket depth. The scribed lines also can be used to check setup after the joint is tacked up or welded.

The second method of ensuring the correct gap is to use a gap ring. A gap ring is a split ring that is formed to the gap required for the pipe diameter being welded. The gap ring is placed in the bottom of the socket end and becomes a permanent part of the joint. Gap rings are the fastest way to fit up socket welds and are available for various pipe sizes in packages of 20 each or more.

Figure 32 shows both methods of preparing socket-weld pipe joints.

2.2.9 Intermittent Welding

Many joints do not require welds along their full length. Often, stiffeners, brackets, and braces can be intermittently welded rather than continuously welded. Using short intermittent welds instead of continuous welds will reduce distortion.

2.2.10 Backstep Welding

Backstep welding (*Figure 33*) is a welding technique in which the general progression of welding is from left to right, but the weld beads are deposited in short increments from right to left.

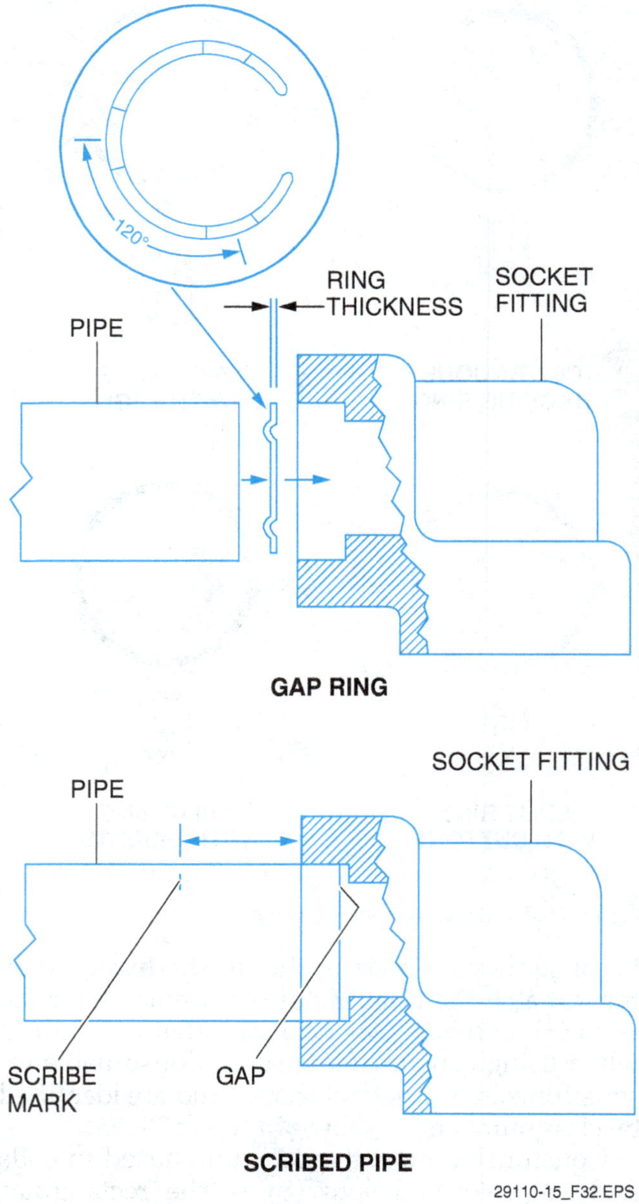

Figure 32 Socket-weld joint preparation.

This technique reduces distortion by minimizing and interrupting the heat input.

2.2.11 Welding Sequence

A welding sequence involves placing welds at different points on a weldment so that shrinkage forces in one location are counteracted by shrinkage forces in another location. Welding sequences can be quite complicated for large complex weldments or quite simple for less complicated weldments. When possible, use a welding sequence because it is a very effective means of controlling distortion. A simple welding sequence is to make short welds on alternating sides of the joint. This can be used for fillet or groove, intermittent, or continuous welding. *Figure 34* shows examples of intermittent welding sequences.

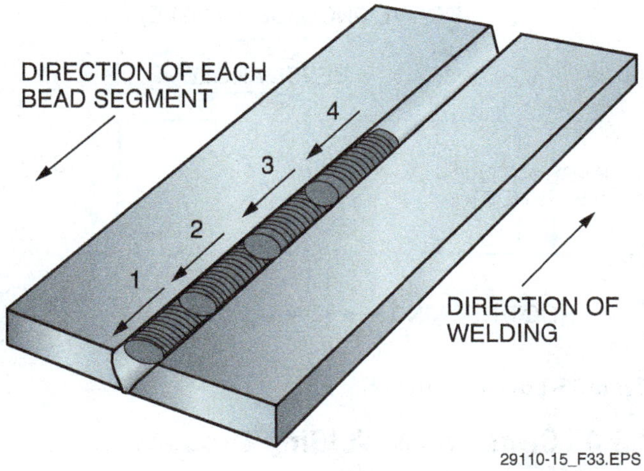

Figure 33 Backstep welding.

Welding sequences are often performed by two individuals welding on opposite sides of a joint at the same time. This type of welding is often called buddy welding and is a very effective way to control distortion.

2.2.12 Heat Treatments

Distortion can be controlled by preheating and postheating. Applying heat before welding is called preheating. Applying heat immediately after welding is called postheating. Preheating and postheating will be covered in detail in another module.

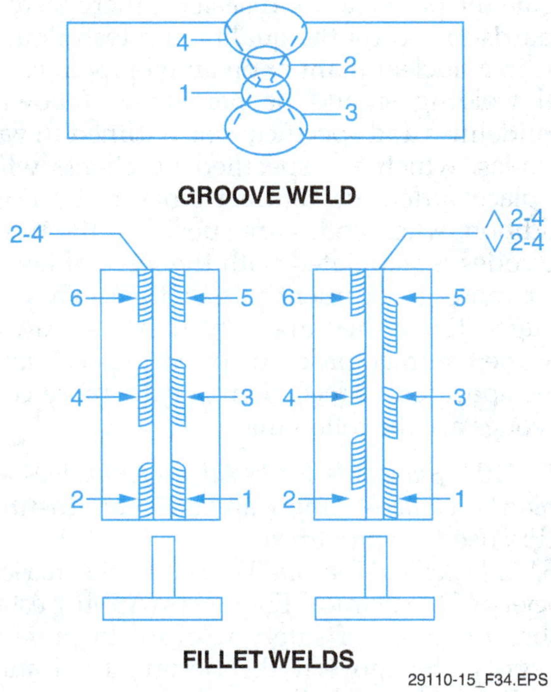

Figure 34 Intermittent welding sequences.

Welding must be stopped periodically to allow inspection of the weld in progress. These hold points should be scheduled strategically. Any work that will be obstructed after welding is complete, and therefore will not be easily inspected, must be inspected before the weld covers it. The following are among the items to be checked:

- Preheat temperature
- Postheat temperature
- Compliance with WPS requirements
- Weld root pass
- Weld layers

2.2.13 Checking Joint Misalignment and Fit-Up

The quality of joint preparation and fit-up directly affects the quality of the completed weld. After a weld is made, it is very costly and time-consuming to repair a defect. By thoroughly checking the joint fit-up, potential problems can be avoided. Follow these steps to check a fit-up:

Step 1 Determine if the weld is covered by a WPS. If the weld has a WPS, obtain a copy and follow it to make the weld.

Step 2 Check that the base metal type and grade are as specified. If either the base metal type or grade is wrong, contact your supervisor.

Step 3 Check, if required, that the welder who will perform the welding is certified/qualified to the WPS.

Alternating Welds

When possible, an alternating weld should be placed directly adjacent to or across from its matching weld on the opposite side. On lengthy welds, however, this sequence may not be advisable because it could leave long runs unwelded in between these welds. In these cases, stagger the welds from side to side.

If welding alternately on either side of the joint is not possible, or if one side must be completed first, a joint preparation may be used that deposits more weld metal on the second side. The greater contraction resulting from depositing more weld metal on the second side will help counteract the distortion on the first side.

Step 4 Check that the joint surfaces are free of contamination such as grease, oil, moisture, and rust. Also, check the surfaces parallel to the root and face of the weld. If there is contamination, clean the joint before continuing.

Step 5 Check the joint surface for cracks or laminations. If you find either, contact your supervisor.

Step 6 Check that the edge preparation is as specified for the joint (*Figure 35*). Check the following:

- Groove type
- Root opening
- Root face
- Included angle
- Bevel angle
- Base metal thickness

Step 7 Check the fit-up of backing strips, backing rings, and consumable inserts if required. Backing strips and backing rings must be tight against the joint along or around its entire length with even root spacing. Consumable inserts must fit tightly at the root of the weld. Correct any fit-up problems before proceeding.

Step 8 Check that the parts of the weldment are square, level, and plumb, and check that any angles are as specified. Correct any problems before proceeding.

Step 9 Check that the tack welds have been cleaned and feathered. Correct any problems before proceeding.

Step 10 Check that the welding process to be used is as specified on the WPS.

Step 11 Check that the consumables (type and size) are as specified on the WPS. Also check that arrangements have been made for the heated storage of consumables if required on the WPS.

Step 12 Check that the welding sequence is available if required.

Step 13 Check that the welding machine settings are as specified on the WPS.

Step 14 Check that arrangements have been made for preheating, interpass temperature control, and postheating if required on the WPS.

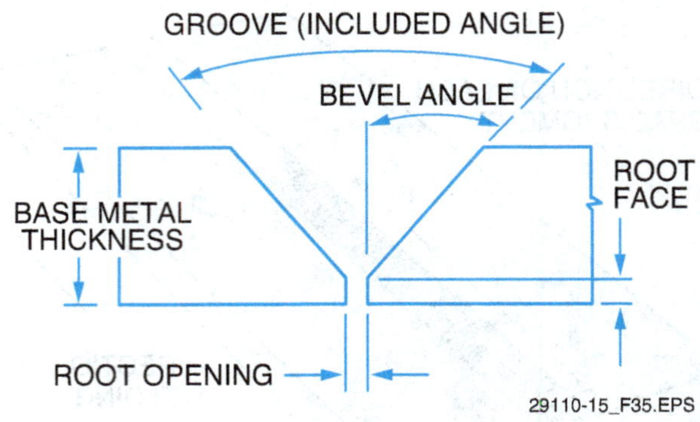

Figure 35 Edge preparation.

2.3.0 Codes and Welding Procedure Specifications

Whenever a bridge, building, ship, or pressure vessel is welded, the manufacturer and the buyer must reach agreement on how each weld will be made. To eliminate the need to write a new code for each job, government agencies, societies, and associations have developed codes. These codes are used universally to ensure safety and quality when welds are made.

2.3.1 Governing Codes and Standards

A welding code is a detailed listing of the rules and principles that apply to specific welded products. Codes ensure that safe and reliable welded products will be produced and that persons associated with the welding operation will be safe. Imagine the potential for disaster if there were no standards to control the quality and technique of welds in a nuclear plant or on an oil pipeline.

All welding should be performed following the guidelines and specifications outlined in various codes, which are specified by clients when they place orders or contract work to be done. In addition, when codes are specified, the use of these codes is mandated with the force of law by one or more government jurisdictions. They are not suggestions; they are requirements. Always check the contract, order, or project specification for the specified code(s). Some of the more common codes are the following:

- *API 1104, Standards for Welding of Pipelines and Related Facilities* – American Petroleum Institute (API); used for pipelines.
- *ASME Boiler and Pressure Vessel Code* – American Society of Mechanical Engineers (ASME); establishes the rules of safety relevant to pressure integrity that govern the design, fabrication, and inspection of boilers and pressure vessels, including nuclear power plant components.

- *ASME B31.1, Power Piping* – American Society of Mechanical Engineers (ASME); used for pressure piping.
- *ASME B31.3, 2006 Process Piping* – American Society of Mechanical Engineers (ASME); used for process piping in refineries, chemical and pharmaceutical, and similar plants.
- *ABS* – American Bureau of Shipping & Coast Guard; sets standards of excellence in Marine and Offshore Classifications. (see: **http://www.eagle.org/company/overview.html**).
- *AWS D1.1, Structural Welding Code* – American Welding Society (AWS); commonly specified for buildings and civil structures of all types for welding standard carbon and low alloy structured steel materials
- *AASHTO/AWS D1.5, 2008 Bridge Welding Code* – American Welding Society (AWS); used for welded bridges.

2.3.2 Code Changes

Periodically, codes are reviewed and changes made. Normally this occurs yearly, but some codes may go more than one year before they are updated. Most codes that are extremely broad in scope may undergo changes one section or category at a time. When a code is updated, the entire code is reissued with an updated year suffix, or **addendum** sheets for the areas of the code affected by the changes are issued. Changes are typically noted to alert readers to what has changed from the prior edition. The AWS code indicates additions to the code with a double vertical line in the margin alongside the changed area. Editorial changes to the AWS code are indicated by a single vertical line.

The ASME updates the entire code every three years. An addendum is issued annually. The annual addendum is identified by placing the letter A in front of the year on the cover of the code. API updates their entire code every five years.

It is important to recognize code changes. When a client specifies a code year, the code for that year, and that year only, is to be used unless otherwise specified by the client. When referring to a code, be sure the year matches the specifications for the job.

2.3.3 Welding Procedure Specifications

A WPS is a written set of instructions for producing sound welds in a prescribed manner. Each WPS is written and tested in accordance with a particular welding code or specification and must be in accordance with industry practice. Note that all welding requires that acceptable industry standards be followed, but not all welds require a WPS. If a weld does require a WPS, the WPS must be followed. The consequences of not following a required WPS are severe. Consequences include producing an unsafe weldment that could endanger life as well as the rejection of the weldment and lawsuits. Even if an unsafe condition is not known to exist, a structure known to contain numerous welds that were not done according to the WPS may have to be destroyed and redone. The economic impact can be devastating and cost welders their jobs.

When it is required, always follow the WPS. The requirement for the use of a WPS is often listed on job blueprints as a note or in the tail of the welding symbol. If you are unsure whether the welding being performed requires a WPS, do not proceed until you check with your supervisor. As a welder, you have a responsibility to follow the WPS and the applicable codes any time you are on the job. Your employer and fellow workers depend on you to carry that responsibility without deviation.

Each WPS is written by an individual who knows welding codes, specifications, and acceptable industry practices. It then becomes the responsibility of each manufacturer or contractor to test and qualify the WPS before using it. The WPS is tested by welding test coupons. Then, the coupons are tested according to the applicable code. For shielded metal arc welding (SMAW) of complete-penetration groove welds, the testing required includes nondestructive testing (NDT), tensile strength tests, and root, face, or side bend tests. The results of the testing are recorded on a Procedure Qualification Record (PQR). The WPS and PQR must be kept on file.

Many different formats are used for WPSs, but they all contain the same essential information. *Figure 36* shows a portion of a typical WPS.

Information typically found on a WPS includes the following:

- *Scope* – The welding process to which the WPS applies as well as the governing code and/or specification to be used.
- *Base metal* – The chemical composition or specification of the applicable base metal using the industry-standard identification; for example, A36 for carbon steel.
- *Welding process* – The welding process to be used to make the weld. For example, SMAW, gas metal arc welding (GMAW), flux-core arc welding (FCAW), or gas tungsten arc welding (GTAW).
- *Filler metal* – The composition, identifying type, or classification of the filler metal to be used.

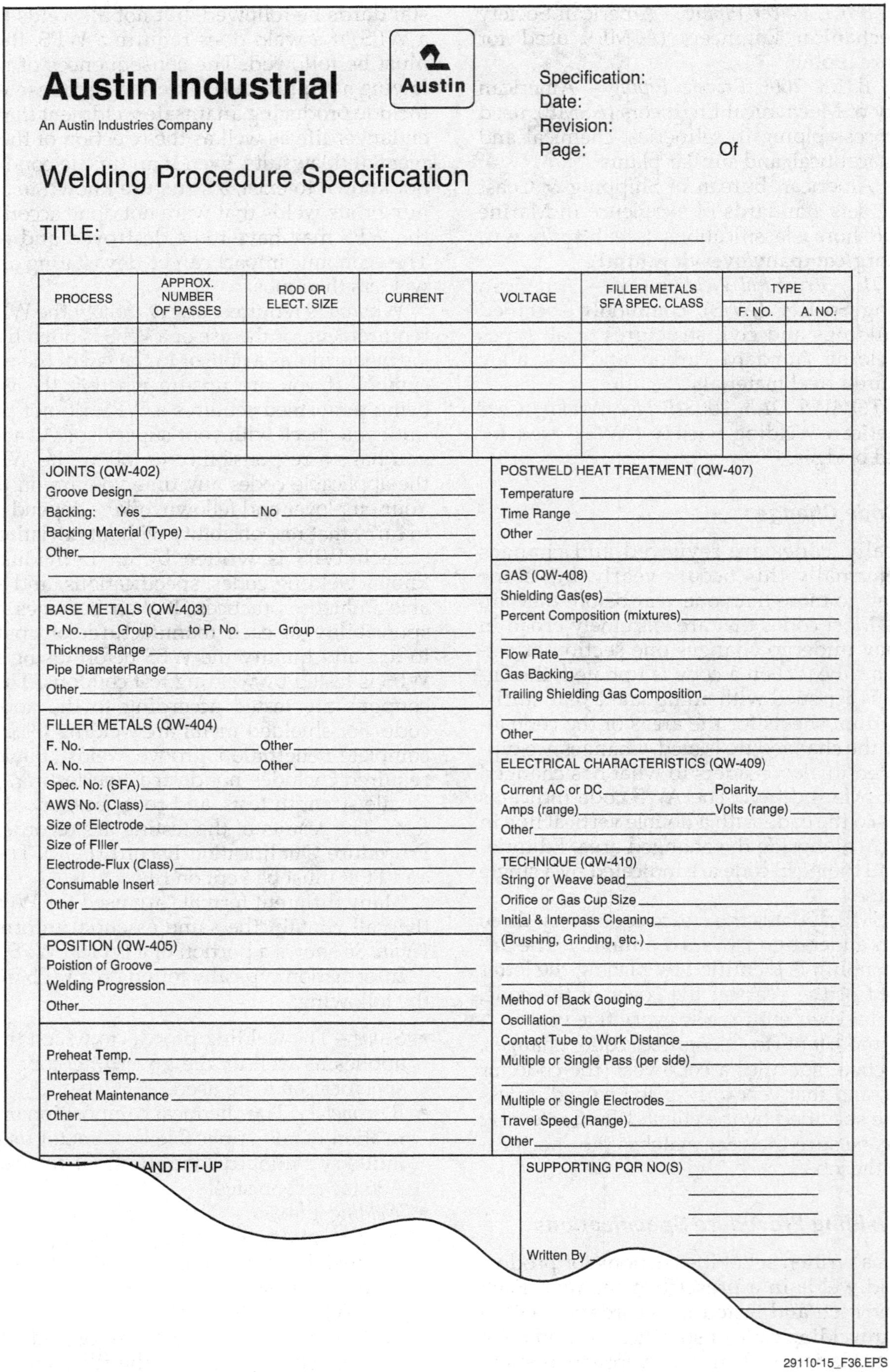

Austin Industrial

An Austin Industries Company

Austin

Specification:
Date:
Revision:
Page: Of

Welding Procedure Specification

TITLE:

PROCESS	APPROX. NUMBER OF PASSES	ROD OR ELECT. SIZE	CURRENT	VOLTAGE	FILLER METALS SFA SPEC. CLASS	TYPE	
						F. NO.	A. NO.

JOINTS (QW-402)
Groove Design
Backing: Yes No
Backing Material (Type)
Other

BASE METALS (QW-403)
P. No. Group to P. No. Group
Thickness Range
Pipe DiameterRange
Other

FILLER METALS (QW-404)
F. No. Other
A. No. Other
Spec. No. (SFA)
AWS No. (Class)
Size of Electrode
Size of FIller
Electrode-Flux (Class)
Consumable Insert
Other

POSITION (QW-405)
Poistion of Groove
Welding Progression
Other

Preheat Temp.
Interpass Temp.
Preheat Maintenance
Other

AND FIT-UP

POSTWELD HEAT TREATMENT (QW-407)
Temperature
Time Range
Other

GAS (QW-408)
Shielding Gas(es)
Percent Composition (mixtures)
Flow Rate
Gas Backing
Trailing Shielding Gas Composition
Other

ELECTRICAL CHARACTERISTICS (QW-409)
Current AC or DC Polarity
Amps (range) Volts (range)
Other

TECHNIQUE (QW-410)
String or Weave bead
Orifice or Gas Cup Size
Initial & Interpass Cleaning
(Brushing, Grinding, etc.)
Method of Back Gouging
Oscillation
Contact Tube to Work Distance
Multiple or Single Pass (per side)
Multiple or Single Electrodes
Travel Speed (Range)
Other

SUPPORTING PQR NO(S)

Written By

29110-15_F36.EPS

Figure 36 Portion of typical WPS.

- *Type of current* – Either alternating current (AC) or direct current (DC). If DC is specified, the polarity must also be given.
- *Arc voltage and travel speed* – An arc voltage range. Ranges for travel speed for automatic processes are mandatory and are recommended for semiautomatic processes.
- *Joint design and tolerances* – Joint design details, tolerances, and welding sequences that are given as a cross-sectional drawing or as a reference to drawings or specifications.
- *Joint and surface preparation* – The methods that can be used to prepare joint faces and the degree of surface cleaning required.
- *Tack welding* – Details pertaining to tack welding. Tack welders must use the WPS.
- *Welding details* – The size of electrodes to use for different portions and positions of the joint, the arrangement of welding passes to fill the joint, and the pass width and weave limitations.
- *Positions of welding* – The welding positions that can be used; for example, 1G, 3F, 5G.
- *Peening* – The details of peening and the type of tool to be used if peening is permitted.
- *Heat input* – The details to control heat input.
- *Second-side preparation* – The method used to prepare the second side when joints are welded from two sides.
- *Postheat treatment* – The details about postheat treatment or a reference to a separate document.

Essential and Nonessential Variables on a WPS

A different WPS is required for each change to an essential variable. Some essential variables for welds made with SMAW include the following:

- Base-metal thickness (within range listed)
- Composition of base metal
- Strength and composition of filler metal
- Base-metal preheat temperature

Nonessential variables for welds made with SMAW include the following:

- Type of groove used
- Omission of backing material in butt joints
- Electrode size

Changes in these nonessential variables would not normally require a different WPS.

Additional Resources

2014 Technical Training Guide. Current Edition. Cleveland, OH. USA: The Lincoln Electric Company. **www.lincolnelectric.com**

Welding Handbook. Current Edition. Miami, FL: The American Welding Society.

Mathey Dearman Product Manuals. The Mathey Dearman Company. **www.matheydearman.com**

2.0.0 Section Review

1. Stainless steel is more likely to be affected by distortion than carbon steel because it has a ____.
   a. higher coefficient of thermal expansion and a lower specific heat per unit volume
   b. lower coefficient of thermal expansion and a lower specific heat per unit volume
   c. lower coefficient of thermal expansion and a higher specific heat per unit volume
   d. higher coefficient of thermal expansion and a higher specific heat per unit volume

2. If a WPS is being followed, it also applies to the ____.
   a. tack weld
   b. welding machine model
   c. ground lead
   d. alignment tool

3. Existing welding codes eliminate the need for ____.
   a. welder training and certification
   b. creating standards for each and every job
   c. Welding Procedure Specifications (WPS)
   d. quality control inspections

Summary

It is very important to perform proper joint fit-up and alignment to ensure an acceptable weld. The proper fit-up measuring devices and tools must be used to accomplish this. Welders must know how to compensate for welding distortion as well as how to check for poor fit-up after the weld has been completed. Even though many welders are not required to fit joints to be welded, especially pipe joints, many others are able to fit as well as they weld, adding to their value and versatility. When the welder is not responsible for fitting the joint, he or she must be familiar with the techniques and tools used to create a reliable joint that meets the specifications.

Government agencies, professional societies, and associations have written guidelines for joint fit-up and alignment. These guidelines not only help ensure quality welds, but also help ensure safe welds and welding environments. All fit-up procedures must embrace these guidelines that have been proven over the years through use and repeated testing. When specified, these codes must be followed under penalty of law.

Review Questions

1. Which tool is often fabricated on the job from small channel or angle iron?
   a. Straightedge
   b. Plumb bob
   c. Square
   d. Level

2. Before welding a 45-degree joint, check the angle of the joint with a _____.
   a. straightedge
   b. Hi-Lo gauge
   c. framing square
   d. combination square

3. The primary use of a Hi-Lo gauge is to check for _____.
   a. pipe joint misalignment
   a. I-beam joint alignment
   b. 45-degree joints
   c. 90-degree joints

4. Secure a chain hoist over a weldment to be raised with an approved sling wrapped around _____.
   a. conduit
   b. piping
   c. a structural member
   d. raceways

5. When using a strongback to hold two weldment parts that have been aligned and fitted up, the tack welds used to hold the strongback to the weldment parts should be placed on only _____.
   a. opposite sides of the strongback
   b. the notched side of the strongback
   c. the vertical sides of the strongback
   d. one side of the strongback

6. If a joint needs to be forced into alignment, yokes and wedges, or bolts, can be used with _____.
   a. strongbacks
   b. cage clamps
   c. peening
   d. a come-along

7. When working with pipe over 10" (DN250), OSHA requires _____.
   a. only specific filler metals be used
   b. a WPS
   c. chain hoists with a minimum capacity of 10 tons
   d. four-legged jack stands

8. The arched crossbar available for cage clamps is designed to _____.
   a. strengthen the clamp
   b. allow welds to be made under the clamp
   c. allow the clamp to be used on a many pipe sizes
   d. hold a section of plate to the pipe

9. Which of the following devices may *not* be allowed on some applications or projects?
   a. Strongbacks
   b. Cage clamps
   c. Chain clamps
   d. Pipe pullers

10. The stress locked in a weldment when it returns to the ambient temperature, and equal to the strength of the base metal, is called _____.
    a. base stress
    b. residual stress
    c. thermal conductivity
    d. coefficient of expansion

11. The more weld metal placed into a joint, the greater the forces of _____.
    a. stress
    b. magnetism
    c. shrinkage
    d. thermal conductivity

12. Socket-welded joints are generally found on _____.
    a. lightweight plate
    b. heavy plate
    c. pipe sizes 5" (DN125) or smaller
    d. pipe sizes larger than 6" (DN150)

13. A detailed listing of the rules and principles that apply to specific welded products is called a _____.
    a. work order
    b. procedure
    c. punch list
    d. welding code

14. The written set of instructions for producing sound welds in a prescribed manner is called a _____.
    a. contract
    b. Welding Procedure Specification (WPS)
    c. welding code
    d. PQR

15. Welding test coupons prepared for a job are tested according to the _____.
    a. applicable code
    b. WPS guidelines
    c. manufacturer's guidelines
    d. PQR

Trade Terms Quiz

Fill in the blank with the correct term that you learned from your study of this module.

1. _______ refers to the difference in the internal alignment of two pipe sections.
2. The stress remaining in a weldment as a result of heating and cooling is called _______.
3. The quantity of heat needed to raise one unit mass of a substance by one unit degree is called _______.
4. Preplaced filler metal called a(n) _______ is completely fused into the root of a joint during welding.
5. The term _______ refers to a vertical line that is perpendicular to the earth's horizontal axis.
6. A(n) _______ is extra information used to identify corrections or revisions made to documents and shared with document users.
7. Mechanically working metal with repeated hammer blows is called _______.
8. The _______ refers to the temperature of air surrounding an object or individual.
9. The term _______ refers to a line that is parallel to the earth's horizontal axis.
10. A weld made to hold weldment parts in alignment until a final weld is made is called a(n) _______.
11. _______ is the change in length, per unit length of material, for each 1°C of change in temperature.

Trade Terms

Addendum
Ambient temperature
Coefficient of thermal expansion (linear)
Consumable insert
High-low
Level
Peening
Plumb
Residual stress
Specific heat per unit volume
Tack weld

Trade Terms Introduced in This Module

Addendum: Supplementary information typically used to describe corrections or revisions to documents.

Ambient temperature: The room temperature of the atmosphere completely surrounding an object on all sides.

Coefficient of thermal expansion (linear): The change in length per unit length of material for a 1°C change in temperature.

Consumable insert: Preplaced filler metal that is completely fused into the root of the joint during welding, becoming part of the weld.

High-low: The discrepancy in the internal alignment of two sections of pipe.

Level: A line on a horizontal axis parallel with the earth's surface (horizon). Also refers to the tool used to check if an object is level.

Peening: The mechanical working of metal using repeated hammer blows, typically using a ball-shaped hammer head such as that found on a ball peen hammer. Peening may be done for both metallurgical and appearance reasons.

Plumb: A line on a vertical axis perpendicular to the earth's surface (horizon).

Residual stress: Stress remaining in a weldment as a result of heat.

Specific heat per unit volume: The quantity of heat required to raise one unit mass of a substance by one unit degree.

Tack weld: A weld made to hold parts of a weldment in proper alignment until the final weld is made.

Additional Resources

This module presents thorough resources for task training. The following resource material is suggested for further study.

2014 Technical Training Guide. Current Edition. Cleveland, OH. USA: The Lincoln Electric Company. **www.lincolnelectric.com**

Welding Handbook. Current Edition. Miami, FL: The American Welding Society.

Mathey Dearman Product Manuals. The Mathey Dearman Company. **www.matheydearman.com**

Figure Credits

Topaz Publications, Inc., Module Opener, Figures 7 (photo), 8, 24 (photo)

Courtesy of Mathey Dearman, Figures 1A, 14–16, 18, 19

Courtesy of The LS Starrett Company, Figure 1B

RIDGID®, Figure 2A

© Stanley Black & Decker, Figure 2B

G.A.L. Gage Co, Figures 3 (photo), 5 (photo)

Courtesy of Sumner Manufacturing Co., Figures 13, 17B

Courtesy of H & M Pipe Beveling Machine Company, Inc., Figure 17A

Section Review Answer Key

Answer	Section Reference	Objective
Section One		
1. c	1.1.3	1a
2. a	1.2.2	1b
3. b	1.3.2	1c
4. c	1.4.4	1d
Section Two		
1. d	2.1.0	2a
2. a	2.2.2	2b
3. b	2.3.0	2c

NCCER CURRICULA — USER UPDATE

NCCER makes every effort to keep its textbooks up-to-date and free of technical errors. We appreciate your help in this process. If you find an error, a typographical mistake, or an inaccuracy in NCCER's curricula, please fill out this form (or a photocopy), or complete the online form at **www.nccer.org/olf**. Be sure to include the exact module ID number, page number, a detailed description, and your recommended correction. Your input will be brought to the attention of the Authoring Team. Thank you for your assistance.

Instructors – If you have an idea for improving this textbook, or have found that additional materials were necessary to teach this module effectively, please let us know so that we may present your suggestions to the Authoring Team.

NCCER Product Development and Revision

13614 Progress Blvd., Alachua, FL 32615

Email: curriculum@nccer.org
Online: www.nccer.org/olf

❑ Trainee Guide ❑ Lesson Plans ❑ Exam ❑ PowerPoints Other ____________

Craft / Level: Copyright Date:

Module ID Number / Title:

Section Number(s):

Description:

Recommended Correction:

Your Name:

Address:

Email: Phone:

Glossary

Addendum: Supplementary information typically used to describe corrections or revisions to documents.

Alloy: A metal that has had other elements added to it that substantially change its mechanical properties.

Ambient temperature: The room temperature of the atmosphere completely surrounding an object on all sides.

Amperage: The unit of measure used for the intensity of an electric current; often abbreviated to amp or amps.

Annealed: To free from internal stress by heating and gradually cooling.

Arc blow: The deflection of the arc from its intended path due to magnetic forces that develop from welding current flow.

Arc burn: Burn to the skin produced by brief exposure to intense radiant heat and ultraviolet light.

Arc strike: A discontinuity consisting of localized melting of the base metal or finished weld caused by the initiation of an arc. It remains visible as a result of striking the arc outside the area to be welded and can lead to rejection of an otherwise good weld.

Austenitic: Containing austenite (a solid solution of carbon or of carbon and other elements in a ferrous alloy), added through heating.

Backfire: A loud snap or pop as a torch flame is extinguished.

Backing: A weldable or nonweldable material used behind a root opening to allow defect-free welding at the open root of a joint.

Base metal: Metal to be welded, cut, or brazed.

Bonded: The permanent joining of metallic parts to form an electrically conductive path that will assure electrical continuity and the capacity to safely conduct any current likely to be imposed on it.

Brazing: A method of joining metal using heat and a filler metal with a melting point above 842°F (450°C). Unlike welding, the base metal is not melted during the brazing process.

Carbon steel: An alloy of iron combining iron and usually less than 1 percent carbon to provide hardness.

Carbon-graphite electrode: An electrode composed of a mixture of soft amorphous carbon and hard graphite carbon that may be coated with copper.

Carburizing flame: A flame burning with an excess amount of fuel; also called a reducing flame.

Castings: Something cast; any article that has been cast in a mold.

Code: A document that establishes the minimum requirements for a product or process. Codes can be, and often are, adopted as laws.

Coefficient of thermal expansion (linear): The change in length per unit length of material for a 1°C change in temperature.

Concentric cable system: A-CAC configuration in which a unique combination fitting is used to connect the torch cable to welding power in order to enable compressed air passage through the power conductor.

Condensation: The process in which atmospheric water vapor condenses (returns to its liquid state) on a cool surface.

Conductor: A material that will support the flow of an electrical current. Copper wire is the most common conductor.

Consumable insert: Preplaced filler metal that is completely fused into the root of the joint during welding, becoming part of the weld.

Defect: A discontinuity or imperfection that renders a part of the product or the entire product unable to meet minimum acceptable standards or specifications.

Discontinuity: A change or break in the shape or structure of a part that may or may not be considered a defect, depending on the code.

Distortion: The expansion and contraction of welded parts caused by the heating and subsequent cooling of the weld joint.

Drag angle: Describes the travel angle when the electrode is pointing in a direction opposite to the welding bead's progression.

Drag lines: The lines on the edge of the material that result from the travel of the cutting oxygen stream into, through, and out of the metal.

Dross: A waste byproduct of molten metal; the material (oxidized and molten metal) that is expelled from the kerf when cutting using a thermal process. It is sometimes called slag.

Ductile: Able to undergo change of form without breaking.

Ductility: Refers to the mechanical property of a material that allows it to be bent or shaped without breaking.

Duty cycle: The percentage of a ten-minute period that a welding machine can continuously produce its rated amperage without overheating; the percentage of time a plasma arc cutting machine can cut without overheating within a ten-minute period.

Electric arc: The flow of an electrical current across an air gap or gaseous space.

Electrically grounded: Connected to Earth or to some conducting body that serves in place of the earth.

Electrode: The point from which a welding arc is produced.

Embrittled: Metal that has been made brittle and that will tend to crack with little bending.

Essential variable: Items in a welding procedure specification (WPS) that cannot be changed without requalifying the WPS.

Ferritic: Steel containing less than 0.10 percent carbon and is magnetic. This steel can't be hardened via heat treatment.

Ferrous: Containing iron.

Ferrous metals: Metals containing iron.

Flash burn: Burns to the eyes sometimes called welder's flash; caused by exposure to intense radiant heat and ultraviolet light.

Flashback: The flame burning back into the tip, torch, hose, or regulator, causing a high-pitched whistling or hissing sound.

Flux: A material used to dissolve or prevent the formation of oxides and other undesirable substances on a weld joint; material used to prevent, dissolve, or facilitate the removal of oxides and other undesirable substances on a weld or base metal.

Fume plume: The fumes, gases, and particles from the consumables, base metal, and base metal coating during the welding process.

Galvanized steel: Carbon steel dipped in zinc to inhibit corrosion.

Gouging: The process of cutting a groove into a surface.

Governor: A device that limits engine speed.

Hardenable materials: Metals that have the ability to be made harder by heating and then cooling.

Heat-affected zone: The part of the base metal that has been altered by heating, but not melted by the heat.

Hermetically sealed: Having an airtight seal.

High-low: The discrepancy in the internal alignment of two sections of pipe.

Homogeneity: The quality or state of having a uniform structure or composition throughout.

Immediate danger to life and health (IDLH): As defined by OSHA, this term refers to an atmosphere that poses an immediate threat to life, would cause irreversible adverse health effects, or would impair an individual's ability to escape from a dangerous atmosphere.

Inclusion: Foreign matter introduced into and remaining in a weld.

Kerf: The gap produced by a cutting process.

Laminations: Cracks in the base metal formed when layers separate.

Level: A line on a horizontal axis parallel with the earth's surface (horizon). Also refers to the tool used to check if an object is level.

Load: The amount of force applied to a material or a structure.

Low-hydrogen electrode: An electrode specially manufactured to contain little or no moisture.

Malleable: Capable of being extended or shaped by hammering or by pressure from rollers.

Martensitic: Steel that shares some characteristics with ferritic, but has a higher levels of carbon, up to a full 1 percent. It can be tempered and hardened and is used where strength is more important than a resistance to oxidation.

Melt-through: Complete joint penetration.

Motor-generator: A combination device in which the motor turns the shaft of a generator, which in turn produces an AC voltage.

Neutral flame: A flame burning with correct proportions of fuel gas and oxygen.

Non-essential variable: Items in a welding procedure specification (WPS) that can be changed without requalifying the WPS.

Nonferrous metal: A metal, such as aluminum, copper, or brass, lacking sufficient quantities of iron to have any effect on its properties.

Notch toughness: The ability of a material to absorb energy in the presence of a flaw such as a notch; the ability of a material to resist breaking at points where stress is concentrated.

Oscillation: A repetitive and consistent side-to-side motion.

Oxide: The scale that forms on metal surfaces when they are exposed to oxygen or air containing oxygen.

Oxidize: To combine with oxygen, such as in burning (rapid oxidation) or rusting (slow oxidation).

Oxidizing flame: A flame burning with an excess amount of oxygen.

Peening: The mechanical working of metal using repeated hammer blows, typically using a ball-shaped hammer head such as that found on a ball peen hammer. Peening may be done for both metallurgical and appearance reasons.

Phase: In a three-phase power supply system, the sine waves of the voltage on each of the three separate conductors is displaced from each of the others by 120 degrees, although all conductors are carrying alternating current at the same frequency and are synchronized. Each of the three sine waves is referred to as a phase of the power source.

Pierce: To penetrate through metal plate with an oxyfuel cutting torch.

Piping porosity: A form of porosity having a length greater than its width and that is approximately perpendicular to the weld face.

Plasma: A fourth state of matter (not solid, liquid, or gas) created by heating a gas to such a high temperature that it boils electrons off the gas molecules (ionization).

Plumb: A line on a vertical axis perpendicular to the earth's surface (horizon).

Polarity: Refers to the direction of electrical current flow in a DC welding circuit; the condition of a system in which it has opposing physical properties at different points, such as an electric charge.

Porosity: Gas pockets, or voids in the weld metal.

Potential: The relative electrical voltage difference between two points of reference; the electromotive force or difference in potential between two points, expressed in volts.

Procedure qualification record (PQR): The document containing the results of the nondestructive and destructive testing required to qualify a WPS.

Purge gas: An inert gas such as nitrogen used to drive oxygen away from a weld site.

Push angle: Describes the travel angle when the electrode is pointing in the same direction as the welding bead's progression.

Quench: To cool suddenly by plunging into a liquid; to rapidly cool a hot component such as a freshly welded coupon. Note that only coupons should be quenched; real welds with a function should never be quenched.

Radiographic: Describes images made by passing X-rays or gamma rays through an object and recording the variations in density on photographic film.

Rectifier: An electronic device that converts AC voltage to DC voltage.

Residual stress: Stress remaining in a weldment as a result of heat.

Restart: The action of restarting a weld bead that is already in progress that requires the crater left from the termination of the previous arc.

Root opening: The space between the base metal pieces at the bottom or root of the joint.

Shielding gas: A gas such as argon, helium, or carbon dioxide used to protect the welding electrode wire from contamination in GMAW and FCAW welding.

Soapstone: Soft, white stone used to mark metal.

Solenoid valve: A valve used to control the flow of gases or liquids that is opened or closed by the action of an energized electromagnet.

Specific heat per unit volume: The quantity of heat required to raise one unit mass of a substance by one unit degree.

Specification: A document that defines in detail the work to be performed or the materials to be used in a product or process.

Stainless steel: An iron-based alloy usually containing at least 11 percent chromium.

Standard: A document that defines how a code is to be implemented.

Step-down transformer: An electrical device that uses two wire coils of different sizes, referred to as the primary and secondary windings, to convert a higher voltage to a lower voltage through induction. Primary windings are connected to the primary power source; the secondary windings provide the reduced voltage to the process.

Stringer bead: A type of weld bead made without any significant weaving motion. With SMAW, stringer beads are not more than three times the diameter of the electrode.

Supplemental essential variable: The variables that must be considered when notch toughness requirements are invoked.

Surfacing: The application by welding, brazing, or thermal spraying of a layer of material to a surface to obtain desired properties or dimensions.

Tack weld: A weld made to hold parts of a weldment in proper alignment until the final weld is made.

Tempering: To impart strength or toughness to (steel or cast iron) by heating and cooling.

Tensile strength: The measure of the ability of a material to withstand a longitudinal stress without breaking.

Traceability: The ability to verify that a procedure has been followed by reviewing the documentation step-by-step.

Transducer: A device that converts one form of energy into another.

Ultraviolet (UV) radiation: Invisible rays capable of causing burns. UV rays from the sun are the causes of sunburn.

Underbead cracking: Cracking in the base metal near the weld, but under the surface.

Undercut: A defect that occurs when the cross-sectional thickness of the base metal is reduced at the edges of the weld. Excessive weld current can cause this as it causes the base metal at the edges to melt and drain into the weld puddle. A downward sloped edge toward the weld bead is the visual evidence.

Ventricular fibrillation: Irregular heart rhythm characterized by a rapid fluttering heartbeat.

Vertical welding: Welding with an upward or downward progression.

Washing: A term used to describe the process of cutting out bolts, rivets, previously welded pieces, or other projections from the metal surface.

Weathering steel: Steel alloy that, under specific conditions, is designed to form a very dense oxide layer on its outer surfaces, which retards further oxidation.

Weave bead: A type of weld bead made by transverse oscillation of the electrode in a particular pattern.

Weld axis: A straight line drawn through the center of a weld along its length.

Weld coupon: Metal pieces to be welded together for test or practice.

Welding procedure qualification: The demonstration that welds made following a specific process can meet prescribed standards.

Welding procedure specification (WPS): The document containing all the detailed methods and practices required to produce a sound weld.

Weldment: An assembly that is fastened together by welded joints.

Work angle: An angle less than 90 degrees between a line perpendicular to the major workpiece surface and a plane determined by the electrode axis and the weld axis. 0-degree work angle is common. In a T-joint or corner joint, the line is perpendicular to the non-butting member. The definition of work angle for a pipe weld is covered in a later module, as it differs from that of plate welding.

Wrought: Produced or shaped by beating with a hammer, as with iron.

Structural Fitter

Level One

Trainee Guide

PEARSON

Boston Columbus Indianapolis New York San Francisco Upper Saddle River
Amsterdam Cape Town Dubai London Madrid Milan Munich Paris Montreal Toronto
Delhi Mexico City São Paulo Sydney Hong Kong Seoul Singapore Taipei Tokyo

NCCER
President: Don Whyte
Director of Product Development: Daniele Dixon
Structural Fitter Project Manager: Chris Wilson
Senior Manager: Tim Davis
Quality Assurance Coordinator: Debie Ness
Desktop Publishing Coordinator: James McKay
Permissions Specialist: Megan Casey
Production Specialist: Megan Casey
Editor: Chris Wilson

Writing and development services provided by Topaz Publications, Liverpool, NY
Lead Writer/Project Manager: Tom Burke
Desktop Publisher: Joanne Hart
Art Director: Alison Richmond
Permissions Editors: Toni Burke, Andrea LaBarge
Writers: Tom Burke, Troy Staton, Veronica Westfall

Pearson Education, Inc.
Editorial Director: Vernon R. Anthony
Executive Editor: Alli Gentile
Editorial Assistant: Douglas Greive
Program Manager: Alexandrina B. Wolf
Operations Supervisor: Deidra M. Skahill
Art Director: Jayne Conte
Director of Marketing: David Gesell
Executive Marketing Manager: Derril Trakalo
Marketing Manager: Brian Hoehl
Marketing Coordinator: Crystal Gonzalez
Composition: NCCER
Printer/Binder: Document Technology Resources, Fredericksburg, VA
Cover Printer: Document Technology Resources, Fredericksburg, VA
Text Fonts: Palatino and Univers

Credits and acknowledgments for content borrowed from other sources and reproduced, with permission, in this textbook appear at the end of each module.

PEARSON

Perfect bound ISBN-13: 978-0-13-294864-7
ISBN-10: 0-13-294864-8

Preface

To the Trainee

Welcome to the structural fitter trade! Structural fitters lay out, fabricate, and position structural steel elements for a variety of vessel types. Bulkheads, plates, frames, stanchions, and other heavy metal parts within the hull of a ship are all assembled by structural fitters.

Structural fitters in the maritime industry need to have an eye for detail. One of the key tasks for any structural fitter is to be able to precisely fabricate structural systems according to specifications and plans provided to them. If you have an attention for detail and mechanical abilities, structural fitting may be for you.

The maritime industry is facing an aging workforce and shortage of skilled workers. This shortage extends to the structural fitter trade. Skilled structural fitters are in high demand in almost every shipyard. By completing this training you are gaining the abilities needed to help fill that demand. We hope you continue your training beyond this textbook to further your knowledge and skills in this field.

If you're training through an NCCER-Accredited Training Program Sponsor and successfully pass the module exams and performance tests in this course, you may be eligible for credentialing through NCCER's National Registry. Check with your instructor or local program sponsor to find out. To learn more, go to **www.nccer.org** or contact us at 1.888.622.3720.

We invite you to visit the NCCER website at **www.nccer.org** for information on the latest product releases and training, as well as online versions of the *Cornerstone* magazine and Pearson's NCCER product catalog.

Your feedback is welcome. You may email your comments to **curriculum@nccer.org** or send general comments and inquiries to **info@nccer.org**.

NCCER Standardized Curricula

NCCER is a not-for-profit 501(c)(3) education foundation established in 1996 by the world's largest and most progressive construction companies and national construction associations. It was founded to address the severe workforce shortage facing the industry and to develop a standardized training process and curricula. Today, NCCER is supported by hundreds of leading construction and maintenance companies, manufacturers, and national associations. The NCCER Standardized Curricula was developed by NCCER in partnership with Pearson, the world's largest educational publisher.

Some features of the NCCER Standardized Curricula are as follows:

- An industry-proven record of success
- Curricula developed by the industry for the industry
- National standardization providing portability of learned job skills and educational credits
- Compliance with the Office of Apprenticeship requirements for related classroom training *(CFR 29:29)*
- Well-illustrated, up-to-date, and practical information

NCCER also maintains a National Registry that provides transcripts, certificates, and wallet cards to individuals who have successfully completed a level of training within a craft in NCCER's Curricula. *Training programs must be delivered by an NCCER Accredited Training Sponsor in order to receive these credentials.*

Special Features

In an effort to provide a comprehensive, user-friendly training resource, we have incorporated many different features for your use. Whether you are a visual or hands-on learner, this book will provide you with the proper tools to get started in the Structural Fitter trade.

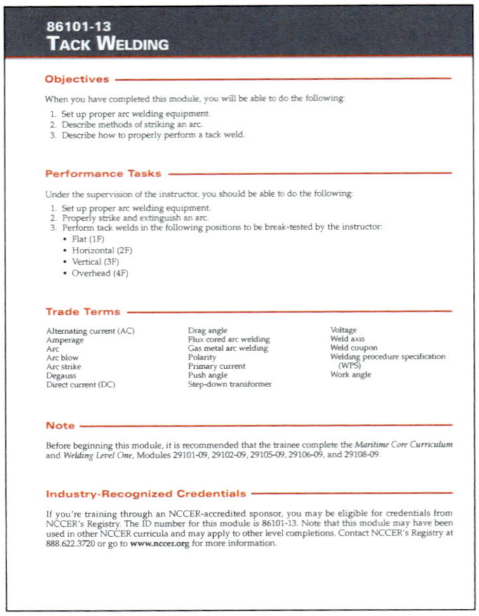

86101-13
TACK WELDING

Objectives

When you have completed this module, you will be able to do the following:

1. Set up proper arc welding equipment.
2. Describe methods of striking an arc.
3. Describe how to properly perform a tack weld.

Performance Tasks

Under the supervision of the instructor, you should be able to do the following:

1. Set up proper arc welding equipment.
2. Properly strike and extinguish an arc.
3. Perform tack welds in the following positions to be break-tested by the instructor:
   - Flat (1F)
   - Horizontal (2F)
   - Vertical (3F)
   - Overhead (4F)

Trade Terms

Alternating current (AC)	Drag angle	Voltage
Amperage	Flux cored arc welding	Weld axis
Arc	Gas metal arc welding	Weld coupon
Arc blow	Polarity	Welding procedure specification (WPS)
Arc strike	Primary current	Work angle
Degauss	Push angle	
Direct current (DC)	Step-down transformer	

Note

Before beginning this module, it is recommended that the trainee complete the *Maritime Core Curriculum* and *Welding Level One*, Modules 29101-09, 29102-09, 29105-09, 29106-09, and 29108-09.

Industry-Recognized Credentials

If you're training through an NCCER-accredited sponsor, you may be eligible for credentials from NCCER's Registry. The ID number for this module is 86101-13. Note that this module may have been used in other NCCER curricula and may apply to other level completions. Contact NCCER's Registry at 888.622.3720 or go to **www.nccer.org** for more information.

Introduction

This page is found at the beginning of each module and lists the Objectives, Performance Tasks, Trade Terms, and Required Trainee Materials for that module. The Objectives list the skills and knowledge you will need in order to complete the module successfully. The Performance Tasks give you an opportunity to apply your knowledge to the real-world duties that structural fitters perform. The list of Trade Terms identifies important terms you will need to know by the end of the module. Required Trainee Materials list the materials and supplies needed for the module.

Special Features

Features provide a head start for those entering the Maritime Pipefitting field by presenting technical tips and professional practices from craftworkers in various disciplines. These features often include real-life scenarios similar to those you might encounter on the job site.

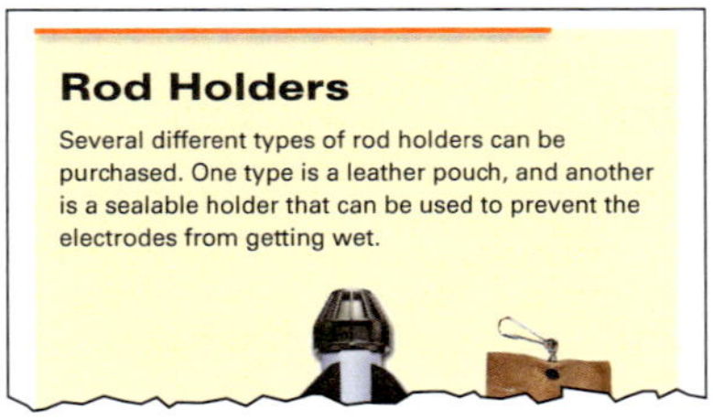

Color Illustrations and Photographs

Full-color illustrations and photographs are used throughout each module to provide vivid detail. These figures highlight important concepts from the text and provide clarity for complex instructions. Each figure reference is denoted in the text in *italics* for easy reference.

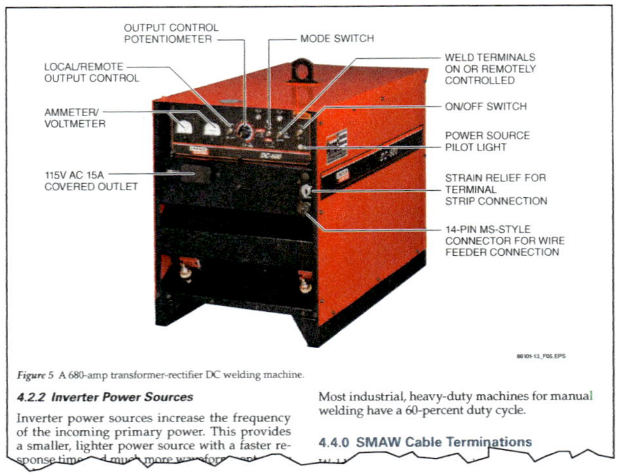

Notes, Cautions, and Warnings

Safety features are set off from the main text in highlighted boxes and are organized into three categories based on the potential danger of the issue being addressed. Notes simply provide additional information on the topic area. Cautions alert you of a danger that does not present potential injury but may cause damage to equipment. Warnings stress a potentially dangerous situation that may cause injury to you or a co-worker.

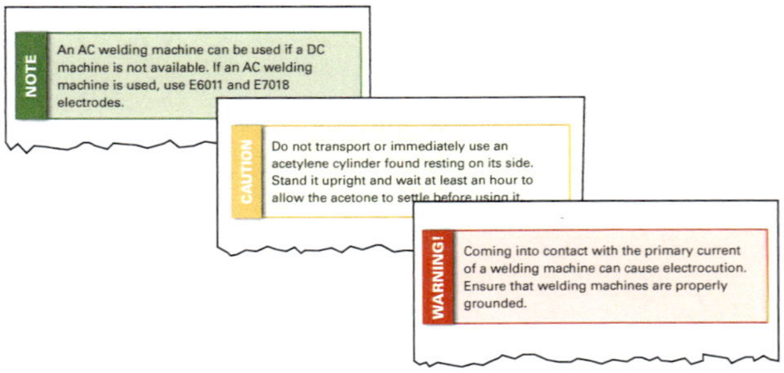

Case History

Case History features emphasize the importance of safety by citing examples of the costly (and often devastating) consequences of ignoring best practices or OSHA regulations.

Case History

Mexico City Refinery

In Mexico City in 1984, a weld on a liquid propane gas (LPG) tank cracked. The escaping gas cloud drifted toward a nearby residential area until it encountered an ignition source. When the gas ignited, the fireball quickly burned its way back to the leak source. The resulting explosion destroyed the refinery and many residential neighborhoods, leaving 503 people dead and more than 4,000 injured.

The Bottom Line:

Welds must be made correctly and inspected properly to prevent injury, loss of property, and sometimes death.

Going Green

Going Green looks at ways to preserve the environment, save energy, and make good choices regarding the health of the planet. Through the introduction of new construction practices and products, you will see how the "greening of America" has already taken root.

GOING GREEN

Test Coupons

Test coupons are routinely taken for testing in the welding industry. When those test coupons are no longer needed, they should be properly recycled, to save the company money and to keep salvageable items out of the landfills.

Step-by-Step Instructions

Step-by-step instructions are used throughout to guide you through technical procedures and tasks from start to finish. These steps show you not only how to perform a task but also how to do it safely and efficiently.

Follow these steps to make a vertical-up fillet tack weld:

Step 1 In the 1F position, tack two plates together to form a T-joint for the fillet tack weld coupon.

Step 2 Clamp or tack-weld the coupon in the vertical position.

Step 3 Run a number of ½" to ¾" long tack welds along the root of the joint (starting at the bottom) using an electrode (work) angle of approximately 45 degrees with a 0-degree to 10-degree push angle (*Figure 25*). Use a whipping motion by quickly raising the electrode about ¼" and then dropping it

Trade Terms

Each module presents a list of Trade Terms that are discussed within the text and defined in the Glossary at the end of the module. These terms are denoted in the text with **bold, blue type** upon their first occurrence. To make searches for key information easier, a comprehensive Glossary of Trade Terms from all modules is located at the back of this book.

Welding machines that produce only DC welding voltage and current can be further classified by the characteristics of the welding current they produce. If the welding current varies between the higher open-circuit voltage and the lower operating voltage, it is classified as a variable-voltage constant-current power source. This type of welding machine is used for SMAW. If open-circuit and operating voltages are nearly the same, the machine is classified as a constant-voltage or constant-potential DC welding machine. This type of machine is used for **gas metal arc welding (GMAW)** and **flux cored arc welding (FCAW)**.

4.2.0 Industrial SMAW Machine Types

Review Questions

Review Questions reinforce the knowledge you have gained and are a useful tool for measuring what you have learned.

Review Questions

1. SMAW is often referred to as _____.
   a. AC welding
   b. stick welding
   c. gap welding
   d. resistive welding

2. When working in a confined space, the oxygen and acetylene tanks must be left outside the space.
   a. True
   b. False

3. During SMAW welding, current jumps across the _____.
   a. workpiece clamp from the tip of the electrode to the surface of the workpiece
   b. gap from the tip of the electrode to the surface of the workpiece
   c. electrode holder from the tip of the electrode to the surface of the workpiece
   d. molten metal from the tip of the electrode to the surface of the workpiece

4. In AC current, the number of cycles completed in one second is called the _____.
   a. polarity
   b. travel
   c. frequency
   d. current value

5. Electrical current that has no frequency is known as _____.
   a. DC current
   b. AC current
   c. AC voltage
   d. 120VAC

6. Which of the following only applies to DC current?
   a. Frequency
   b. Polarity
   c. Amperage
   d. Current flow

7. The voltage in a welding machine is reduced or stepped down by a(n) _____.
   a. limiter
   b. generator
   c. alternator
   d. transformer

8. The duty cycle of a welding machine is based on a _____.
   a. 10-minute period
   b. 15-minute period
   c. 20-minute period
   d. 30-minute period

9. To prevent the lead from pulling loose from the welding machine terminal, welding cable lugs are generally _____.
   a. slotted
   b. soldered in place
   c. ringed
   d. insulated

10. When preparing to practice welding of fillet tack welds, the metal should be cut into _____.
    a. 3" × 12" rectangles for the base
    b. 4" × 12" rectangles for the base
    c. 4" × 4" squares for the base
    d. 12" × 12" squares for the base

11. The easiest method of striking an arc is the _____.
    a. scratching method
    b. grounding method
    c. weaving method
    d. tapping method

12. Arc blow is caused by _____.
    a. strong winds
    b. magnetic fields
    c. fluctuating weld current
    d. damaged electrodes

NCCER Standardized Curricula

NCCER's training programs comprise more than 80 construction, maintenance, pipeline, and utility areas and include skills assessments, safety training, and management education.

Boilermaking
Cabinetmaking
Carpentry
Concrete Finishing
Construction Craft Laborer
Construction Technology
Core Curriculum:
  Introductory Craft Skills
Drywall
Electrical
Electronic Systems Technician
Heating, Ventilating, and
  Air Conditioning
Heavy Equipment Operations
Highway/Heavy Construction
Hydroblasting
Industrial Coating and Lining
  Application Specialist
Industrial Maintenance
  Electrical and Instrumentation
  Technician
Industrial Maintenance
  Mechanic
Instrumentation
Insulating
Ironworking
Masonry
Millwright
Mobile Crane Operations
Painting
Painting, Industrial
Pipefitting
Pipelayer
Plumbing
Reinforcing Ironwork
Rigging
Scaffolding
Sheet Metal
Signal Person
Site Layout
Sprinkler Fitting
Tower Crane Operator
Welding

Maritime

Maritime Industry Fundamentals
Maritime Pipefitting
Structural Fitter

Green/Sustainable Construction

Building Auditor
Fundamentals of Weatherization
Introduction to Weatherization
Sustainable Construction
  Supervisor
Weatherization Crew Chief
Weatherization Technician
Your Role in the Green
  Environment

Energy

Alternative Energy
Introduction to the Power
  Industry
Introduction to Solar
  Photovoltaics
Introduction to Wind Energy
Power Industry Fundamentals
Power Generation Maintenance
  Electrician
Power Generation I&C
  Maintenance Technician
Power Generation Maintenance
  Mechanic
Power Line Worker
Power Line Worker: Distribution
Power Line Worker: Substation
Power Line Worker:
  Transmission
Solar Photovoltaic Systems
  Installer
Wind Turbine Maintenance
  Technician

Pipeline

Control Center Operations,
  Liquid
Corrosion Control
Electrical and Instrumentation
Field Operations, Liquid
Field Operations, Gas
Maintenance
Mechanical

Safety

Field Safety
Safety Orientation
Safety Technology

Management

Fundamentals of Crew
  Leadership
Project Management
Project Supervision

Supplemental Titles

Applied Construction Math
Careers in Construction
Tools for Success

Spanish Translations

Basic Rigging
  (Principios Básicos de
  Maniobras)
Carpentry Fundamentals
  (Introducción a la
  Carpintería, Nivel Uno)
Carpentry Forms
  (Formas para Carpintería,
  Nivel Trés)
Concrete Finishing, Level One
  (Acabado de Concreto,
  Nivel Uno)
Core Curriculum:
  Introductory Craft Skills
  (Currículo Básico:
  Habilidades Introductorias del
  Oficio)
Drywall, Level One
  (Paneles de Yeso, Nivel Uno)
Electrical, Level One
  (Electricidad, Nivel Uno)
Field Safety
  (Seguridad de Campo)
Insulating, Level One
  (Aislamiento, Nivel Uno)
Ironworking, Level One
  (Herrería, Nivel Uno)
Masonry, Level One
  (Albañilería, Nivel Uno)
Pipefitting, Level One
  (Instalación de Tubería
  Industrial, Nivel Uno)
Reinforcing Ironwork, Level One
  (Herreria de Refuerzo,
  Nivel Uno)
Safety Orientation
  (Orientación de Seguridad)
Scaffolding
  (Andamios)
Sprinkler Fitting, Level One
  (Instalación de Rociadores,
  Nivel Uno)

Acknowledgments

This curriculum was revised as a result of the farsightedness and leadership of the following sponsors:

Ameri-Force Industrial Services, Inc.
Bollinger Shipyards, Inc.
Huntington Ingalls
Newport Area CTC
Newport News Shipbuilding
Signal International
Tecnico Corporation
VT Halter Marine, Inc.

This curriculum would not exist were it not for the dedication and unselfish energy of those volunteers who served on the Authoring Team. A sincere thanks is extended to the following:

Frank Cunningham
Hector DeLeon
Bill Docalovich
Sterling Dolese
Brian Folse
Paul Hennis
Jeff McKenzie
Ronnie Parrish
Joe Sabol
Randolph Scott
Michelle Shirley
Terry Stead
Wesley Lane

NCCER Partners

American Fire Sprinkler Association
Associated Builders and Contractors, Inc.
Associated General Contractors of America
Association for Career and Technical Education
Association for Skilled and Technical Sciences
Carolinas AGC, Inc.
Carolinas Electrical Contractors Association
Center for the Improvement of Construction Management and Processes
Construction Industry Institute
Construction Users Roundtable
Construction Workforce Development Center
Design Build Institute of America
GSSC – Gulf States Shipbuilders Consortium
Manufacturing Institute
Mason Contractors Association of America
Merit Contractors Association of Canada
NACE International
National Association of Minority Contractors
National Association of Women in Construction
National Insulation Association
National Ready Mixed Concrete Association
National Technical Honor Society
National Utility Contractors Association
NAWIC Education Foundation
North American Technician Excellence
Painting & Decorating Contractors of America
Portland Cement Association
SkillsUSA®
Steel Erectors Association of America
U.S. Army Corps of Engineers
University of Florida, M. E. Rinker School of Building Construction
Women Construction Owners & Executives, USA

NCCER Business Partners

Contents

Module Seven

Fire Watch

Prepares a worker to perform fire watch duties in support of welding and flame cutting activities. Describes the classes of fires and the methods used to extinguish them, as well as the responsibilities of a person assigned as a fire watch. (Module ID 86102-14; 5 Hours)

Module Eight

Introduction to Structural Fitter Drawings

Covers fundamental skills needed to read fabrication drawings that are commonly used by structural fitters. Focuses on basic drawing elements such as title blocks, revision blocks, and drawing lines and introduces plan, elevation, and detail drawings. (Module ID 86103-14; 10 Hours)

Module Nine

Fitting I

Introduces layout tools, fitting tools, and fitting aids used to fit up and align plate joints. Incorporates hands-on tasks through which the beginning fitter will learn how to perform basic layout, alignment, and fit-up tasks. (Module ID 86104-14; 40 Hours)

Glossary

Index

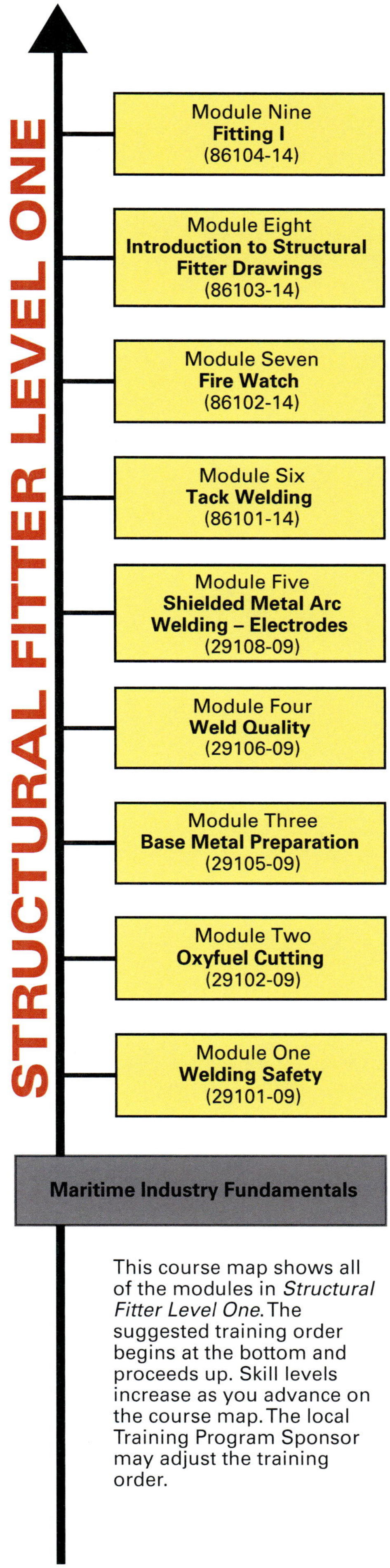

This course map shows all of the modules in *Structural Fitter Level One*. The suggested training order begins at the bottom and proceeds up. Skill levels increase as you advance on the course map. The local Training Program Sponsor may adjust the training order.

29101-09

Welding Safety

Module One

Trainees with successful module completions may be eligible for credentialing through NCCER's National Registry. To learn more, go to **www.nccer.org** or contact us at **1.888.622.3720**. Our website has information on the latest product releases and training, as well as online versions of our *Cornerstone* magazine and Pearson's product catalog.

Your feedback is welcome. You may email your comments to **curriculum@nccer.org**, send general comments and inquiries to **info@nccer.org**, or fill in the User Update form at the back of this module.

This information is general in nature and intended for training purposes only. Actual performance of activities described in this manual requires compliance with all applicable operating, service, maintenance, and safety procedures under the direction of qualified personnel. References in this manual to patented or proprietary devices do not constitute a recommendation of their use.

29101-09
Welding Safety

Objectives

When you have completed this module, you will be able to do the following:

1. Identify some common hazards in welding.
2. Explain and identify proper personal protection used in welding.
3. Describe how to avoid welding fumes.
4. Explain some of the causes of accidents.
5. Identify and explain uses for material safety data sheets.
6. Explain safety techniques for storing and handling cylinders.
7. Explain how to avoid electric shock when welding.
8. Describe proper material handling methods.

Performance Tasks

This is a knowledge-based module. There are no Performance Tasks.

Trade Terms

Bonded
Electrically grounded
Fume plume
Immediate danger to life and health (IDLH)
Purging gas
Ventricular fibrillation

Required Trainee Materials

1. Pencil and paper
2. Appropriate personal protective equipment

Prerequisites

Before you begin this module, it is recommended that you successfully complete *Core Curriculum.*

Industry-Recognized Credentials

If you're training through an NCCER-accredited sponsor, you may be eligible for credentials from NCCER's Registry. The ID number for this module is 29101-09. Note that this module may have been used in other NCCER curricula and may apply to other level completions. Contact NCCER's Registry at 888.622.3720 or go to **www.nccer.org** for more information.

Contents

Topics to be presented in this module include:

Figures and Tables

1.0.0 Introduction

Qualified welders are in high demand all over the world. The work of welders includes joining pipe sections for oil and natural gas pipelines, building ships, and working in a variety of industrial environments such as power plants, refineries, chemical plants, and manufacturing facilities.

Welding is the process of joining similar metals using a high-intensity electric arc at temperatures from 1,500°F to 3,000°F. The arc melts and fuses the base metals. In most cases a filler metal is melted along with the base metal in order to strengthen the joint. The finished joint is as strong as, or even stronger than, the base metals that have been joined. Welding is commonly done on carbon steel, stainless steel, and aluminum.

There is a lot more to welding than simply applying an electric arc to the base metals. Welders must know the characteristics of the metals being welded and the special materials and techniques that apply to the different metals. They must also know how to properly handle and store base metals, how to prepare metals for welding, and how to interpret welding specifications.

The most common type of welding is shielded metal arc welding, or SMAW. In the SMAW method, a flux-coated, consumable electrode is used to place the weld (*Figure 1*). SMAW is used primarily for welding iron and steel, including stainless steel.

Gas tungsten arc welding (GTAW), sometimes called TIG (tungsten inert gas) welding, uses a non-consumable tungsten electrode to produce the weld. GTAW is commonly used in welding stainless steel and light metals such as aluminum and magnesium.

Gas metal arc welding (GMAW), sometimes called MIG, or metal inert gas welding, uses a consumable wire that is fed through the welding gun to make the weld. GMAW is used on steel and other metals. It is widely used in industrial applications, especially in the manufacture of automobiles where robotic welding is common.

In addition to welding, welders are often required to cut steel using oxyfuel, air carbon arc, or plasma arc equipment (*Figure 2*). Oxyfuel cutting is the most common method. In the oxyfuel method, oxygen and a fuel gas are mixed in a cutting torch to produce a high-intensity flame that can cut through thick steel. Acetylene is the most common fuel gas because it produces the most intense heat. Propane, MAPP® gas, and butane are also used for cutting work. Plasma arc cutting relies on a jet of superheated gas that is passed through an electric

SMAW

101F01A.EPS

GMAW

101F01B.EPS

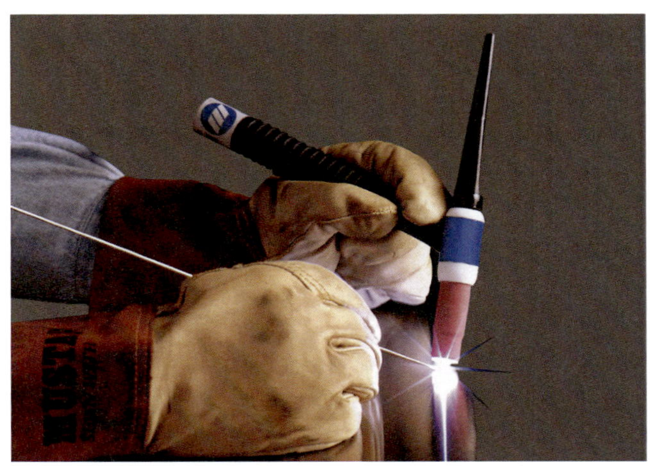

GTAW

101F01C.EPS

Figure 1 Welding methods.

OXYFUEL CUTTING

101F02A.EPS

MANUAL PLASMA ARC CUTTING

101F02B.EPS

CARBON ARC CUTTING

101F02C.EPS

Figure 2 Cutting methods.

arc. It is used to cut steel, as well as other metals. Carbon arc cutting uses an electric arc formed between a carbon electrode and the base metal to melt the base metal. Carbon arc cutting is often used in preparing a base metal for welding.

Welders work on critical structures such as bridges, high-rise buildings, ships, pipelines, and automobiles. As you can imagine, defective welds in any of these applications can lead to a disaster. For that reason, welding work is subjected to intense quality control inspections. The more critical the weld, the more intense the inspection is. Inspectors use a variety of methods to examine welds, including X-ray inspection.

In order to qualify as a welder, a person must undergo extensive training that requires many hours of practice. The training provided by this program will prepare you to weld plate and pipe in a variety of positions and to cut carbon steel plate and pipe using common cutting equipment and techniques.

There are numerous opportunities for welders in heavy construction, manufacturing, ship building, and auto racing. Once you have completed your training and have become a journey-level welder, you can advance in the trade in a variety of ways (*Figure 3*). The more training and education you complete, the greater the opportunities.

1.1.0 Standardized Training by NCCER

The National Center for Construction Education and Research (NCCER) is a not-for-profit education foundation established by the nation's leading construction companies. NCCER was created to provide the industry with standardized construction education materials, the NCCER Standardized Curricula (the welding modules are part of this series), and a system for tracking and recognizing students' training accomplishments—NCCER's National Registry.

NCCER also offers accreditation, instructor certification, and skills assessments. NCCER is committed to developing and maintaining a training process that is internationally recognized, standardized, portable, and competency-based.

Working in partnership with industry and academia, NCCER has developed a system for program accreditation that is similar to those found in institutions of higher learning. NCCER's accreditation process ensures that students receive quality training based on uniform standards and criteria. These standards are outlined in NCCER's *Accreditation Guidelines* and must be adhered to by NCCER Accredited Training Sponsors.

More than 550 training and assessment centers across the United States and eight other countries are proud to be NCCER Accredited Training Sponsors. Millions of craft professionals and construction managers have received quality construction education through NCCER's network of Accredited Training Sponsors and the thousands of Training Units associated with the Sponsors. Every year the number of NCCER Accredited Training Sponsors increases significantly.

A craft instructor is a journeyman craft professional or career and technical educator trained and certified to teach the NCCER Standardized Curricula. This network of certified instructors ensures that NCCER training programs will meet the standards of instruction set by the industry. There are more than 3,350 master trainers and 33,000 craft instructors within the NCCER instructor network. More information is available at **www.nccer.org**.

1.2.0 Apprenticeship Training

Formal apprenticeship programs conform to federal and state requirements under *CFR Titles 29:29 and 29:30*. All approved apprenticeship programs provide OJT as well as classroom instruction. The related training requirement is fulfilled by all NCCER craft training programs. The main difference between NCCER training and registered ap-

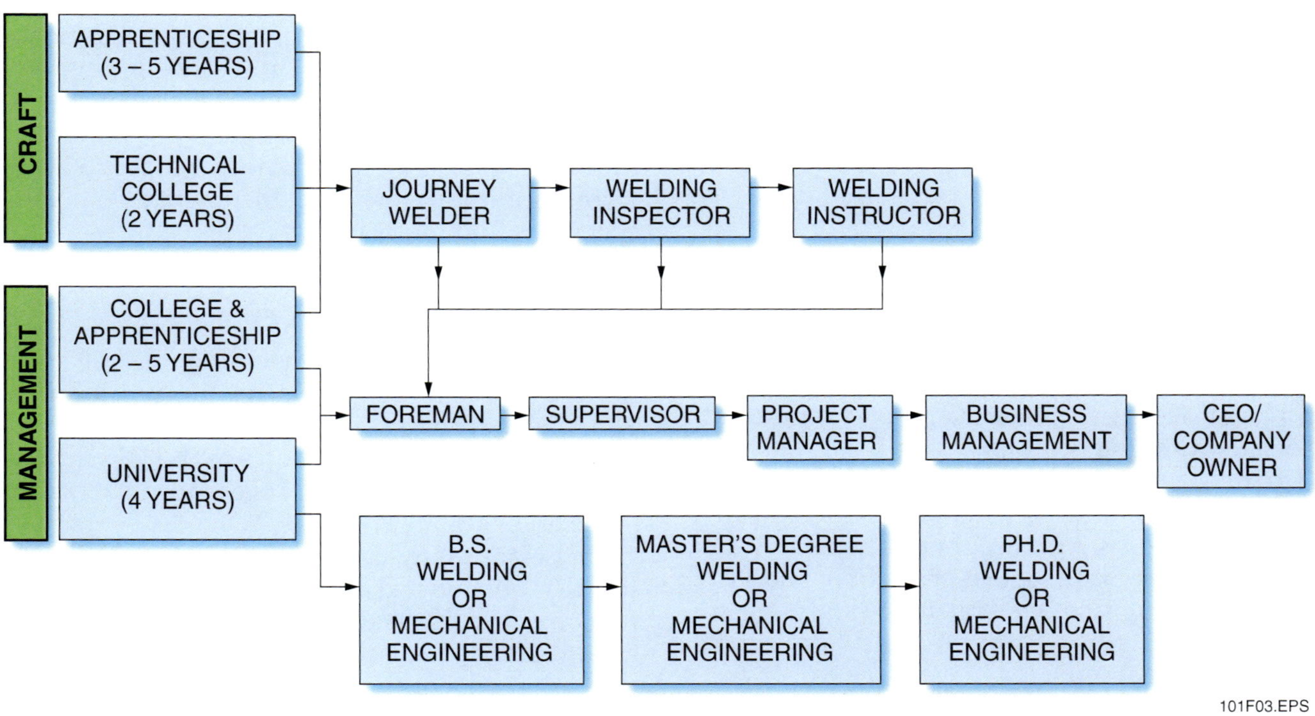

Figure 3 Opportunities in the welding trade.

prenticeship programs is that apprenticeship has specific time limits in which the training must be completed. Apprenticeship standards set guidelines for recruiting and outreach, and a specific time limit for each of a variety of OJT tasks. Additionally, there are reporting requirements and audits to ensure adherence to the apprenticeship standards. Companies and employer associations register their individual apprenticeship programs with the Office of Apprenticeship within the US Department of Labor, and in some instances, with state apprenticeship councils (SAC). OJT of 2,000 hours per year and a minimum of 144 hours of classroom-related training are required. Apprenticeship programs vary in length from 2,000 hours to 10,000 hours.

1.2.1 Youth Training and Apprenticeship Programs

Youth apprenticeship programs are available that allow students to begin their apprenticeship or craft training while still in high school. A student entering the program in the 11th grade may complete as much as one year of the NCCER training program by high school graduation. In addition, programs (in cooperation with local construction industry employers) allow students to work in the craft and earn money while still in school. Upon graduation, students can enter the industry at a higher level and with more pay than someone just starting in a training program.

Students participating in the NCCER or youth apprenticeship training are recognized through official transcripts and can enter the second level or year of the program wherever it is offered. They may also have the option of applying credits at two-year or four-year colleges that offer degree or certificate programs in their selected field of study.

2.0.0 Welding Safety Practices

Many hazards are associated with welding, cutting, and related processes. Some of these are electric shock, infrared rays, ultraviolet (UV) rays, hot metal, slag, sparks, and associated welding fumes.

Trainees must understand the hazards of welding and develop the proper attitude toward safety. This begins with an understanding of the need for proper clothing, eye protection, and face protection, along with general shop safety. *Figure 4* shows a welder using proper protective equipment.

The industry is fortunate to have a well-developed standard for safety: *American National Standards Institute (ANSI) Z49.1-2005, Safety in Welding, Cutting, and Allied Processes*. Everyone involved with welding, cutting, and related processes should be familiar with this document. The document discusses personal protective equipment, ventilation, and fire prevention as well as welding in confined spaces. It contains precautionary information that is important to welders, supervisors, and managers. The document also details specific welding process safety procedures.

101F04.EPS

Figure 4 Welder using proper protective equipment.

This module is an extension of the *Core Curriculum* safety module. Its purpose is to identify general safety considerations that apply to various aspects of the field of welding and the steps needed to avoid job-related deaths and injuries. Specific equipment or process hazards are covered in more detail in modules that deal with the use of the equipment or process in a particular application.

There are many causes of accidents. They can usually be divided into two broad categories: personal and physical. You must be aware of the factors that can cause accidents so you can understand the consequences of your actions on the job site.

2.1.0 Personal Factors That Cause Accidents

Accidents can often be traced to personal factors such as poor health, lack of experience, and the improper use of alcohol and medications.

People who are ill or injured may not be able to concentrate on their work. In addition, they may be physically weakened and unable to handle strenuous work. A person doing work that is inherently dangerous is more likely to cause an accident when he or she is sick or injured. Mental stress can also play a role in accidents. Again, concentration is the issue. An employee who is worried about a serious personal problem is likely to be distracted and unable to focus on work. Workers need to realize that they endanger themselves and those around them when they are not 100 percent effective. As a worker, you need to be able to recognize when others may not be up to par and take appropriate precautions.

Age and inexperience often play a role in accidents. Insurance company studies show that a person who lacks experience is more likely to take risks that cause accidents. Sometimes, accidents occur just because someone hasn't had enough experience to learn how to avoid them. In fact, insurance company statistics show that more accidents occur in the under-18 age group than in any other. An inexperienced person often is not able to foresee the outcome of an action or just lacks the knowledge to know what works and what doesn't.

People who consume alcohol or use illegal drugs while working risk their lives and the lives of their co-workers. A drug or alcohol hangover is nearly as bad. Legal prescription and over-the-counter drugs can also cause problems. Some cold remedies and cough medicines contain alcohol or other substances that will make people drowsy. Alcohol and drugs affect coordination, alertness, and decision-making ability. Never use these substances while you are working, and don't work with people who use them on the job.

2.2.0 Physical Factors That Cause Accidents

Accidents are often caused by conditions at the job site. For example, foot and vehicle traffic congestion increases at starting and quitting times. As quitting time approaches, people are more likely to hurry, possibly taking risks they might not otherwise take. People often slow down right after a meal, so workers may be less attentive after lunch than at other times.

People who learned to put their toys away when they were young probably keep a neat, safe work site as adults. The others are likely to have tools, equipment, scrap material, and other clutter lying around their work site that will trip people. If you keep a messy work site, eventually you or someone else will be injured because of it. Put away your tools, dispose of scrap materials, and secure your equipment and cables when not in use. If you have flammable or hazardous material at your work site, keep it properly contained and covered. When you're done with it, return it to the designated location for disposal or storage.

Damaged or defective welding equipment is dangerous. For example, a frayed or damaged welding cable could give someone a lethal shock or start a fire. Any damaged or defective equipment should be so tagged and either repaired or replaced. All equipment should be periodically inspected for damage. Make sure that periodic maintenance procedures recommended by the manufacturer are performed on schedule.

3.0.0 Appropriate Personal Protective Equipment

Each person in the shop wears general work clothing. Extra protection is needed for each person in direct contact with hot materials. Depending on the specific job and conditions, protective equipment can include:

- Body protection
- Foot protection
- Hand protection
- Ear protection
- Eye, face, and head protection

3.1.0 Body Protection

Basic clothing should offer protection from flying sparks, heat, and ultraviolet radiation from an electric arc. Shirts should be made of a tight-weave fabric, have long sleeves and pocket flaps, and be worn with the collar buttoned. Pants must not have cuffs and should fit so they hang straight down the leg, touching the shoe tops without creases. Cuffs and creases can catch sparks, which can cause fires. Never wear polyester or other synthetic fibers; sparks will melt these materials, causing serious burns. Wear wool and cotton, which are more resistant to sparks, instead of synthetic fibers. Cotton clothing, if worn, should be treated with a fire-retardant.

For cutting operations, out-of-position or overhead welding/cutting, or welding operations that produce spattering molten metal and sparks

Terms

Don't confuse the terms *fire-retardant* and *fireproof.* A product impregnated with a fire-retardant delays ignition. A fireproof product is not supposed to burn. Clothing treated with a fire-retardant may lose some of its protective ability if washed, so it should be retreated after washing.

or large amounts of heat, additional protective coverings are required. *Figure 5* shows a welder in protective clothing and equipment. Protective coverings, usually made of leather because of its durability, are often just referred to as leathers. However, because leather is heavy and hot to wear, lightweight fire- and heat-resistant clothing is also available. *Figure 6* shows a variety of leather and fire-resistant items that can be worn in combinations to provide various degrees of covering. Full jackets offer the most protection. A cape is cooler but offers only shoulder, chest, and arm protection. Leather pants, in combination with a jacket, will protect a welder's lap and legs. If the weather is hot and full leather pants are uncomfortable, leggings, sometimes called chaps, are available. They are strapped to the legs, leaving the back open. However, an apron alone is cooler and will protect a welder's lap and most of the leg area when squatting, sitting, or bending over a table. In some cases, if the welding is at or below the waist level, only full- or half-sleeve arm covers are required in combination with an apron.

3.2.0 Foot Protection

The Occupational Safety and Health Administration (OSHA) requires that protective footwear be worn when working where falling, rolling, or sharp objects pose a danger of foot injuries and where feet are exposed to electrical hazards. Footwear with leather-reinforced soles or innersoles of flexible metal are recommended. High-top safety shoes or boots 8 inches or taller should be worn. Ensure that the pant leg covers the tongue and lace area of the footwear. If the tongue and lace area is exposed, wear leather spats (*Figure 6*) under the pants or leggings and over the front top of the footwear to protect them from sparks or falling molten metal. Spats will prevent sparks from burning through the front of lace-up boots. Protective footwear must comply with *ANSI Z49.1-2005.* Sneakers, tennis shoes, and similar types of footwear must never be worn on the job site.

3.3.0 Hand Protection

Gloves are the primary type of hand protection. Gloves must be selected on the basis of the hazards involved in doing the work. Gauntlet-type welding gloves must be worn when welding or cutting to protect against UV rays from an electric arc and heat from any thermal welding/cutting process. The most common glove worn is the standard leather welding glove shown in *Figure 7.* For heavy-duty welding or cutting, special heat-resistant or heat-reflective gloves are also available.

WARNING!

Don't weld or cut metal with gloves that have been used to handle petroleum products, especially when oxygen is being used. Also consider that the transfer of petroleum residue to the welding rod can affect the weld quality.

3.4.0 Ear Protection

Welding areas can be very noisy. In addition, if overhead work is being performed, hot sparks can cause burns to the ears and ear canals unless a leather hood is used. Such burns are extremely painful, so it is important that you also wear ear plugs when welding (*Figure 8*). Disposable earplugs are the most common form of hearing protection used in the industry. These devices usually have an outer layer of pliable foam and a core layer of acoustical fiber that filters out harmful noise yet allows you to hear normal conversation. To use disposable earplugs, simply roll each plug into a cone and insert the tapered end into the ear canal while pulling up on the upper portion of your ear. The earplugs will expand, filling the ear canal and creating a proper fit. Reusable earplugs that can be cleaned and worn repeatedly are also commonly used. These are typically cleaned with boiling water or alcohol. Plain cotton placed in the ear is not an acceptable protective device.

3.5.0 Eye, Face, and Head Protection

The heat and light produced by cutting or welding operations can damage the skin and eyes. Injury to the eyes may result in permanent loss of vision. Oxyfuel cutting and welding can cause eye fatigue and mild burns to the skin because of the infrared heat radiated by the process. Welding or cutting operations involving an electric arc of any kind produce UV radiation, which can cause severe burns to the eyes and exposed skin and permanent damage to the retina. A flash burn can

Figure 5 Personal protective equipment.

harm unprotected eyes in just seconds. Welders should never view an electric arc directly or indirectly without wearing a properly tinted lens designed for electric arc use.

As specified in *ASTM Z49.1*, protective spectacles with side shields, arc goggles, or other approved eye protection must be worn by welders in addition to a helmet or hand shield with filter lens. The spectacles or goggles may have either clear or filtered lenses, depending on the amount of exposure to welding or cutting radiation (see *Table 1*). Welding helmets will not protect against flying slag chips, grinding fragments, and similar hazards, so other eye protection is essential. For overhead oxyfuel operations, a leather hood may be used in place of the face shield to obtain protection from sparks and molten metal (*Figure 9*).

For electric arc operations, a leather hood or welding helmet with a properly tinted lens (shades 9 to 14) must be worn over safety glasses to provide proper protection. Many varieties of helmets are available; typical styles are shown in *Figure 10*. Some of the helmets are available with additional side-view lenses in a lighter tint so that welders can sense, by peripheral vision, any activities occurring beside them.

Most welding and cutting tasks require the use of safety glasses, chemical-resistant goggles, dust goggles, or face shields. Always check your company's policy for the welding or cutting product being used to find out what type of eye protection is needed.

Figure 6 Leather and fire-resistant protective coverings.

Figure 7 Gauntlet-type welding gloves.

4.0.0 Fumes and Gases

The gases, fumes, and dust produced by welding and cutting processes can be hazardous. Adequate ventilation must be provided to prevent workers from breathing these products. There are three methods available to protect personnel against fumes and gases:

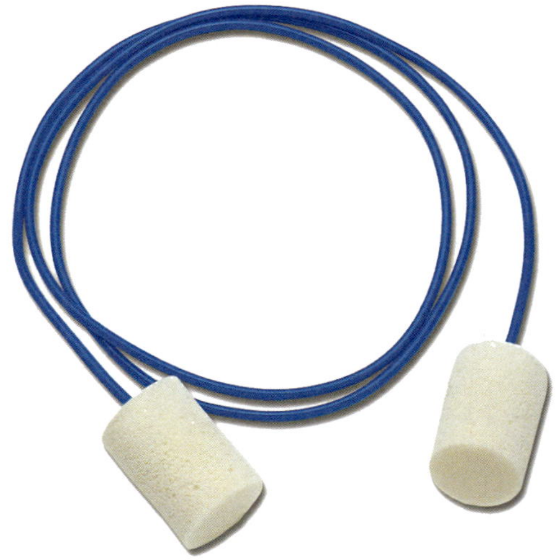

Figure 8 Ear plugs.

- *Natural ventilation* – The movement of air through the workplace caused by natural forces is often sufficient to remove fumes and gases.
- *Mechanical ventilation* – If natural ventilation is not adequate, portable or fixed fans can be used to provide the necessary ventilation.
- *Source extraction* – This method uses a mechanical device to capture welding fumes at or near the arc.

Lightweight Leather Gloves

Soft, flexible, leather gloves are used for light-duty work and for operations involving gas metal arc welding (GMAW), gas tungsten arc welding (GTAW), brazing, soldering, and oxyfuel welding where free hand movement is required.

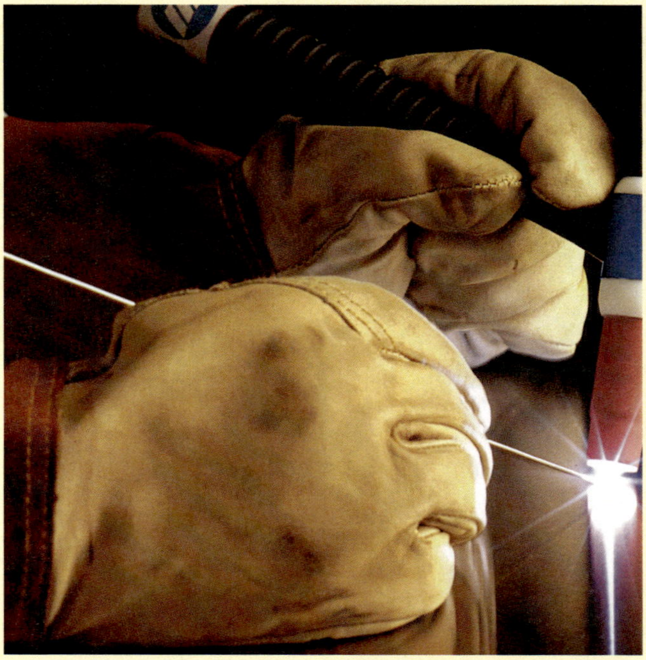

Table 1 Guide for shade numbers.

From *AWS F2.2 Lens Shade Selector.* Shade numbers are given as a guide only and may be varied to suit individual needs.

Process	Electrode Size in (mm)	Arc Current (Amperes)	Minimum Protective Shade	Suggested* Shade No. (Comfort)
Shielded Metal Arc Welding (SMAW)	Less than 3/32 (2.4)	Less than 60	7	–
	3/32–5/32 (2.4–4.0)	60–160	8	10
	5/32–1/4 (4.0–6.4)	160–250	10	12
	More than 1/4 (6.4)	250–550	11	14
Gas Metal Arc Welding (GMAW) and Flux Cored Arc Welding (FCAW)		Less than 60	7	–
		60–160	10	11
		160–250	10	12
		250–500	10	14
Gas Tungsten Arc Welding (GTAW)		Less than 50	8	10
		50–150	8	12
		150–500	10	14
Air Carbon Arc Cutting (CAC-A)	(Light)	Less than 500	10	12
	(Heavy)	500–1000	11	14
Plasma Arc Welding (PAW)		Less than 20	6	6 to 8
		20–100	8	10
		100–400	10	12
		400–800	11	14
Plasma Arc Cutting (PAC)		Less than 20	4	4
		20–40	5	5
		40–60	6	6
		60–80	8	8
		80–300	8	9
		300–400	9	12
		400–800	10	14
Torch Brazing (TB)		–	–	3 or 4
Torch Soldering (TS)		–	–	2
Carbon Arc Welding (CAW)		–	–	14

	Plate Thickness		Suggested* Shade No. (Comfort)
	in	mm	
Oxyfuel Gas Welding (OFW)			
Light	Under 1/8	Under 3	4 or 5
Medium	1/8 to 1/2	3 to 13	5 or 6
Heavy	Over 1/2	Over 13	6 or 8
Oxygen Cutting (OC)			
Light	Under 1	Under 25	3 or 4
Medium	1 to 6	25 to 150	4 or 5
Heavy	Over 6	Over 150	5 or 6

*As a rule of thumb, start with a shade that is too dark to see the weld zone. Then go to a lighter shade which gives sufficient view of the weld zone without going below the minimum. In oxyfuel gas welding, cutting, or brazing where the torch and/or the flux produces a high yellow light, it is desirable to use a filter lens that absorbs the yellow or sodium line of the visible light spectrum.

101T01.EPS

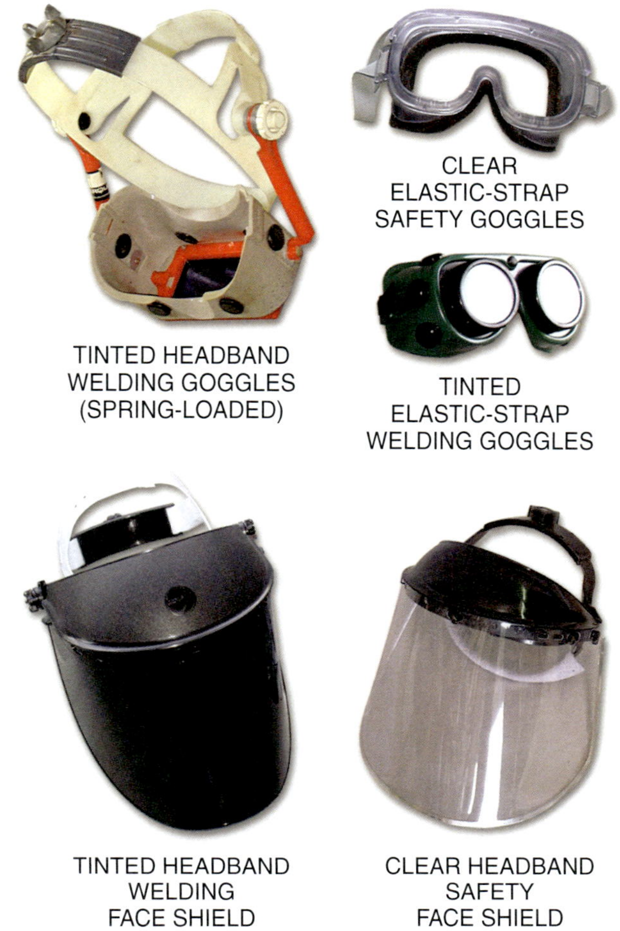

Figure 9 Oxyfuel welding/cutting goggles and face shield combinations.

Regardless of the ventilation method used, it is important to avoid breathing in the welding fumes. This means keeping your head up and away from the **fume plume**, which is the term used to describe the fumes, gases, and particles

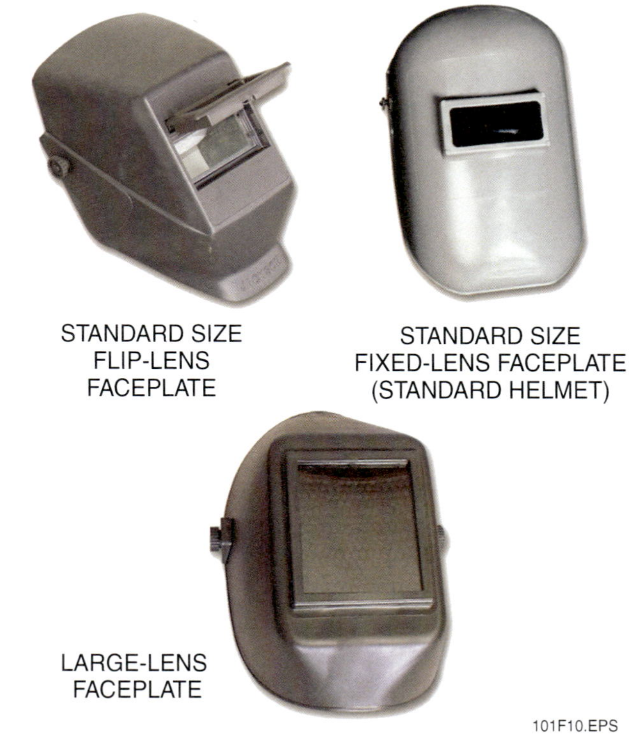

Figure 10 Electric arc welding helmets.

created by the welding process. The heated fumes and smoke generally rise straight up. Observe the column of smoke and position yourself to avoid it. A small fan may also be used to divert the smoke, but take care to keep the fan from blowing directly on the work area; the fumes and gases must be present at an electric arc in order to protect the molten metal from the air.

In addition to the fume plume itself, some consumables and base metals contain toxic materials that require special ventilation. These materials include:

Auto-Darkening Helmets

Many new helmets are available with an adjustable auto-darkening lens that changes from a shade 3 or 4 to a shade 10, 11, or 12 the instant an electric arc starts to occur. A sensor just above the lens detects the start of an electric arc and electronically causes the shade of the lens to change in less than 1/25,000 of a second. The use of this type of lens eliminates the need for the welder to raise the helmet or flip up the faceplate to view the work before striking an arc. The helmet is expensive, and most require periodic battery replacement and testing to ensure proper operation.

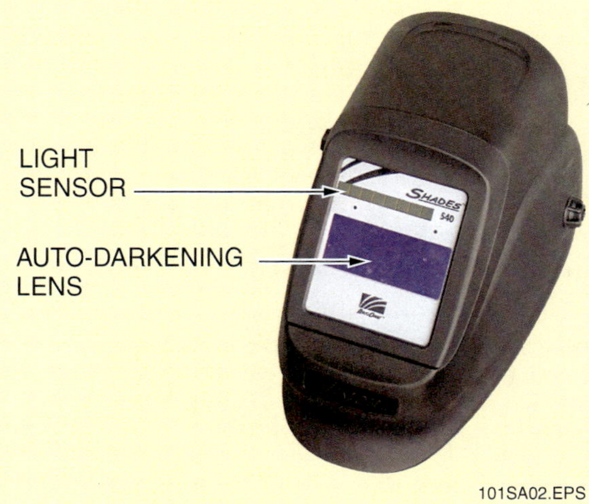

- Barium
- Cadmium (found in some steel and fasteners)
- Chromium (found in stainless steel)
- Base metal coatings such as paint
- Cobalt
- Copper
- Manganese
- Nickel
- Silica
- Zinc (found in galvanized metals)

WARNING!

It is essential to be aware of the materials being cut or welded. Some of the materials just listed produce fumes that are known to cause cancer or can lead to serious respiratory illness. Long-term exposure can even result in death. Some consumables are designated as Special Ventilation products for these reasons. When welding or cutting these materials, wear an approved full-face supplied-air respirator (SAR). For occasional short-term exposure to fumes from zinc- or copper-coated materials, a high-efficiency particulate arresting (HEPA)-rated or metal-fume filter may be used on a standard respirator.

In many cases, natural ventilation provides the required level of protection. However, it may be necessary to have a specialist determine if the ventilation is adequate. Some guidelines used in making this determination are:

- The welding area must contain at least 10,000 cubic feet. This represents an area of 22' × 22' × 22' for each welder.
- The ceiling height must be no less than 16 feet.
- Cross-ventilation must not be blocked by partitions, equipment, or other barriers.
- The area must not be designated a confined space.

4.1.0 Source Extraction Equipment

Source extraction, also called local exhaust, can be provided by portable, mobile, or stationary systems (*Figure 11*). As a general rule, the farther the extraction hose is from the arc, the greater the volume of air required. There are two classes of source extraction equipment: low vacuum/high volume and high vacuum/low volume.

Low vacuum/high volume equipment can be used in situations where the extraction hose is 6 to 15 inches from the arc. It uses extraction hoses, or hose and tubing combinations, with a diameter of 6 to 8 inches to accommodate the higher air volume of 560 to 860 cubic feet per minute. The mobile unit shown in *Figure 11* is an example of a low vacuum/high volume system.

High vacuum/low volume systems are designed for close-proximity (2 to 4 inch) work and is effective in difficult to reach or confined work spaces. There are two types of high vacuum/low volume systems. The fume extraction gun type has a fume extraction nozzle built into the welding gun. Suction nozzles—the second type—are positioned near the weld, usually within 4 inches of the work. Adequate ventilation can be a problem in tight or cramped working quarters. To ensure adequate room ventilation in such cases, local exhaust ventilation should be used to capture fumes.

The exhaust hood should be kept 4 to 6 inches away from the source of the fumes. Welders should recognize that fumes of any type, regardless of their source, should not be inhaled. The best way to avoid problems is to provide adequate ventilation. If this is not possible, breathing protection must be used. Respirators for use in poorly ventilated or confined spaces are shown in *Figures 12* and *13*.

If respirators are used, your employer should offer worker training on respirator fitting and use. Medical screenings should also be given.

WARNING!

Fumes and gases can be dangerous. Overexposure can cause nausea, headaches, dizziness, metal fume fever, and severe toxic effects that can be fatal. Studies have demonstrated irritation to eyes, skin, and the respiratory system and even more severe complications. In confined spaces, fumes and gases may cause asphyxiation.

4.2.0 Respirators

Special metals require the use of respirators to protect welders from harmful fumes. Respirators are grouped into three main types based on how they work to protect the wearer from contaminants. The types are:

- Air-purifying respirators
- Supplied-air respirators (SARs)
- Self-contained breathing apparatus (SCBA)

4.2.1 Air-Purifying Respirators

Air-purifying respirators provide the lowest level of protection. They are made for use only in atmospheres that have enough oxygen to sustain life (at least 19.5 percent). Air-purifying respira-

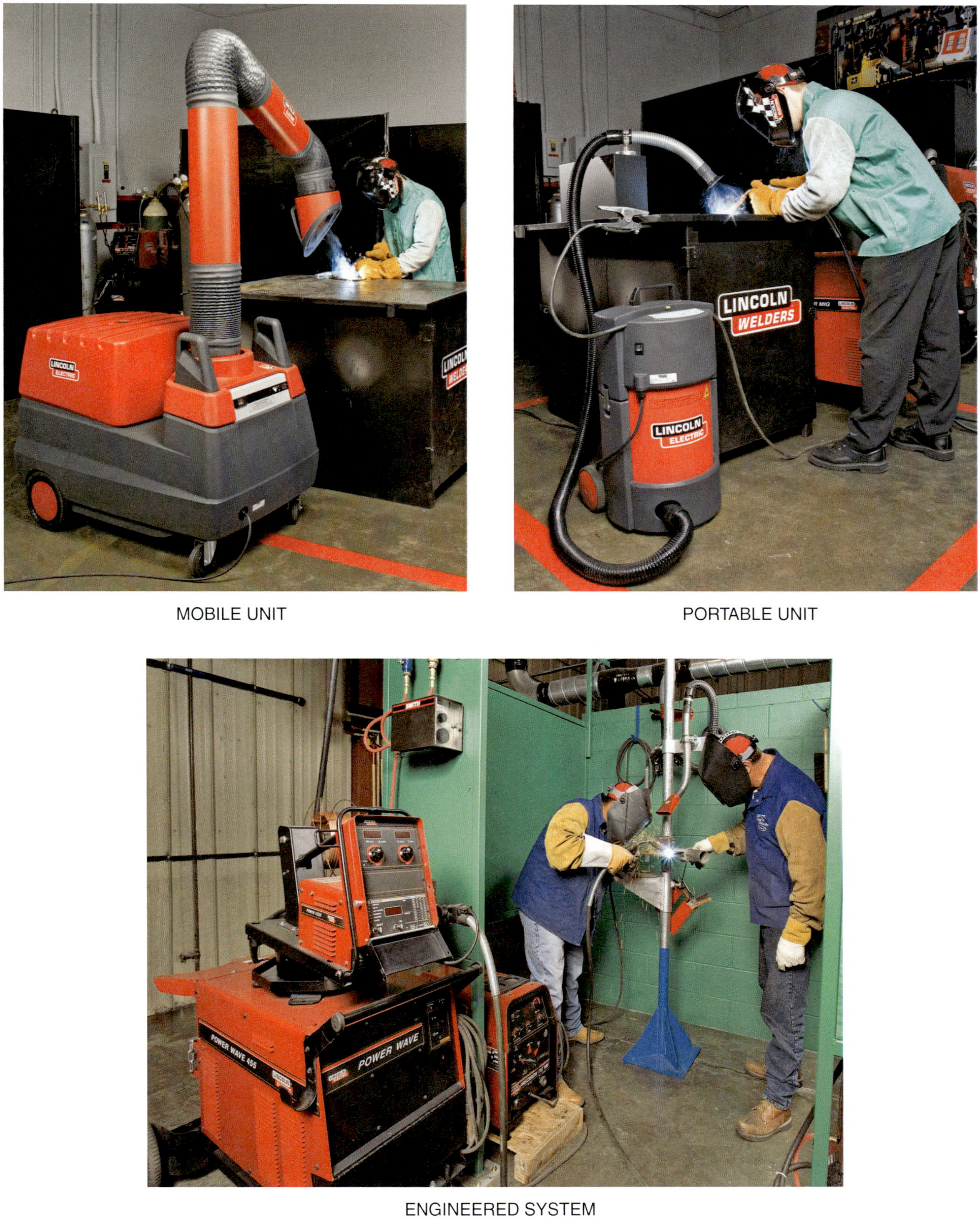

Figure 11 Source extraction equipment.

101F12.EPS

Figure 12 Typical respirator.

tors use special filters and cartridges to remove specific gases, vapors, and particles from the air. The respirator cartridges contain charcoal, which absorbs certain toxic vapors and gases. When the wearer detects any taste or smell, the charcoal's absorption capacity has been reached and the cartridge can no longer remove the contaminant. The respirator filters remove particles such as dust, mists, and metal fumes by trapping them within the filter material. Filters should be changed when it becomes difficult to breathe. Depending on the contaminants, cartridges can be used alone or in combination with a filter/pre-filter and filter cover. Air-purifying respirators should be used for protection only against the types of contaminants listed on the filters and cartridges and on the National Institute for Occupational Safety and Health (NIOSH) approval label affixed to each respirator carton and replacement filter/cartridge carton. Respirator manufacturers typically classify air-purifying respirators into four groups:

- No maintenance
- Low maintenance
- Reusable
- Powered air-purifying respirators (PAPRs)

No-maintenance and low-maintenance respirators are typically used for residential or light commercial work that does not call for constant and heavy respirator use. No-maintenance respirators are typically half-mask respirators with permanently attached cartridges or filters. The entire respirator is discarded when the cartridges

101F13.EPS

Figure 13 A belt-mounted respirator.

or filters are spent. Low-maintenance respirators generally are also half-mask respirators that use replaceable cartridges and filters. However, they are not designed for constant use.

Reusable respirators (*Figure 14*) are made in half-mask and full facepiece styles. These respirators require the replacement of cartridges, filters, and respirator parts. Their use also requires a complete respirator maintenance program. Air respirator maintenance is discussed later.

Think About It

Ventilation

Areas that have less than 10,000 cubic feet (283 cubic meters) per welder or that have ceilings less than 16 feet (4.9 meters) high require forced ventilation. This is the equivalent of a room 25 feet by 25 feet with a ceiling of 16 feet. What is the cubic area of a 90-foot long by 25-foot wide room with 8-foot ceilings?

101F14.EPS

Figure 14 Reusable half-mask air-purifying respirator.

101F15.EPS

Figure 15 Powered air-purifying respirator (PAPR).

Powered air-purifying respirators (PAPRs) are made in half-mask, full facepiece, and hood styles. They use battery-operated blowers to pull outside air through the cartridges and filters attached to the respirator. The blower motors can be either mask- or belt-mounted. Depending on the cartridges used, they can filter particulates, dusts, fumes, and mists along with certain gases and vapors. PAPRs like the one shown in *Figure 15* have a belt-mounted, powered air-purifier unit connected to the mask by a breathing tube. Many models also have an audible and visual alarm that is activated when airflow falls below the required minimum level. This feature gives an immediate indication of a loaded filter or low battery charge condition. Units with the blower mounted in the mask do not use a belt-mounted powered air purifier connected to a breathing tube.

4.2.2 Supplied-Air Respirators

Supplied-air respirators (*Figure 16*) provide a supply of air for extended periods of time via a high-pressure hose that is connected to an external source of air, such as a compressor, compressed-air cylinder, or pump. They provide a higher level of protection in atmospheres where air-purifying respirators are not adequate. Supplied-air respirators are typically used in toxic atmospheres. Some can be used in atmospheres that are **immediately dangerous to life and health (IDLH)** as long as they are equipped with an air cylinder for emergency escape. An atmosphere is considered IDLH if it poses an immediate hazard to life or produces immediate, irreversible, and debilitating effects on health. There are two types of supplied-air respirators: continuous-flow and pressure-demand.

101F16.EPS

Figure 16 Supplied-air respirator.

The continuous-flow supplied-air respirator provides air to the user in a constant stream. One or two hoses are used to deliver the air from the air source to the facepiece. Unless the compressor or pump is especially designed to filter the air or a portable air-filtering system is used, the unit must be located where there is breathable air (grade D

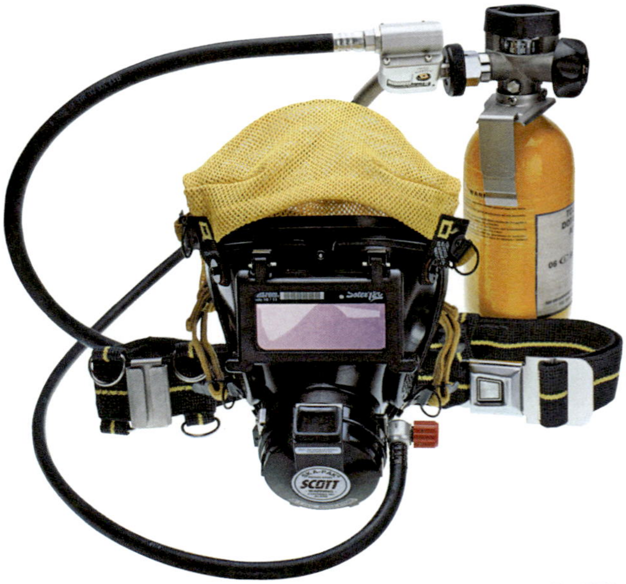

Figure 17 Self-contained breathing apparatus.

or better as described in *Compressed Gas Association [CGA] Commodity Specification G-7.1*). Continuous-flow respirators are made with tight-fitting half-masks or full facepieces. They are also made with hoods. The flow of air to the user may be adjusted either at the air source (fixed flow) or on the unit's regulator (adjustable flow). Pressure-demand supplied-air respirators are similar to the continuous-flow type except that they supply air to the user's facepiece via a pressure-demand valve as the user inhales and fresh air is required. They typically have a two-position exhalation valve that allows the worker to switch between pressure-demand and negative-pressure modes to facilitate entry into, movement within, and exit from a work area.

4.2.3 Self-Contained Breathing Apparatus (SCBA)

SCBAs (*Figure 17*) can be used in oxygen-deficient atmospheres (below 19.5 percent oxygen), in poorly ventilated or confined spaces, and in IDLH atmospheres. These respirators provide a supply of air for 30 to 60 minutes from a compressed-air cylinder worn on the user's back. Note that the emergency escape breathing apparatus (EEBA) is a smaller version of an SCBA cylinder. EEBA units are used for escape from hazardous environments and generally provide a five- to ten-minute supply of air.

4.3.0 Respiratory Program

Local and OSHA procedures must be followed when selecting the proper type of respirator for a particular job (*Figure 18*). A respirator must be

Figure 18 Use the right respirator for the job.

properly selected (based on the contaminant present and its concentration level), properly fitted, and used in accordance with the manufacturer's instructions. It must be worn during all times of exposure. Regardless of the kind of respirator needed, OSHA regulations require employers to have a respirator protection program consisting of:

- Standard operating procedures for selection and use
- Employee training
- Regular cleaning and disinfecting
- Sanitary storage
- Regular inspection
- Annual fit testing
- Pulmonary function testing

As an employee, you are responsible for wearing respiratory protection when needed. When it comes to vapors or fumes, both can be eliminated in certain concentrations by the use of air-purifying devices as long as oxygen levels are acceptable. Examples of fumes are smoke billowing from a fire or the fumes generated when welding. Always check the cartridge on your respirator to make sure it is the correct type to use for the air conditions and contaminants found on the job site.

When selecting a respirator to wear while working with specific materials, you must first determine the hazardous ingredients contained in the material and their exposure levels, then choose the proper respirator to protect yourself at these levels. Always read the product's MSDS. It identifies the hazardous ingredients and should list the type of respirator and cartridge recommended for use with the product.

Limitations that apply to all half-mask (air-purifying) respirators are as follows:

- These respirators do not completely eliminate exposure to contaminants, but they will reduce the level of exposure to below hazardous levels.
- These respirators do not supply oxygen and must not be used in areas where the oxygen level is below 19.5 percent.
- These respirators must not be used in areas where chemicals have poor warning signs, such as no taste or odor.

If your breathing becomes difficult, if you become dizzy or nauseated, if you smell or taste the chemical, or if you have other noticeable effects, leave the area immediately, return to a fresh air area, and seek any necessary assistance.

4.3.1 Positive and Negative Fit Checks

All respirators are useless unless properly fit-tested to each individual. To obtain the best protection from your respirator, you must perform positive and negative fit checks each time you wear it. These fit checks must be done until you have obtained a good face seal.

To perform the positive fit check, do the following:

Step 1 Adjust the facepiece for the best fit, then adjust the head and neck straps to ensure good fit and comfort.

WARNING!

Do not overtighten the head and neck straps. Tighten them only enough to stop leakage. Overtightening can cause facepiece distortion and dangerous leaks.

Step 2 Block the exhalation valve with your hand or other material.

Step 3 Breathe out into the mask.

Step 4 Check for air leakage around the edges of the facepiece.

Step 5 If the facepiece puffs out slightly for a few seconds, a good face seal has been obtained.

To perform a negative fit check, do the following:

Step 1 Block the inhalation valve with your hand or other material.

Step 2 Attempt to inhale.

Step 3 Check for air leakage around the edges of the facepiece.

Step 4 If the facepiece caves in slightly for a few seconds, a good face seal has been obtained.

Respirator Inspection, Care, and Maintenance

A respirator must be clean and in good condition, and all of its parts must be in place for it to give you proper protection. Respirators must be cleaned every day. Failure to do so will limit their effectiveness and offer little or no protection. For example, suppose you wore the respirator yesterday and did not clean it. The bacteria from breathing into the respirator, plus the airborne contaminants that managed to enter the facepiece, will have made the inside of your respirator very unsanitary. Continued use may cause you more harm than good. Remember, only a clean and complete respirator will provide you with the necessary protection. Follow these guidelines:

- Inspect the condition of your respirator before and after each use.
- Do not wear a respirator if the facepiece is distorted or if it is worn and cracked. You will not be able to get a proper face seal.
- Do not wear a respirator if any part of it is missing. Replace worn straps or missing parts before using.
- Do not expose respirators to excessive heat or cold, chemicals, or sunlight.
- Clean and wash your respirator after every time you use it. Remove the cartridge and filter, hand wash the respirator using mild soap and a soft brush, and let it air dry overnight.
- Sanitize your respirator each week. Remove the cartridge and filter, then soak the respirator in a sanitizing solution for at least two minutes. Thoroughly rinse with warm water and let it air dry overnight.
- Store the clean and sanitized respirator in its resealable plastic bag. Do not store the respirator face down. This will cause distortion of the facepiece.

5.0.0 Confined Space Permits

A confined space refers to a relatively small or restricted space, such as a storage tank, boiler, or pressure vessel or small compartments, such as underground utility vaults, small rooms, or the unventilated corners of a room.

OSHA 29 CFR 1910.146 defines a confined space (*Figure 19*) as a space that:

- Is large enough and so configured that an employee can bodily enter and perform assigned work
- Has a limited or restricted means of entry or exit; for example, tanks, vessels, silos, storage bins, hoppers, vaults, and pits
- Is not designed for continuous employee occupancy

OSHA 29 CFR 1910.146 further defines a permit-required confined space as a space that:

- Contains or has the potential to contain a hazardous atmosphere
- Contains a material that has the potential for engulfing an entrant
- Has an internal configuration such that an entrant could be trapped or asphyxiated by inwardly converging walls or by a floor that slopes downward and tapers to a smaller cross section
- Contains any other recognized serious safety or health hazard

For safe working conditions, the oxygen level in a confined space atmosphere must range between 19.5 percent and 23.5 percent by volume, with 21 percent being considered the normal level (*Table 2*). Oxygen concentrations below 19.5 percent by volume are considered deficient.

WARNING!

If too much oxygen is ventilated into a confined space, it can be absorbed by the welder's clothing and ignite. If too little oxygen is present, it can lead to the welder's death in minutes. For this reason, make sure confined spaces are ventilated properly for cutting or welding purposes and *never* use oxygen in confined spaces for ventilation purposes.

When welding or cutting is being performed in any confined space, the gas cylinders and welding machines must be left on the outside. Before operations are started, the wheels of heavy portable equipment must be securely blocked to prevent accidental movement. Where a welder must enter a confined space through a manhole or other opening, all means must be provided for quickly removing the worker in case of emergency. When safety belts and lifelines are used for this purpose, they must be attached to the welder's body so that he or she cannot be jammed in a small exit opening. An attendant with a pre-planned rescue procedure must be stationed outside to observe the welder at all times and must be capable of putting the rescue operations into effect.

When welding or cutting operations are suspended for any substantial period of time, such as during lunch or overnight, all electrodes must be removed from the holders, and the holders must be carefully located so that accidental contact cannot occur. The welding machines must also be disconnected from the power source.

In order to eliminate the possibility of gas escaping through leaks or improperly closed valves when gas welding or cutting, the gas and oxygen supply valves must be closed, the regulators released, the gas and oxygen lines bled, and the valves on the torch shut off when the equipment will not be used for a substantial period of time. Where practical, the torch and hose must also be removed from the confined space. After welding operations are completed, the welder must mark the hot metal or provide some other means of warning other workers.

WARNING!

Nitrogen and argon are used as backing and purge gases for some welding applications. These gases will displace all the oxygen in the space where they are used.

At a refinery in Corpus Christi, Texas, while the welder was at lunch, a helper entered a pipe in which argon had been used as a backing gas. When the welder returned, he found the helper dead.

Nitrogen is lighter than air and, if allowed, will float upward. Argon, on the other hand, is heavier than air and will settle in low-lying areas. When removing nitrogen from a confined space, open the top to allow it to float out. When removing argon from a confined space, provide some way for it to drain out the bottom.

6.0.0 Area Safety

An important factor in area safety is good housekeeping. The work area should be picked up and swept clean. The floors and workbenches should be free of dirt, scrap metal, grease, oil, and anything that is not essential to accomplishing the given task. Collections of steel, welding electrode stubs, wire, hoses, and

101F19.EPS

Figure 19 Worker entering a confined space with a restricted opening for entry and exit.

cables are difficult to work around and easy to trip over. An electrode caddy can be used to hold the electrodes and stubs. Hooks can be made to hold hoses and cables, and scrap steel should be thrown into scrap bins.

The ideal welding shop should have bare concrete floors and **electrically grounded** and **bonded** bare metal walls and ceilings to reduce the possibility of fire. Never weld or cut over wood floors, as this increases the possibility of fire.

It is important to keep flammable liquids as well as rags, wood scraps, piles of paper, and other combustibles out of the welding area.

If you must weld in an enclosed building, make every effort to eliminate anything that could trap a spark. Sparks can smolder for hours and then burst into flames. Regardless of where you're welding, be sure to have a fire extinguisher nearby. Also, keep a five-gallon bucket of water handy to cool off hot metal and quickly douse small fires. If a piece of hot metal must be left un-

Table 2 Effects of an Increase or Decrease in Oxygen Levels

Oxygen Level	Effects
> 23.5%	Easy ignition of flammable material such as clothes
19.5% – 23.5%*	Normal
17%	Deterioration of night vision, increased breathing volume, accelerated heartbeat
14% – 16%	Very poor muscular coordination, rapid fatigue, intermittent respiration
6% – 10%	Nausea, vomiting, inability to perform, unconsciousness
< 6%	Spasmodic breathing, convulsive movements, and death in minutes

* Above 23.5% is considered enriched.

attended, use soapstone to write the word HOT on it before leaving. This procedure can also be used to warn people of hot tables, vises, firebricks, and tools.

Never use a cutting torch inside your workshop unless a proper cutting area is available. Take whatever you're going to cut outside, away from flammables. Also be aware that welding sparks can ignite gasoline fumes in a confined space.

Whenever welding must be done outside a welding booth, use portable screens to protect other personnel from the arc or reflected glare (see *Figure 20*). The portable screen also prevents drafts of air from interfering with the stability of the arc.

The most common welding accident is burned hands and arms. Keep first-aid equipment nearby to treat burns in the work area. Eye injuries can also occur if you are careless. Post emergency phone numbers in a prominent location.

The following are some work-area reminders:

- Eliminate tripping hazards by coiling cables and keeping clamps and other tools off the floor.
- Don't get entangled in cables, loose wires, or clothing while you work. This allows you to move freely, especially should your clothing ignite or some other accident occur.
- Clean up oil, grease, or other agents that may ignite and splatter off surfaces while welding.
- Shut off the welder and disconnect the power plug before performing any service or maintenance.

101F20.EPS

Figure 20 A typical welding screen.

- Keep the floor free of electrodes once you begin to weld. They could cause a slip or fall.
- Work in a dry area, booth, or other shielded area whenever possible.
- Make sure there are no open doors or windows through which sparks may travel to flammable materials.
- Don't coil electrical leads over your body while working on scaffolding.

7.0.0 Hot Work Permits and Fire Watches

A hot work permit (*Figure 21*) is an official authorization from the site manager to perform work that may pose a fire hazard. The permit includes information such as the time, location, and type of work being done. The hot work permit system promotes the development of standard fire safety guidelines. Permits also help managers keep records of who is working where and at what time. This information is essential in the event of an emergency or at other times when personnel need to be evacuated.

During a fire watch, a person other than the welder or cutting operator must constantly scan the work area for fires. Fire watch personnel must have ready access to fire extinguishers and alarms and know how to use them. Cutting operations must never be performed without a fire watch. Whenever oxyfuel cutting equipment is used, there is a great danger of fire. Hot work permits and fire watches are used to minimize this danger. Most sites require the use of hot work permits and fire watches. When they are violated, severe penalties are imposed.

HOT WORK PERMIT

For Cutting, Welding, or Soldering with Portable Gas or ARC Equipment

(References: 1997 Uniform Fire Code Article 49 & National Fire Protection Association Standard NFPA 51B.)

Job Date____________ Start Time____________ Expiration____________ WO #____________

Name of Applicant__________________________ Company____________ Phone____________

Supervisor__________________________ Phone____________

Location / Description of work ____

____

IS FIRE WATCH REQUIRED?

1. ________ (yes or no) Are combustible materials in building construction closer than 35 feet to the point of operation?
2. ________ (yes or no) Are combustibles more than 35 feet away but would be easily ignited by sparks?
3. ________ (yes or no) Are wall or floor openings within a 35 foot radius exposing combustible material in adjacent areas, including concealed spaces in floors or walls?
4. ________ (yes or no) Are combustible materials adjacent to the other side of metal partitions, walls, ceilings, or roofs which could be ignited by conduction or radiation?
5. ________ (yes or no) Does the work necessitate disabling a fire detection, suppression, or alarm system component?

YES to any of the above indicates that a qualified fire watch is required.

Fire Watcher Name(s) ______________________________ Phone ________________

NOTIFICATIONS

Notify the following groups at least 72 hours prior to work and 30 minutes after work is completed.
Write in names of persons contacted.

Notify in person OR by phone ONLY if question #5 above is answered "yes":

- Facilities Management Fire Alarm Supervisor

Notify by phone or in person: (If by phone, write down name of person and send them a completed copy of this permit.)

- Facilities Management Fire Protection Group
- Environmental Health & Safety Industrial Hygiene Group

SIGNATURES REQUIRED

University Project Manager______________________________ Date __________ Phone____________

I understand and will abide by the conditions described in this permit. I will implement the necessary precautions which are outlined on both sides of this permit form. Thirty minutes after each hot work session, I will reinspect work areas and adjacent areas to which spark and heat might have spread to verify that they are fire safe, and contact Facilities Management Alarm Technicians to have any disabled fire protection systems reactivated.

____________________ ____________________ Date __________ Phone ____________

Permit Applicant Company or Department

1/17/03

101F21.EPS

Figure 21 Hot work permit.

> **WARNING!**
>
> Never perform any type of heating, cutting, or welding until you have obtained a hot work permit and established a fire watch. If you are unsure of the procedure, check with your supervisor. Violation of hot work permit and fire watch procedures can result in serious injury or death. OSHA requires that the fire watch be maintained for one-half hour after welding or cutting is finished.

8.0.0 Oxyfuel Gas Welding and Cutting Safety

Most welding environment fires occur during oxyfuel gas welding or cutting. Welders should be well trained in the function and operation of each part of an oxyfuel gas welding or cutting station. Welders should also be trained and tested in the correct methods of starting, testing for leaks, and shutting down an oxyfuel gas welding or cutting station. They must be aware of the hazards involved in the use of fuel gas, oxygen, and shielding gas cylinders and how these cylinders are stored safely.

It is recommended that at least one fire watch be posted with an extinguisher to watch for possible fires.

9.0.0 Cutting Containers

Cutting and welding activities present unique hazards depending upon the material being cut or welded and the fuel used to power the equipment. All cutting and welding should be done in designated areas of the shop if possible. These areas should be made safe for welding and cutting operations with concrete floors, arc filter screens, protective drapes, curtains or blankets (*Figure 22*), and fire extinguishers. No combustibles should be stored nearby.

> **WARNING!**
>
> Welding or cutting must never be performed on drums, barrels, tanks, vessels, or other containers until they have been emptied and cleaned thoroughly, eliminating all flammable materials and all substances (such as detergents, solvents, greases, tars, or acids) that might produce flammable, toxic, or explosive vapors when heated.

Containers must be cleaned by steam cleaning, flushing with water, or washing with detergent until all traces of the material have been removed.

Oxyfuel Gas Welding Safety Precautions

101SA03.EPS

- Always light the oxyfuel gas torch flame using an approved torch lighter to avoid burning your fingers.
- Never point the torch tip at anyone when lighting it or using it.
- Never point the torch at the cylinders, regulators, hoses, or anything else that may be damaged and cause a fire or explosion.
- Never lay a lighted torch down on the bench or workpiece, and do not hang it up while it is lighted. If the torch is not in the welder's hands, it must be off.
- To prevent possible fires and explosions, check valves and flashback arrestors must be installed in all oxyfuel gas welding and cutting outfits.
- When cutting with oxyfuel gas equipment, clear the area of all combustible materials.
- Skin contact with liquid oxygen can cause frostbite. Be careful when handling liquid oxygen.
- Never use oxygen as a substitute for compressed air.
- All oxygen cylinders with leaky valves or safety fuse plugs and discs should be set aside and marked for the attention of the supplier. Do not tamper with or attempt to repair oxygen cylinder valves. Do not use a hammer or wrench to open the valves.

3,000°F INTERMITTENT, 1,500°F CONTINUOUS
SILICON DIOXIDE CLOTH

101F22.EPS

Figure 22 Welding blanket.

> **WARNING!**
>
> Clean containers only in well-ventilated areas. Vapors can accumulate during cleaning, causing explosions or injury.

After cleaning the container, fill it with water or a **purging gas,** such as carbon dioxide, argon, or nitrogen to displace the explosive fumes (*Figure 23*). Air, which contains oxygen, is displaced from inside the container by the water or inert gas. Without oxygen, combustion cannot take place. When using water, position the container to minimize the air space. When using an inert gas, provide a vent hole so the inert gas can purge the air.

> **WARNING!**
>
> Do not assume that a container that has held combustibles is clean and safe until proven so by proper tests. Do not weld in places where dust or other combustible particles are suspended in air or where explosive vapors are present. Removal of flammable materials from vessels/containers may be done by steaming or boiling.

Proper procedures for cutting or welding hazardous containers are described in the *American Welding Society (AWS) F4.1-2005, Recommended Safe Practices for the Preparation for Welding and Cutting of Containers and Piping.*

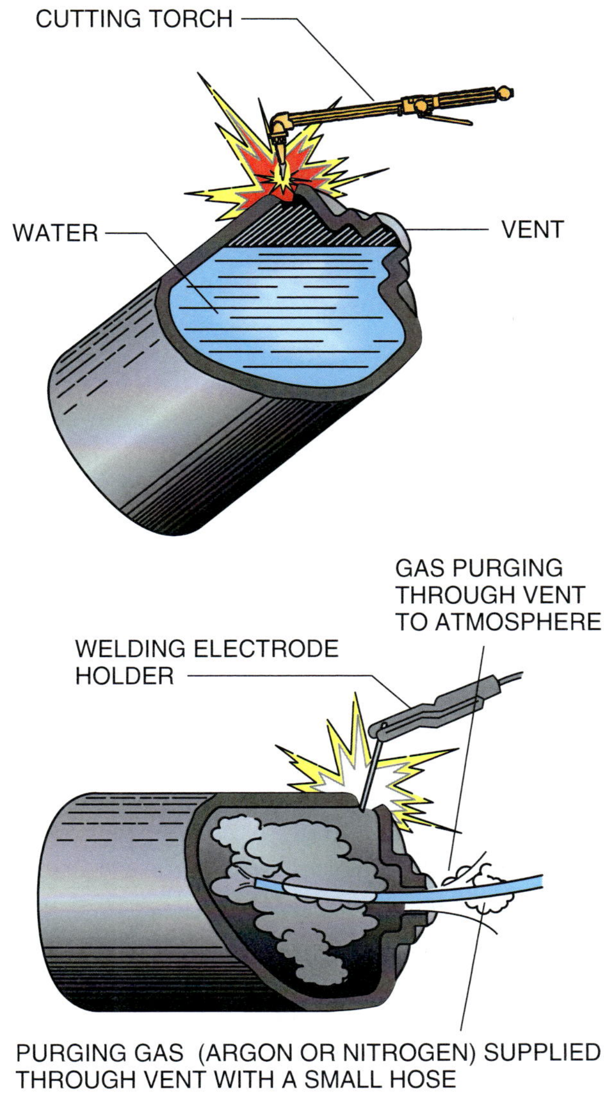

101F23.EPS

Figure 23 Purging containers.

10.0.0 Cylinder Storage and Handling

Oxygen and fuel gas cylinders or other flammable materials must be stored separately. The storage areas must be separated by 20 feet or by a wall 5 feet high with at least a 30-minute burn rating. The purpose of the distance or wall is to keep the heat of a small fire from causing the oxygen cylinder safety valve to release. If the safety valve releases the oxygen, a small fire would become a raging inferno.

Inert gas cylinders may be stored separately or with either fuel cylinders or oxygen cylinders. Empty cylinders must be stored separately from full cylinders, although they may be stored in the same room or area. All cylinders must be stored vertically and have the protective caps screwed on firmly.

10.1.0 Securing Gas Cylinders

Cylinders must be secured with a chain or other device so that they cannot be knocked over accidentally. Even though they are more stable, cylinders attached to a manifold should be chained, as should cylinders stored in a special room used only for cylinder storage.

10.2.0 Storage Areas

Cylinder storage areas must be located away from halls, stairwells, and exits so that in case of an emergency they will not block an escape route. Storage areas should also be located away from heat, radiators, furnaces, and welding sparks. The location of storage areas should be where unauthorized people cannot tamper with the cylinders. A warning sign that reads *Danger—No Smoking, Matches, or Open Lights*, or similar wording, should be posted in the storage area.

10.3.0 Cylinders with Valve Protection Caps

Cylinders equipped with a valve protection cap must have the cap in place unless the cylinder is in use. The protection cap prevents the valve from being broken off if the cylinder is knocked over. If the valve of a full high-pressure cylinder (argon, oxygen, CO_2, or mixed gases) is broken off, the cylinder can fly around the shop like a missile if it has not been secured properly. Never lift a cylinder by the safety cap or valve. The valve can easily break off or be damaged. When moving cylinders, the valve protection cap must be replaced (*Figure 24*), especially if the cylinders are mounted on a truck or trailer. Cylinders must never be dropped or handled roughly.

WARNING!

Using a wrench inserted through the cap can open the valve. If the cap is stuck, use a strap wrench to remove, or call the gas supplier.

10.4.0 General Precautions

Use warm water (not boiling) to loosen cylinders that are frozen to the ground. Any cylinder that leaks, has a bad valve, or has damaged threads must be identified and reported to the supplier. Use a piece of soapstone or a tag to write the problem on the cylinder. If closing the cylinder valve

101F24.EPS

Figure 24 Handle cylinders with care.

cannot stop the leak, move the cylinder outdoors to a safe location, away from any source of ignition, and notify the supplier. Post a warning sign, then slowly release the pressure.

In its gaseous form, acetylene is extremely unstable and explodes easily. For this reason it must remain at pressures below 15 pounds per square inch (psi). If an acetylene cylinder is tipped over, stand it upright and wait at least an hour before using it. If liquid acetone is withdrawn from a cylinder, it will gum up the safety check valves and regulators and decrease the stability of the acetylene stored in the cylinder. For this reason, acetylene must never be withdrawn at a per-hour rate that exceeds one-seventh of the volume of the cylinder(s) in use. Acetylene cylinders in use should be opened no more than one and one-half turns and, preferably, no more than three-fourths of a turn.

Other precautions include:

- Use only compressed gas cylinders containing the correct gas for the process used and properly operating regulators designed for the gas and pressure used.

- Make sure all hoses, fittings, and other parts are suitable and maintained in good condition.
- Keep cylinders in the upright position and securely chained to an undercarriage or fixed support.
- Keep combustible cylinders in one area of the building for safety. Cylinders must be at a safe distance from arc welding or cutting operations and any other source of heat, sparks, or flame.
- Never allow the electrode, electrode holder, or any other electrically hot parts to come in contact with the cylinder.
- When opening a cylinder valve, stand with the valve stem between yourself and the regulator.
- Always use valve protection caps on cylinders when they are not in use or are being transported. Make sure the cap is on tight.

11.0.0 Power Tool Safety

All power tools must be properly grounded to prevent accidental electric shock. If you feel even a slight tingle while using a power tool, stop and have the tool checked by a technician. Power tools should never be used with force or allowed to overheat from excessive or incorrect use. If an extension cord is used, it should have a large enough current rating to carry the load. An extension cord that is too small will cause the tool and the cord to overheat.

> **WARNING!**
> Grinders create sparks that can cause a fire.

12.0.0 Electrical Safety

Electric shock from welding and cutting equipment can kill or cause severe burns by coming in contact with bare skin. Serious injury can also result if a fall occurs because of the shock. The amount of current that passes through the human body determines the outcome of an electric shock. The higher the voltage, the greater the chance for a fatal shock. Electric current flows along the path of least resistance to return to its source. If you come in contact with a live conductor, you become a load. *Figure 25* shows how much resistance the human body presents under various circumstances and how this converts into amps or milliamps when the voltage is 110V. Note that the potential for shock increases dramatically if the skin is damp. A cut will also reduce resistance and increase shock potential. Currents of less than 1 amp can severely injure and even kill a person. Under the right conditions, even a small current can cause serious injury or death. The Construction Employers Association has issued the following guidelines regarding the seriousness of electric shock:

- *Greater than 3 milliamps (mA)* – Indirect accident
- *Greater than 10 mA* – Muscle contraction
- *Greater than 30 mA* – Lung paralysis, usually temporary
- *Greater than 50 mA* – Possible **ventricular fibrillation**
- *100 mA to 4A* – Certain ventricular fibrillation
- *Greater than 4A* – Heart paralysis, severe burns

Each welder and operator of the equipment must be trained and able to recognize the dangers associated with each particular type of equipment to avoid injuries, fatalities, and other electrical accidents.

Power Tool Safety Precautions

- Wear heavy protective clothing, a hard hat, goggles, and work gloves to guard against injury from flying debris.
- Use adequate hearing protection.
- Use the appropriate respiratory protection when dust hazards are produced.
- To avoid shock, ensure that electrically powered tools are adequately grounded and used in a dry environment. For added protection, use tools that are double insulated and equipped with ground fault circuit interrupters.
- Do not use power tools in confined spaces where sparks could cause explosions.
- Use nonsparking tools around combustible and flammable materials.
- Do not exceed the rated speed of the tool.
- Use recommended tool guards.

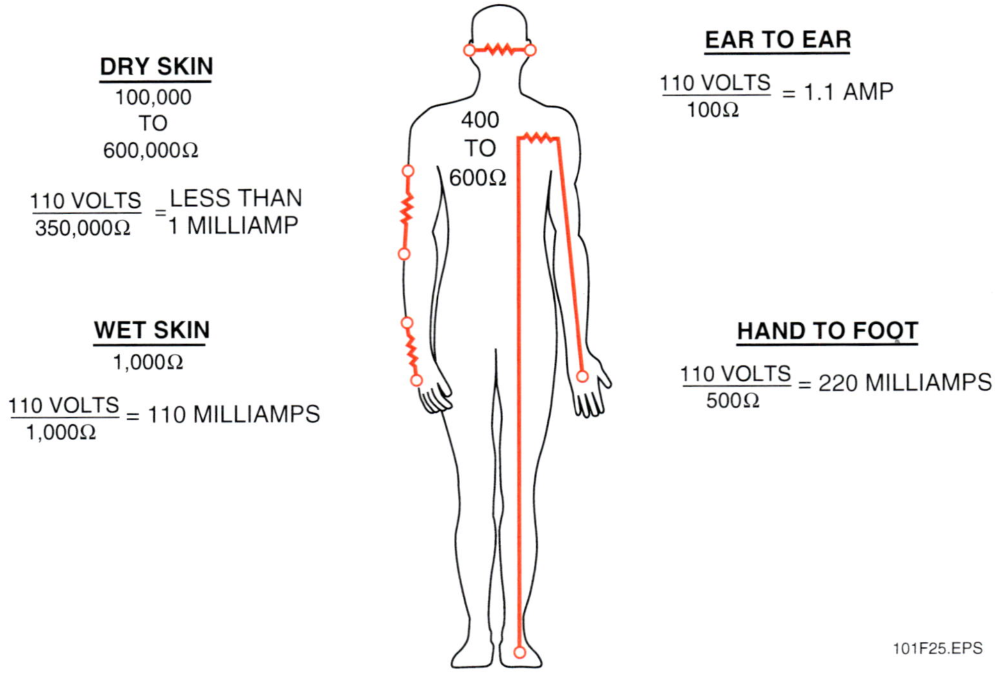

Figure 25 Typical body resistances and current.

It is a good idea for all involved to know first-aid procedures, such as cardiopulmonary resuscitation (CPR) for treating electric shock. Treat electrical burns like any other type of burn by applying ice or a cold compress over the burned area. Prevent infection and cover with a clean, dry dressing. Seek medical attention if necessary.

Welding-related electrical hazards come from two sources. The first, which we will call primary hazards, come from the source voltage that powers the welding machine. This voltage is usually 230 or 460 volts. These are potentials that create the possibility of serious, even fatal, shocks. It is important to know that this voltage exists inside the welding machine any time the machine is plugged in. The only way to disable it is to unplug the machine or turn off the power at the safety disconnect switch.

WARNING!

Be sure to follow applicable lockout/tagout procedures when disabling the power to the unit. Also, have a qualified electrician install your welding machine so that it will be correctly wired and grounded. If your welder does not work properly, have a qualified technician repair it.

Secondary hazards exist in the welding process itself. This type of shock can occur if a person touches part of the electrode circuit, such as a bare spot on the electrode cable, and touches the metal work material with another part of the body at the same time. Such a condition can occur if the welder is leaning against the work while welding. Here are some things to remember that will help keep you safe:

Electrical Safety Precautions

- Never operate arc-welding equipment on a wet or damp floor.
- Never touch an electrode to a grounded surface because these surfaces will become electrically live. The electrode and work circuit is electrically energized when the output is on.
- Never place electrode holders in contact with a grounded metal surface because it could short circuit the welding machine.

- The voltage is highest at the electrode when you are not welding.
- The jaws and screws on the electrode holder, as well as the stick electrode itself, are electrically hot. Make sure they do not touch your skin or clothing, especially if the clothing is damp or wet.
- Wear dry gloves that are in good condition when welding. Keep an extra pair on hand.
- Keep dry insulation between your body, including your arms and legs, and the metal being welded or a ground such as a metal floor or wet earth. Use plywood or a rubber mat between you and the ground or floor.
- Keep the welding cables and electrode holder in good condition.
- If you must work in conditions where you might come into contact with the work while welding, use one of the following types of machines:
  - Semiautomatic DC constant-voltage welder
  - DC manual (stick) welder
  - AC welder with reduced voltage control
- Use a GFCI.

13.0.0 Material Safety Data Sheets

All manufacturers of potentially hazardous materials must provide detailed information regarding possible hazards resulting from the use of their product in the form of a material safety data sheet (MSDS) to the users of their products. An MSDS must be provided to anyone using the product or anyone working in the area where the product is in use. Often, companies will post these sheets on a bulletin board or put them in a convenient place near the work area.

All MSDSs must contain the same basic information in at least nine required sections. However, some MSDSs may have additional sections covering transportation, environmental issues, and other subjects, as deemed necessary by the manufacturer. Depending on the manufacturer, MSDSs are often formatted in different ways. *Figure 26* shows a portion of an MSDS produced by one major manufacturer.

14.0.0 Material Handling

Proper lifting, moving, and handling of large, heavy, welded assemblies is important to the safety of the workers and the weldment. Improper work habits can cause serious personal injury and/or damage to equipment and materials.

14.1.0 Lifting

When lifting a heavy object, distribute the weight of the object evenly between both hands. Use your legs to lift, not your back. Do not try to lift a large or bulky object without help if the object is heavier than you can lift with one hand.

14.2.0 Hoists or Cranes

Hoists or cranes can be overloaded with welded assemblies. Check the capacity of the equipment before trying to lift a load. Keep any load as close to the ground as possible while it is being moved. Pushing a load on an overhead crane is better than pulling a load. If it is necessary to pull a load, use a rope. When moving or lifting a load with ropes, chains, and cables, stand to one side of the load to prevent injury if the equipment or material holding the load should break or snap back.

15.0.0 Safety Planning and Emergency Action Plans

Planning is critical to safety. Accidents occur if you let them, but if you anticipate and plan ways to avoid them, accidents can be reduced or eliminated. Hazards in welding can arise from conditions at the work site and from the tools, equipment, and materials being used. Before work begins, the site should be surveyed

Shock Prevention

- Floors must be dry at all times. Use wooden platforms or rubberized carpet/floor coverings or other insulated material.
- Only experienced electricians may work on electric arc welding machine power connections.
- Workbenches must be grounded.
- With the power off, cable connections must be checked for cracked insulators, loose contacts, and worn or cut hoses.
- A machine must never be operated above its rated capacity.

LINCOLN® ELECTRIC

Date:	1/708	MSDS No.:	US-M292
Trade Name:	Excalibur 7018-1 MR		
Sizes:	All		
Supersedes:	1/23/2004		

MATERIAL SAFETY DATA SHEET

For Welding Consumables and Related Products

Conforms to Hazard Communication Standard 29CFR 1910.1200 Rev. October 1988

SECTION I - IDENTIFICATION

Manufacturer/ Supplier: The Lincoln Electric Company
22801 St. Clair Avenue
Cleveland, OH 44117-1199
(216) 481-8100

Product Type: Covered Electrode

Classification: AWS E7018-1H4R

SECTION II - HAZARDOUS MATERIAL (1)

IMPORTANT!
This section covers the materials from which this product is manufactured. The fumes and gases produced during welding with the normal use of this product are covered by Section V; see it for industrial hygiene information.
CAS Number shown is representative for the ingredients listed. All ingredients listed may not be present in all sizes.
(1) The term 'hazardous' in 'Hazardous Materials' should be interpreted as a term required and defined in the Hazards Communication Standard and does not necessarily imply the existence of any hazard.

Ingredients:	CAS No.	Wt.%	TLV mg/m^3	PEL mg/m^3
Iron	7439-89-6	15	10*	10*
Limestone and/or calcium carbonate	1317-65-3	10	10	15
Fluorides (as F)	7789-75-5	< 5	2.5	2.5
Silicates and other binders	1344-09-8	< 5	10*	10*
Titanium dioxides	13463-67-7	< 5	10	15
Manganese and/or manganese alloys and compounds (as Mn)****	7439-96-5	< 5	0.2	5 (c)
Mineral silicates	1332-58-7	< 5	5**	5**
Silicon and/or silicon alloys and compounds (as Si)	7440-21-3	1	10*	10*
Zirconium alloys and compounds (as Zr)	12004-83-0	1	5	5
Molybdenum alloys (as Mo)	7439-98-7	< 0.5	10	10
Lithium compounds (as Li)	554-13-2	< 0.5	10*	10*
Carbon steel core wire	7439-89-6	55	10*	10*

Supplemental Information: (*) Not listed. Nuisance value maximum is 10 milligrams per cubic meter. PEL value for iron oxide is 10 milligrams per cubic meter. TLV value for iron oxide is 5 milligrams per cubic meter.

(**) As respirable dust.

(*****) Subject to the reporting requirements of Sections 311, 312, and 313 of the Emergency Planning and Community Right-to-Know Act of 1986 and of 40CFR 370 and 372.

(c) Values are for manganese fume. STEL (Short Term Exposure Limit) is 3.0 milligrams per cubic meter.

SECTION III - FIRE AND EXPLOSION HAZARD DATA

Non Flammable; Welding arc and sparks can ignite combustibles and flammable products. See Z49.1 referenced in Section VI.

(CONTINUED ON SIDE TWO)

101F26A.EPS

Figure 26 Sample MSDS (1 of 2).

Product: **Excalibur 7018-1 MR**

Date: **1/7/2008**

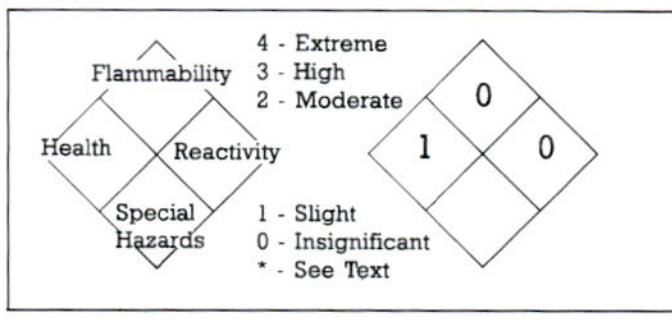

SECTION IV - HEALTH HAZARD DATA

Threshold Limit Value: The ACGIH recommended general limit for Welding Fume NOS - (Not Otherwise Specified) is 5 mg/m^3. ACGIH-1999 preface states that the TLV-TWA should be used as guides in the control of health hazards and should not be used as fine lines between safe and dangerous concentrations. See Section V for specific fume constituents which may modify this TLV. Threshold Limit Values are figures published by the American Conference of Government Industrial Hygienists. Units are milligrams per cubic meter of air.

Effects of Overexposure: Electric arc welding may create one or more of the following health hazards:
Fumes and Gases can be dangerous to your health. Common entry is by inhalation. Other possible routes are skin contact and ingestion.

Short-term (acute) overexposure to welding fumes may result in discomfort such as metal fume fever, dizziness, nausea, or dryness or irritation of nose, throat, or eyes. May aggravate pre-existing respiratory problems (e.g. asthma, emphysema).

Long-term (chronic) overexposure to welding fumes can lead to siderosis (iron deposits in lung) and may affect pulmonary function. Manganese overexposure can affect the central nervous system, resulting in impaired speech and movement. Bronchitis and some lung fibrosis have been reported. WARNING: This product, when used for welding or cutting, produces fumes or gases which contain chemicals known to the State of California to cause birth defects and, in some cases, cancer. (California Health & Safety Code Section 25249.5 et seq.)

Arc Rays can injure eyes and burn skin. *Skin cancer has been reported.*
Electric Shock can kill. If welding must be performed in damp locations or with wet clothing, on metal structures or when in cramped positions such as sitting, kneeling or lying, or if there is a high risk of unavoidable or accidental contact with workpiece, use the following equipment: Semiautomatic DC Welder, DC Manual (Stick) Welder, or AC Welder with Reduced Voltage Control.

Emergency and First Aid Procedures: Call for medical aid. Employ first aid techniques recommended by the American Red Cross.
IF BREATHING IS DIFFICULT give oxygen. IF NOT BREATHING employ CPR (Cardiopulmonary Resuscitation) techniques.
IN CASE OF ELECTRICAL SHOCK, turn off power and follow recommended treatment. In all cases call a physician.

SECTION V - REACTIVITY DATA

Hazardous Decomposition Products: Welding fumes and gases cannot be classified simply. The composition and quantity of both are dependent upon the metal being welded, the process, procedure and electrodes used.

Other conditions which also influence the composition and quantity of the fumes and gases to which workers may be exposed include: coatings on the metal being welded (such as paint, plating, or galvanizing), the number of welders and the volume of the worker area, the quality and amount of ventilation, the position of the welder's head with respect to the fume plume, as well as the presence of contaminants in the atmosphere (such as chlorinated hydrocarbon vapors from cleaning and degreasing activities).

When the electrode is consumed, the fume and gas decomposition products generated are different in percent and form from the ingredients listed in Section II. Decomposition products of normal operation include those originating from the volatilization, reaction, or oxidation of the materials shown in Section II, plus those from the base metal and coating, etc., as noted above.

Reasonably expected fume constituents of this product would include: Primarily iron oxide; secondarily complex oxides of manganese, silicon and sodium.

Maximum fume exposure guideline and PEL for this product is 5.0 milligrams per cubic meter.

Gaseous reaction products may include carbon monoxide and carbon dioxide. Ozone and nitrogen oxides may be formed by the radiation from the arc.

Determine the composition and quantity of fumes and gases to which workers are exposed by taking an air sample from inside the welder's helmet if worn or in the worker's breathing zone. Improve ventilation if exposures are not below limits. See ANSI/AWS F1.1, F1.2, F1.3, F1.4, and F1.5, available from the American Welding Society, 550 N.W. LeJeune Road, Miami, FL 33126.

SECTION VI AND VII
CONTROL MEASURES AND PRECAUTIONS FOR SAFE HANDLING AND USE

Read and understand the manufacturer's instruction and the precautionary label on the product. Request Lincoln Safety Publication E205. See American National Standard Z49.1, 'Safety in Welding, Cutting and Allied Processes' published by the American Welding Society, 550 N.W. LeJeune Road, Miami, FL, 33126 and OSHA Publication 2206 (29CFR1910), U.S. Government Printing Office, Superintendent of Documents, P.O. Box 371954, Pittsburgh, PA 15250-7954 for more details on many of the following:

Ventilation: Use enough ventilation, local exhaust at the arc, or both to keep the fumes and gases from the worker's breathing zone and the general area. Train the welder to keep his head out of the fumes. *Keep exposure as low as possible.*

Respiratory Protection: Use respirable fume respirator or air supplied respirator when welding in confined space or general work area when local exhaust or ventilation does not keep exposure below TLV.

Eye Protection: Wear helmet or use face shield with filter lens shade number 12 or darker. Shield others by providing screens and flash goggles.

Protective Clothing: Wear hand, head, and body protection which help to prevent injury from radiation, sparks and electrical shock. See Z49.1. At a minimum this includes welder's gloves and a protective face shield, and may include arm protectors, aprons, hats, shoulder protection, as well as dark substantial clothing. Train the welder not to permit electrically live parts or electrodes to contact skin . . . or clothing or gloves if they are wet. Insulate from work and ground.

Disposal Information: Discard any product, residue, disposable container, or liner as ordinary waste in an environmentally acceptable manner according to Federal, State and Local Regulations unless otherwise noted.

101F26B.EPS

Figure 26 Sample MSDS. (2 of 2).

for hazardous materials and unsafe conditions. Regardless of the size of the job, the site should always be evaluated to reveal any hazard such as confined spaces, unguarded openings, possible electrical hazards, and environmental hazards.

Once all potential hazards have been identified, a comprehensive emergency action plan should be developed to deal with all emergencies. Depending on prevailing OSHA standards and the number of people involved, the plan may be required to be in writing, or it may be communicated orally. As appropriate, an emergency action plan must include the following elements:

- Emergency escape procedures and emergency route assignments
- Procedures to be followed by employees who remain to take corrective actions
- Procedures that account for all employees after any emergency evacuation has been completed
- Rescue and medical duties, along with the necessary training for those individuals who are to perform them
- The preferred means for reporting fires and other emergencies, including all phone numbers
- Names and/or job titles of persons or departments to be contacted for further information or explanation of duties under the plan

Safety Meetings

It is recommended that brief, informal safety meetings be held by a crew supervisor or safety specialist to review safety concerns and identify specific hazards that may be encountered in upcoming work. These meetings can be held at the beginning of a shift.

Summary

Welders join metal using a high-intensity electric arc to melt and fuse the base metals. Welders work on critical structures such as pipelines, high-rise buildings, ships, and bridges so the trade requires a high level of skill and concentration, which can only be achieved through extensive training and practice. Welders also are called upon to cut and gouge metal using equipment that produces a high-intensity flame. The most common cutting method is oxyfuel cutting, in which oxygen is mixed with a fuel gas such as acetylene to produce the cutting flame.

Safety is everyone's responsibility. Proper clothing, footwear, and eye protection are essential for safe welding and cutting. Workers who fail to comply with safety rules are subject to dismissal. All welding shops must have established plans for dealing with accidents. Take the time to learn the proper procedures for accident response and reporting before you need to respond to an emergency.

Accidents are very harmful for both employees and employers, and they are often caused by careless behavior and unsafe conditions. Most accidents can be prevented. By knowing and avoiding the behaviors that cause accidents and by keeping working conditions safe, it may be possible to avoid injuries and reduce hazards.

Trying to bluff your way through a job you do not understand is asking for trouble. Even if you think you know the correct procedures, a review may bring out an important part of the job that you may have forgotten. Don't be afraid to ask questions. The responses you receive may help new or less experienced co-workers get answers to questions they may be reluctant to ask.

Materials, fumes, welding radiation exposure, and storing and handling cylinders present particular hazards for the welder. Developing an attitude of safety is an excellent way for every worker to avoid or reduce all of these hazards. Practicing good safety attitudes means that you:

- Report all unsafe conditions and behaviors immediately.
- Keep work areas clean and orderly at all times.
- Immediately report all accidents and injuries, no matter how minor.
- Be certain you completely understand the instructions given before starting work.
- Know how and where needed medical help may be obtained.
- Wear the required protective devices when working in a hazardous operation area.
- Do not use alcohol or drugs. If you are ill and must take prescribed medication, notify your supervisor immediately.

Review Questions

1. Accidents are more likely for employees in the age group _____.
   a. 18 and under
   b. 19 to 25
   c. 26 to 45
   d. 60 and over

2. Body protection while welding is best provided by wearing _____.
   a. pants with cuffs
   b. polyester clothing
   c. cotton clothing
   d. loose-fitting shirts

3. Which of the following is a correct statement?
   a. It is not necessary to wear other eye protection if you are wearing a welding helmet.
   b. The same shade of lens can be used for almost all welding and cutting tasks.
   c. The recommended shade for SMAW welding is 10 – 14.
   d. Safety glasses with side shields should not be used because they block peripheral vision.

4. The potentially toxic coating that is used in galvanized metal is _____.
   a. zinc
   b. chromium
   c. barium
   d. manganese

5. To make sure your respirator provides proper protection, you must _____ each time you wear it.
   a. perform a positive fit check
   b. perform both positive and negative fit checks
   c. perform a negative fit check
   d. tighten the respirator head and neck straps as much as possible

6. A confined space is one that _____.
   a. has a flammable atmosphere
   b. has unrestricted means of entry or exit
   c. is designed for continuous employee occupancy
   d. is large enough that an employee can bodily enter and perform tasks

7. A space with an oxygen level below _____ percent is considered oxygen deficient.
   a. 19.5
   b. 20
   c. 21.5
   d. 23.5

8. A good area safety practice is _____.
   a. welding near an open window
   b. welding over wooden floors
   c. to use cardboard to deflect welding/grinding sparks away from others
   d. writing HOT on hot metal before leaving it unattended

9. A hot work permit _____.
   a. authorizes the performance of work potentially posing a fire hazard
   b. promotes development of standard fire safety guidelines
   c. records unsafe conditions at a job site
   d. helps the manager keep records of hazardous spaces

10. Cutting operations should never be performed without a _____ in the area.
    a. bucket of sand
    b. bucket of water
    c. fire watch
    d. fire hose

11. Before performing oxyfuel cutting on a cleaned container, fill the container with an inert gas or _____ and position the container to displace as much air as possible from inside the container.
    a. air
    b. sand
    c. water
    d. solvent

12. Because of its instability, the acetylene gas used in oxyfuel cutting must remain at pressures less than _____ psi when in its gaseous form.

    a. 10
    b. 15
    c. 25
    d. 35

13. Acetylene cylinders in use should be opened no more than _____ turn(s).

    a. ½
    b. ¾
    c. 1½
    d. 2

14. The potential for electric shock _____.

    a. decreases when the skin is damp
    b. remains the same when the skin is damp
    c. increases when the skin is damp
    d. decreases when the skin is cut

15. An MSDS is a form used to _____.

    a. file a worker's compensation claim with an insurance company
    b. list the contents, hazards, and precautions that pertain to a chemical or material
    c. record unsafe conditions that exist at a job site
    d. recommend changes to your employer's safety program

Trade Terms Quiz

Fill in the blank with the correct key trade term that you learned from your study of this module.

1. An electric circuit that is connected to the earth is said to be ________.
2. Nitrogen or argon is used as a(n) ________ in some welding and cutting operations.
3. The cloud of gases and other materials that rises during the welding process is called the ________.
4. An irregular heart rhythm characterized by a fluttering heartbeat is called ________.
5. A condition that poses an immediate or delayed threat to life or health is referred to as ________.
6. When metal parts are joined together to form an electrically conductive path, they are said to be ________.

Trade Terms

Bonded
Electrically grounded
Fume plume
Immediate danger to life and health (IDLH)
Purging gas
Ventricular fibrillation

Cornerstone of Craftmanship

Barry Breeden

Lincoln College of Technology Instructor

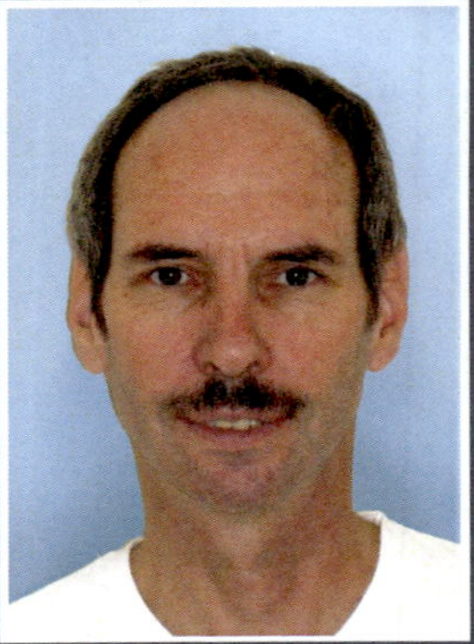

Barry Breeden has had a highly successful career in welding, and he claims the secrets to his success are simple. First, don't ever give up. Second, always keep educating yourself and learning new things. Barry has been an aircraft mechanic for NASA, done welding procedures for the Tomahawk cruise missile and the Trident launch tube, welded in shipyards throughout the U.S., and helped build the Super Collider for Lawrence Livermore National Laboratory. He is currently teaching welding to some very fortunate students at Lincoln College of Technology.

How did you choose a career in the field?
I started out in aeronautics, but when we got to the welding part, I did extremely well. I took an Introduction to Welding class, and did very well in that too. After taking only one class, I was able to get a job in welding. I wanted to learn more, so I went to night school. During the day I practiced what I had learned in that night school class. The more I learned about welding, the more I wanted to learn. I went back to school full time and got an A.S. degree in Welding Technology. Eventually I got my teaching credentials in welding.

What types of training have you been through?
I have been trained as an aircraft mechanic. I have also been through training for most of the welding processes including SMAW, GMAW, GTAW, Pulsed GTAW, FCAW (flux-cored arc welding), SAW (submerged arc welding), and electric arc stud welding, among others. In addition, I have completed training as an OSHA 500 instructor.

What kinds of work have you done in your career?
I am fortunate to have had an interesting and varied career. I was an aircraft mechanic for NASA, worked for Westinghouse Marine Division doing welding procedures for the Tomahawk cruise missile and the Trident launch tube, have welded in most of the shipyards in the United States., and helped build the Super Collider for Lawrence Livermore National Laboratory. I have done lots of welding inspection involving many different codes. Currently, I am teaching at Lincoln College of Technology.

Tell us about your present job and what you like about it.
As a teacher, I really like seeing the faces of new welders when they understand what I'm talking about and grasp what it means to make a good weld. This makes me feel very good.

What factors have contributed most to your success?
I never give up when things get tough. If I got laid off because work was slow, I just went back to school and learned something new. Always make the best use of your time, and never give up.

What advice would you give to those new to the field?
You will have days when everything you do is great, and days when nothing you do turns out right. Whatever you do, don't give up. Education is the key to your growth.

Tell us some interesting career-related facts or accomplishments.
I have many certifications in welding and nondestructive testing (NDT), and have worked in NDT for the aerospace and many other industries. I've served on the board of directors of the local chapter of the AWS for the past 13 years. It's important that the welding industry continue to grow, and I want to do whatever I can to help accomplish that.

Trade Terms Introduced in This Module

Bonded: The permanent joining of metallic parts to form an electrically conductive path that will assure electrical continuity and the capacity to safely conduct any current likely to be imposed on it.

Electrically grounded: Connected to earth or to some conducting body that serves in place of the earth.

Fume plume: The fumes, gases, and particles from the consumables, base metal, and base metal coating during the welding process.

Immediate danger to life and health (IDLH): As defined by OSHA, this term refers to any condition that poses an immediate or delayed threat to life or that would cause irreversible adverse health effects or that would interfere with an individual's ability to escape unaided from a permit space.

Purging gas: A gas such as nitrogen or argon that is classified as inert because it will not react with other material.

Ventricular fibrillation: Irregular heart rhythm characterized by a rapid fluttering heartbeat.

Additional Resources

This module is intended to present thorough resources for task training. The following reference works are suggested for further study. These are optional materials for continued education rather than for task training.

Arc Welding Safety E205, Latest Edition. Cleveland, OH: Lincoln Electric.

Ventilation Guide for Weld Fumes AWS F32M/F32, Latest Edition. Miami, FL: American Welding Society.

Figure Credits

Lincoln Electric Company, 101F01A, 101F01B, 101F02A, 101F04, 101F11, 101F26
Miller Electric Mfg. Co., 101F01C, 101F02B, 101SA01
Rex Ball, 101F02C
Topaz Publications, Inc., 101F05–101F07, 101F09, 101F10, 101SA02, 101SA03, 101F22
North Safety Products USA, 101F08
Table 1 courtesy of American Welding Society *AWS F2.2:2001, Lens Shade Selector*, Reproduced with permission from the American Welding Society (AWS), Miami, FL USA.
Scott Health and Safety, 101F12–101F17
Photo Courtesy of Capital Safety (DBI-SALA and PROTECTA), 101F19
Sellstrom Manufacturing, 101F20
J. R. Yochum, 101F24

NCCER CURRICULA — USER UPDATE

NCCER makes every effort to keep its textbooks up-to-date and free of technical errors. We appreciate your help in this process. If you find an error, a typographical mistake, or an inaccuracy in NCCER's curricula, please fill out this form (or a photocopy), or complete the online form at **www.nccer.org/olf**. Be sure to include the exact module ID number, page number, a detailed description, and your recommended correction. Your input will be brought to the attention of the Authoring Team. Thank you for your assistance.

Instructors – If you have an idea for improving this textbook, or have found that additional materials were necessary to teach this module effectively, please let us know so that we may present your suggestions to the Authoring Team.

NCCER Product Development and Revision

13614 Progress Blvd., Alachua, FL 32615

Email: curriculum@nccer.org

Online: www.nccer.org/olf

❑ Trainee Guide ❑ Lesson Plans ❑ Exam ❑ PowerPoints Other ______________________

Craft / Level: Copyright Date:

Module ID Number / Title:

Section Number(s):

Description:

Recommended Correction:

Your Name:

Address:

Email: Phone:

29102-09

Oxyfuel Cutting

Module Two

Trainees with successful module completions may be eligible for credentialing through NCCER's National Registry. To learn more, go to **www.nccer.org** or contact us at **1.888.622.3720**. Our website has information on the latest product releases and training, as well as online versions of our *Cornerstone* magazine and Pearson's product catalog.

Your feedback is welcome. You may email your comments to **curriculum@nccer.org**, send general comments and inquiries to **info@nccer.org**, or fill in the User Update form at the back of this module.

This information is general in nature and intended for training purposes only. Actual performance of activities described in this manual requires compliance with all applicable operating, service, maintenance, and safety procedures under the direction of qualified personnel. References in this manual to patented or proprietary devices do not constitute a recommendation of their use.

29102-09 Oxyfuel Cutting

Objectives

When you have completed this module, you will be able to do the following:

1. Identify and explain the use of oxyfuel cutting equipment.
2. Set up oxyfuel equipment.
3. Light and adjust an oxyfuel torch.
4. Shut down oxyfuel cutting equipment.
5. Disassemble oxyfuel equipment.
6. Change cylinders.
7. Perform oxyfuel cutting:
   - Straight line and square shapes
   - Piercing and slot cutting
   - Bevels
   - Washing
   - Gouging
8. Operate a motorized, portable oxyfuel gas cutting machine.

Performance Tasks

Under the supervision of your instructor, you should be able to do the following:

1. Set up oxyfuel equipment.
2. Light and adjust an oxyfuel torch.
3. Shut down oxyfuel cutting equipment.
4. Disassemble oxyfuel equipment.
5. Change empty cylinders.
6. Cut shapes from various thicknesses of steel, emphasizing:
   - Straight line
   - Square shape
   - Piercing
   - Bevel
   - Slot
7. Perform washing.
8. Perform gouging.

Trade Terms

Backfire
Carburizing flame
Drag lines
Dross
Ferrous metals
Flashback
Kerf
Neutral flame
Oxidizing flame
Pierce
Soapstone

Prerequisites

Before you begin this module, it is recommended that you successfully complete the following: *Core Curriculum* and *Welding Level One*, Module 29101-09.

Industry-Recognized Credentials

If you're training through an NCCER-accredited sponsor, you may be eligible for credentials from NCCER's Registry. The ID number for this module is 29102-09. Note that this module may have been used in other NCCER curricula and may apply to other level completions. Contact NCCER's Registry at 888.622.3720 or go to **www.nccer.org** for more information.

Contents

Topics to be presented in this module include:

Figures and Tables

Figures and Tables (continued)

1.0.0 INTRODUCTION

Oxyfuel cutting (OFC), also called flame cutting or burning, is a process that uses the flame and oxygen from a cutting torch to cut **ferrous metals.** The flame is produced by burning a fuel gas mixed with pure oxygen. The flame heats the metal to be cut to the kindling temperature (a cherry-red color); then a stream of high-pressure pure oxygen is directed from the torch at the metal's surface. This causes the metal to instantaneously oxidize or burn. The cutting process results in oxides that mix with molten iron and produce **dross**, which is blown from the cut by the jet of cutting oxygen. This oxidation process, which takes place during the cutting operation, is similar to a greatly sped-up rusting process. *Figure 1* shows oxyfuel cutting.

The oxyfuel cutting process is usually used only on ferrous metals such as straight carbon steels, which oxidize rapidly. This process can be used to quickly cut, trim, and shape ferrous metals, including the hardest steel.

Oxyfuel cutting can be used for certain metal alloys, such as stainless steel; however, the process requires higher preheat temperatures (white heat) and about 20 percent more oxygen for cutting. In addition, sacrificial steel plate or rod may have to be placed on top of the cut to help maintain the burning process. Other methods, such as carbon arc cutting, powder cutting, inert gas cutting, and plasma arc cutting, are much more practical for cutting steel alloys and nonferrous metals. These other methods will be covered in a later module.

2.0.0 OXYFUEL SAFETY SUMMARY

The proper safety equipment and precautions must be used when working with oxyfuel equipment because of the potential danger from the high-pressure flammable gases and high temperatures used. The following is a summary of safety procedures and practices that must be observed while cutting or welding. Keep in mind that this is just a summary. Complete safety coverage is provided in the *Level One* module, *Welding Safety*. If you have not completed that module, do so before continuing. Above all, be sure to wear appropriate protective clothing and equipment when welding or cutting.

2.1.0 Protective Clothing and Equipment

- Always use safety glasses with a full face shield or a helmet (*Figure 2*). The glasses, face shield, or helmet lens must have the proper light-reducing tint for the type of welding or cutting to be performed.

102F01.EPS

Figure 1 Oxyfuel cutting.

- Wear proper protective leather and/or flame retardant clothing along with welding gloves that will protect you from flying sparks and molten metal, as well as heat.
- Wear 8" or taller high-top safety shoes or boots. Make sure that the tongue and lace area of the footwear will be covered by a pant leg. If the tongue and lace area is exposed or the footwear must be protected from burn marks, wear leather spats under the pants or chaps and over the top of the footwear.
- Wear a solid material (nonmesh) hat with a bill pointing to the rear or, if much overhead cutting or welding is required, a full leather hood with a welding face plate and the correctly tinted lens. If a hard hat is required, use a hard hat that allows the attachment of rear deflector material and a face shield.
- If a full leather hood is not worn, wear a face shield over safety glasses for gas welding or cutting. Either the face shield or the lenses of the safety glasses must be an approved shade. A shade 3 to 6 filter is recommended, depending on the thickness of the metal being cut, as required by *ANSI Z49.1*.
- If a full leather hood is not worn, wear earmuffs, or at least earplugs, to protect your ear canals from sparks.

2.2.0 Fire/Explosion Prevention

- Never carry matches or gas-filled lighters in your pockets. Sparks can cause the matches to ignite or the lighter to explode, resulting in serious injury.

Figure 2 Personal protective equipment for oxyfuel cutting.

- Never perform any type of heating, cutting, or welding until a hot work permit is obtained and an approved fire watch is established. Most work site fires caused by these types of operations are started by cutting torches.
- Never use oxygen to blow off clothing. The oxygen can remain trapped in the fabric for a time. If a spark hits the clothing during this time, the clothing can burn rapidly and violently out of control.
- Make sure that any flammable material in the work area is moved or shielded by a fire-resistant covering. Approved fire extinguishers must be available before attempting any heating, welding, or cutting operations.
- Always comply with any site requirement for a hot-work permit and/or a fire watch.
- Never release a large amount of oxygen or use oxygen as compressed air. Its presence around flammable materials or sparks can cause rapid and uncontrolled combustion. Keep oxygen away from oil, grease, and other petroleum products.
- Never release a large amount of fuel gas, especially acetylene. Methane and propane tend to concentrate in and along low areas and can ignite at a considerable distance from the release point. Acetylene is lighter than air but is even more dangerous. When mixed with air or oxygen, it will explode at much lower concentrations than any other fuel.
- To prevent fires, maintain a neat and clean work area and make sure that any metal scrap or dross is cold before disposal.
- Before cutting containers such as tanks or barrels, check to see if they have contained any explosive, hazardous, or flammable materials, including petroleum products, citrus products, or chemicals that decompose into toxic fumes when heated. As a standard practice, always clean and then fill any tanks or barrels with water, or purge them with a flow of inert gas to displace any oxygen (*Figure 3*).

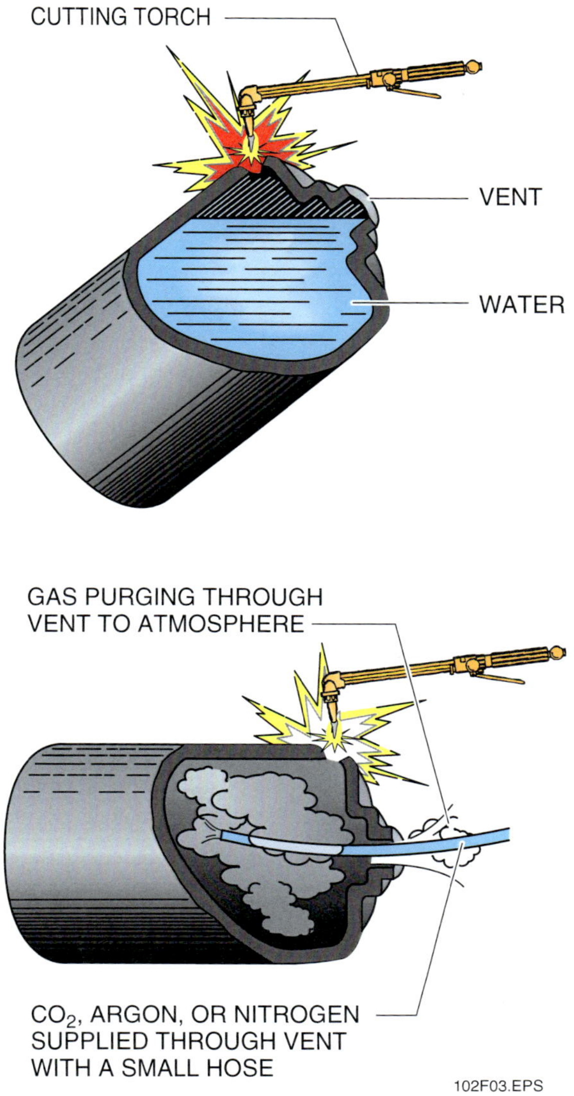

Figure 3 Eliminating/minimizing oxygen in a container.

2.3.0 Work Area Ventilation

- Make sure confined-space procedures are followed before conducting any welding or cutting in the confined space.
- Always perform cutting operations in a well-ventilated area. Cutting operations involving materials or coatings containing cadmium, mercury, lead, zinc, chromium, and beryllium result in toxic fumes. For long-term cutting of such materials, always wear an approved full face, supplied-air respirator (SAR) that uses breathing air supplied externally of the work area. For occasional, very short-term exposure to zinc or copper fumes, a high-efficiency particulate arresting (HEPA)-rated or metal-fume filter may be used on a standard respirator.
- Make sure confined spaces are ventilated properly for cutting or welding purposes.
- Never use oxygen in confined spaces for ventilation purposes.

WARNING!

Respect Explosive Mixtures

An accumulation of an oxygen and fuel-gas mixture can result in a dangerous explosion when ignited. A 2" balloon filled with an oxygen and acetylene mixture has the explosive power of an M80 firecracker (one-quarter stick of dynamite). The mixed gases from 100' of ¼" twin hose will fill a 9" balloon. A 9" balloon filled with mixed oxygen and acetylene gases has an explosive power that is 90 times more powerful than an M80 firecracker.

3.0.0 Oxyfuel Cutting Equipment

The equipment used to perform oxyfuel cutting includes oxygen and fuel gas cylinders, oxygen and fuel gas regulators, hoses, and a cutting torch. A typical movable oxyfuel (oxyacetylene) cutting outfit is shown in *Figure 4*.

Figure 4 Typical oxyacetylene welding/cutting outfit.

3.1.0 Oxygen

Oxygen (O_2) is a colorless, odorless, tasteless gas that supports combustion. Combined with burning material, pure oxygen causes a fire to flare and burn out of control. When mixed with fuel gases, oxygen produces the high-temperature flame required in order to flame cut metals.

3.1.1 Oxygen Cylinders

Oxygen is stored at more than 2,000 psi in hollow steel cylinders. The cylinders come in a variety of sizes based on the cubic feet they hold. The smallest standard cylinder holds about 85 cubic feet of oxygen, and the largest ultra-high-pressure cylinder holds about 485 cubic feet. The most common size oxygen cylinder used for welding and cutting operations is the 227 cubic foot cylinder. It is more than 4' tall and 9" in diameter. *Figure 5* shows standard high-pressure oxygen cylinder markings and sizes. The cylinders must be tested every 10 years.

Oxygen cylinders have bronze cylinder valves on top (*Figure 6*). Turning the cylinder valve handwheel controls the flow of oxygen out of the cylinder. A safety plug on the side of the cylinder valve allows oxygen in the cylinder to escape if the pressure in the cylinder rises too high. Oxygen cylinders are usually equipped with Compressed Gas Association (CGA) 540 valves for service up to 3,000 per square inch gauge (psig). Some cylinders are equipped with CGA 577 valves for up to 4,000 psig service or CGA 701 valves for up to 5,500 psig service. Use care when handling oxygen cylinders because oxygen is stored at such high pressures. When it is not in use, always cover the cylinder valve with the protective steel safety cap and tighten it securely (*Figure 7*).

WARNING!

Do not remove the protective cap unless the cylinder is secured. If the cylinder falls over and the nozzle breaks off, the cylinder will be propelled like a rocket, causing severe injury or death to anyone in its way.

3.2.0 Acetylene

Acetylene gas (C_2H_2), a compound of carbon and hydrogen, is lighter than air. It is formed by dissolving calcium carbide in water. It has a strong, distinctive, garlic-like odor, which is added to the gas so that it can be detected. In its gaseous form, acetylene is extremely unstable and explodes easily. Because of this instability, it cannot be compressed at pressures of more than 15 psi when in its gaseous form. At higher pressures, acetylene gas breaks down chemically, producing heat and

Oxygen Consumption

You should be aware of the fact that two-thirds of the oxygen in a neutral flame comes from the breathing air inside the work space. This could be an issue in a tight work space.

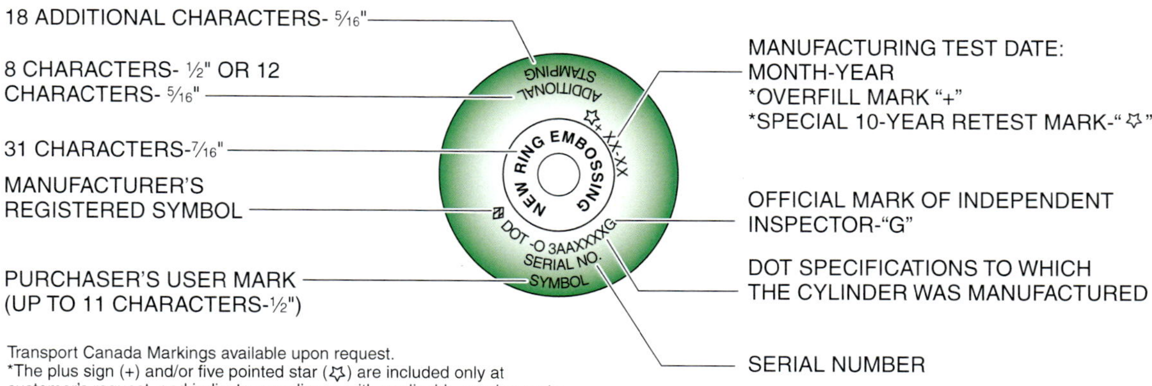

Transport Canada Markings available upon request.
*The plus sign (+) and/or five pointed star (✩) are included only at customer's request, and indicate compliance with applicable requirements of the Code of Federal Regulations, Title 49, Transportation.

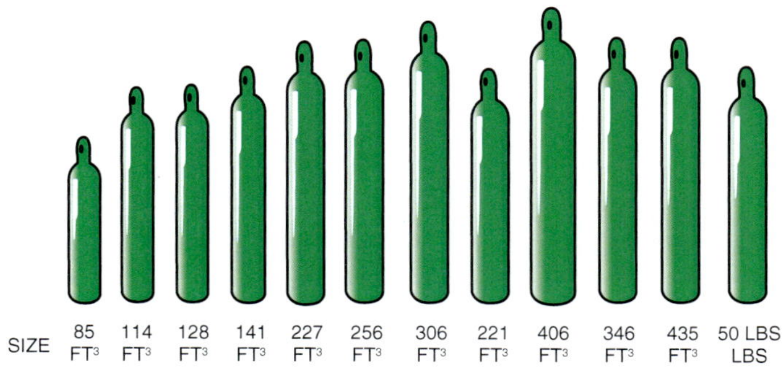

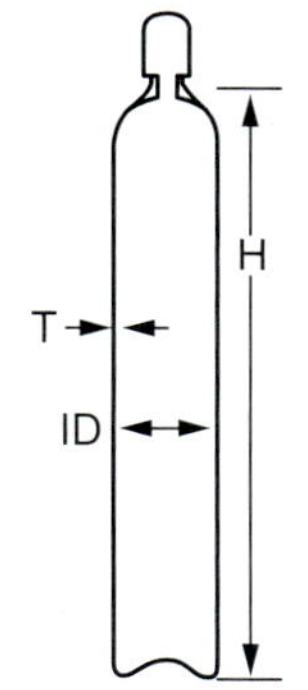

HIGH PRESSURE CYLINDER MARKINGS

DOT SPECIFICATIONS	O_2 CAPACITY (FT³)		WATER CAPACITY (IN³)		NOMINAL DIMENSIONS (IN)			NOMINAL WEIGHT (LB)	PRESSURE (PSI)	
	AT RATED SERVICE PRESSURE	AT 10% OVERCHARGE	MINIMUM	MAXIMUM	AVG. INSIDE DIAMETER "ID"	HEIGHT "H"	MINIMUM WALL "T"		SERVICE	TEST
STANDARD HIGH PRESSURE CYLINDERS[1]										
3AA2015	85	93	960	1040	6.625	32.50	0.144	48	2015	3360
3AA2015	114	125	1320	1355	6.625	43.00	0.144	61	2015	3360
3AA2265	128	140	1320	1355	6.625	43.00	0.162	62	2265	3775
3AA2015	141	155	1630	1690	7.000	46.00	0.150	70	2015	3360
3AA2015	227	250	2640	2710	8.625	51.00	0.184	116	2015	3360
3AA2265	256	281	2640	2710	8.625	51.00	0.208	117	2265	3775
3AA2400	306	336	2995	3060	8.813	55.00	0.226	140	2400	4000
3AA2400	405	444	3960	4040	10.060	56.00	0.258	181	2400	4000
ULTRALIGHT® HIGH PRESSURE CYLINDERS[1]										
E-9370-3280	365	NA	2640	2710	8.625	51.00	0.211	122	3280	4920
E-9370-3330	442	NA	3181	3220	8.813	57.50	0.219	147	3330	4995
ULTRA HIGH PRESSURE CYLINDERS[2]										
3AA3600	347[3]	374	2640	2690	8.500	51.00	0.336	170	3600	6000
3AA6000	434[3]	458	2285	2360	8.147	51.00	0.568	267	6000	10000
E-10869-4500	435[3]	NA	2750	2890	8.813	51.00	0.260	148	4500	6750
E-10869-4500	485[3]	NA	3058	3210	8.813	56.00	0.260	158	4500	6750

1. Regulators normally permit filling these cylinders with 10% overcharge, provided certain other requirements are met.
2. Under no circumstances are these cylinders to be filled to a pressure exceeding the marked service pressure at 70°F.
3. Nitrogen capacity at 70°F.

All cylinders normally furnished with 3/4" NGT internal threads, unless otherwise specified.
Nominal weights include neck ring but exclude valve and cap, add 2 lbs. (.91 kg) for cap and 1 1/2 lb. (.8 kg) for valve.
Cap adds approximately 5 in. (127 mm) to height.
Cylinder capacities are approximately 5 in. (127 mm) to height.
Cylinder capacities are approximately at 70°F. (21°C).

102F05.EPS

Figure 5 High-pressure oxygen cylinder markings and sizes.

Figure 6 Oxygen cylinder valve.

pressure that could result in a violent explosion. When combined with oxygen, acetylene creates a flame that burns hotter than 5,500°F, one of the hottest gas flames. Acetylene can be used for flame cutting, welding, heating, flame hardening, and stress relieving.

Tank Identification Transducers

Many gas suppliers mount transducers on their tanks so that they can readily identify the tanks. The transducer is electronically scanned, and a coded number is matched against the supplier's records for identification. This aids in quickly determining the purchaser or user of the tank as well as the required retesting date.

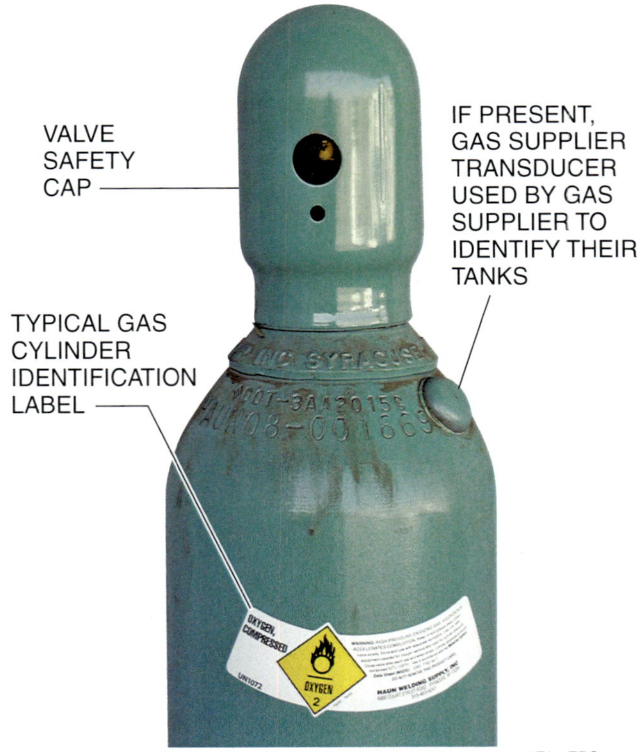

Figure 7 Oxygen cylinder with standard safety cap.

3.2.1 Acetylene Cylinders

Because of the explosive nature of acetylene gas, it cannot be stored above 15 psi in a hollow cylinder. To solve this problem, acetylene cylinders are specially constructed to store acetylene at higher pressures. The acetylene cylinder is filled with a porous material that creates a solid, instead of a hollow, cylinder. The porous material is soaked with acetone, which absorbs the acetylene, stabilizing it for storage at higher pressures. Because of the liquid acetone inside the cylinder, acetylene cylinders must always be used in an upright

Alternate High-Pressure Cylinder Valve Cap

High-pressure cylinders can also be equipped with a clamshell cap that can be closed to protect the cylinder valve with or without a regulator installed on the valve. This enables safe movement of the cylinder after the cylinder valve is closed. This type of cap is usually secured to the cylinder body cap threads when it is installed so that it cannot be removed. When the clamshell is closed, it can also be padlocked to prevent unauthorized operation of the cylinder valve.

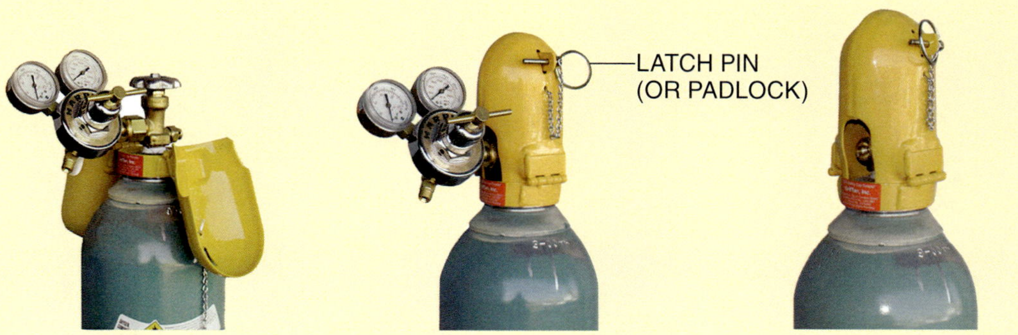

102SA01.EPS

position. If the cylinder is tipped over, stand the cylinder upright and wait at least one hour before using it. If liquid acetone is withdrawn from a cylinder, it will gum up the safety check valves and regulators. Always take care to withdraw acetylene gas from a cylinder at pressures less than 15 psig and at hourly rates that do not exceed one-seventh of the cylinder capacity. Higher rates may cause liquid acetone to be withdrawn along with the acetylene.

Acetylene cylinders have safety fuse plugs in the top and bottom of the cylinder that melt at 220°F (*Figure 8*). In the event of a fire, the fuse plugs will release the acetylene gas, preventing the cylinder from exploding.

Acetylene cylinders are available in a variety of sizes based on the cubic feet of acetylene that they can hold. The smallest standard cylinder holds about 10 cubic feet of gas. The largest standard cylinder holds about 420 cubic feet of gas. A cylinder that holds about 850 cubic feet is also available. *Figure 9* shows standard acetylene cylinder markings and sizes. Like oxygen cylinders, acetylene cylinders must be tested every 10 years.

Acetylene cylinders are usually equipped with a standard CGA 510 brass cylinder valve (see *Figure 9*). The handwheel of the valve controls the flow of acetylene from the cylinder to a regulator. Some acetylene cylinders are equipped with an alternate standard CGA 300 valve. Some obsolete valves still in use require a special long-handled wrench with a square socket end to operate the valve.

The smallest standard acetylene cylinder, which holds 10 cubic feet, is equipped with a CGA 200 small series valve, and 40-cubic-foot cylinders use a CGA 520 small series valve. As with oxygen cylinders, place a protective valve cap on the acetylene cylinders during transport (*Figure 10*).

WARNING!

Do not remove the protective cap unless the cylinder is secured. If the cylinder falls over and the nozzle breaks off, the cylinder will release highly explosive gas.

102F08.EPS

Figure 8 Standard acetylene cylinder valve and safety plugs.

Alternate Acetylene Cylinder Safety Cap

Acetylene cylinders can be equipped with a ring guard cap that protects the cylinder valve with or without a regulator installed on the valve. This enables safe movement of the cylinder after the cylinder valve is closed. This type of cap is usually secured to the cylinder body cap threads when it is installed so that it cannot be removed.

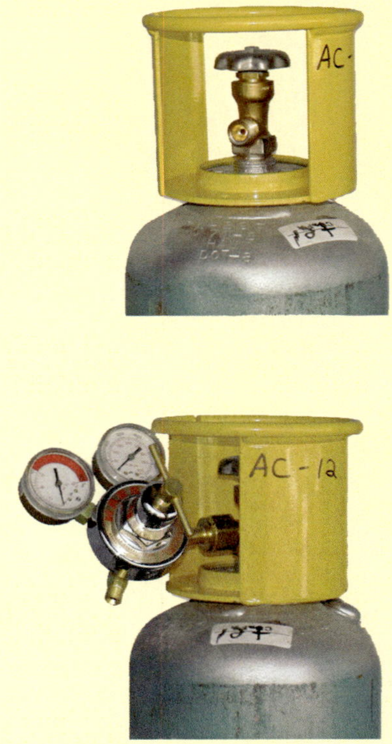

102SA02.EPS

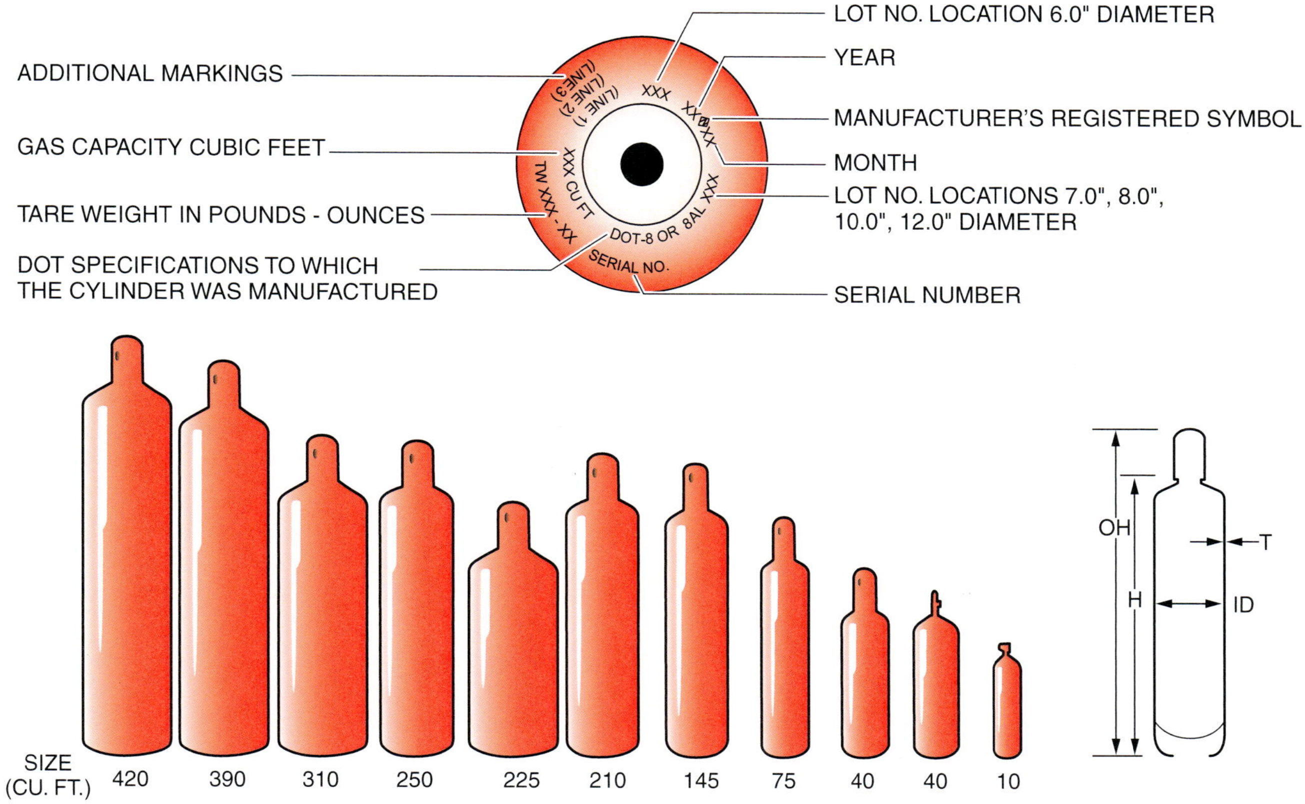

ACETYLENE CYLINDER MARKINGS

DOT SPECIFICATIONS	CAPACITY ACETYLENE (FT³)	CAPACITY MIN. WATER (IN³)	CAPACITY MIN. WATER (LB.)	NOMINAL DIMENSIONS (IN) AVG. INSIDE DIAMETER "ID"	HEIGHT W/OUT VALVE OR CAP "H"	HEIGHT W/VALVE AND CAP "OH"	MINIMUM WALL "T"	ACETONE (LB - OZ)	APPROXIMATE TARE WEIGHT WITH VALVE WITHOUT CAP (LB)
8 AL[1]	10	125	4.5	3.83	13.1375	14.75	0.0650	1-6	8
8[1]	40	466	16.8	6.00	19.8000	23.31	0.0870	5-7	25
8[2]	40	466	16.8	6.00	19.8000	28.30	0.0870	5-7	28
8[3]	75	855	30.8	7.00	25.5000	31.25	0.0890	9-8	45
8	100	1055	38.0	7.00	30.7500	36.50	0.0890	12-2	55
8	145	1527	55.0	8.00	34.2500	40.00	0.1020	18-10	76
8	210	2194	79.0	10.00	32.2500	38.00	0.0940	25-13	105
8AL	225	2630	94.7	12.00	27.5000	32.75	0.1280	29-6	110
8	250	2606	93.8	10.00	38.0000	43.75	0.0940	30-12	115
8AL	310	3240	116.7	12.00	32.7500	38.50	0.1120	39-5	140
8AL	390	4151	150.0	12.00	41.0000	46.75	0.1120	49-14	170
8AL	420	4375	157.5	12.00	43.2500	49.00	0.1120	51-14	187
8	60	666	24.0	7.00	25.79 OH		0.0890	7-11	40
8	130	1480	53.3	8.00	36.00 OH		0.1020	17-2	75
8AL	390	4215	151.8	12.00	46.00 OH		0.1120	49-14	180

1. Tapped for 3/8" valve but are not equipped with valve protection caps.
2. Includes valve protection cap.
3. Can be tared to hold 60 ft³ (1.7 m³) of acetylene gas.

Standard tapping (except cylinders tapped for 3/8") 3/4"-14 NGT.

Weight includes saturation gas, filler, paint, solvent, valve, fuse plugs.
Does not include cap of 2 lb. (.91 kg.)
Cylinder capacities are based upon commercially pure acetylene gas at 250 psi (17.5 kg/cm²), and 70°F (15°C).

102F09.EPS

Figure 9 Acetylene cylinder markings and sizes.

102F10.EPS

Figure 10 Acetylene cylinder with standard valve safety cap.

3.3.0 Liquefied Fuel Gases

Many fuel gases other than acetylene are used for cutting. They include natural gas and liquefied fuel gases such as methylacetylene propadiene (MAPP®), propylene, and propane. Their flames are not as hot as acetylene, but they have higher British thermal unit (Btu) ratings and are cheaper and safer to use. The supervisor at your job site will determine which fuel gas to use. *Table 1* compares the flame temperatures of oxygen mixed with various fuel gases.

MAPP® is a Dow Chemical Company product that is a chemical combination of acetylene and propane gases. MAPP® gas burns at temperatures almost as high as acetylene and has the stability of propane. Because of this stability, it can be used at pressures over 15 psi and is not as likely as acetylene to **backfire** or **flashback**. MAPP® also has an offensive odor that can be detected easily. MAPP® gas can be used for flame cutting, heating, stress relieving, brazing, soldering, and scarfing (cleaning cutting dross or other material from the workpiece).

Propylene mixtures are hydrocarbon-based gases that are stable and shock-resistant, making them relatively safe to use. They are purchased under trade names such as High Purity Gas (HPG™), Apachi™, and Prestolene™. These gases and others have distinctive odors to make leak detection easier. They burn at temperatures around 5,193°F, hotter than natural gas and propane. Propylene gases are used for flame cutting, scarfing, heating, stress relieving, brazing, and soldering.

Propane is also known as liquefied petroleum (LP) gas. It is stable and shock-resistant, and it has a distinctive odor for easy leak detection. It burns at 4,580°F, which is the lowest temperature of any fuel gas. It has a slight tendency toward backfire and flashback and is used quite extensively for cutting procedures.

Natural gas is delivered by pipeline rather than by cylinders. It burns at about 4,600°F. Natural gas is relatively stable and shock-resistant and has a slight tendency toward backfire and flashback. Because of its recognizable odor, leaks are easily detectable. Natural gas is used primarily for cutting on job sites with permanent cutting stations.

Table 1 Flame Temperatures of Oxygen with Various Fuel Gases

Type of Gas	Flame Temperature
Acetylene	More than 5,500°F
MAPP®	5,300°F
Propylene	5,130°F
Natural gas	4,600°F
Propane	4,580°F

3.3.1 Liquefied Fuel Gas Cylinders

Liquefied fuel gases are shipped in hollow steel cylinders (*Figure 11*). When empty, they are much lighter than acetylene cylinders.

The liquefied fuel gas is stored in hollow steel cylinders of various sizes. They can hold from 30 to 225 pounds of fuel gas. As the cylinder valve is opened, the vaporized gas is withdrawn from the cylinder. The remaining liquefied gas absorbs heat and releases additional vaporized gas. The

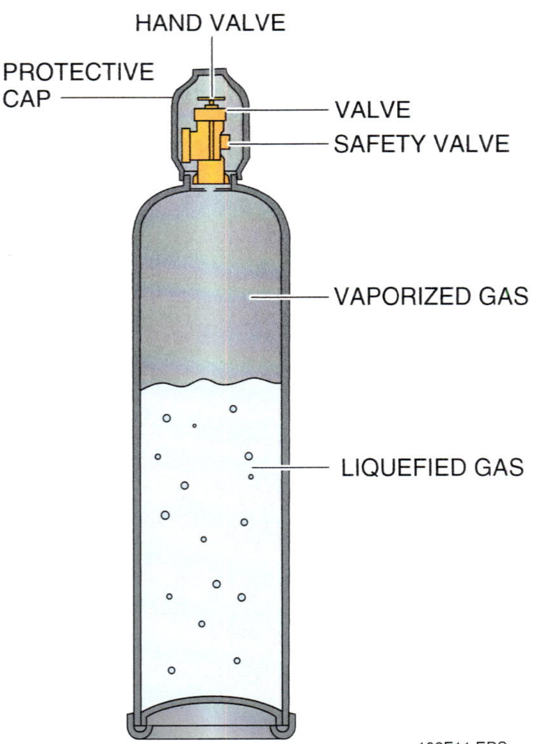

102F11.EPS

Figure 11 Liquefied fuel gas cylinder.

pressure of the vaporized gas varies with the outside temperature. The colder the outside temperature, the lower the vaporized gas pressure will be. If high volumes of gas are removed from a liquefied fuel gas cylinder, the pressure drops, and the temperature of the cylinder will also drop. A ring of frost can form around the base of the cylinder. If high withdrawal rates continue, the regulator may also start to ice up. If high withdrawal rates are required, special regulators with electric heaters should be used.

WARNING!

Never apply heat directly to a cylinder or regulator. This can cause excessive pressures, resulting in an explosion.

The pressure inside a liquefied fuel gas cylinder is not an indicator of how full or empty the cylinder is. The weight of a cylinder determines how much liquefied gas is left. Liquefied fuel gas cylinders are equipped with CGA 510, 350, or 695 valves, depending on the fuel and storage pressures.

WARNING!

Do not remove the protective cap on liquefied fuel gas cylinders unless the cylinder is secured. If the cylinder falls over and the nozzle breaks off, the cylinder will release highly explosive gas. Cylinders containing a liquid such as propane must be kept in an upright position. If the valve is broken or is opened with the cylinder horizontal, the fuel can emerge as a liquid that will shoot a long distance before it vaporizes. If it is ignited, it will act like a flamethrower.

3.4.0 Regulators

Regulators (*Figure 12*) are attached to the cylinder valve. They reduce the high cylinder pressures to the required lower working pressures and maintain a steady flow of gas from the cylinder.

Handling and Storing Liquefied Gas Cylinders

Liquefied fuel gas cylinders have a safety valve built into the valve at the top of the cylinder. The safety valve releases gas if the pressure begins to rise. Use care when handling fuel gas cylinders because the gas in cylinders is stored at such high pressures. Cylinders should never be dropped or hit with heavy objects, and they should always be stored in an upright position. When not in use, the cylinder valve must always be covered with the protective steel cap.

The pressure-adjusting screw controls the gas pressure. Turned clockwise, it increases the pressure of gas. Turned counterclockwise, it reduces the pressure of gas. When turned counterclockwise until loose (released), it stops the flow of gas.

Most regulators contain two gauges. The high-pressure or cylinder-pressure gauge indicates the actual cylinder pressure; the low-pressure or working-pressure gauge indicates the pressure of the gas leaving the regulator.

Oxygen regulators differ from fuel gas regulators. Oxygen regulators may have green markings and always have right-hand threads on all connections. The oxygen regulator's high-pressure gauge generally reads up to 3,000 psi and includes a second scale that shows the amount of oxygen in the cylinder in terms of cubic feet. The low-pressure or working-pressure gauge may read 100 psi or higher.

Fuel gas regulators may have red markings and usually have left-hand threads on all the connections. As a reminder that the regulator has left-hand threads, a V-notch may be cut around the nut. The fuel gas regulator's high-pressure gauge usually reads up to 400 psi. The low-pressure or working-pressure gauge may read up to 40 psi. Acetylene gauges, however, are always red-lined at 15 psi as a reminder that acetylene pressure should not be increased over 15 psi.

There are two types of regulators: single-stage and two-stage.

MAPP® Gas

Unlike acetylene, which must be used at a pressure less than 15 psi, MAPP® gas can be used at the full cylinder pressure of 95 psi. For that reason, it can be used for underwater cutting in deep water. Because it is stored as a liquid, MAPP® is sold by the pound, rather than the cubic foot like acetylene. MAPP® cylinders, unlike acetylene cylinders, do not require acetone or other packing. Therefore, the same size cylinder can hold much more MAPP® than acetylene.

Source: **www.free-ed.net**

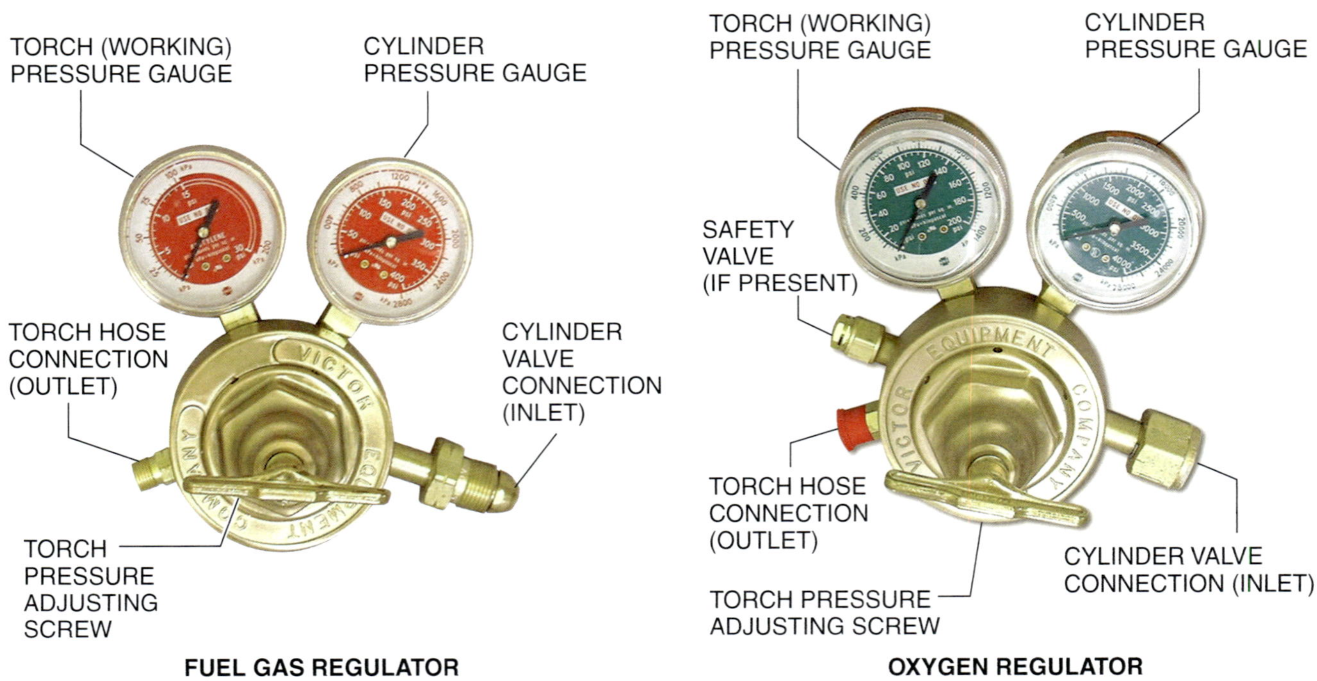

Figure 12 Oxygen and acetylene regulators.

WARNING!

To prevent injury and damage to regulators, always follow these guidelines:

- Never subject regulators to jarring or shaking, as this can damage the equipment beyond repair.
- Always check that the adjusting screw is fully released before the cylinder valve is turned on and when the welding has been completed.
- Always open cylinder valves slowly and stand with the valve stem between you and the regulator.
- Never use oil to lubricate a regulator. This can result in an explosion when the regulator is in use.
- Never use fuel gas regulators on oxygen cylinders or oxygen regulators on fuel gas cylinders.
- Never work with a defective regulator. If it is not working properly, shut off the gas supply and have the regulator repaired by someone who is qualified to work on it.
- Never use large wrenches, pipe wrenches, pliers, or slipjoint pliers to install or remove regulators.

3.4.1 Single-Stage Regulators

Single-stage, spring-compensated regulators reduce pressure in one step. As gas is drawn from the cylinder, the internal pressure of the cylinder decreases. A single-stage, spring-compensated regulator is unable to automatically adjust for this decrease in internal cylinder pressure. Therefore, it becomes necessary to adjust the spring pressure to periodically adjust the output gas pressure as the gas in the cylinder is consumed. These regulators are the most commonly used because of their low cost and high flow rates.

3.4.2 Two-Stage Regulators

The two-stage, pressure-compensated regulator reduces pressure in two steps. It first reduces the input pressure from the cylinder to a predetermined intermediate pressure. The intermediate pressure is then adjusted by the pressure-adjusting screw. With this type of regulator, the delivery pressure to the torch remains constant, and no readjustment is necessary as the gas in the cylinder is consumed. Standard two-stage regulators are more expensive than single-stage regulators and have lower flow rates. There are also heavy-duty types with higher flow rates that are usually preferred for thick material and/or continuous-duty cutting operations.

3.4.3 Check Valves and Flashback Arrestors

Check valves and flashback arrestors (*Figure 13*) are safety devices for regulators, hoses, and torches. Check valves allow gas to flow in one direction only. Flashback arrestors stop fire.

Check valves consist of a ball and spring that open inside a cylinder. The valve allows gas to move in one direction but closes if the gas attempts to flow in the opposite direction. When a torch is first pressurized or when it is being shut off, back-pressure check valves prevent the entry and mixing of acetylene with oxygen in the oxygen hose or the entry and mixing of oxygen with acetylene in the acetylene hose.

Flashback arrestors prevent flashbacks from reaching the hoses and/or regulator. They have a flame-retarding filter that will allow heat, but not flames, to pass through. Most flashback arrestors also contain a check valve.

Add-on check valves and flashback arrestors are designed to be attached to the torch handle connections and to the regulator outlets. As a minimum, flashback arrestors with check valves should be attached to the torch handle connections. Both devices have arrows on them to indicate flow direction. When installing add-on check valves and flashback arrestors, be sure the arrow matches the desired gas flow direction.

3.5.0 Hoses

Hoses transport gases from the regulators to the torch. Oxygen hoses are usually green with right-hand threaded connections. Hoses for fuel gas are usually red and have left-hand threaded connections. The fuel gas connections may also be grooved as a reminder that they have left-hand threads.

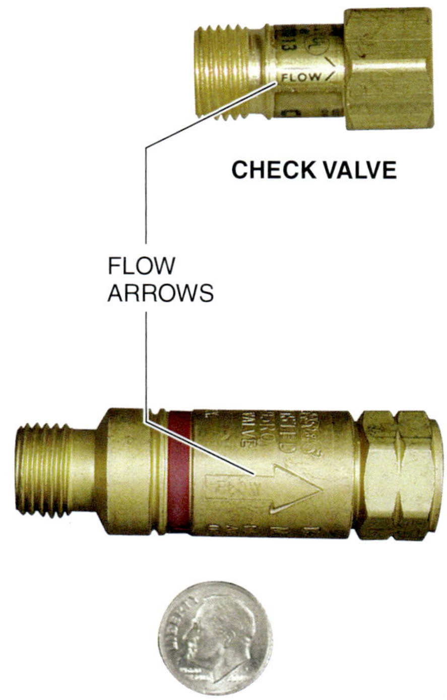

Figure 13 Torch handle add-on check valve and flashback arrestor.

Proper care and maintenance of the hose is important for maintaining a safe, efficient work area. Remember the following guidelines for hoses:

- Protect the hose from molten dross or sparks, which will burn the exterior. Although some hoses are flame retardant, they will burn.
- Remove the hoses from under the metal being cut. If the hot metal falls on the hose, the hose will be damaged.
- Frequently inspect and replace hoses that show signs of cuts, burns, worn areas, cracks, or damaged fittings.

Torch Wrenches

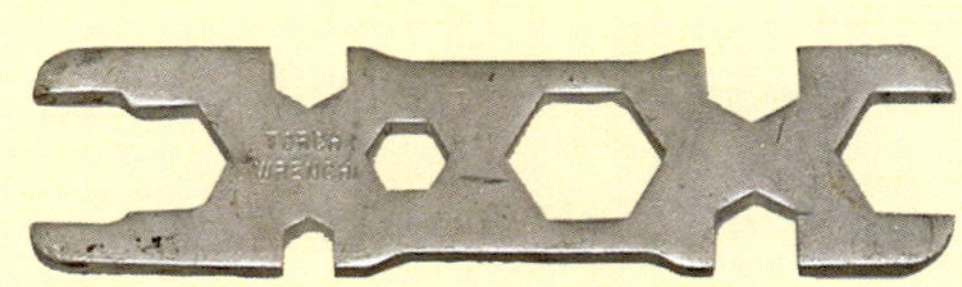

Only a torch wrench, sometimes called a gang wrench, should be used to install regulators, hose connections, check valves, flashback arrestors, torches, and torch tips. The universal torch wrench shown is equipped with various size wrench cutouts for use with a variety of equipment and standard CGA components. The fittings for oxyfuel equipment are brass or bronze, and certain components are often fitted with soft, flexible, O-ring seals. The seal surfaces of the fittings or O-rings can be easily damaged by overtightening with standard wrenches. The length of a torch wrench is limited to reduce the chances of damage to fittings because of excessive torque. In some cases, manufacturers specify only hand-tightening for certain fitting connections of a torch set (tips or cutting/welding attachments). In any event, follow the manufacturer's specific instructions when connecting the components of a torch set.

Flashback Arrestors

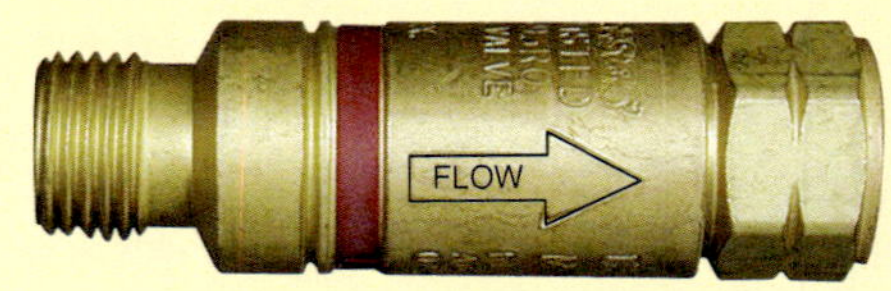

102SA04.EPS

It is highly recommended that add-on flashback arrestors with check valves, rather than just check valves, be installed at both the torch handle and regulator connections to prevent serious injury or property damage. If the flame front of a flashback, sometimes called the backburn, gets by the check valves into the hoses or regulators, the hoses could burn through, and a large, very dangerous, uncontrolled fire could start. Newer torches are available with check valves and flashback arrestors built into the torch handle.

- Never use pipe-fitting compounds or lubricants around hose connections. These compounds often contain oil or grease, which ignite and burn or explode in the presence of oxygen.

NOTE

MAPP®, propane, and propylene require hoses designed for the mixture of hydrocarbons present in these fuels. Check with a hose manufacturer to obtain their recommendations for hoses suitable for use with these gases.

3.6.0 Cutting Torches

Cutting torches mix oxygen and fuel gas for the torch flame and control the stream of oxygen necessary for the cutting jet. Depending on the job site, you may use either a one-piece or a combination cutting torch.

3.6.1 One-Piece Hand Cutting Torch

The one-piece hand cutting torch, sometimes called a demolition torch, contains the fuel gas and oxygen valves that allow the gases to enter the chambers and then flow into the part of the torch where they are mixed. The main body of the torch is called the handle. The torch valves control the fuel gas and oxygen used for preheating the metal to be cut. The cutting oxygen lever, which is spring-loaded, controls the jet of cutting oxygen. Hose connections are located at the end of the torch body behind the valves. *Figure 14* shows a three-tube one-piece positive-pressure hand cutting torch in which the preheat fuel and oxygen are mixed in the tip. These torches are designed for heavy-duty cutting. They have long supply tubes from the torch handle to the torch head to reduce radiated heat to the operator's hands. The torches are generally available in capacities for cutting steel up to 120 thick. Larger-capacity torches, with the ability to cut steel up to 360 thick, can also be obtained.

Two different types of oxyfuel cutting torches are in general use. The positive-pressure torch is designed for use with fuel supplied through a regulator from pressurized fuel storage cylinders. The injector torch is designed to use a vacuum created by oxygen flow to draw the necessary amount of fuel from a very-low-pressure fuel source, such as a natural gas line or acetylene generator. The injector torch, when used, is most often found in continuous-duty high-volume manufacturing applications. Both types may employ one of two different fuel-mixing methods:

- Torch-handle or supply-tube mixing
- Torch-head or tip mixing

The two methods can normally be distinguished by the number of supply tubes from the torch handle to the torch head. Torches that use three tubes (see *Figure 14*) from the handle to the head mix the preheat fuel and oxygen at the torch head or tip. This method tends to help eliminate any flashback damage to the torch head supply tubes and torch handle. Torches with two tubes usually mix the preheat fuel and oxygen in a mixing chamber in the torch body or in one of the supply tubes. Injector torches usually have the injector located in one of the supply tubes, and the mixing occurs in that tube from the injector to the torch head. Some older torches that have only two visible tubes are actually three-tube torches that mix the preheat fuel and oxygen in the torch head or tip. This is accomplished by using a separate preheat fuel tube inside a larger preheat oxygen tube.

3.6.2 Combination Torch

The combination torch consists of a cutting torch attachment that fits onto a welding torch handle. These torches are normally used in light-duty or medium-duty applications. Fuel gas and oxygen valves are on the torch handle. The cutting attachment has a cutting oxygen lever and another oxygen valve to control the preheat flame. When the cutting attachment is screwed onto the torch handle, the torch handle oxygen valve is opened all the way, and the preheat oxygen is controlled by an oxygen valve on the cutting attachment. When the

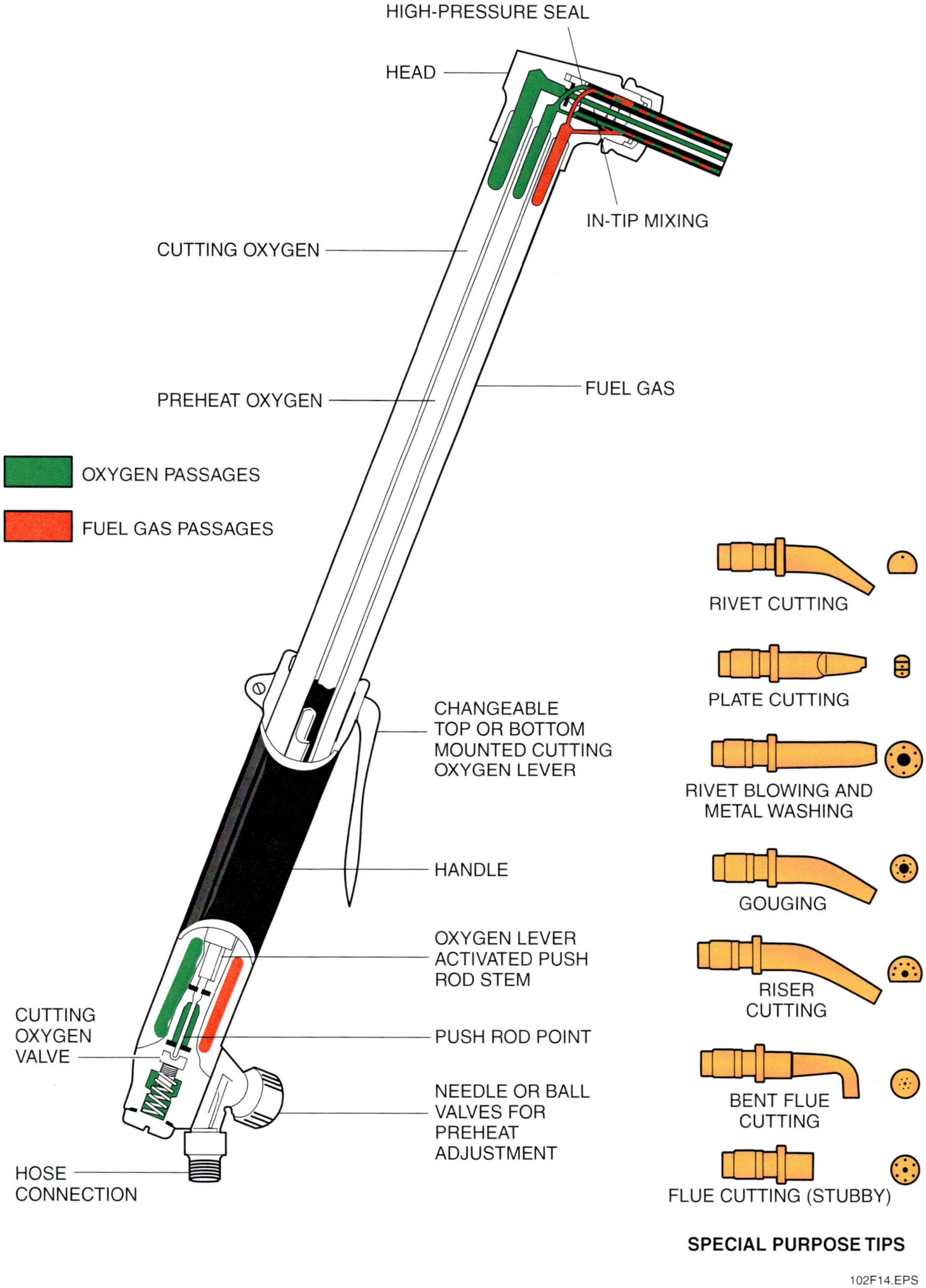

Figure 14 Heavy-duty three-tube one-piece positive-pressure hand cutting torch.

Built-In Flashback Arrestors

Newer one-piece hand cutting torches and combination torches with built-in check valves and/or flashback arrestors are available from most manufacturers.

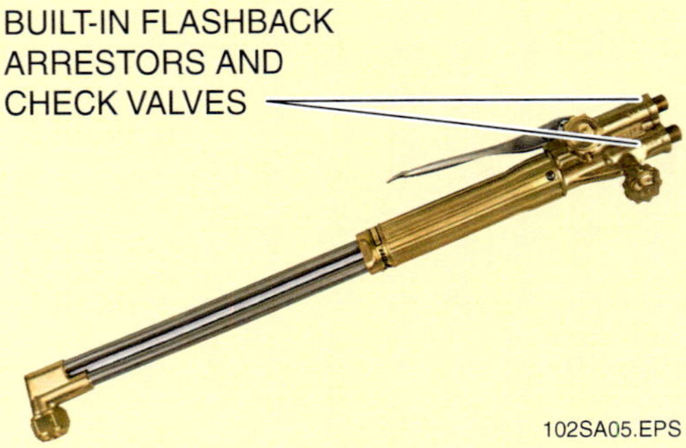

cutting attachment is removed, welding and heating tips can be screwed onto the torch handle. *Figure 15* shows a two-tube combination torch in which the preheat mixing is accomplished in a supply tube. These torches are usually positive-pressure torches with mixing occurring in the attachment body, supply tube, head, or tip. These torches may be equipped with built-in flashback arrestors and check valves.

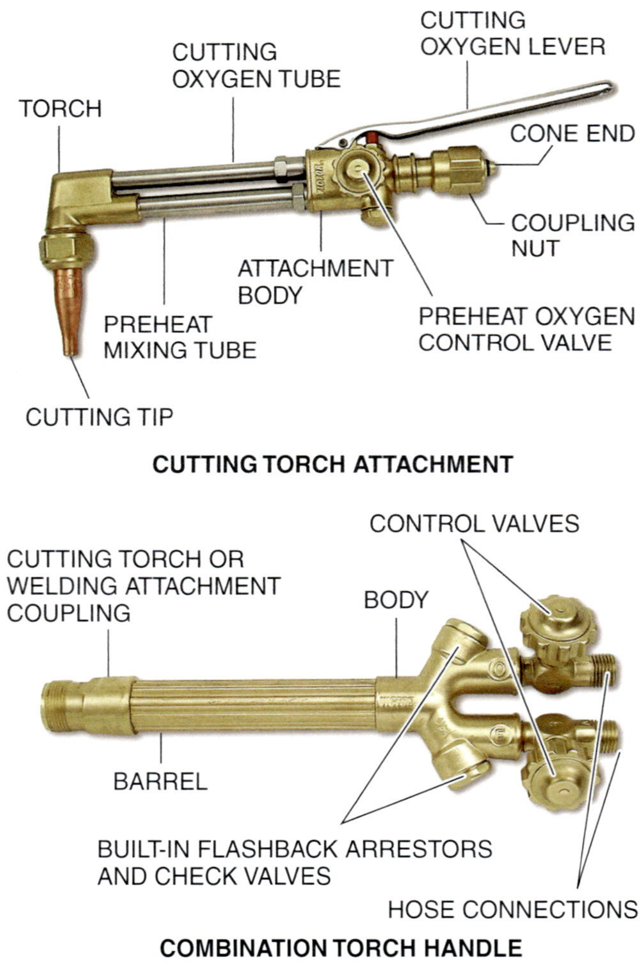

Figure 15 Combination cutting torch.

3.7.0 Cutting Torch Tips

Cutting torch tips, or nozzles, fit into the cutting torch and are either screwed in or secured with a tip nut. There are one- and two-piece cutting tips (*Figure 16*).

One-piece cutting tips are made from a solid piece of copper. Two-piece cutting tips have a separate external sleeve and internal section.

Torch manufacturers supply literature explaining the appropriate torch tips and gas pressures to be used for various applications. *Table 2* shows a sample cutting tip chart that lists recommended tip sizes and gas pressures for use with acetylene fuel gas and a specific manufacturer's torch and tips.

WARNING!

Do not use the cutting tip chart from one manufacturer for the cutting tips of another manufacturer. The gas flow rate of the tips may be different, resulting in excessive flow rates. Different gas pressures may also be required. The cutting torch tip to be used depends on the base metal thickness and fuel gas being used. Special-purpose tips are also available for use in such operations as gouging and grooving.

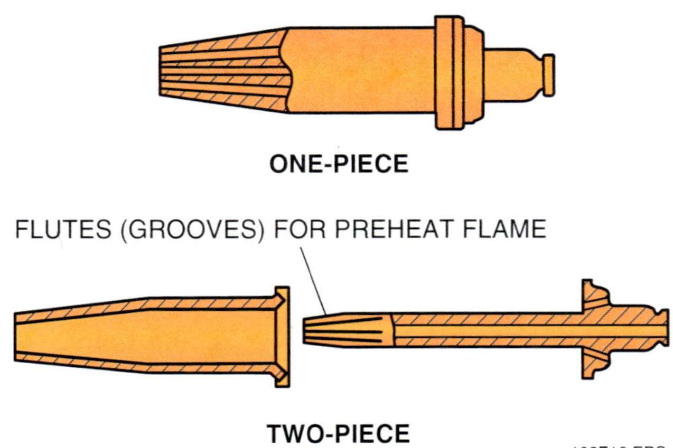

Figure 16 One- and two-piece cutting tips.

Table 2 Sample Acetylene Cutting Tip Chart

Cutting Tip Series 1-101, 3-101, and 5-101						
Metal Thickness	Tip Size	Cutting Oxygen Pressure (psig)*	Preheat Oxygen (psig)*	Acetylene Pressure (psig)	Speed (in./min.)	Kerf Width
⅛"	000	20/25	3/5	3/5	20/30	0.04
¼"	00	20/25	3/5	3/5	20/28	0.05
⅜"	0	25/30	3/5	3/5	18/26	0.06
½"	0	30/35	3/6	3/5	16/22	0.06
¾"	1	30/35	4/7	3/5	15/20	0.07
1"	2	35/40	4/8	3/6	13/18	0.09
2"	3	40/45	5/10	4/8	10/12	0.11
3"	4	40/50	5/10	5/11	8/10	0.12
4"	5	45/55	6/12	6/13	6/9	0.15
6"	6	45/55	6/15	8/14	4/7	0.15
10"	7	45/55	6/20	10/15	3/5	0.34
12"	8	45/55	7/25	10/15	3/4	0.41

*The lower side of the pressure listing is for hand cutting and the higher side is for machine cutting.

The cutting torch tip to be used depends on the base metal thickness and fuel gas being used. Special-purpose tips are also available for use in such operations as gouging and grooving.

3.7.1 Cutting Tips for Acetylene

One-piece torch tips are usually used with acetylene cutting because of the high temperatures involved. They can have four, six, or eight preheat holes in addition to the single cutting hole. *Figure 17* shows the arrangement of typical acetylene torch cutting tips.

3.7.2 Cutting Tips for Liquefied Fuel Gases

Tips used with liquefied fuel gases must have at least six preheat holes. Because fuel gases burn at lower temperatures than acetylene, more holes are necessary for preheating. Tips used with liquefied fuel gases can be one- or two-piece cutting tips. *Figure 18* shows typical cutting tips used with liquefied fuel gases.

3.7.3 Special-Purpose Cutting Tips

Special-purpose tips are available for special cutting jobs. These jobs include cutting sheet metal, rivets, risers, and flues, as well as washing and

Torch Tip Styles

Nearly all manufacturers use different tip-to-torch mounting designs, sealing surfaces, and diameters. In addition, tip sizes and flow rates are usually not the same between manufacturers even though the number designations may be the same. This makes it impossible to safely interchange cutting tips between torches from different manufacturers. Even though some tips from different manufacturers may appear to be the same, do not interchange them. The sealing surfaces are very precise, and serious leaks may occur that could result in a dangerous fire or flashback.

102SA06.EPS

gouging. *Figure 19* shows special-purpose torch cutting tips.

- The sheet metal cutting tip has only one preheat hole. This minimizes the heat and prevents distortion in the sheet metal. These tips are normally used with a motorized carriage but can also be used for hand cutting.
- Rivet cutting tips are used to cut off rivet heads, bolt heads, and nuts.
- Riser cutting tips are similar to rivet cutting tips and can also be used to cut off rivet heads, bolt heads, and nuts. They have extra preheat holes to cut risers, flanges, or angle legs faster. They can be used for any operation that requires a cut close to and parallel to another surface, such as in removing a metal backing.
- Rivet blowing and metal washing tips are heavy-duty tips designed to withstand high heat. They are used for coarse cutting and for removing such items as clips, angles, and brackets.
- Gouging tips are used to groove metal in preparation for welding.
- Flue cutting tips are designed to cut flues inside boilers. They also can be used for any cutting operation in tight quarters where it is difficult to get a conventional tip into position.

3.8.0 Tip Cleaners and Tip Drills

With use, cutting tips become dirty. Carbon and other impurities build up inside the holes, and molten metal often sprays and sticks onto the surface of the tip. A dirty tip will result in

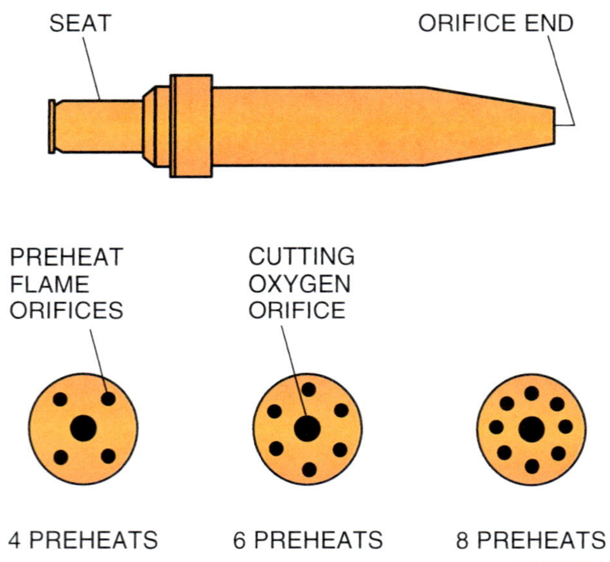

Figure 17 Orifice end views of one-piece acetylene torch cutting tips.

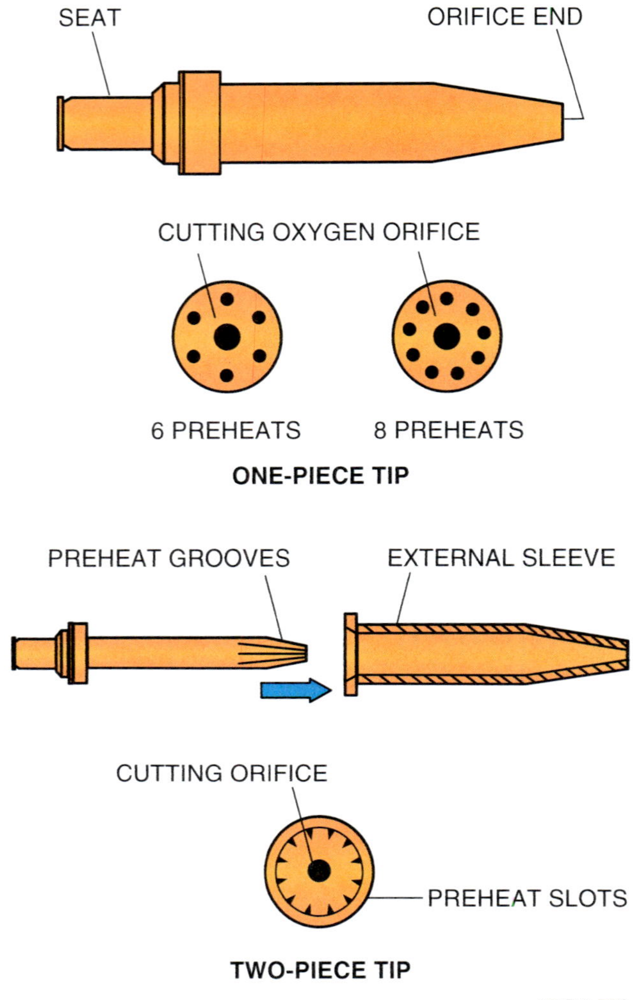

Figure 18 Orifice end views of cutting tips for liquefied fuel gases.

Acetylene Flow Rates

Manufacturers provide listings of the maximum fuel flow rate for each acetylene tip size in addition to recommended acetylene pressures. When selecting a tip, make sure that its maximum flow rate (in cubic feet per hour) does not exceed one-seventh of the total fuel capacity (in cubic feet) for the acetylene cylinder in use. Multiple cylinders must be manifolded together if the flow rate exceeds the cylinder(s) in use in order to prevent withdrawal of acetone along with acetylene.

a poor-quality cut with an uneven **kerf** and excessive dross buildup. To ensure good cuts with straight kerfs and minimal dross buildup, clean cutting tips with tip cleaners or tip drills (*Figure 20*).

Tip cleaners are small round files. They usually come in a set with files to match the diameters of the various tip holes. In addition, each set usually includes a file that can be used to lightly recondition the face of the cutting tip. Tip cleaners are inserted into the tip hole and moved back and forth a few times to remove deposits from the hole.

Tip drills are used for major cleaning and for holes that are plugged. Tip drills are tiny drill bits that are sized to match the diameters of tip holes. The drill fits into a drill handle for use. The handle is held, and the drill bit is turned carefully inside the hole to remove debris. They are more brittle than tip cleaners, making them more difficult to use.

CAUTION

Tip cleaners and tip drills are brittle. If you are not careful, they may break off inside a hole. Broken tip cleaners are difficult to remove. Improper use of tip cleaners or tip drills can enlarge the tip, causing improper burning of gases. If this occurs, tips must be discarded. If the end of the tip has been partially melted or deeply gouged, do not attempt to cut it off or file it flat. The tip should be discarded and replaced with a new tip. This is because some tips have tapered preheat holes, and if a significant amount of metal is removed from the end of the tip, the preheat holes will become too large.

3.9.0 Friction Lighters

Always use a friction lighter (*Figure 21*), also known as a striker or spark-lighter, to ignite the cutting torch. The friction lighter works by rubbing a piece of flint on a steel surface to create sparks.

WARNING!

Do not use a match or a gas-filled lighter to light a torch. This could result in severe burns and/or could cause the lighter to explode.

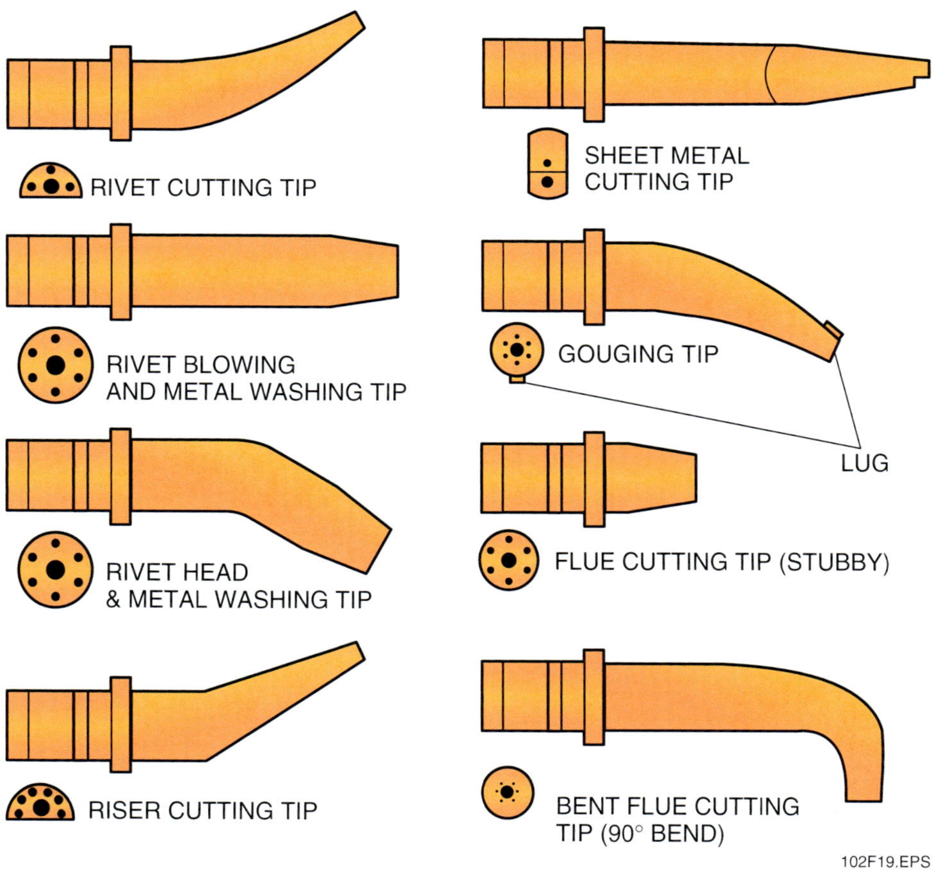

Figure 19 Special-purpose torch cutting tips.

Alternate Design Solid Copper Tip

A manufacturer of replacement torch tips for various brands of torches has a two-piece design for use in place of a standard solid copper tip. It consists of a single tip-to-head sealing adapter and a number of small interchangeable tip sizes that screw onto the adapter. The small screw-on tips are faster to change and somewhat less costly than a full standard tip.

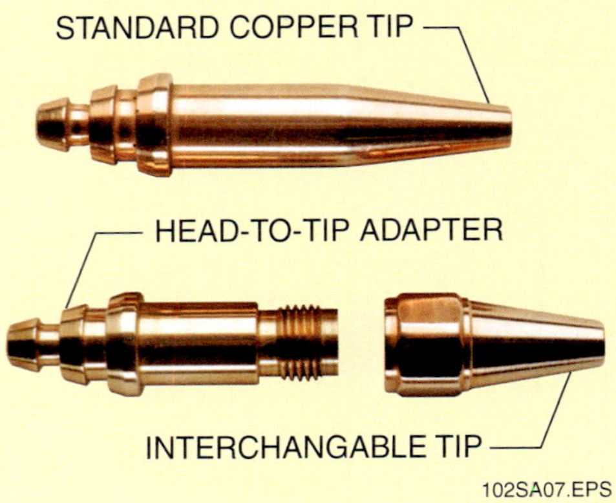

3.10.0 Cylinder Cart

The cylinder cart, or bottle cart, is a modified hand truck that has been equipped with seats and chains to hold cylinders firmly in place. Bottle carts help ensure the safe transportation of gas cylinders. *Figure 22* shows a typical cylinder cart for a cutting rig. Some carts are equipped with tool/accessory trays or boxes as well as rod holders.

3.11.0 Soapstone Markers

Because of the heat involved in cutting operations, along with the tinted lenses that are required, ordinary pen or pencil marking for cutting lines or welding locations is not effective. The oldest and most common material used for marking is **soapstone** in the form of sticks or cylinders (*Figure 23*). Soapstone is soft and feels greasy and slippery. It is actually steatite, a dense, impure form of talc that is heat resistant. It also shows up well through a tinted lens under the illumination of an electric

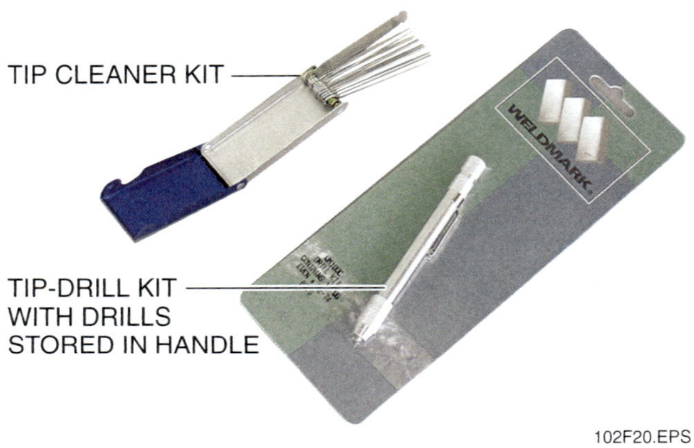

Figure 20 Tip cleaner and drill kits.

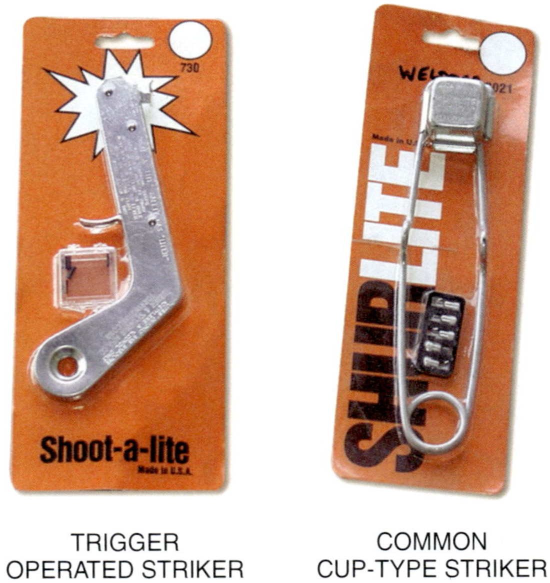

Figure 21 Friction lighters.

Figure 22 Cylinder cart.

Sharpening Soapstone Sticks

The most effective way to sharpen a soapstone stick marker is to shave it on one side with a file. By leaving one side flat, accurate lines can be drawn very close to a straightedge or a pattern.

102SA09.EPS

arc or gas cutting flame. Some welders prefer to use silver-graphite pencils for marking dark materials and red-graphite pencils for aluminum or other bright metals. Graphite is also highly heat resistant (*Figure 23*). A few manufacturers also market heat-resistant paint/dye markers for cutting and welding.

Cup-Type Striker

When using a cup-type striker to ignite a welding torch, hold the cup of the striker slightly below and to the side of the tip, parallel with the fuel gas stream from the tip. This prevents the ignited gas from deflecting back toward you from the cup and reduces the amount of carbon soot in the cup. Note that the flint in a striker can be replaced.

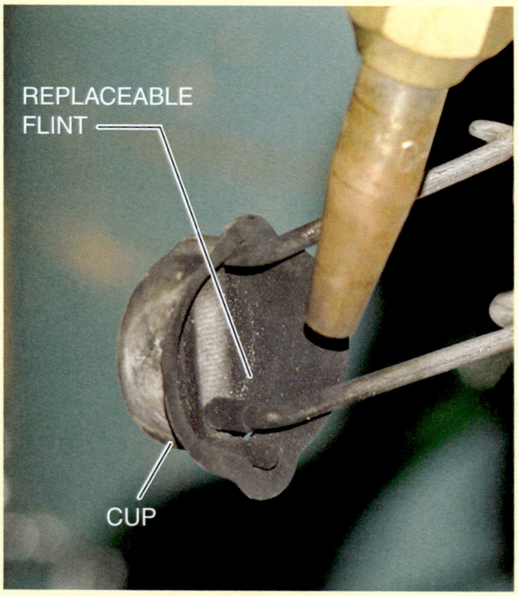

102SA08.EPS

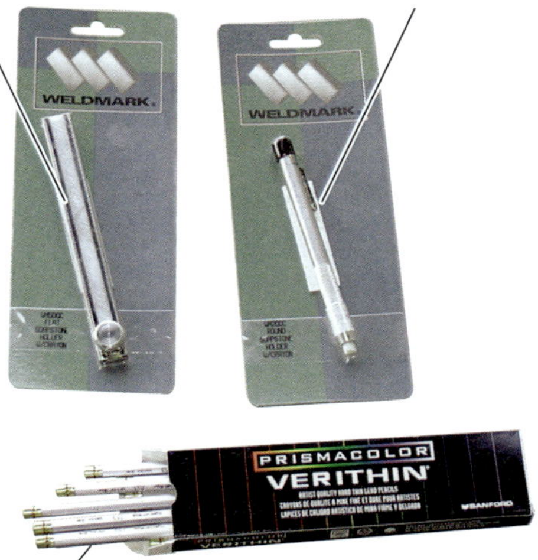

102F23.EPS

Figure 23 Soapstone and graphite markers.

3.12.0 Specialized Cutting Equipment

In addition to the common hand cutting torches, other types of equipment are used in oxyfuel cutting applications. This equipment includes mechanical guides used with a hand cutting torch, various types of motorized cutting machines, and oxygen lances. All of the motorized units use special straight body machine cutting or welding torches with a gear rack attached to the torch body to set the tip distance from the work.

3.12.1 Mechanical Guides

On long, circular, or irregular cuts, it is very difficult to control and maintain an even kerf with a hand cutting torch. Mechanical guides can help maintain an accurate and smooth kerf along the cutting line. For straight line or curved cuts, use a one- or two-wheeled accessory that clamps on the torch tip in a fixed position. The wheeled accessory maintains the proper tip distance while the tip is guided by hand along the cutting line.

Perform arc or circular cuts with the circle cutting accessory shown in *Figure 24*. The torch tip fits through and is secured to a rotating mount between the two small metal wheels. The wheel heights are adjustable so that the tip distance from the work can be set. The radius of the circle is set by moving the pivot point on a radius bar. After a starting hole is cut (if needed), the torch tip is placed through and secured to the circle cutter rotating mount. Then the pivot point is placed in a drilled hole or a magnetic holder at the center of the circle. When the cut is restarted, the torch can be moved in a circle around the cut, guided by the circle cutter. The magnetic guide shown in *Figure 24* is used for straight line cuts. The magnets hold it securely in place during cutting operations.

When large work with an irregular pattern must be cut, a template is often used. The torch is drawn around the edges of the template to trace the pattern as the cut is made. If multiple copies must be cut, a metal pattern, held in place and spaced for tip distance from the work by stacked magnets, is usually used. For a one- or two-time copy, a heavily weighted Masonite or aluminum template that is spaced off the workpiece could be carefully used and discarded.

3.12.2 Motor-Driven Equipment

A variety of fixed and portable motorized cutting equipment is available for straight and curved cutting/welding. The computer-controlled gantry cutting machine (*Figure 25*) and the optical pattern-tracing machine (*Figure 26*) are fixed-location machines used in industrial manufacturing applications. The computer-controlled machine can be programmed to **pierce** and then cut any pattern from flat metal stock. The optical pattern-tracing machine follows lines on a drawing using a light beam and an optical detector. They both can be rigged to cut multiple items using multiple torches operated in parallel. Both units have a motor-driven gantry that travels the length of

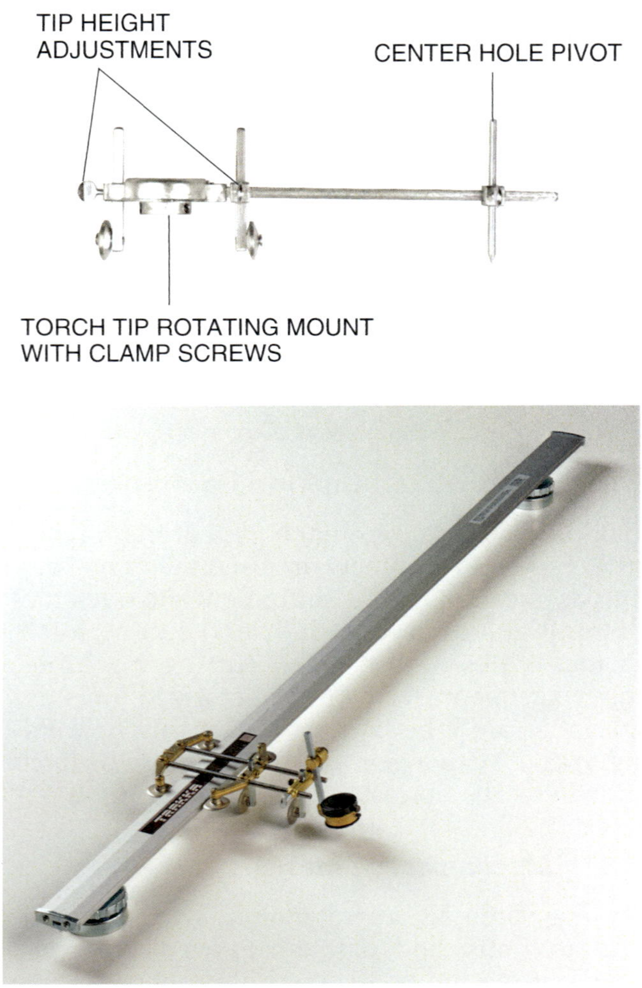

Figure 24 Circle cutting guide.

Figure 25 Computer-controlled gantry cutting machine.

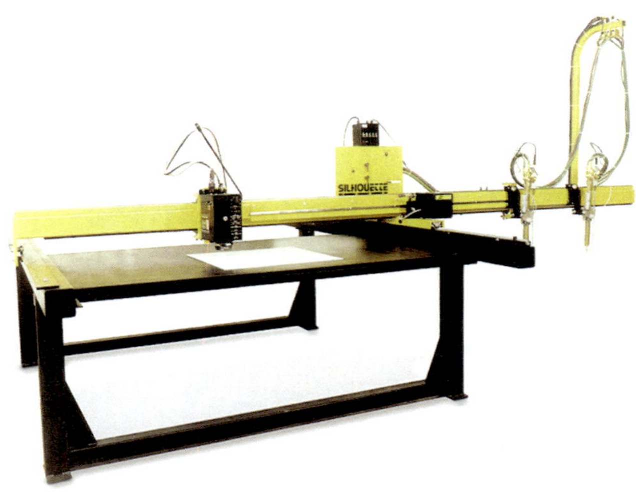

Figure 26 Optical pattern-tracing machine.

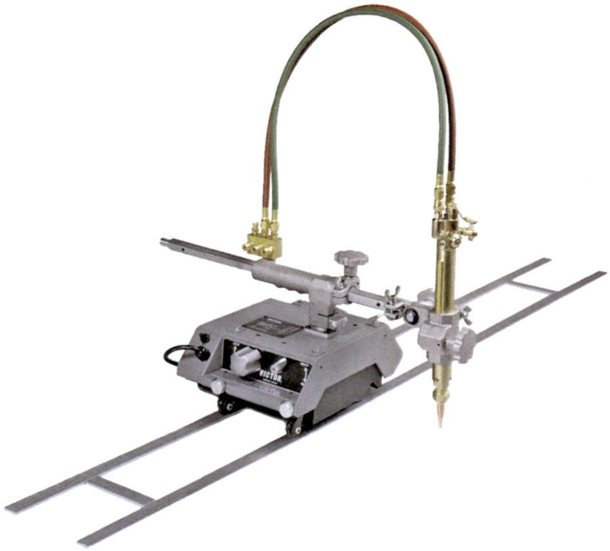

102F27.EPS

Figure 27 Track burner with oxyfuel machine torch.

a table and a transverse motor-driven beam or torch head that moves back and forth across the table. Both units are also equipped to use both oxyfuel cutting and plasma cutting torches. The size of the patterns can also be adjusted.

Other types of pattern-tracing machines use metal templates that are clamped in the machine. A follower wheel traces the pattern from the template. The pattern size can be increased or decreased by electrical or mechanical linkage to a moveable arm holding one or more cutting torches that cut the pattern from flat metal stock.

Portable track cutting machines (track burners) can be used in the field for straight or curved cutting and beveling. *Figures 27* and *28* show units driven by a variable-speed motor. The unit shown in *Figure 27* is available with track extensions for any length of straight cutting or beveling, along with a circle cutting attachment. The unit shown in *Figure 28* uses a somewhat flexible magnetic track for both flat straight-line or large-diameter object cutting or beveling. It is shown equipped with an optional plasma machine torch. Both units can be adapted to metal inert gas (MIG) or tungsten inert gas (TIG) welding.

A portable, motor-driven band track or hand-cranked ring gear cutter/beveler can be set up in the field for cutting and beveling pipe with oxyfuel or plasma machine torches (*Figures 29* and *30*). The stainless steel band track cutter uses a chain and motor sprocket drive to rotate the machine cutting torch around the pipe a full 360 degrees. The all-aluminum ring gear type of cutter/beveler is positioned on the pipe, and then the saddle is clamped in place. In operation, the ring gear and the cutting torch rotate at different rates around the saddle for a full 360-degree cut.

102F28.EPS

Figure 28 Track burner with a plasma machine torch.

Shop-Made Straight-Line Cutting Guide

A simple solution for straight-line cutting is to clamp a piece of angle iron to the work and use a band clamp around the cutting torch tip to maintain the cutting tip distance from the work. When the cut is started, the band clamp rests on the top of the vertical leg of the angle iron, and the torch is drawn along the length of the angle iron at the correct cutting speed.

102SA10.EPS

102F29.EPS

Figure 29 Band-track pipe cutter/beveler.

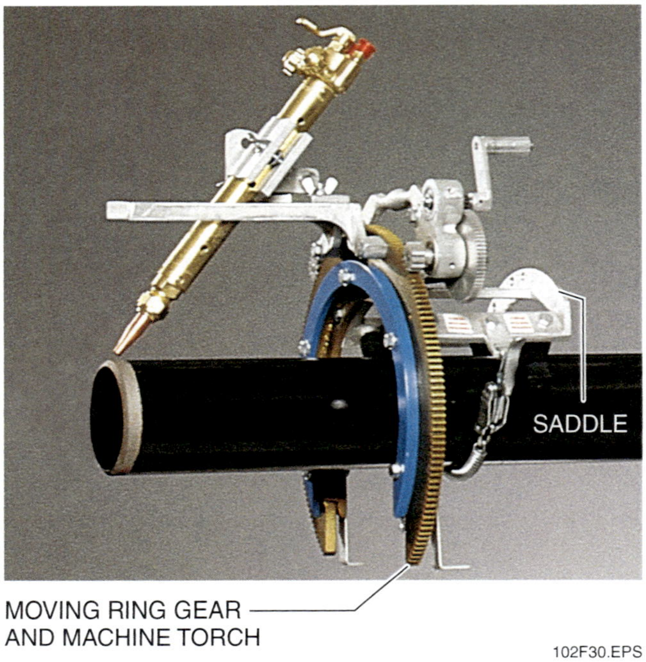

102F30.EPS

Figure 30 Ring gear pipe cutter/beveler.

A hand-guided oxyfuel cutting torch with an integral-precision, variable-speed motor drive (*Figure 31*), which can be used for straight line and curved cutting to achieve machine-quality cuts, is available. A circle cutting accessory is also available for the unit.

3.12.3 Exothermic Oxygen Lances

Exothermic (combustible) oxygen lances are a special oxyfuel cutting tool usually used in heavy industrial applications and demolition work. The lance is a steel pipe that contains magnesium- and

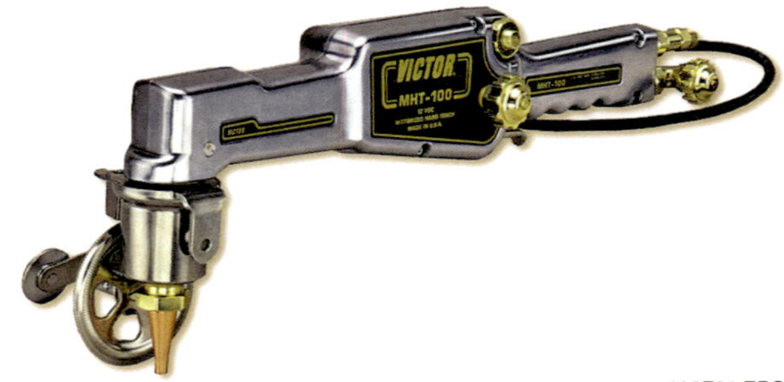

102F31.EPS

Figure 31 Hand-guided motorized oxyfuel cutting torch.

aluminum-cored powder or rods (fuel). In operation, the lance is clamped into a holder (*Figure 32*) that seals the lance to a fitting that supplies oxygen to the lance through a hose at pressures of 75 to 80 psi. With the oxygen turned on, the end of the lance is ignited with an acetylene torch or flare. As long as the oxygen is applied, the lance will burn and consume itself. The oxygen-fed flame of the burning magnesium, aluminum, and steel pipe creates temperatures approaching 10,000°F. At this temperature, the lance will rapidly cut or pierce any material, including steel, metal alloys, and cast iron, even under water. The lances for the holder shown in *Figure 32* are 10' long and range in size from ⅜" to 1" in diameter. The larger sizes can be coupled to obtain a longer

Cutting through Difficult Materials

Fire brick, reinforced concrete, and large, thick iron castings or steel objects are some of the many materials that can quickly be cut or pierced with an exothermic oxygen lance. With the larger lances, use specially insulated heat-reflective suits and full head shields because of the amount of heat generated.

102SA11.EPS

Figure 32 Oxygen lance holder.

lance. A small pistol-grip heat-shielded unit that can, if desired, be used with an electric welder is also available. The arc, combined with an oxygen lance, can create temperatures exceeding 10,000°F. This small unit uses lances from ¼" to ⅜" in diameter that are 22" to 36" long and that cut very rapidly at a maximum burning time of 60 to 70 seconds. The small unit is primarily used to burn out large frozen pins and frozen headless bolts or rivets. Like a large lance, it can be used to cut any material, including concrete-lined pipe. It also can be used to remove hard-surfacing material that has been applied to wear surfaces. Both units are relatively inexpensive and can be set up in the field with only an oxygen cylinder, hose, and ignition device.

4.0.0 Setting Up Oxyfuel Equipment

When setting up oxyfuel equipment, you must follow certain procedures to ensure that the equipment operates properly and safely. The following sections explain the procedures for setting up oxyfuel equipment.

4.1.0 Transporting and Securing Cylinders

Follow these steps to transport and secure cylinders:

WARNING!

Always handle cylinders with care. They are under high pressure and should never be dropped, knocked over, rolled, or exposed to heat in excess of 140°F. When moving cylinders, always be certain that the valve caps are in place. Use a cylinder cage to lift cylinders. Never use a sling or electromagnet.

Step 1 Transport cylinders to the workstation in the upright position on a hand truck or bottle cart (*Figure 33*).

Step 2 Secure the cylinders at the workstation.

Step 3 Remove the protective cap from each cylinder and inspect the outlet nozzles to make sure that the seat and threads are not damaged. Place the protective caps where they will not be lost and where they will be available when the cylinders are empty.

CAUTION

Do not transport or immediately use an acetylene cylinder found resting on its side. Stand it upright and wait at least an hour to allow the acetone to settle before using it.

4.2.0 Cracking Cylinder Valves

Follow these steps to crack cylinder valves:

Step 1 Crack open the cylinder valve momentarily to remove any dirt from the valves (*Figure 34*).

New Cutting Technologies

Manufacturers of welding and cutting equipment are constantly working on new products and methods. Two new cutting technologies were emerging at the time this module was being written.

One of these technologies uses water as the heat source. A special generator separates the water into hydrogen and oxygen using electrolysis. The two gases are then recombined in the cutting torch in the same manner as oxygen and acetylene. This generator can be used in welding, cutting, soldering, and brazing applications. It also has potential for use as a fuel source for automobiles. The product is called Brown's Gas in honor of Yull Brown, the man who devoted his professional life and fortune to the development of this process (**www.brownsgas.com**).

Another emerging technology uses regular gasoline in place of acetylene. Like an oxyacetylene setup, the gasoline and oxygen are combined in the torch to produce the cutting flame (**www.petrogen.com**).

SINGLE-CYLINDER HAND CART

HEAVY-DUTY TWIN-CYLINDER HAND CART

102F33.EPS

Figure 33 Carts for transporting cylinders.

102F34.EPS

Figure 34 Cracking a cylinder valve.

Hoisting Cylinders

Never attempt to lift a cylinder using the holes in a safety cap. Always use a lifting cage. Make sure that the cylinder is secured in the cage. Various size cages are available for high-pressure cylinders and cylinders containing liquids.

102SA12.EPS

WARNING!

Always stand with the valve stem between you and the regulator when opening valves to avoid injury from dirt that may be lodged in the valve.

Step 2 Wipe out the connection seat of the valves with a clean cloth. Dirt frequently collects in the outlet nozzle of a cylinder valve and must be cleaned out to keep it from entering the regulator when pressure is turned on.

WARNING!

Be sure the cloth used does not have any oil or grease on it. Oil or grease mixed with compressed oxygen will cause an explosion.

4.3.0 Attaching Regulators

Follow these steps to attach the regulators:

Step 1 Check that the regulator is closed (adjustment screw loose).

CAUTION

Use care not to overtighten connections. The brass connections used will strip if overtightened. Repair or replace equipment that does not seal properly. If the torch valves leak, try lightly tightening the packing nut at the valves. If a cutting attachment or torch tip leaks, disassemble and check the sealing surfaces and/or O-rings for damage. If the equipment or torch tip is new and the sealing surfaces do not appear damaged, overtighten the connection slightly to seat the sealing surfaces; then loosen and retighten normally. If the leaks persist or the equipment is visibly damaged, replace the torch, cutting attachment (if used), or cutting tip as necessary.

Fuel and Oxygen Cylinder Separation for Fixed Installations

For fixed installations involving one or more cylinders coupled to a manifold, fuel and oxygen cylinders must be separated by at least 20' or be divided by a wall 5' or more high with a 30-minute burn rating (*American National Standards Institute Z49.1*).

Step 2 Check the regulator fittings to ensure that they are free of oil and grease (*Figure 35*).

Step 3 Connect and tighten the oxygen regulator to the oxygen cylinder using a torch wrench (*Figure 36*).

WARNING!

Do not work with a regulator that shows signs of damage, such as cracked gauges, bent thumbscrews, or worn threads. Set it aside for repairs. Never attempt to repair regulators yourself. When tightening the connections, always use a torch wrench.

Step 4 Connect and tighten the fuel gas regulator to the fuel gas cylinder. Remember that most fuel gas fittings have left-hand threads.

Step 5 Crack the cylinder valve slightly and open the regulator to expel any debris from the outlet. Shut the cylinder valve and close the regulator (*Figure 37*).

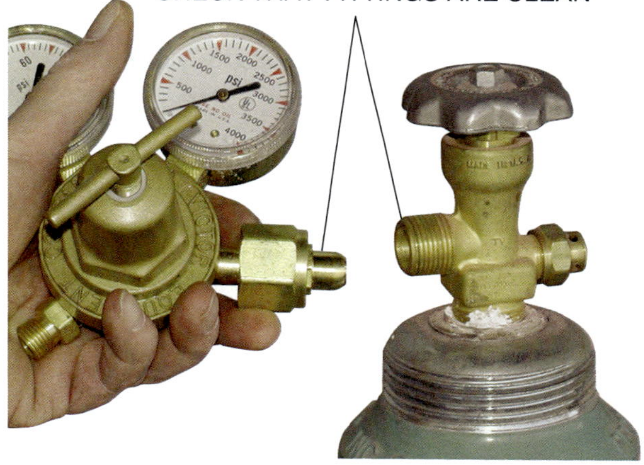

102F35.EPS

Figure 35 Checking connection fittings.

102F36.EPS

Figure 36 Tightening regulator connection

Figure 37 Cleaning the regulator.

4.4.0 Installing Flashback Arrestors or Check Valves

Follow these steps to install flashback arrestors or check valves:

> **WARNING!** At least one flashback arrestor must be used with each hose. The absence of a flashback arrestor could result in flashback. Flashback arrestors can be attached either to the regulator, the torch, or both; however, flashback arrestors installed at the torch handle are preferred if only one is being used.

Step 1 Attach a flashback arrestor or check valve to the hose connection on the oxygen regulator (*Figure 38*) and tighten with a torch wrench.

Step 2 Attach and tighten a flashback arrestor or check valve to the hose connection on the fuel gas regulator. Keep in mind that most fuel gas fittings have left-hand threads.

4.5.0 Connecting Hoses to Regulators

New hoses contain talc and loose bits of rubber. These materials must be blown out of the hoses before the torch is connected. If they are not blown out, they will clog the torch needle valves.

Follow these steps to connect the hoses to the regulators:

Step 1 Inspect both the oxygen and fuel gas hoses for any damage, burns, cuts, or fraying.

Step 2 Replace any damaged hoses.

Figure 38 Attaching a flashback arrestor.

Step 3 Connect the oxygen hose to the oxygen regulator flashback arrestor or check valve (*Figure 39*).

Step 4 Connect the fuel gas hose to the fuel gas regulator flashback arrestor or check valve. Keep in mind that most fuel gas fittings have left-hand threads.

Step 5 Open the regulator and purge the hose until it is clear.

4.6.0 Attaching Hoses to the Torch

Follow these steps to attach the hoses to the torch:

Step 1 Attach flashback arrestors to the oxygen and fuel gas hose connections on the torch body unless the torch has built-in flashback arrestors and check valves. Keep in mind that most fuel gas fittings have left-hand threads.

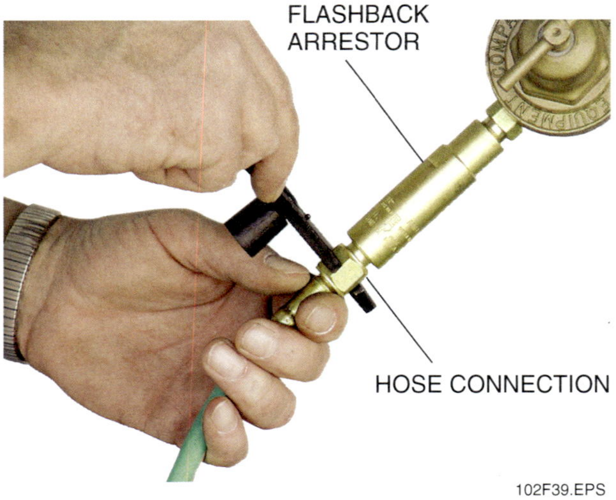

Figure 39 Connecting hose to regulator flashback arrestor.

Step 2 Attach and tighten the oxygen hose to the oxygen fitting on the flashback arrestor or torch (*Figure 40*).

Step 3 Attach and tighten the hose to the fuel gas fitting on the flashback arrestor or torch. Remember that all hose fittings have left-hand threads.

4.7.0 Connecting Cutting Attachments (Combination Torch Only)

If a combination torch is being used, connect cutting attachments as follows:

Step 1 Check the torch manufacturer's instructions for the correct method of installing the attachment.

Step 2 Connect the attachment and tighten by hand as required.

4.8.0 Installing Cutting Tips

Follow these steps to install a cutting tip in the cutting torch:

Step 1 Identify the thickness of the material to be cut.

> **WARNING!**
>
> If acetylene fuel is being used, make sure that the maximum fuel flow rate per hour of the tip does not exceed one-seventh of the fuel cylinder capacity. If a purplish flame is observed when the torch is operating, the fuel rate is too high and acetone is being withdrawn from the acetylene cylinder along with the acetylene gas.

Step 2 Identify the proper size cutting tip from the manufacturer's recommended tip size chart for the fuel being used.

Step 3 Inspect the cutting tip sealing surfaces and orifices for damage or plugged holes. If the sealing surfaces are damaged, discard the tip. If the orifices are plugged, clean them with a tip cleaner or drill.

Step 4 Check the torch manufacturer's instructions for the correct method of installing cutting tips.

Step 5 Install the cutting tip, securing it with a torch wrench or by hand as required (*Figure 41*).

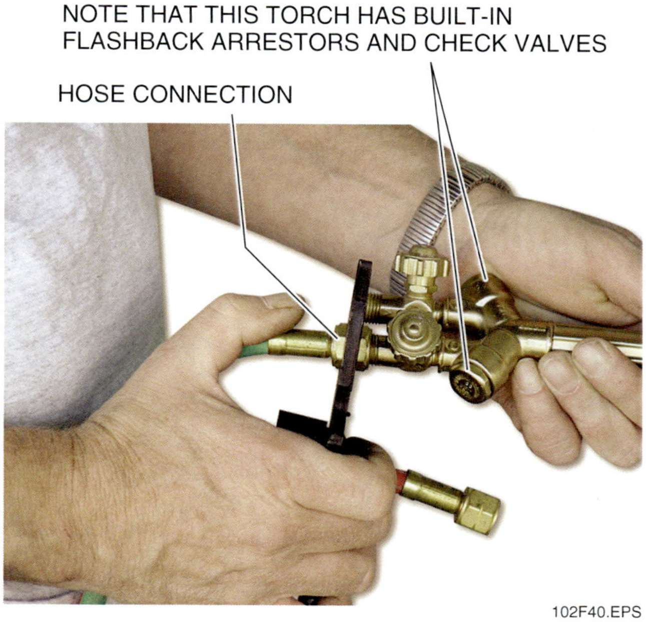

Figure 40 Connecting hoses to torch body.

Figure 41 Installing a cutting tip.

4.9.0 Closing Torch Valves and Loosening Regulator Adjusting Screws

Follow these steps to close the torch valves and loosen the regulator adjusting screws (*Figure 42*):

Step 1 Check the fuel and oxygen valves on the torch to be sure they are closed. Closing the torch gas valves prevents gases from backing up inside the torch.

> **CAUTION**
>
> Loosening regulator adjusting screws closes the regulators and prevents damage to the regulator diaphragms when the cylinder valves are opened.

Step 2 Check the oxygen regulator adjusting screw to be sure it is loose (backed out).

Figure 42 Torch valves and regulator adjusting screws.

Step 3 Check the fuel gas regulator adjusting screw to be sure it is loose.

4.10.0 Opening Cylinder Valves

Follow these steps to open cylinder valves (*Figure 43*):

> **WARNING!**
>
> Never stand directly in front of or behind a regulator. The regulator adjusting screw can blow out, causing serious injury. Always open the cylinder valve gradually. Quick openings can damage a regulator or gauge or even cause a gauge to explode.
>
> Oxygen cylinder valves must be opened all the way until the valve seats at the top. Seating the valve at the fully open position prevents high-pressure leaks at the valve stem.

Step 1 Standing on the cylinder valve side of the oxygen regulator, slowly open the oxygen cylinder valve until the desired pressure is reached. Allow the pressure in the cylinder pressure gauge to rise gradually until the gauge indicates the oxygen cylinder pressure, then open it all the way.

Figure 43 Cylinder valves and gauges.

> **WARNING!**
>
> When opening a cylinder valve, keep the valve stem between you and the regulator.

Step 2 Slowly open the fuel gas cylinder valve until the cylinder pressure gauge indicates the cylinder pressure, but no more than 1½ turns. This allows it to be quickly closed in case of a fire.

4.11.0 Purging the Torch and Setting the Working Pressures

Follow these steps to purge the torch and set the working pressures (*Figure 44*):

> **WARNING!**
>
> The working pressure gauge readings on single-stage regulators will rise after the torch valves are turned off. This is normal. However, if acetylene is being used as the fuel gas, make sure that the static pressure does not rise above 15 psig. Make sure that equipment is purged and leak tested in a well-ventilated area to avoid creating an explosive concentration of gases.

102F44.EPS

Figure 44 Typical points for purging the equipment and setting the working pressures.

Step 1 Fully open the oxygen valve on the torch. Then depress and hold or lock open the cutting oxygen lever.

Step 2 Tighten the oxygen regulator adjusting screw until the working pressure gauge shows the correct oxygen gas working pressure with the gas flowing. Allow the gas to flow for five to ten seconds to purge the torch and hoses of air or fuel gas.

Step 3 At the torch, release the cutting lever and close the oxygen valve.

Step 4 Open the fuel valve on the torch about ⅛ of a turn.

Step 5 Tighten the fuel regulator adjusting screw until the working pressure gauge shows the correct fuel gas working pressure with the gas flowing. Allow the gas to flow for five to ten seconds to purge the hoses and torch of air.

Step 6 At the torch, close the fuel valve. If acetylene is used, check that the acetylene static pressure does not rise above 15 psig. If it does, immediately open the torch fuel valve and reduce the regulator output pressure as required.

4.12.0 Testing for Leaks

Equipment must be tested for leaks immediately after it is set up and periodically thereafter. The torch should be checked for leaks each time be-

Portable Oxyacetylene or Oxyfuel Equipment

The equipment shown in *Figure 43* is mounted on a hand truck and is very heavy. This type of equipment is typically used in a shop or on a job site when extensive cutting must be accomplished. For small jobs, portable equipment that can be hand carried by one person may be used.

102SA13.EPS

Fuel Cylinder Wrench

If the fuel cylinder is equipped with a valve requiring a T-wrench, always leave the wrench in place on the valve so that the fuel can be quickly turned off. This type of valve is obsolete but still in use.

102SA14.EPS

fore use. Leaks could cause a fire or explosion if undetected. To test for leaks, brush a commercially prepared leak-testing formula or a solution of detergent and water on the following points. If bubbles form, a leak is present.

> **WARNING!**
> Make sure the detergent contains no oil. In the presence of oxygen, oil can cause fires or explosions.

Leak points include:

- Oxygen cylinder valve
- Fuel gas cylinder valve
- Oxygen regulator and regulator inlet and outlet connections
- Fuel gas regulator and regulator inlet and outlet connections
- Hose connections at the regulators, check valves/flashback arrestors, and torch
- Torch valves and cutting oxygen lever valve
- Cutting attachment connection (if used)
- Cutting tip

If there is a leak at the fuel gas cylinder valve stem, attempt to stop it by tightening the packing gland. If this does not stop the leak, mark and remove the cylinder and notify the supplier. For other leaks, tighten the connections slightly with a wrench. If this does not stop the leak, turn off the gas pressure, open all connections, and inspect the screw threads.

> **WARNING!**
> Do not use Teflon® tape or pipe dope on these fittings because you will not get a good seal. Make sure that equipment is purged and leak tested in a well-ventilated area to avoid creating an explosive concentration of gases.

4.12.1 Initial and Periodic Leak Testing

Perform the following steps at initial equipment setup and periodically thereafter:

Step 1 Set the equipment to the correct working pressures with the torch valves turned off.

Step 2 Using a leak-test solution, check for leaks at the cylinder valves, regulator relief ports, and regulator gauge connections (*Figure 45*). Also, check for leaks at hose connections, regulator connections, and check valve/flame arrestor connections up to the torch.

4.12.2 Leak-Down Testing of Regulators, Hoses, and Torch

Perform the following steps to quickly leak test the regulators, hoses, and torch before the torch is ignited.

Step 1 Set the equipment to the correct working pressures with the torch valves turned off. Then loosen both regulator adjusting screws. Check the working pressure gauges after a minute or two to see if the pressure drops. If the pressure drops, check the hose connection and regulators for leaks; otherwise, proceed to Step 2.

Step 2 Place a thumb over the cutting tip orifices and press to block the orifices.

Step 3 Turn on the torch oxygen valve and then depress and hold the cutting oxygen lever down.

Step 4 After the gauge pressure drops slightly, observe the oxygen working pressure gauge for a minute to see if the pressure continues to drop. If the pressure keeps dropping, perform the leak test described in the following section to determine the source of the leak. If the pressure does not change, close the torch oxygen valve and release the pressure at the cutting tip.

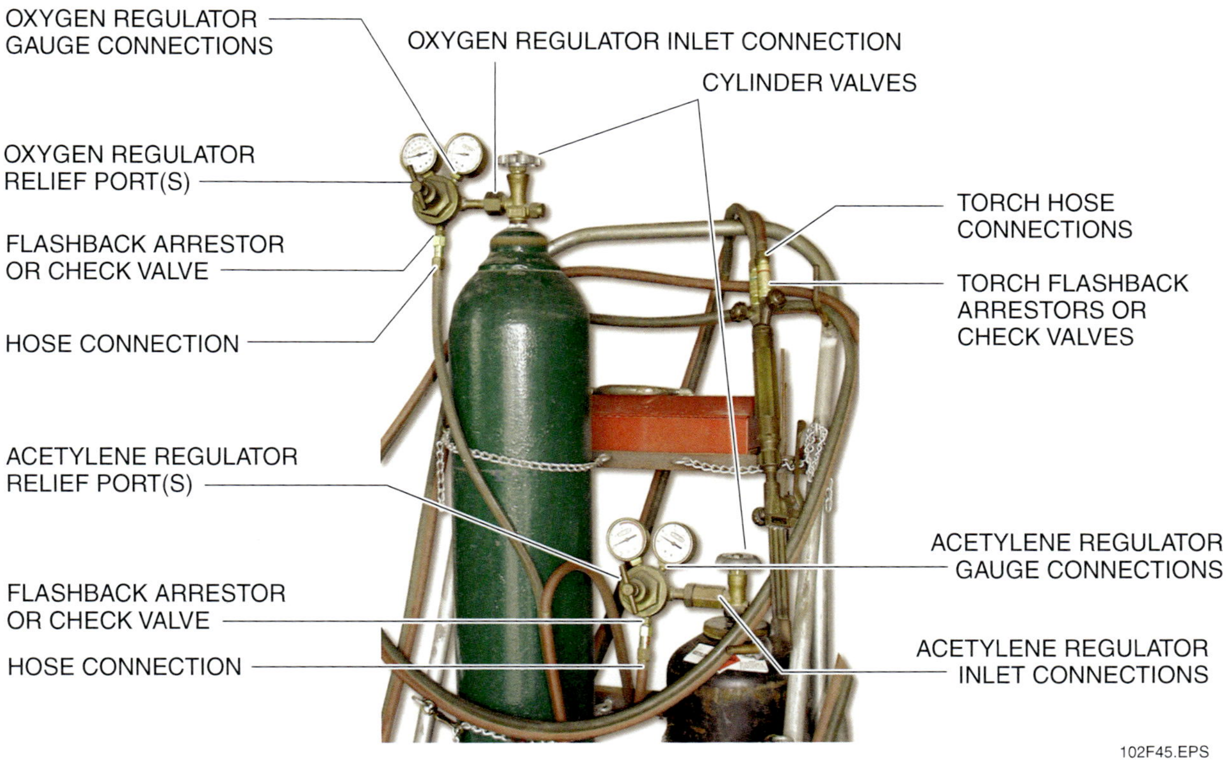

Figure 45 Typical initial and periodic leak-test points.

Step 5 With the tip blocked, turn on the torch fuel valve. After the gauge pressure drops slightly, carefully observe the fuel working pressure gauge for a minute. If the pressure continues to drop, perform the leak test described in the following section to determine the source of the leak. If the pressure does not change, close the torch fuel valve and release the pressure at the cutting tip.

Step 6 If no leaks are apparent, set the equipment to the correct working pressures.

4.12.3 Full Leak Testing of a Torch

Perform the following steps to test for and isolate torch leaks (*Figure 46*).

Step 1 Set the equipment to the correct working pressures with the torch valves turned off.

Step 2 Place a thumb over the cutting tip orifices and press to block the orifices.

Step 3 Turn on the torch oxygen valve and then depress and lock the cutting oxygen lever down.

Step 4 With the cutting tip blocked, check for leaks using a leak-test solution at the torch oxygen valve, cutting oxygen lever valve, cutting attachment connection (if used), preheat oxygen valve (if used), and cutting tip seal at the torch head.

Step 5 Release the cutting oxygen lever and close the torch oxygen valve. Release the pressure at the cutting tip.

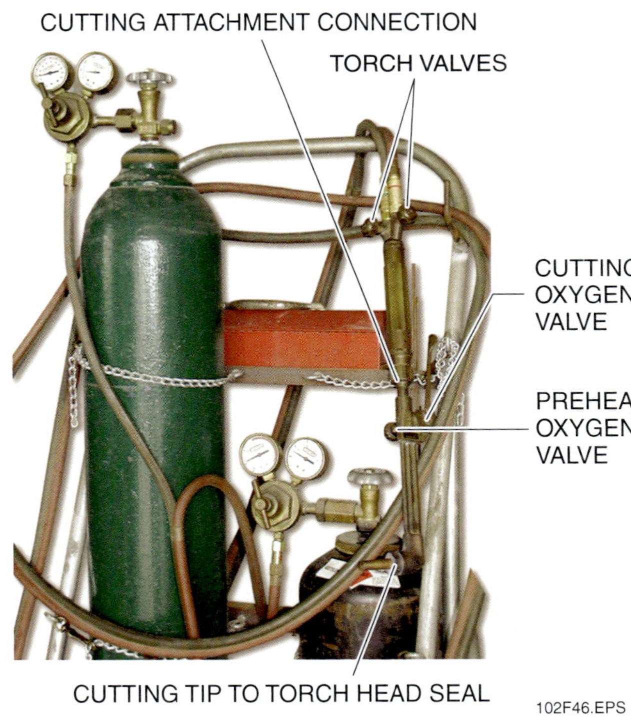

Figure 46 Torch leak-test points.

Step 6 With the cutting tip blocked, open the torch fuel valve.

Step 7 Using a leak-test solution, check for leaks at the torch fuel valve, cutting attachment (if used), and cutting tip seal at the torch head.

Step 8 Close the torch fuel valve and remove thumb from the cutting tip.

5.0.0 Controlling the Oxyfuel Torch Flame

To be able to safely use a cutting torch, the operator must understand the flame and be able to adjust it and react to unsafe conditions. The following sections will explain the oxyfuel flame and how to control it safely.

5.1.0 Oxyfuel Flames

There are three types of oxyfuel flames: **neutral flame**, **carburizing flame**, and **oxidizing flame**.

- *Neutral flame* – A neutral flame burns proper proportions of oxygen and fuel gas. The inner cones will be light blue in color, surrounded by a darker blue outer flame envelope that results when the oxygen in the air combines with the super-heated gases from the inner cone. A neutral flame is used for all but special cutting applications.
- *Carburizing flame* – A carburizing flame has a white feather created by excess fuel. The length of the feather depends on the amount of excess fuel present in the flame. The outer flame envelope is longer than that of the neutral flame, and it is much brighter in color. The excess fuel in the carburizing flame (especially acetylene) produces large amounts of carbon. The carbon will combine with red-hot or molten metal, making the metal hard and brittle. The carburizing flame is cooler than a neutral flame and is never used for cutting. It is used for some special heating applications.
- *Oxidizing flame* – An oxidizing flame has an excess of oxygen. The inner cones are shorter, much bluer in color, and more pointed than a neutral flame. The outer flame envelope is very short and often fans out at the ends. An oxidizing flame is the hottest flame. A slightly oxidizing flame is recommended with some special fuel gases, but in most cases it is not used. The excess oxygen in the flame can combine with many metals, forming a hard, brittle, low-strength oxide. However, the preheat flames of a properly adjusted cutting torch will be slightly oxidizing when the cutting oxygen is shut off.

Figure 47 shows the various flames that occur at a cutting tip for both acetylene and LP gas.

5.2.0 Backfires and Flashbacks

When the torch flame goes out with a loud pop or snap, a backfire has occurred. Backfires are usually caused when the tip or nozzle touches the work surface or when a bit of hot dross briefly interrupts the flame. When a backfire occurs, you can relight the torch immediately. Sometimes the torch even relights itself. If a backfire recurs without the tip making contact with the base metal, shut off the torch and find the cause. Possible causes are:

- Improper operating pressures
- A loose torch tip
- Dirt in the torch tip seat or a bad seat

When the flame goes out and burns back inside the torch with a hissing or whistling sound, a flashback is occurring. Immediately shut off the oxygen valve on the torch; the flame is burning inside the torch. If the flame is not extinguished quickly, the end of the torch will melt off. The flashback will stop as soon as the oxygen valve is closed. Therefore, quick action is crucial. Flashbacks can cause fires and explosions within the cutting rig and, therefore, are very dangerous. Flashbacks can be caused by:

- Equipment failure
- Overheated torch tip

Always Leak Test a Cutting/Welding Torch

Always take the time to perform a leak test on regulators, hoses, and the torch before you use them the first time that day. Always perform a leak test on a torch after you change tips or if you convert the torch from welding to cutting or vice versa. Leaks, especially fuel leaks, can cause a fire or explosion after a cutting or welding operation begins. A dangerous fire occurring at or near the torch may not be immediately noticed by the welder due to limited visibility caused by the tinted lenses in the welding goggles or face shield.

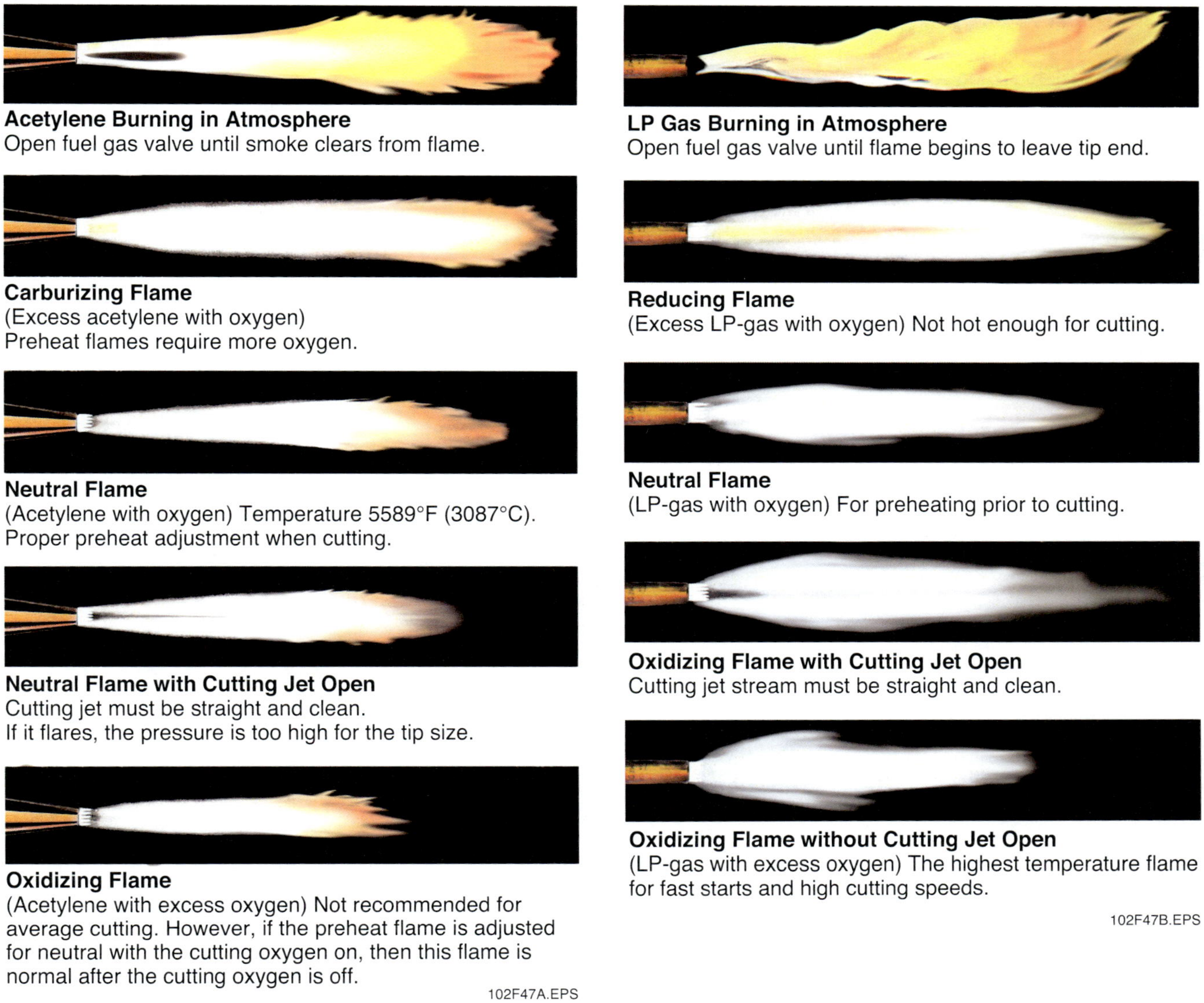

Figure 47 Acetylene and LP gas flames.

- Dross or spatter hitting and sticking to the torch tip
- Oversized tip (tip is too large for the gas flow rate being used)

After a flashback has occurred, wait until the torch has cooled. Then, blow oxygen (not fuel gas) through the torch for several seconds to remove soot that may have built up in the torch during the flashback before relighting it. If you hear the hissing or whistling after the torch is reignited or if the flame does not appear normal, shut off the torch immediately and have the torch serviced by a qualified technician. If the torch or tip has been damaged, replace the flashback arrestors.

5.3.0 Igniting the Torch and Adjusting the Flame

After the cutting equipment has been properly set up and purged, the torch can be ignited and the flame adjusted for cutting. Follow these steps to ignite the torch:

Step 1 Choose the appropriate cutting torch tip according to the base metal thickness you will be cutting and fuel gas you are using.

> **NOTE**
> Refer to the manufacturer's charts. You may have to readjust the oxygen and fuel gas pressure depending on the tip selected.

Step 2 Inspect the tip sealing surfaces and orifices. Attach the tip to the cutting torch or cutting attachment by placing it on the end of the torch and tightening the nut.

> **NOTE**
>
> Some manufacturers recommend tightening the nut with a torch wrench. Others recommend tightening the nut by hand. Check the tip manual to see if the manufacturer recommends that the nut be tightened manually or with a torch wrench.

Step 3 Put on proper protective clothing, gloves, and eye/face protection.

Step 4 Release the oxygen cutting lever. If present, close the preheat oxygen valve and open the torch oxygen valve fully.

Step 5 Open the fuel gas valve on the torch handle about one-quarter turn.

Step 6 Holding the friction lighter near the side and to the front of the torch tip, ignite the torch.

> **WARNING!**
>
> Hold the friction lighter near the side of the tip when igniting the torch to prevent deflecting the ignited gas back toward you. Always use a friction lighter. Never use matches or cigarette lighters to light the torch because this could result in severe burns and/or could cause the lighter to explode. Always point the torch away from yourself, other people, equipment, and flammable material.

Step 7 Once the torch is lit, adjust the torch fuel gas flame by adjusting the flow of fuel gas with the fuel gas valve. Increase the flow of fuel gas until the flame stops smoking or pulls slightly away from the tip. Decrease the flow until the flame returns to the tip.

Step 8 Open the preheat oxygen valve (if present) or the oxygen torch valve very slowly and adjust the torch flame to a neutral flame.

Step 9 Press the cutting oxygen lever all the way down and observe the flame. It should have a long, thin, high-pressure oxygen cutting jet up to 8" long extending from the cutting oxygen hole in the center of the tip. If it does not:

- Check that the working pressures are set as recommended on the manufacturer's chart.
- Clean the cutting tip. If this does not clear up the problem, change the cutting tip.

Step 10 With the cutting oxygen on, observe the preheat flame. If it has changed slightly to a carburizing flame, increase the preheat oxygen until the flame is neutral. After this adjustment, the preheat flame will change slightly to an acceptable oxidizing flame when the cutting oxygen is shut off.

5.4.0 Shutting Off the Torch

Follow these steps to shut off the torch after a cutting operation is completed:

> **WARNING!**
>
> Always turn off the oxygen flow first to prevent a possible flashback into the torch.

Step 1 Release the cutting oxygen lever. Then close the torch or preheat oxygen valves.

Step 2 Close the torch fuel gas valve quickly to extinguish the flame.

6.0.0 Shutting Down Oxyfuel Cutting Equipment

When a cutting job is completed and the oxyfuel equipment is no longer needed, it must be shut down. Follow these steps to shut down oxyfuel cutting equipment (*Figure 48*):

Step 1 Close the fuel gas and oxygen cylinder valves.

Step 2 Open the fuel gas valve to allow gas to escape, then close it. Open the oxygen valve to allow gas to escape, then close it. Do not proceed to the next step until all pressure is released and all regulator gauges read zero.

> **Obtaining Maximum Fuel Flow**
>
> Increasing the fuel flow until the flame pulls away from the tip and then decreasing the flow until the flame returns to the tip sets the maximum fuel flow for the tip size in use.

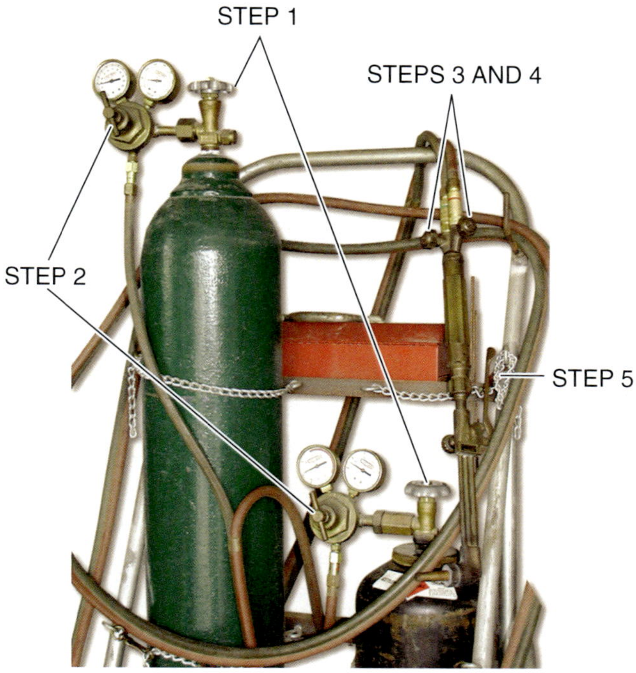

Figure 48 Shutting down oxyfuel cutting equipment.

Step 3 Back out the fuel gas and oxygen regulator adjusting screws until they are loose.

Step 4 Coil the hose and secure the torch to prevent damage.

7.0.0 Disassembling Oxyfuel Equipment

Follow these steps if the oxyfuel equipment must be disassembled after use:

Step 1 Check to be sure the equipment has been properly shut down. This includes checking that the cylinder valves are closed and all pressure gauges read zero.

Step 2 Remove both hoses from the torch.

Step 3 Remove both hoses from the regulators.

Step 4 Remove both regulators from the cylinder valves.

Step 5 Replace the protective caps on the cylinders.

Marking and Tagging Cylinders

Do not use permanent markers on cylinders; use soapstone or another temporary marker. If a cylinder is defective, place a warning tag on it.

Step 6 Return the oxygen cylinder to its proper storage place.

WARNING!

Always transport and store gas cylinders in the upright position. Be sure they are properly secured (chained) and capped.

Step 7 Return the fuel gas cylinder to its proper storage place.

WARNING!

Regardless if the cylinders are empty or full, never store fuel gas cylinders and oxygen cylinders together. Storing cylinders together is a violation of OSHA and local fire regulations and could result in a fire and explosion.

8.0.0 Changing Cylinders

Empty is a relative term when discussing gas cylinders. These cylinders should never be completely emptied because reverse flow could occur. Oxygen tanks should never get below the required working pressure or 25 psi. Follow these procedures to change a cylinder:

WARNING!

When moving cylinders, always be certain that they are in the upright position and the valve caps are secured in place. Never use a sling or electromagnet to lift cylinders. To lift cylinders, use a cylinder cage.

Step 1 Check to be sure the equipment has been properly shut down. This includes checking that the cylinder valves are closed and all pressure gauges read zero.

Step 2 Remove the regulator from the empty cylinder.

Step 3 Replace the protective cap on the empty cylinder.

Step 4 Transport the empty cylinder from the workstation to the storage area.

Step 5 Mark MT (empty) and the date (or the accepted site notation for indicating an empty cylinder) near the top of the cylinder using soapstone (*Figure 49*).

Step 6 Place the empty cylinder in the empty cylinder section of the cylinder storage area for the type of gas in the cylinder.

102F49.EPS

Figure 49 Typical empty cylinder marking.

9.0.0 Performing Cutting Procedures

The following sections explain how to recognize good and bad cuts, how to prepare for cutting operations, and how to perform straight-line cutting, piercing, bevel cutting, washing, and gouging.

9.1.0 Inspecting the Cut

Before attempting to make a cut, you must be able to recognize good and bad cuts and know what causes bad cuts. This is explained in the following list and illustrated in *Figure 50*:

- A good cut features a square top edge that is sharp and straight, not ragged. The bottom edge can have some dross adhering to it but not an excessive amount. What dross there is should be easily removable with a chipping hammer. The **drag lines** should be near vertical and not very pronounced.
- When preheat is insufficient, bad gouging results at the bottom of the cut because of slow travel speed.
- Too much preheat will result in the top surface melting over the cut, an irregular cut edge, and an excessive amount of dross.
- When the cutting oxygen pressure is too low, the top edge will melt over because of the resulting slow cutting speed.
- Using cutting oxygen pressure that is too high will cause the operator to lose control of the cut, resulting in an uneven kerf.
- A travel speed that is too slow results in bad gouging at the bottom of the cut and irregular drag lines.
- When the travel speed is too fast, there will be gouging at the bottom of the cut, a pronounced break in the drag line, and an irregular kerf.
- A torch that is held or moved unsteadily across the metal being cut can result in a wavy and irregular kerf.
- When a cut is lost and then not restarted carefully, bad gouges will result at the point where the cut is restarted.

> **NOTE**
> The following tasks are designed to develop your skills with a cutting torch. Practice each task until you are thoroughly familiar with the procedure. As you complete each task, take it to your instructor for evaluation. Do not proceed to the next task until your instructor tells you to continue.

9.2.0 Preparing for Oxyfuel Cutting with a Hand Cutting Torch

Before metal can be cut, the equipment must be set up and the metal prepared. One important step is to properly lay out the cut by marking it with soapstone or punch marks. The few minutes this takes will result in a quality job, reflecting craftsmanship and pride in your work. Follow these steps to prepare to make a cut:

Step 1 Prepare the metal to be cut by cleaning any rust, scale, or other foreign matter from the surface.

Step 2 If possible, position the work so you will be comfortable when cutting.

Step 3 Mark the lines to be cut with soapstone or a punch.

Step 4 Select the correct cutting torch tip according to the thickness of the metal to be cut, the type of cut to be made, the amount of preheat needed, and the type of fuel gas to be used.

Step 5 Ignite the torch.

Step 6 Use the procedures outlined in the following sections for performing particular types of cutting operations.

9.3.0 Cutting Thin Steel

Thin steel is $\frac{3}{16}$" thick or less. A major concern when cutting thin steel is distortion caused by

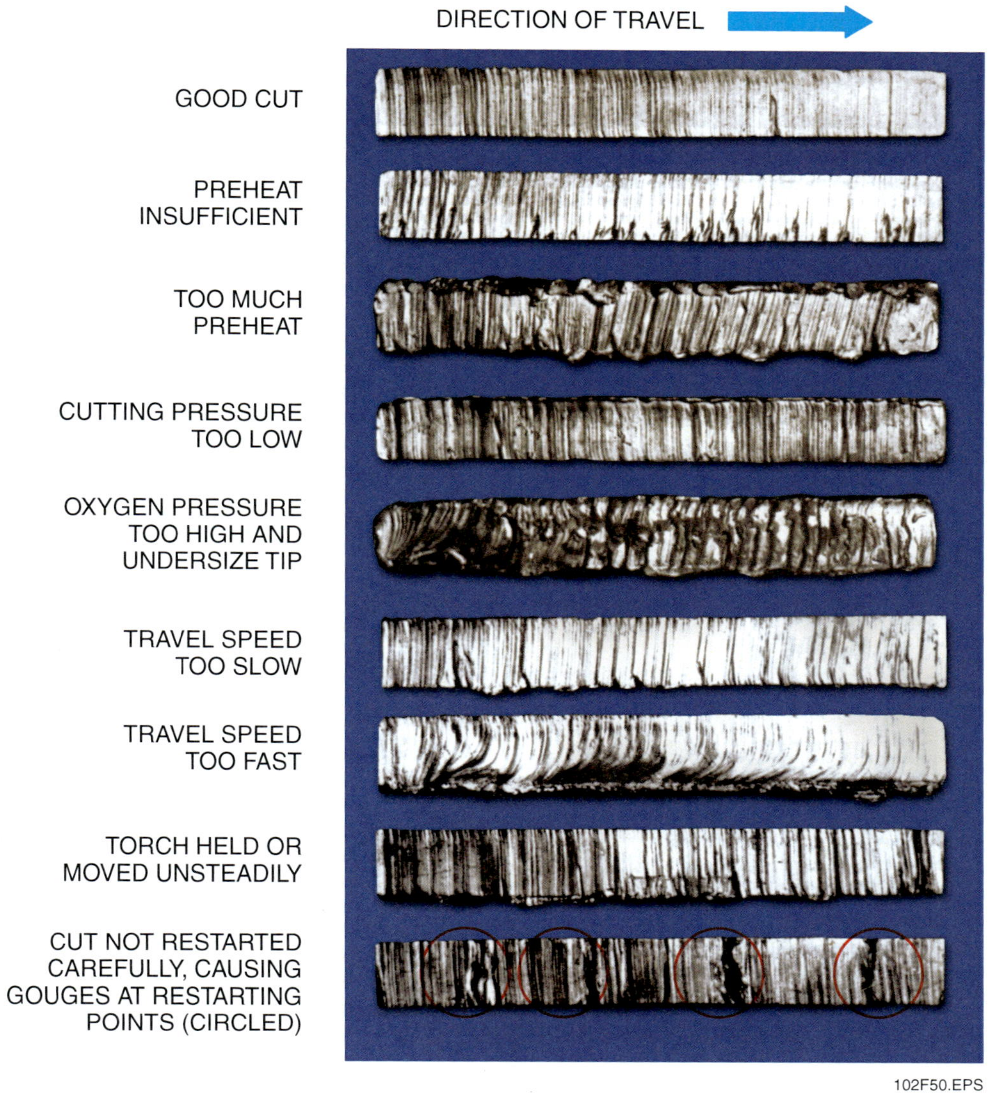

Figure 50 Examples of good and bad cuts.

the heat of the torch and the cutting process. To minimize distortion, move as quickly as you can without losing the cut. Follow these steps to cut thin steel:

Step 1 Prepare the metal surface.

Step 2 Light the torch.

Step 3 Hold the torch so that the tip is pointing in the direction the torch is traveling at a 15- to 20-degree angle. Make sure that a preheat orifice and the cutting orifice are centered on the line of travel next to the metal (*Figure 51*).

CAUTION

Holding the tip upright when cutting thin steel will overheat the metal, causing distortion.

Step 4 Preheat the metal to a dull red. Use care not to overheat thin steel because this will cause distortion.

NOTE

The edge of the tip can be rested on the surface of the metal being cut and then slid along the surface when making the cut.

Step 5 Press the cutting oxygen lever to start the cut, and then move quickly along the line to be cut. To minimize distortion, move as quickly as you can without losing the cut.

9.4.0 Cutting Thick Steel

Most oxyfuel cutting will be on steel that is more than 3⁄16" thick. Whenever heat is applied to metal, distortion is a problem, but as the steel gets

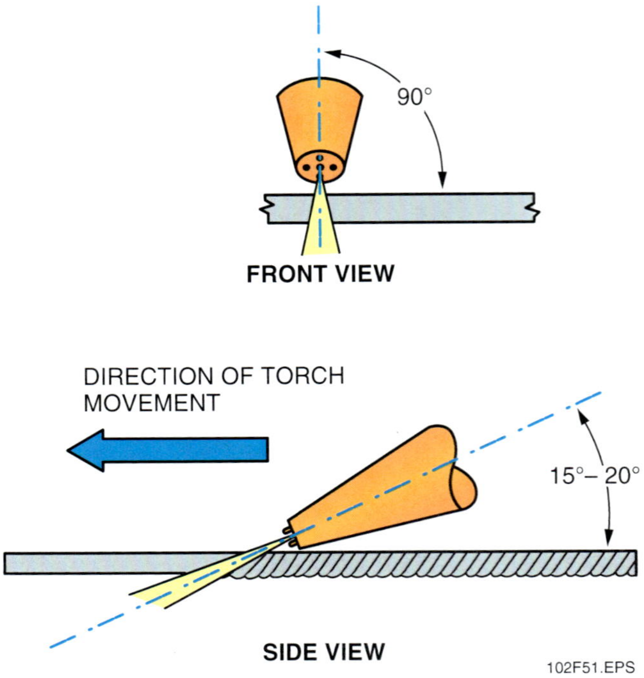

Figure 51 Cutting thin steel.

thicker, it becomes less of a problem. Follow these steps to cut thick steel with a hand cutting torch:

Step 1 Prepare the metal surface and torch. Light the torch.

Step 2 Ignite and adjust the torch flame.

> **NOTE**
> The torch can be moved from either right to left or left to right. Choose the direction that allows the best visibility of the cut.

Step 3 Follow the number sequence shown in *Figure 52* to perform the cut.

> **NOTE**
> When cutting begins, the tips of the preheat flame should be held 1⁄16" to 1⁄8" above the workpiece. For steel up to 3⁄8" thick, the first and third procedures can usually be omitted.

9.5.0 Piercing a Plate

Before holes or slots can be cut in a plate, the plate must be pierced. Piercing puts a small hole through the metal where the cut can be started. Because more preheat is necessary on the surface of a plate than at the edge, choose the next-larger cutting tip than is recommended for the thickness to be pierced. When piercing steel that is more than 3" thick, it may help to preheat the bottom side of the plate directly under the spot to be pierced. Follow these steps to pierce a plate for cutting:

Step 1 Prepare the metal surface and torch.

Step 2 Ignite the torch and adjust the flame.

Step 3 Hold the torch tip 1⁄4" to 5⁄16" above the spot to be pierced until the surface is a bright cherry red (*Figure 53*).

Step 4 Slowly press the cutting oxygen lever. As the cut starts, raise the tip about 1⁄2" above the metal surface and tilt the torch slightly so that molten metal does not blow back onto the tip. The tip should be raised and tipped before the cutting oxygen lever is fully depressed.

Step 5 Maintain the tipped position until a hole burns through the plate. Then rotate the tip vertically.

Step 6 Lower the torch tip to about 3⁄16" above the metal surface and continue to cut outward from the original hole to the edge of the line to be cut.

9.6.0 Cutting Bevels

Bevel cutting is often performed to prepare the edge of steel plate for welding. Follow these steps to perform bevel cutting (*Figure 54*):

Step 1 Prepare the metal surface and the torch.

Step 2 Ignite the torch and adjust the flame.

Step 3 Hold the torch so that the tip faces the metal at the desired bevel angle.

> **NOTE**
> An angle iron can be used as a guide.

Step 4 Preheat the edge to a bright cherry red.

Step 5 Press the cutting oxygen lever to start the cut.

Step 6 As cutting begins, move the torch tip at a steady rate along the line to be cut. Pay particular attention to the torch angle to ensure it is uniform along the entire length of the cut.

9.7.0 Washing

Washing is a term used to describe the process of cutting out bolts or rivets. Washing operations use a special tip with a large cutting hole that produces a low-velocity stream of oxygen. The low-velocity oxygen stream helps prevent cutting into the surrounding base metal. Washing tips can

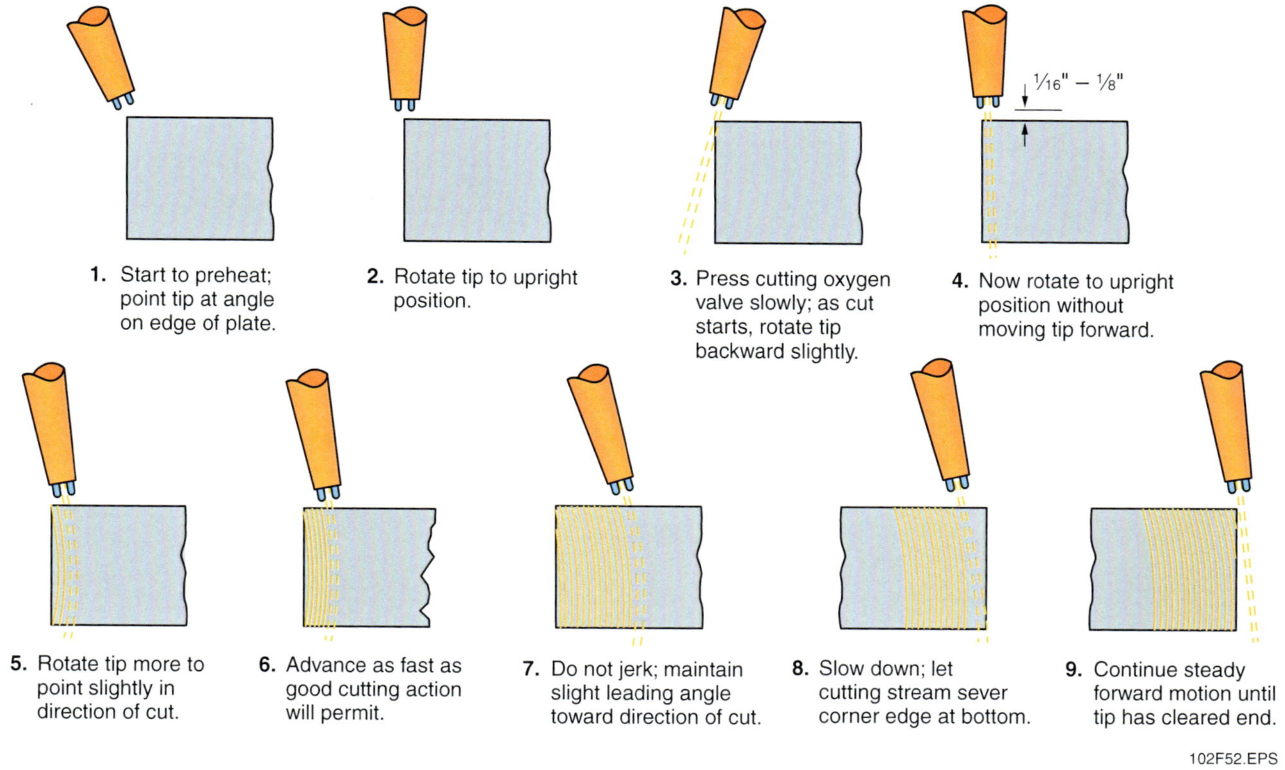

Figure 52 Flame cutting with a hand torch.

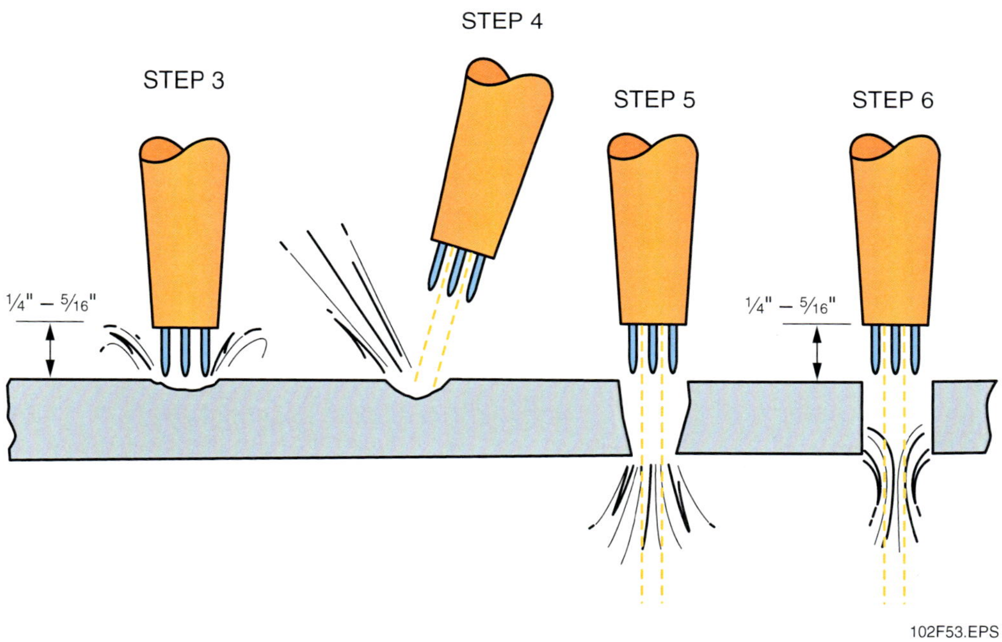

Figure 53 Steps for piercing steel.

also be used to remove items such as blocks, angles, or channels that are welded onto a surface. Follow these steps to perform washing (*Figure 55*):

Step 1 Prepare the metal surface and torch.

Step 2 Ignite the torch and adjust the flame.

Step 3 Preheat the metal to be cut until it is a bright cherry red.

Step 4 Move the cutting torch at a 55-degree angle to the metal surface.

Step 5 At the top of the material, press the cutting oxygen lever to cut the material to

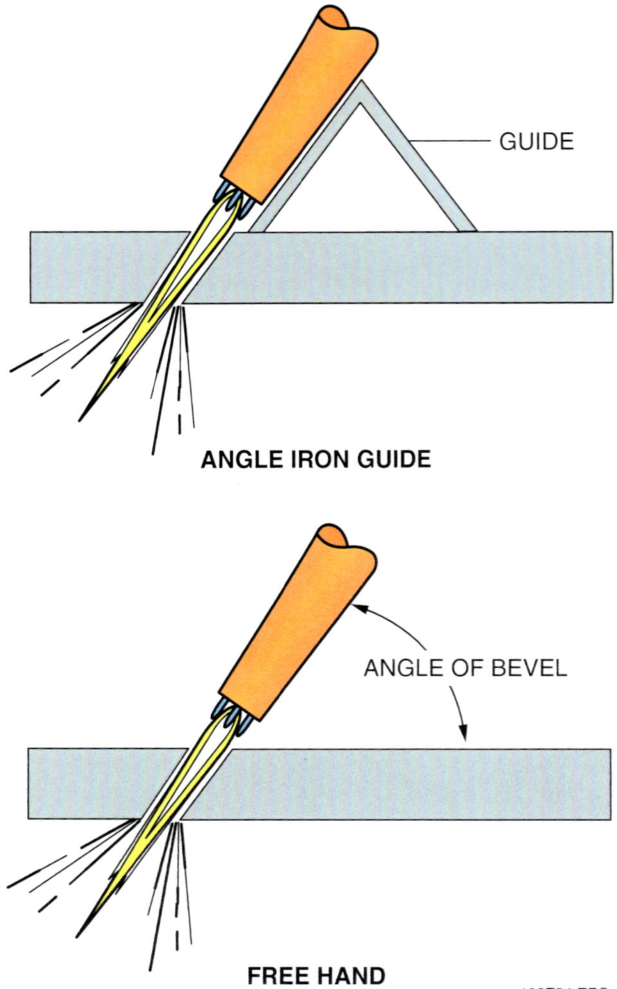

Figure 54 Cutting a bevel.

be removed. Continue moving back and forth across the material while rotating the tip to a position parallel with the material. Move the tip back and forth and down to the surrounding metal. Use care not to cut into the surrounding metal.

> **CAUTION**
>
> As the surrounding metal heats up, there is a greater danger of cutting into it. Try to complete the washing operation as quickly as possible. If the surrounding metal gets too hot, stop and let it cool down.

9.8.0 Gouging

Gouging is the process of cutting a groove into a surface. Gouging operations use a special curved tip that produces a low-velocity stream of oxygen that curves up, allowing the operator to control the depth and width of the groove. It is an effective means to gouge out cracks or weld defects for welding. Gouging tips can also be used to remove

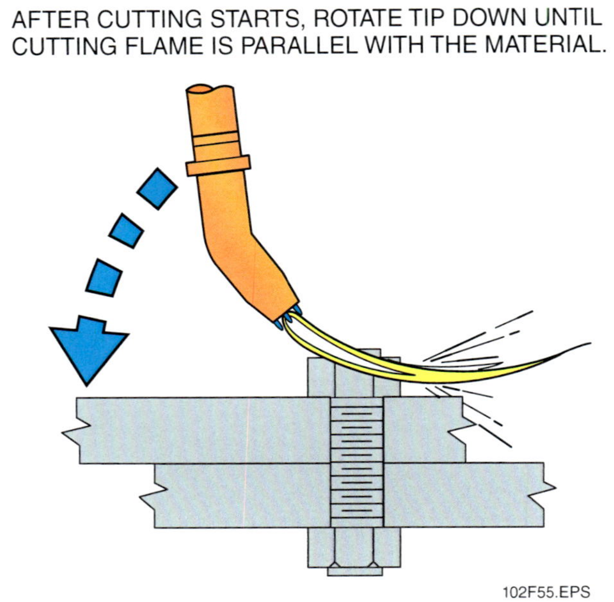

Figure 55 Washing.

steel backing from welds or to wash off bolt or rivet heads. Gouging tips are not as effective as washing tips for removing the shank of the bolt or rivet. Follow these steps to gouge (*Figure 56*):

Step 1 Prepare the metal surface and torch.

Step 2 Ignite the torch and adjust the flame.

Step 3 Holding the torch so that the preheat holes are pointed directly at the metal, preheat the surface until it becomes a bright cherry red.

Step 4 When the steel has been heated to a bright cherry red, slowly roll the torch away from the metal so that the holes are at an angle that will enable you to cut the gouge to the correct depth. As you roll the torch away, press the cutting oxygen lever gradually.

Step 5 Move the cutting torch along the line of the gouge. As you move the torch, rock it back and forth as necessary to create a gouge of the required depth and width.

> **NOTE**
>
> The travel speed and torch angle are very important when gouging. If the travel speed or torch angle is incorrect, the gouge will be irregular and there will be a buildup of dross inside the gouge. Practice until the gouge is clean and even.

Figure 56 Gouging.

10.0.0 Portable Oxyfuel Cutting Machine Operation

As explained previously, machine oxyfuel gas cutters or track burners are basic guidance systems driven by a variable speed electric motor to enable the operator to cut or bevel straight lines at any desired speed. The device (*Figure 57*) is usually mounted on a track or used with a circle-cutting attachment to enable the operator to cut various diameters from 4" to 96". It consists of a heavy-duty tractor unit fitted with an adjustable torch mount and gas hose attachments. It is also equipped with an On/Off switch, a reversing switch, a clutch, and a speed-adjusting dial calibrated in feet/meters per minute.

The device shown in *Figure 57* offers the following operational features:

- Makes straight-line cuts of any length
- Makes circle cuts up to 96" in diameter
- Makes bevel or chamfer cuts
- Has an infinitely variable cutting speed from 1" to 60" per minute
- Has dual speed and clutch controls to enable operation of the machine from either end

Figure 58 shows the location of the following controls:

- *Directional control* – Set the machine direction by toggling the Fwd-Off-Rev toggle switch located next to the power cord.
- *Speed control* – Turn the large knob on either end of the machine to position the speed indicator at the desired cutting speed.

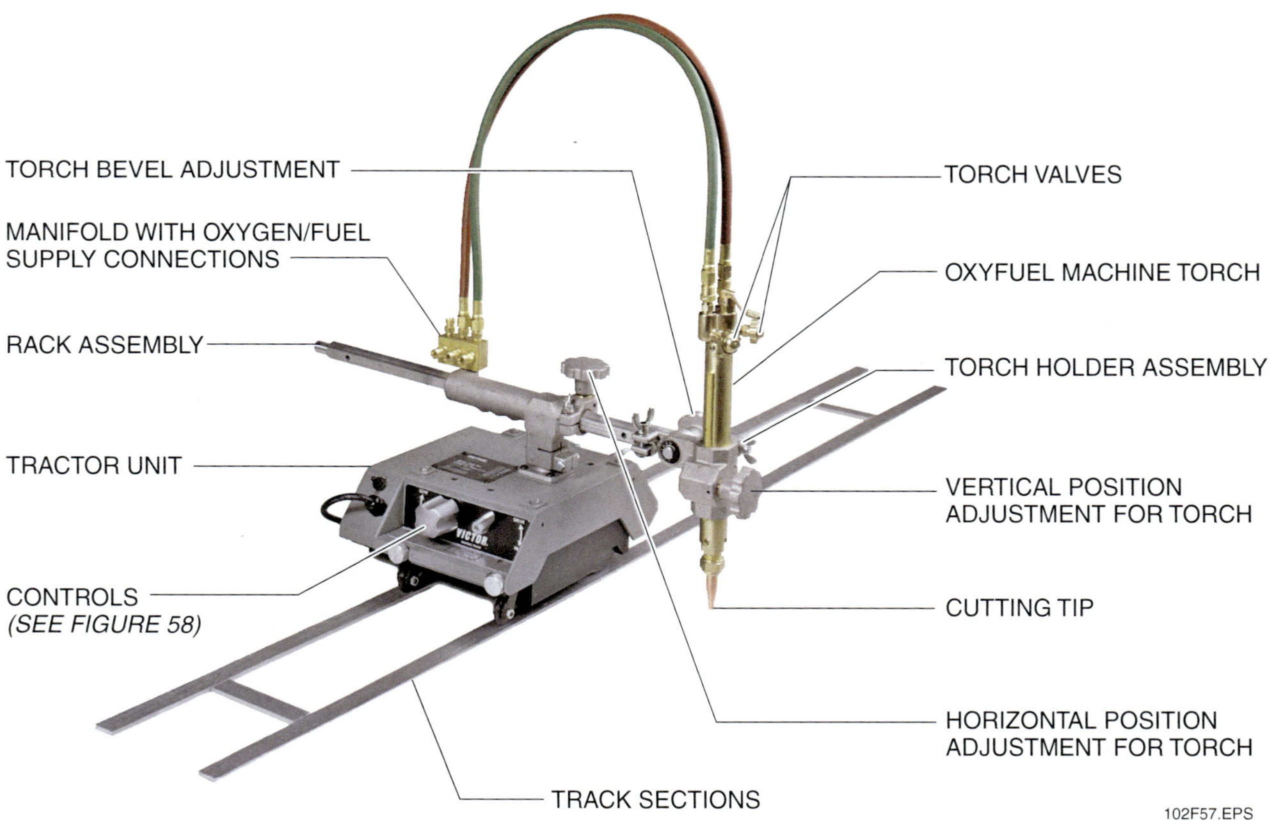

Figure 57 Victor track burner.

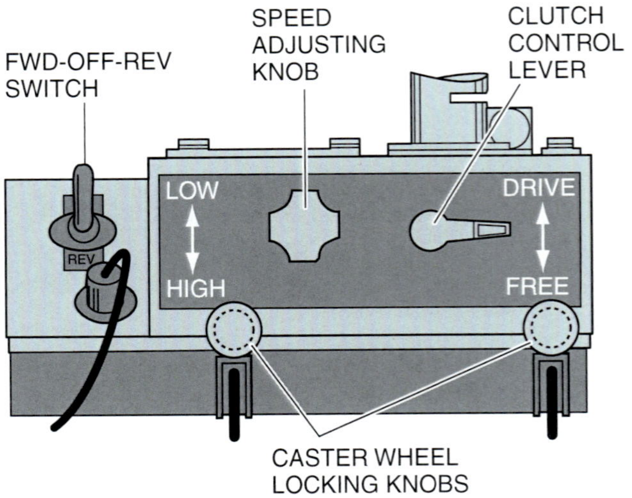

Figure 58 Victor track burner controls.

- *Clutch operation* – Engage the clutch by rotating one of the two clutch levers, located on either end of the machine, to the Drive position. Place the clutch lever in the Free position to permit easy manual positioning of the machine prior to or after the actual cutting operation.

10.1.0 Torch Adjustment

The rack assembly permits the torch holder assembly to move toward or away from the tractor unit. The torch holder allows vertical positioning of the torch. The torch bevel adjustment allows torch positioning at any angle from 190 to 290 degrees in a plane perpendicular to the track. After adjusting the torch to the desired position, tighten all clamping screws to prevent the torch from making any unexpected movements.

10.2.0 Straight-Line Cutting

Cut straight lines using the following procedure:

> **WARNING!** Most cutting machines are not designed to detect the end of their track or workpiece. Take care that an unattended machine does not fall from an elevated workpiece.

Step 1 Place the machine track on the workpiece and line it up before placing the machine on the track.

Step 2 Be sure the track is long enough for the cut to be made. If not, install additional track. Connect track sections carefully. Extend the track on both sides of the cut and support the track. When properly connected, the machine should travel smoothly from one track section to the next. If the cut is long, the track may have to be clamped at both ends beyond the cut to keep the track from moving during the cut.

Step 3 Place the machine on the track. Place the clutch lever in the Free position. Be sure that the supply gas hoses and the power lines are long enough and free to move with the machine so that it can complete the cut properly.

Step 4 Move the machine to the approximate point where the cut will start. Set the drive speed control to the desired cutting speed. Set the Fwd-Off-Rev switch to the Off position. Plug the power cord into a 115 alternating current (AC), 60 Hertz (Hz) power outlet.

Step 5 Ensure that all clamping screws are properly tightened. Ignite and properly adjust the torch, then preheat the start of the cut. Set the Fwd-Off-Rev switch to the desired direction of travel. Simultaneously turn on the cutting oxygen and set the clutch lever to the Drive position.

Step 6 When the cut is completed, stop the machine and shut off the torch.

10.3.0 Bevel Cutting

Perform bevel cutting operations using the following procedure:

> **WARNING!** Most cutting machines are not designed to detect the end of their track or workpiece. Take care that an unattended machine does not fall from an elevated workpiece.

Step 1 Place the machine track on the workpiece and line it up before placing the machine on the track.

Step 2 Be sure the track is long enough for the cut to be made. If not, install additional track. Connect track sections carefully. Extend the track on both sides of the cut and support the track. When properly connected, the machine should travel smoothly from one track section to the next. If the cut is long, the track may have to be clamped at both ends beyond the cut to keep the track from moving during the cut.

Step 3 Place the machine on the track. Place the clutch lever in the Free position. Be sure that the supply gas hoses and the power lines are long enough and free to move with the machine so that it can complete the cut properly.

Step 4 Loosen the bevel adjusting knob, set the torch angle to the desired bevel angle, and then tighten the bevel adjusting knob.

Step 5 Move the machine to the approximate point where the cut will start. Set the drive speed control to the desired cutting speed. Set the Fwd-Off-Rev switch to the Off position. Plug the power cord into a 115AC, 60Hz power outlet.

Step 6 Ensure that all clamping screws are properly tightened. Ignite and properly adjust the torch, then preheat the start of the cut. Set the Fwd-Off-Rev switch to the desired direction of travel. Simultaneously turn on the cutting oxygen and set the clutch lever to the Drive position.

Step 7 When the cut is completed, stop the machine and shut off the torch.

SUMMARY

Oxyfuel cutting has many uses on job sites. It can be used to cut plate and shapes to size, prepare joints for welding, clean metals or welds, and disassemble structures. Because of the high pressures and flammable gases involved, there is a danger of fire and explosion when using oxyfuel equipment. However, these risks can be minimized when the oxyfuel cutting operator is well trained and knowledgeable. Be sure you know and understand the safety precautions and equipment presented in this module before using oxyfuel equipment.

Review Questions

1. For gas welding, either the face shield or the lenses of the safety glasses must be an approved shade _____ filter.
   a. 2
   b. 3 to 6
   c. 7 to 8
   d. 9 to 10

2. Most worksite fires caused by welding or cutting are started by _____.
   a. improper disposal of dross
   b. escape of acetylene
   c. welding torches
   d. cutting torches

3. When mixed with air or oxygen, _____ will explode at much lower concentrations than any other fuel.
   a. propylene
   b. propane
   c. acetylene
   d. MAPP®

4. When pure oxygen is combined with fuel gases, the pure oxygen produces _____.
   a. argon
   b. hydrogen
   c. a colorless, odorless, and tasteless gas
   d. a high-temperature flame needed for flame cutting

5. Oxygen has a strong garlic-like odor so leaks can be readily detected.
   a. True
   b. False

6. The smallest standard oxygen cylinder holds about _____ cubic feet of oxygen.
   a. 40
   b. 65
   c. 80
   d. 85

7. Acetylene gas must be withdrawn from a cylinder at an hourly rate that does *not* exceed _____ of the cylinder capacity.
   a. ½
   b. ⅓
   c. ⅕
   d. ⅐

8. Of the fuel gases listed, _____ burns with the hottest flame.
   a. propane
   b. propylene
   c. butane
   d. MAPP®

9. The amount of liquefied gas remaining in a cylinder is determined by the _____ of the cylinder.
   a. color
   b. heat
   c. weight
   d. pressure

10. The regulators used on fuel gas cylinders usually have red markings and usually have _____ on all connections.
    a. right-hand threads
    b. left-hand threads
    c. metric threads
    d. safety latches

11. The attachment on the top of a fuel gas cylinder that allows the gas to flow only in one direction is called a _____.
    a. single-stage regulator
    b. two-stage regulator
    c. flashback arrestor
    d. check valve

12. The cutting tips used with liquefied fuel gases must have at least _____ preheat holes.
    a. four
    b. five
    c. six
    d. seven

13. When lifting oxyfuel cutting cylinders, always use a(n) _____.
    a. sling cable
    b. sling strap
    c. cylinder cage
    d. electromagnet

14. If an acetylene cylinder is found lying on its side, you should _____ before using it.

  a. contact the supplier for instructions
  b. release a small amount to the air
  c. stand it upright and wait at least one hour
  d. roll it around to stabilize the gas

15. Most fuel gas fittings have _____ threads.

  a. left-hand
  b. right-hand

16. The first step in selecting a cutting tip is to _____.

  a. inspect the cutting tip sealing surfaces and orifices for damage
  b. determine the size of cutting tip to use
  c. determine the kind of gas being used
  d. identify the thickness of the material to be cut

17. Before opening cylinder valves, verify that the adjusting screws on the oxygen and fuel gas regulators are _____.

  a. tight
  b. fully clockwise
  c. loose
  d. fully counterclockwise

18. When a cutting flame has an excess of fuel, the flame is called a(n) _____ flame.

  a. cold
  b. neutral
  c. oxidizing
  d. carburizing

19. When disassembling oxyfuel equipment, verify that all pressure gauges read _____ before starting to take the equipment apart.

  a. 21.0
  b. 0
  c. within 0.3 of 0
  d. within 0.2 of 0

20. When inspecting a completed cut made with oxyfuel cutting equipment, the drag lines of the cut should be near _____ and not very pronounced.

  a. 30 degrees
  b. 45 degrees
  c. horizontal
  d. vertical

Trade Terms Quiz

Fill in the blank with the correct key trade term that you learned from your study of this module.

1. The gap produced by a cutting process is called a(n) ________.
2. A loud snap or pop as a torch flame is extinguished is called a(n) ________.
3. The soft, white material used to mark metal is ________.
4. ________ contain iron.
5. A flame burning with too much fuel is called a(n) ________.
6. A flame burning with too much oxygen is called a(n) ________.
7. When correct portions of fuel gas and oxygen are fed to a flame, the flame is said to be a(n) ________.
8. When a thermal process is used for cutting, the material expelled from the kerf is called ________.
9. When a flame burning back into the tip of a torch causes a high-pitched whistling sound, the condition is called ________.
10. When metal plate is penetrated with an oxyfuel cutting torch, the action is called a(n) ________.
11. On the edge of a cut, the lines resulting from the cutting oxygen stream into, through, and out of the metal are called ________.

Trade Terms

Backfire	Dross	Kerf	Pierce
Carburizing flame	Ferrous metals	Neutral flame	Soapstone
Drag lines	Flashback	Oxidizing flame	

Cornerstone of Craftmanship

Ray French

Execution Coordinator,
Special Projects, Flint Hills Resource

As a young man, Ray French got off on the wrong foot, but eventually he found his niche in the welding trade.

How did you get your start in the welding trade?
I received a college football scholarship but flunked out of college in my freshman year. I then spent a year loading ships and barges until my father suggested that I learn a trade. My uncle had a welding shop, so I decided to try welding. I really liked it! I went to welding school for two years while working part time at my uncle's welding shop.

How did your career progress after welding school?
I was hired by an oil refinery and worked in the welding shop until I received my first-class welder classification. Then I was assigned to a hot-shot team made up of welders, pipefitters, boilermakers, pump mechanics, and electricians. During my time on that team, I had the opportunity to learn a lot about each of these trades. I supervised several shutdowns—that's where a unit is taken out of service and completely overhauled—and eventually was promoted to maintenance foreman. Today, my job title is Execution Coordinator, and I supervise large projects at a major oil refinery.

What do you like most about your job?
I like the pride that comes from doing a good job and being able to see the results of my work every day. I also like the respect I get because of my skills and my knowledge of the trade.

What do you think it takes to be a success?
You have to take pride in your work. Do a quality job every time and don't take shortcuts. Your reputation is everything. It will follow you from job to job. No matter where you go, someone will know you or know someone who knows you.

Drew Fontenot

Calcasieu Parish School Systems
Welding Instructor, Lake Charles, Louisiana

The seeds of a welding career took root when Drew was just a boy as he watched his dad and uncle repair and build farm equipment. As a teenager, Drew studied welding in high school. Then he went off into the world to work in various jobs—from maintenance welder in oil refineries and the petrochemical industry, to welder for general contractors on various projects, to sharing his knowledge with young welders through teaching. Drew's philosophy for a successful career in welding is simple: have diverse skills and get as much training as you can. According to Drew, this will make you competitive in the job market and an excellent candidate for advancement.

How did you choose a career in the field?
As a youngster, I was intrigued with welding, and I liked to watch my dad and uncle repair and build farm equipment. Later on in high school, I heard about a welding program that was offered starting in the tenth grade. I enrolled and took welding for the next three years. Near the end of my senior year, I got a job as a maintenance welder for Cities Service Oil Company, which later became CITGO. After graduation, I kept working for Cities Service evenings and nights, and I studied business agriculture at college during the day. By the time graduation rolled around, jobs in my major were few and far between, so I kept on working in welding. Eventually I got a job teaching welding back at my old high school. Ironically, this was the same position that had been held by the teacher who taught me welding in the first place.

What types of training have you been through?
I got my formal welding training in a high school welding program, but I also had lots of on-the-job experience. With respect to training and certifications, I am a Certified Welding Educator and Certified Welding Inspector through the AWS, and I've completed the NCCER Craft Instructor and Master Trainer programs.

What kinds of work have you done in your career?
Over the course of my career, I've worked as a maintenance welder in oil refineries and in the petrochemical industry, worked as a welder for several general contractors on a variety of projects, and done contract welding. I'm currently teaching welding at the high school level. I've served as an officer for my local AWS section, and am presently chairman of the Educational Committee for the section. I also serve as an advisor on the National AWS Educators Committee. With everything else that's going on, I still like to find the time to do welding repair work on the family farm.

Tell us about your present job and what you like about it.
In my present job, I teach the NCCER Core Curriculum, NCCER Welding Level One and Welding Level Two, and the AWS SENSE (Schools Excelling through National Skill Education) Entry Level Welder. I am also the lead SkillsUSA advisor for our school. Something I really enjoy is talking with former students and having them tell me about how successful they are in their welding careers.

What factors have contributed most to your success?
I've made it a point to participate in professional organizations such as the AWS and SkillsUSA. I've always found a lot of value in participating in these organizations—not only in the benefits offered, but especially in the opportunity to actively promote the goals of organizations like the AWS and SkillsUSA.

What advice would you give to those new to the field?

Get as much training in as many processes as you can before you start working. Once you start to work full time, you may find it difficult to go back and get additional training. The more diverse your skills are, the easier it will be for you to find work. Diverse skills and additional training will also give you a greater opportunity for advancement.

Tell us some interesting career-related facts or accomplishments.

When I was a high school senior, I had the honor of winning first place in the Louisiana State Vocational Welding Contest. I've been recipient of the AWS Section and District Educator Awards. In 2005, I was among those chosen to attend the AWS Instructors Institute in Miami, Florida. Most recently, in 2007, the Louisiana Department of Education selected me to serve as Subject Matter Expert (SME) on NCCER's Welding Committee.

Trade Terms Introduced in This Module

Backfire: A loud snap or pop as a torch flame is extinguished.

Carburizing flame: A flame burning with an excess amount of fuel; also called a reducing flame.

Drag lines: The lines on the edge of the material that result from the travel of the cutting oxygen stream into, through, and out of the metal.

Dross: The material (oxidized and molten metal) that is expelled from the kerf when cutting using a thermal process. It is sometimes called slag.

Ferrous metals: Metals containing iron.

Flashback: The flame burning back into the tip, torch, hose, or regulator, causing a high-pitched whistling or hissing sound.

Kerf: The gap produced by a cutting process.

Neutral flame: A flame burning with correct proportions of fuel gas and oxygen.

Oxidizing flame: A flame burning with an excess amount of oxygen.

Pierce: To penetrate through metal plate with an oxyfuel cutting torch.

Soapstone: Soft, white stone used to mark metal.

Additional Resources

This module is intended to present thorough resources for task training. The following reference work is suggested for further study. This is optional material for continued education rather than for task training.

ANSI Z49.1, Safety in Welding, Cutting, and Allied Processes, American Welding Society, Miami, FL.

Figure Credits

Lincoln Electric Company, 102F01 (left)
J. R. Yochum, 102F01 (right)
Topaz Publications, Inc., 102F02, 102F04, 102F06–102F08, 102SA01–102SA04, 102F10, 102F12, 102F13, 102F20–102F23, 102SA08–102SA10, 102F32–102F46, 102SA13, 102SA14, 102F48
Thermadyne, Inc., 102SA05, 102F15, 102F24B, 102F27, 102F31, 102F52, 102F57, 102F58
American Torch Tip Company, 102SA06, 102SA07
Lenco d/b/a NLC, Inc., 102F24A
Koike Aronson, Inc., 102F25
ESAB Welding and Cutting Products, 102F26
Bug-O Systems, Inc., 102F28
H&M Pipe Beveling Machine Company, Inc., 102F29, 102F30
Figure SA11 courtesy of American Welding Society *Welding Handbook Volume 2*,
Figure 14.26, Reproduced with permission from the American Welding Society,
Miami, FL, Eighth Edition.
Saf-T-Cart, 102SA12
Smith Equipment, 102F47
Figure 50 courtesy of American Welding Society *Welding Handbook Volume 2*, Figure 8.16, Reproduced with permission from the American Welding Society, Miami, FL, Eighth Edition.

Appendix

Performance Accreditation Tasks

The Performance Accreditation Tasks (PATs) correspond to and support the learning objectives in *AWS EG2.0, Guide for the Training and Qualification of Welding Personnel: Entry-Level Welder*.

Note that in order to satisfy all learning objectives in *AWS EG2.0*, the instructor must also use the PATs contained in the second level in the NCCER Welding curriculum.

PATs provide specific acceptable criteria for performance and help to ensure a true competency-based welding program for students.

The following task is designed to test your competency with an oxyfuel cutting torch. Do not perform this cutting task until directed to do so by your instructor.

SETTING UP, IGNITING, ADJUSTING, AND SHUTTING DOWN OXYFUEL EQUIPMENT

Using oxyfuel equipment that has been completely disassembled, demonstrate how to:

- Set up oxyfuel equipment
- Ignite and adjust the flame
  - Carburizing
  - Neutral
  - Oxidizing
- Shut off the torch
- Shut down the oxyfuel equipment

Criteria for Acceptance:

- Set up the oxyfuel equipment in the correct sequence ____________
- Demonstrate that there are no leaks ____________
- Properly adjust all three flames ____________
- Shut off the torch in the correct sequence ____________
- Shut down the oxyfuel equipment ____________

CUTTING A SHAPE FROM THIN STEEL

Using a carbon steel plate, lay out and cut the shape and holes shown in the figure. If available, use a machine track cutter to straight cut the 6" dimensions.

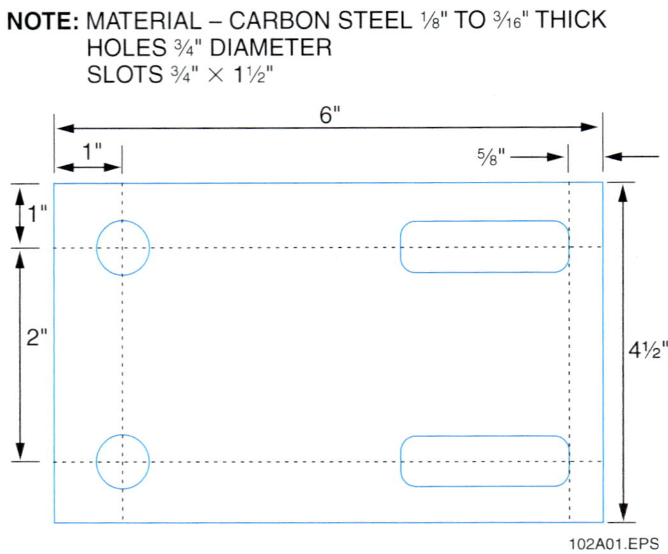

Criteria for Acceptance:

- Outside dimensions ±1/16" ________
- Inside dimensions (holes and slots) ±1/8" ________
- Square ±5° ________
- Minimal amount of dross sticking to plate which can be easily removed ________
- Square kerf face with minimal notching not exceeding 1/16" deep ________

CUTTING A SHAPE FROM THICK STEEL

Using a carbon steel plate, lay out and cut the shape and holes shown in the figure. If available, use a machine track cutter to bevel and straight cut the 6" dimensions.

NOTE: MATERIAL – CARBON STEEL OVER ¼" THICK OR GREATER
HOLES ¾" DIAMETER
SLOTS ¾" × 1½"

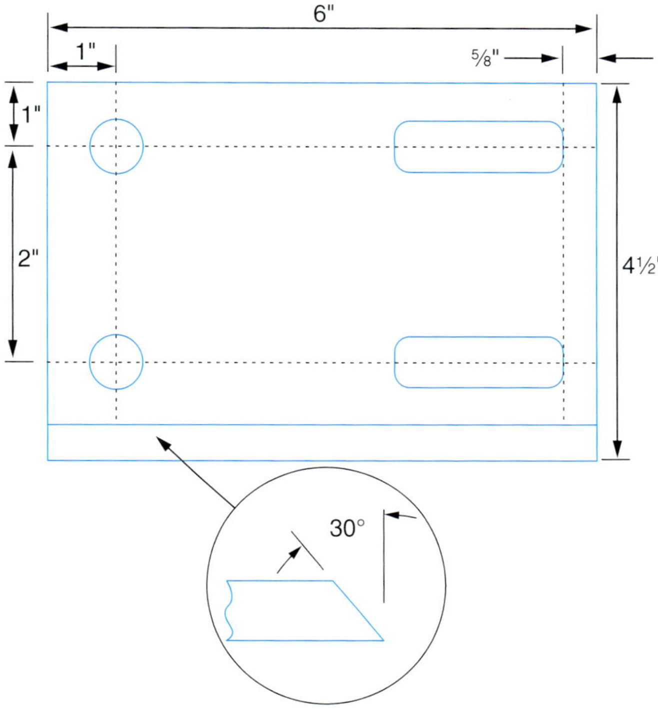

102A02.EPS

Criteria for Acceptance:

- Outside dimensions ±1⁄16" ________
- Inside dimensions (holes and slots) ±⅛" ________
- Square ±5° ________
- Bevel ±2° ________
- Minimal amount of dross sticking to plate which can be easily removed ________
- Square kerf face with minimal notching not exceeding 1⁄16" deep ________

NCCER CURRICULA — USER UPDATE

NCCER makes every effort to keep its textbooks up-to-date and free of technical errors. We appreciate your help in this process. If you find an error, a typographical mistake, or an inaccuracy in NCCER's curricula, please fill out this form (or a photocopy), or complete the online form at **www.nccer.org/olf**. Be sure to include the exact module ID number, page number, a detailed description, and your recommended correction. Your input will be brought to the attention of the Authoring Team. Thank you for your assistance.

Instructors – If you have an idea for improving this textbook, or have found that additional materials were necessary to teach this module effectively, please let us know so that we may present your suggestions to the Authoring Team.

NCCER Product Development and Revision

13614 Progress Blvd., Alachua, FL 32615

Email: curriculum@nccer.org

Online: www.nccer.org/olf

❑ Trainee Guide ❑ Lesson Plans ❑ Exam ❑ PowerPoints Other ________________________

Craft / Level: Copyright Date:

Module ID Number / Title:

Section Number(s):

Description:

Recommended Correction:

Your Name:

Address:

Email: Phone:

29105-09

Base Metal Preparation

Module Three

Trainees with successful module completions may be eligible for credentialing through NCCER's National Registry. To learn more, go to **www.nccer.org** or contact us at **1.888.622.3720**. Our website has information on the latest product releases and training, as well as online versions of our *Cornerstone* magazine and Pearson's product catalog.

Your feedback is welcome. You may email your comments to **curriculum@nccer.org**, send general comments and inquiries to **info@nccer.org**, or fill in the User Update form at the back of this module.

This information is general in nature and intended for training purposes only. Actual performance of activities described in this manual requires compliance with all applicable operating, service, maintenance, and safety procedures under the direction of qualified personnel. References in this manual to patented or proprietary devices do not constitute a recommendation of their use.

29105-09
Base Metal Preparation

Objectives

When you have completed this module, you will be able to do the following:

1. Clean base metal for welding or cutting.
2. Identify and explain joint design.
3. Explain joint design considerations.
4. Mechanically bevel the edge of a mild steel plate.
5. Thermally bevel the end of a mild steel plate.
6. Select the proper joint design based on a welding procedure specification (WPS) or instructor direction.

Performance Tasks

Under the supervision of the instructor, you should be able to do the following:

1. Mechanically bevel the edge of a mild steel plate ¼" to ¾" thick at 22½ degrees (or 30 degrees, depending on the equipment available).
2. Thermally prepare a bevel.

Trade Terms

Backing
Base metal
Code
Distortion
Load
Melt-through
Nonferrous metal
Oxide
Peen
Piping porosity
Porosity
Root face
Root opening
Surfacing
Transverse
Weathering steel
Welding procedure specification (WPS)
Weldment

Prerequisites

Before you begin this module, it is recommended that you successfully complete the following: *Core Curriculum* and *Welding Level One*, Modules 29101-09 through 29104-09.

Industry-Recognized Credentials

If you're training through an NCCER-accredited sponsor, you may be eligible for credentials from NCCER's Registry. The ID number for this module is 29105-09. Note that this module may have been used in other NCCER curricula and may apply to other level completions. Contact NCCER's Registry at 888.622.3720 or go to **www.nccer.org** for more information.

Contents

Topics to be presented in this module include:

Figures

1.0.0 Introduction

To ensure that the highest quality welds are produced, and to comply with welding **codes**, **base metals** must be properly prepared prior to welding. The type of preparation required depends on the governing code requirements, the base metal type, the condition of the base metal, the welding process to be used, and the equipment available. This module explains how to properly prepare various base metals to conform to the appropriate welding codes.

2.0.0 Basic Welding Safety

Welders routinely use manual and powered tools to clean and shape metals. *Figure 1* shows welders cutting metal with oxyfuel torches, and *Figure 2* shows welders grinding metal. Everyone entering the field of welding starts by doing basic tasks, which includes preparing the metals for welding.

When using tools to clean the metals prior to welding, wires from the rotating wire brushes and particles from the sanding discs become flying objects. Metal particles ground from the workpieces are usually hot, and will burn your skin or anything combustible. Welding screens or similar protective devices are used to keep the hot particles from igniting combustible materials near the work. Most companies require a fire extinguisher and a fire watch (person) posted near the cutting or grinding work. To prevent injury from the flying particles and debris, always wear protective gloves, arm and chest protectors, and face shields when cutting and preparing metals. Also, wear hearing protection to protect your ears from the noise generated by the grinders.

While much of the preparation work is performed in the shop where parts can be secured in clamping devices, some of it must be performed in the field on structures or equipment. Welders must sometimes perform such work at heights requiring the wearing of fall protection devices. Always make sure that fall protection devices are undamaged and up-to-date before using them.

The following is a summary of safety procedures and practices that must be observed while cutting, cleaning, shaping, and welding. Keep in mind that this is a summary. Complete safety coverage is provided in *Welding Safety*. If you have not completed that module, do so before continuing.

105F01.EPS

Figure 1 Cutting metal with oxyfuel torches.

2.1.0 Protective Clothing and Equipment for Preparing Metals

To maintain safety and prevent injury, it is essential that you wear the appropriate protective clothing and equipment when preparing metals. Be sure to follow these guidelines during all phases of metal preparation:

- Always use safety glasses with a full face shield or a welding helmet when preparing metals. Wear welding glasses, a face shield, or a welding helmet with the proper light-reducing tint for the type of welding or cutting being performed. Never directly or indirectly view an electric arc without using the properly tinted lenses.
- Wear the proper protective leather and/or flame-retardant clothing, along with protective gloves, when cutting, grinding, or welding (see *Figure 3*).
- Wear 8" or taller high-top safety shoes or boots. Make sure that the tongue and lace area of the footwear will be covered by a pant leg. If the tongue and lace area is exposed, or if the footwear must be protected from burn marks, wear leather spats under your pants or chaps, and over the top front area of the footwear.
- Wear a solid material (non-mesh) hat with a bill pointing to the rear, or—if much overhead cutting or welding is required—wear a full leather hood with a welding face plate and the correct

105F02.EPS

Figure 2 Welders grinding metal.

105F03.EPS

Figure 3 Appropriate personal protective equipment.

tinted lens. If a hard hat is required, use one that allows the attachment of rear deflector material and a face shield.

- If a full leather hood is not worn, wear a face shield and snug-fitting safety glasses for gas welding or cutting. Either the face shield or the lenses of the welding glasses must be an approved filter. For electric arc welding, wear safety glasses and a welding hood with the correct tinted lens.
- If a full leather hood is not worn, wear earplugs to protect ear canals from sparks.

2.2.0 Fire/Explosion Prevention

Welding activities usually involve the use of fire or extreme heat to melt metal. Whenever fire is used, it must be controlled and contained. Welding activities are often performed on vessels that may once have contained flammable or explosive materials. Residues from those materials can catch fire or explode when a welder begins work on such a vessel. The following are fire and explosion prevention guidelines associated with welding:

- Never carry matches or gas-filled lighters in your pockets. Sparks can cause the matches to ignite or the lighter to explode, causing serious injury.
- Never perform any type of heating, cutting, or welding until a hot-work permit has been obtained and an approved fire watch established. Most work site fires caused by these types of operations are started by cutting torches.
- Never use oxygen to blow dirt or dust off clothing. The oxygen can remain trapped in the fabric for a time. If a spark hits the oxygen in the fabric, the clothing can burn rapidly and violently out of control.
- Make sure that any flammable material in the work area is moved or shielded by a fire-resistant covering. Approved fire extinguishers must be available before any heating, welding, or cutting operations are attempted.
- Never release a large amount of oxygen or use oxygen as compressed air. The presence of oxygen around flammable materials or sparks can cause rapid and uncontrolled combustion. Keep oxygen away from oil, grease, and other petroleum products.
- Never release a large amount of fuel gas, especially acetylene. Methane and propane tend to concentrate in and along low areas and can be ignited at a considerable distance from the release point. Acetylene is lighter than air; when mixed with air or oxygen, it will explode at much lower concentrations than any other fuel.
- To prevent fires, maintain a neat and clean work area, and make sure that any metal scrap or dross is cold before disposal.
- Before cutting or welding containers such as tanks or barrels, find out if they have contained any explosive, hazardous, or flammable materials including petroleum products, citrus products, or chemicals that decompose into toxic fumes when heated. As a standard practice, always clean and then fill any tanks or barrels with water, or fill them with a purging gas such as nitrogen or argon to displace any oxygen.

2.3.0 Work Area Ventilation

Vapors and fumes tend to rise in the air from their sources, and welders often have to work above welding areas where fumes are being created. Welding fumes can be harmful. Good work area ventilation helps to remove the vapors and protect the welder. The following is a list of work area ventilation guidelines to consider before and during welding activities:

- Make sure confined space procedures are followed before conducting any welding or cutting in the confined space.
- Never use oxygen in a confined space for the purpose of ventilation.
- Always perform cutting or welding operations in a well-ventilated area. Cutting or welding operations involving zinc or cadmium materials or coatings result in toxic fumes. For long-term cutting or welding of such materials, always wear an approved full face, supplied-air respirator that uses breathing air supplied from outside of the work area. For occasional, very short-term exposure, a HEPA-rated or metal-fume filter may be used on a standard respirator.
- Make sure confined spaces are ventilated properly for cutting or welding purposes.

GOING GREEN

Welding Water

To better protect the environment, send any water used in a welding process, or used to clean tanks, to a waste treatment facility instead of simply allowing it to run into a storm drain or onto the ground.

3.0.0 Base Metal Cleaning

All base metals should be cleaned before welding. Even new materials that may look clean often are not. They pick up contaminants and surface corrosion during shipping and handling. When performing maintenance welding, a welder often comes across components that have been exposed to surface contamination such as corrosion, paint, oil, and grease. To ensure quality welds and to conform to code requirements, surface contaminants and **oxides** must be removed prior to welding. In addition, the heat generated by the welding process could cause the contaminants to give off toxic fumes, endangering the worker performing the welding.

Because of the defects caused by surface contaminants and the possibility of toxic fumes, all welding codes require surface cleaning. In general, the codes state that the surface to be thermally cut or welded must be clean and free from paint, oil, rust, scale, and other material that would be harmful either to the weld or to the base metal when heat is applied. Cleaning is typically performed by mechanical and/or chemical means.

3.1.0 Surface Corrosion

All metal has surface corrosion. Corrosion occurs when metal is exposed to air. Some corrosion exists as a thin, hard film that merely stains the metal and acts as a retardant to further corrosion. Examples of this type of corrosion are found on stainless steel, copper, and aluminum. The thin, hard layer of corrosion bonds tightly to the surface and protects the metal. A more recognizable type of corrosion that is found on mild steel is called rust (*Figure 4*). Rust is a very coarse type of corrosion that tends to flake easily, exposing more of the base metal to corrosion. Alloys such as chromium and copper are often added to mild steel to protect it from this coarse type of corrosion. The alloys cause the corrosion (rust) to be very fine and to bond to the surface, protecting the base metal beneath. These types of steels are called alloy steels. **Weathering steel** is an example of a copper-alloyed steel that protects itself from corrosion by forming a hard, tough layer of brown corrosion (rust) on its surface. Weathering steels are used outdoors and require no painting.

Regardless of the type of corrosion, it must be removed prior to welding. Any type of corrosion can cause weld defects.

105F04.EPS

Figure 4 Examples of rust and corrosion.

3.2.0 Defects Caused by Surface Contamination

The most common defect caused by surface contamination is **porosity**. Examples of porosity are shown in *Figure 5*. Porosity occurs when gas pockets or voids appear in the weld metal. When the porosity has a length greater than its width and is approximately perpendicular to the weld face, it is called **piping porosity**. Piping porosity is formed as the gas pocket floats toward the surface of the weld, leaving a void behind. The gas pocket is trapped as the weld metal solidifies, but as the next layer of weld is deposited, it continues to float up. Porosity may not be visible on the surface of the weld.

Surface contamination can result from any of several causes, including the following:

- Use of gloves contaminated with oil or other substances
- Allowing alloys to come into contact with carbon steel objects, including chain falls, jack stands, and fork lifts
- The use of chloride-based markers to write on the base metal

GOING GREEN

Welding Fumes

To better protect the environment, perform any welding activities under a vapor extraction system that can filter the welding fumes before they reach the atmosphere.

Because of the risk of contamination, metals such as carbon steel, stainless steel, and aluminum must be stored separately. Aluminum must be stored indoors.

3.3.0 Mechanical Cleaning

Mechanical cleaning is the most common method of removing surface contamination. Tools used for mechanical cleaning include hand tools, power tools, and special sandblasting equipment. When performing mechanical cleaning, be sure to wear safety glasses and a face shield for protection from the flying particles produced during the cleaning operation. In addition, special clothing is required for sandblasting.

3.3.1 Hand Tools

Flexible scrapers are used to remove dirt and grease from **weldments** or weldment components. Rigid scrapers can be used to remove hardened paint or dirt.

Wire brushes are used to remove paint and light-to-medium surface corrosion. Wire brushes with stainless steel bristles must be used to clean stainless steel and aluminum. When brushing stainless steel, use stainless steel brushes that have been used only on stainless steel. When brushing aluminum, use stainless steel brushes that have been used only on aluminum; if a standard wire brush with steel bristles is used, it will contaminate the surface with mild steel, causing weld defects.

Files are another way to remove surface corrosion. When filing aluminum or stainless steel, do not use files that have been used on other metals. Filings often stick to the file and could contaminate the metal being filed, resulting in weld defects. When using a file, be sure it has a handle on the file tang.

> **CAUTION**
>
> Using a file without a handle can result in serious injury from the pointed file tang.

Use a fine file and keep it clean to prevent scratching the base metal surface. Take care not to damage or scratch the base metal outside the weld zone. Clean files by using stainless steel brushes or file cards that either are new or have only been used to brush the same type of metal (*Figure 6*).

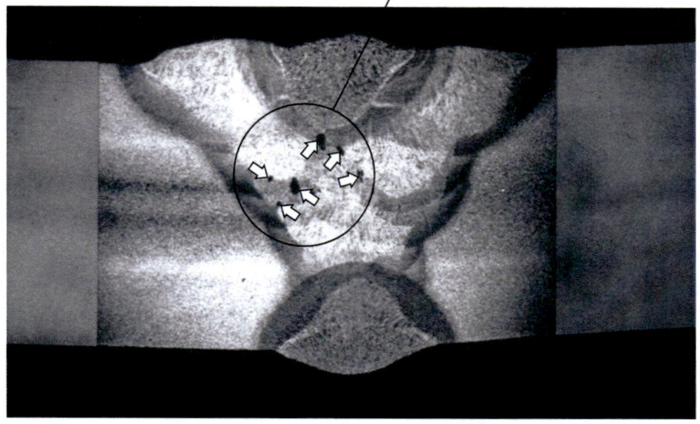

UNIFORMLY DISTRIBUTED POROSITY

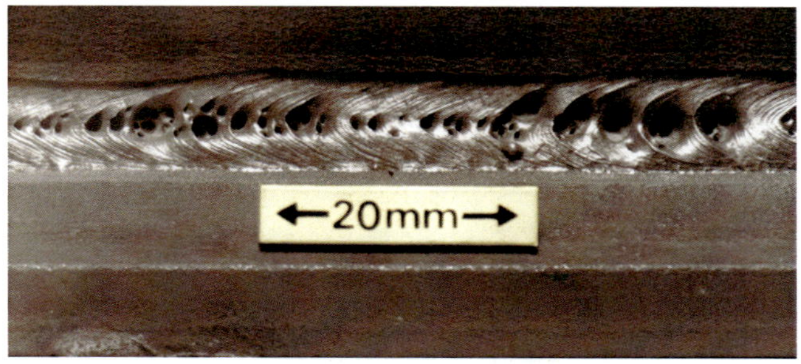

SURFACE-BREAKING POROSITY

Reproduced by permission TWI Ltd. 105F05.EPS

Figure 5 Porosity.

3.3.2 Power Tools

Tight corrosion must be removed by filing, grinding, or sandblasting. For large jobs or jobs where speed is important, power tools are best. Power tools may be electrically or pneumatically powered.

> **NOTE**
>
> Sandblasting may not be used on aluminum.

> **WARNING!**
>
> To contain sparks when cleaning metal with power tools, use a welding screen. Also, remember to wear appropriate personal protection equipment (earplugs, safety glasses and face shield, and gloves) when using power tools.

Appearances Can Be Deceiving

Do not assume a metal's surface is corrosion-free just because it is shiny. Corrosion on stainless steel and aluminum is silver. The metal may appear shiny, but to obtain high-quality welds, the corrosion must be removed.

Figure 6 File and cleaning tools.

Angle grinders and end grinders are very effective in removing large areas of surface contamination. Die grinders and small angle grinders work very well for small areas such as weld grooves and bevel edges. Grinders have attachments that can be used for special applications (*Figure 7*). These include grinding disks and wheels, wire brushes, rotary files, flapper wheels, and cutoff wheels.

Grinding disks and wheels are made for specific types of metals. Be sure you have the correct disk or wheel for the type of metal being ground. Always use aluminum oxide disks for grinding aluminum or stainless steel. For stainless steel or aluminum, do not use wheels that have been used on other metals. They will contaminate the surface of the aluminum or the stainless steel.

When using wire brush attachments on stainless steel or aluminum, brushes with stainless steel bristles must be used. Use stainless steel brushes that have only been used on stainless steel for brushing stainless steel. Use stainless steel brushes that have only been used on aluminum for brushing aluminum. Mild steel bristles will contaminate stainless steel or aluminum, causing weld defects.

When grinding, care must be taken to prevent grinding the base metal below the minimum allowable base metal thickness. If this should happen, the base metal will have to be discarded or, if allowable, built up with weld. Both of these alternatives are expensive.

Cutoff wheels are used to cut metals and other materials. They are made for specific types of metals, or other materials such as stone. Be sure you have the correct cutoff wheel for the type of material being cut.

Weld flux chippers, needle scalers, and chipping hammers are air (pneumatically) powered. These tools are used by welders to clean surfaces and remove dross from cuts and welds. Weld

Think About It

Hammers in Welding

Welders often use versions of the hammers shown in the photo. What do you think they are used for in welding operations?

105SA01.EPS

Grinding Wheel and Disk Safety

If a grinding wheel or disk shatters while in use, the pieces of the wheel or disk become deadly projectiles as the weights of the pieces are multiplied by the centrifugal force and the revolutions per minute (rpm) of the grinding motor. The force with which a piece of wheel or disk is released is equivalent to that of an armor-piercing bullet. Never grind without full face protection. Avoid standing in the orbital path of the wheel or disk in case it disintegrates.

Figure 7 Grinders and attachments.

flux chippers, needle scalers, and chipping hammers are also excellent for removing paint, heavy scale, or hardened dirt, but they are not very effective for removing surface corrosion. Weld flux chippers have a single chisel; needle scalers have about 18 to 20 blunt steel needles about 10" long (*Figure 8*). Most weld flux chippers can be converted to needle scalers with the needle scaler attachment. Pneumatic chipping hammers, which hit much harder than weld flux chippers, are used for **peening** or gouging weld joints.

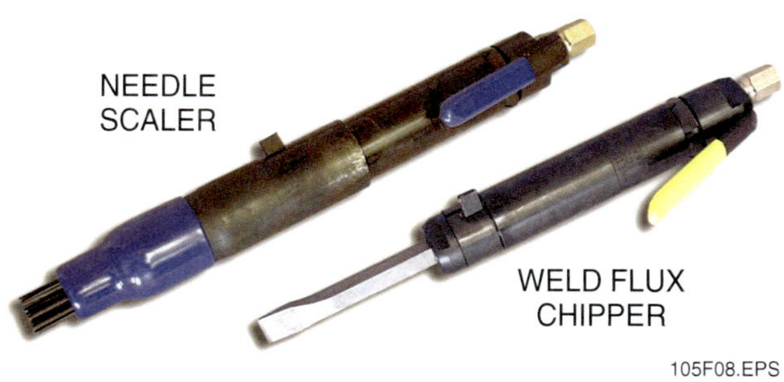

Figure 8 Weld flux chipper and needle scaler.

3.4.0 Chemical Cleaning

There will be times when mechanical cleaning is not enough. Mechanical cutting, grinding, and brushing may leave residue on a metal that will contaminate the weld. The oil from the skin of the welder handling the metal can also contaminate a weld. For these reasons, welders may have to resort to some form of chemical cleaner to ensure that the metals being prepared are very clean. The two chemical cleaners most often used by welders are acetone and butanone. Butanone is better known as methyl ethyl ketone, or MEK.

3.4.1 Acetone

Acetone is a flammable, colorless liquid. Acetone is a manufactured product, but it can also be found in some natural environments. It is one of the major ingredients in nail polish remover. Anyone who has worked in an auto repair shop has probably used acetone to remove over-sprayed paint from car windows. It will dissolve resins, epoxies, and many plastics. It is a great degreaser that is often used to prepare metals for painting.

From a safety standpoint, acetone is an irritant. Because it can cause permanent eye damage (corneal clouding), workers handling it or working near it must wear safety goggles. Acetone vapors should be avoided, and a suitable respirator is recommended. Prolonged exposure may eventually cause liver damage. Check the acetone MSDS closely to ensure that you understand its hazards.

3.4.2 MEK

MEK is a manufactured organic chemical. It is a colorless liquid that dissolves substances such as gums, resins, vinyl film coatings, and other similar materials that may still be on the metals. MEK is highly flammable, but it presents much less of a health risk than acetone. Anyone working with MEK needs to review the MSDS to make sure all safety precautions are taken.

3.4.3 Applications

After the metal parts have been mechanically cleaned, the chemical cleaner is applied, usually with a brush or spray. The chemical is allowed to work for a short time before being removed. A clean cloth is used to remove the cleaning agent along with any contamination the cleaner may have loosened from the metal. When cleaning metals with chemicals, be aware that the contaminants being removed may give off vapors that could be dangerous. Be sure to use personal protective equipment as required.

4.0.0 Joint Design

Welded joints are selected primarily for the safety and strength required for the conditions that will be encountered. When selecting the joint for a particular application, many factors must be taken into account, including **load** considerations, the environment in which it will be used, materials, processes, and cost.

4.1.0 Load Considerations

The loads in a welded steel structure or component are transferred through the welded joint. The welded joint must be designed to withstand the stresses caused by the loads. *Figure 9* shows how force creates stress on welded joints.

Any or all of these forces may act on welded parts. When two parts are welded together, in whatever manner, they are supposed to become one. If the surface areas of the two parts being welded are not properly prepared, the welds may fail when one or more of these forces is applied. The following are brief explanations of load terms as they relate to welds:

- *Tensile* – Capable of stretching or being stretched. When metals are pulled in opposite directions by forces (F_1 vs. F_2), they are expected to stretch some, but if stretched too far, they will eventually break. Their strength to remain one is referred to as their tensile strength. When metals are heated, as when they are welded, the molecules in the metals change. When the metals are cooled, their molecules change again. These heating and cooling cycles can weaken the tensile qualities of metals. When two pieces of metal are welded end-to-end, they become one piece. When they are pulled at each end, the tensile strength of the two pieces, and the weld, are tested. If the weld is a good weld, one of the welded pieces will usually break before the weld breaks.
- *Compression* – The reduction of volume or mass. Metals can be compressed. If you strike a steel nail or bolt with a hammer, you can eventually compress the nail or bolt into a different shape. When two pieces of metal are welded end-to-end and forces (F_1 and F_2) are applied inwardly on each end, the pieces and their weld are compressed.
- *Bending* – To make curved. For example, you may be able to manually bend a section of straight lightweight flat bar around a 6" pipe. The pipe in this case is a force (F_2) pushing against the center of the bar as you pull or ap-

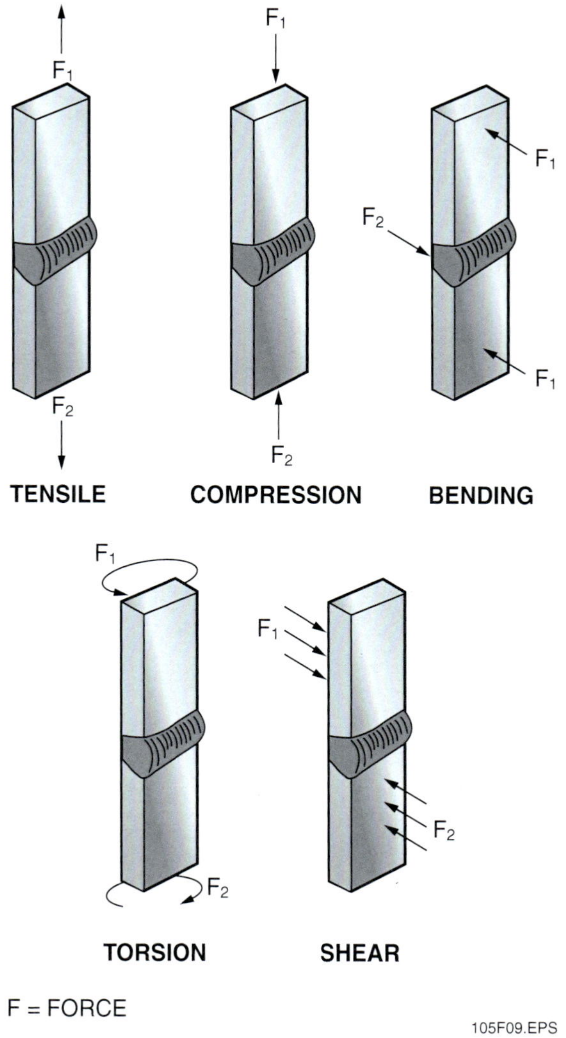

Figure 9 Forces that act on welded joints.

ply force (F_1) in the opposite direction at each end of the bar. The result of such action is that the flat bar becomes curved. When two pieces of metal are welded end-to-end, they may be tested in what is known as a bend test. The purpose of a bend test is to see if the weld will hold when force is applied against the ends of the pieces.

- *Torsion* – Twisting. Torsion is the amount of stress the welded bars can withstand as a counterclockwise force (F_1) is applied at one end and a clockwise force (F_2) is applied at the other end. For example, the driveshaft of a truck is twisted as it transmits the rotational movements of the drive motor and transmission to the drive axle.
- *Shear* – To cut off or trim. When a piece of sheet metal is placed into a press brake machine, one side of the sheet metal is resting against a stationary surface of the machine. That stationary surface is applying some force to that side of the sheet metal. That stationary force is shown as F_1 (see *Figure 9*). When the brake is activated, a moveable part of the machine applies force to the opposite side of the sheet metal. F_2 represents the forced created by that moveable part. When the moveable force (F_2) is increased, it eventually shears or cuts the metal because the stationary surface (F_1) is applying force to the opposite side of the sheet metal.

4.2.0 Types of Joints

There are five basic types of joints. They are the butt joint, lap joint, corner joint (inside and outside), T-joint, and edge joint (*Figure 10*). Each of these joints may be accomplished using various types of welds.

4.3.0 Types of Welds

Most types of welds require some degree of base metal edge preparation depending on the type of joint being welded. Common types of welds include the following: **surfacing**, plug or slot, fillet, square groove, bevel groove, V-groove, J-groove, and U-groove. If a particular type of weld requires a specified **root opening**, it is typically shown on the welding drawing or **welding procedure specification (WPS)**. Base metals may require that some type of groove is cut on one or both edges. Edge preparations will be discussed for each type of weld as it is covered in this section.

Think About It

Using a File

Should the file shown in Figure 6 be used in the condition shown? Why or why not?

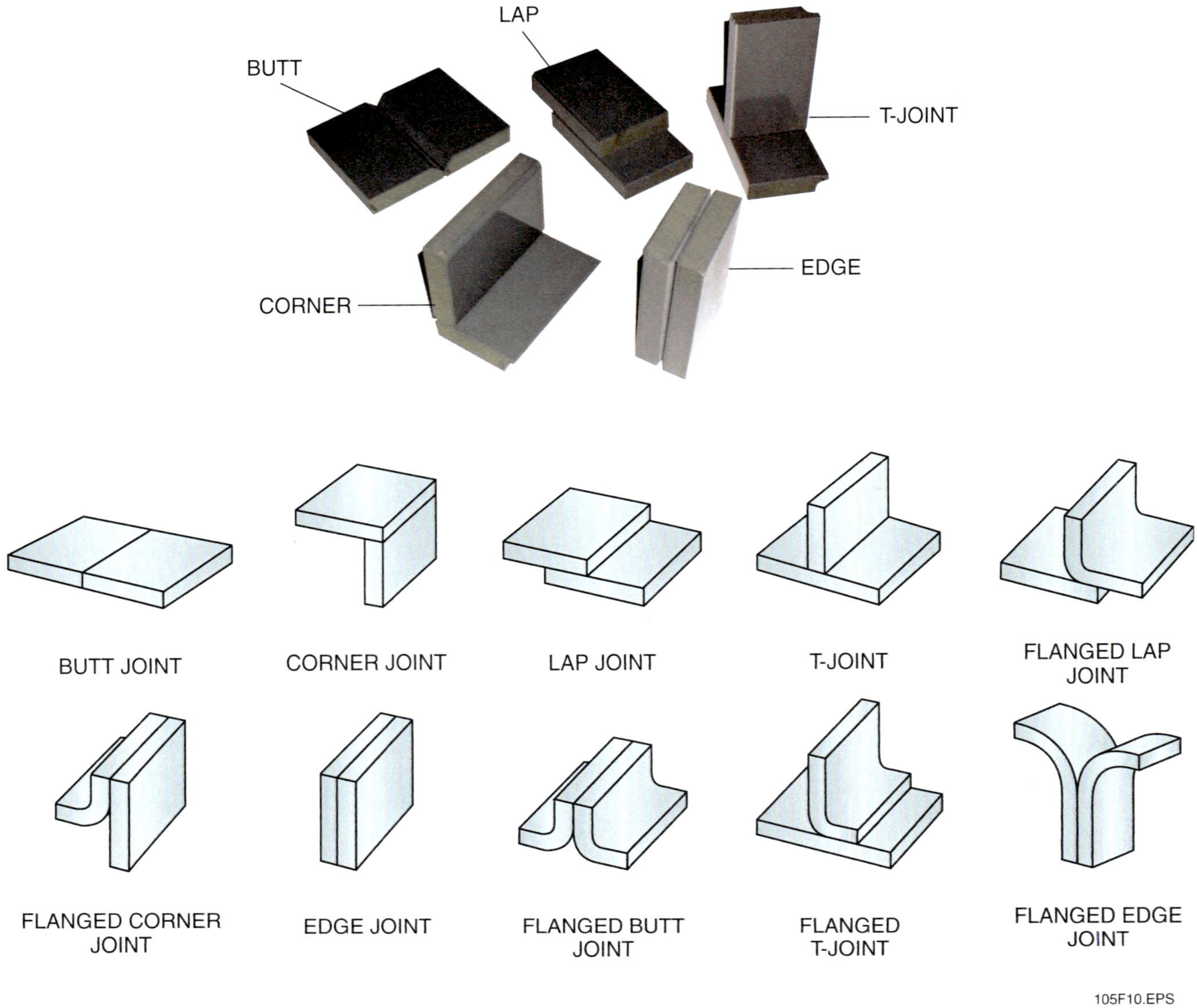

Figure 10 Five basic joint types with variations.

4.3.1 Surfacing Welds

Surfacing is the only type of weld that is applied to just one base metal surface. It is used either to build up a base surface that has become worn below the desired thickness or dimension, or to add material to a base to increase its strength and stability. An example of surfacing is shown in *Figure 11*.

Before applying the first layer, the base metal surface must be cleaned to remove contaminants, such as oxidation and dirt, using manual or power cleaning tools as previously described. When applying the surfacing weld using an oxyfuel process, the surface should be evenly preheated before surfacing to eliminate warping of the base material.

For proper penetration and layering to take place, each layer of surfacing must be properly chipped, scraped, and cleaned before applying additional layers of surfacing.

Surfacing is often applied in the tool and die industry to rebuild relatively expensive tools and dies that have worn down. It is also used in the repair of heavy equipment to build up areas that

Did You Know?

Surfacing

Surfacing, as defined by the American Welding Society, is the application by welding, brazing, or thermal spraying of a layer of material to a surface to obtain desired properties or dimensions, as opposed to making a joint.

105F11.EPS

Figure 11 Surfacing welding on an inclined vertical surface.

are exposed to various excessive stress factors such as shear, tensile stress, and torsion.

4.3.2 Plug and Slot Welds

Plug and slot welds are used to join metal pieces when the edges cannot be welded. The plug or slot weld may be applied on lap, corner, and T-joints. The hole or slot can either be completely or partially filled with weld material when joining it to the other base metal. If the hole is round, the weld is referred to as a plug weld. If the weld is slotted or elongated, it is called a slot weld. *Figure 12* illustrates how a plug or slot weld may be used on a lap joint. A typical application is to overfill the hole or slot, then grind the excess material to a level, finished surface.

Preparation of the base metals for this type of weld requires proper cleaning and removal of any oxides or dirt from both pieces. The hole or slot is then drilled to receive the weld. The walls of the hole or slot must also be cleaned of any oxides, dirt, and oils prior to welding.

Plug and slot welds are commonly used where a finished surface is required, or where overall dimensions are not permitted to exceed those of the plate thickness.

4.3.3 Fillet Welds

Fillet welds may be applied to lap joints, T-joints, or corner joints. The fillet weld does not require any base metal edge preparation other than removing contaminants by cleaning with a wire brush or other appropriate cleaning tools. The two base metals are positioned together with or without a root opening, and one or more passes (welding beads) are applied at the intersection of the two base metal edges (*Figure 13*). Whenever more than one pass is made on a fillet weld—or on any weld—the previous pass must be cleaned of all contaminants such as dross, dirt, oxides, and porosity before the next pass is applied.

When using a fillet weld on outside corner welds, a corner-to-corner or a half-lap fit can be used (*Figure 14*). The corner-to-corner joint is difficult to assemble, because neither plate can support the other, and care must be used when welding to prevent burning through the corner. The half-lap joint is easier to assemble, requires less welding material, and is less likely to burn through the corner. The half-lap does require a second weld on the inside of the corner. If a half-lap fit is used, allowances in the plate dimensions must be made for the lap.

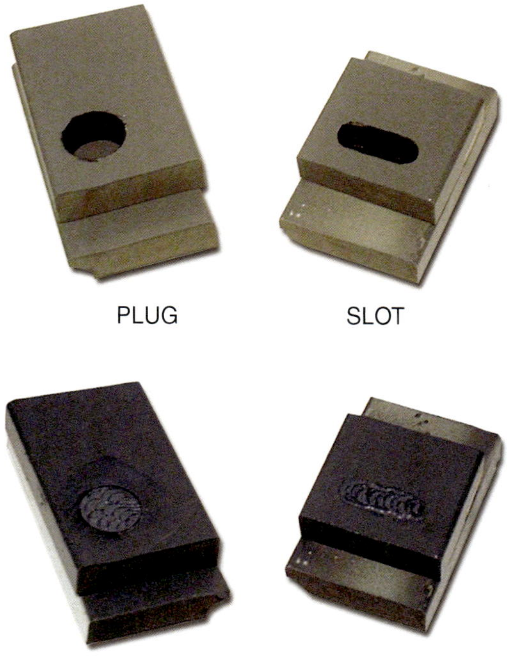

105F12.EPS

Figure 12 Plug and slot welds.

Preheating Surface Welds

For the shielded metal arc welding (SMAW) process, obtain a higher current setting than would normally be used for joint welding with the designated electrode diameter. This preheats the metal ahead of the arc and allows the surface to remain in a molten state long enough for any impurities to surface.

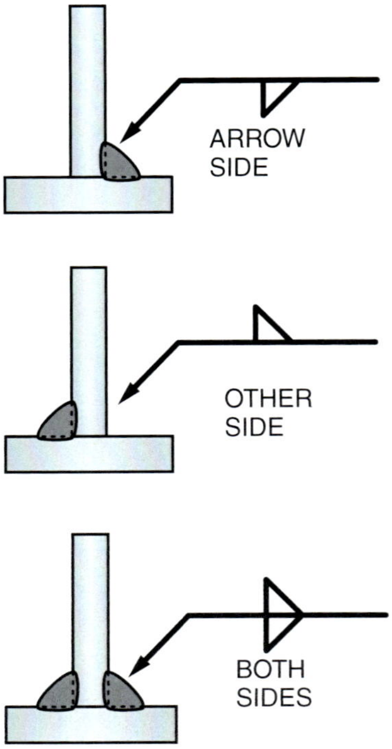

Figure 13 Fillet welds and drawing symbols.

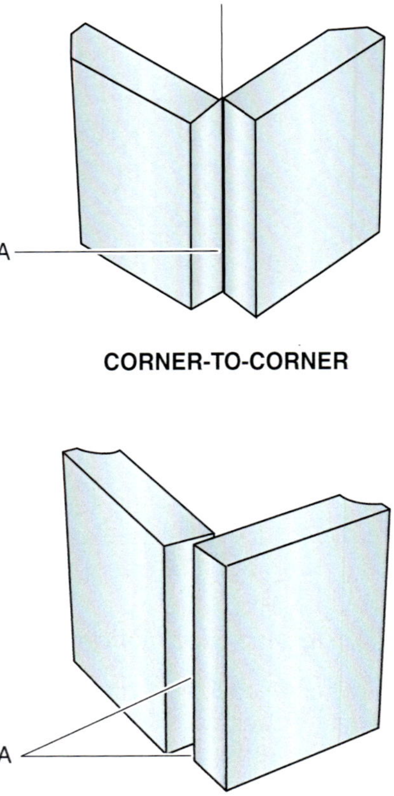

Figure 14 Fit-up of corner-to-corner and half-lap fillet welds.

4.3.4 Square-Groove Welds

Square-groove welds can be used with butt joints, corner joints (inside and outside), T-joints, and edge joints. The difference between the fillet and the square groove is that a square groove requires a root opening to be set or prepared between the two base pieces prior to welding, as shown in *Figure 15*. This allows penetration into a greater portion of the surface area of the two base metals. A partial joint-penetration weld will have much less strength than a complete joint-penetration weld.

Welding codes impose restrictions for complete joint-penetration welds. With shielded metal arc welding (SMAW), the maximum base metal thickness is ¼", and welding from both sides is required. In addition, a root opening of half the thickness of the base metal is required, and the root of the first weld must be gouged before the second weld is made. For gas metal arc welding (GMAW) and flux-cored arc welding (FCAW), the maximum base metal thickness is ⅜", with the same requirements for welding from both sides and for back gouging. Welding codes and specifications will be discussed later in this module.

4.3.5 Bevel-Groove Welds

Bevel-groove welds are commonplace in the welding industry because the base metal preparation is relatively simple and provides greater surface penetration than the square groove. The bevel-groove weld can be applied to all five of the basic weld joints. Typically, a single bevel is cut on the edge of one of the base metal pieces. The

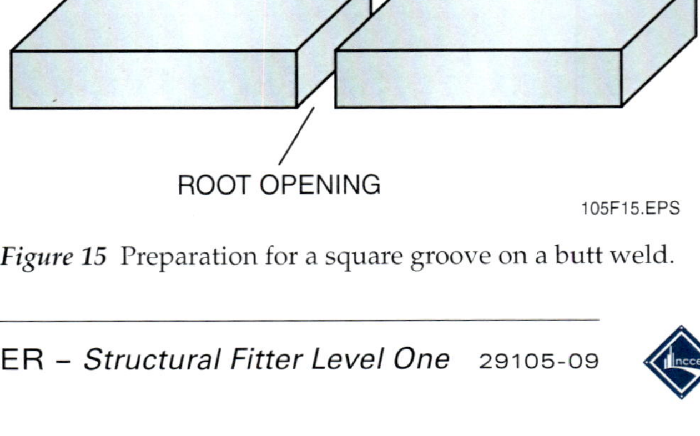

Figure 15 Preparation for a square groove on a butt weld.

preparation may include a specified root opening along with the beveled cut, as shown in the butt weld arrangement in *Figure 16*. Cutting the bevel may be accomplished by various cutting methods that will be discussed later in this module.

4.3.6 V-Groove Welds

V-groove welds are generally applied to inside or outside corner joints, butt joints, or edge joints.

V-grooves and bevel grooves are more economical to prepare than other types of welds. *Figure 17* shows a prepared V-groove on a butt joint to be welded. Typically, a 45-degree angle must be cut on both edges to prepare the base metals to receive the weld. In preparing inside or outside corner joints or butt joints for V-groove welds, the angle may not extend from the top of the bevel

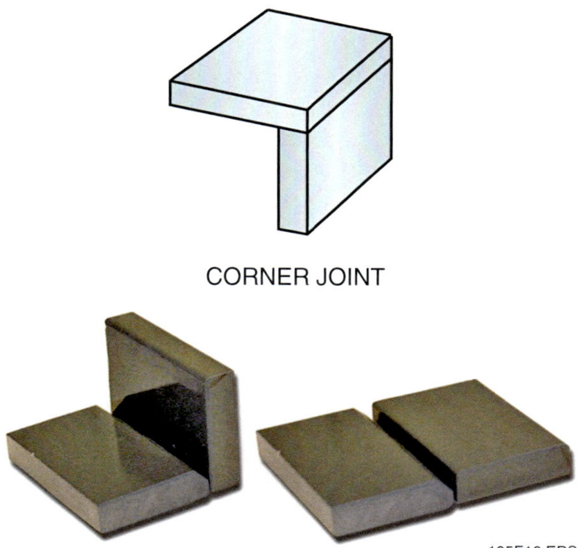

Figure 16 Preparations for corner and butt bevel-groove welds.

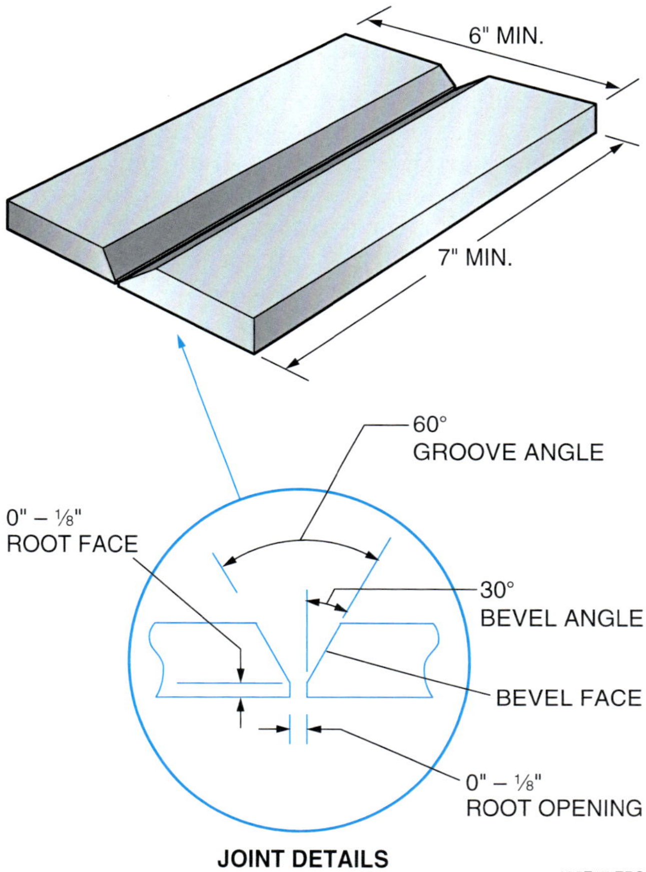

Figure 17 Open root V-groove weld.

Surface Welding

Certain shafts located within gearboxes used in the marine industry turn at relatively high rpms and will corrode and wear rapidly if subjected to salt water. It is not always cost effective to discard and replace each shaft with a new one. Surface welding may be used to bring the shafts back to the desired dimensions by building them up and re-machining them. They would then be classified as rebuilt shafts. All rotating parts must be balanced before being installed. Otherwise, they will vibrate, which in turn will destroy bearings and may cause the part itself to break.

edge to the bottom of the bevel edge on each piece, but may be cut short of the bottom corner if a **root face** is required. In this edge preparation, the flat surface from the point where the angle stops to the bottom of the piece is referred to as the root face.

4.3.7 Single Versus Double V-Groove Welds

When possible, a double V-groove weld should be used in place of a single V-groove (*Figure 18*). The double V-groove requires half the weld metal compared to the single V-groove. Welding from both sides also reduces **distortion**, because the forces of distortion work against each other.

4.3.8 J- and U-Groove Welds

J- and U-groove welds (*Figure 19*) are similar. The main difference is that a J-groove weld requires only one base metal to have its edge grooved in the form of a J. A U-groove is formed by preparing two matching J-grooves, one on each base metal edge. Even though both types require much less weld material than bevel or V-groove welds, they require much more preparation time because of the shape of the grooves. A J-groove weld can be applied to all five types of weld joints. The U-groove can only be used on butt, inside and outside corner, and edge joints.

4.3.9 Combination Fillet and Groove Welds

A combination groove and fillet weld is sometimes used in place of a fillet weld alone. A groove weld requires more preparation time than a fillet weld, but a fillet weld requires more welding time and material than a groove weld. The combination weld requires more preparation time than a fillet weld alone, but helps to reduce distortion.

4.3.10 Groove Angles and Root Openings

The purpose of the groove angle is to allow access to the root of the weld. The root preparation is sized to control **melt-through.** Increasing or decreasing the root preparation (root opening and root face) will result in excess melt-through or insufficient root penetration. As a general rule, as the groove angle decreases, the root opening increases to compensate (*Figure 20*). Root faces are

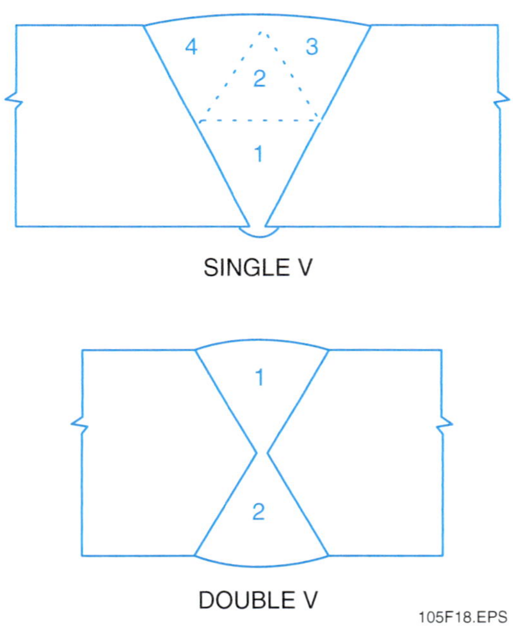

Figure 18 Comparison of double and single V-grooves.

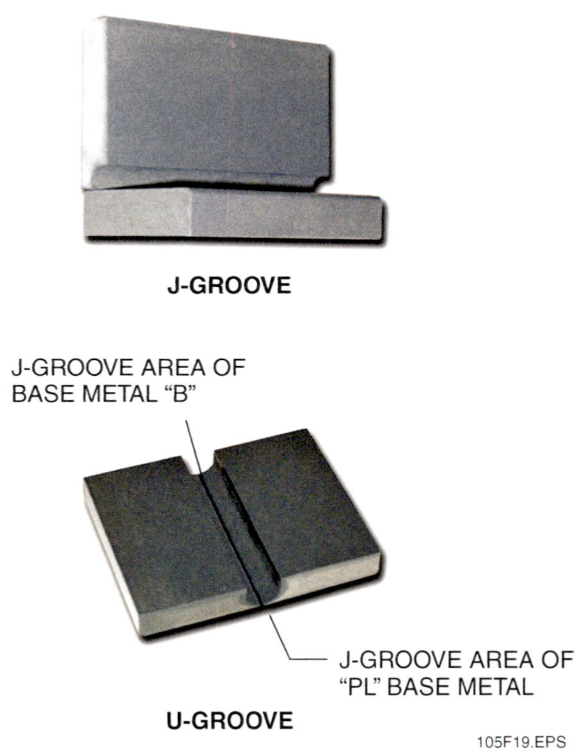

Figure 19 J- and U-groove welds.

Groove Angles

If the groove angle is larger than necessary, it will require additional weld metal to fill. This will increase the time and cost to complete the weld, and also increase distortion. If the groove angle is too small, it may result in weld defects.

used with open root joints, but not when metal **backing** strips are used.

4.3.11 Open Root Welds

For open root welds on plate, the groove angle should be 60 degrees, and the maximum size of the root opening and root face is ⅛". For open root welds on pipe, the groove angle is 60 degrees or 75 degrees, depending on the code or specifications used. The maximum size of the root opening and root face is ⅛". *Figure 21* shows groove angles for open root welds on plate and pipe.

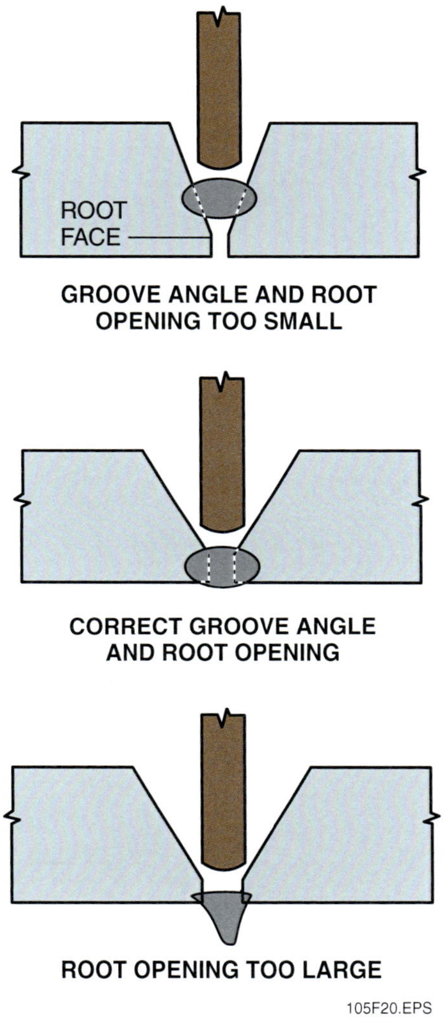

Figure 20 Groove angles and root openings.

4.3.12 Welds with Backing on Plate

Backing for plate can be strips made from the same material as the base metal, flux-coated tape, fiberglass-coated tape, ceramic tape, or gas (*Figure 22*). When gas is used as a backing, the same groove angles and root preparation used for open root welds are used. When flux-coated tape, fiberglass-coated tape, or ceramic tape is used as a backing, the groove angle should be 60 degrees, the maximum root opening should be 3⁄16", and the maximum root face should be 3⁄16". When applying tape, be sure to clean the surface to ensure that the tape adheres tightly, and make sure that the tape is centered on the weld joint.

> **NOTE**
> Check the welding procedure specification (WPS) or manufacturer's recommendations for joint preparation before using flux-coated, fiberglass-coated, or ceramic tape.

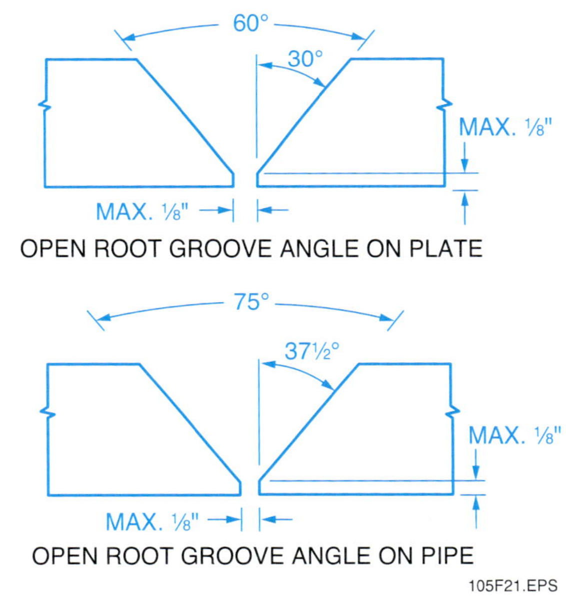

Figure 21 Open root joint preparation.

Backing Gas Hazards

Nitrogen and argon are used as backing and purge gases for some welding applications. These gases will displace all the oxygen in the space where they are used.

At a refinery in Corpus Christi, Texas, while the welder was at lunch, a helper entered a pipe in which argon had been used as a backing gas. When the welder returned, he found the helper dead.

Nitrogen is lighter than air and, if allowed, will float upward. Argon, on the other hand, is heavier than air and will settle in low-lying areas. When removing nitrogen from a confined space, open the top to allow it to float out. When removing argon from a confined space, provide some way for it to drain out the bottom.

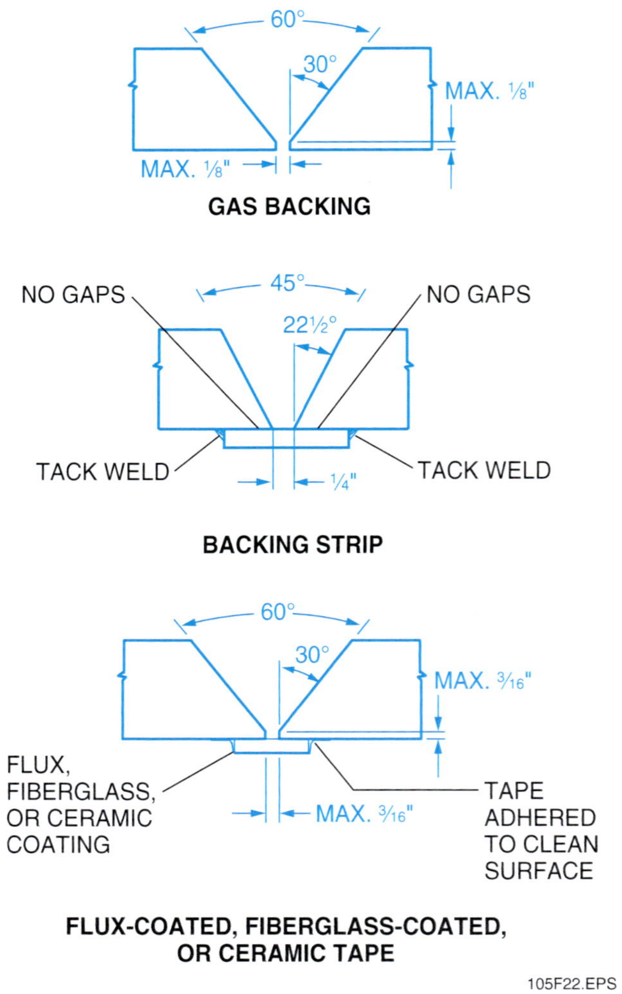

Figure 22 Plate joint preparation for backing.

When strips made from the same material as the base metal are used, a groove angle of 45 degrees, a root opening of ¼", and a feather edge (no root face) are used. The backing strip must be thick enough and wide enough to prevent burnthrough at the root and to absorb and dissipate the heat of the root pass. The recommendation for a backing strip for welding mild steel up to ¾" thick is ⅜" thick by 3" wide.

> **NOTE**
> Check the WPS or site procedures for the backing size requirements for your site.

The backing strip must be tack welded so that it will be held securely in place during welding. It must also be centered on the groove and be in close contact (no gaps) with the back of the plate being welded.

4.4.0 Welding Position

It is easier and faster to weld groove welds in the flat position and fillets in the flat or horizontal position than it is to weld out of position. Always try to position the weldment so that welding is performed in the flat position. If a weldment is an assembly of parts, look for welds that can be made on the bench before tacking the assembly together. When welding pipe, weld as many fittings as possible before tacking the pipe in position, and try to leave welds that will be the most accessible and the easiest to perform. When preparing weld joints, try to prepare the joint so that welding takes place from the most accessible position. *Figure 23* shows the positioning of weld joints for welding.

4.5.0 Codes and Welding Procedure Specifications

A welding code is a detailed listing of the rules and principles that apply to specific welded products. Codes ensure that safe and reliable welded products will be produced and that persons associated with the welding operation will be safe. Clients may specify which welding codes should

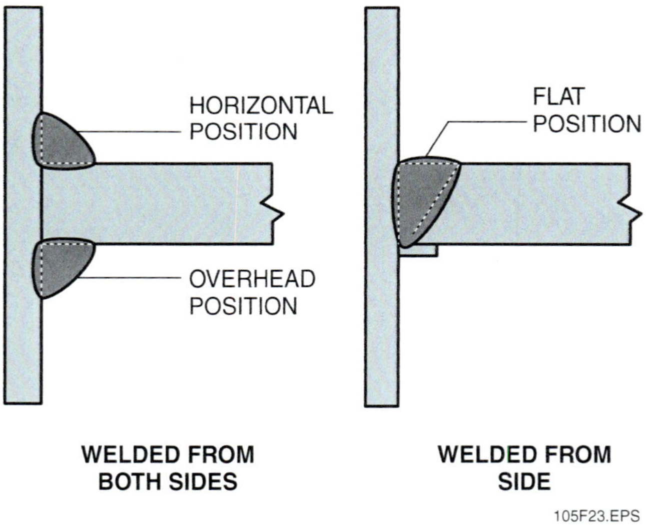

Figure 23 Positioning welds.

Out-of-Position Welding

Out-of-position welding generally requires smaller electrodes with more passes to fill the joint. Because out-of-position welding is more difficult, there is also a greater likelihood of weld defects.

be followed when they place their orders or award contracts.

If codes are mandated, a welding procedure specification (WPS) must be written for each weld. A WPS (*Figure 24*) is a written set of instructions for producing reliable welds. It includes the type of joint to be used, as well as the type of weld and any groove preparation that may be required for that particular weld. Each WPS is written and tested in accordance with a particular welding code or specification. All welding requires that acceptable industry standards be followed, but not all welds are mandated by codes. The requirement for use of a WPS is often listed on blueprints as a note or in the tail of the welding symbol.

> **NOTE**
> Welding symbols and how to read them are covered in another module.

Weld Procedure Number	30 P1 TIG 01 Issue A
Qualifying Welding Procedure (WPAR)	WP T17/A

Manufacturer:	National Fabs Ltd 25 Lane End Birkenshaw Leeds	Method of Preparation and Cleaning:	Machine and Degrease
		Parent Metal Specification:	Grade 304L Stainless Steel
		Parent Metal Thickness	3 to 8mm Wall
		Pipe Outside Diameter	25 to 100mm
Location:	Workshop	Welding Position:	All Positions
Welding Process:	Manual TIG	Welding Progression:	Upwards
Joint Type:	Single Sided Butt Weld		

Joint Design

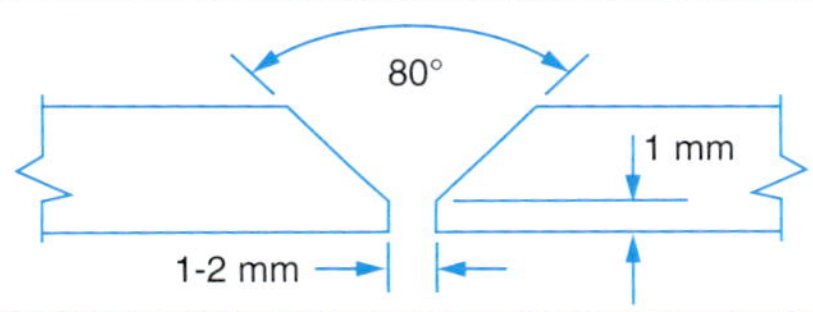

Welding Sequences

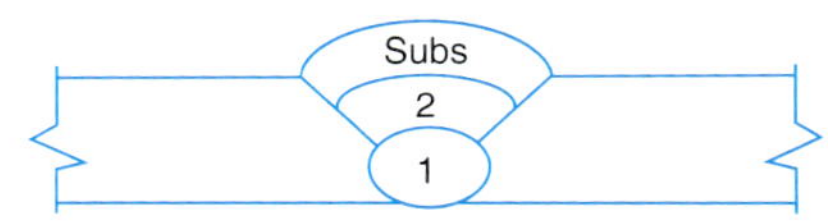

Run	Process	Size of Filler Metal	Current A	Voltage V	Type of Current/Polarity	Wire Feed Speed	Travel Speed	Heat Input
1 2 And Subs	TIG TIG	1.2 mm 1.6 mm	70 - 90 80 - 140	N/A	DC– DC–	N/A	N/A	N/A

Welding Consumables:-	
Type, Designation Trade Name:	BS 2901 Part 2 : 308S92
Any Special Baking or Drying:	No
Gas Flux:	Argon 99.99% Purity
Gas Flow Rate - Shield:	8 - 12 LPM
- Backing:	5 LPM
Tungsten Electrode Type/ Size:	2% Thoriated 2.4mm Dia
Details of Back Gouging/Backing:	Gas Backing
Preheat Temperature:	5°C Min
Interpass temperature:	200°C Max
Post Weld Heat Treatment Time, temperature, method: Heating and Cooling Rates*:	Not Required

Production Sequence

1. Clean weld and 25mm borders to bright metal using approved solvent.
2. Position items to be welded ensuring good fit up and apply purge.
3. Tack weld parts together using TIG, tacks to at least 5mm min length.
4. Deposit root run using 1.2mm dia. wire.
5. Inspect root run internally.
6. Complete weld using 1.6mm dia wire using stringer beads as required.
7. 100% Visual inspection of completed weld

Revision History

Date	Issue	Changes	Authorization
26/11/2000	A	First changes	Jack Straw *Jack Straw*

105F24.EPS

Figure 24 Example of a WPS.

If you are unsure whether or not the welding being performed requires a WPS, you should not proceed until you verify this information.

5.0.0 Welding Joint Preparation

There are many different ways to prepare a joint for welding. The joint can be prepared mechanically using nibblers, grinders, or cutters. It can also be prepared thermally using an oxyfuel, plasma arc, or carbon arc cutting torch. The method used to prepare the joint will generally be based on the base metal type, ease of use, and code or procedure specifications.

5.1.0 Identify Joint Specification

Before starting a joint preparation, determine if the joint is covered by a WPS or site quality standard. If the joint is covered by a WPS or site quality standard, the joint type and welding parameters will be specified. These specifications must be followed. If you are unsure whether or not a joint is covered by a WPS or site quality standard, check with your supervisor before proceeding.

5.2.0 Mechanical Joint Preparation

Mechanical joint preparation is most often used on alloy steel, stainless steel, and **nonferrous metal** piping or plate. It is slower than thermal methods such as oxyfuel, carbon arc, and plasma arc, but has the advantages of high precision, low heat input, and the absence of oxides commonly left by thermal methods.

5.2.1 Grinders

Handheld electric or air-operated grinders are used in welding shops, and even more often in the field, to cut and bevel pipe and plate to prepare them for welding. *Figure 25* shows a handheld electric grinder being used to prepare a root bead on a test pipe.

WARNING!

Cutoff and grinding wheels must be selected based on the material to be cut or beveled. Using an improper wheel may damage the wheel or workpiece, and can sometimes create hazardous conditions for the operator should the wheel shatter during operation. Refer to the manufacturer's recommendations and warnings when selecting grinding wheels.

105F25.EPS

Figure 25 Handheld grinder.

5.2.2 Pipe Beveling Machines

Welded piping is found extensively in the utilities industry and petrochemical industries. Nearly every piece of pipe that is welded to a fitting, to another piece of pipe, or to any other connection requires that the edge be cut square and beveled according to specifications before welding. If a mechanical cutting and beveling process is used, the procedure may be done using an electrically or pneumatically powered beveling machine (*Figure 26*). Many of these are portable machines that operate much like a lathe. They have mandrels that hold various cutting tools, as well as numerous adjusting mechanisms that make it easy to set the bevel angle and depth. Various models are available to cut and bevel 2" to 60" pipe. When specifications call for tubing to be welded, machines similar to pipe beveling machines are available to face, square, and chamfer the ends of tubes in preparation for welding. Boiler tube ends are typically prepared using these smaller machines before welding the tubes in place.

Special cutoff machines are also made to be mounted on the outside of the pipe. They can be

Thermally Sensitive Material Joint Preparation

Mechanical preparation is often required for materials that could be affected by the heat of thermal processes.

105F26.EPS

Figure 26 Pipe-beveling machine.

mounted with a ring or a special chain with rollers. An electrically or pneumatically operated grinder with a cutoff blade is mounted and manually or electrically powered around the pipe to cut it off.

WARNING!

Always follow all manufacturers' safety procedures when using grinders. Failure to follow the manufacturer's safety recommendations could result in serious personal injury.

5.2.3 Nibblers and Cutters

Nibblers prepare the edge of a plate or pipe with a reciprocal punch that cuts off a chip with each stroke. The bevel angle is set by adjusting the nibbler. A nibbler must have access to an edge on which it will be used (*Figure 27*).

Cutters use round cutting tools similar to mill cutting tools. The bevel angle is set by adjusting the cutter or by changing the cutter blade. Cutters leave the surface much smoother than nibblers, and can be used for cutoff operations. Another advantage of cutters is that they can easily be used to prepare J- or compound bevels by changing the shape of the cutter. Cutters made for pipe are sometimes called pipe-end-prep lathes.

WARNING!

Always follow all manufacturers' safety procedures when using nibblers and cutters. Failure to follow the manufacturer's safety recommendations could result in serious personal injury.

5.3.0 Thermal Joint Preparation

Thermal joint preparation includes preparing a joint with the oxyfuel, plasma arc, or carbon arc cutting process. Although the carbon arc cutting process can be used for beveling plate and pipe, the results are not as satisfactory as beveling performed with oxyfuel or plasma arc cutting. Because of carbon deposits, the carbon arc cutting process is best for gouging seams, cracks, or making repair welds.

NOTE

Other modules in this curriculum provide more detailed coverage of the specifics of oxyfuel cutting, plasma arc cutting, and carbon arc cutting.

The torch used for oxyfuel or plasma arc cutting can be handheld or mounted on a motorized or manual carriage. The motorized carriage for

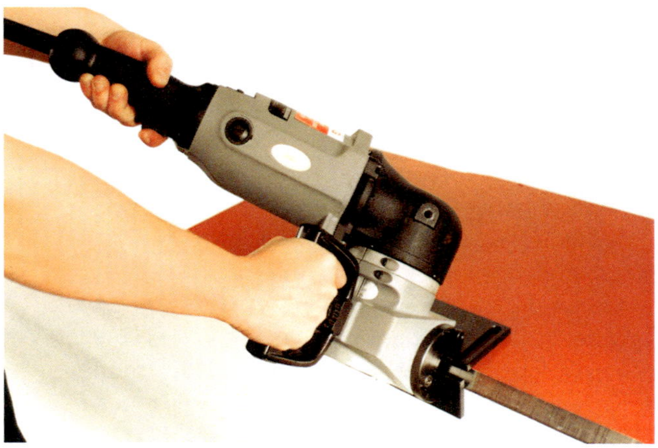

105F27.EPS

Figure 27 Nibbler.

Portable Pipe-Beveling Machine Floor Stands

Floor stands are available to convert portable pipe-beveling machines to more permanent stationary devices.

plate cutting runs on flat tracks positioned on the surface of the plate to be cut (*Figure 28*). The carriage has On/Off and forward/reverse switches and a speed adjustment. A handwheel on top of the carriage adjusts the torch holder **transversely**, and a handwheel at the torch holder adjusts the torch vertically. The bevel angle is set by pivoting the torch holder. The carriage can carry a special oxyfuel or plasma arc cutting torch designed to fit into the torch holder.

Special equipment is used for cutting pipe. A steel ring or special chain with rollers is attached to the outside of the pipe. The torch is carried around the pipe on a torch holder that is powered by electricity, air, or a hand crank. The bevel angle is set by pivoting the torch holder, and the torch is adjusted vertically with a handwheel on the torch holder. An out-of-round attachment can be used to compensate for pipe that is out-of-round. For large-diameter pipe (54" or larger), special internal equipment is available. With this special equipment, the torch mechanism is mounted on the inside of the pipe. Special plasma or oxyfuel torches can be mounted in the torch holders of the external or internal pipe cutting equipment (*Figure 29*).

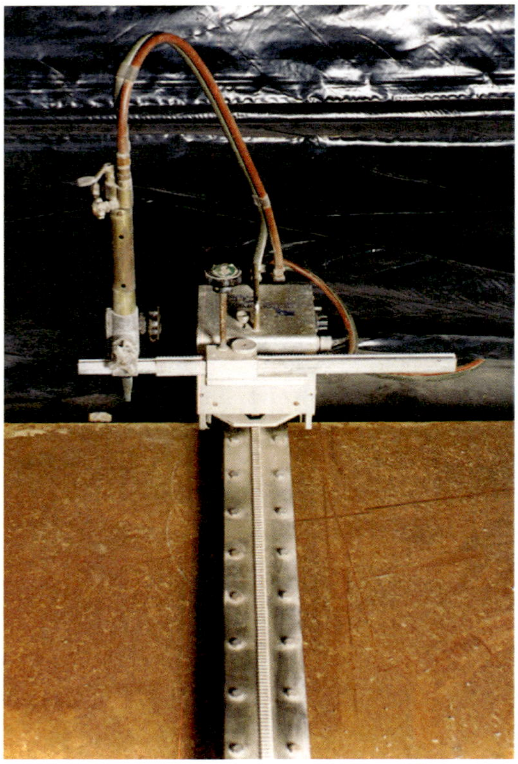

105F28.EPS

Figure 28 Motorized carriage for cutting and beveling plate.

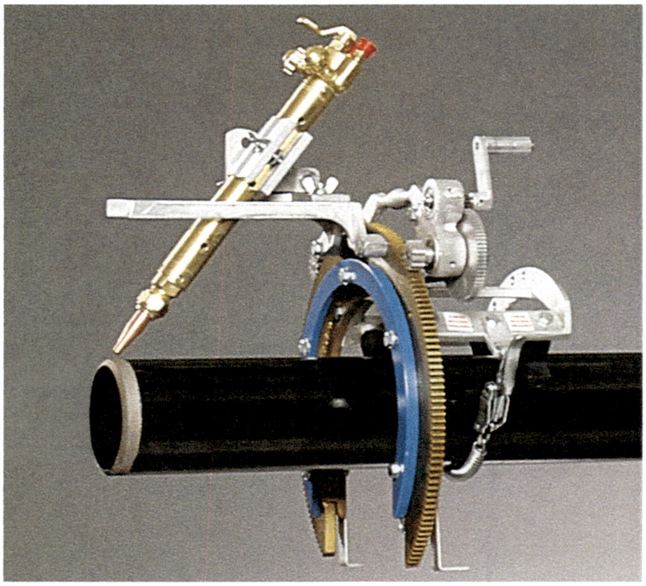

105F29.EPS

Figure 29 Pipe cutting and beveling equipment.

Thermal Joint Preparation Precautions

When preparing a joint with the oxyfuel, plasma arc, or carbon arc cutting process, all dross must be removed prior to welding. Any dross remaining on the joint during welding will cause porosity in the weld. Joints prepared with the carbon arc cutting torch must be carefully inspected for carbon deposits. It is common for small carbon deposits to be left behind as the carbon electrode is consumed during the cutting or gouging operation. These carbon deposits will cause defects such as hard spots, loss of ductility, and cracking in the weld. Before welding, use a grinder to clean surfaces prepared with the carbon arc torch.

Automatic Pipe Bevelers

Automatic versions of pipe bevelers and cutters are an improvement over pattern cutters and other similar equipment. This is because there is no need to reset the preheat flame before each cut. Once an automatic pipe beveler has been initially set up, all subsequent cuts can be made with the same settings. These systems also save time and cutting gases, because all gases are extinguished immediately when the system is switched off.

SUMMARY

The importance of proper joint preparation cannot be overemphasized. If a joint is not properly prepared, the resulting weld will not perform as designed. This will cause an unsafe weldment. Chemical or mechanical cleaning can be performed to remove base metal surface contamination prior to welding. It is important to select the proper type of joint and use the correct joint preparation methods to ensure acceptable welds. Welding codes set guidelines that must be adhered to during joint preparation. Following the procedures presented in this module will ensure that the joints you prepare will follow established industry practices and meet code requirements.

Review Questions

1. Corrosion of metals occurs when they are exposed to _____.
   a. shielding gas
   b. welding
   c. air
   d. polishing
2. A type of steel made from copper-alloyed steel that protects itself from corrosion is _____ steel.
   a. spring
   b. weathering
   c. mild
   d. stainless
3. The most common method for cleaning contaminated surfaces is _____ cleaning.
   a. chemical
   b. ultrasonic
   c. thermal
   d. mechanical
4. A pneumatic tool that uses a number of blunt steel needles to remove surface corrosion is called a _____.
   a. weld flux chipper
   b. needle scaler
   c. needle nose chipper
   d. corrosion needler
5. The factors that must be considered when selecting a joint for a particular application include load considerations, environment, materials, cost, and _____.
   a. processes
   b. personnel qualifications
   c. preheat
   d. personal protective equipment
6. The five basic types of joints are butt, lap, corner, T-joint, and _____.
   a. chamfer
   b. edge
   c. bevel
   d. square
7. If a type of weld requires a specific root opening, it is shown on the _____.
   a. title block
   b. welding drawing
   c. sectional view
   d. elevation view
8. Lap joints, T-joints, or corner joints are welded using _____ welds.
   a. surface
   b. butt
   c. fillet
   d. spot
9. Before using flux-coated, fiberglass-coated, or ceramic tape as a backer, the welder must check the _____.
   a. specification book
   b. WPS
   c. field drawings
   d. revision block
10. The written set of specifications for producing sound welds is a(n) _____.
    a. WPS
    b. PQR
    c. WRS
    d. SWS
11. Grinding wheels that are used improperly or are poorly selected can cause injury due to _____.
    a. radiation
    b. reversing
    c. shattering
    d. air-locking
12. Before a pipe can be welded to a fitting or other connection, its edge usually must be _____.
    a. burnished
    b. hardened
    c. ground to fit
    d. cut square and beveled

13. Nibblers must have access to an _____ to be used.

a. inverter
b. excess pipe flange
c. edge
d. axis

14. Thermal preparation of joints includes the use of oxyfuel, plasma arc, and _____ cutting.

a. electric
b. low heat
c. continuous-arc
d. carbon arc

15. The best cutting process for gouging seams, cracks, or making repair welds is the _____ cutting process.

a. plasma arc
b. carbon steel
c. grinding
d. carbon arc

Trade Terms Quiz

Fill in the blank with the correct key trade term that you learned from your study of this module.

1. ________ is an assembly that is fastened together by weld joints.
2. ________ is a weldable or nonweldable material used behind a root opening to allow defect-free welding at the open root of a joint.
3. A(n) ________ is the document containing all the detailed methods and practices required to produce a sound weld.
4. Metal to be welded, cut, or brazed is ________.
5. ________ is steel alloy that, under specific conditions, is designed to form a very dense oxide layer on its outer surfaces, which retards further corrosion.
6. A(n)________ is a set of regulations covering permissible materials, service limitations, fabrication inspections, testing procedures, and qualifications of welders.
7. ________ means placed or running crosswise.
8. ________ is the expansion and contraction of welded parts caused by the heating and subsequent cooling of the weld joint.
9. The space between the base metal pieces at the bottom or root of a joint is called a(n) ________.
10. The amount of force applied to a material or structure is called a(n) ________.
11. A small facing on the end of a bevel for a groove weld is called a(n) ________.
12. Complete joint penetration is ________.
13. Gas pockets, inclusions, or voids in the weld metal are called ________.
14. A metal, such as aluminum, copper, or brass, lacking sufficient quantities of iron to have any effect on its properties is called a(n) ________.
15. ________ is a form of porosity having a length greater than its width and is approximately perpendicular to the weld face.
16. The scale that forms on metal surfaces when they are exposed to air is called ________.
17. ________ is the rounded end of a ball peen hammer used to strike a weld as it cools to reduce stress on the weld.
18. An object's surface properties or dimensions can be modified using a process known as ________.

Trade Terms

Backing	Load	Peen	Root opening	Welding procedure specification (WPS)
Base metal	Melt-through	Piping porosity	Surfacing	Weldment
Code	Nonferrous metal	Porosity	Transverse	
Distortion	Oxide	Root face	Weathering steel	

Cornerstone of Craftmanship

Larry Thurston

Trainer and Quality Assurance/Quality Control
Gulf States, Inc.
Midland, Michigan

Larry credits his father with being his early role model for a career in welding. His father, a boilermaker/welder, was always working on projects in his shop at home. Larry watched his dad, and he was soon working with him on the projects. In 1966 he got his first real job as an operator. That was the start of a long and successful career in the trades, construction, and maintenance, and most recently, in teaching and training.

How did you choose a career in the welding field?
I was very fortunate to have the influence of my father. He was a boilermaker/welder at the Dow Chemical Company. There were constant projects going on in his shop at home that I was always involved in. Eventually I was hired by the Dow Chemical Company as an operator. A few years later, I was awarded an apprenticeship as a pipefitter, working and training in capital construction projects. From that came a career in construction and maintenance. I had always been interested in mechanical projects, so this was a great start for me.

What types of training have you been through?
I am a journeyman insulator, pipefitter, and welder. I have been through apprenticeships in all three trades, not to mention the training I have received from some great journeymen.

What kinds of work have you done in your career?
I started out as an operator, learning how processes worked, and from there I started my trades. During my career, I have worked in various phases of the construction and maintenance field including takeoff, fabrication, installation, and helping with design. I retired from Dow Chemical Company in 1998. I decided to return to work and was subsequently employed by Gulf States, Inc. in Midland, Michigan as a pipefitter/welder. While at Gulf States, I was asked if I would start a training program for people in the trades. Training in several trades has been both a challenge and a learning experience for me. Passing on to younger journeymen what I was taught over the years has been more than fun and very rewarding. I am also an instructor at Associated Builders and Contractors, where I teach apprentices all levels of pipefitting.

Tell us about your present job.
After spending many years in the trades and learning my crafts, it is very rewarding to share with others the knowledge that was given to me. Training is a never-ending opportunity to coach other craftspersons as new technologies present themselves. I work for a great company with a vision to train and add to the skill levels of craft professionals. This gives me a chance not only to provide better trained people for the company, but to provide those in the trades with the opportunity to improve their skills and their earning capacity.

What factors have contributed the most to your success?
I have thought about this many times, and I always come up with the same answer. I worked for a company that saw value in training its people from within the workforce. I was rewarded with three apprenticeships, and the power of learning cannot be replaced.

What advice would you give to those new to the welding field?

First, have a good attitude. A good attitude makes people want to be around you, and that makes your learning and work experience much easier. Second, have a good work ethic. Be at work on time every day. A good attendance record is very important to your employer. Knowing that you can be counted on every day makes you a more valuable employee. Third, learn your craft well. Your knowledge cannot be replaced or taken away from you. The knowledge you have will gain you the respect of your peers. Fourth, always work safely so you and your fellow workers can be injury free and accomplish what you need to do each day. A can-do attitude will take you a long way. The construction field is wide open for advancement to those who learn their trade well and have a willingness to lead through example.

Tell us some interesting career-related facts or accomplishments.

I have made a very good living for myself and my family for many years. In the future, I believe those in the trades will do even better because of fewer people entering the field.

Trade Terms Introduced in This Module

Backing: A weldable or nonweldable material used behind a root opening to allow defect-free welding at the open root of a joint.

Base metal: Metal to be welded, cut, or brazed.

Code: A set of regulations covering permissible materials, service limitations, fabrication, inspection, testing procedures, and qualifications of welders.

Distortion: The expansion and contraction of welded parts caused by the heating and subsequent cooling of the weld joint.

Load: The amount of force applied to a material or a structure.

Melt-through: Complete joint penetration.

Nonferrous metal: A metal, such as aluminum, copper, or brass, lacking sufficient quantities of iron to have any effect on its properties.

Oxide: The scale that forms on metal surfaces when they are exposed to air.

Peen: The rounded end of a ball peen hammer used to strike a weld as it cools to reduce stress on the weld.

Piping porosity: A form of porosity having a length greater than its width and that is approximately perpendicular to the weld face.

Porosity: Gas pockets, inclusions, or voids in the weld metal.

Root face: A small facing on the end of a bevel for a groove weld.

Root opening: The space between the base metal pieces at the bottom or root of the joint.

Surfacing: The application by welding, brazing, or thermal spraying of a layer of material to a surface to obtain desired properties or dimensions.

Transverse: Placed or running crosswise.

Weathering steel: Steel alloy that, under specific conditions, is designed to form a very dense oxide layer on its outer surfaces, which retards further corrosion.

Welding procedure specification (WPS): The document containing all the detailed methods and practices required to produce a sound weld.

Weldment: An assembly that is fastened together by welded joints.

Additional Resources

This module is intended to present thorough resources for task training. The following reference works are suggested for further study. These are optional materials for continued education rather than for task training.

Welding Handbook, Volume 5, 2001. Miami, FL: The American Welding Society.

The Procedure Handbook of Arc Welding, 2000. Cleveland, OH: The Lincoln Electric Company.

OSHA Standard 1926.351, Arc Welding and Cutting

www.lincolnelectric.com

Figure Credits

Lincoln Electric Company, 105F01A, 105F02, 105F03

J. R. Yochum, 105F01B, 105F25

Topaz Publications, Inc., 105F04, 105F06–105F08, 105SA01, 105F10 (photos), 105F11, 105SA02, 105F12, 105F16 (photo), 105F19, 105F26, 105F28

Reproduced by permission TWI Ltd., 105F05

Heck Industries, Inc., 105F27

H&M Pipe Beveling Machine Company, Inc., 105F29

Performance Accreditation Tasks

The Performance Accreditation Tasks (PATs) correspond to and support the learning objectives in *AWS EG2.0, Guide for the Training and Qualification of Welding Personnel: Entry-Level Welder*.

Note that in order to satisfy all learning objectives in *AWS EG2.0*, the instructor must also use the PATs contained in the second level in the NCCER Welding curriculum.

PATs provide specific acceptable criteria for performance and help to ensure a true competency-based welding program for students.

The following task is designed to test your competency with an oxyfuel cutting torch. Do not perform this cutting task until directed to do so by your instructor.

PREPARE PLATE JOINTS MECHANICALLY

Using a nibbler or cutter, mechanically prepare the edge of a ¼" to ¾" carbon steel plate with a bevel of 22½° (or 30° bevel depending on equipment available).

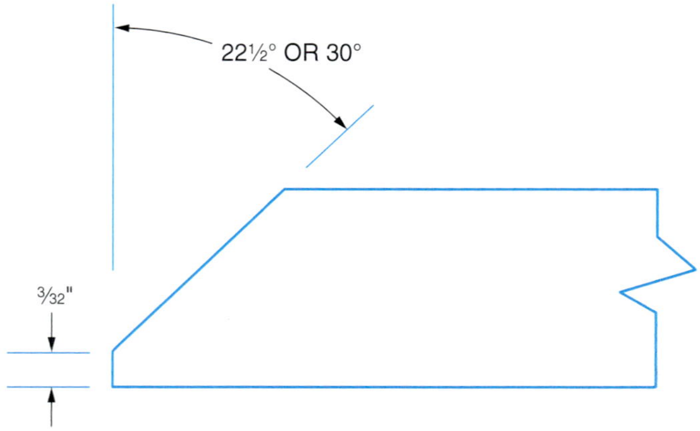

NOTE: BASE METAL = CARBON STEEL PLATE

105A01.EPS

Criteria for Acceptance:

- Bevel angle ±2½° ________
- Bevel face smooth and uniform to 1⁄16" ________
- Root face ±1⁄32" ________

PREPARE PLATE JOINTS THERMALLY

Using oxyfuel or plasma arc cutting equipment, thermally prepare the edge of a ¼" to ¾" carbon steel plate with a bevel of 22½° (or 30° depending on equipment available).

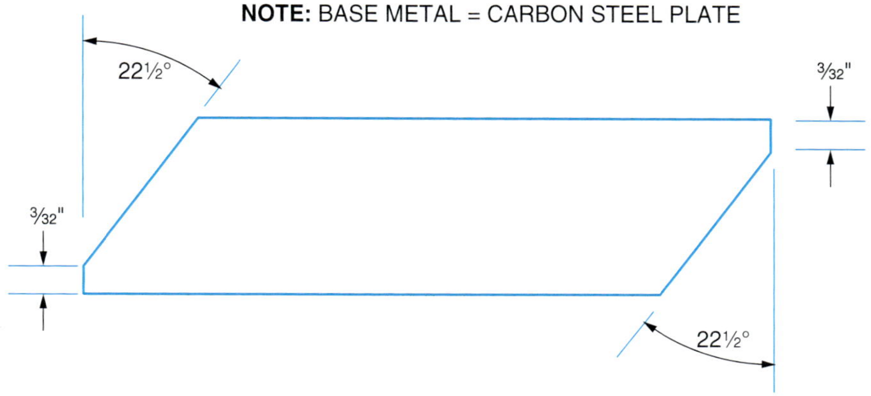

105A02.EPS

Criteria for Acceptance:

- Bevel angle ±2½° ________
- No dross ________
- Minimal notching not exceeding 1/16" deep on the kerf face ________
- Minimum of 1/16" radius at the top edge and bottom edge of the kerf ________
- Root face ±1/32" ________

NCCER CURRICULA — USER UPDATE

NCCER makes every effort to keep its textbooks up-to-date and free of technical errors. We appreciate your help in this process. If you find an error, a typographical mistake, or an inaccuracy in NCCER's curricula, please fill out this form (or a photocopy), or complete the online form at **www.nccer.org/olf**. Be sure to include the exact module ID number, page number, a detailed description, and your recommended correction. Your input will be brought to the attention of the Authoring Team. Thank you for your assistance.

Instructors – If you have an idea for improving this textbook, or have found that additional materials were necessary to teach this module effectively, please let us know so that we may present your suggestions to the Authoring Team.

NCCER Product Development and Revision

13614 Progress Blvd., Alachua, FL 32615

Email: curriculum@nccer.org

Online: www.nccer.org/olf

❑ Trainee Guide ❑ Lesson Plans ❑ Exam ❑ PowerPoints Other ______________________

Craft / Level: Copyright Date:

Module ID Number / Title:

Section Number(s):

Description:

Recommended Correction:

Your Name:

Address:

Email: Phone:

29106-09

Weld Quality

Module Four

Trainees with successful module completions may be eligible for credentialing through NCCER's National Registry. To learn more, go to **www.nccer.org** or contact us at **1.888.622.3720**. Our website has information on the latest product releases and training, as well as online versions of our *Cornerstone* magazine and Pearson's product catalog.

Your feedback is welcome. You may email your comments to **curriculum@nccer.org**, send general comments and inquiries to **info@nccer.org**, or fill in the User Update form at the back of this module.

This information is general in nature and intended for training purposes only. Actual performance of activities described in this manual requires compliance with all applicable operating, service, maintenance, and safety procedures under the direction of qualified personnel. References in this manual to patented or proprietary devices do not constitute a recommendation of their use.

29106-09
Weld Quality

Objectives

When you have completed this module, you will be able to do the following:

1. Identify and explain codes governing welding.
2. Identify and explain weld imperfections and their causes.
3. Identify and explain nondestructive examination practices.
4. Identify and explain welder qualification tests.
5. Explain the importance of quality workmanship.
6. Identify common destructive testing methods.
7. Perform a visual inspection of fillet welds.

Performance Tasks

This is a knowledge-based module. There are no Performance Tasks.

Trade Terms

Defect
Discontinuity
Embrittled
Hardenable materials
Homogeneity
Inclusion
Laminations
Procedure qualification
Procedure qualification record (PQR)
Radiographic
Underbead cracking

Prerequisites

Before you begin this module, it is recommended that you successfully complete the following: *Core Curriculum* and *Welding Level One*, Modules 29101-09 through 29105-09.

Industry-Recognized Credentials

If you're training through an NCCER-accredited sponsor, you may be eligible for credentials from NCCER's Registry. The ID number for this module is 29106-09. Note that this module may have been used in other NCCER curricula and may apply to other level completions. Contact NCCER's Registry at 888.622.3720 or go to **www.nccer.org** for more information.

Contents

Topics to be presented in this module include:

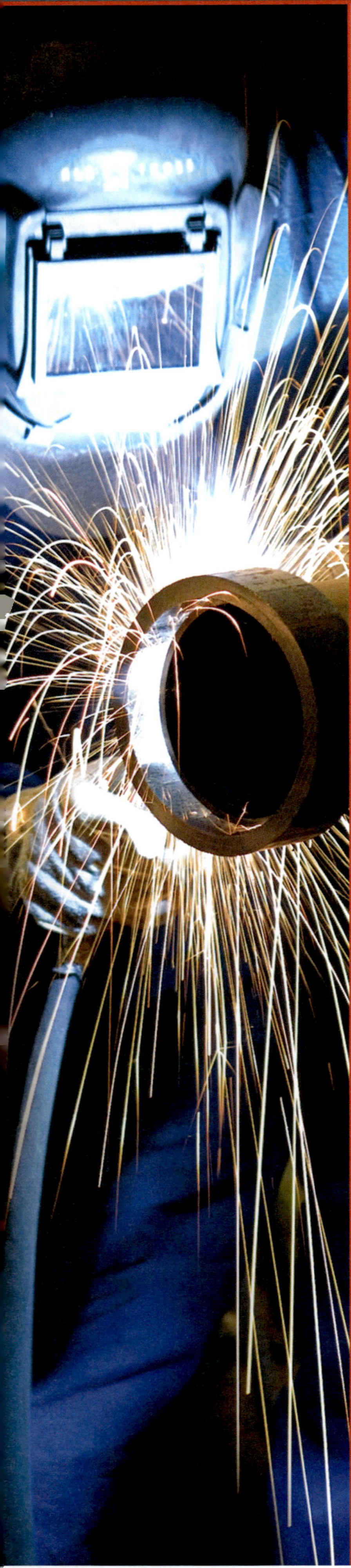

Figures and Tables

1.0.0 Introduction

Quality is an important aspect of a welder's job. A weld that can pass or be acceptable for one welding application may not be acceptable for another. Acceptable welding criteria have been established in codes and standards. Several codes govern welding activities, qualification requirements, and tests that can be performed on weldments to determine and identify various weld imperfections.

The following sections will identify and explain applicable codes, various weld imperfections and their causes, types of weldment testing, and welder qualification test requirements. The importance of quality workmanship will also be discussed, along with typical site quality organizational structures for ensuring that quality welds are achieved.

2.0.0 Codes Governing Welding

A code is a set of requirements covering permissible materials, service limitations, fabrication, inspection, testing procedures, and qualifications of welders. Welding codes ensure that safe and reliable welded products are produced and that persons associated with the welding operation are safe. Clients specify in the contract the codes that will be used on the project. All welding must then be performed according to the guidelines and specifications outlined in that code. Since there are a number of codes and each is updated periodically, you should know which welding codes and code year apply to your job.

Codes and specifications that apply to welding safety and quality have been developed and are published by a number of nationally recognized agencies. To eliminate the necessity of writing a code for each new job, sections of these existing codes are referenced by the project contract. Agencies and societies that have established codes include the following:

- American Society of Mechanical Engineers (ASME)
- American Welding Society (AWS)
- American Petroleum Institute (API)
- American National Standards Institute (ANSI)

This module will not cover all the detailed requirements of each individual code or code section. Instead, general considerations that apply to weld quality will be covered.

2.1.0 American Society of Mechanical Engineers

The American Society of Mechanical Engineers has two codes: the *ASME Boiler and Pressure Vessel Code* and *ASME B31, Code for Pressure Piping.* Both of these codes are endorsed by the American National Standards Institute.

The *ASME Boiler and Pressure Vessel Code (BPVC)* contains eleven sections. The sections most frequently referenced by welders are as follows:

- *Section II, Material Specifications* – This section contains the specifications for acceptable ferrous (Part A) and nonferrous (Part B) base metals and for acceptable welding and brazing filler metals and fluxes (Part C). Many of these specifications are identical to and have the same number designation as AWS specifications for welding consumables. This section is used to match base metals and filler metals.
- *Section V, Nondestructive Examination* – This section covers the methods and standards for nondestructive examination of boilers and pressure vessels.
- *Section IX, Welding and Brazing Qualifications* – This section covers the qualification of welders, welding operators, brazers, and brazing operators. It also covers the welding and brazing procedures that must be used for welding or brazing boilers or pressure vessels. This section of the code is often cited in other codes and standards as the welding qualification standard.

The Importance of Welding Inspection

It is important that you learn to recognize the various types of weld discontinuities and identify their causes so that you can be the first line of defense against problem welds.

Welding is used to join metals in many critical applications, including buildings, bridges, pipelines, motor vehicles, and heavy machinery. Failure of a weld in any of these applications could mean loss of life or major property damage. For this reason, welding inspectors will frequently check your work to make sure your welds meet the standards established for the project. The greater the potential hazard or cost risk, the more frequent and intense the inspections will be.

ASME B31, Code for Pressure Piping, consists of eight sections. Each section gives the minimum requirements for the design, materials, fabrication, erection, testing, and inspection of a particular type of piping system. In particular, *B31.1, Power Piping,* covers power and auxiliary service systems for electric generation stations. *B31.3* covers chemical plant and petroleum refining piping.

All sections of *ASME B31, Code for Pressure Piping,* require qualification of the welding procedures and testing of welders and welding operators. Some sections require these qualifications to be performed in accordance with Section IX of the *ASME Boiler and Pressure Vessel Code,* while in others it is optional.

2.2.0 American Welding Society

The American Welding Society publishes numerous documents covering welding. These documents include codes, standards, specifications, recommended practices, and guides. *AWS D1.1, Structural Welding Code – Steel,* is the code most frequently referenced. It covers welding and qualification requirements for welded structures of carbon and low-alloy steels. It is not intended to apply to pressure vessels, pressure piping, or base metals less than ⅛" thick.

2.3.0 American Petroleum Institute

The American Petroleum Institute publishes documents in all areas related to petroleum production. *API 1104, Standard for Welding of Pipelines and Related Facilities,* applies to arc and oxyfuel gas welding of piping, pumping, transmission, and distribution systems for petroleum. It presents methods for making acceptable welds by qualified welders using approved welding procedures, materials, and equipment. It also presents suitable methods to ensure proper analysis of weld quality.

2.4.0 American National Standards Institute

The American National Standards Institute is a private organization that does not actually prepare standards. Instead, it adopts standards that it feels are of value to the public interest. ANSI standards deal with dimensions, ratings, terminology and symbols, test methods, and performance, as well as safety specifications for materials, equipment, components, and products in many fields, including construction. Many codes used today have been adopted as ANSI standards.

2.5.0 Marine Codes

The AWS website lists the codes associated with marine welds and provides the following summary statements for each code:

- *AWS D3.5-93R, Guide for Steel Hull Welding* – This guide provides information to users in the marine construction industry about the best and most practical methods to weld steel hulls for ships, barges, mobile offshore drilling units, and other marine vessels. It provides information on steel plates, shapes, castings, and forging, as well as their selection and weldability. It discusses welding processes and proper design for welding. Hull construction is presented in terms of preparation of materials, erection and fitting, and control of distortion. Qualification of procedures and personnel are outlined, and inspection methods are discussed. A common shipyard problem, stray current protection, is discussed as is the health and safety of the workforce. Supplementary non-mandatory appendices are provided for informational purposes.
- *AWS D3.6M, Specification for Underwater Welding* – This specification covers the requirements for welding structures or components under the surface of the water. It includes welding in both dry and wet environments. Sections 1 through 6 constitute the general requirements for underwater welding, while Sections 7 through 10 contain the special requirements applicable to four individual classes of weld: Class A – Comparable to above-water welding; Class B – For less critical applications; Class C – Where load bearing is not a primary consideration; and Class O – To meet the requirements of another designated code or specification.
- *AWS D3.7, Guide for Aluminum Hull Welding* – This guide provides information on the welding of seagoing aluminum hulls and other structures in marine construction. Included are sections on hull materials, construction preparation, welding equipment and processes, qualification requirements, welding techniques, and safety precautions.

GOING GREEN

Weld Leaks

Maintaining weld quality is very important. When welds fail, leaks can occur that could be potentially harmful to the environment.

Anyone involved in marine welding must be knowledgeable of these welding codes. When in doubt, find a current copy of the applicable code and review it.

3.0.0 Basic Elements of Welding Codes

A major function of welding codes is the qualification of operators to perform a particular operation. The operational qualification usually entails specific aspects of the procedure in question. All welding codes provide detailed information about qualification in the following general areas:

- Welding procedure qualification
- Welder performance qualification
- Welding operator qualification

> **NOTE**
> The information in this module is provided as a general guideline only. Check with your supervisor if you are unsure of the codes and specification requirements for your project.

Machine welding is covered in some codes but is not common to all codes. Each type of qualification mentioned is different and subject to different requirements.

3.1.0 Welding Procedure Qualification

A welding procedure is a written document that contains materials, methods, processes, electrode types, techniques, and all other necessary and relevant information about the weldment. Welding procedures must be qualified before they can be used. **Procedure qualification** has nothing to do with the skills of the individual welder.

Welding procedure qualifications are limiting instructions written to explain how welding will be done. These limiting instructions are listed in a document known as a welding procedure specification (WPS). *Figure 1* show a sample WPS. The purpose of the WPS is to define and document in detail the variables involved in welding a certain base metal. The WPS lists the following information in detail:

- Various base metals to be joined by welding
- Filler metal to be used
- Range of preheat and postheat treatment
- Thickness and other variables described for each welding process

WPS variables are identified either as essential or nonessential variables. Essential variables are items in the welding procedure specification that cannot be changed without requalifying the welding procedure. Nonessential variables are items in the WPS that may be changed within a range identified by the code, but that do not affect the qualification status. The following can be considered essential variables in a welding procedure:

- Filler metal classification
- Material thickness
- Joint design
- Type of base metal
- Welding process

Examples of nonessential variables that may be changed without having to requalify the welding procedure include:

- Amperage
- Travel speed
- Shielding gas flow (if applicable)
- Electrode and filler wire size

> **NOTE**
> Do not change any variable (essential or nonessential) without discussing it with your supervisor.

The WPS is qualified for use by welding test coupons (samples of the welded material) and by testing the coupons in accordance with the applicable code. A test weld is made and test coupons are cut from it. The test coupons are used to make tensile tests, root bends, and face bends as required by the code. The test results are then

Inspection Standards

Among the many documents and standards published by the American Welding Society are documents covering weld inspection. These include:

- *AWS B1.10, Guide for the Nondestructive Inspection of Welds*
- *AWS B1.11, Guide for the Visual Inspection of Welds*
- *Certification Manual for Welding Inspectors*

ANNEX B—SAMPLE FORMS

WELDING PROCEDURE SPECIFICATION (WPS) Yes ☐
PREQUALIFIED __________ QUALIFIED BY TESTING X
or PROCEDURE QUALIFICATION RECORDS (PQR) Yes ☐

Identification # PQR 231
Revision 1 Date 12-1-07 By W. Lye
Authorized by J. Jones Date 1-18-08
Type—Manual ☐ Semi-Automatic ☐
Machine ☐ Automatic ☐

Company Name Red Inc.
Welding Process(es) FCAW
Supporting PQR No.(s) –

JOINT DESIGN USED
Type: Butt
Single ☒ Double Weld ☐
Backing: Yes ☒ No ☐
Backing Material: ASTM A131A
Root Opening 1/4" Root Face Dimension –
Groove Angle: 35° Radius (J–U) –
Back Gouging: Yes ☐ No ☒ Method –

BASE METALS
Material Spec. ASTM A131
Type or (Grade) A
Thickness: Groove 1" Fillet –
Diameter (Pipe) –

FILLER METALS
AWS Specification A5.20
AWS Classification E71T-1

SHIELDING
Flux – Gas CO_2
Composition ______
Electrode-Flux (Class)______ Flow Rate 30-40 cfh
Gas Cup Size 5/8" or 3/4"

PREHEAT
Preheat Temp., Min 75° Ambient
Interpass Temp., Min 75°F Max 350° F

POSITION
Position of Groove: 4G Fillet: –
Vertical Progression: Up ☐ Down ☐

ELECTRICAL CHARACTERISTICS
Transfer Mode (GMAW) Short-Circuiting ☐
Globular ☒ Spray ☐
Current: AC ☐ DCEP ☒ DCEN ☐ Pulsed ☐
Other ______
Tungsten Electrode (GTAW)
Size: ______
Type: ______

TECHNIQUE
Stringer or Weave Bead: Stringer
Multi-pass or Single Pass (per side) Multipass
Number of Electrodes 1
Electrode Spacing Longitudinal –
Lateral –
Angle –

Contact Tube to Work Distance 3/4-1"
Peening None
Interpass Cleaning: Wire Brush

POSTWELD HEAT TREATMENT
Temp. N.A.
Time ______

WELDING PROCEDURE

Pass or Weld Layer(s)	Process	Filler Metals		Current		Volts	Travel Speed	Joint Details
		Class	Diam.	Type & Polarity	Amps or Wire Feed Speed			
1	FCAW	E71T-1	.045"	DC+	180	26	8	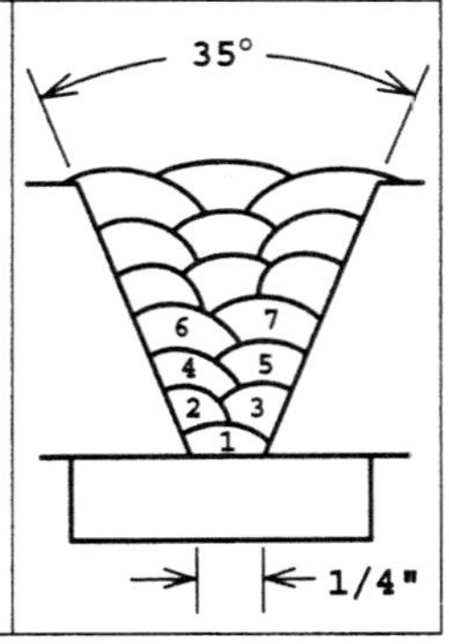
2-8	"	"	"	"	200	27	10	
9-11	"	"	"	"	200	27	11	
12-15	"	"	"	"	200	27	9	
16	"	"	"	"	200	27	11	

Form E-1 (Front)

106F01.EPS

Figure 1 Example of a WPS.

recorded on a document known as a **procedure qualification record (PQR).** If the weldment produced by the particular procedure meets the standards of the code, the procedure becomes qualified. Under most codes, each WPS must have a PQR to document the quality of the weld produced. *Figure 2* shows a sample PQR.

The methods used to qualify procedures are more detailed and thorough than those used to qualify either welders or welding operators. This is because procedures must qualify physical and metallurgical properties.

3.2.0 Welder Performance Qualification

Once a procedure has been qualified, the welder using it must be qualified to use that procedure by passing a welding performance qualification test. Because no single performance test can qualify welders for all the different types of welding that must be done, a welder may be required to pass additional performance qualification tests. Performance tests used to qualify welders are covered later in this module.

3.3.0 Welder Operator Qualification

When fully automatic welding equipment is used, the operators of the equipment must demonstrate their ability to set up and monitor the equipment so that it will produce acceptable welds. The codes also contain qualification tests for these operators.

4.0.0 Weld Discontinuities and Their Causes

Codes and standards define the quality requirements necessary to achieve the integrity and reliability of the weldment. These quality requirements help ensure that welded joints are capable of serving their intended function for the anticipated life of the weldment. Weld discontinuities can prevent a weld from meeting the minimum quality requirements.

AWS defines a **discontinuity** as an interruption of the typical structure of a weldment, such as a lack of **homogeneity** in the mechanical, metallurgical, or physical characteristics of the material or weldment. A discontinuity is not necessarily a **defect.** A defect found during inspection will require the weld to be rejected. A single excessive discontinuity or a combination of discontinuities can make the weldment defective (unable to meet minimum quality requirements). However, a weld can have one or more discontinuities and still be acceptable.

The welder should be able to identify discontinuities and understand the effect they have on weld integrity. Some can be determined from visual inspection. Those that are internal to the weldment can only be detected through other testing methods. The most common weld discontinuities are the following:

- Porosity
- **Inclusions**
- Cracks
- Incomplete joint penetration
- Incomplete fusion
- Undercuts
- Arc strikes
- Spatter
- Unacceptable weld profiles

4.1.0 Porosity

Porosity is the presence of voids or empty spots in the weld metal. It is the result of gas pockets being trapped in the weld as it is being made. As the molten metal hardens, the gas pockets form voids. Unless the gas pockets work up to the surface of the weld before it hardens, porosity cannot be detected through visual inspection.

Porosity can be grouped into the following major types:

- *Uniformly scattered porosity* – May be located throughout single-pass welds or throughout several passes in multiple-pass welds (*Figure 3B*).
- *Clustered porosity* – A localized grouping of pores that results from improperly starting or stopping the welding.
- *Linear porosity* – May be aligned along a weld interface, the root of a weld, or a boundary between weld beads (*Figure 3A*).

Avoiding Porosity

Excessive porosity has a serious effect on the mechanical properties of the joint. Although some codes permit a certain amount of porosity in welds, it is best to have as little as possible. This can be accomplished by properly cleaning the base metal, avoiding excessive moisture in the electrode covering, and using proper welding techniques.

Procedure Qualification Record (PQR) # 231
Test Results

TENSILE TEST

Specimen No.	Width	Thickness	Area	Ultimate Tensile Load, lb	Ultimate Unit Stress, psi	Character of Failure and Location
231-1	.75"	1.00"	.75"	52 500	70 000	Ductile
231-3	.75"	1.00"	.75"	52 275	69 700	Ductile

GUIDED BEND TEST

Specimen No.	Type of Bend	Result	Remarks
231-2	Side	Pass	
231-4	Side	Pass	Small (< 1/16") opening acceptable
231-6	Side	Pass	
231-5	Side	Pass	

VISUAL INSPECTION
Appearance acceptable
Undercut acceptable
Piping porosity none
Convexity none
Test date 12-3-2002
Witnessed by D. Davis

Radiographic-ultrasonic examination
RT report no.: D231 Result passed
UT report no.: ______ Result ______

FILLET WELD TEST RESULTS
Minimum size multiple pass Macroetch
1. ______ 3. ______
2. ______

Maximum size single pass Macroetch
1. ______ 3. ______
2. ______

Other Tests

All-weld-metal tension test
Tensile strength, psi 83,100
Yield point/strength, psi 72,600
Elongation in 2 in., % 28
Laboratory test no. PW 231

Welder's name W. T. Williams Clock no. 261 Stamp no. ______

Tests conducted by RED Inc. & ABC Testing Laboratory

Test number PQR 231
Per D. Miller

We, the undersigned, certify that the statements in this record are correct and that the test welds were prepared, welded, and tested in conformance with the requirements of Section 4 of AWS D1.1/D1.1M, 2002 (year) Structural Welding Code—Steel.

Signed RED Inc.
Manufacturer or Contractor
By R. M. Boncrack
Title Q.C. Mgr.
Date 12-15-2002

106F02.EPS

Figure 2 Example of a PQR.

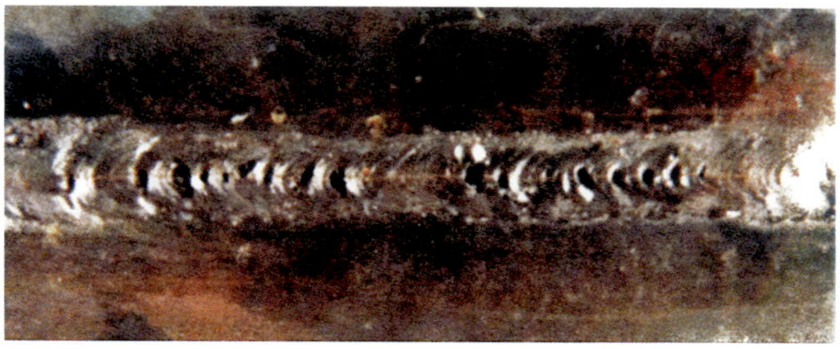

(A) LINEAR POROSITY

(B) SCATTERED SURFACE POROSITY

(C) PIPING POROSITY

106F03.EPS

Figure 3 Examples of porosity.

- *Piping porosity* – Normally extends from the root of the weld toward the face. These elongated gas pores are also called wormholes. They do not often extend to the surface, and the porosity cannot be visually detected (*Figure 3C*).

Most porosity is caused by improper welding techniques or contamination. Improper welding techniques may cause inadequate shielding gas to be formed. As a result, parts of the weld are left unprotected. Oxygen in the air, or moisture in the flux or on the base metal that dissolves in the weld pool can become trapped and produce porosity.

The intense heat of the weld can decompose paint, dirt, oil, or other contaminants, producing hydrogen. This gas can become trapped in the solidifying weld pool and produce porosity.

4.2.0 Inclusions

Inclusions are foreign matter trapped in the weld metal (*Figure 4*), between weld beads, or between the weld metal and the base metal. Inclusions are sometimes jagged and irregularly shaped. Sometimes they form in a continuous line. This concentrates stresses in one area and reduces the structural integrity (strength) of the weld.

Inclusions generally result from faulty welding techniques, improper access to the joint for welding, or both. A typical example of an inclusion is slag, which normally forms over a deposited weld. If the electrode is not manipulated correctly, the force of the arc will cause some of the slag particles to be blown into the molten pool. If the pool solidifies before the inclusions can float to the top, they become lodged in the metal, producing a discontinuity. Sharp notches in joint boundaries or between weld passes also can result in slag entrapment.

SURFACE SLAG INCLUSIONS

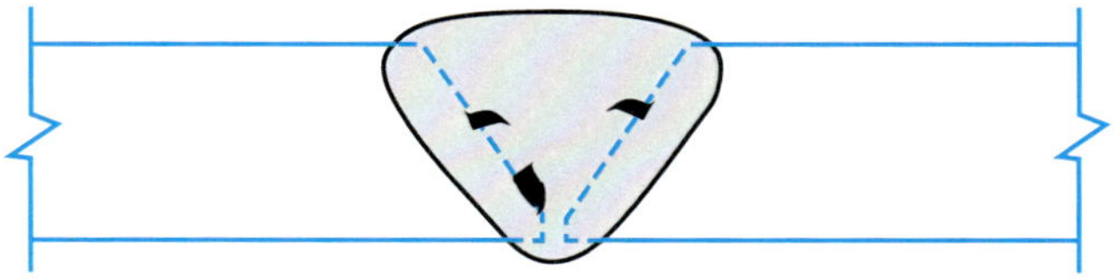

106F04.EPS

Figure 4 Examples of nonmetallic inclusions.

Discontinuities

Ideally, a weld should not have any discontinuities, but most will have one or more. When evaluating a weld, it is important to note the type, size, and location of the discontinuity. Any one of these factors, or all three, can change a discontinuity to a defect, requiring the weld to be rejected during the inspection process. For example, discontinuities located where stresses exist tend to expand and thus are more detrimental than those in other locations. Surface or near-surface discontinuities may be more detrimental than similarly shaped internal discontinuities.

Inclusions are more likely to occur in out-of-position welding because the tendency is to keep the molten pool small and allow it to solidify rapidly to prevent it from sagging.

With proper welding techniques and the correct electrode used with the proper setting, inclusions can be avoided or kept to a minimum. Other remedies include the following:

- Position the work to maintain slag control
- Change the electrode to improve control of molten slag
- Thoroughly remove slag between weld passes
- Grind the weld surface if it is rough and likely to entrap slag
- Remove heavy mill scale or rust on weld preparations
- Avoid the use of electrodes with damaged coverings

4.3.0 Cracks

Cracks are narrow breaks that occur in the weld metal, in the base metal, or in the crater formed at the end of a weld bead (*Figure 5*). They are caused when localized stresses exceed the ultimate strength of the metal. Cracks are generally located near other weld or base metal discontinuities.

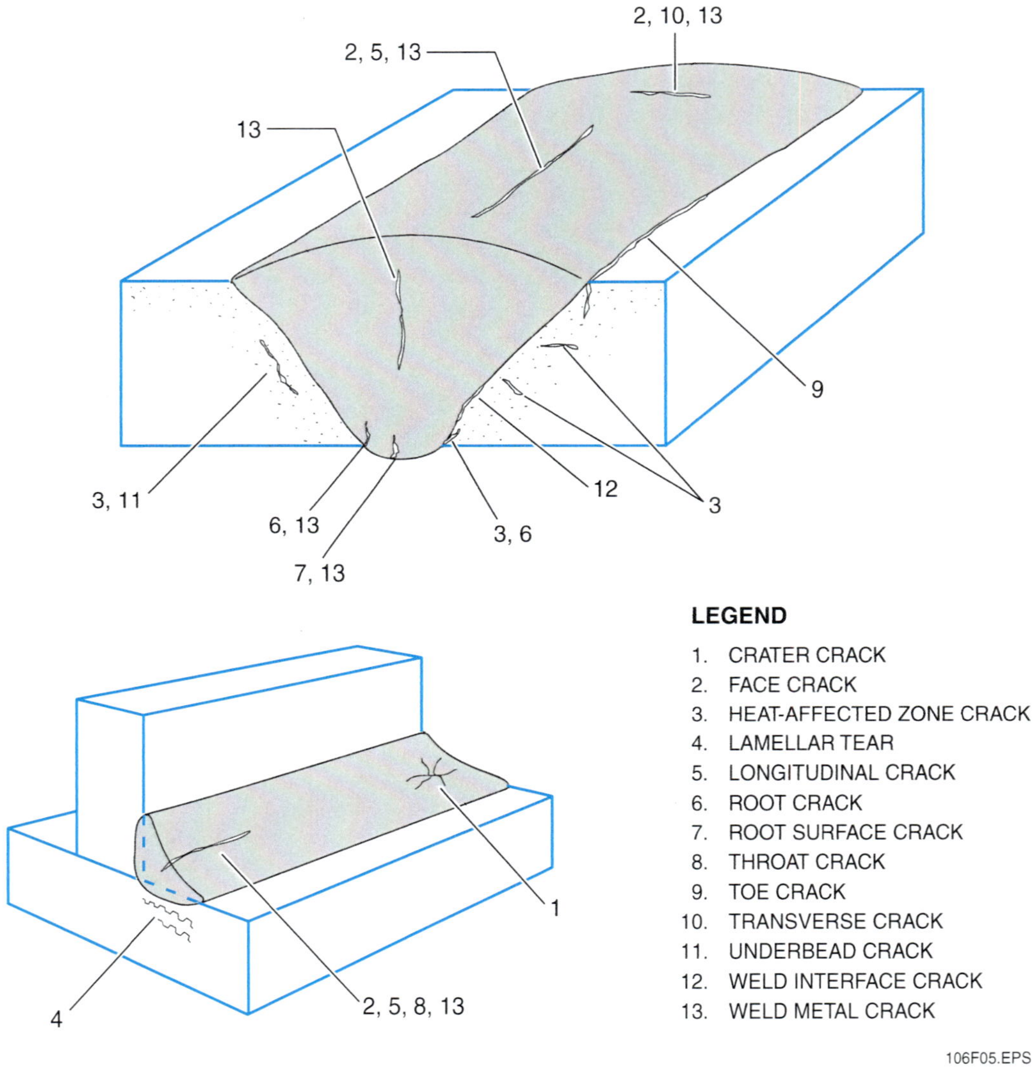

Figure 5 Types of weld metal and base metal cracks.

Hot and Cold Cracks

Hot cracks occur while the weld is solidifying. They can be caused by insufficient ductility at high temperature. Cold cracks occur after the weld has solidified. They are often caused by improper welding techniques.

4.3.1 Weld Metal Cracks

Three basic types of cracks can occur in weld metal: transverse, longitudinal, and crater. As seen in *Figure 5*, weld metal cracks are named to correspond with their location and direction.

Transverse cracks run across the face of the weld and may extend into the base metal. They are more common in joints that have a high degree of restraint.

Longitudinal cracks are usually located in the center of the weld deposit. They may be the continuation of crater cracks or cracks in the first layer of welding. Cracking of the first pass is likely to occur if the bead is thin. If this cracking is not eliminated before the other layers are deposited, the crack will progress through the entire weld deposit.

Crater cracks have a tendency to form in the crater whenever the welding operation is interrupted. These cracks usually proceed to the edge of the crater and may be the starting point for longitudinal weld cracks. Crater cracks can be minimized or prevented by filling craters to a slightly convex shape prior to breaking the welding arc.

Figure 6 shows various kinds of weld metal cracks.

Weld metal cracking can usually be reduced by taking one or more of the following actions:

- Improve the contour or composition of the weld deposit by changing the electrode manipulation or electrical conditions
- Increase the thickness of the deposit and provide more weld metal to resist the stresses by decreasing the travel speed
- Reduce thermal stress by preheating
- Use low-hydrogen electrodes
- Balance shrinkage stress by sequencing welds
- Avoid rapid cooling conditions

4.3.2 Base Metal Cracks

Base metal cracking usually occurs within the heat-affected zone of the metal being welded. The possibility of cracking increases when working with **hardenable materials.** These cracks usually occur along the edges of the weld and through the heat-affected zone into the base metal. Types of base metal cracking include **underbead cracking** and toe cracking.

Underbead cracks are limited mainly to steel. They are usually found at regular intervals under the weld metal and usually do not extend to the surface. Because of this, they cannot be detected by visual methods of inspection.

Toe cracks are generally the result of strains caused by thermal shrinkage acting on a heat-

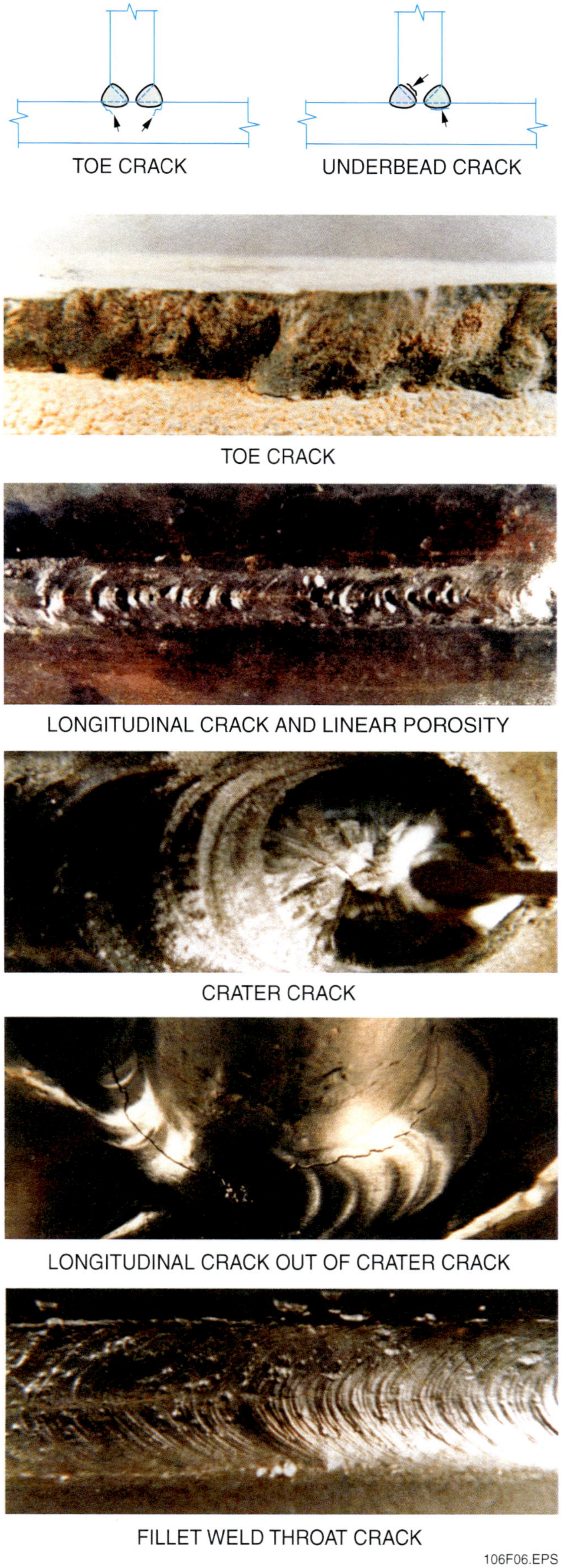

Figure 6 Types of weld metal cracks.

affected zone that has been **embrittled**. They sometimes occur when the base metal cannot accommodate the shrinkage strains that are imposed by welding.

Base metal cracking can usually be reduced or eliminated by one of the following methods:

- Control the cooling rate by preheating
- Control heat input
- Use the correct electrode
- Control welding materials

4.4.0 Incomplete Joint Penetration

Incomplete joint penetration (*Figure 7*) occurs when the filler metal fails to penetrate and fuse with an area of the weld joint. This incomplete penetration will cause weld failure if the weld is subjected to tension or bending stresses.

Insufficient heat at the root of the joint is a frequent cause of incomplete joint penetration. If the metal being joined first reaches the melting point at the surfaces above the root of the joint, molten metal may bridge the gap between these surfaces and screen off the heat source before the metal at the root melts.

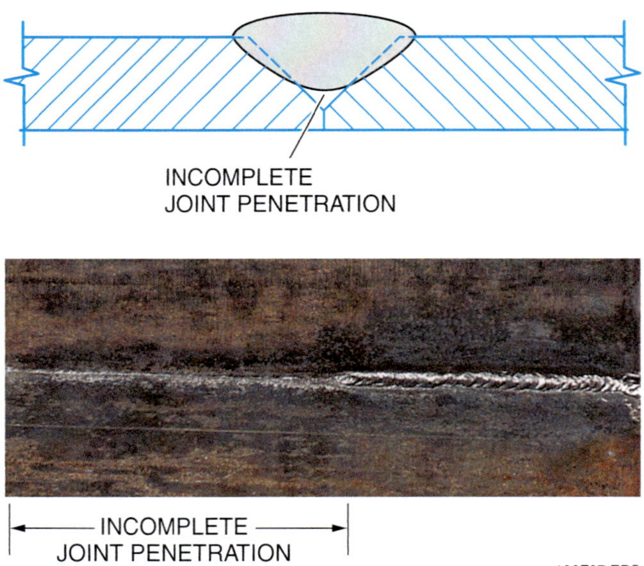

Figure 7 Incomplete joint penetration.

Improper joint design is another leading cause of incomplete joint penetration. If the joint is not prepared or fitted accurately, an excessively thick root face or an insufficient root gap may cause incomplete penetration. Incomplete joint penetration is likely to occur under the following conditions:

- If the root face dimension is too big, even though the root opening is adequate
- If the root opening is too small
- If the included angle of a V-groove is too small

Figure 8 is a labeled diagram showing correct and incorrect joint designs.

Even if the welding heat is correct and the joint design is adequate, incomplete penetration can result from poor control of the welding arc. Examples of poor control include the following:

- Using an electrode that is too large
- Traveling too fast
- Using a welding current that is too low

Incomplete penetration is always undesirable in welds, especially in single-groove welds where the root of the weld is subject either to tension or bending stresses. It can lead directly to weld failure or can cause a crack to start at the unfused area.

4.5.0 Incomplete Fusion

Many welders confuse incomplete joint penetration with incomplete fusion. It is possible to have good penetration without complete root fusion. Incomplete fusion is the failure of a welding process to fuse, or join together, layers of weld metal or weld metal and base metal.

Incomplete fusion may occur at any point in a groove or fillet weld, including the root of the weld. Often the weld metal simply rolls over onto the plate surface. This is generally referred to as overlap. In many cases, the weld has good fusion

Case History

Mexico City Refinery

In Mexico City in 1984, a weld on a liquid propane gas (LPG) tank cracked. The escaping gas cloud drifted toward a nearby residential area until it encountered an ignition source. When the gas ignited, the fireball quickly burned its way back to the leak source. The resulting explosion destroyed the refinery and many residential neighborhoods, leaving 503 people dead and more than 4,000 injured.

The Bottom Line: Welds must be made correctly and inspected properly to prevent injury, loss of property, and sometimes death.

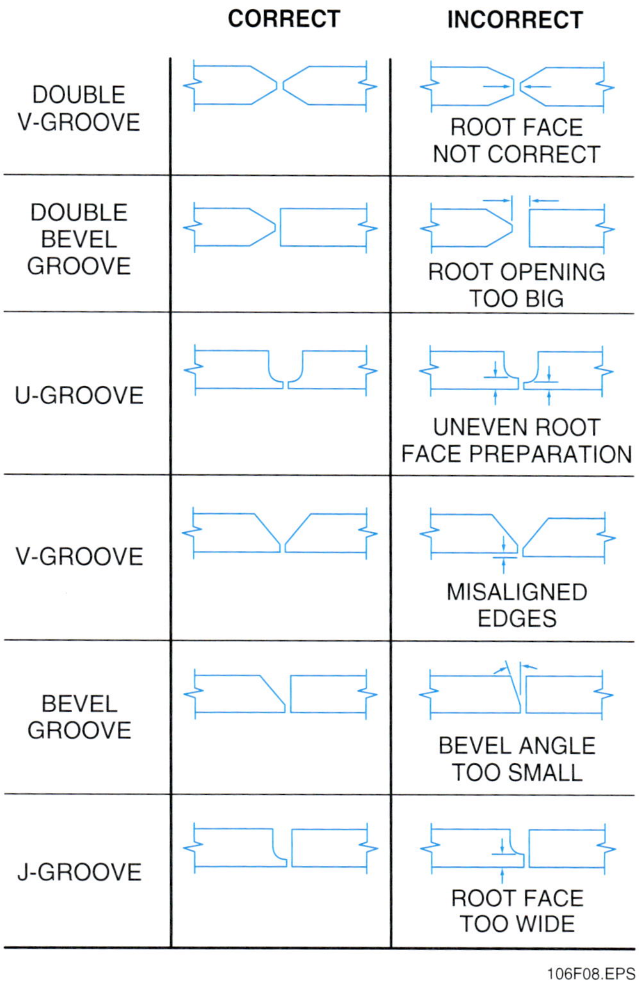

Figure 8 Correct and incorrect joint designs.

at the root and at the plate surface, but because of poor technique and insufficient heat, the toe of the weld does not fuse. *Figure 9* includes diagrams and photos showing incomplete fusion and overlap. Causes for incomplete fusion include the following:

- Insufficient heat as a result of low welding current, high travel speeds, or an arc gap that is too close
- Wrong size or type of electrode
- Failure to remove oxides or slag from groove faces or previously deposited beads
- Improper joint design
- Inadequate gas shielding

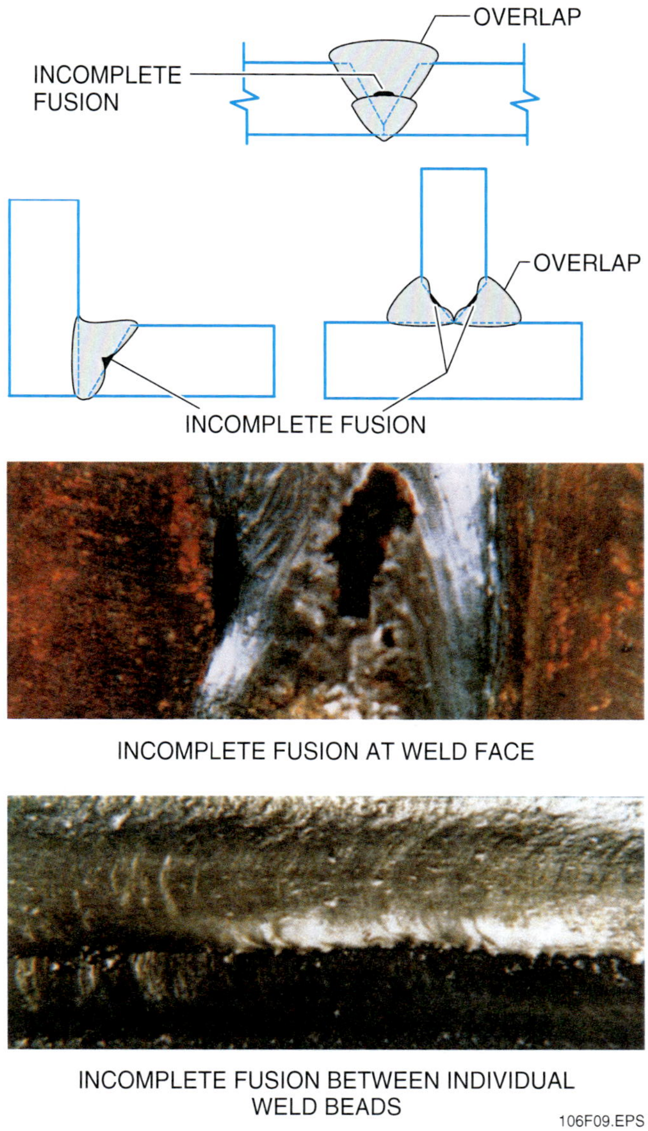

Figure 9 Incomplete fusion and overlap.

Incomplete fusion discontinuities affect weld joint integrity in much the same way as porosity and slag inclusion.

4.6.0 Undercut

Undercut is a groove melted into the base metal beside the weld. It is the result of the arc removing more metal from the joint face than is replaced by weld metal. On multilayer welds, it may also occur at the point where a layer meets the wall of a groove (*Figure 10*).

Causes of Incomplete Joint Penetration

Incomplete joint penetration is generally associated with groove welds. It may result from insufficient welding heat, improper joint design (too much metal for the welding arc to penetrate), or poor control of the welding arc.

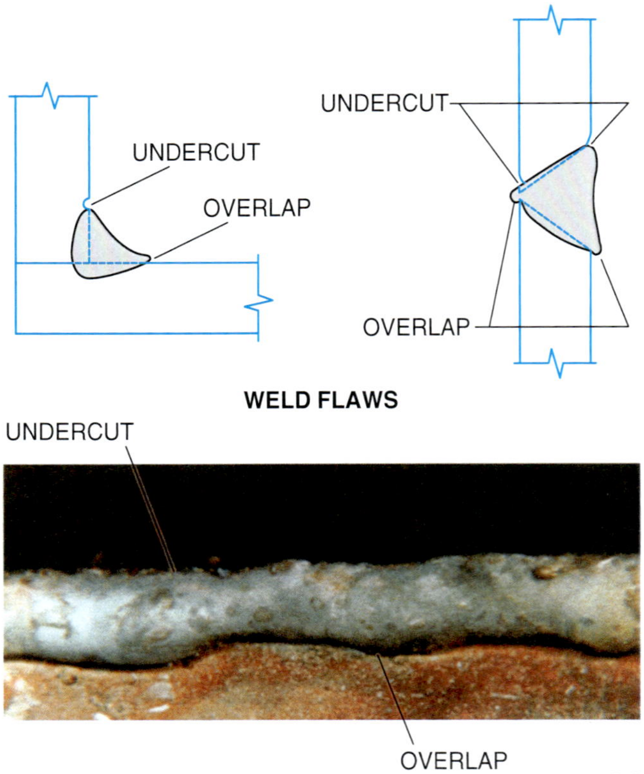

Figure 10 Undercut and overlap.

Undercutting is usually due to improper electrode manipulation. Other causes of undercutting include the following:

- Using a current adjustment that is too high
- Having an arc gap that is too long
- Failing to fill up the crater completely with weld metal

Most welds have some undercut that can be found upon careful examination. When it is controlled within the limits of the specifications and does not create a sharp or deep notch, undercut is usually not considered a weld defect. However, when it exceeds the limits, undercutting can be a serious defect because it reduces the strength of the joint.

4.7.0 Arc Strikes

Arc strikes (*Figure 11*) are small, localized points where surface melting occurs away from the joint. These spots may be caused by accidentally strik-

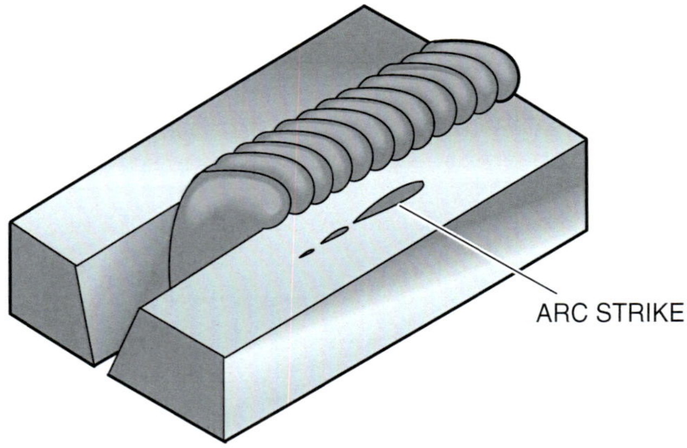

Figure 11 Arc strike.

ing the arc in the wrong place or by faulty ground connections.

Striking an arc on base metal that will not be fused into the weld metal should be avoided. Arc strikes can cause hardness zones in the base metal and can become the starting point for cracking. Arc strikes can cause a weld to be rejected.

4.8.0 Spatter

Spatter is made up of very fine particles of metal on the plate surface adjoining the weld area. It is usually caused by high current, a long arc, an irregular and unstable arc, or improper shielding. Spatter makes a poor appearance on the weld and base metal and can make it difficult to inspect the weld. Spatter can also cause coating failure.

4.9.0 Acceptable and Unacceptable Weld Profiles

The profile of a finished weld can affect the performance of the joint under load as much as other discontinuities affect it. This applies to the profile of a single-pass and to a layer of a multiple-pass weld. An unacceptable profile for a single-pass or multiple-pass weld could lead to the formation of discontinuities such as incomplete fusion or slag inclusions as the other layers are deposited.

4.9.1 Fillet Welds

A fillet weld is a weld that is approximately triangular in cross section and is used with T-, lap,

Preferred Fillet Weld Contours

In single-pass, weave-bead fillet welds where two workpieces are being joined at an angle (not lap joints), flat or slightly convex faces are usually preferred because weld stresses are more uniformly distributed through the fillet and workpieces.

and corner joints. The sizes and locations of fillet welds are given as welding symbols. The two types of fillet welds are convex and concave (*Figure 12*). A convex fillet weld has its surface bowed out like the outside surface of a ball. A concave fillet weld has its surface bowed in like the inside surface of a bowl.

The following terms are used to describe a fillet weld:

- *Weld face* – The exposed surface of the weld
- *Leg* – The distance from the root of the joint to the toe of a fillet weld
- *Weld toe* – The junction between the face of a weld and the base metal
- *Weld root* – The point shown in cross section at which the weld metal intersects with the base metal and extends farthest into the weld joint

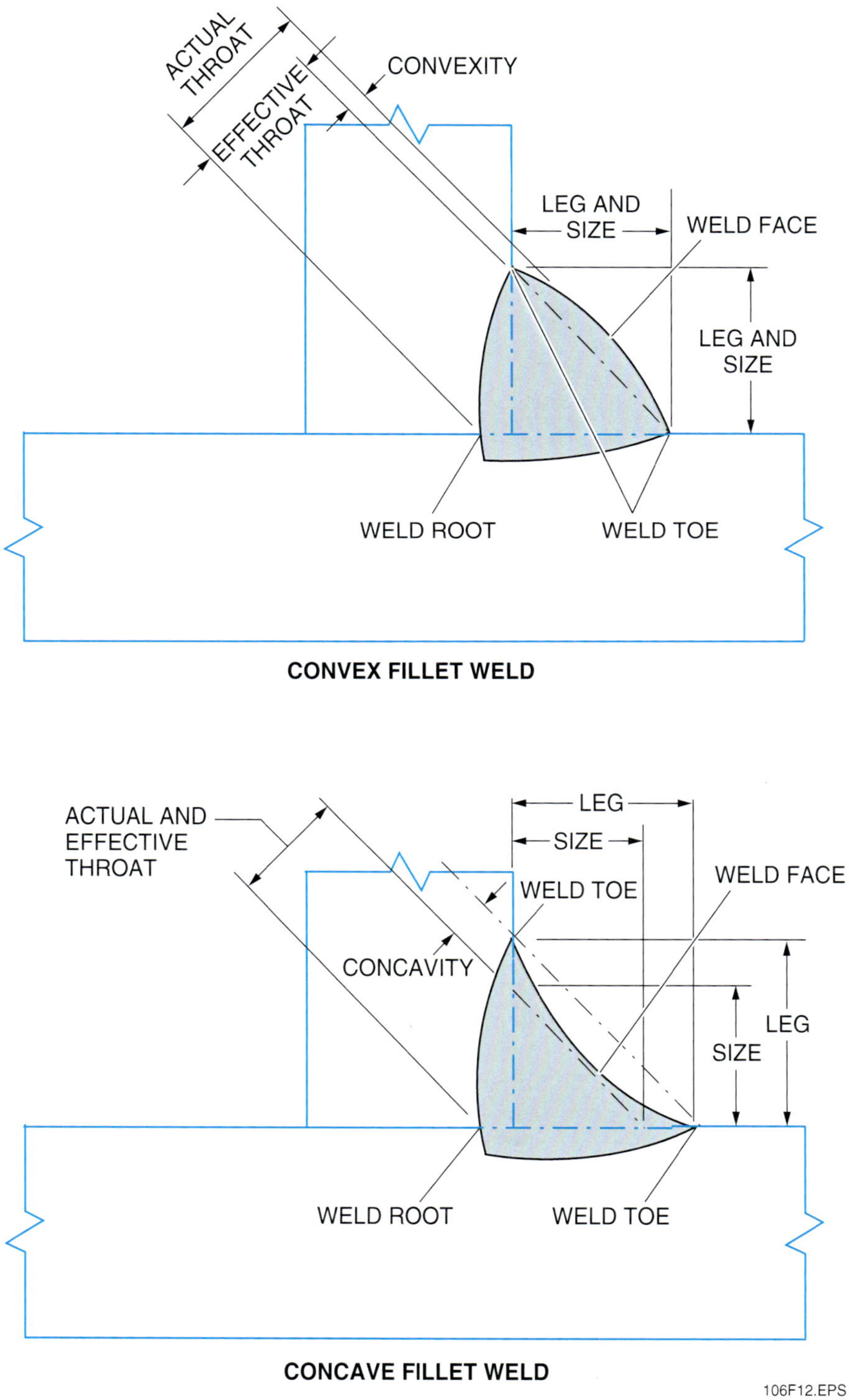

Figure 12 Convex and concave fillet welds.

- *Size* – The leg lengths of the largest right triangle that can be drawn within the cross section of a fillet weld
- *Actual throat* – The shortest distance from the root of the weld to its face
- *Effective throat* – The minimum distance, minus any convexity, from the root of the weld to its face
- *Theoretical throat* – The distance from the beginning of the joint root (with a zero opening) that is perpendicular to the hypotenuse of the largest right triangle that can be inscribed within the cross section of a fillet weld.

Fillet welds may be either equal leg or unequal leg (*Figure 13*). The face may be slightly convex, flat, or slightly concave. Welding codes require that fillet welds have a uniform concave or convex face, although a slightly nonuniform face is acceptable. The convexity of a fillet weld or individual surface bead should not exceed 0.07 times the actual face width or individual surface bead plus 0.06".

The ideal fillet has a uniform concave or convex face, although a slightly nonuniform face is acceptable. The convexity of a fillet weld or individual surface bead will be approximately 0.07 times the actual face width or the width of the individual surface bead, plus 1⁄16". *Figure 14* shows acceptable and unacceptable weld profiles for both fillet and groove welds.

> **NOTE**
> Refer to your site's WPS for specific requirements on fillet and butt welds. The information provided here is only a general guideline. The site WPS or quality specifications must be followed for all welds. Check with your supervisor if you are unsure of the specifications for your application.

A fillet weld is unacceptable and must be repaired if the profile has insufficient throat, excessive convexity, excessive undercut, overlap, insufficient leg, or incomplete fusion. Fillet welds require little base metal preparation except for cleaning the weld area and removing all dross from cut surfaces. Any dross from oxyfuel, plasma arc, or carbon arc cutting will cause porosity in the weld. For this reason, the codes require that all dross be removed prior to welding.

4.9.2 Butt Welds

Butt welds should be made with slight reinforcement (not exceeding 1⁄8") and a gradual transition to the base metal at each toe. Butt welds should not have excess reinforcement, insufficient throat, excessive undercut, or overlap. If a butt weld has any of these defects, it should be repaired. The bead width should not exceed the groove width by more than 1⁄8".

> **NOTE**
> Refer to your site's WPS for specific requirements on fillet and butt welds. The information provided here is only a general guideline. The site WPS or quality specifications must be followed for all welds. Check with your supervisor if you are unsure of the specifications for your application.

5.0.0 Nondestructive Examination (NDE) Practices

Nondestructive examination (NDE), sometimes referred to as nondestructive testing or nondestructive inspection, is a term used for those inspection methods that allow materials to be examined without changing or destroying them. NDE methods can usually detect the discontinuities and defects previously described.

> **NOTE**
> Various codes and specifications often require similar methods for qualifying welders. The applicable code or specification should be consulted for specific details and requirements. Ask your supervisor if you are unsure about which codes or specifications apply to your project.

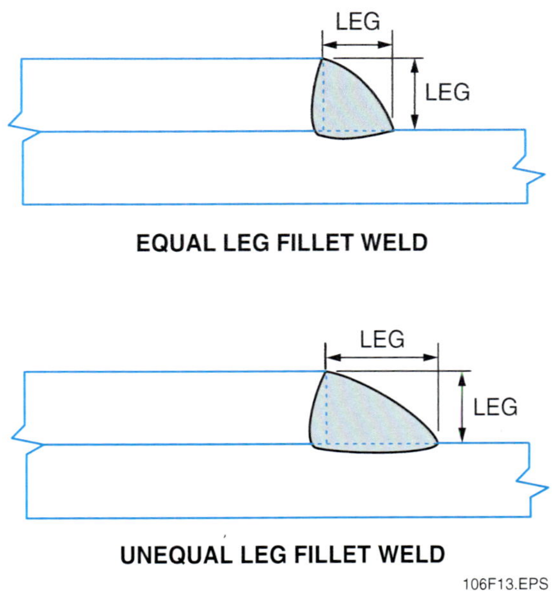

Figure 13 Equal leg and unequal leg fillet welds.

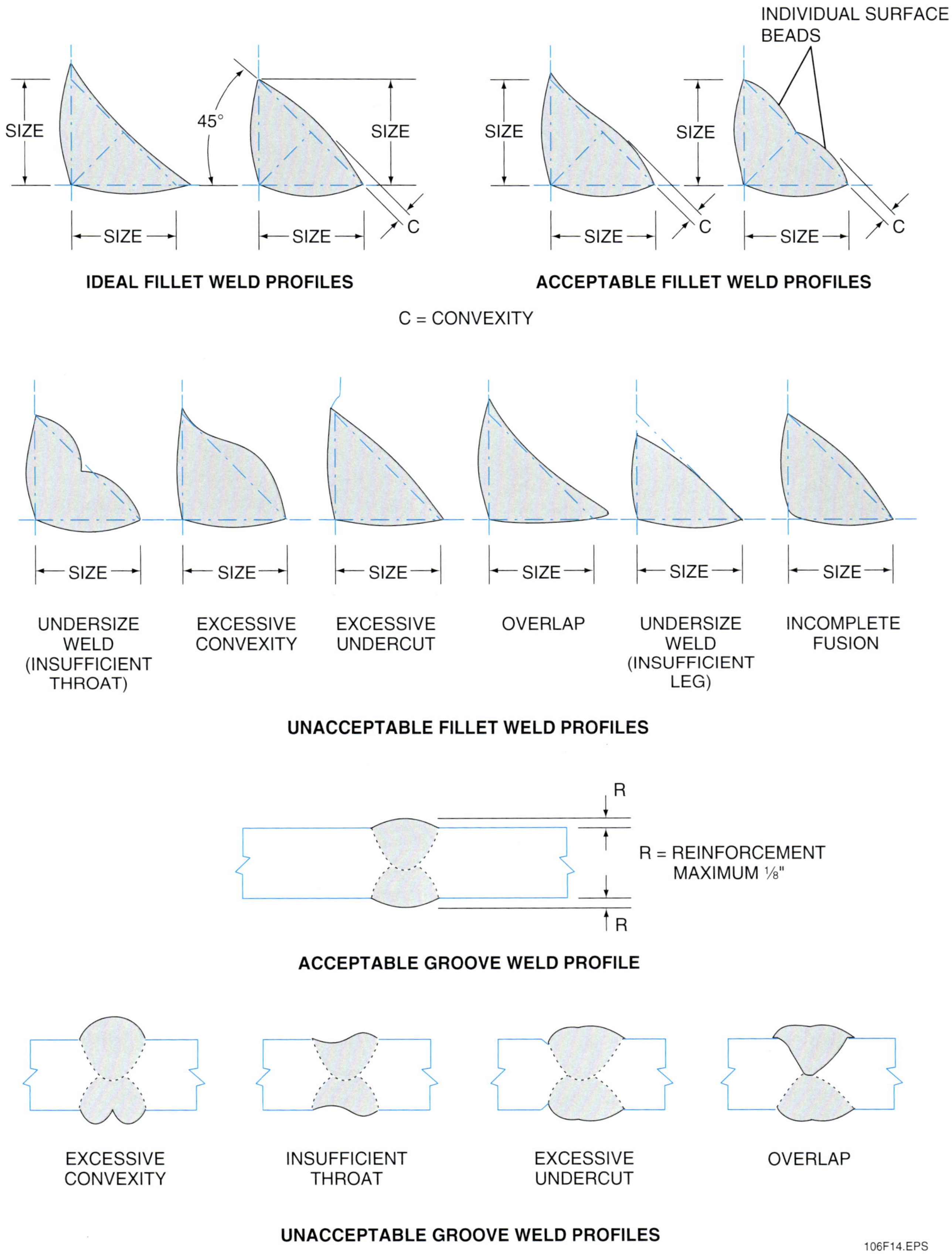

Figure 14 Acceptable and unacceptable weld profiles.

Nondestructive examination is usually performed by the site quality group as part of the site quality program. Inspectors trained in the proper test methods conduct the examinations.

The welder should be familiar with the following basic nondestructive examination practices:

- Visual inspection (VT)
- Liquid penetrant inspection (PT)
- Magnetic particle inspection (MT)
- Radiographic inspection (RT)
- Ultrasonic inspection (UT)
- Electromagnetic (eddy current) inspection (ET)
- Leak testing (LT)

5.1.0 Visual Inspection

In visual inspection, the surface of the weld and the base metal are observed for visual imperfections. Certain tools and gauges may be used during the inspection. Visual inspection is the examination method most commonly used by welders and inspectors. It is the fastest and most inexpensive method for examining a weld. However, it is limited to what can be detected by the naked eye or through a magnifying glass.

Properly done before, during, and after welding, visual inspection can detect more than 75 percent of discontinuities before they are found by more expensive and time-consuming nondestructive examination methods.

Prior to welding, the base metal should be examined for conditions that may cause weld defects. Dimensions, including edge preparation, should also be confirmed by measurements. If problems or potential problems are found, corrections should be made before proceeding any further.

After the parts are assembled for welding, the weld joint should be visually checked for a proper root opening and any other aspects that might affect the quality of the weld. Visually check for the following conditions:

- Proper cleaning
- Joint preparation and dimensions
- Clearance dimensions for backing strips, rings, or consumable inserts
- Alignment and fit-up of the pieces being welded
- Welding procedures and machine settings
- Specified preheat temperature (if applicable)
- Tack weld quality

During the welding process, visual inspection is the primary method for controlling quality. Some of the aspects that should be visually examined include the following:

- Quality of the root pass and the succeeding weld layers
- Sequence of weld passes
- Interpass cleaning
- Root preparation prior to welding a second side
- Conformance to the applicable procedure

After the weld has been completed, the weld surface should be thoroughly cleaned. A complete visual examination may disclose weld surface defects such as cracks, shrinkage cavities, undercuts, incomplete penetration, nonfusion, overlap, and crater deficiencies before they are discovered using other nondestructive inspection methods.

An important aspect of visual examination is checking the dimensional accuracy of the weld after it has been completed. Dimensional accuracy is determined by conventional measuring gauges. The purpose of using the gauges is to determine if the weld is within allowable limits as defined by the applicable codes and specifications.

Some of the more common welding gauges are the following:

- Undercut gauge
- Butt weld reinforcement gauge
- Fillet weld blade gauge set

5.1.1 Undercut Gauge

An undercut gauge is used to measure the amount of undercut on the base metal. Typically, codes allow for undercut to be no more than 0.01" deep when the weld is transverse to the primary stress in the part that is undercut. Several types of gauges can be used to check the amount of undercut. These gauges have a pointed end that is pushed into the undercut. The reverse side of the gauge indicates the measurement in either inches or millimeters. Two types of undercut gauges currently used are the bridge cam gauge and the V-WAC gauge (*Figure 15*). These gauges can be used for measuring undercut and for many other measurements.

5.1.2 Butt Weld Reinforcement Gauge

The butt weld reinforcement gauge has a sliding pointer calibrated to several different scales that are used to measure the size of a fillet weld or the reinforcement of a butt weld. To use the gauge for a fillet weld, position it as shown in *Figure 16*, and slide the pointer to contact the base metal or weld metal. Be sure to read the correct scale for the measurement being taken. The other end of the gauge is used for butt welds.

5.1.3 Fillet Weld Blade Gauge Set

The fillet weld blade gauge set has seven individual blade gauges for measuring convex and concave fillet welds. The individual gauges are held together by a screw secured with a knurled nut. The seven individual blade gauges can measure eleven concave and convex fillet weld sizes: ⅛", $^{3}/_{16}$", ¼", $^{5}/_{16}$", ⅜", $^{7}/_{16}$", ½", ⅝", ¾", ⅞", and 1" and their metric equivalents. The same blade size cannot be used for measuring both concave and convex fillet welds.

To use the fillet weld blade gauge set, identify the type of fillet weld to be measured (concave or convex) and the size. Select the appropriate blade and position it. Be sure the gauge blade is flush to the base metal with the tip touching the vertical member.

Figure 17 shows an application of a fillet weld blade gauge.

5.2.0 Liquid Penetrant Inspection

Liquid penetrant inspection (PT) is a nondestructive method for locating defects that are open to the surface. It cannot detect internal defects. The technique is based on the ability of a penetrating liquid, which is usually red in color, to wet the surface opening of a discontinuity and be drawn into it. A liquid or dry powder developer, which is usually white in color, is then applied over the metal. If the flaw is significant, red penetrant bleeds through the white developer to indicate a discontinuity or defect.

The dye, cleaner, and developer are available in spray cans for convenience. Some solvents used in the cleaners and developers contain high amounts of chlorine, a known health hazard, to make the liquids nonflammable.

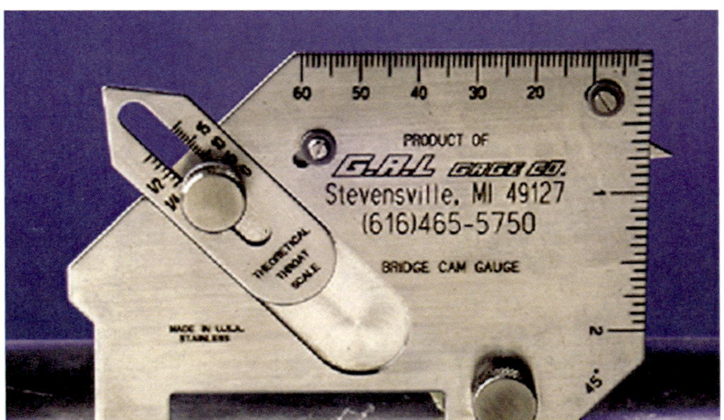

BRIDGE CAM GAUGE

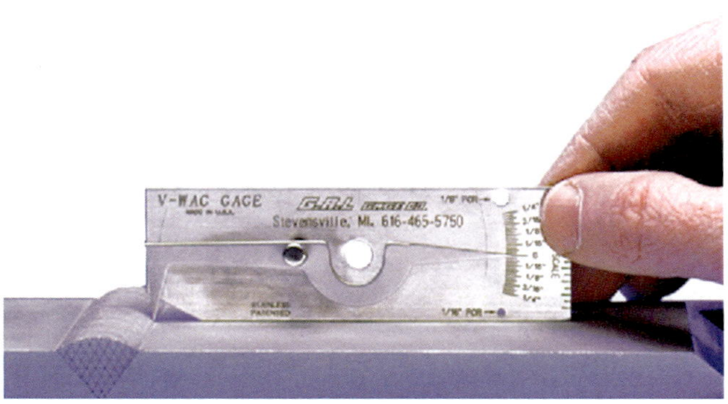

V-WAC GAUGE

106F15.EPS

Figure 15 Undercut gauges.

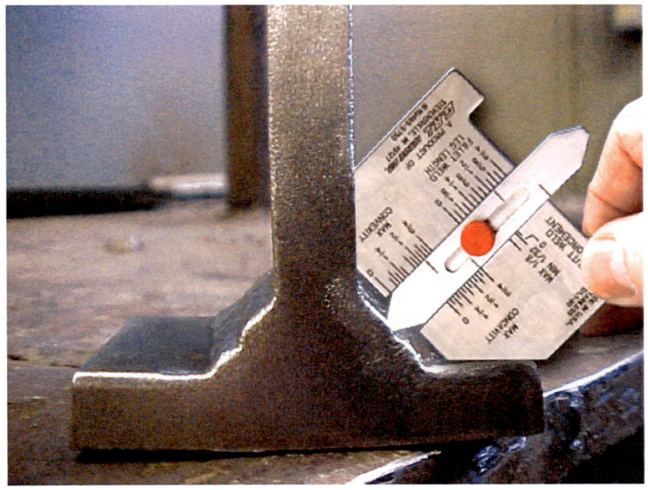

106F16.EPS

Figure 16 Automatic weld size gauge (AWS).

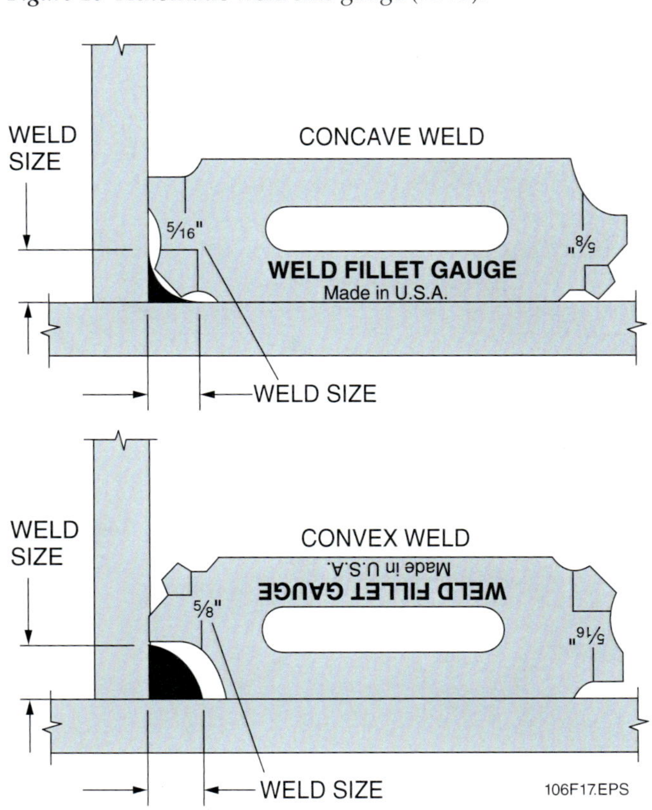

106F17.EPS

Figure 17 Fillet weld blade gauge.

WARNING!

Refer to the material safety data sheet (MSDS) for hazards associated with the liquid penetrant solvent.

The most common defects found using this process are surface cracks. Most cracks exhibit an irregular shape. The width of the bleed-out (the red dye bleeding through the white developer) is a relative measure of the depth of a crack.

Surface porosity, metallic oxides, and slag will also hold penetrant and cause bleed-out. These indications are usually more circular and have less width than a crack. *Figure 18* shows liquid penetrant materials and an example of the results of liquid penetrant inspection.

5.2.1 Advantages of Liquid Penetrant Inspection

The advantages of liquid penetrant inspection are that it can find small defects not visible to the naked eye, it can be used on most types of metals, it is basically inexpensive, and it is fairly easy to use and interpret. PT is most useful to examine welds that are susceptible to surface cracks. Except for visual inspection, it is perhaps the most commonly used nondestructive examination method for surface inspection of nonmagnetic parts.

5.2.2 Disadvantages of Liquid Penetrant Inspection

The disadvantages of liquid penetrant inspection are that it takes more time to use than visual inspection, and it can only find surface defects. The presence of weld bead ripples and other irregularities can also hinder the interpretation of indicators. Because chemicals are used, care must be taken when performing the inspection. When testing the rough, irregular surfaces produced by welding, the presence of nonrelevant indicators may also make interpretation difficult.

5.3.0 Magnetic Particle Inspection

Magnetic particle inspection (MT) is a nondestructive examination method that uses electricity to magnetize the weld that will be examined (*Figure 19*). After the metal has been magnetized, metal particles are sprinkled onto the weld surface. If there are defects in the surface or just below it, the metal particles will be grouped into a pattern around the defect. The defect can be iden-

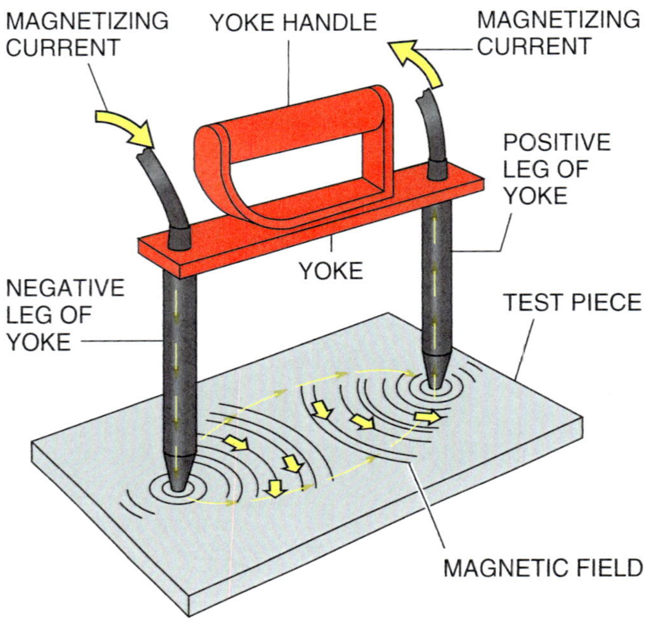

106F18.EPS

Figure 18 Liquid penetrant materials and inspection example.

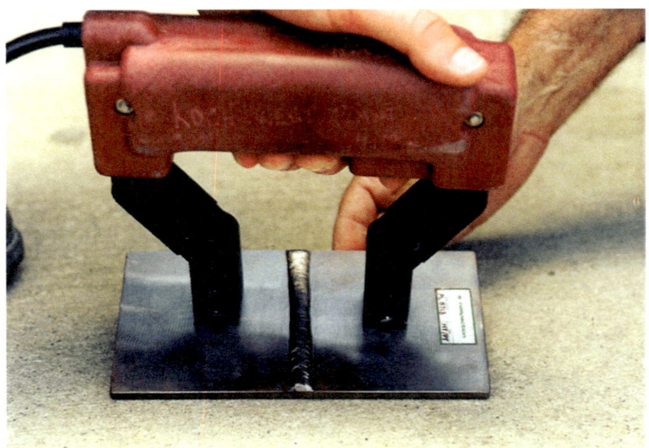

106F19.EPS

Figure 19 Magnetic particle examination with electromagnetic yokes.

tified by the shape, width, and height of the particle pattern.

Magnetic particle inspection is used to test welds for such defects as surface cracks, nonfusion, porosity, and slag inclusion. It can also be used to inspect plate edges for surface imperfections prior to welding. Defects can be detected only at or near the surface of the weld. Defects much deeper than this are not likely to be found. Certain discontinuities exhibit characteristic powder patterns that can be identified by a skilled inspector.

For magnetic particle examination, the part to be inspected must be ferromagnetic (made of steel or a steel alloy), smooth, clean, dry, and free from oil, water, and excess slag. The part is magnetized by using an electric current to set up a magnetic field within the material. The magnetized surface is covered with a thin layer of magnetic powder. If there is a defect, the powder is held to the surface at the defect because of the powerful magnetic field.

When this examination method is used, there is normally a code or standard that governs both the method and the acceptance/rejection criteria of indications.

5.3.1 Advantages and Disadvantages of Magnetic Particle Inspection

The advantages of magnetic particle inspection are that it can find small defects not visible to the naked eye, and it is faster than liquid penetrant inspection.

Disadvantages of magnetic particle inspection are that the materials must be capable of being magnetized, the inspector must be skilled in interpreting indications, rough surfaces can interfere with the results, the method requires an electrical power source, and it cannot find internal discontinuities located deep in the weld.

5.4.0 Radiographic Inspection

Radiography (RT) is a nondestructive examination method that uses radiation (X-rays or gamma rays) to penetrate the weld and produce an image on film (*Figure 20*). When a joint is radiographed, the radiation source is placed on one side of the weld and the film on the other. The joint is then exposed to the radiation source. The radiation penetrates the metal and produces an image on the film. The film is called a radiograph and provides a permanent record of the weld quality. Radiography should only be used and interpreted by trained, qualified personnel.

Radiographic inspection can produce a visible image of weld discontinuities, both surface and subsurface, when they are different in density from the base metal and different in thickness parallel to the radiation. Surface discontinuities are better identified by visual, penetrant, or magnetic particle examination.

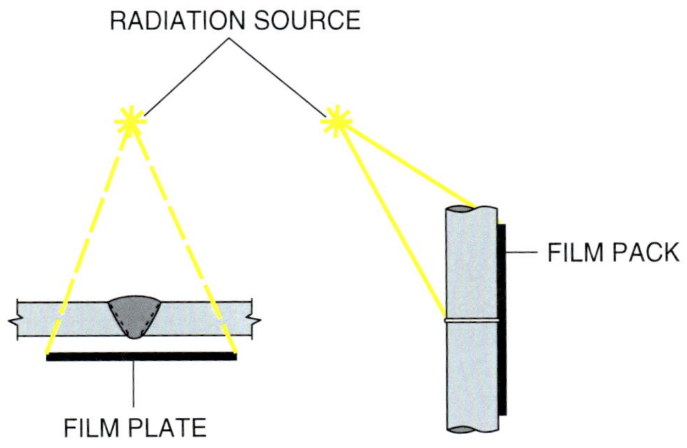

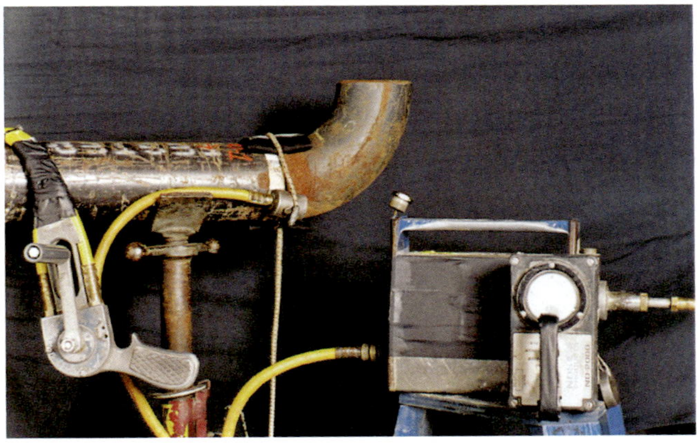

106F20.EPS

Figure 20 Examination using radiography.

Fixed Weld Fillet Gauge

This fixed weld fillet gauge can be used to measure 11 fillet weld sizes.

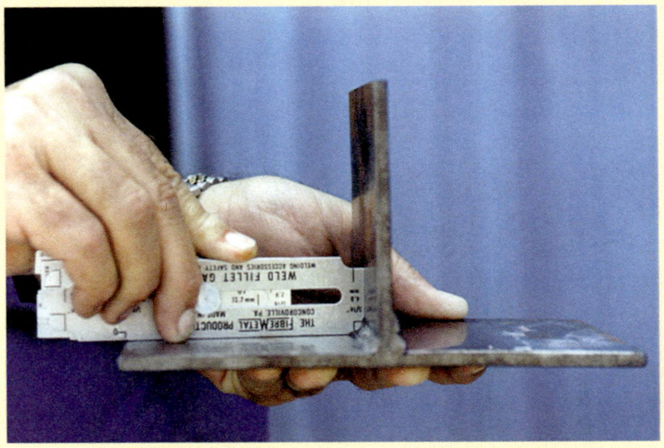

106SA01.EPS

5.4.1 Advantages and Disadvantages of Radiographic Inspection

The advantages of radiographic inspection are that the film gives a permanent record of the weld quality, the entire thickness can be examined, and it can be used on all types of metals.

Disadvantages of radiographic inspection are that it is a slow and expensive method for inspecting welds, some joints are inaccessible to radiography, and radiation of any type is very hazardous to humans. Cracks can frequently be missed if they are very small or are not aligned with the radiation beam.

5.5.0 Ultrasonic Inspection

Ultrasonic inspection (UT) is a relatively low-cost, nondestructive examination method that uses soundwave vibrations to find surface and subsurface defects in the weld material. Ultrasonic waves are passed through the material being tested and are reflected back by any density change caused by a defect. The reflected signal is shown on the screen display of the instrument (*Figure 21*).

The term *ultrasonic* indicates frequencies above those heard by the human ear. Ultrasonic devices operate very much like depth sounders or fish finders.

Ultrasonic examination can be used to detect and locate cracks, **laminations**, shrinkage cavities, pores, slag inclusions, incomplete fusion, and incomplete joint penetration, as well as other discontinuities in the weld. A qualified inspector can interpret the signal on a screen to determine the approximate position, depth, and size of the discontinuity.

Ultrasonic beams can deeply penetrate weld material with high sensitivity and accuracy. The output is readily digitized, and operation of the monitor is entirely nonhazardous. Some of the different types of ultrasonic techniques are pulse-echo, pulse-transmission, ultrasonic attenuation, continuous wave resonance, and ultrasonic spectroscopy. Ultrasonic monitors are very portable, but they require experienced technicians to monitor the equipment. Reading an ultrasonic flaw detector requires some practice because the screen image shows the peaks of back reflection. *Figure 22* shows an ultrasonic flaw detector and an angle-beam transducer.

Figure 23 shows the effects of moving an angle-beam transducer back and forth across a weld.

Figure 24 shows the effects of moving an angle-beam transducer in an arc around a weld.

5.5.1 Advantages and Disadvantages of Ultrasonic Inspection

The advantages of ultrasonic inspection are the following: it can find defects throughout the material being examined, it can be used to check materials that cannot be radiographed, it is nonhazardous to personnel and equipment, and it can detect even small defects from one side of the material.

The disadvantages of ultrasonic inspection are that it requires a high degree of skill to properly interpret the patterns, and very small or thin weldments are difficult to inspect.

5.6.0 Electromagnetic (Eddy Current) Inspection

Like magnetic particle testing, electromagnetic, or eddy current, inspection (ET) uses electromagnetic energy to detect defects in the joint. A coil, which produces a magnetic field, is placed on or around the part being tested. After being

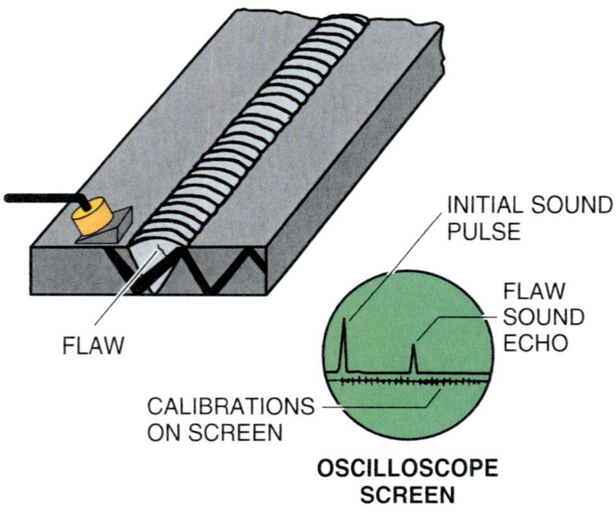

106F21.EPS

Figure 21 Portable ultrasonic device.

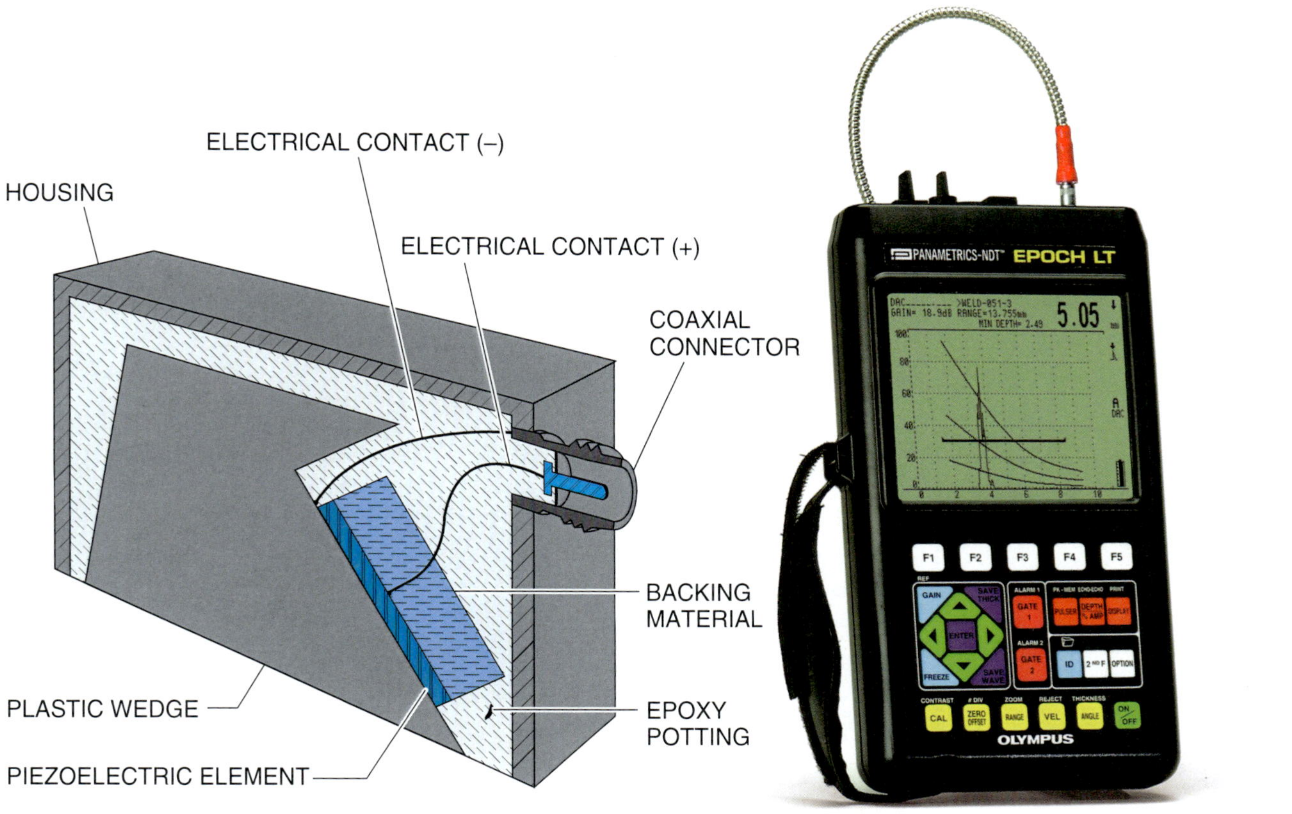

Figure 22 Ultrasonic flaw detector and angle-beam transducer.

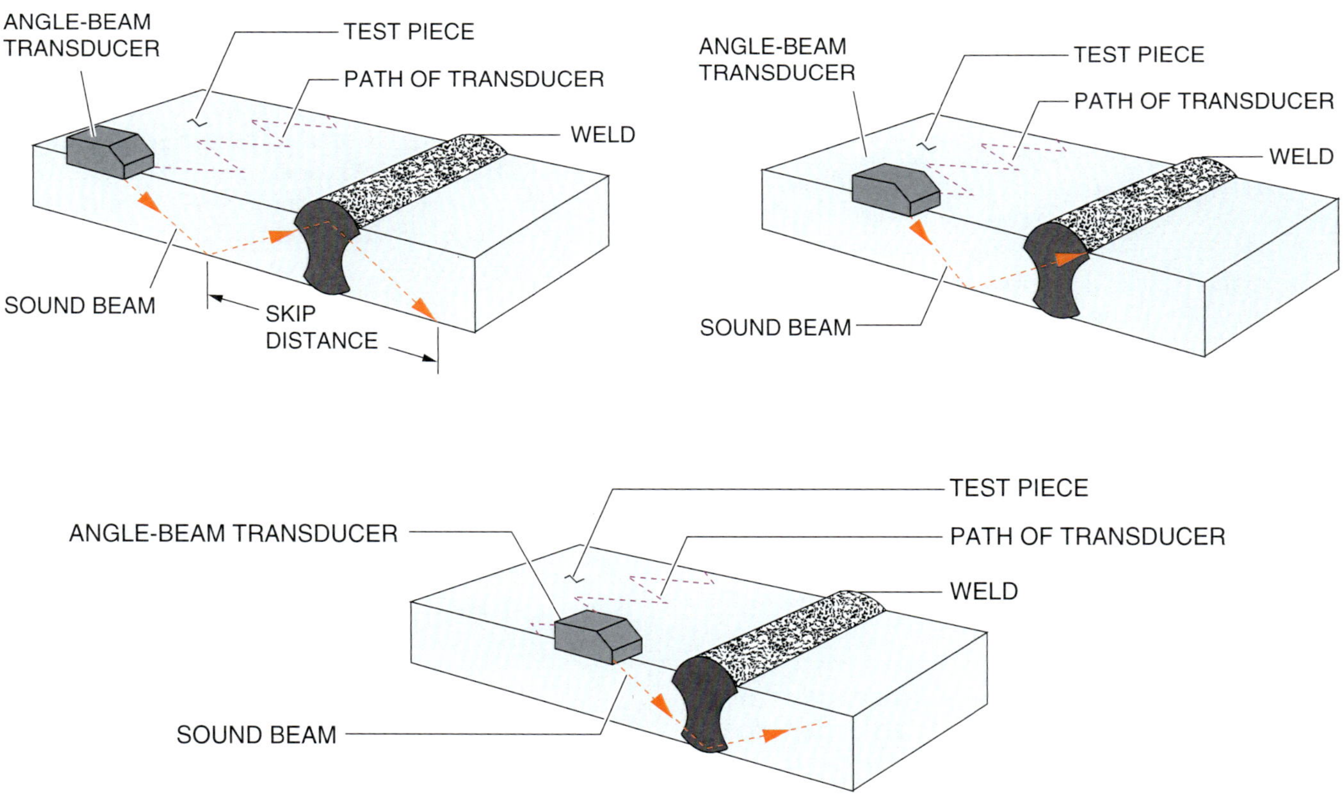

Figure 23 Moving a transducer back and forth.

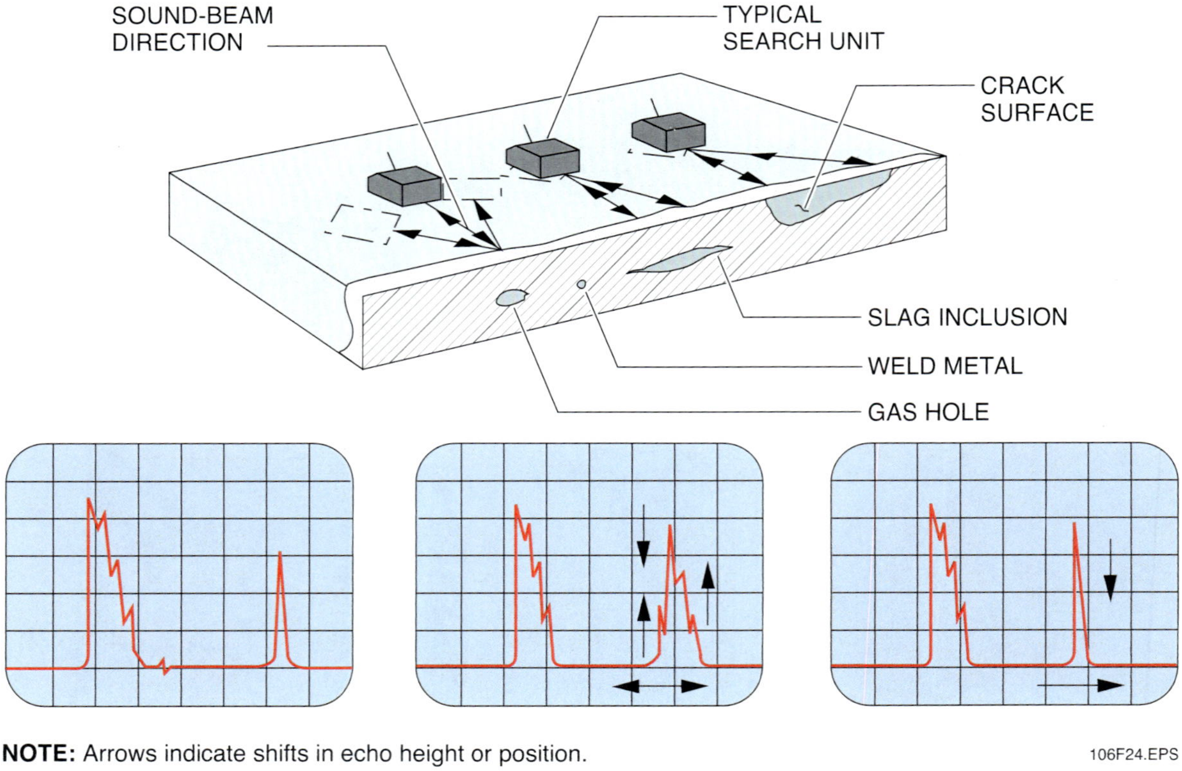

Figure 24 Moving a transducer in an arc.

calibrated, the coil is moved over the part to be inspected. The coil produces a current in the metal through induction. The induced current is called an eddy current. If a discontinuity is present in the test part, it will interrupt the flow of the eddy currents. This change can be observed on the oscilloscope.

Eddy currents only detect discontinuities near the surface of the part. This method is suitable for both ferrous and nonferrous materials, and it is used in testing welded tubing and pipe. It can determine the physical characteristics of a material and the wall thickness in tubing. It can also check for porosity, pinholes, slag inclusions, internal and external cracks, and nonfusion.

5.6.1 Advantage and Disadvantage of Eddy Current Inspection

The advantage of eddy current inspection is that it can detect surface and near-surface weld defects. It is particularly useful in inspecting circular parts like pipes and tubing.

The disadvantage of eddy current inspection is that eddy currents decrease with depth, so defects farther from the surface may go undetected. The accuracy of the examination depends in large part on the calibration of the instrument and the qualification of the inspector.

5.7.0 Leak Testing

Leak testing (*Figure 25*) is used to determine the ability of a pipe or vessel to contain a gas or liquid under pressure. Testing methods vary depending on the application of the weldment. In some cases, the vessel is pressurized and tested by immersing it in water or by applying a soap bubble solution to the weld. An open tank can be tested using water that contains fluorescein, which can be detected by ultraviolet light.

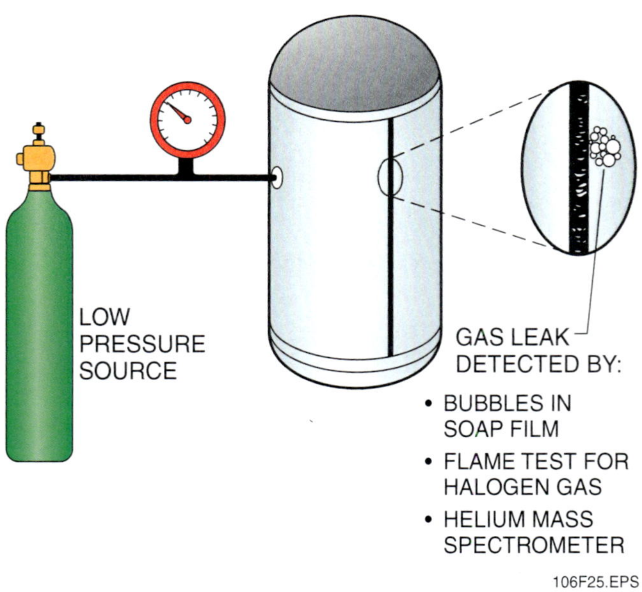

Figure 25 Example of a leak test.

A method called the vacuum box test is used to test a vessel when only one side of the weld is accessible. The base of a storage tank is an example.

The vacuum box is a transparent box with a soft rubber seal. A vacuum pump is used to extract all the air from the box. A leak is indicated by the presence of bubbles.

In the helium spectrometer leak test, helium is used as a tracer gas. Because of the small size of helium atoms, they can pass through an opening so small that it might not be detectable by other test methods. Sensitive instruments are used to detect the presence of helium.

6.0.0 DESTRUCTIVE TESTING

There is considerable overlap in the application of nondestructive and destructive tests. Destructive tests, which destroy the weld, are frequently used on several sample weldments to supplement, confirm, or establish the limits of nondestructive tests. Destructive testing is also used to provide supporting information. Once this information has been established, nondestructive examinations can be made on similar welds to locate all discontinuities above the critical defect size that was determined by the destructive tests.

In a destructive test, the test sample is destroyed or damaged in the testing process and is no longer suitable for use. Destructive testing is often done in a laboratory setting using machines designed to apply different stresses to the weld or the base metal, depending upon which is being tested. Some tests can be accomplished by striking the sample with a hammer. Examples of destructive tests commonly used to examine welds and base metals include the following:

- *Tensile test* – In this test, a sample is placed in a tensile testing machine and is pulled until it breaks. The tensile test is sometimes used to determine if the weld performs as well as the base metal. In most cases, however, the test is performed to determine specific characteristics of the weldment, such as strength and ductility (the amount the sample will stretch before breaking).
- *Hardness test* – Whereas strength testing examines the ability of the sample to transmit a load, hardness testing examines its ability to resist penetration. Hardness testing is usually done using a penetrating device that leaves an indentation in the sample. The depth or diameter of the indentation, depending on the type of hardness test being conducted, is then measured to determine the hardness.
- *Impact test* – The ability of a weld to withstand an impact, such as a hammer strike, is measured by this test. The characteristic being tested is referred to as toughness. In a laboratory environment, a pendulum-type machine that simulates a heavy hammer blow is used. A notch of a specified size is made in the sample. Then the sample is placed in the jaws of the testing machine and struck with a pendulum blow.
- *Soundness test* – The three types of soundness tests include bend, nick-break, and fillet weld break tests.

  – Bend testing is the most commonly used soundness test used in determining the qualifications of a welder or welding procedures. Bend tests are performed by placing the sample in a special fixture and applying stress in a way that causes the sample to bend 180 degrees. The bend is then inspected for weld defects. Guided bend tests (*Figure 26*) are used to evaluate groove test welds on both plates and pipes. In this method, specimens are bent into a U-shape with a device called a jig. The bending action places stress on the weld metal and reveals any discontinuities in the weld.

  The three types of tests performed on the jig are root bends, face bends, and side bends. These bends test for penetration and fusion throughout the root joint. They also test for porosity, inclusions, or other defects, as well as the quality of the fusion to the side walls and the face of the weld joint. Face and root bend tests are used for materials up to 3/8" thick. For materials thicker than 3/8", side bend testing may be used.

  – Nick-break testing is used primarily in the pipeline industry. The specimen is saw-cut so it will break in a specific place. Then it is broken, and the weld zone is examined for defects.

  – In a fillet weld break test, a fillet weld is made on one side of a T-joint. Stress is then applied to the T-joint until the weld fractures (*Figure 27*). The weld is then examined for defects. This is not a weld strength test; it is intended to break the weld so that it can be examined for discontinuities.

GOING GREEN

Test Coupons

Test coupons are routinely taken for testing in the welding industry. When those test coupons are no longer needed, they should be properly recycled, to save the company money and to keep salvageable items out of the landfills.

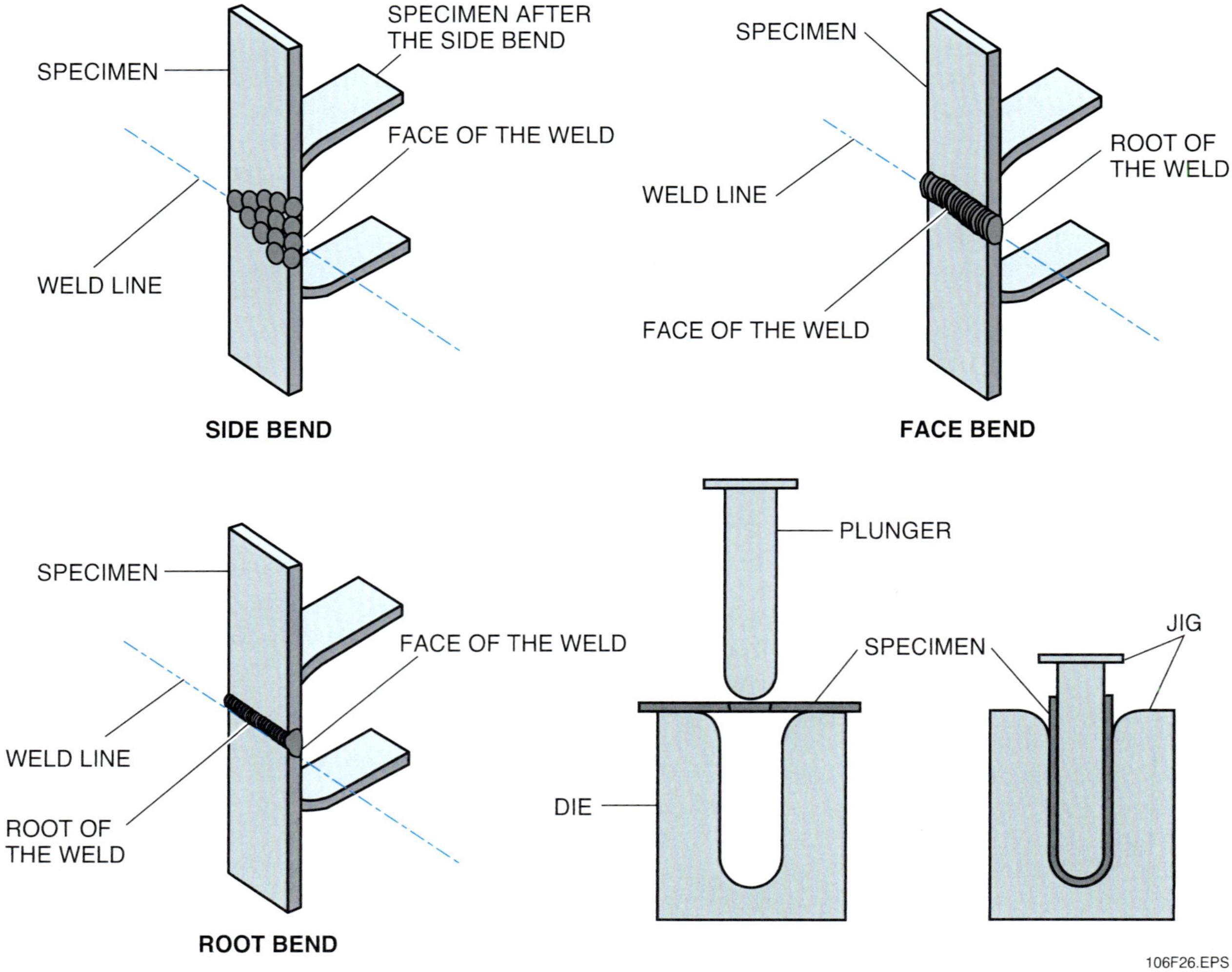

Figure 26 Guided bend test method and samples.

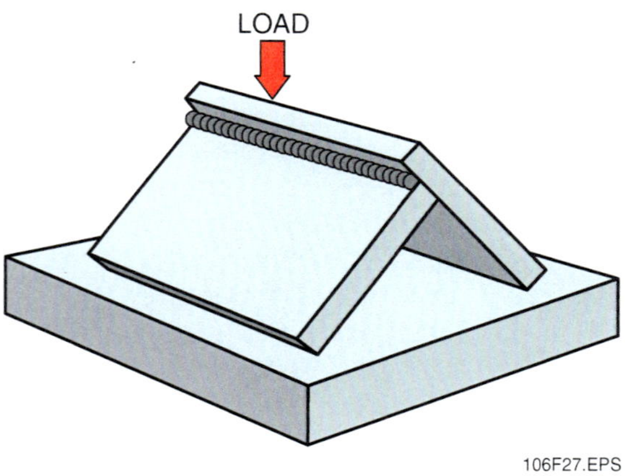

Figure 27 Fillet weld break test.

7.0.0 Welder Performance Qualification Tests

The purpose of welder performance qualification is to measure the proficiency of individual welders. As previously discussed, codes require that welders take a test to qualify for performing a welding procedure.

> **NOTE**
> Various codes and specifications often require similar methods for qualifying welders. The applicable code or specification should be consulted for specific details and requirements. Ask your supervisor if you are unsure about which codes or specifications apply to your project.

7.1.0 Welding Positions Qualification

The welder or welding operator is qualified by welding position. Welders may be qualified to perform a welding procedure in only one, or possibly all, position(s) by taking a welding test. The qualification tests are designed to measure the welder's ability to make groove and fillet welds in different positions on plate, pipe, or both in accordance with the applicable code. Each welding position is designated by a number and a letter, such as 1G. These designations are standard for all codes.

The letter G designates a groove weld. The letter F designates a fillet weld. For plate welding, the positions are designated by the following numbers:

- *1* – Flat position welding
- *2* – Horizontal position welding
- *3* – Vertical position welding
- *4* – Overhead position welding

Figure 28 shows the plate welding positions for both fillet welds and groove welds.

For pipe welding, there are these additional positions: 5F, 5G, and 6G. The numbers 5 and 6 indicate that multiple-position welds are required. Also, in the 1G (flat groove) and 1F (flat fillet) positions, the pipe is rotated during welding. *Figure 29* shows the pipe welding positions for both fillet welds and groove welds.

A welder who qualifies in one position does not automatically qualify to weld in all positions. However, in most cases qualification for groove welds will qualify the welder for fillet welds; qualification for pipe will qualify the welder for plate. Qualification in one code may or may not qualify the welder in other codes. The qualification requirements between codes may not match. Refer to your site requirements/code for qualifying requirements.

7.2.0 AWS Structural Steel Code

The AWS structural steel code provides information concerning the qualification of welding procedures, welders, and welding operators for the types of welding done by contractors and fabricators in building and bridge construction. Qualification for plate welding also qualifies the welder for rectangular tubing.

The mild steel electrodes used with shielded metal arc welding (SMAW) are classified by F numbers: F1, F2, F3, and F4. Qualification with an electrode in a particular F-number classification will qualify the welder with all electrodes identified in that classification and in lower F-number classifications. *Table 1* shows AWS F-number electrode classifications.

Material thickness is an essential variable in qualification tests under AWS code. Some of the tests in the code qualify the welder only up to twice the thickness of the test piece. Others qualify the welder for unlimited thicknesses.

A typical AWS welder qualification test is a V-groove weld with metal backing in the 3G and 4G positions using an F4 electrode. Passing this test qualifies the welder to weld with F4 or lower electrodes and to make groove and fillet welds in all positions. *Figure 30* shows an example of an AWS plate test.

7.3.0 ASME Code

Individual welders and welding operators who are required to weld to ASME code must qualify in accordance with Section IX of the *ASME Boiler and Pressure Vessel Code* on either plate or pipe. Qualification on pipe also qualifies the welder to weld plate, but not vice versa. Qualification with groove welds also qualifies the welder for fillet welds, but not vice versa. It is possible under the code to qualify for fillet welds only.

The typical ASME welder qualification test is to weld pipe in the 6G position using an open root. Passing this test qualifies the welder to weld pipe in all positions and plate in all positions (fillet and groove). If F3 electrodes are used all the way out, the welder only qualifies on F3 electrodes on both plate and pipe. If F3 electrodes are used for welding the root and F4 electrodes are used for filler, the welder qualifies to weld pipe or plate with F3 electrodes and pipe or plate with F4 electrodes as long as backing is used. The drawing in *Figure 31* shows a typical ASME pipe test.

7.4.0 Welder Qualification Tests

The welder becomes qualified by successfully completing a weld made in accordance with the WPS. It is general practice for code welding to qualify welders on the groove weld tests. Passing these tests also permits the welder to weld fillet welds.

7.4.1 Making the Test Weld

Although the qualification tests are designed to determine the capability of welders, some welders have failed for reasons not related to their welding ability. This is due principally to carelessness in the application of the weld and in the preparation of the test specimen. It is important to note prior to welding where the test strips will be cut from the weld coupon. By doing this, you can avoid potential problems such as restarts in the area of the test strips. The following sections explain how to prepare a test specimen.

7.4.2 Removing Test Specimens

After making the qualification test weld, the test specimens are cut from the test pipe or plate by any suitable means. There are specific locations where the test specimen is cut from the pipe or plate.

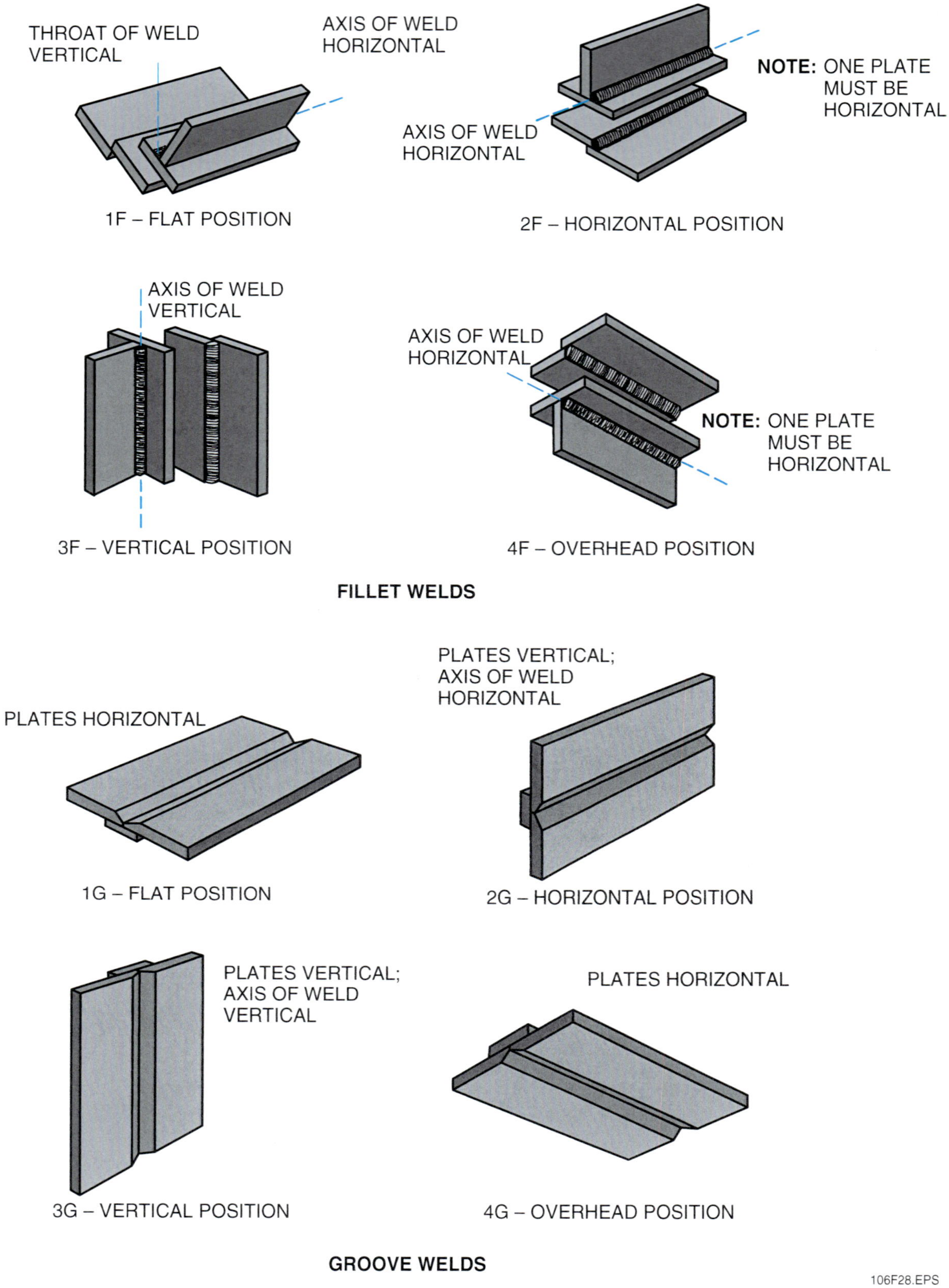

Figure 28 Welding positions for plate.

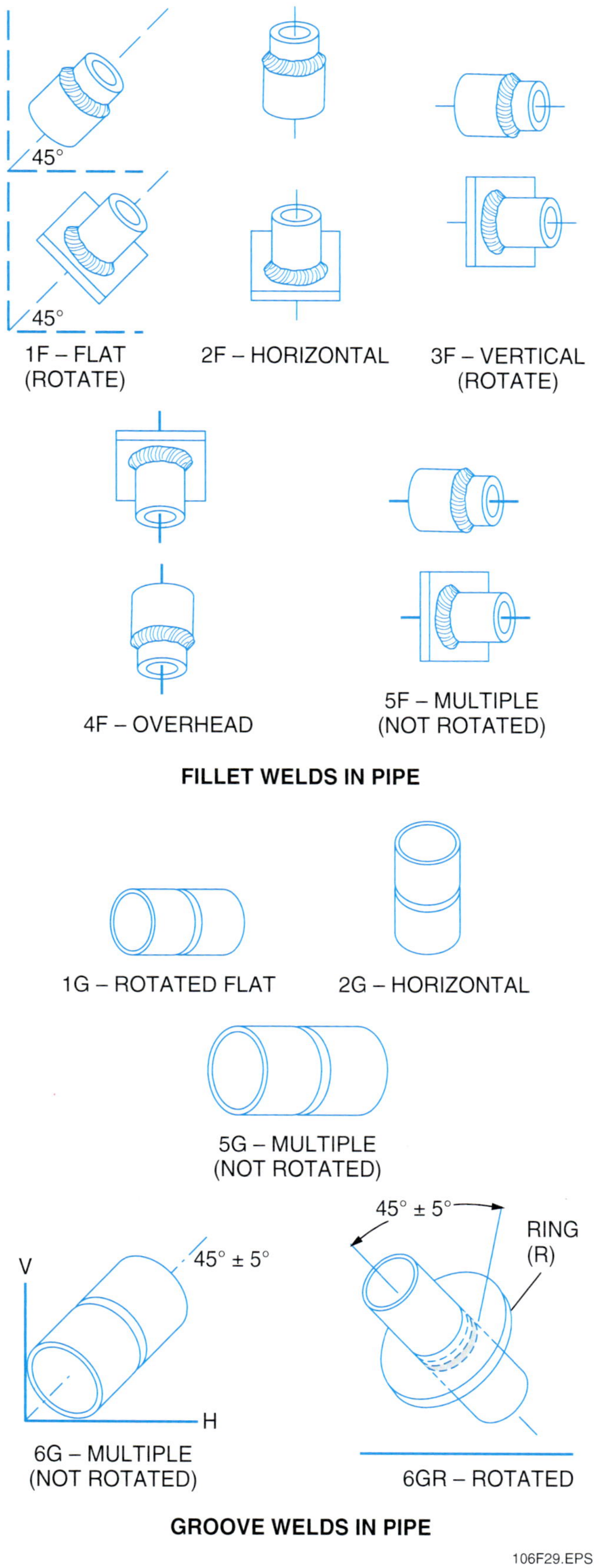

Figure 29 Welding positions for pipe.

For pipe welded in the 1G or 2G positions, two specimens are required. For material 3/8" thick and under, a face bend and a root bend are required. For 3/8" only, side bends can be substituted. For material thicker than 3/8", two side bends are required. *Figure 32* shows where the specimens are cut from the test weld.

For pipe welded in the 5G or 6G positions, four specimens are required. For material 1/16" thick up to 3/8" thick, two face bends and two root bends are required. For material more than 3/8" thick, four side bends are required. *Figure 33* shows where the specimens are cut from the test weld.

Typical specimen locations for plate welds are shown in *Figure 34*. For material 3/8" thick, a face bend and a root bend are required. For material thicker than 3/8", two side bends are required.

NOTE

Tests are usually given on plate thicknesses of 3/8" for limited thickness and 1" for unlimited thickness qualifications.

7.4.3 Preparing the Specimens for Testing

After the specimen has been cut from the test piece, it must be properly prepared for testing (*Figure 35*). Poor specimen preparation can cause a sound weld metal to fail. For example, a slight nick may open up under the severe bending stress of the test, causing the specimen to fail. To properly prepare the test specimen, do the following:

- Grind or machine the surface to a smooth finish. All grinding and machining marks must be lengthwise on the sample. Otherwise, they produce a notch effect, which may cause failure.
- Remove any face or root reinforcement from the weldment. This is part of the test requirement and, more important to the welder, failure to do so can cause the failure of a good weld. Grinding of the weld reinforcement may not exceed 1/32" or 5 percent of the thickness of the base material (*AWS D1.1: 2006*).
- Round the edges to a smooth 1/16" radius. This can be done with a file. Rounded edges help prevent failure caused by cracks starting at a sharp corner.
- Refrain from water-quenching specimens you are grinding when they are hot. Quenching may create small surface cracks that become larger during the bend test.

Table 1 AWS F-Number Electrode Classification

Group	AWS Electrode Classification				Type
F4	EXX15	EXX16	EXX18		Low-Hydrogen
F3	EXX10	EXX11			Fast-Freeze
F2	EXX12	EXX13	EXX14		Fill-Freeze
F1	EXX20	EXX24	EXX27	EXX28	Fast-Fill

Figure 30 Typical AWS plate test.

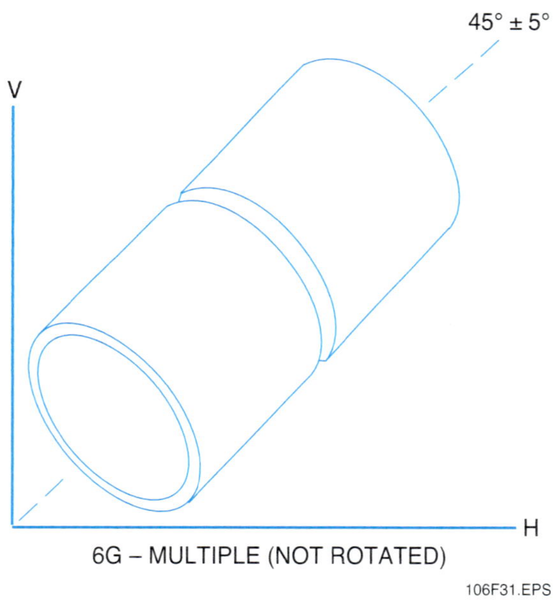

Figure 31 Typical ASME pipe test.

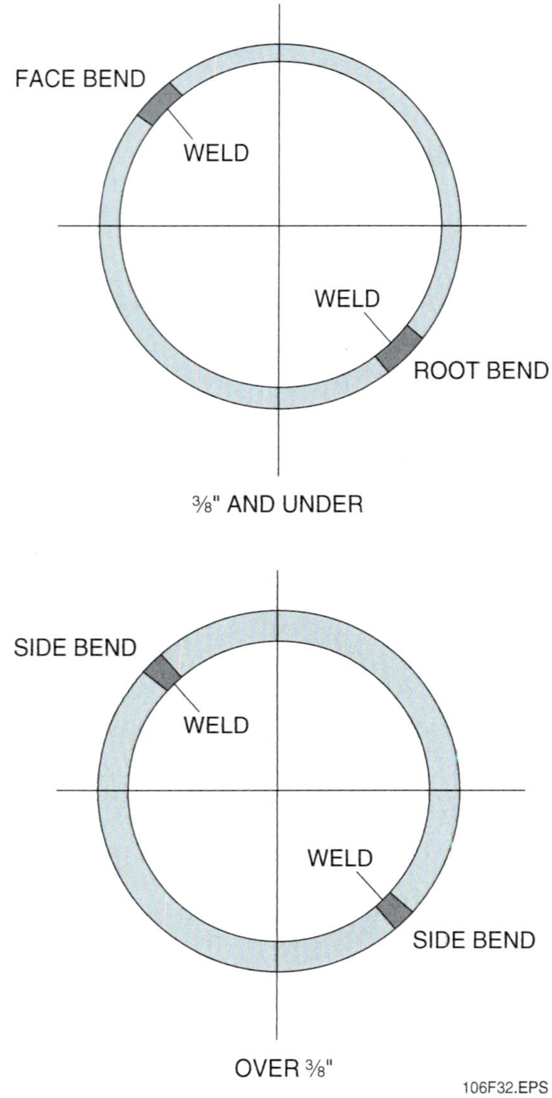

Figure 32 Locations of pipe specimens for 1G and 2G positions.

After bending, the specimen is evaluated by measuring the discontinuities that are exposed. The criteria for acceptance can vary by code or site quality standards. *AWS D1.1* states that for acceptance, the surface shall contain no discontinuities exceeding the following dimensions:

- *1/8" (3.2 mm)* – Measured in any direction on the surface.
- *3/8" (9.5 mm)* – Sum of the greatest dimensions of all discontinuities exceeding 1/32" (0.8 mm) but less than or equal to 1/8" (3.2 mm).
- *1/4" (6.4 mm)* – Maximum corner crack, except when the corner crack results from visible slag inclusion or other fusion-type discontinuities, then a 1/8" (3.2 mm) maximum shall apply. A

specimen with corner cracks exceeding ¼" (6.4 mm), with no evidence of slag inclusions or other fusion-type discontinuities may be discarded, and a replacement test specimen from the original weldment shall be tested.

> **NOTE**
> In some cases, a radiographic inspection will be used instead of the guided bend test. This allows the entire weld to be examined and can detect small discontinuities at any location within the weld.

When the welder passes the qualification tests, the test results and the procedures that the welder may weld to are listed on a record that is kept by the company. This record becomes part of the quality documentation, and the welder becomes qualified to weld to that procedure.

7.4.4 Welder Qualification Limits

Welders may retest if they initially fail the test. An immediate retest consists of two test welds of each type of test that was failed. All the test specimens must pass this retest. A complete retest may be made if the welder has had further training or practice since the last test.

> **NOTE**
> Retest requirements may vary depending on site quality standards. Check your site's quality standards for specific retest requirements.

Machines Used in Destructive Testing

A wide variety of specialized machines, such as those shown here, are used in destructive testing.

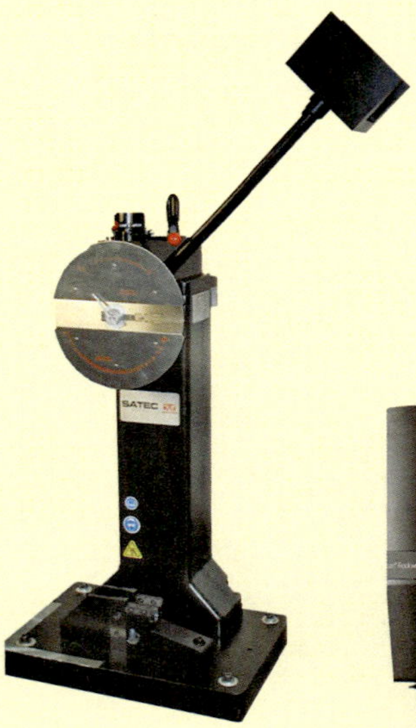

IMPACT TEST MACHINE

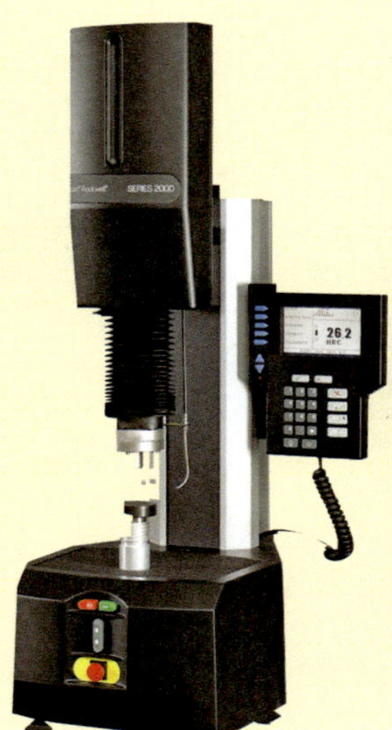

ROCKWELL HARDNESS TESTER

PHOTOS COURTESY OF INSTRON®

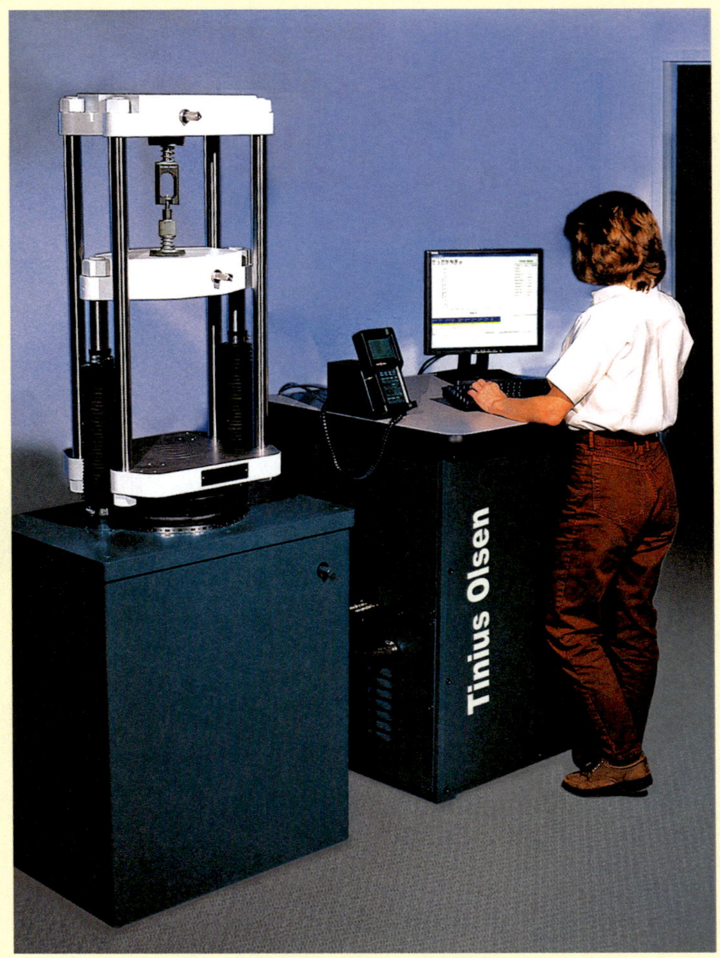

DESTRUCTIVE TESTING MACHINE

106SA02.EPS

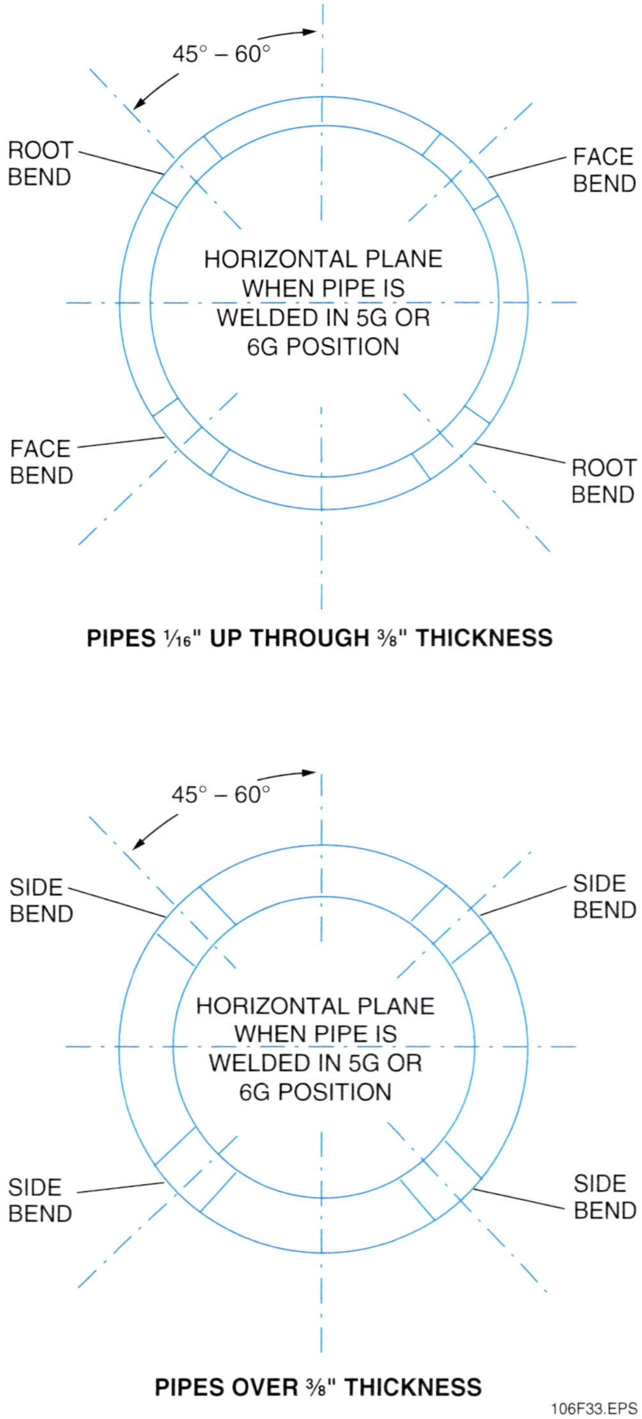

Figure 33 Locations of pipe specimens for 5G and 6G positions.

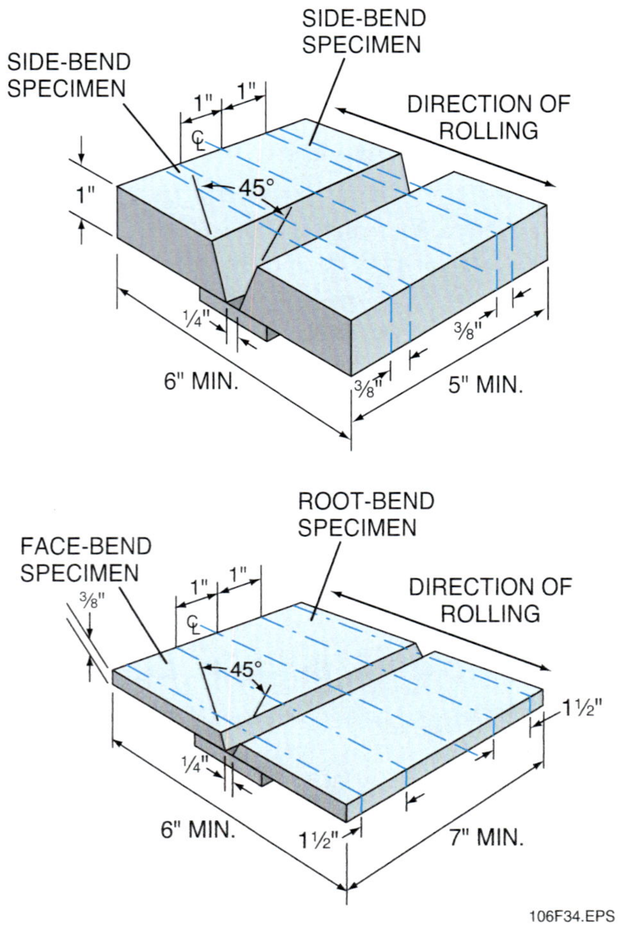

Figure 34 Specimen locations for plate welds.

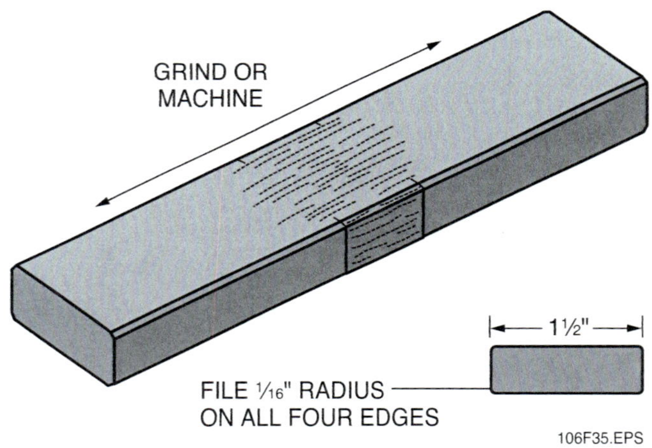

Figure 35 Example of a prepared test specimen.

Jig Plunger and Die

The jig has a plunger and die that are dimensioned for the thickness of the specimen being bent. Refer to the welding code being used for testing for the required dimensions of the bending jig's plunger and die.

After welders have qualified, they may have to requalify if they have not used the specific process for a certain time period. This time period varies depending on the codes. Welders may also be required to requalify if there is a reason to question their ability to make welds that meet the WPS. Also, since welder performance qualification is limited to the essential variables of a particular procedure, any change in one or more of the essential variables requires the welder to requalify with the new procedure.

8.0.0 Quality Workmanship

The codes and standards discussed in this module were written to ensure that welders consistently make quality welds. Although many weldments will be examined and tested, due to time and cost, not all will be. Nevertheless, quality workmanship is expected in every weld a welder makes.

8.1.0 Typical Site Organization

The welder should be able to work with appropriate site representatives to ensure that quality work is achieved. To do this, the site organizational structure needs to be understood. If quality problems arise, the welder needs to follow the appropriate chain of command to eliminate any problems (*Figure 36*).

> **NOTE**
> Your site may differ in its organizational structure. Check with your immediate supervisor to find out the structure for your site.

8.2.0 Chain of Command

A welder should always follow the site chain of command. However, there may be instances when the welder should bypass the chain of command. Examples of such instances include the following:

- You have been directed to perform an unsafe act. If you cannot resolve the matter with your immediate supervisor, it is your responsibility to go to the general foreman, superintendent, project manager, or safety engineer.
- You have been directed to perform a weld that requires certification, and you are not certified to perform it. If you cannot resolve the matter with your immediate supervisor, it is your responsibility to go to the general foreman, superintendent, project manager, or quality engineer.

> **NOTE**
> Always try to resolve problems with your immediate supervisor before bypassing the chain of command.

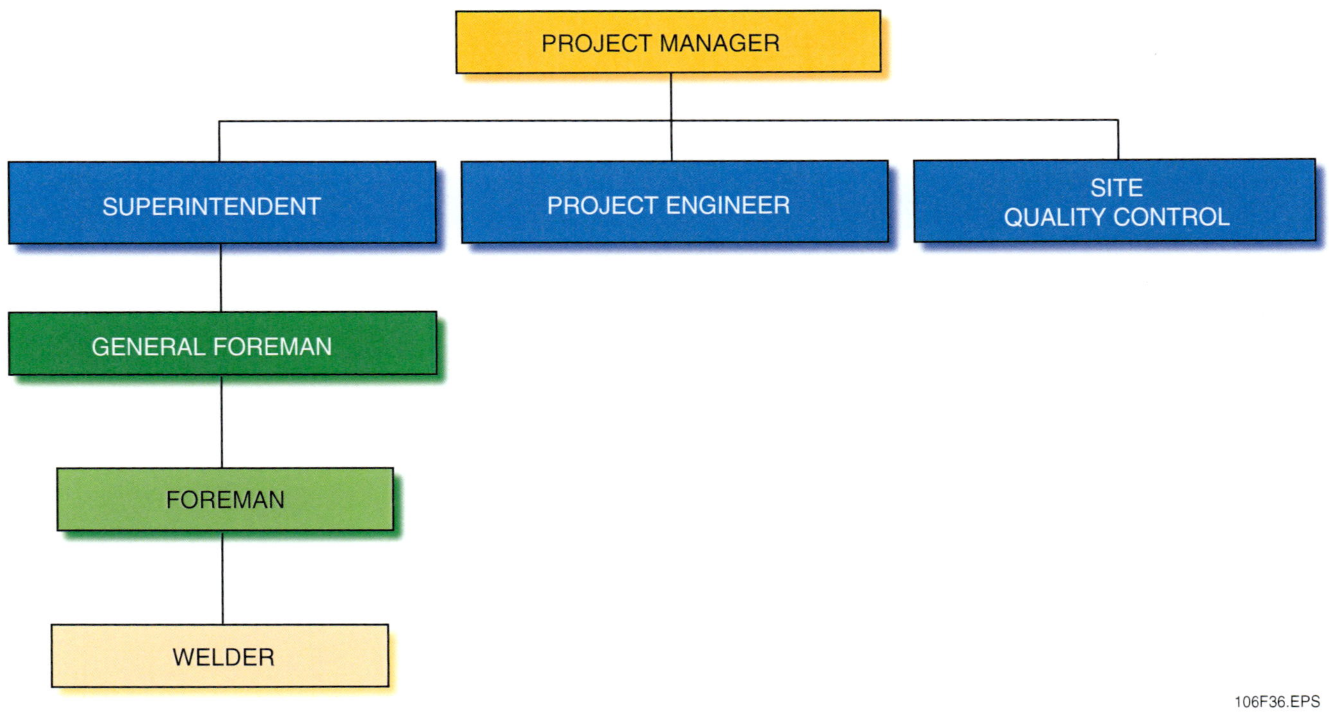

Figure 36 Typical site organizational structure.

SUMMARY

Quality is everyone's responsibility. If the work being done cannot be defined as quality work, it reflects on all those involved in the process. One essential trait of a craftsperson is a sense of quality workmanship. The craftsperson is generally closest to the work and will therefore have a major impact on product quality. Keeping quality in mind as you perform each step of your job will help you identify and correct small problems before they become major ones. This will make everyone's job easier and instill a sense of pride in what has been accomplished.

Safety, quality, and production each has a cost of its own. On the project, each of these factors should have proper guidelines. At the completion of the project, when all records for safety, cost, planning, scheduling, and effectiveness have been evaluated, the papers are usually filed away, but quality remains indefinitely for all eyes to see. Quality is, perhaps, the major reason for repeat business. How well the craftsperson performed will be noticed long after the project has been completed.

Review Questions

1. The specifications for acceptable ferrous (Part A) and nonferrous (Part B) base metals are contained in Section _____ of the *American Society of Mechanical Engineers (ASME) Boiler and Pressure Vessel Code (BPVC)*.
   a. I
   b. II
   c. V
   d. IX

2. The minimum requirements for the design, materials, fabrication, erection, testing, and inspection of various types of piping systems are covered by _____.
   a. *ASME B31*
   b. *AWS D1.1*
   c. *AWS B1.11*
   d. *API Std. 1104*

3. The code that covers the requirements for welding structures or components under the surface of water is _____.
   a. *ASME B31*
   b. *AWS D1.1*
   c. *AWS D3.6M*
   d. *API 1104*

4. The presence of voids or empty spots in the weld metal is called _____.
   a. porosity
   b. inclusion
   c. defect
   d. discontinuity

5. The type of porosity in which the weld exhibits a localized grouping of pores is known as _____.
   a. uniformly scattered porosity
   b. clustered porosity
   c. linear porosity
   d. piping porosity

For Questions 6 through 8, match a picture in *Figure 1* to each discontinuity listed by putting the correct letter in the blank.

6. _____ Linear porosity

7. _____ Surface slag inclusion

8. _____ Toe crack

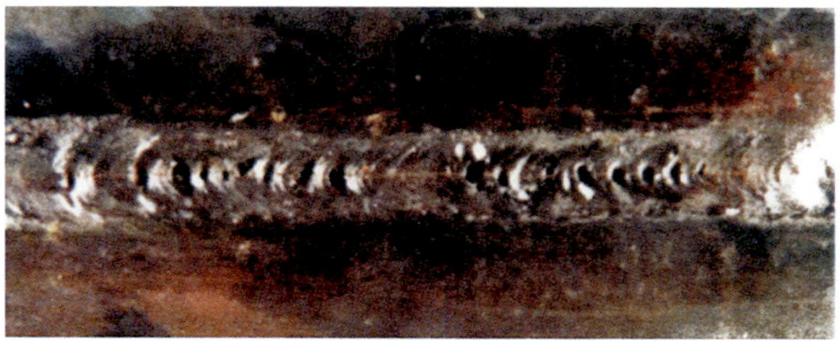

(A)

(B)

(C)

(D)

Figure 1

106RQ01.EPS

9. One method of reducing weld metal cracking is to use _____ electrodes.

   a. fast-fill
   b. fill-freeze
   c. low-hydrogen
   d. fast-freeze

10. The test in which a vacuum box is used is done to check for _____.

   a. inclusions
   b. porosity
   c. leaks
   d. incomplete fusion

11. The test method shown in *Figure 2* is the _____.

   a. nick-break test
   b. fillet weld break test
   c. impact test
   d. hardness test

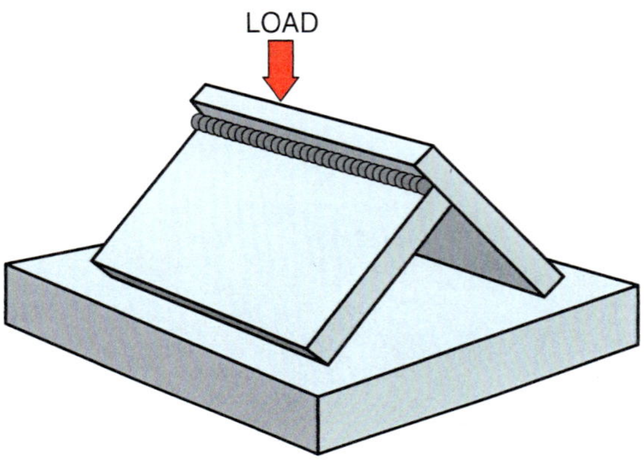

Figure 2

106RQ02.EPS

12. A typical American Welding Society (AWS) welder qualification test is a V-groove weld with metal backing in the 3G and 4G positions using an _____ electrode.

   a. F4
   b. F3
   c. F2
   d. F1

13. An ASME welder qualification test that qualifies a welder to make all types of welds is to weld pipe in the _____ position.

   a. 1F
   b. 6G
   c. 4F
   d. 4G

14. To properly prepare a test specimen, it is important to _____.

   a. water-quench the specimen while it is hot
   b. grind it widthwise to produce a notch effect
   c. remove any face or root reinforcement from the weldment
   d. make sure the edges are sharp

15. If you are directed to perform a weld for which you are not qualified, you should _____.

   a. bypass the chain of command
   b. report directly to the safety engineer as soon as possible
   c. try to resolve the matter with your immediate supervisor
   d. perform the weld to the best of your ability

Trade Terms Quiz

Fill in the blank with the correct key trade term that you learned from your study of this module.

1. _______ is the demonstration that welds made following a specific procedure can meet prescribed standards.
2. A(n) _______ is a discontinuity or imperfection that renders a part of the product or the entire product unable to meet minimum acceptable standards or specifications.
3. _______ is cracking in the base metal near the weld, but under the surface.
4. Metals that have the ability to be made harder by heating and then cooling are called _______.
5. The document containing the results of the nondestructive and destructive testing required to qualify a welding procedure specification (WPS) is called a(n) _______.
6. _______ is a change or break in the shape or structure of a part that may or may not be considered a defect, depending on the code.
7. _______ is the quality or state of having a uniform structure or composition throughout.
8. _______ describes images made by passing X-rays or gamma rays through an object and recording the variations in density on photographic film.
9. Cracks in the base metal formed when layers separate are called _______.
10. _______ metal has been made brittle and will tend to crack with little bending.
11. Foreign matter introduced into and remaining in a weld is called a(n) _______.

Trade Terms

Defect
Discontinuity
Embrittled
Hardenable materials
Homogeneity
Inclusion
Laminations
Procedure qualification
Procedure qualification record (PQR)
Radiographic
Underbead cracking

Rod Hellyer

Industrial Liaison Workforce
Education Department
Lee College Baytown, TX

Rod was already working as a mechanic when two of his younger brothers persuaded him to try a areer in construction. Rod credits much of his success to his parents, who set very high standards for their children. According to Rod, his dad was a hardworking man who put pride and craftsmanship into everything he did. Inspired by his father, Rod adopted these same values and continues to practice them today as he celebrates nearly 25 years as a college welding instructor. Rod's advice to new welders entering the field is this—take pride in yourself and in your work, do everything to the best of your ability, and be a worker who can be trusted to get the job done.

How did you choose a career in the field?
I owe it all to my two younger brothers who were welders and pipefitters. I was working as a mechanic at the time. My brothers convinced me to switch gears and go into construction.

What types of training have you been through?
When I first entered construction, I received some on-the-job training. Then I enrolled in a local college and started my formal training in welding. I took classes in SMAW, GTAW, GMAW, and metallurgy, along with enough academic courses to complete my A.A.S. in Welding Technology.

What kinds of work have you done in your career?
I started out doing structural welding and fabricating. I worked on anything and everything—ship loaders, conveyors, oil storage tanks, towers, pipelines, boilers—you name it. I've performed work in accordance with a wide variety of welding codes.

Tell us about your present job and what you like about it.
I've been a welding instructor at Lee College going on 25 years now. This includes 15 years working at a local high school teaching students interested in welding. I have a deep admiration for welding and cutting, and I try to communicate this to my students. We teach all classes in SMAW, GTAW, and GMAW. I've recently accepted a position as Industrial Liaison in the Workforce Education Department. I'll be in charge of industrial training for our local businesses and contractors. That training will comply with the training and recordkeeping guidelines of the NCCER's Construction & Maintenance Education Foundation (CMEF) program.

What factors have contributed most to your success?
I attribute most of my success to my parents, who had very high moral standards. My dad worked hard all his life. He put a lot of pride and craftsmanship into everything he did, and he instilled the same values in us. Dad believed that anything worth doing was worth doing well. He required that even the least important jobs be done to the best of our ability. Today, I continue to practice those values instilled in me by my father.

What advice would you give to those new to the field?
I would stress the importance of taking pride in yourself and in your work. Do everything to the best of your ability. You will be judged on your work by your peers and your employers. Don't worry about the work of others. Just focus on the job you are doing. Be one of those workers who will be called upon to tackle the jobs that no one else can get done.

Tell us some interesting career-related facts or accomplishments.
Some of the most interesting work I've done was on pipelines inside the plants making tie-ins on live units. I also enjoyed the manufacturing of products. And what an experience it is to see a new plant built and then started up for operation—knowing that you were a part of building something and transforming it from a bare spot of ground to a large industrial unit.

Cornerstone of Craftmanship

John J. Knapp

Industrial Coordinator/Trainer, Welding
Tulsa Technology Center,
Tulsa, OK

Like others before him, John Knapp was influenced by a family member who was in the welding trade. Also, like others before him, he started at the bottom as a helper and worked his way up. John now serves on the welding committee that steers the NCCER Welding curriculum.

How did you choose a career in the field?
During the summers of my junior and senior years of high school, I worked as a welder's helper on a power plant in Nebraska. My father worked in quality control and inspected welding work. I saw the potential that welders had and decided that I wanted to learn the trade. I always wanted to teach, but did not think that welding was going to be my subject matter. Certain people believed that I could teach and it just happened. Most of my career has been in the education of welding.

What types of training have you been through?
As a welder's helper, I was "old-schooled" by the experienced welders that I worked with on that first power plant job. They would show me a trick or two and would let me weld a little on non-critical projects. I have since received more formal training and experience from Southeast Community College, Milford, NE—Welding and Metallurgy degree; Tulsa Welding School, Tulsa, OK—Welding Combo Pipe; Oklahoma State University, Stillwater, OK—Trade and Industrial Education; American Welding Society Certified Inspector, Educator, Welder; National Occupational Competency Testing Institute—Exam Welding; and numerous welding courses from Hobart Institute of Welding Technology in Troy, OH, and from Lincoln Electric Welding in Cleveland, OH. I worked seven years in the field as a structural and pipe welder before getting into a teaching job. If you are in the welding field, you are always learning. As a welding instructor, you must constantly learn as much as you can about the trade.

What kinds of work have you done in your career?
I've been a structural welder on power plant projects, a pipe and tube welder on chemical and power plant projects, a boiler tube welder and quality control inspector in paper mills, and a welding instructor for private and public schools. I have customized and coordinated welding training and testing for manufacturing companies, and served as an AWS welding inspector at the AWS-Accredited Test Facility in Oklahoma.

Tell us about your present job and what you like about it.
Currently, I hold the position of Coordinator of Business and Industry Training. My duties include helping companies custom train new welders or upgrade the skills of their current welders. I manage our AWS-Accredited Test Facility in Tulsa. To help offset the cost of training, I also help companies qualify for local and state funding set aside for those purposes. It is a rewarding challenge to help a company get their procedures, training, and testing up-to-date.

What factors have contributed most to your success?
My parents were my role models. They both worked hard and displayed great leadership skills. Being polite and getting along with people is critical, and you must have passion for your craft. You must be a lifelong learner; most good organizations give time and financial support to help us stay current in our field. Doing your job in a professional manner is very important, along with mentoring those who are newer to the trade.

What advice would you give to those new to the field?

Be willing to learn and apply what you have learned. Have a good attitude, and work will never feel like work. I tell trainees that if they take pride in their work and are up for the tough physical challenges of a career in welding the opportunities are endless.

Tell us some interesting career-related facts or accomplishments:

I have done many things in the field of welding. I have won my share of awards and have had students compete on a national level. It amazes me to have students who could barely strike an arc turn into competent welders. I have been fortunate to have a couple of retired instructors help me along the way—Barry Lawrence, a welding engineer out of Ohio State, and Bill Conrad, a trade and industrial educator out of Oklahoma State. They are both committed and well qualified to serve the trade in all areas from beginning welding to engineering and testing.

Trade Terms Introduced in This Module

Defect: A discontinuity or imperfection that renders a part of the product or the entire product unable to meet minimum acceptable standards or specifications.

Discontinuity: A change or break in the shape or structure of a part that may or may not be considered a defect, depending on the code.

Embrittled: Metal that has been made brittle and that will tend to crack with little bending.

Hardenable materials: Metals that have the ability to be made harder by heating and then cooling.

Homogeneity: The quality or state of having a uniform structure or composition throughout.

Inclusion: Foreign matter introduced into and remaining in a weld.

Laminations: Cracks in the base metal formed when layers separate.

Procedure qualification: The demonstration that welds made following a specific process can meet prescribed standards.

Procedure qualification record (PQR): The document containing the results of the nondestructive and destructive testing required to qualify a welding procedure specification (WPS).

Radiographic: Describes images made by passing X-rays or gamma rays through an object and recording the variations in density on photographic film.

Underbead cracking: Cracking in the base metal near the weld, but under the surface.

Additional Resources

This module is intended to present thorough resources for task training. The following reference works are suggested for further study. These are optional materials for continued education rather than for task training.

AWS B1.10: Guide for the Nondestructive Inspection of Welds, 1999. Miami, FL: The American Welding Society.

AWS B1.11: Guide for the Visual Inspection of Welds, 2000. Miami, FL: The American Welding Society.

OSHA Standard 1926.351, Arc Welding and Cutting.

Welding Handbook, Volume 1 (2001) and Volume 2 (2004). Miami, FL: The American Welding Society (AWS)

AWS D3.5-93R: Guide for Steel Hull Welding, 1993. Miami, FL: The American Welding Society (AWS)

AWS D3.6M: Specification for Underwater Welding, 1999. Miami, FL: The American Welding Society (AWS)

AWS D3.7: Guide for Aluminum Hull Welding, 2004. Miami, FL: The American Welding Society (AWS)

The Procedure Handbook of Arc Welding, 2000. Cleveland, OH: The Lincoln Electric Company. **www.lincolnelectric.com**

Figure Credits

Reprinted from ASME BPVC Edition, Section IX by permission of The American Society of Mechanical Engineers. All rights reserved, 106A01A, 106A01B

American Welding Society, F106F01

Figure 2 courtesy of American Welding Society *AWS D1.1/D1.1M:2008*, Procedure Qualification Record #231, Reproduced with permission from the American Welding Society (AWS), Miami, FL USA.

Figure 3 courtesy of American Welding Society *AWS B1.11-2000*, Figures 1, 2, and 3, Reproduced with permission from the American Welding Society (AWS), Miami, FL USA.

Figure 4 courtesy of American Welding Society *AWS B1.11-2000*, Figure 30, Reproduced with permission from the American Welding Society (AWS), Miami, FL USA.

Figure 5 courtesy of American Welding Society *AWS B1.11-2000*, Figure 20, Reproduced with permission from the American Welding Society (AWS), Miami, FL USA.

Figure 6 courtesy of American Welding Society *AWS B1.11-2000*, Figures 3, 24, 25, 26, and 28, Reproduced with permission from the American Welding Society (AWS), Miami, FL USA.

Figure 9 courtesy of American Welding Society *AWS B1.11-2000*, Figures 7 and 8, Reproduced with permission from the American Welding Society (AWS), Miami, FL USA.

Figure 10 courtesy of American Welding Society *AWS B1.11-2000*, Figure 18, Reproduced with permission from the American Welding Society (AWS), Miami, FL USA.

Figure 25 courtesy of American Welding Society *Certification Manual for Welding Inspectors*, Figure 12.11, Reproduced with permission from the American Welding Society,

Miami, FL, Eighth Edition.

Topaz Publications, Inc., 106F07, 106SA01, 106F18, 106F19 (photo)–106F21

G.A.L. Gage Co., 106F15, 106F16

Olympus NDT, 106F22

Photo courtesy of Instron®, 106SA02 (left)

J. R. Yochum, 106F30

PROCEDURE QUALIFICATION RECORD

QW-483 SUGGESTED FORMAT FOR PROCEDURE QUALIFICATION RECORD (pqr)
(See QW-200.2, Section IX, ASME Boiler and Pressure Vessel Code)
Record Actual Conditions Used to Weld Test Coupon.

Company Name ______________________________

Procedure Qualification Record No. ______________ Date ______________

WPS No. ______________________________

Welding Process(es) ______________________________

Types (Manual, Automatic, Semi-Auto.) ______________________________

JOINTS (QW-402)

Groove Design of Test Coupon

(For combination qualifications, the deposited weld metal thickness shall be recorded for each filler metal or process used.)

BASE METALS (QW-403)

Material Spec. ______________

Type or Grade ______________

P-No. ______________ to P-No. ______________

Thickness of Test Coupon ______________

Diameter of Test Coupon ______________

Other ______________

POSTWELD HEAT TREATMENT (QW-407)

Temperature ______________

Time ______________

Other ______________

GAS (QW-408)

	Gas(es)	Percent Composition (Mixture)	Flow Rate
Shielding			
Trailing			
Backing			

FILLER METALS (QW-404)

SFA Specification ______________

AWS Classification ______________

Filler Metal F-No. ______________

Weld Metal Analysis A-No. ______________

Size of Filler Metal ______________

Other ______________

Weld Metal Thickness ______________

ELECTRICAL CHARACTERISTICS (QW-409)

Current ______________

Polarity ______________

Amps ______________ Volts ______________

Tungsten Electrode Size ______________

Other ______________

POSITION (QW-405)

Position of Groove ______________

Weld Progression (Uphill, Downhill) ______________

Other ______________

TECHNIQUE (QW-410)

Travel speed ______________

String or Weave Bead ______________

Oscillation ______________

Multipass or Single Pass (per side) ______________

Single or Multiple Electrodes ______________

Other ______________

PREHEAT (QW-406)

Preheat Temp. ______________

Interpass Temp. ______________

Other ______________

This form (E00007) may be obtained from the Order Dept., ASME, 22 Law Drive, Box 2300, Fairfield, NJ 07007-2300

(Source ASME B31.1, SEC IX)

106A01A.EPS

QW-483 (Back)

Tensile Test (QW-150) **PQR No.** ________________

Specification No.	Width	Thickness	Area	Ultimate Total Load lb	Ultimate Unit Stress psi	Type of Failure & Location

Guided-Bend Tests (QW-160)

Type and Figure No.	

Toughness Tests (QW-170)

Specimen No.	Notch Location	Notch Type	Test Temp.	Impact Values	Lateral Exp.		Drop Weight	
					% Shear	Mils	Break	No Break

Fillet-Weld Test (QW-180)

Result - Satisfactory: Yes ________ No ________ Penetration into Parent Metal: Yes ________ No ________

Macro - Results ________________

Other Tests

Type of Test ________________

Deposit Analysis ________________

Other ________________

- -

Welder's Name ________________ Clock No. ________ Stamp No. ________

Tests Conducted by: ________________ Laboratory Test No. ________

We certify that the statements in this record are correct and that the test welds were prepared, welded, and tested in accordance with the requirements of Section IX of the ASME Code.

Manufacturer ________________

Date ________________ By ________________

(Detail of record of tests are illustrative only and may be modified to conform to the type and number of tests required by the Code.)

(Source ASME B31.1, SEC IX)

106A01B.EPS

NCCER CURRICULA — USER UPDATE

NCCER makes every effort to keep its textbooks up-to-date and free of technical errors. We appreciate your help in this process. If you find an error, a typographical mistake, or an inaccuracy in NCCER's curricula, please fill out this form (or a photocopy), or complete the online form at **www.nccer.org/olf**. Be sure to include the exact module ID number, page number, a detailed description, and your recommended correction. Your input will be brought to the attention of the Authoring Team. Thank you for your assistance.

Instructors – If you have an idea for improving this textbook, or have found that additional materials were necessary to teach this module effectively, please let us know so that we may present your suggestions to the Authoring Team.

NCCER Product Development and Revision

13614 Progress Blvd., Alachua, FL 32615

Email: curriculum@nccer.org

Online: www.nccer.org/olf

❑ Trainee Guide ❑ Lesson Plans ❑ Exam ❑ PowerPoints Other ________________________

Craft / Level: Copyright Date:

Module ID Number / Title:

Section Number(s):

Description:

Recommended Correction:

Your Name:

Address:

Email: Phone:

29108-09

Shielded Metal Arc Welding – Electrodes

Module Five

Trainees with successful module completions may be eligible for credentialing through NCCER's National Registry. To learn more, go to **www.nccer.org** or contact us at **1.888.622.3720**. Our website has information on the latest product releases and training, as well as online versions of our *Cornerstone* magazine and Pearson's product catalog.

Your feedback is welcome. You may email your comments to **curriculum@nccer.org**, send general comments and inquiries to **info@nccer.org**, or fill in the User Update form at the back of this module.

This information is general in nature and intended for training purposes only. Actual performance of activities described in this manual requires compliance with all applicable operating, service, maintenance, and safety procedures under the direction of qualified personnel. References in this manual to patented or proprietary devices do not constitute a recommendation of their use.

29108-09

Shielded Metal Arc Welding – Electrodes

Objectives

When you have completed this module, you will be able to do the following:

1. Identify factors that affect electrode selection.
2. Explain the American Welding Society (AWS) and the American Society of Mechanical Engineers (ASME) filler metal classification system.
3. Identify different types of filler metals.
4. Explain the storage and control of filler metals.
5. Explain filler metal traceability requirements and how to use applicable code requirements.
6. Identify and select the proper electrode for a specified welding task.

Performance Tasks

This is a knowledge-based module. There are no Performance Tasks.

Trade Terms

Alloy
Condensation
Ductile
Flux
Heat-affected zone
Hermetically sealed
Low-hydrogen electrode
Notch toughness
Traceability
Vertical welding

Prerequisites

Before you begin this module, it is recommended that you successfully complete the following: *Core Curriculum* and *Welding Level One*, Modules 29101-09 through 29107-09.

Industry-Recognized Credentials

If you're training through an NCCER-accredited sponsor, you may be eligible for credentials from NCCER's Registry. The ID number for this module is 29108-09. Note that this module may have been used in other NCCER curricula and may apply to other level completions. Contact NCCER's Registry at 888.622.3720 or go to **www.nccer.org** for more information.

Contents

Topics to be presented in this module include:

Figures and Tables

1.0.0 Introduction

Shielded metal arc welding cannot take place without electrodes. SMAW has often been referred to as stick welding, and the electrodes have been referred to as stick electrodes. The welder must be able to distinguish different types of electrodes and select the correct electrode for the job. In addition, the welder must know how to handle and store electrodes, some of which require special considerations.

This module explains the electrode classification system, electrodes, and the considerations for selecting the correct electrode for the job. It also explains how to properly store electrodes in compliance with the appropriate welding codes.

2.0.0 Shielded Metal Arc Welding Electrodes

A shielded metal arc welding electrode has a wire core coated with **flux.** The wire core transfers the welding current from the electrode holder to the work. The arc at the end of the electrode melts the metal core, the flux coating, and the base metal at a temperature exceeding 6,000°F. The wire core mixes with the melted base metal, forming the weld. As the flux coating is melted, it does the following:

- Gives off a gaseous shield of carbon dioxide that protects the molten weld by pushing away harmful oxygen and hydrogen present in the atmosphere
- Acts as a fluxing agent to clean and deoxidize the molten metal
- Stabilizes the arc and reduces spatter
- Forms a slag covering over the weld to further protect it from the atmosphere and slow the cooling rate

Figure 1 shows an electrode in use.

Some flux coatings also have powdered metal and alloying elements in them. The powdered metal adds filler metal and helps improve the appearance of the weld. The alloying elements change the chemical composition and strength of the deposited weld metal. **Alloys** are normally added to electrodes so the deposited weld metal will match the base metal chemical composition and strength.

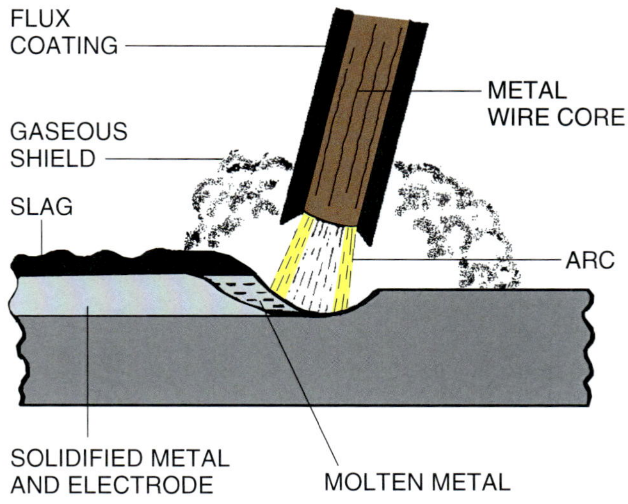

Figure 1 Electrode in use.

3.0.0 AWS Filler Metal Specification System

The American Welding Society writes specifications for welding consumables, such as electrodes, filler rods, and fluxes, used with all the various welding processes. There are a series of different specifications identified as AWS A5.XX. These specifications are used in all major welding codes and accepted in all industries except where other approvals are specified, such as the American Bureau of Shipping (ABS), the U.S. Coast Guard, the Federal Highway Administration, Military Specifications (MIL), and Lloyds of London. The American Society of Mechanical Engineers Boiler Code electrode specifications closely follow the AWS specifications.

The purpose of the specifications is to set standards that all manufacturers must follow when manufacturing welding consumables. This ensures consistency for the user regardless of which company manufactured the product. The specifications set standards for the following:

- Classification system, identification, and marking
- Chemical composition of the deposited weld metal
- Mechanical properties of the deposited weld metal

The following are some examples of AWS specifications that pertain to shielded metal arc welding electrodes:

- *A5.1 – Specification for Carbon Steel Electrodes for Shielded Metal Arc Welding*
- *A5.3 – Specification for Aluminum and Aluminum Alloy Electrodes for Shielded Metal Arc Welding*
- *A5.4 – Specification for Stainless Steel Electrodes for Shielded Metal Arc Welding*
- *A5.5 – Specification for Low-Alloy Steel Electrodes for Shielded Metal Arc Welding*
- *A5.6 – Specification for Covered Copper and Copper Alloy Arc Welding Electrodes*
- *A5.11 – Specification for Nickel and Nickel-Alloy Welding Electrodes for Shielded Metal Arc Welding*
- *A5.15 – Specification for Welding Electrodes and Rods for Cast Iron*
- *A5.16 – Specification for Titanium and Titanium Alloy Welding Electrodes and Rods*
- *A5.21 – Specification for Bare Electrodes and Rods for Surfacing*
- *A5.24 – Specification for Zirconium and Zirconium Alloy Welding Electrodes and Rods*

The most common electrodes are discussed in the *A5.1* and *A5.5* specifications. The electrodes described in the *A5.1* specification are used to weld carbon (mild) steel. The electrodes in the *A5.5* specification are used to weld low-alloy, high-strength steels. These two specifications will be explained later in this module.

NOTE

The specification number is generally followed by the year of publication, such as *A5.1:2004* or *A5.5:2006*. The year of publication is the year that the specification was last revised. These documents are normally on a five-year revision cycle. Always use the latest version.

3.1.0 Classification System

The classification system used to identify SMAW A5.1 carbon steel electrodes and A5.5 low-alloy electrodes uses the prefix E followed by four or five numbers, such as E6010 or E10018. This is printed on each electrode near the end where the electrode holder grasps the electrode. *Figure 2* shows electrode markings.

The meanings of the markings are as follows (refer to *Figure 3*):

- *E* stands for electrode to signify use with electric welding.
- The first two digits, or three if 100 or over, multiplied by 1,000 designate the minimum tensile strength in pounds per square inch of the deposited weld metal:
  - E6010 has a minimum tensile strength of 60,000 psi.
  - E7018 has a minimum tensile strength of 70,000 psi.
  - E12018 has a minimum tensile strength of 120,000 psi.

Sizing and Packaging of SMAW Electrodes

SMAW electrodes are sized according to the wire core diameter, and are commonly packed in 10- and 50-pound cans, as shown in the photo.

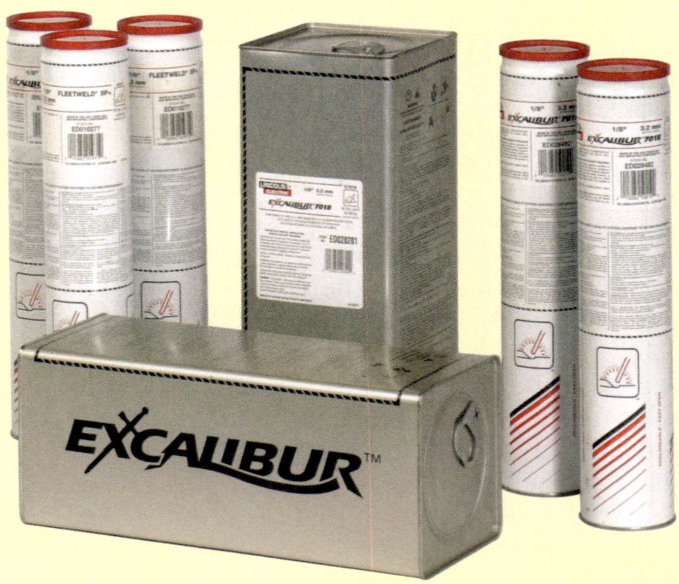

108SA01.EPS

108F02.EPS

Figure 2 Electrode markings.

- The third or fourth digit indicates the possible welding positions:
  - *1* can be used to weld in all positions.
  - *2* can be used to weld in flat and horizontal fillet weld positions only.
  - *4* can be used to weld in flat, horizontal, overhead, and vertical positions.

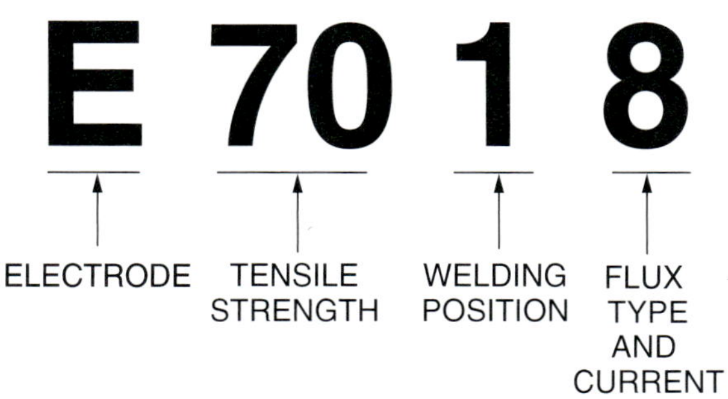

108F03.EPS

Figure 3 Electrode marking system.

- The fourth or fifth digit indicates special characteristics of the flux coating and the type of welding current required, such as alternating current, direct current electrode negative (DCEN), or direct current electrode positive (DCEP).

Table 1 shows the characteristics of mild steel electrodes covered by the *A5.1* specification.

The low-alloy steel electrodes that are covered by *A5.5* have a suffix in addition to the standard classification numbers. The suffix has a letter and a number. The letter indicates the chemical composition of the deposited weld metal, and the number indicates the composition of the chemical classification. *Figure 4* shows an electrode classification marking from the *A5.5* specification.

The suffix letters indicate the following:

- *A* – carbon-molybdenum alloy steel
- *B* – chromium-molybdenum alloy steel
- *C* – nickel steel alloy

Table 1 Mild Steel Electrode Characteristics

AWS Class	Flux Coating	Current Requirements	Position	Characteristics
EXX 10 EXX 20	CELLULOSE SODIUM	DCEP	All FLAT, HOR. FILLET	DEEP PENETRATION, FLAT BEADS
EXX 11	CELLULOSE POTASSIUM	AC, DCEP	All	DEEP PENETRATION, FLAT BEADS
EXX 12	TITANIA SODIUM	AC, DCEN	All	MEDIUM PENETRATION
EXX 13	TITANIA POTASSIUM	AC, DCEP, DCEN	All	SHALLOW PENETRATION
EXX 14 EXX 24	TITANIA IRON POWDER	AC, DCEP, DCEN	All FLAT, HOR. FILLET	MEDIUM PENETRATION, FAST DEPOSIT
EXX 15	LOW-HYDROGEN SODIUM	DCEP	All	MODERATE PENETRATION
EXX 16	LOW-HYDROGEN POTASSIUM	AC, DCEP	All	MODERATE PENETRATION
EXX 27	IRON POWDER, IRON OXIDE	AC, DCEP, DCEN	FLAT, HOR. FILLET	MEDIUM PENETRATION
EXX 18 EXX 28	IRON POWDER LOW-HYDROGEN	AC, DCEP	All FLAT, HOR. FILLET	SHALLOW TO MEDIUM PENETRATION

108T01.EPS

- *D* – manganese-molybdenum alloy steel
- *G* – other alloys with minimal elements

The meaning of the numbers varies by alloy type, but always indicates the percentages of the following alloying elements:

- Carbon (C)
- Manganese (Mn)
- Phosphorus (P)
- Sulphur (S)
- Silicon (Si)
- Nickel (Ni)
- Chromium (Cr)
- Molybdenum (Mo)
- Vanadium (V)

Figure 5 shows the electrode marking system for low-alloy, high-strength electrodes.

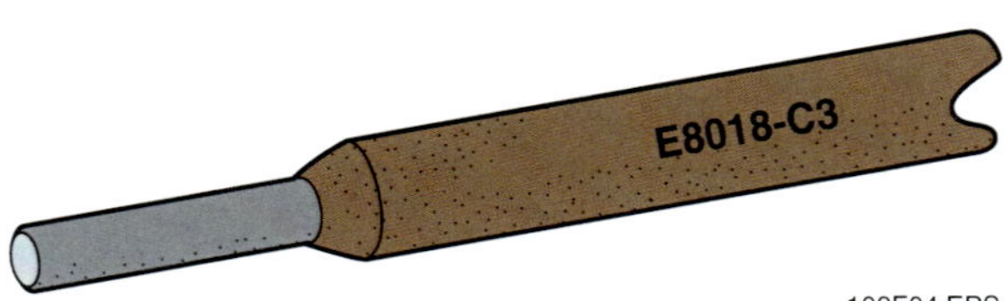

108F04.EPS

Figure 4 Marking on a low-alloy, high-strength electrode.

CAUTION

Be very careful when selecting electrodes. For example, a 7018-B2 cannot be used in place of a 7018 electrode because the additional alloy will cause a significant metallurgical change in the weld.

Figure 6 shows electrodes listed by alloy content.

3.2.0 Manufacturers' Classification

All electrodes used for code work must have the AWS classification number printed on them as well as on the container in which they are shipped. In addition, electrode manufacturers generally print their own unique classification name or number on the electrode and/or on the container in which the electrodes are shipped. Manufacturers may also make more than one electrode within the same AWS classification. For example, Lincoln Electric manufactures two electrodes with the AWS classification E6010. One is called Fleetweld 5P, and the other is called Fleetweld 5P+.

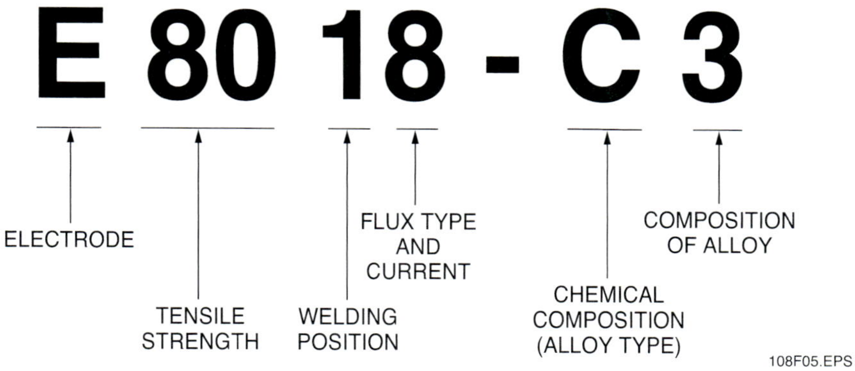

Figure 5 Marking system for low-alloy, high strength electrodes.

Lincoln Electric recommends Fleetweld 5P for general fabrication and maintenance welding and states that it is good for pipe. However, Lincoln Electric states that Fleetweld 5P+ is the best choice for pipe. The major difference between the two electrodes is the mechanical properties of the deposited weld metal. Even though both fall into the E6010 classification, the 5P+ electrode has improved characteristics to decrease cracking possibilities in **heat-affected zones.** It is also better suited for open root joints. When 5P+ is used for the root pass, its greater elongation results in improved resistance to heat-affected zone cracking. Also, the 5P+ has higher tensile and yield strength.

When there is more than one electrode in an AWS classification and you are unsure about which electrode to use, check with your supervisor.

Table 2 shows some examples of manufacturers' and AWS classifications for common mild-steel electrodes.

CARBON-MOLYBDENUM STEEL

E7010-A1
E7011-A1
E7015-A1
E7016-A1
E7018-A1
E7020-A1
E7027-A1

NICKEL STEEL

E8016-C1
E8018-C1
E8016-C2
E8018-C2
E8016-C3
E8018-C3

MANGANESE-MOLYBDENUM STEEL

E9015-D1
E9018-D1
E10015-D2
E10016-D2
E10018-D2

CHROMIUM-MOLYBDENUM STEEL

E8016-B1
E8018-B1
E8015-B2L
E8016-B2
E8018-B2
E8018-B2L
E9015-B3L
E9015-B3
E9016-B3
E9018-B3
E9018-B3L
E8015-B4L
E8016-B5

OTHER LOW-ALLOY STEEL

EXX10-G
EXX11-G
EXX13-G
EXX15-G
EXX16-G
EXX18-G
E7020-G
E9018-M
E10018-M
E11018-M
E12018-M

108F06.EPS

Figure 6 Electrodes listed by alloy.

3.3.0 Electrode Sizes

Mild steel electrodes for SMAW are available in standard sizes, usually ranging from 3⁄32" to 1⁄4". They are typically packaged in 10- and 50-pound cans, but also come in smaller containers, as shown in *Figure 7*. The amperage application for which the electrodes are used is proportional to the diameter size of the electrode. The number of electrodes in a 50-pound can varies with the diameter and length of the electrodes. The length of the electrode may vary from manufacturer to manufacturer. *Table 3* is taken from the McKay® electrode catalog, and lists some mild steel electrodes and the number of electrodes in one pound. Lincoln Electric electrodes and electrodes from other manufacturers will be of very similar weights.

Think About It

Determining Tensile Strength

What factor determines the selection of an electrode with the proper tensile strength? How is this tensile strength determined?

Table 2 Manufacturers' and AWS Classifications for Common Mild-Steel Electrodes

	AWS Classification				
Manufacturer	**E6010**	**E6011**	**E6013**	**E7014**	**E7018**
Hobart	Pipemaster 60	335A, 335C	447A, 447C	14A	718, 718C
Airco	Pipecraft	6011, 6011C	6013, 6013D	7014	Easy Arc 7018MR
ESAB	AP100, SW-610, SW-10P	SW-14	SW-15	SW-15P	Atom Arc 7018
Lincoln	Fleetwood 5P, 5P+	Fleetwood 35, 35LS, 180	Fleetwood 37, 57	Fleetwood 47	Jetweld LH-70
McKay	6010M	6011	6013	7014	–

108T02.EPS

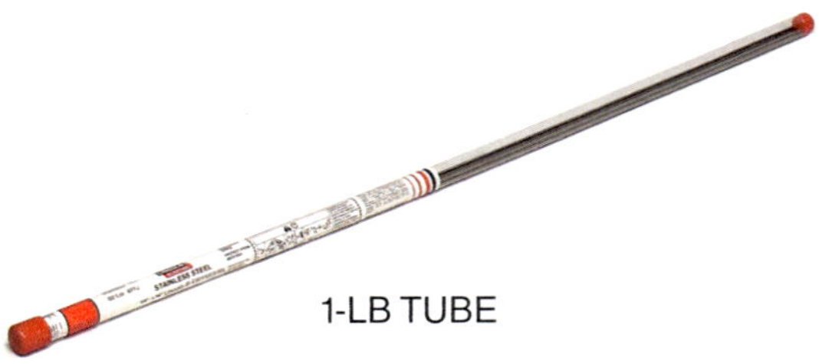

1-LB TUBE

50-LB BOX

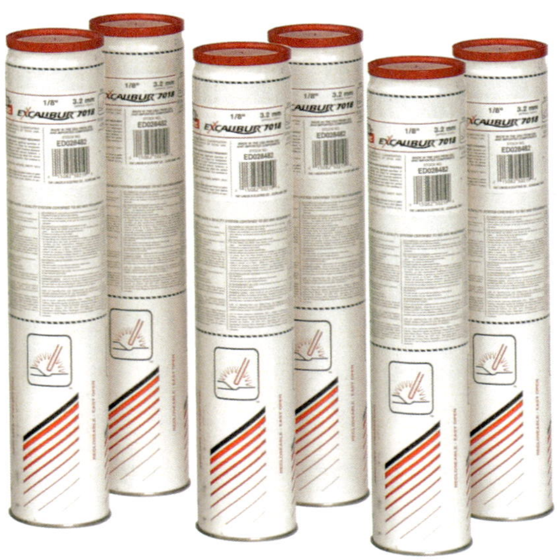

10-LB EASY-OPEN CANS

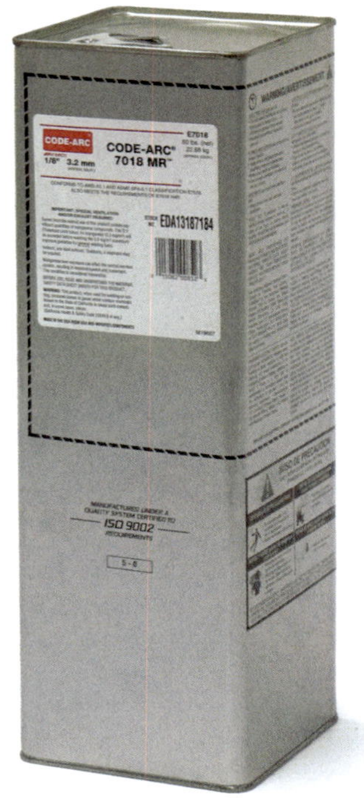

50-LB CAN

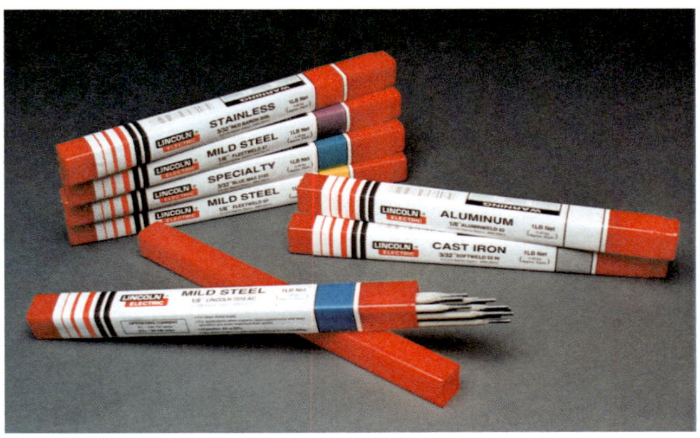

1-LB BOX TUBES

108F07.EPS

Figure 7 Electrode containers.

Table 3 Number of Electrodes per Pound

McKay Type	Diameter: Length:	3/32" (2.4mm) 14"	1/8" (3.2mm) 14"	5/32" (4.0mm) 14"	3/16" (4.8mm) 14"	7/32" (5.6mm) 18"	1/4" (6.4mm) 18"
6010 PM		30	17	12	8	—	—
6011, Soft-Arc 6011		25	15	11	7	—	—
6013		25	15	10	7	—	—
7014		24	13	9	6	—	—
7024		—	10	7	—	4	2
7018 XLM, Soft-Arc 7018-1		32	15	10	7	—	3
7018 AC, Millennia 7018		32	15	10	7	4	3

108T03.EPS

4.0.0 Selecting Electrodes

Because so many different electrodes are available, the welder must be able to distinguish between them and select the correct electrode for the job. The following sections will explain how to select electrodes.

4.1.0 Electrode Groups

Electrodes are often classified into one of four groups based on their general characteristics:

- Fast-freeze
- Fast-fill
- Fill-freeze
- Low-hydrogen

4.1.1 Fast-Freeze Electrodes

Fast-freeze electrodes are all-position electrodes that provide deep penetration. However, the arc has a lot of fine weld spatter. The weld bead is flat with distinct ripples and a light slag coating that can be difficult to remove. The arc is easy to control, making fast-freeze electrodes good for overhead and **vertical welding**. Electrodes in the fast-freeze group include the following:

- E6010
- E7010
- E6011

Typical applications for fast-freeze electrodes include the following:

- Vertical-up and overhead plate welding
- Pipe welding (cross-country and in-plant)
- Welds made on galvanized, plated, or painted surfaces
- Joints requiring deep penetration, such as square butts
- Sheet metal welds

WARNING!

Many surface coatings, such as galvanized coatings, give off very toxic fumes when heated, cut, or welded. Fumes from certain types of electrodes may also be hazardous. Breathing these fumes can cause damage to the lungs and other organs. Identify any hazard for the coatings with which you are working by referring to the MSDS for that particular coating. Also, read the MSDS contained in the electrode package. Use proper ventilation and take the necessary precautions to protect yourself.

4.1.2 Fast-Fill Electrodes

Fast-fill electrodes have powdered iron in the flux, which gives them high deposition rates but requires that they only be used for flat welds and horizontal fillet welds. These electrodes have shallow penetration with excellent weld appearance and almost no spatter. The heavy slag covering is easy to remove. Electrodes in the fast-fill group will also have a 2 as the next-to-last number, and include the following:

- E6027
- E7024
- E7028 (also in the low-hydrogen group)

Typical applications for fast-fill electrodes are production welds on plate more than ¼" thick and flat and horizontal fillets and lap welds.

4.1.3 Fill-Freeze Electrodes

Fill-freeze electrodes, also called fast-follow electrodes, are all-position electrodes that are commonly used for vertical and flat welding. They have medium deposition rates and penetration and are excellent for sheet metal. The weld bead ranges from smooth and ripple-free to even with distinct ripples. The medium slag covering is easy to remove.

Electrodes in the fill-freeze group include the following:

- E6012
- E6013
- E7014

Typical applications for fill-freeze electrodes include the following:

- Vertical fillet or lap welds
- Sheet metal lap and fillet welds
- Joints having poor fit-up in the flat position
- General purpose welding

4.1.4 Low-Hydrogen Electrodes

Low-hydrogen electrodes are designed for welding high-sulphur, phosphorus, and medium- to high-carbon steels, which have a tendency to develop porosity and cracks under the weld bead. These problems are caused by hydrogen that is absorbed during welding. Low-hydrogen electrodes are available with fast-fill and fill-freeze characteristics. They produce dense, X-ray-quality welds with excellent **notch toughness** and **ductility**. Electrodes in the low-hydrogen group include the following:

- E7015
- E7016
- E7018
- E7028
- E7048

Typical applications for low-hydrogen electrodes include the following:

- X-ray quality welds
- Welds requiring high mechanical properties
- Crack-resistant welds in medium- to high-carbon steels
- Welds that resist hot-short cracking in phosphorus steels
- Welds that minimize porosity in sulphur steels
- Welds in thick or highly restrained mild or alloy steels to minimize the danger of weld cracking
- Welds in alloy steels requiring a strength of 70,000 psi or more
- Multiple-pass welds in all positions

If hydrogen is present in the weld zone, it will create defects. The greatest source of hydrogen is water moisture, which is composed of oxygen and hydrogen. Low-hydrogen electrodes are carefully manufactured to eliminate moisture in the flux coating. Low hydrogen is defined as being less than 16 milliliters (ml) per 100 grams of weld metal. The amount of hydrogen in these electrodes varies from 4 to 16 ml per 100 grams of metal weld.

Low-hydrogen electrodes have two unique suffixes. The M suffix is for compliance. Most military codes require an M suffix. When the suffix H and a number are added, the numbers relate to the amount of hydrogen in the electrodes. As a result, the electrodes will have suffixes such as H4, H8, and H16. These three suffixes indicate the dryness of the electrodes. An additional suffix may also be added to the dryness suffixes. A suffix such as H4R would be one such additional suffix. The R in this case indicates that the electrode was designed to meet specific low-moisture pickup limits under controlled humidification tests. *Figure 8* shows a 7018 electrode with the H4R suffix.

If low-hydrogen electrodes are exposed to the atmosphere, they will start to absorb moisture and will no longer be low-hydrogen. For this reason, all low-hydrogen electrodes are shipped in **hermetically sealed** metal containers. Once opened, the electrodes must be stored in a heated oven. *Figure 9* shows low-hydrogen electrodes properly stored. Guidelines for the storage and handling of low-hydrogen electrodes are covered later in this module.

CAUTION

The welding codes establish strict guidelines for the storage, handling, and redrying of low-hydrogen electrodes. These procedures will be referenced in your site quality standards and welding procedure specifications (WPSs). Severe penalties will result from violation of these standards. Improperly storing these electrodes will lead to poor-quality welds. Be sure you read and understand your site standards before working with low-hydrogen electrodes.

4.2.0 Electrode Selection Considerations

Many factors enter into the proper selection of electrodes for the job. The available welding equipment and its current ranges limit the choice of electrodes, and base metal properties play an

108F08.EPS

Figure 8 Electrode with suffix.

108F09.EPS

Figure 9 Low-hydrogen electrode storage.

important role in the selection. Also, the position in which the weld must be made and design factors affect the choice of electrodes for a particular application. Taken from Hobart's welding catalog, *Table 4* indicates position, polarity, and common usages for some of the common steel electrodes, which are listed by AWS classification.

4.2.1 Welding Procedure Specification

If there is a WPS for the weld to be made, the filler metal (electrode) will be specified. It will be given as an AWS classification and/or manufacturer's standard and will sometimes include a manufacturer's name. The filler metal specified in the WPS must be used.

4.2.2 Base Metal Type

The filler metal should match the base metal's chemical composition and mechanical properties. The chemical composition and mechanical properties of the base metal can be obtained by referring to the base metal mill specifications. The mill specifications can then be compared to various tables that list base metal and filler metal acceptability. These tables are available in a number of codes, such as *the AWS Structural Welding Code – Steel D1.1, Table 3.1 Prequalified Base Metal – Filler Metal Combinations for Matching Strength*, and the *ASME Boiler and Pressure Vessel Code, Section II, Parts A, B, and C, Material Specifications.* Tables are also available from the electrode or base metal manufacturers. If the filler metal and base metal are not compatible, the weld produced will be defective.

A weld made on mild steel with a high-tensile electrode such as E12018 will be weaker than one made with E7018. The high-tensile weld will be so strong that when stress is placed on the weld, it will not give, transferring all of the stress to the heat-affected zone along the weld. This will cause cracking. A weld made with E7018 will give along with the base metal, distributing the stress across the weld, making a stronger joint. The opposite occurs when mild steel electrodes are used to weld high-strength steel. All the stress occurs in the weld because the surrounding base metal is so much stronger, causing the weld to fail prematurely.

Table 4 Common Steel Electrode Selection Factors

AWS Class	Position	Polarity	Usage
E6011	All	AC, DCEP	Good for dirty, rusty steel
E6013	All	AC, DCEN, DCEP	Good for thin steel and poor fit-up
E7014	All	AC, DCEN, DCEP	Good for thin steel
E7018	All	AC, DCEP	Good for high deposition; good for medium- or high-carbon steels

DCEN – DC Electrode Negative (Straight Polarity)
DCEP – DC Electrode Positive (Reverse Polarity)

108T04.EPS

4.2.3 Base Metal Thickness

Normally, an electrode that is smaller in diameter than the thickness of the metal to be welded is used. For sheet metal, a 3⁄32" E6013 electrode may be recommended because it has shallow penetration. For very thick sections, always use a low-hydrogen electrode for its **ductile** qualities. To minimize joint preparation, a deep penetration E6010 or E7010 electrode can be used for the root pass.

4.2.4 Base Metal Surface Condition

When possible, the base metal surface should always be cleaned before welding. Surface corrosion, plating, coatings, and paint should be removed. If this is not possible, use a deep penetrating electrode, such as an E6010.

> **WARNING!**
>
> Many surface coatings, such as galvanized coatings, give off very toxic fumes when heated, cut, or welded. Breathing these fumes can cause damage to the lungs and other organs. Identify any hazard for the coatings with which you are working by referring to the MSDS for that particular coating. Use proper ventilation and take the necessary precautions to protect yourself.

4.2.5 Welding Position

The position of welding is important because some electrodes, such as E7024 and E7028, can only be used in the flat and horizontal fillet weld positions. In addition, welding in vertical and overhead positions should never be attempted with electrodes larger than 3⁄16", except for low-hydrogen electrodes. The limitation for low-hydrogen electrodes and EXX14 electrodes is 5⁄32". As a general rule, the joint should be positioned for flat welding so you can use the largest electrode possible.

4.2.6 Joint Design

The type of joint to be welded is an important consideration when selecting an electrode. When T- and lap joints can be positioned flat, they are usually welded with E7024 or E7028 electrodes. Butt joints that need deep penetration should be welded with E6010 electrodes. If the joint has poor fit-up, use a shallow penetrating electrode such as an E6012 or E6013. If low-hydrogen electrodes are used for open root joints, back gouging may be required. To eliminate the need for back gouging and to make the root pass easier to run, use metal backing or put the root in with an E6010 electrode and then fill out with a low-hydrogen electrode.

4.2.7 Welding Current

The most common type of welding current is DC. Because all electrodes can be run on DC, there is no problem when DC current is available. Welding machines that can produce only AC current cannot use E6010 or E7010 electrodes. Although most other electrodes will run on AC, they will run better using DC current.

The amount of current produced by the welding machine is also a consideration. Running large diameter electrodes that require high amperages near the maximum output of the welding machine will cause the welding machine to overheat and cut out. Check the welding machine current type, rated output, and duty cycle before selecting the electrode size.

5.0.0 Filler Metal Storage and Control

Electrodes used for shielded metal arc welding must be dry. The dry mineral flux coatings on electrodes start to absorb moisture as soon as they are exposed to the atmosphere. In as little as one-half hour, some electrodes can pick up enough moisture from the atmosphere to cause weld defects. Some of these defects, such as underbead cracking, may not be visible to the naked eye.

Low-hydrogen electrodes pick up moisture faster than any other type of electrode. Also, because of the nature of the base metal welded with low-hydrogen electrodes, a smaller amount of moisture will cause defects. For these reasons, special care must be used when handling and storing low-hydrogen electrodes.

> **Material Safety Data Sheet for Electrodes**
>
> Every box or can of welding electrodes must contain a copy of the MSDS for that product. The complete MSDS for a Fleetweld 5P electrode is shown in the *Appendix*.

5.1.0 Code Requirements

All welding codes require electrodes with a low-hydrogen coating to receive special handling. They must be shipped in hermetically sealed containers. Immediately after opening the sealed containers, low-hydrogen electrodes must be stored in special ovens at a temperature of 250°F (120°C). When low-hydrogen electrodes are removed from the oven, they can be exposed to the atmosphere only for a designated period of time. This time period varies with the classification of the electrode, but generally ranges from one-half hour to ten hours. *AWS D.1.1, Table 5.1 Allowable Atmospheric Exposure of Low-Hydrogen Electrodes* gives allowable exposure times (see *Table 5*). The electrodes must then be returned to the oven and generally require redrying at higher temperatures. The exact procedures for handling filler metal at your site are typically written out in a quality standard. These procedures must be followed. If you are unsure of the requirements at your site, check with your supervisor.

CAUTION

The exposure times shown are general guidelines only. Excessive exposure times will cause weld defects. Check your site quality standards for the actual exposure times for your location.

CAUTION

The consequences of not following your site's standards on electrode handling and storage may be severe. Be sure you understand your site's standards before handling electrodes. Improper handling of electrodes can cause weld defects. The information provided in this module is of a general nature. Your site's particular quality standards must be followed.

5.2.0 Receiving Filler Metal

Inspect all filler metal as it is received. All low-hydrogen electrodes (EXX15, 16, or 18), stainless steel electrodes, and nickel alloy coated electrodes must be supplied in hermetically sealed containers. If the seal has been broken, reject the electrodes. Mark them "Not to Be Used" and notify your supervisor.

All electrodes other than low-hydrogen electrodes must be shipped in moisture-resistant containers. Inspect these containers for damage. The electrodes can be used if their coatings are not damaged and they have not come into direct contact with water or oil. Mark electrodes that cannot be used and notify your supervisor.

Table 5 Allowable Atmospheric Exposure of Low-Hydrogen Electrodes

Electrode	Column A (hours)	Column B (hours)
A5.1		
E70XX	4 max	Over 4 to 10 max
E70XXR	9 max	
E70XXHXR	9 max	
E7018M	9 max	
A5.5		
E70XX-X	4 max	Over 4 to 10 max
E80XX-X	2 max	Over 2 to 10 max
E90XX-X	1 max	Over 1 to 5 max
E100XX-X	½ max	Over ½ to 4 max
E110XX-X	½ max	Over ½ to 4 max

Notes:

1. Column A: Electrodes exposed to atmosphere for longer periods than shown shall be redried before use.
2. Column B: Electrodes exposed to atmosphere for longer periods than those established by testing shall be redried before use.
3. Electrodes shall be issued and held in quivers, or other small open containers. Heated containers are not mandatory.
4. The optional supplemental designator, R, designates a low-hydrogen electrode which has been tested for covering moisture content after exposure to a moist environment for 9 hours and has met the maximum level allowed in *AWS A5.1, Specification for Carbon Steel Electrodes for Shielded Metal Arc Welding.*

108T05.EPS

Each container must be marked with at least the following information:

- ASME, AWS, or other electrode specification number and electrode classification designation
- Electrode size
- Weight

If the containers are not properly marked, mark them "Not to Be Used" and notify your supervisor.

5.3.0 Storing Filler Metal

All filler metal must be stored in a warm (40°F minimum), dry storage area. Filler metals are not to be stored directly on the floor. Electrodes that are stored on the floor will form **condensation.** Use pallets or shelves to store electrodes off the floor. They must also be stored in such a manner that they will not be damaged. Once low-hydrogen, stainless steel, or nickel alloy electrode containers have been opened, the electrodes must immediately be stored in an oven.

5.4.0 Storage Ovens

Electrode storage ovens are typically electrical and are controlled by a thermostat. Storage ovens for large shops can hold several hundred pounds of electrodes and can reach maximum temperatures of more than 800°F. This maximum temperature is too high for most electrodes. Always follow the electrode manufacturer's guidelines for storing electrodes in ovens. Smaller portable ovens that are designed to be used on the job may hold only 25 or 30 pounds of electrodes. Depending on the WPS or job quality standards, portable ovens may be required to hold electrodes at the point of use. The electrodes remaining in the portable ovens after performing the task must be returned to the holding oven. *Figure 10* shows examples of electrode storage ovens.

Storage ovens should be set between 250°F and 300°F. The oven must remain on at all times. Never unplug an oven or plug it into an outlet that could accidentally be shut off.

5.4.1 Exposure Times

The time that a low-hydrogen, stainless steel, or nickel alloy electrode can be exposed to the atmosphere is limited and depends on the electrode. It is the welder's responsibility to take out of the oven only the number of electrodes that can be used before the maximum exposure time is reached. The exposure time is defined as the period of time that the electrode is not in a heated holding or portable oven.

Samples of electrodes being used at a site are often collected by the quality control group for moisSamples of electrodes being used at a site are often collected by the quality control group for moisture testing. Depending on the results of this testing, the maximum exposure times can be adjusted. In addition, if high moisture content is found in the electrodes, the welds made with the electrodes may have to be replaced. If high-moisture electrodes were used to make welds on critical equipment, the welds will have to be replaced.

Figure 10 Electrode storage ovens.

Manufacturers' Welding Guides

Most electrode manufacturers publish paperback welding guides that include information on their electrodes, such as recommended amperage settings and uses. These booklets can be obtained through your local distributor for free or at a nominal charge.

Depending on the site quality standards, electrodes that exceed the maximum exposure time must either be destroyed or dried. Check your site quality standards to see if you can dry the electrodes or if they must be destroyed. Low-hydrogen electrodes that are exposed to water must be destroyed.

5.4.2 Drying Electrodes

Electrodes that exceed the maximum exposure time can be dried once. They are dried by placing them in a holding oven at an elevated temperature for a period of time. The amount of time and temperature depend on the nature of the moisture pickup and the electrode.

CAUTION

Do not attempt to dry electrodes before determining whether your site quality standards allow drying. Improperly drying electrodes will cause weld defects. Check with your supervisor.

6.0.0 Filler Metal Traceability Requirements

Welding codes and the WPS on a project or job site normally specify the type of electrode to be used. Once a weld is made, there must be a means to trace and document the type of electrode that was used to make the weld or welds in order to verify that the correct welding codes and WPS were followed.

Traceability requirements vary according to client requirements and specifications. On many job sites, these traceability requirements may be found in the quality control documentation mandating the job. One of the more common requirements followed may be found in the *ASME Boiler and Pressure Vessel Code, Section II, Material Specifications.* This section of the ASME code lists acceptable base metal and filler metal combinations.

NOTE

The traceability requirements in site quality control manuals must be followed. The consequences for violation of the requirements can be severe. Understand traceability requirements before proceeding with a welding job. Verify this information with your supervisor, project engineer, or welding inspector.

Cellulose Electrodes

Do not put cellulose electrodes into ovens. Doing so will destroy the operating characteristics of the electrode coating.

Electrode Mix-Up

Do not mix electrode types in an oven. If electrodes of various types must be stored in the same oven, keep them organized by placing them on separate shelves.

Summary

Many different electrodes are produced by many different manufacturers. To ensure quality and uniformity, the American Welding Society has established a classification system for electrodes. Understanding this classification system and why electrodes are placed in a particular classification is essential for the welder to correctly identify and/or select electrodes for the weld to be made.

Welders must be able to select proper types of filler metals and electrodes to make welds on specific materials. It is also important for welders to understand how to properly store electrodes to prevent damage or deterioration.

Review Questions

1. Welding current is transferred from the electrode holder, through the _____, and into the work.
   a. base metal
   b. flux coating
   c. welding ground
   d. wire core of the electrode

2. The harmful oxygen and hydrogen present in the atmosphere surrounding a molten weld is pushed away from the weld by the _____ created during the welding process.
   a. gaseous shield of carbon dioxide
   b. gaseous shield of nitrogen
   c. welding spatter
   d. heat

3. The purpose of the specifications written by the American Welding Society (AWS) is to set standards that all manufacturers must follow when manufacturing _____.
   a. welding consumables
   b. welding machines
   c. chemicals
   d. metals

4. The number to the right of an AWS specification number indicates the _____.
   a. type of electrodes discussed in the specification
   b. year the specification was last revised
   c. number of chapters in the document
   d. year the specification was written

5. All-position electrodes that provide deep penetration belong to the _____ electrode group.
   a. fast-fill
   b. fill-freeze
   c. fast-freeze
   d. low-hydrogen

6. Electrodes designed for welding high-sulphur, phosphorus, and medium- to high-carbon steels belong to the _____ electrode group.
   a. fast-fill
   b. fill-freeze
   c. fast-freeze
   d. low-hydrogen

7. Of all electrodes, the ones that collect moisture faster than any other type are the _____ electrodes.
   a. fast-fill
   b. fill-freeze
   c. fast-freeze
   d. low-hydrogen

8. Typical exposure time for low-hydrogen electrodes ranges from _____ hours.
   a. ½ to 4
   b. ½ to 10
   c. 2 to 5
   d. 3 to 6

9. All filler metal must be stored in a dry storage area at a minimum temperature of _____.
   a. 30°F
   b. 40°F
   c. 50°F
   d. 60°F

10. The traceability requirements of filler metals vary according to the _____ requirements.
    a. base metal
    b. welder type
    c. site and client
    d. state and local

Trade Terms Quiz

Fill in the blank with the correct key trade term that you learned from your study of this module.

1. Welding from top to bottom is called ________.
2. A(n) ________ is a metal that has had other elements added to it that substantially change its mechanical properties.
3. The part of a base metal that has been altered but not melted by the heat is called the ________.
4. ________ is the ability of a material to resist breaking at points where stress is concentrated.
5. The process in which water from the atmosphere forms on a cool surface is called ________.
6. A container that is sealed airtight is said to be ________.
7. ________ is the ability to verify that a procedure has been followed by reviewing the documentation step by step.
8. ________ is the mechanical property of a material that allows the material to be bent or shaped without breaking.
9. An electrode that is specially manufactured to contain little or no moisture is called a(n) ________.
10. ________ is the material used to prevent, dissolve, or facilitate the removal of oxides and other undesirable substances on a weld.

Trade Terms

Alloy
Condensation
Ductile
Flux
Heat-affected zone
Hermetically sealed
Low-hydrogen electrode
Notch toughness
Traceability
Vertical welding

Cornerstone of Craftmanship

Jeff Hyde

Project Manager – Engineering Lead

Flint Hills Resources

As a young man, Jeff Hyde learned to weld on his grandparents' farm in southeastern Kansas. Jeff's career demonstrates that life has a lot to offer to someone with ambition and perseverance.

How did you get your start?

As a teenager I spent summers on my grandparents' farm. My grandfather taught me to weld when I was 14. I also learned to repair farm equipment and did a lot of that work, but it wasn't as much fun as welding. When I finished high school, the oil fields were booming. Because of my ability to weld, I got a job as an oil rig welder in Odessa, Texas, and took welding classes at Odessa College at night.

Going from being an oilfield welder to a maintenance engineer is a big jump. How did that happen?

The oil field boom didn't last forever. After a couple of years of welding oil rigs, I decided to join the Navy and learn about electronics as a sonar technician. When my Navy hitch was up, I took advantage of GI education assistance funds to go to college, where I majored in mechanical engineering. When I finished college, I was hired as a project engineer in the petrochemical industry. Because of my background in welding and electronics, and the knowledge of metallurgy I gained from my welding background, the move into maintenance and reliability engineering was natural.

What do you do in your job?

I review the maintenance needs for plant mechanical equipment and determine the safest, most reliable, and most cost-effective maintenance approach. I also troubleshoot equipment failures and perform failure analysis to determine what caused the equipment to fail. An important part of my job is designing methods to prevent failures from reoccurring.

What do you like about the work you do?

I'm given a lot of freedom to do what I think is necessary in my area of responsibility. Also, the job gives me the opportunity to do the things I really like to do. I have always liked to build things and to tinker with anything new. And I've always enjoyed taking broken machines apart to figure out why they failed and to see if I could fix them.

What do you think it takes to be a success?

To be a success at anything, you can't give up when the going gets tough. During my life, I've encountered many difficulties that could have prevented me from reaching my goals. Any of them would have given me a good excuse to quit, but I didn't let them. I'm not a genius—ask any of my former teachers. I learned by seeking out the people who could teach me what I needed to know. I always kept my goals in view so that problems wouldn't defeat me.

What advice would you give someone just starting out?

Find something you truly like to do and then do that to the best of your ability. It may take a while, and you may have to try different things before you find what you're looking for. If it takes more training or education, go after it, and don't let anything stand in your way.

Trade Terms Introduced in This Module

Alloy: A metal that has had other elements added to it that substantially change its mechanical properties.

Condensation: The process in which water from the atmosphere forms on a cool surface.

Ductile: The mechanical property of a material that allows it to be bent or shaped without breaking.

Flux: Material used to prevent, dissolve, or facilitate the removal of oxides and other undesirable substances on a weld.

Heat-affected zone: The part of the base metal that has been altered but not melted by the heat.

Hermetically sealed: Having an airtight seal.

Low-hydrogen electrode: An electrode specially manufactured to contain little or no moisture.

Notch toughness: The ability of a material to resist breaking at points where stress is concentrated.

Traceability: The ability to verify that a procedure has been followed by reviewing the documentation step by step.

Vertical welding: Welding with a downward progression from top to bottom.

Additional Resources

This module is intended to present thorough resources for task training. The following reference works are suggested for further study. These are optional materials for continued education rather than for task training.

OSHA Standard 1926.351, Arc Welding and Cutting.

The Procedure Handbook of Arc Welding, 2000. Cleveland, OH: The Lincoln Electric Company.

Stick Electrode Welding Guide, 2004. Cleveland, OH: The Lincoln Electric Company.

Stick Electrode Product Catalog, 2008. Cleveland, OH: The Lincoln Electric Company. **www.lincolnelectric.com**

Figure Credits

Lincoln Electric Company, 108F01, 108F02, 108SA01, 108SA02, 108F07, 108F08, 108A01, 108A02

Topaz Publications, Inc., 108F09

Table 5 courtesy of American Welding Society *AWS D1.1/D1.1M:2008*, Table 5.1, Allowable atmospheric exposure of low-hydrogen electrodes, reproduced with permission from the American Welding Society (AWS) Miami, FL USA.

Phoenix International, Inc., 108F10

Sample Material Safety Data Sheet

LINCOLN® ELECTRIC

Date:	12/1/01	MSDS No.:	US-M205
Trade Name:	Fleetweld 5P		
Sizes:	All		
Supersedes:	2/10/99		

MATERIAL SAFETY DATA SHEET

For Welding Consumables and Related Products

Conforms to Hazard Communication Standard 29CFR 1910.1200 Rev. October 1988

SECTION I - IDENTIFICATION

Manufacturer/ Supplier: The Lincoln Electric Company
22801 St. Clair Avenue
Cleveland, OH 44117-1199
(216) 481-8100

Product Type: Covered Electrode

Classification: AWS E6010

SECTION II - HAZARDOUS MATERIAL ([1])

IMPORTANT!
This section covers the materials from which this product is manufactured. The fumes and gases produced during welding with the normal use of this product are covered by Section V; see it for industrial hygiene information.
CAS Number shown is representative for the ingredients listed. All ingredients listed may not be present in all sizes.
(1) The term 'hazardous' in 'Hazardous Materials' should be interpreted as a term required and defined in the Hazards Communication Standard and does not necessarily imply the existence of any hazard.

Ingredients:	CAS No.	Wt.%	TLV mg/m^3	PEL mg/m^3
Cellulose and other carbohydrates	65996-61-4	< 5	10*	10*
Silicates and other binders	1344-09-8	< 5	10*	10*
Titanium dioxides (as Ti)*****	13463-67-7	< 5	10	15
Iron	7439-89-6	< 5	10*	10*
Manganese and/or manganese alloys and compounds (as Mn)****	7439-96-	1	0.2	1.0(c)
Magnesite	1309-48-4	1	10	15
Mineral silicates	1332-58-7	0.5	5**	5**
Iron oxides (as Fe)	65996-74-9	< 0.5	5	10
Limestone and/or calcium carbonate	1317-65-3	< 0.5	10	15
Graphite	7782-42-5	< 0.5	2.0	2.5
Carbon steel core wire	7439-89-6	85	10*	10*

Supplemental Information: (*) Not listed. Nuisance value maximum is 10 milligrams per cubic meter. PEL value for iron oxide is 10 milligrams per cubic meter. TLV value for iron oxide is 5 milligrams per cubic meter.

(c) Values are for manganese fume. STEL (Short Term Exposure Limit) is 3.0 milligrams per cubic meter.

(**) As respirable dust.

(*****) Subject to the reporting requirements of Sections 311, 312, and 313 of the Emergency Planning and Community Right-to-Know Act of 1986 and of 40CFR 370 and 372.

SECTION III - FIRE AND EXPLOSION HAZARD DATA

Non Flammable; Welding arc and sparks can ignite combustibles and flammable products. See Z49.1 referenced in Section VI.

(CONTINUED ON SIDE TWO)

108A01A.EPS

Product:	**Fleetweld 5P**
Date:	**12/1/01**

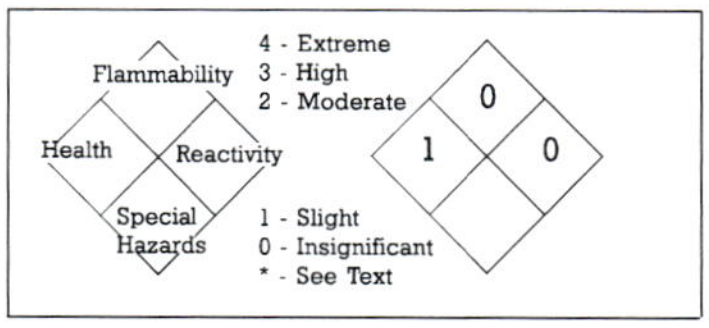

SECTION IV - HEALTH HAZARD DATA

Threshold Limit Value: The ACGIH recommended general limit for Welding Fume NOC - (Not Otherwise Classified) is 5 mg/m^3. ACGIH-1999 preface states that the TLV-TWA should be used as guides in the control of health hazards and should not be used as fine lines between safe and dangerous concentrations. See Section V for specific fume constituents which may modify this TLV. Threshold Limit Values are figures published by the American Conference of Government Industrial Hygienists. Units are milligrams per cubic meter of air.

Effects of Overexposure: Electric arc welding may create one or more of the following health hazards:
Fumes and Gases can be dangerous to your health. Common entry is by inhalation. Other possible routes are skin contact and ingestion.

Short-term (acute) overexposure to welding fumes may result in discomfort such as metal fume fever, dizziness, nausea, or dryness or irritation of nose, throat, or eyes. May aggravate pre-existing respiratory problems (e.g. asthma, emphysema).

Long-term (chronic) overexposure to welding fumes can lead to siderosis (iron deposits in lung) and may affect pulmonary function. Manganese overexposure can affect the central nervous system, resulting in impaired speech and movement. Bronchitis and some lung fibrosis have been reported. WARNING: This product, when used for welding or cutting, produces fumes or gases which contain chemicals known to the State of California to cause birth defects and, in some cases, cancer. (California Health & Safety Code Section 25249.5 et seq.)

Arc Rays can injure eyes and burn skin. *Skin cancer has been reported.*
Electric Shock can kill. If welding must be performed in damp locations or with wet clothing, on metal structures or when in cramped positions such as sitting, kneeling or lying, or if there is a high risk of unavoidable or accidental contact with workpiece, use the following equipment: Semiautomatic DC Welder, DC Manual (Stick) Welder, or AC Welder with Reduced Voltage Control.

Emergency and First Aid Procedures: Call for medical aid. Employ first aid techniques recommended by the American Red Cross.
IF BREATHING IS DIFFICULT give oxygen. IF NOT BREATHING employ CPR (Cardiopulmonary Resuscitation) techniques.
IN CASE OF ELECTRICAL SHOCK, turn off power and follow recommended treatment. In all cases call a physician.

SECTION V - REACTIVITY DATA

Hazardous Decomposition Products: Welding fumes and gases cannot be classified simply. The composition and quantity of both are dependent upon the metal being welded, the process, procedure and electrodes used.

Other conditions which also influence the composition and quantity of the fumes and gases to which workers may be exposed include: coatings on the metal being welded (such as paint, plating, or galvanizing), the number of welders and the volume of the worker area, the quality and amount of ventilation, the position of the welder's head with respect to the fume plume, as well as the presence of contaminants in the atmosphere (such as chlorinated hydrocarbon vapors from cleaning and degreasing activities.)

When the electrode is consumed, the fume and gas decomposition products generated are different in percent and form from the ingredients listed in Section II. Decomposition products of normal operation include those originating from the volatilization, reaction, or oxidation of the materials shown in Section II, plus those from the base metal and coating, etc., as noted above.

Reasonably expected fume constituents of this product would include: Primarily iron oxide; secondarily complex oxides of manganese, silicon and sodium.

Maximum fume exposure guideline and PEL for this product is 5.0 milligrams per cubic meter.

Gaseous reaction products may include carbon monoxide and carbon dioxide. Ozone and nitrogen oxides may be formed by the radiation from the arc.

Determine the composition and quantity of fumes and gases to which workers are exposed by taking an air sample from inside the welder's helmet if worn or in the worker's breathing zone. Improve ventilation if exposures are not below limits. See ANSI/AWS F1.1, F1.2, F1.3, F1.4, and F1.5, available from the American Welding Society, 550 N.W. LeJeune Road, Miami, FL 33126.

SECTION VI AND VII
CONTROL MEASURES AND PRECAUTIONS FOR SAFE HANDLING AND USE

Read and understand the manufacturer's instruction and the precautionary label on the product. Request Lincoln Safety Publication E205. See American National Standard Z49.1, 'Safety In Welding, Cutting and Allied Processes' published by the American Welding Society, 550 N.W. LeJeune Road, Miami, FL, 33126 and OSHA Publication 2206 (29CFR1910), U.S. Government Printing Office, Superintendent of Documents, P.O. Box 371954, Pittsburgh, PA 15250-7954 for more details on many of the following:

Ventilation: Use enough ventilation, local exhaust at the arc, or both to keep the fumes and gases from the worker's breathing zone and the general area. Train the welder to keep his head out of the fumes. *Keep exposure as low as possible.*

Respiratory Protection: Use respirable fume respirator or air supplied respirator when welding in confined space or general work area when local exhaust or ventilation does not keep exposure below TLV.

Eye Protection: Wear helmet or use face shield with filter lens shade number 12 or darker. Shield others by providing screens and flash goggles.

Protective Clothing: Wear hand, head, and body protection which help to prevent injury from radiation, sparks and electrical shock. See Z49.1. At a minimum this includes welder's gloves and a protective face shield, and may include arm protectors, aprons, hats, shoulder protection, as well as dark substantial clothing. Train the welder not to permit electrically live parts or electrodes to contact skin . . . or clothing or gloves if they are wet. Insulate from work and ground.

Disposal Information: Discard any product, residue, disposable container, or liner as ordinary waste in an environmentally acceptable manner according to Federal, State and Local Regulations unless otherwise noted.

108A01B.EPS

NCCER CURRICULA — USER UPDATE

NCCER makes every effort to keep its textbooks up-to-date and free of technical errors. We appreciate your help in this process. If you find an error, a typographical mistake, or an inaccuracy in NCCER's curricula, please fill out this form (or a photocopy), or complete the online form at **www.nccer.org/olf**. Be sure to include the exact module ID number, page number, a detailed description, and your recommended correction. Your input will be brought to the attention of the Authoring Team. Thank you for your assistance.

Instructors – If you have an idea for improving this textbook, or have found that additional materials were necessary to teach this module effectively, please let us know so that we may present your suggestions to the Authoring Team.

NCCER Product Development and Revision

13614 Progress Blvd., Alachua, FL 32615

Email: curriculum@nccer.org
Online: www.nccer.org/olf

❏ Trainee Guide ❏ Lesson Plans ❏ Exam ❏ PowerPoints Other ______________________

Craft / Level: Copyright Date:

Module ID Number / Title:

Section Number(s):

Description:

Recommended Correction:

Your Name:

Address:

Email: Phone:

86101-14

Tack Welding

Module Six

Trainees with successful module completions may be eligible for credentialing through NCCER's National Registry. To learn more, go to **www.nccer.org** or contact us at **1.888.622.3720**. Our website has information on the latest product releases and training, as well as online versions of our *Cornerstone* magazine and Pearson's product catalog.

Your feedback is welcome. You may email your comments to **curriculum@nccer.org**, send general comments and inquiries to **info@nccer.org**, or fill in the User Update form at the back of this module.

This information is general in nature and intended for training purposes only. Actual performance of activities described in this manual requires compliance with all applicable operating, service, maintenance, and safety procedures under the direction of qualified personnel. References in this manual to patented or proprietary devices do not constitute a recommendation of their use.

86101-14
Tack Welding

Objectives

When you have completed this module, you will be able to do the following:

1. Set up proper arc welding equipment.
2. Describe methods of striking an arc.
3. Describe how to properly perform a tack weld.

Performance Tasks

Under the supervision of the instructor, you should be able to do the following:

1. Set up proper arc welding equipment.
2. Properly strike and extinguish an arc.
3. Perform tack welds in the following positions to be break-tested by the instructor:
   - Flat (1F)
   - Horizontal (2F)
   - Vertical (3F)
   - Overhead (4F)

Trade Terms

Alternating current (AC)
Amperage
Arc
Arc blow
Arc strike
Degauss
Direct current (DC)
Drag angle
Flux cored arc welding
Gas metal arc welding
Polarity
Primary current
Push angle
Step-down transformer
Voltage
Weld axis
Weld coupon
Welding procedure specification (WPS)
Work angle

Note

Before beginning this module, it is recommended that the trainee complete the *Maritime Core Curriculum* and *Welding Level One*, Modules 29101-09, 29102-09, 29105-09, 29106-09, and 29108-09.

Industry-Recognized Credentials

If you're training through an NCCER-accredited sponsor, you may be eligible for credentials from NCCER's Registry. The ID number for this module is 86101-14. Note that this module may have been used in other NCCER curricula and may apply to other level completions. Contact NCCER's Registry at 888.622.3720 or go to **www.nccer.org** for more information.

Contents

Topics to be presented in this module include:

Figures

1.0.0 Introduction

In the maritime industries, tack welds are used to hold the parts of steel weldments or frames in proper alignment during fit-up by a structural fitter. After fit-up, the final welds are made by a finish welder. For the tack welds to be effective, their length and the distance between them depend on the weight and thickness of the materials being joined. The same type of electrode that will be used for the finish weld must be used for tack welds. The size and quality of the tack welds must be such that they can be incorporated in the finish weld without the need for repairs by the finish welder. In some cases, preheating is required for tack welds. It is important that the **welding procedure specification (WPS)** and project drawings for tack welds be followed during fit-up. *Figure 1* shows an example of an acceptable tack weld and tack welds that require rework. The structural fitter is responsible for the following during fit-up:

- Use of proper electrodes and proper preheat, if required.
- Removal of slag over the tack welds and any adjacent weld spatter.
- Repair of tack welds that require rework.
- Elimination of weld burrs after removal of temporarily welded support attachments.
- Repair of any visible damage caused during the process of making tack welds, including **arc strikes**. This must be done to prevent possible future stress failure of the weldment or frame parts.

Tack welds, as well as finish welds on steel, are made using shielded metal arc welding (SMAW) with electrodes. This is often referred to as stick welding. This welding process can use either **alternating current (AC)** or **direct current (DC)** with adjustments made on the welding machine to regulate the level of current. This module provides an overview of some SMAW industrial welders used in the maritime industries, along with some basic weld cleaning tools. The module also provides information on welding machine setup, striking an **arc**, maintaining an arc to run tack welds, and performing fillet tack welds in all positions.

2.0.0 Safety Summary

The following is a summary of safety procedures and practices that must be observed when cutting or welding. Keep in mind that this is a summary. Complete safety coverage is provided in Welding Level One, *Welding Safety*. If you have not completed that module, do so before continuing. Above all, be sure to wear appropriate protective clothing and equipment when welding or cutting.

2.1.0 Protective Clothing and Equipment

To maintain safety and prevent injury, it is essential that you wear the appropriate protective clothing and equipment when cutting or welding metals. Be sure to follow these guidelines during all phases of cutting or welding:

- Always use safety glasses with a full-face shield or a helmet. The glasses, face shield, or helmet lens must have the proper light-reducing tint for the type of welding or cutting being performed. Never directly or indirectly view an electric arc without using a properly tinted lens.

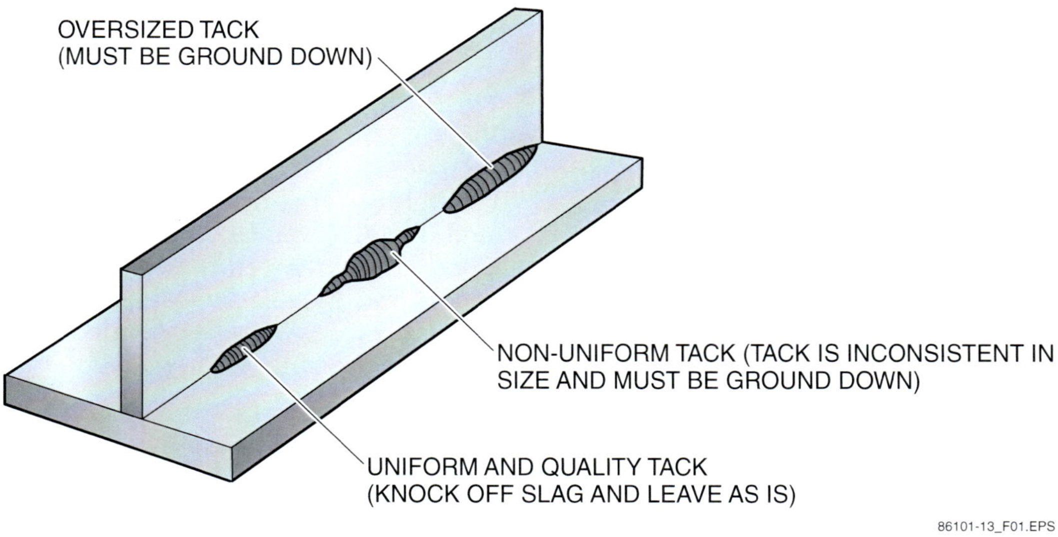

Figure 1 Example of a uniform quality tack, a non-uniform tack, and an oversized tack.

- Wear proper protective leather and/or flame-retardant clothing, along with welding gloves that will protect you from flying sparks and molten metal as well as heat.
- Wear 8" or taller high-top safety shoes or boots. Make sure that the tongue and lace area of the footwear will be covered by a pant leg. If the tongue and lace area is exposed, or if the footwear must be protected from burn marks, wear leather spats under the pants or chaps and over the front of the footwear.
- Wear a solid material (non-mesh) hat with a bill pointing to the rear or, if much overhead cutting or welding is required, wear a full leather hood with a welding face plate and the correct tinted lens. If a hard hat is required, use one that allows the attachment of rear deflector material and a face shield or welding hood.
- If a full leather hood is not worn, wear a face shield and snug-fitting welding goggles over safety glasses for gas welding or cutting. Either the face shield or the lenses of the welding goggles must be an approved shade 5 or 6 filter. For electric arc welding, wear safety goggles and a welding hood with the correct tinted lens (shade 9 to 14).
- If a full leather hood is not worn, wear earmuffs, or at least earplugs, to protect ears and ear canals from sparks from overhead operations.
- If working at heights, remember to use appropriate fall protection equipment.

2.2.0 Fire/Explosion Prevention

Welding activities usually involve the use of fire or extreme heat to melt metal. Whenever fire is used, it must be controlled and contained. Welding activities are often performed on vessels that may once have contained flammable or explosive materials. Residues from those materials can catch fire or explode when a welder begins work on such a vessel. The following are fire and explosion prevention guidelines associated with welding:

- Never carry matches or gas-filled lighters in your pockets. Sparks can cause the matches to ignite or the lighter to explode, causing serious injury.
- Never perform any type of heating, cutting, or welding until a hot work permit has been obtained and an approved fire watch is established during and after the operation. Most job site fires caused by these types of operations are started by cutting torches.
- Never use oxygen to blow dust or dirt off of clothing. The oxygen can remain trapped in the fabric for a period of time. If a spark hits the clothing during this time, the clothing can burn rapidly and violently.
- Make sure that any flammable material in the work area is moved or shielded by a fire-resistant covering. Approved fire extinguishers must be available before any heating, welding, or cutting operations are attempted.
- Never release a large amount of oxygen or use oxygen as compressed air. The presence of oxygen around flammable materials or sparks can cause rapid and uncontrolled combustion. Keep oxygen away from oil, grease, or other petroleum products, as oxygen can explode on contact with these substances.
- Never release a large amount of fuel gas, especially acetylene. Methane and propane tend to concentrate in and along low areas and can be ignited at a considerable distance from the release point. Acetylene is lighter than air but is even more dangerous; when mixed with air or oxygen, it will explode at much lower concentrations than any other fuel.
- To prevent fires, maintain a neat and clean work area, and make sure that any metal scrap or slag is cold before disposal.
- When working in a confined space, leave the tanks outside of the space.
- Before cutting or welding containers, such as tanks or barrels, find out if they have contained any explosive, hazardous, or flammable materials, including petroleum products, citrus products, or chemicals that decompose into toxic fumes when heated. As a standard practice, always clean and then fill any tanks or barrels with water or purge them with a flow of inert gas such as nitrogen or argon to displace any oxygen.

2.3.0 Work Area Ventilation

Vapors and fumes tend to rise in the air from their sources. Welders often have to work above the welding area where the fumes are being created. Welding fumes can cause personal injuries. Good work area ventilation helps to remove the vapors and protect the welder. The following is a list of work area ventilation guidelines to consider before and during welding activities:

- Make sure confined space procedures are followed before conducting any welding or cutting in the confined space.
- Make sure confined spaces are ventilated properly for cutting or welding purposes.

- Never use oxygen in confined spaces for purposes of ventilation.
- Always perform cutting or welding operations in a well-ventilated area. Cutting or welding operations involving materials, coatings, or electrodes containing cadmium, mercury, lead, zinc, chromium, and beryllium result in toxic fumes. For cutting or welding of such materials, always use proper area ventilation and wear an approved full-face, supplied-air respirator (SAR) that uses breathing air supplied from outside of the work area. For occasional, very short-term exposure to fumes from zinc or copper coated materials, a high-efficiency, particulate arresting (HEPA)-rated or metal-fume filter may be used on a standard respirator.

2.4.0 Lifting Hazards

Welders sometimes forget the basics of lifting as they transport their welding machines, cables, and even the parts to be welded. Use rigging or lifting devices whenever possible. Get help if you must lift a heavy object by hand. Always remember to use tag lines when hoisting heavy objects that tend to spin or swing.

2.5.0 Working at Heights

Welders may have to be suspended from the side of a structure to weld something. Remember to wear all the approved fall protection devices when performing such work. Make sure to inspect your fall protection equipment items each time you use them, and be certain that they are current and up-to-date.

3.0.0 Welding Current

In welding, there must be current flow between the workpiece and the welding machine. The arc that is produced during SMAW is an electrical current that jumps across the gap from the tip of the electrode to the surface of the workpiece. The current originates in the transformer or generator of the welding machine. It is routed through one welding lead, the electrode gap, through the workpiece, and finally back to the transformer or generator through the other welding lead (*Figure 2*). If the electrode tip is touched and held against the workpiece instead of allowing for a gap between the two, the electrode will pass the electricity directly to the workpiece without creating an arc. No welding will take place in this situation because an arc across the gap is needed to heat the workpiece and the electrode to a molten state.

WARNING!

Never remove your welding helmet if the electrode is stuck or frozen to the workpiece. You must free the electrode with your helmet in place because the electrode will flash when it breaks free. First remove the holder, then remove the electrode with pliers.

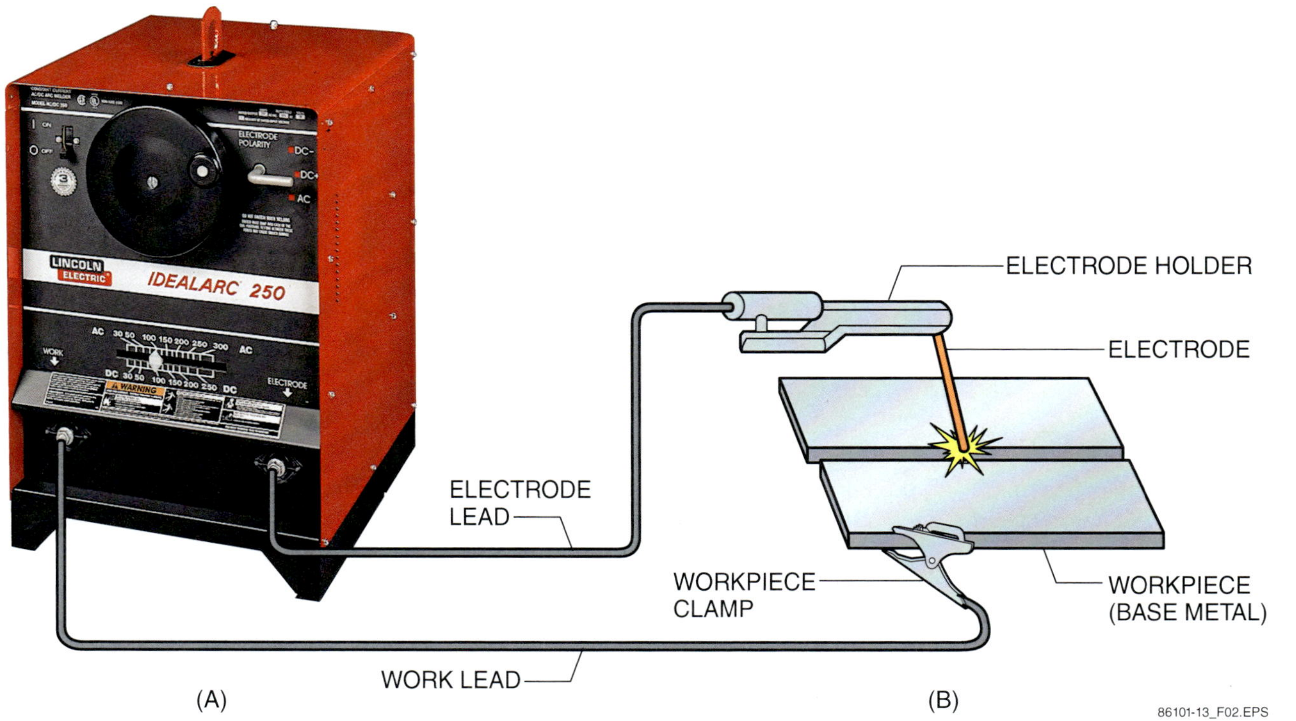

Figure 2 A welding setup.

3.1.0 Types of Welding Current

SMAW machines are designed to produce an output that has a nearly constant current. The currents may be either AC or DC. The type of current used depends on the type of welding to be done and/or the welding equipment available.

3.1.1 AC Welding Current

Alternating current in welding machines is derived from either a transformer-type welding machine or an alternator-type machine. The transformer-type machine changes high-voltage, low-current commercial AC power to the low-voltage, high-current power required for welding. An alternator-type machine uses an electric or fuel-driven motor to turn a rotor inside a number of electromagnets to produce the current needed for welding.

AC current alternates between positive and negative values only because the **voltage** changes from positive to negative values. *Figure 3* represents a graph of 120VAC commercial power. The top half of the cycle of the voltage path is considered positive. The current follows the voltage. In the bottom half of the cycle, the voltage is considered negative, with the current continuing to follow the voltage. In one complete cycle, the wave crosses the axis or zero point twice. The peaks or extreme high or low points on the wave indicate the maximum level of the voltage, either positive (on the top) or negative (on the bottom). The number of cycles completed in one second is called the frequency. Frequency varies depending on the number of poles (magnetic field coils) installed in the generator during the generation stages of the power and on the speed of the rotor rotation. The standard frequency in the United States for commercial power is set at 60 cycles per second.

3.1.2 DC Welding Current

Direct current is electrical current that has no frequency. It does not travel from positive to negative and negative to positive within the same operation. The waveform for DC power resembles a straight line; therefore, it is not really a wave. Remember, frequency is associated with AC power only because frequency refers to the change in direction during one cycle, and DC power does not change direction during its flow. Direct current in welding is derived either from a transformer-rectifier type of system or from a generator. In a transformer-rectifier type of system, AC voltage first is reduced (transformed) to a lower voltage level through the use of a transformer, then it is converted (rectified) into DC voltage. Most gasoline-powered SMAW welding machines produce DC current through a generator. SMAW units that plug into 60-cycle AC power generally produce DC current using a transformer and a rectifier.

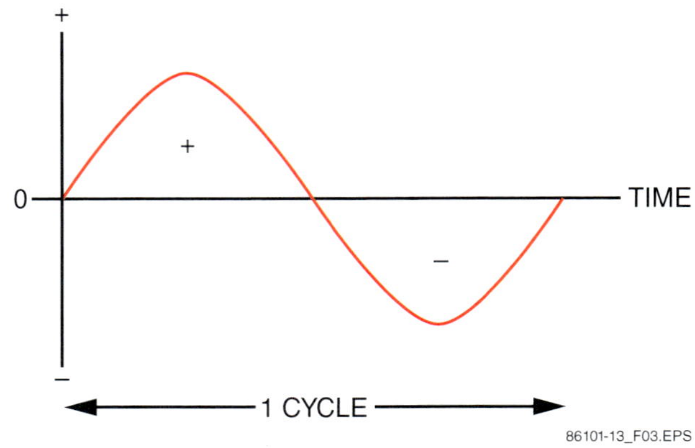

Figure 3 AC current waveform.

3.2.0 Polarity

Polarity only applies to DC current. Polarity in welding is determined either by the way the welding leads are connected to the welding machine and workpiece or by a polarity switch on the DC welding machine. The universally accepted theory is that DC current always travels from the negative pole to the positive pole.

Most, if not all, welding leads are configured with one end of one lead holding an electrode or

Striking an Arc

Two general methods of striking the initial arc are scratching the electrode on the workpiece and tapping the electrode on the workpiece.

The scratching method is generally used by most beginners and even by many experienced welders. However, it is usually not acceptable for code or finished-product welding unless the burn trails from scratching are within the area covered by the weld, or are repaired. Some of the newer inverter power sources do not activate the open-circuit voltage until after an electrode touches a surface. With these machines, a scratching motion must be used in order to activate the required open-circuit voltage and start the arc unless an adjustable starting current capability exists on the machine.

rod and the other lead equipped on one end with a clamping device to attach to the workpiece. Both of the other ends of these two cables typically are equipped with ring-type or plug-in connectors for attaching the leads to the welding machine. When the electrode-holder lead is connected to the plus or positive terminal on the machine and the lead with the clamping device is connected to the minus or negative terminal on the welding machine, direct current electrode positive (DCEP) polarity is established. If these leads are manually switched at the welding machine terminals or through the operation of the manual polarity switch located on the machine, the electrode-holder lead becomes the negative lead and the workpiece clamping device lead becomes the positive lead. This is referred to as direct current electrode negative (DCEN) polarity.

CAUTION

On machines that have polarity switches, do not switch the polarity during operation. This may damage the machine.

3.3.0 Characteristics of Welding Current

The voltage in a welding machine, whether the machine is connected to commercial power or driven by an engine, is reduced or stepped down by a transformer called a **step-down transformer**. The step-down transformer reduces the voltage level and provides a high current level. The high current allows the arc to occur with a voltage potential that carries less of a risk of electrocution. However, electrical shock in welding remains a hazard that can be reduced or eliminated only by using proper personal protective equipment and applying safe welding practices. Some newer machines have automatic open-circuit voltage reduction capability for added safety.

3.3.1 Voltage

Voltage is the measure of the electromotive force or pressure that causes current to flow in a circuit. Two types of voltage are associated with welding current: open-circuit voltage and operating voltage. Open-circuit voltage is the voltage present when the welding machine is on, but no arc is present. For SMAW, open-circuit voltage is usually between 50 and 80 volts.

Operating voltage, or arc voltage, is the voltage measured after the arc has been struck. This voltage is lower than open-circuit voltage and usually measures between 18 and 45 volts. The higher, open-circuit voltage is required to establish the arc because the air gap between the electrode and the work has high resistance to current flow. Once the arc has been established, less voltage is needed, and the welding machine automatically compensates by lowering the voltage. This characteristic of an SMAW machine is called variable-voltage, constant-current power.

3.3.2 Amperage

Amperage is a measurement of the electric current flowing in a circuit. The unit of measurement for current is the ampere (*amp* for short). The number of amps selected at the welding machine determines the amount of heat available to melt the work and the electrode. The current is increased or decreased according to the size of electrode being used and the position in which the welding is being performed.

4.0.0 SMAW Machines

Welding machines, like other tools, have specific classifications, different special types, and rated capacities. A lathe, for example, might be rated by its throw, speeds, and special toolings. SMAW welding machines are broken down by classifications, types, and ratings.

4.1.0 SMAW Machine Classifications

SMAW welding machines are classified by the type of welding current they produce: AC, DC, or AC/DC.

An SMAW welding machine classified as an AC machine will produce only AC welding voltage and current, and a DC machine will produce only DC welding voltage and current. If a welding machine produces both AC and DC welding voltage and current, it is classified as an AC/DC welding machine.

Properties of an Arc

An arc produces a temperature in excess of 6,000°F at the tip of an electrode. This extreme heat is what melts the base metal and electrode, producing a pool of molten metal often referred to as a crater. As the crater solidifies behind the weld, it makes the bond.

Welding machines that produce only DC welding voltage and current can be further classified by the characteristics of the welding current they produce. If the welding current varies between the higher open-circuit voltage and the lower operating voltage, it is classified as a variable-voltage constant-current power source. This type of welding machine is used for SMAW. If open-circuit and operating voltages are nearly the same, the machine is classified as a constant-voltage or constant-potential DC welding machine. This type of machine is used for **gas metal arc welding (GMAW)** and **flux cored arc welding (FCAW)**.

4.2.0 Industrial SMAW Machine Types

Several different types of basic industrial SMAW machines are available to produce the current necessary for welding. They include transformer rectifiers and inverter power sources.

Transformer rectifiers and electric motor generators require electrical power from commercial power lines to operate. This high-voltage, low-amperage current from commercial power lines coming into the welding machine is called the **primary current**. The primary current first flows from the power lines to an electrical circuit breaker distribution panel and from the panel to a disconnecting fuse box. From the disconnecting device, the current flows by cable to the machine or to a receptacle. The disconnecting device can be used to shut off the flow of current to the machine or the receptacle if the welding machine is plugged into the receptacle. The primary current required for welding machines can be 208V, 240V, or 460V single-phase or 240V, 380V, 415V, 460V, 500V, or 575V three-phase.

WARNING!

Coming into contact with the primary current of a welding machine can cause electrocution. Ensure that welding machines are properly grounded.

4.2.1 Transformer-Rectifier Welding Machines

A rectifier is a device that converts AC current to DC current. A transformer-rectifier welding machine uses a transformer to reduce the primary voltage to welding voltage and then rectifies the current to change it from AC to DC. Transformer-rectifier welding machines can be designed to produce AC and DC welding current or DC current only. Transformer rectifiers that produce both AC and DC welding current are usually intended for lighter duty than those that produce DC current only.

Transformer rectifiers can also be designed to produce variable-voltage constant-current or constant-voltage (constant-potential) DC welding current. Some are even designed to provide both variable and constant voltage; those are referred to as multi-process power sources, since they can be used for any welding process. Refer to the manufacturer's documentation to determine how the machine is configured.

Transformer rectifiers have an On/Off switch and an output amperage control. The welding cables (electrode lead and workpiece lead) are connected to terminals. They often have selector switches to select DCEP or DCEN or to select AC welding current if it is available. If there is no selector switch, the cables must be manually changed on the machine terminals to select the type of current desired. *Figures 4* and *5* show the various parts of typical transformer-rectifier welding machines. These machines can typically be stacked three high to save floor space.

The 680-amp machine can be used for welding with electrode sizes up to and including ⅜ inch. The 680-amp unit is also a multi-process welding machine. Besides stick welding, it can be used for various other welding processes, including carbon arc gouging. Carbon arc gouging is used to remove temporary attachments.

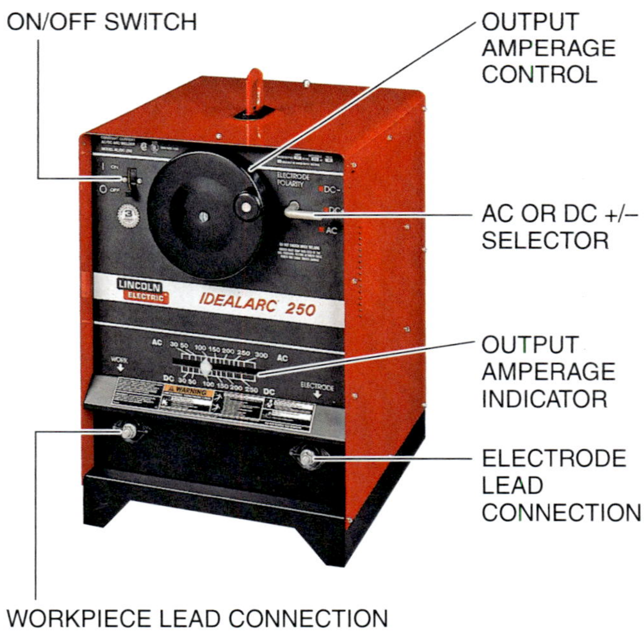

Figure 4 A 250-amp transformer-rectifier AC/DC welding machine.

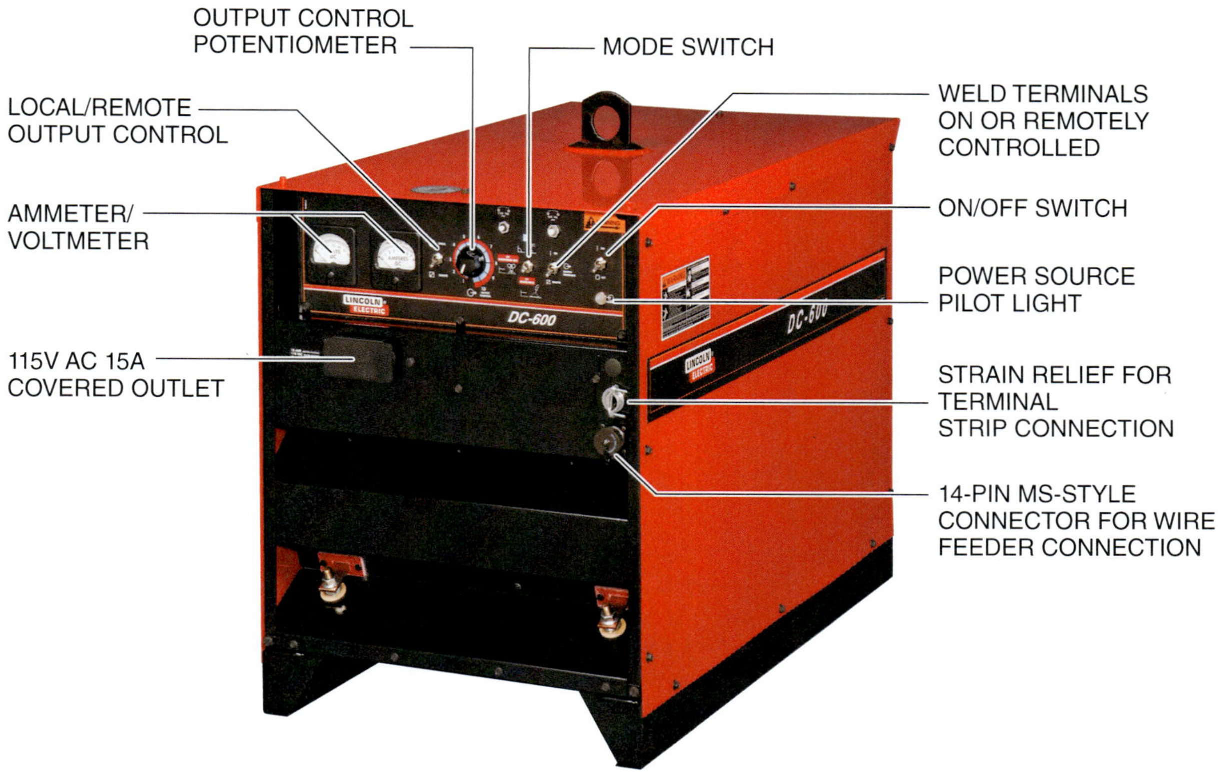

Figure 5 A 680-amp transformer-rectifier DC welding machine.

4.2.2 Inverter Power Sources

Inverter power sources increase the frequency of the incoming primary power. This provides a smaller, lighter power source with a faster response time and much more waveform control for pulse welding. Inverter power sources are used where limited space and portability are important factors. The controls on these machines vary according to size and application.

4.3.0 SMAW Machine Ratings

The size of a welding machine is determined by the amperage output of the machine at a given duty cycle. The duty cycle of a welding machine is based on a ten-minute period. It is the percentage of ten minutes that the machine can continuously produce its rated amperage without overheating. For example, a machine with a rated output of 300A at 60 percent duty cycle can deliver 300A of welding current for six minutes out of every ten (60 percent of ten minutes) without overheating. Most industrial, heavy-duty machines for manual welding have a 60-percent duty cycle.

4.4.0 SMAW Cable Terminations

Welding cables must be equipped with the proper end connections. End connections used on welding cables include lugs, quick disconnects, ground clamps, and electrode holders.

CAUTION

If the end connection is not tightly secured to the cable, the connection will overheat. An overheated connection will cause variations in the welding current and permanent damage to the connector and/or cable. Repair any connectors that overheat.

4.4.1 Lugs and Quick Disconnects

Lugs are used at the end of the welding cable to connect the cable to the welding machine terminals. Lugs come in various sizes to match the

AC/DC Machines

Transformer-rectifier welding machines that can provide either AC or DC current do not utilize the rectifier circuit when welding with AC current. In this situation, the welding machine functions like a transformer welding machine.

welding cable size and are mechanically crimped onto the welding cable. Lugs are normally ringed to prevent the welding lead from pulling loose from the stud connector on the machine.

Quick disconnects are also mechanically connected to the cable ends. They are insulated and serve as cable extensions for splicing two lengths of cable together. Quick disconnects are connected or disconnected with a half twist. When using quick disconnects, make sure that they are tightly connected to prevent overheating or arcing in the connector. They must be kept free of dirt, sand, mud, and other contaminants to ensure a good connection.

4.4.2 Workpiece Clamps

Workpiece clamps establish the connection between the end of the workpiece and the workpiece lead. Workpiece clamps are mechanically connected to the welding cable and come in a variety of shapes and sizes. The size of a workpiece clamp is the rated amperage that it can carry without overheating. When selecting a workpiece clamp, be sure that it is rated at least as high as the rated capacity of the power source on which it will be used. *Figure 6* shows several examples of workpiece clamps.

4.4.3 Electrode Holders

Electrode holders grasp the electrode and provide the electrical contact between the electrode and the end of the welding cable. Electrode holders are also mechanically connected to the welding cable and come in a variety of styles and sizes. The size of an electrode holder is the rated amperage it will carry without overheating. The physical size of an electrode holder is also proportional to its amperage capacity. A small 200A electrode holder will be light and easy to handle but will not hold large electrodes that require more than 200A. When selecting an electrode holder, select the size for the amperage and electrodes being used. *Figure 7* shows two styles of electrode holders.

5.0.0 Tools for Cleaning Welds

Hand tools such as files, chipping hammers, and wire brushes, as well as small power tools such as slag chippers and scalers, are common tools used by welders to prepare surfaces before welding and to clean welds after they are completed.

Figure 6 Workpiece clamps.

WARNING!

Always wear eye and hand protection when preparing and cleaning weldment surfaces. Remember that finished welds remain extremely hot for extended periods and can cause serious burns. Chipping slag and other coatings from finished welds is very hazardous and requires proper eye protection, including goggles or face shields.

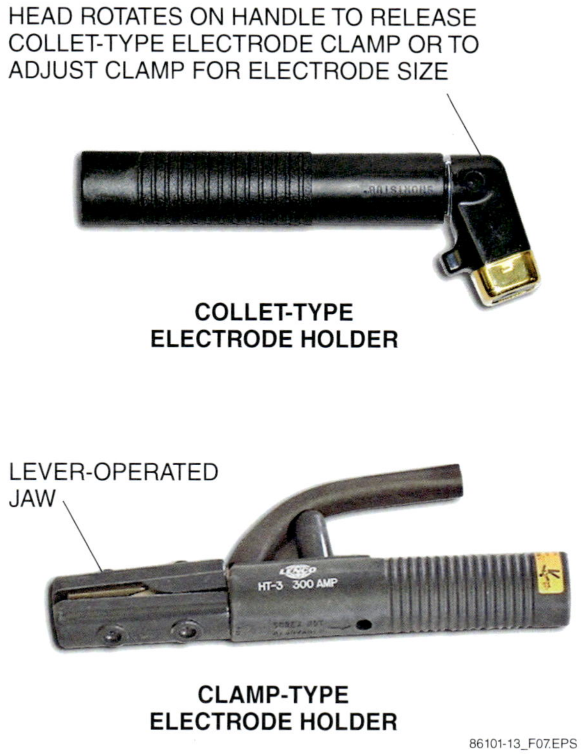

Figure 7 Electrode holders.

5.1.0 Hand Tools

It is extremely important that the base metals are thoroughly cleaned and prepared before welding. Dirty or corroded surfaces are very difficult, if not impossible, to weld properly. Also, before any additional weld beads can be applied over previous beads, the previous pass must be cleaned of any slag that was deposited along with the weld bead. Many mechanical methods have been developed to assist in these cleaning and removing processes, including sandblasting and ultrasonic cleaning. However, many welders prefer simple manual procedures using tools such as files, wire brushes, and chipping hammers. *Figure 8* shows some of the hand tools used by welders to prepare base metals prior to welding and to remove slag after the bead has been laid.

5.2.0 Pneumatic Cleaning and Slag Removal Tools

Weld slag chippers and needle scalers (*Figure 9*) are pneumatically powered. They are used by welders to clean surfaces and to remove slag from cuts and welds. Weld slag chippers and needle scalers are also excellent for removing paint or hardened dirt, but they are not very effective for removing surface corrosion. Weld slag chippers have a single chisel, and needle scalers have 18 to 20 blunt steel needles approximately 10 inches long. Most weld slag chippers can be converted to needle scalers with a needle scaler attachment.

6.0.0 Arc Welding Equipment Setup

Before welding can take place, the area has to be made ready, the welding equipment must be set up, and the metal to be welded must be prepared. The following sections explain how to set up equipment for welding.

6.1.0 Preparing the Welding Area

To practice welding, a welding station with a table, bench, or stand is usually needed (*Figure 10*). The welding surface must be steel, and provisions must be made for placing **weld coupons** out of position.

To set up the area for welding, follow these steps:

Step 1 Ensure that the area is properly ventilated. Make use of doors, windows, and fans.

Figure 8 Typical hand tools used in welding.

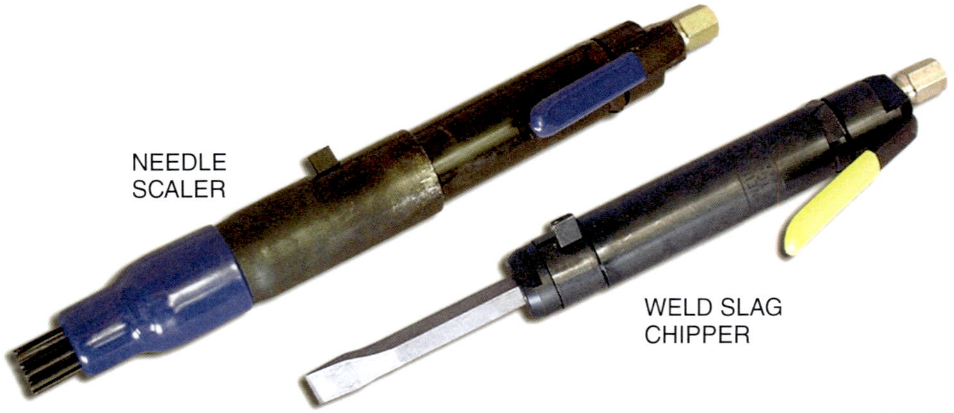

Figure 9 Pneumatic needle scaler and weld slag chipper.

Step 2 Check the area for fire hazards. Remove any flammable materials before proceeding.

Step 3 Check the location of the nearest fire extinguisher. Do not proceed unless the extinguisher is charged and you know how to use it.

Step 4 Set up flash shields around the welding area.

6.2.0 Preparing Tack Weld Coupons

Tack weld coupons should be carbon steel, ¼" to ¾" thick. Use a wire brush or grinder to remove heavy mill scale or corrosion. Prepare weld coupons to practice the welds indicated as follows:

- Striking an arc – The coupons can be any size or shape that is easily handled.
- Running tack beads – The coupons can be any size or shape that is easily handled.
- Fillet tack welds – Cut the metal into 4" × 12" rectangles for the base and 3" × 12" rectangles for the web.

Figure 11 shows the weld coupons for tack fillet welding.

Steel for practice welding is expensive and difficult to obtain. Every effort should be made to conserve the material that is available. Reuse weld coupons until all surfaces have been welded upon. Weld on both sides of the joint, and then cut the weld coupon apart and reuse the pieces. Use material that cannot be cut into weld coupons to practice striking an arc and running beads.

6.3.0 Electrodes

Obtain a small quantity of the electrodes to be used. Electrodes are sometimes referred to by the nonstandard term of *rods*. For the welding exer-

Figure 10 Welding station.

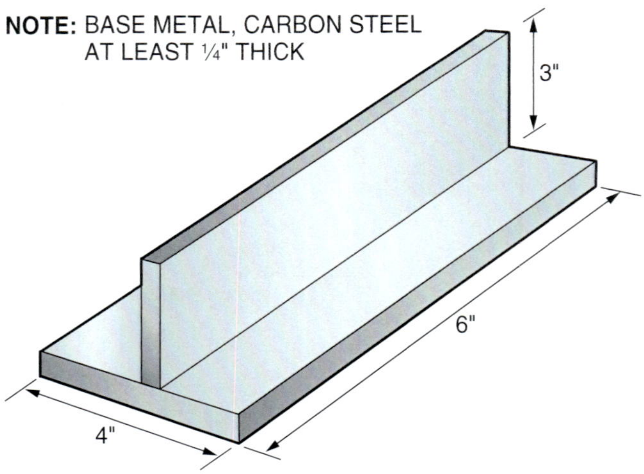

Figure 11 Fillet tack weld coupons.

cises in this module, 3/32" or 1/8" E6010 and E7018 electrodes can be used. Obtain only the electrodes to be used for a particular welding exercise at one time. Have some type of pouch or rod holder in which to store the electrodes, to prevent them from becoming damaged. Never store electrodes loose on a table. They may fall on the floor and be damaged, or create a tripping hazard. Some type of metal container, such as a pail, must also be available to discard hot electrode stubs. Return unused electrodes to the proper storage location.

WARNING!

Do not throw electrode stubs on the floor. They easily roll, and could cause someone to slip and fall. In some locations, stubs must be returned to a supervisor.

6.4.0 Preparing the Welding Machine

Welders can expect to find different types and makes of welding machines in welding shops and on jobs in the field. Most machines will have no operator manuals with them. As a trainee, and even as a trained tack welder, you must be able to recognize the different types of welding machines and to figure out how to safely set up and operate them. Select a welding machine to use and then follow these steps to set it up for welding:

Step 1 Verify that the welding machine can be used for DC welding.

NOTE

An AC welding machine can be used if a DC machine is not available. If an AC welding machine is used, use E6011 and E7018 electrodes.

Step 2 Verify the location of the primary current disconnect to the machine.

Step 3 Check the area for proper ventilation.

Step 4 Select the polarity of the welding leads for the type of electrode to be used.

Step 5 Connect the clamp of the workpiece lead to the workpiece.

Step 6 Set the amperage for the electrode type and size to be used.

Step 7 Ensure that the electrode holder is not touching the workpiece lead clamp, the workpiece, or a grounded object.

Step 8 Turn on the welding machine.

Step 9 Insert the bare end of the electrode into the electrode holder.

WARNING!

Even though workpiece leads and clamps are sometimes called ground leads or clamps, they may not actually be grounded. In this case, the full open-circuit voltage of the welding machine may exist between the workpiece lead or clamp and any grounded object.

7.0.0 Striking an Arc

The first step in starting a weld is striking the arc. After an arc is established, the general rule is that the arc length should be the diameter of the electrode being used; for example, if a 1/8" electrode is being used, the arc length should be 1/8". The arc length is measured from the end of the electrode core to the base metal. If the electrode has a heavier flux coating, such as E7018 or E6013, the end of the electrode core will be recessed in the coating. For these electrodes, the visible arc length (from the end of the coating to the base metal) should be slightly less to compensate for the part of the arc that is recessed. This will ensure that the actual arc length is correct. *Figure 12* shows arc lengths for a 1/8" rod. There are two ways to strike an arc: the scratching method and the tapping method.

7.1.0 Scratching Method

The scratching method (*Figure 13*) is the easiest way to strike an arc; it is used by trainees who are

GOING GREEN

Recycling Metal

Salvage any coupons or scrap metal that is no longer usable for welding practice and place it into a recycling bin. Almost all metals are recyclable and valuable. One technical school took in donated steel, used it for cutting and welding practice, and then sold all their recyclable metals to fund their school's annual picnic. The bottom line is that metal that is no longer usable for cutting or welding practice can be recycled and kept out of the landfills.

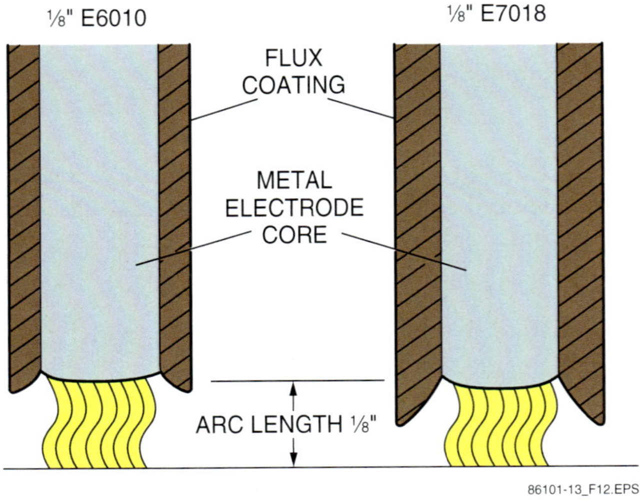

Figure 12 Arc lengths.

learning to weld. It is similar to striking a match. The end of the electrode is simply scratched along the base metal to establish an arc. When the arc has been established, the electrode is raised to establish the correct arc length.

Rod Holders

Several different types of rod holders can be purchased. One type is a leather pouch, and another is a sealable holder that can be used to prevent the electrodes from getting wet.

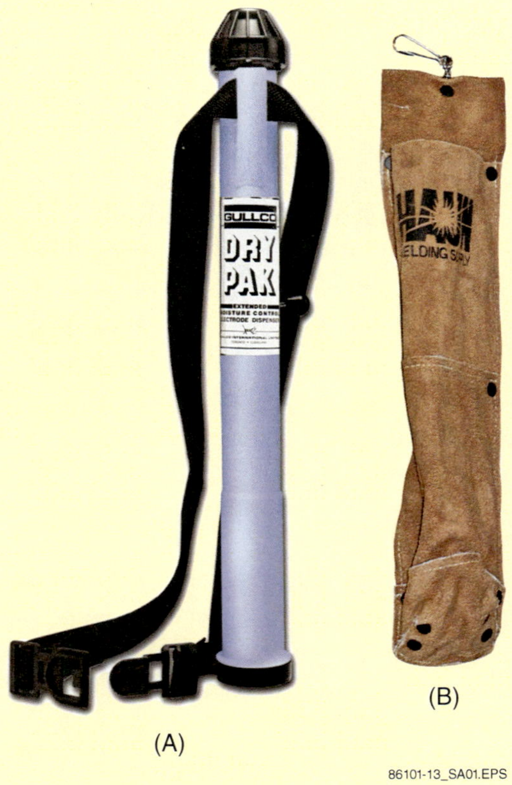

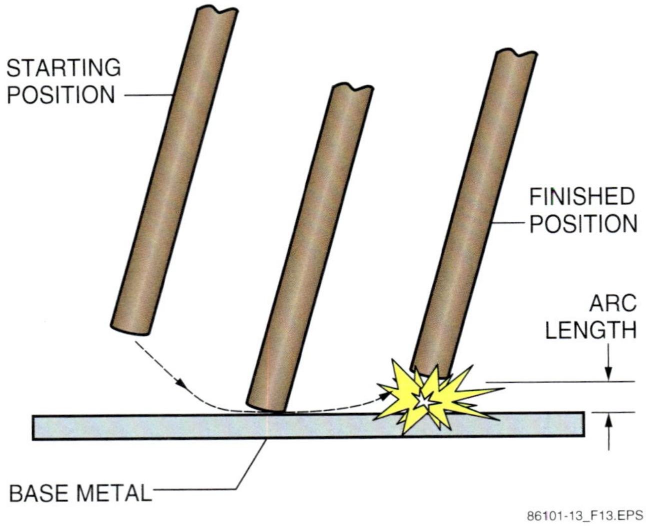

Figure 13 Scratching method of striking an arc.

7.2.0 Tapping Method

The tapping method (*Figure 14*) is the best method of establishing an arc when using a transformer DC welding machine. It is more difficult and takes more practice to perfect, but as welders develop their skills, it is the method that should be used. The scratching method leaves arc strikes on the base metal that are not allowed by the welding codes unless they occur within the welded area or are repaired. With the tapping method, the arc is usually established just after the point where the welding should begin (¼" to ½"). The arc is then moved back to the correct starting point as it stabilizes. To strike an arc with the tapping method, move the electrode quickly to the base metal, lightly tap it, and then raise the electrode to establish the correct arc length.

7.3.0 Practicing Striking and Extinguishing an Arc

Striking the arc correctly is an important step in learning to weld. Practice both methods using 3/32" and 1/8" E6010 or E6011 electrodes until you can strike and maintain an arc consistently. When striking an arc, place the base metal flat on the welding table. To extinguish the arc at the end of a weld, quickly lift the electrode away from the work.

NOTE

Striking an arc with E7018 (low hydrogen) electrodes is more difficult than with most other electrodes because they tend to stick to the base metal more easily. Practice striking an arc with E7018 after you become more skilled at running tack beads with E6010 or E6011 electrodes.

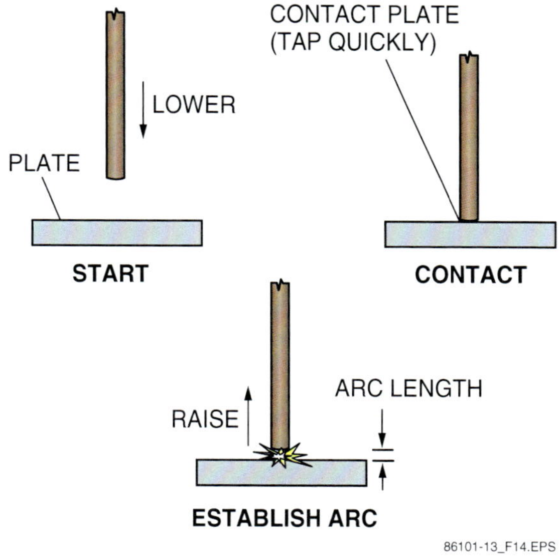

Figure 14 Tapping method of striking an arc.

When practicing striking an arc, you will generally experience the following two problems:

- The end of the electrode will weld itself to the base metal. This is caused by keeping the electrode in contact with the base metal too long or by trying to maintain too short an arc length. If this condition occurs, free the electrode by quickly moving the electrode holder from side to side. If this fails, pull the holder from the electrode and remove the electrode from the base metal with pliers. When the electrode is released from the holder, arcing generally occurs between the holder jaws and the electrode. If arcing has occurred, inspect the holder jaws for any heavy pitting or gouging damage, and replace them if necessary.
- The arc will break (go out). This is caused by raising the end of the electrode too far above the base metal.

These problems will be eliminated as you gain experience in striking and controlling the arc.

8.0.0 Arc Blow

When current flows, especially DC current, strong magnetic fields may be created. The magnetic fields tend to concentrate in corners, in deep grooves, or in the ends of the base metal. When the arc approaches these concentrated magnetic fields, it is deflected. This phenomenon is known as **arc blow**. In some cases, ferrous metal welding fixtures or jigs that are part of a DC welding current path and subjected to repeated current flow can become magnetized and contribute to arc blow. In AC welding, arc blow is rarely a problem because the magnetic field is constantly reversing at twice the frequency of the primary power source. This effectively cancels any strong magnetic field effects. DC arc blow can cause defects such as excessive weld spatter and porosity. If arc blow occurs, try one or more of these methods to control it:

- Change the position of the workpiece lead clamp. This will change the flow of welding current, affecting the way the magnetic fields are created.
- Shorten the arc length. With a shorter arc length, the magnetic field will have less effect on the arc.
- Change the angle of the electrode (*Figure 15*). The normal electrode angle is 10 degrees to 15 degrees in the direction of travel. Raising the electrode angle toward 90 degrees or, in extreme cases, up to 20 degrees in the opposite direction of travel will compensate for the arc blow.
- If possible, tack-weld the workpieces together at the ends and in the middle.
- Periodically **degauss** welding fixtures or jigs with an AC-powered degaussing coil or probe.

Constant-Current Machines

The AC or DC machines desired for SMAW or GTAW are basically constant-current types of machines, some of which are referred to as droop-current machines or droopers. In a true constant-current machine, the output current varies little over a relatively wide range of circuit voltage. Because of this, raising and lowering the electrode from the workpiece has little effect on the weld quality. In addition, striking an arc is easier with a true constant-current machine because of the high open-circuit voltage. On the other hand, droopers can be used to control the molten weld pool temperature more easily. This is because the current and resultant weld heat change much more with a smaller change in voltage that is caused by raising or lowering the electrode in relation to the workpiece.

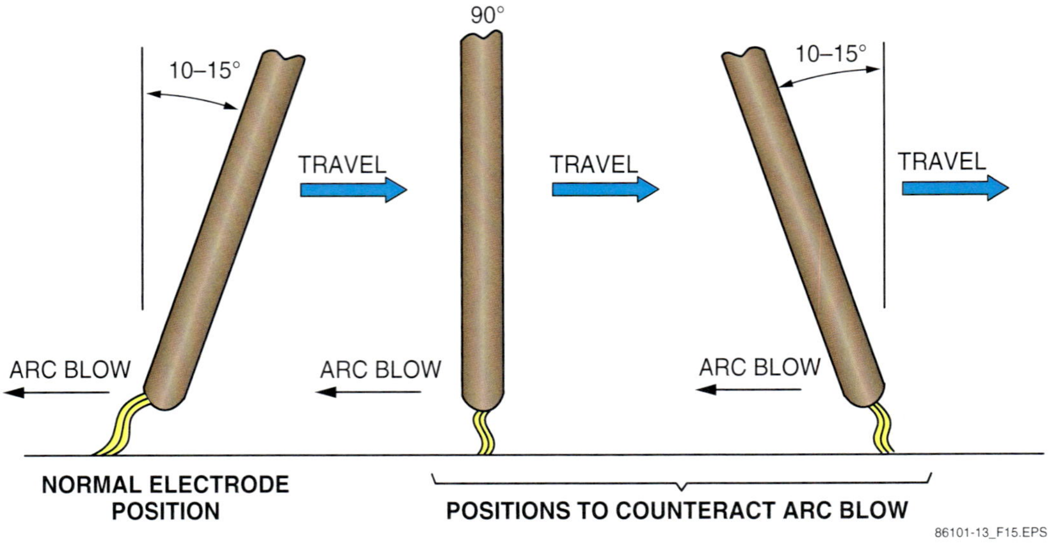

Figure 15 Controlling arc blow.

9.0.0 Tack Weld Beads

A tack weld bead (*Figure 16*) is a weld bead that is made with very little or no side-to-side motion of the electrode. The width of a tack weld bead will vary with the electrode type and size. The actual length and depth of a tack weld bead depends on the size, weight, and thickness of the materials being tacked together. This is usually dictated by the specific instructions or drawings of the WPS.

Tack welds must be placed in accordance with applicable drawings or specifications. The width of tack weld beads is usually specified in the drawing welding symbols. Do not exceed the widths or depths specified for the work. Finished tack welds must meet established quality standards. Poor welds will be rejected during inspection. *Figure 17* shows examples of good and bad tack welds.

9.1.0 Practicing Tack Weld Beads with E6010

Practice running tack weld beads in the flat (F1) position using 3⁄32" or 1⁄8" E6010 or E6011 electrodes. After striking the arc, the electrode axis angle

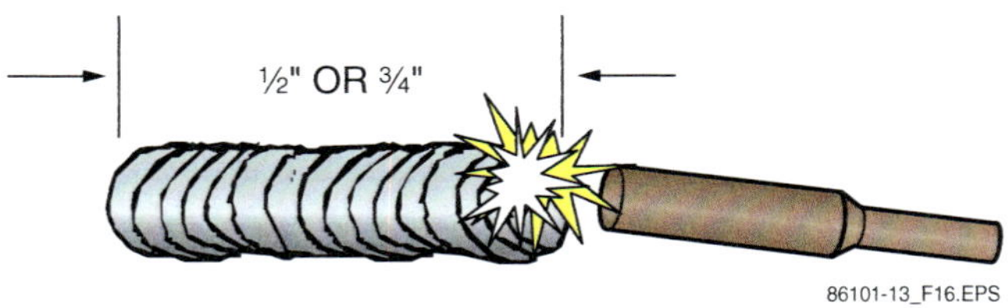

Figure 16 A tack weld bead.

Restriking a Cold Electrode

When attempting to strike an arc using a cold, partially consumed electrode, the cup formed by the hardened flux sheath extending beyond the metal core may not allow the core to contact the workpiece to establish an arc. Scratch striking or hard tapping can be used to break off the sheath. Unfortunately, this sometimes causes large pieces of the flux coating to break off the sides, exposing part of the core. Electrodes with missing flux are difficult to strike. If an arc is struck, it will be unstable and will blow or wander until the core is consumed up to where the entire flux sheath cup is reestablished. During this time, poor welds will occur. One method of removing the flux sheath without chipping flux away from the sides of the rod is to draw the end of the rod across the face of a medium-coarse file. If code-requirement welding is being performed, electrodes with chipped, missing, or cracked flux coatings must be discarded.

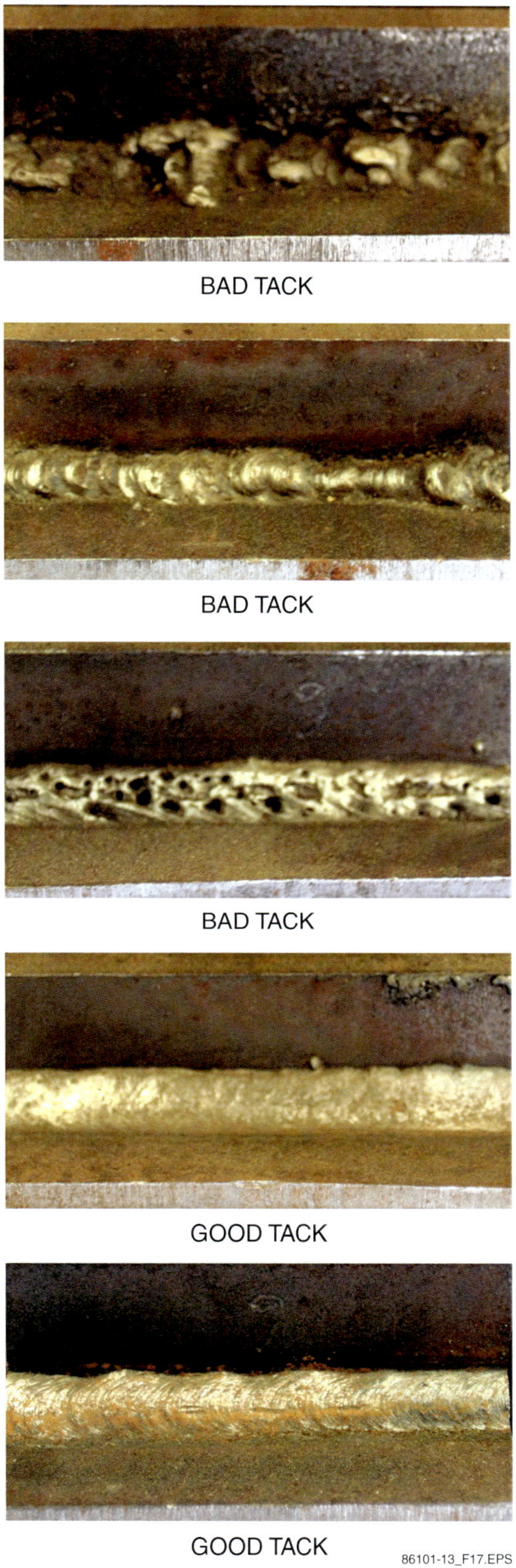

Figure 17 Examples of good and bad tack welds.

should be a 10-degree to 15-degree **drag angle** in the direction of travel for the weld and at a zero **work angle** that is perpendicular (90 degrees) to the base metal across the weld. *Figure 18* shows electrode axis angles.

Travel angles used for SMAW and other types of welding are categorized as either drag or **push angles**. An electrode drag angle is a travel angle in which the electrode axis points at the weld bead during the running of a tack weld bead. A push angle is the opposite condition; it occurs when the electrode axis points away from the weld bead and toward the direction of travel when running a weld bead. Except for vertical welds, SMAW is usually accomplished with an electrode drag angle. Electrode work angle is the side-to-side tilt of a plane containing the electrode axis from a perpendicular line to the major work surface. The plane is formed by the intersection of the **weld axis** and the electrode axis. Work angles typically range from 0 to 25 degrees. In the case of a single tack bead being run on a flat plate, the work angle is zero. The work angle for a fillet weld in a T- or corner joint is always taken from a perpendicular line to the non-butting member.

The whipping motion (*Figure 19*), also called a stepping motion, can be used when depositing the tack weld bead to control the weld puddle. It is performed by moving the electrode up and forward about 1/4", then down and backward about 3/16". A short pause at the end of the backward travel deposits the weld metal. The exact length of travel can vary. By momentarily lengthening and advancing the arc, the molten weld pool is allowed to cool. Lengthening the arc lowers the arc temperature (amperage) and reduces metal transfer from the rod core. Advancing the arc preheats the base metal ahead of the weld and burns off contaminants.

Carefully observe and listen to the weld as it is being made. If the arc length, speed, angles, current selection, and motions are correct, a distinctive frying sound will be heard. Experiment by

Reduced Open Circuit Voltage Welding Machines and Inverter Power Sources

When not actively welding, some welding machines or inverter power sources used in the SMAW mode deactivate open-circuit voltage at the electrode. Only a very low sensing voltage exists. For these machines, the scratch method must be used to activate the welding voltage and current and start the arc unless an adjustable starting current capability exists on the machine.

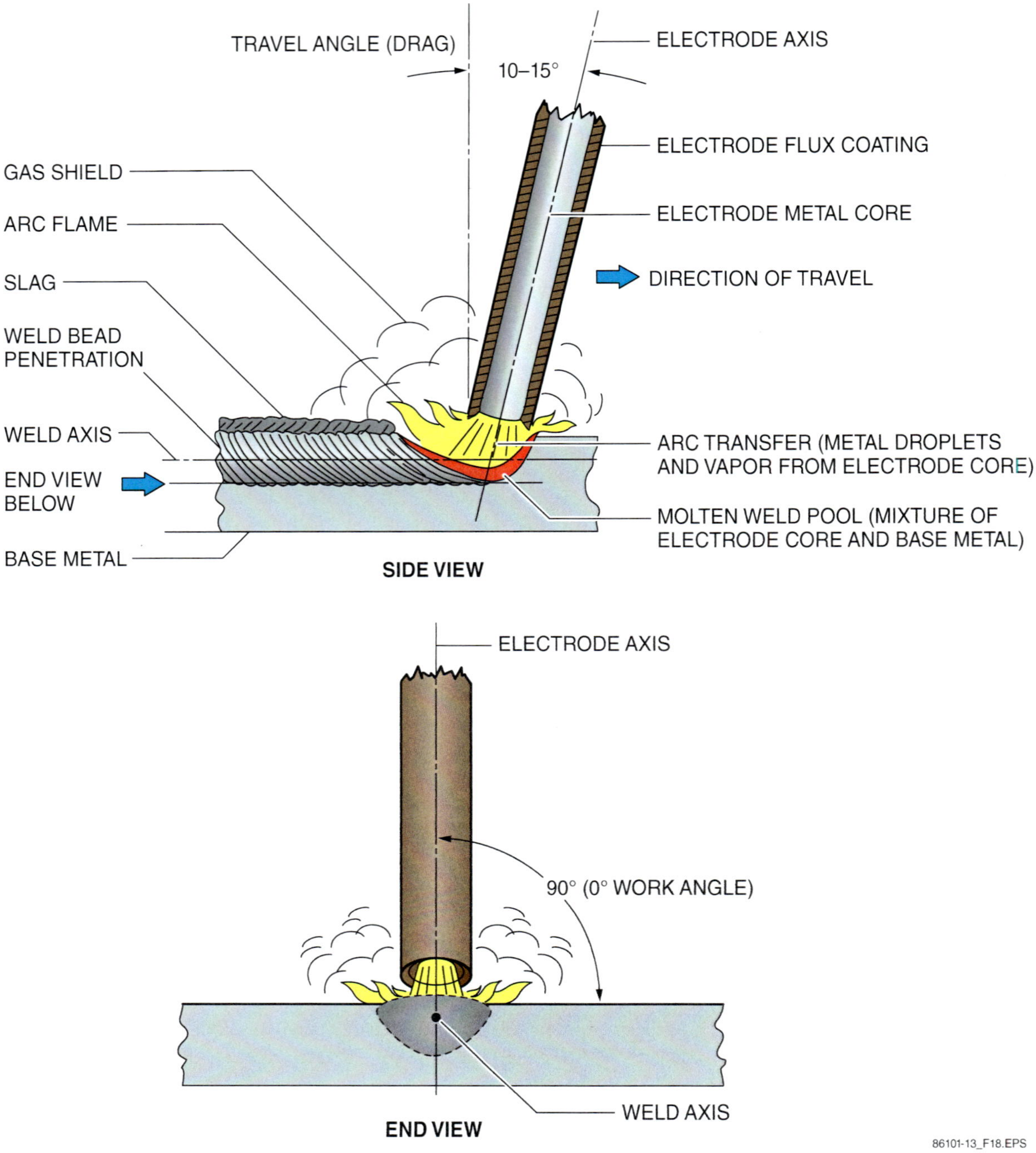

Figure 18 SMAW characteristics with proper travel and work angle in a flat position.

making minor changes in these factors and notice the effects.

Follow these steps to run tack weld beads:

Step 1 Shield your eyes, strike the arc, and move the electrode back slightly to the correct starting point. Then position the electrode for the travel angle and perpendicular to the base metal.

Step 2 Hold the arc in place until the weld puddle widens to about two times the diameter of the electrode.

Step 3 Slowly move the arc forward, keeping a constant arc length. As necessary, use the whipping motion and adjust the forward speed to control buildup and preheat the weld zone.

WARNING!

Always wear face protection to prevent hot slag from striking your face.

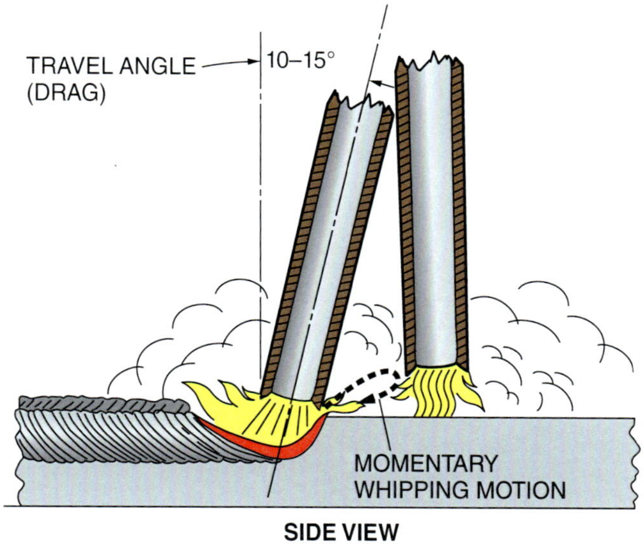

Figure 19 Whipping motion.

Step 4 Continue to weld until a bead about ½" to ¾" long is formed and then break the arc by quickly lifting the electrode straight up. A crater will be left at the point where the arc was broken.

Step 5 Chip the weld slag with a chipping hammer.

Step 6 Clean the bead with a wire brush.

Step 7 Have your instructor inspect the bead for the following:

- Straightness
- A uniform appearance on the bead face
- A smooth, flat transition with complete fusion at the toes (edges) of the weld
- Lack of porosity
- No undercut at the toes (edges)
- Freedom from inclusions
- No cracks
- Freedom from overlap

Step 8 Repeat steps 1 through 7 until you can make acceptable welds every time. *Figure 20* shows proper and improper weld beads.

9.2.0 Practicing Tack Weld Beads with E7018

Continue running tack weld beads, this time using 3⁄32" and ⅛" E7018 electrodes. Use the same electrode angles, but do not whip the electrode. When running all low-hydrogen electrodes, the arc should never leave the weld puddle, and the visible arc should be shorter than when using the E6010, due to the thicker, longer sheath of the flux coating. Weld defects, such as porosity or hydrogen embrittlement, can occur if the arc leaves the weld puddle or if the arc is too long. The arc can be moved within the weld puddle to control the bead shape.

10.0.0 Fillet Tack Welds

A fillet tack weld is a weld that is approximately triangular in cross section and is used with T-, lap, and corner joints. The sizes and locations of fillet tack welds are given as welding symbols.

Fillet tack welds should have a uniform concave or convex face (*Figure 21*), although a slightly non-uniform face is acceptable. The convexity of a fillet weld or individual surface bead should not exceed 0.07 times the actual face width or individual surface bead plus 0.06". A fillet tack weld is unacceptable and must be repaired if the profile has insufficient throat (depth) or excessive con-

Accurate Tap Striking of an Electrode

An accurate method of tap striking an electrode at an exact spot is to rest the electrode on the finger of your gloved, free hand, similar to the way you would support a pool cue. Angle the rod in the direction of travel. Then move the electrode quickly down and up to strike the arc. This method is feasible if your helmet can be lowered by nodding your head, or if you are using an auto-darkening lens helmet.

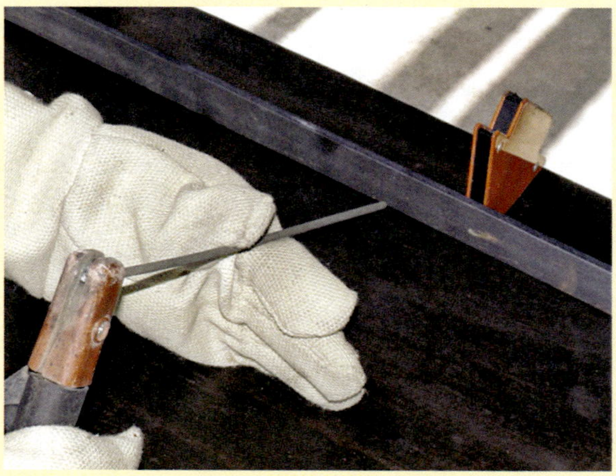

86101-13_SA03.EPS

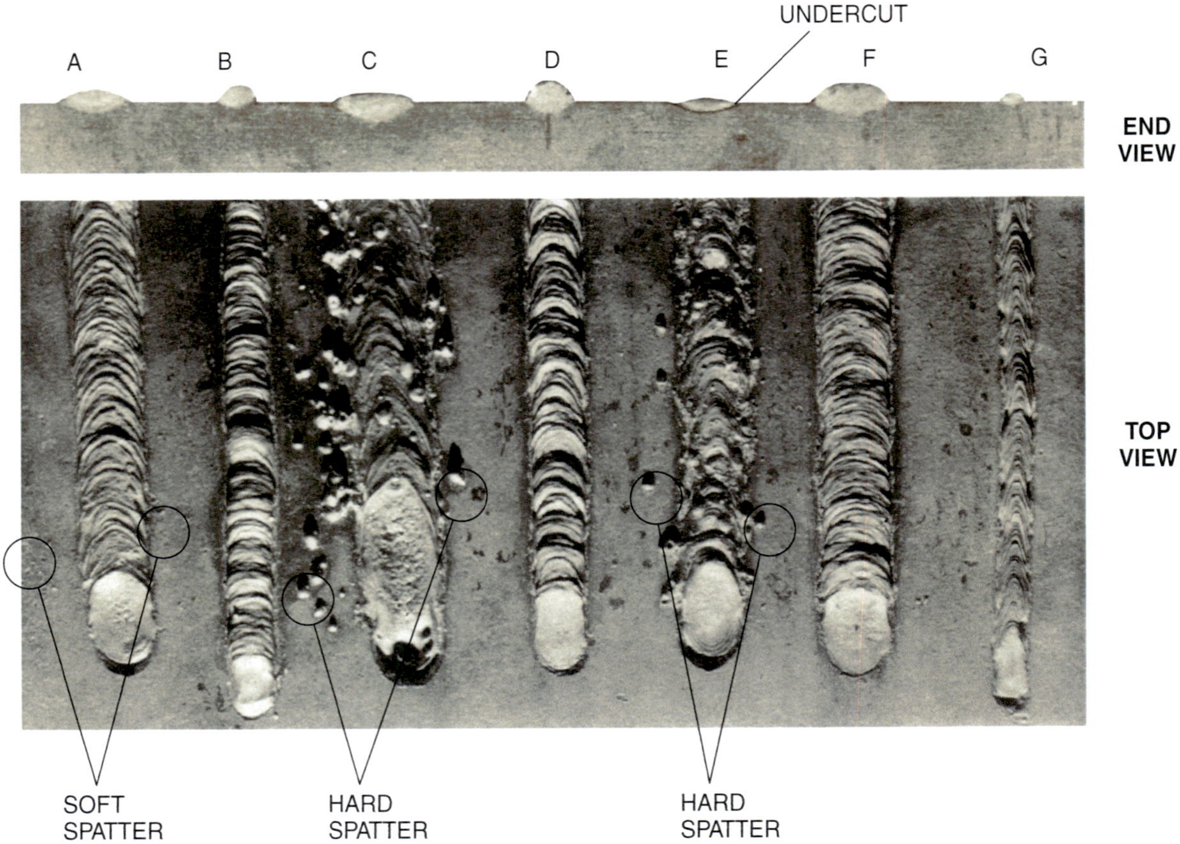

A = Correct current, arc length, and travel speed. Note the easily removed spatter (soft spatter).

B = Current set too low.

C = Current set too high. Note the hard spatter (spatter firmly bonded to base material that must be ground or chiseled off). Note the pointed ends of the bead indicating the weld pool was too hot and cooled too slowly. Impurities are usually trapped in the weld due to the slow cooling.

D = Arc length too short (narrow, high bead caused by arc pressure).

E = Arc length too long. Note the hard spatter and bead undercut of the edges.

F = Travel speed too slow (wide high bead).

G = Travel speed too fast. Note the pointed ends of bead.

© American Welding Society (AWS) *Welding Handbook.*

86101-13_F20.EPS

Figure 20 Effect of current, arc length, and travel speed on SMAW beads.

Reusing a Stuck Electrode

Carefully inspect the end of any electrode that has been stuck to the workpiece and freed by bending. If any of the flux coating is missing, cracked, or badly burned, discard the electrode.

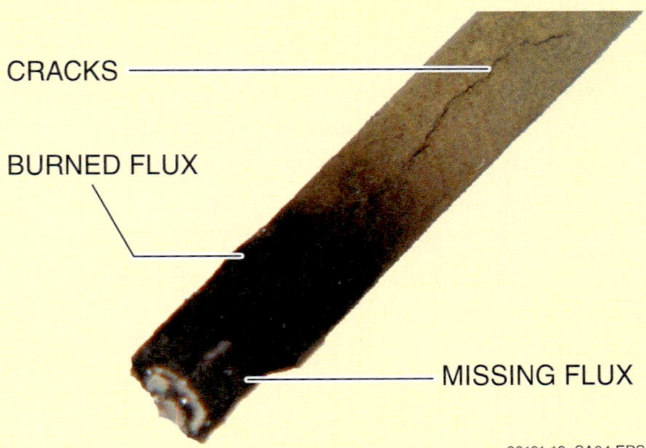

86101-13_SA04.EPS

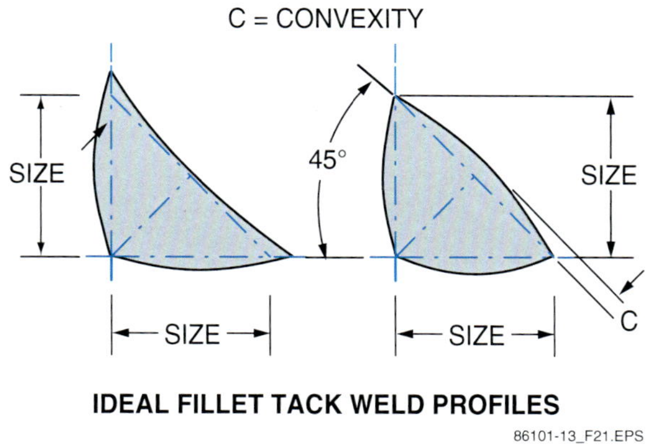

Figure 21 Ideal fillet tack weld profiles.

vexity (C). In all cases, the fillet tack weld must be smaller in all respects to the final specified fillet weld for the joint.

> **NOTE**
>
> Refer to your site's WPS for specific requirements on fillet welds. The information in this module is provided as a general guideline only. The site WPS or quality specifications must be followed for all welds. Check with your supervisor if you are unsure of the specifications for your application.

Fillet tack welds require little base metal preparation except for cleaning the weld area and removing all dross from cut surfaces. Any dross from oxyfuel, plasma arc, or carbon arc cutting will cause porosity in the weld. For this reason, the codes require that all dross be removed prior to welding.

10.1.0 Fillet Weld Positions

The most common fillet welds are made in lap and T-joints. The weld position for plate is determined by the weld axis and the orientation of the workpiece. The positions for fillet welding on plate are flat, or 1F (the letter F stands for fillet); horizontal, or 2F; vertical, or 3F; and overhead, or 4F (*Figure 22*). In the 1F and 2F positions, the weld axis can be inclined up to 15 degrees. Any weld axis inclination for the other positions varies with the rotational position of the weld face as specified in AWS standards.

10.2.0 Practicing Horizontal Fillet Tack Welds with E6010 (2F Position)

Practice horizontal fillet (2F) tack welding by placing fillet tack welds in a T-joint using 3⁄32" or 1⁄8" E6010 electrodes. When making horizontal fillet tack welds, pay close attention to the electrode an-

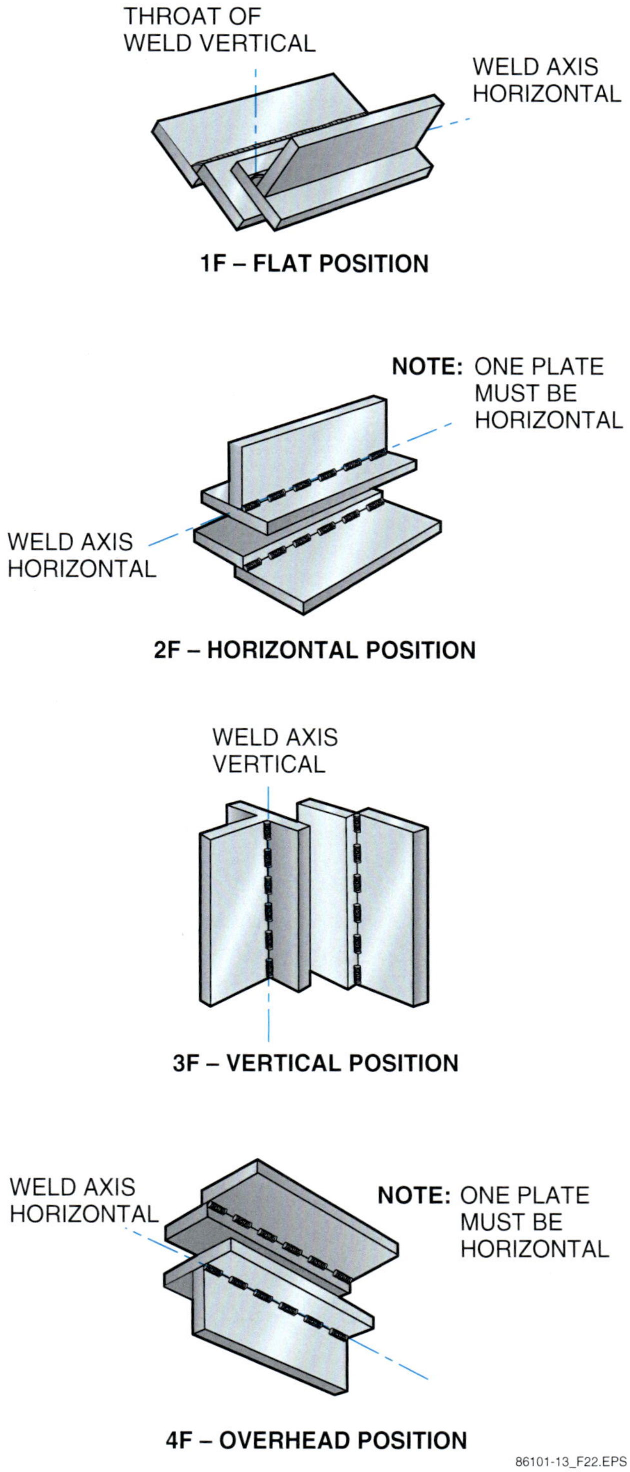

Figure 22 Fillet welding positions for plate.

gle and travel speed. The electrode work angle is 45 degrees. Increase or decrease the travel speed to control the amount of weld-metal buildup.

Follow these steps to make a horizontal fillet tack weld:

Step 1 In the 1F position, tack two plates together to form a T-joint for the fillet tack weld coupon (*Figure 23*).

Step 2 Position the coupon on the welding table in the 2F position.

Step 3 Run a number of ½" to ¾" long tack weld beads along the root of the joint using an electrode work angle of approximately 45 degrees with a 10- to 15-degree drag angle (*Figure 24*). Use a stepping motion and push the arc into the root. Keep the root of the joint fusing together or else a notch will appear on the leading edge of the bead.

Step 4 Clean all the tack welds by chipping the slag and using a wire brush.

Step 5 Have the instructor inspect the tack welds. The welds are acceptable if they have the following characteristics:

- Uniform appearance on the bead face
- Uniform weld size ±1⁄16"
- Acceptable weld profile
- No porosity
- No undercut
- No inclusions

10.3.0 Practicing Horizontal Fillet Tack Welds with E7018 (2F Position)

Repeat horizontal fillet (2F) tack welding using ⅛" E7018 electrodes. Use the same procedure and electrode angle that was used for the horizontal fillet tack welds with E6010 electrodes. Use a short arc and do not whip the electrode.

NOTE: BASE METAL, CARBON STEEL AT LEAST ¼" THICK

Figure 23 Fillet tack weld coupon.

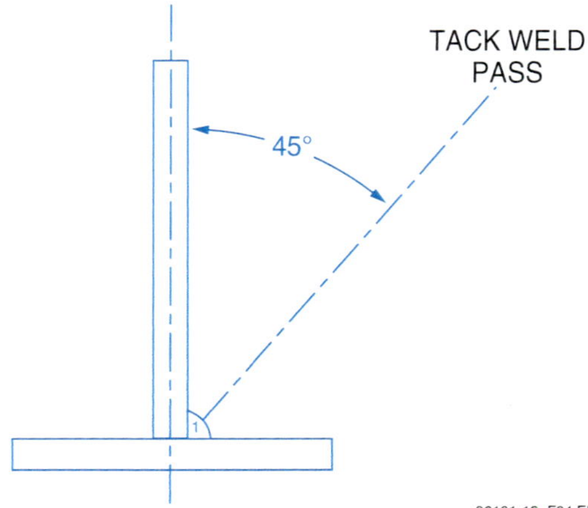

Figure 24 2F tack weld pass and work angle.

10.4.0 Practicing Vertical Fillet Tack Welds with E6010 (3F Position)

Practice vertical fillet (3F) tack welding by placing fillet welds in a T-joint using 3⁄32" or ⅛" E6010 electrodes. Normally, vertical welds are accomplished by welding uphill from the bottom to the top using an electrode push angle (up-angle). Because of the uphill welding and push angle, this type

Stabilize Your Welding Position, Watch the Immediate Welding Area, and Listen to the Arc

A welder should be in a relaxed, comfortable position when welding. Because of the limited view of the welding area through a dark helmet lens, a beginning welder may sway due to a loss of sense of balance. To counteract this, a welder should sit or lean against something to achieve and maintain a stable and relaxed position. This will reduce fatigue and ensure personal safety.

Beginning welders must learn to view the entire electric arc welding work area through the shaded lens of a helmet and to listen to the sound of the arc. At first, a beginning welder's focus is normally concentrated only on the electric arc, because of the difficulty of striking and keeping an arc established. Once beginners have managed to strike and maintain an arc, however, the tremendous light and heat energy generated within the arc are usually fascinating and distracting for a time. With practice, the beginning welder will gradually be able to shift focus and see all sides of a molten weld pool, the weld buildup at the trailing edge of the pool, the cooling slag cover over the weld buildup, and the adjacent welding area on the workpiece. The welder will become attuned to the correct sound of the arc for the rod and work being welded, and the arc itself may not even be noticed. Once the welder is able to view the entire weld area and listen to the arc, running straight and correct welds is relatively easy.

of weld is sometimes called vertical-up fillet tack welding. On the job, the site WPS or quality standard will specify which technique to use.

Follow these steps to make a vertical-up fillet tack weld:

Step 1 In the 1F position, tack two plates together to form a T-joint for the fillet tack weld coupon.

Step 2 Clamp or tack-weld the coupon in the vertical position.

Step 3 Run a number of ½" to ¾" long tack welds along the root of the joint (starting at the bottom) using an electrode (work) angle of approximately 45 degrees with a 0-degree to 10-degree push angle (*Figure 25*). Use a whipping motion by quickly raising the electrode about ¼" and then dropping it back into the weld puddle. Pause in the weld puddle to fill the crater.

Step 4 Clean the weld by chipping the slag and wire brushing the weld.

Step 5 Have the instructor inspect the tack welds. The welds are acceptable if they have the following characteristics:

- Uniform appearance on the bead face
- Uniform weld size ±1⁄16"
- Acceptable weld profile
- No porosity
- No overlap
- No undercut
- No inclusions

10.5.0 Practicing Vertical Fillet Tack Welds with E7018 (3F Position)

Repeat vertical fillet (3F) tack welding using ⅛" E7018 electrodes. Make the tack welds without removing the arc from the weld puddle. The electrode work angle should be approximately 45 degrees with a 0- to 10-degree upward push angle (*Figure 26*). Remember to use a short arc. Do not whip the electrode.

10.6.0 Practicing Overhead Fillet Tack Welds with E6010 (4F Position)

Practice overhead fillet (4F) tack welding by welding convex fillet tack welds in a T-joint using 3⁄32" or ⅛" E6010 electrodes. When making overhead fillet tack welds, pay close attention to the electrode angles and travel speed. Increase or decrease the travel speed to control the amount of weld metal buildup.

Arc Shifting or Wandering

When two workpieces are first being joined, the arc may deposit material on one piece only or wander back and forth from one piece to the other. Both of these occurrences are usually the result of too long an arc. If this is the case, the arc will always jump to the closest workpiece. This is especially noticeable with electrodes that are ⅛" or smaller when welding gapped-butt or T-joints. If the arc initially deposits material on only one workpiece, it is possible that only one of the workpieces is electrically connected to the workpiece lead from the welder and the other piece is not in the circuit. In this case, weaving the electrode back and forth across the workpiece at the start of the weld will usually bridge the gap between the workpieces and tack the pieces together so that they are both in the circuit. Other solutions are to clamp a piece of metal across both pieces or to perform the welding on a metal tabletop.

Lash

To ensure good welding results, remember LASH:

L (length of arc) – The distance between the electrode and the base metal (usually one times the electrode diameter).

A (angle) – Two angles are critical:

- *Travel angle* – The longitudinal angle of the electrode in relation to the axis of the weld joint
- *Work angle* – The traverse angle of the electrode in relation to the axis of the weld joint

S (speed) – Travel speed is measured in inches per minute (IPM). The width of the weld will determine if the travel speed is correct.

H (heat) – Controlled by the amperage setting and dependent upon the electrode diameter, base metal type, base metal thickness, and the welding position.

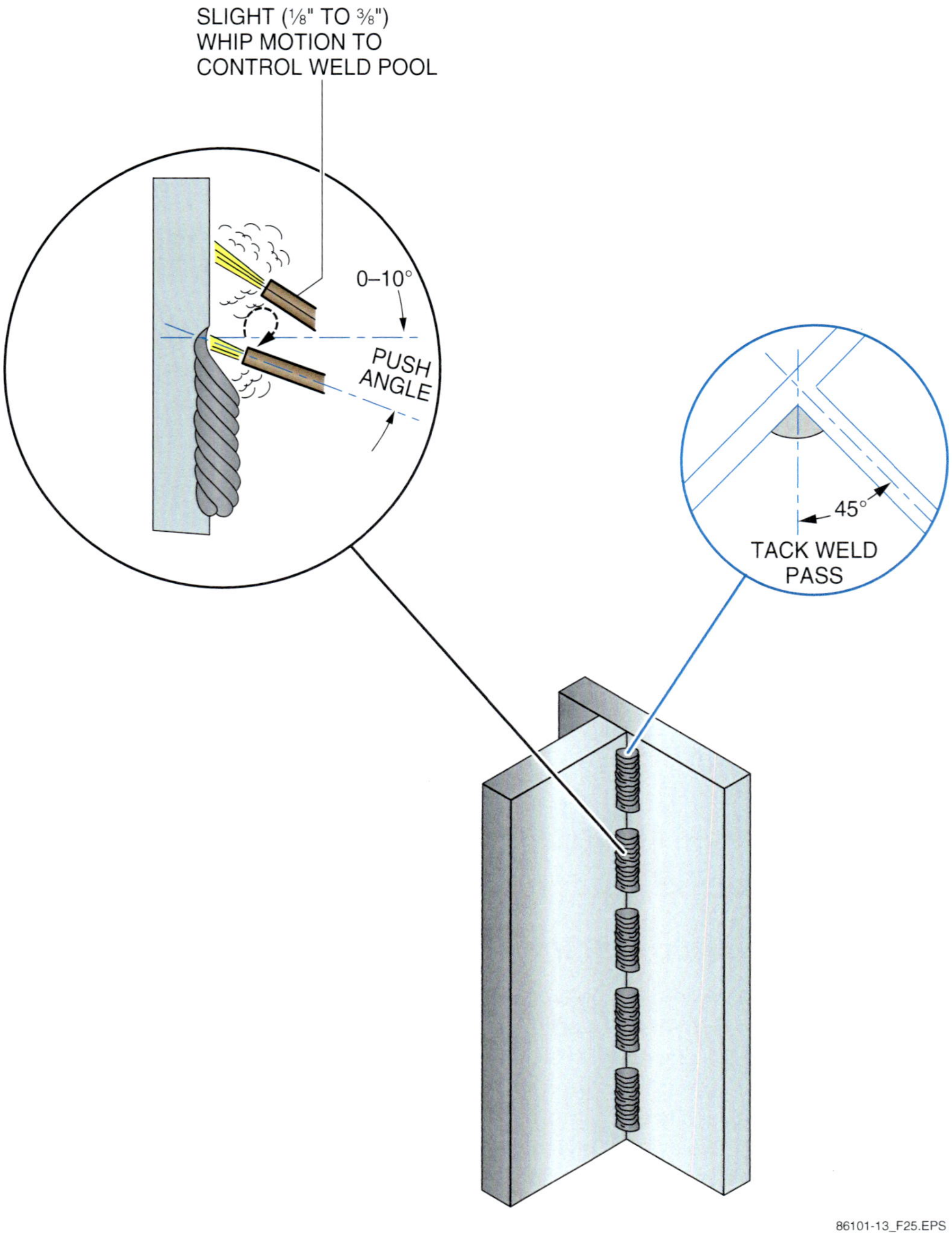

Figure 25 3F tack weld pass and work angle (E6010).

Follow these steps to make an overhead fillet weld:

Step 1 In the 1F position, tack two plates together to form a T-joint for the fillet tack weld coupon.

Step 2 Clamp or tack-weld the coupon so it is in the overhead position.

Step 3 Run a number of using ½" to ¾" long tack welds along the root of the joint using an electrode angle of approximately 45 degrees with a 10-degree to 15-degree drag angle (*Figure 27*).

Step 4 Clean the weld.

Step 5 Have the instructor inspect the tack welds. The welds are acceptable if they have the following characteristics:

- Uniform appearance on the bead face
- Uniform weld size ±1/16"
- Acceptable weld profile
- No porosity
- No overlap
- No undercut
- No inclusions

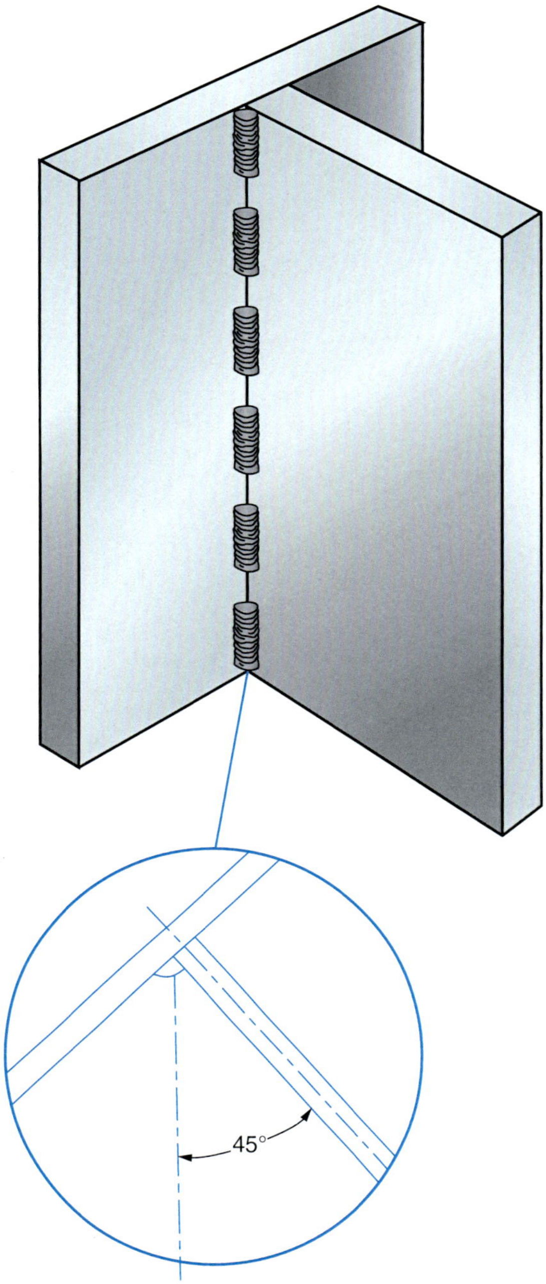

Figure 26 3F tack weld pass and work angle (E7018).

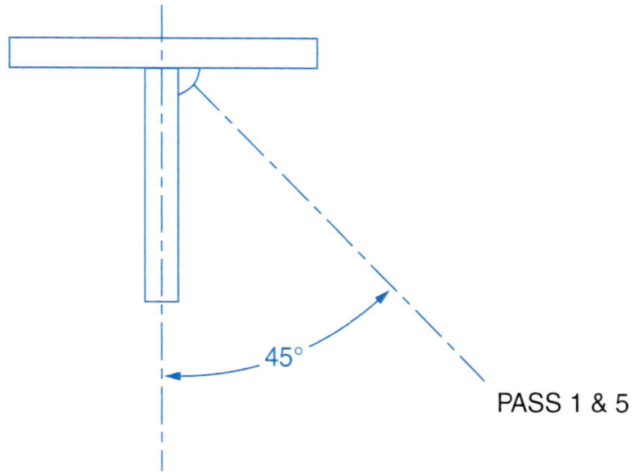

Figure 27 4F tack weld sequence and work angle.

10.7.0 Practicing Overhead Fillet Tack Welds with E7018 (4F Position)

Repeat overhead fillet (4F) tack welding using ⅛" E7018 electrodes. Use the same procedure and electrode angles that were used for the overhead fillet tack weld with E6010 electrodes. Use a short arc and do not whip the electrode.

11.0.0 Practice Weldments

Maritime training facilities may have structural fitter trainees assemble practice weldments or frames as part of the training for fit-up and tacking. This is done after the fitters become proficient at using welding machines and running tack welds in all positions. *Figure 28* shows an example of a fitted-up practice weldment and *Figure 29* shows a structural fitter trainee tacking a vertical part of a practice weldment.

T-Joint Vertical Plate Undercut

The most common defect for T-joints is undercut on the vertical plate of the joint. However, angling the arc slightly toward the vertical plate will force more metal into the bead at the top edge of the weld.

Tacking and Aligning Workpieces

When tacking workpieces together, both sides of the workpieces are usually tacked with welds that are about ½" long, to position the workpieces and minimize distortion when the practice fillet tack welds are made. After the first tack weld, use a hammer or other tool to align the workpieces side to side and end to end; then tack the opposite side. Tack the far ends of the workpieces in the same manner.

86101-13_F28.EPS

Figure 28 Example of a practice weldment.

Vertical Fillet Welds

When making vertical fillet welds, pay close attention to the electrode angles and travel speed. For the first bead, the electrode work angle is approximately 45 degrees. The work angle is adjusted for all other welds. Increase or decrease the travel speed to control the amount of weld metal buildup.

86101-13_F29.EPS

Figure 29 Structural fitter trainee tacking vertical parts of a practice weldment.

Test Joint Heat Dissipation

In T-joints, the welding heat dissipates more rapidly in the thicker or the non-butting member. During a tack weld pass, the arc may have to be concentrated slightly more on the thicker or the non-butting member to compensate for the heat loss.

Summary

Striking an arc, running short tack weld beads, and making fillet tack welds in all positions are essential skills a structural fitter must learn to perform using SMAW electrodes. It is important to practice these welds until acceptable welds are produced. In addition, structural fitters must understand various industrial welding machines, the SMAW process, and their responsibilities during fit-ups. They must also be able to set up the welding machine and use some basic accessories to achieve a reliable weld while maintaining a safe working environment.

Review Questions

1. SMAW is often referred to as _____.
   a. AC welding
   b. stick welding
   c. gap welding
   d. resistive welding

2. When working in a confined space, the oxygen and acetylene tanks must be left outside the space.
   a. True
   b. False

3. During SMAW welding, current jumps across the _____.
   a. workpiece clamp from the tip of the electrode to the surface of the workpiece
   b. gap from the tip of the electrode to the surface of the workpiece
   c. electrode holder from the tip of the electrode to the surface of the workpiece
   d. molten metal from the tip of the electrode to the surface of the workpiece

4. In AC current, the number of cycles completed in one second is called the _____.
   a. polarity
   b. travel
   c. frequency
   d. current value

5. Electrical current that has no frequency is known as _____.
   a. DC current
   b. AC current
   c. AC voltage
   d. 120VAC

6. Which of the following only applies to DC current?
   a. Frequency
   b. Polarity
   c. Amperage
   d. Current flow

7. The voltage in a welding machine is reduced or stepped down by a(n) _____.
   a. limiter
   b. generator
   c. alternator
   d. transformer

8. The duty cycle of a welding machine is based on a _____.
   a. 10-minute period
   b. 15-minute period
   c. 20-minute period
   d. 30-minute period

9. To prevent the lead from pulling loose from the welding machine terminal, welding cable lugs are generally _____.
   a. slotted
   b. soldered in place
   c. ringed
   d. insulated

10. When preparing to practice welding of fillet tack welds, the metal should be cut into _____.
    a. 3" × 12" rectangles for the base
    b. 4" × 12" rectangles for the base
    c. 4" × 4" squares for the base
    d. 12" × 12" squares for the base

11. The easiest method of striking an arc is the _____.
    a. scratching method
    b. grounding method
    c. weaving method
    d. tapping method

12. Arc blow is caused by _____.
    a. strong winds
    b. magnetic fields
    c. fluctuating weld current
    d. damaged electrodes

13. A short weld bead with very little or no side-to-side motion of the electrode is called a(n) _____.

   a. arc bead
   b. weave bead
   c. tack weld bead
   d. whipping bead

14. After striking the arc when practicing tack welds in the flat position, the electrode axis angle should be 10 to 15 degrees in the direction of travel and at a 0 work angle that is _____.

   a. 30 degrees to the base metal
   b. 45 degrees to the base metal
   c. 60 degrees to the base metal
   d. 90 degrees to the base metal

15. The part of a fillet weld that can be convex or concave is called the _____.

   a. leg
   b. throat
   c. face
   d. root

Trade Terms Quiz

Fill in the blank with the correct term that you learned from your study of this module.

1. ________ is electrical current received from conventional power lines used to energize welding machines.
2. ________ is electrical current that reverses its flow at set intervals.
3. ________ is the measurement of electromotive force or pressure that causes current to flow in a circuit.
4. ________ is electrical current that flows in one direction only.
5. An electrical device that uses two wire coils of different sizes to convert a high-voltage, low-current power source into a lower voltage, higher current output is known by the term ________.
6. A measurement of the rate of flow of electric current is called ________.
7. The direction of flow of electrical current in a direct current welding circuit is called ________.
8. The flow of electricity through an air gap or gaseous space is called a(n) ________.
9. When magnetic forces deflect the arc from its intended path, the action is called ________.
10. A straight line drawn through the center of a weld and along its length is called the ________.
11. ________ is a discontinuity consisting of localized melting of the base metal or finished weld caused by the arc.
12. ________ is an angle that is less than 90 degrees between a line perpendicular to the surface of the workpiece and a plane determined by the electrode axis and the weld axis.
13. ________ is the travel angle at which the electrode is pointing in the same direction as the welding progression.
14. ________ is the travel angle in which the electrode is pointing in a direction opposite to the welding progression.
15. A piece of metal to be welded upon or welded together with another piece as a test or practice is called a(n) ________.
16. A welding method that uses a consumable wire electrode fed through a welding gun is ________.
17. A welding method that uses a non-consumable electrode is ________.
18. The document that contains detailed methods and practices required to produce a weld is the ________.
19. To neutralize a magnetic field using equipment that produces an opposing magnetic field is to ________ it.

Trade Terms

Alternating current (AC)
Amperage
Arc
Arc blow
Arc strike
Degauss
Direct current (DC)
Drag angle
Flux-cored arc welding (FCAW)
Gas metal arc welding (GMAW)
Polarity
Primary current
Push angle
Step-down transformer
Voltage
Weld axis
Weld coupon
Welding procedure specification (WPS)
Work angle

Trade Terms Introduced in This Module

Alternating current (AC): Electrical current that reverses its flow at set intervals.

Amperage: A measurement of the rate of flow of electric current.

Arc blow: The deflection of the arc from its intended path by magnetic forces.

Arc strike: A discontinuity consisting of localized melting of the base metal or finished weld caused by the arc. It occurs as a result of striking the arc outside the area to be welded.

Arc: The flow of electricity through an air gap or gaseous space.

Degauss: To neutralize a magnetic field by using equipment that produces an opposing magnetic field.

Direct current (DC): Electrical current that flows in one direction only.

Drag angle: The travel angle when the electrode is pointing in a direction opposite to the welding progression.

Gas metal arc welding (GMAW): In gas metal arc welding, sometimes called MIG, a consumable wire fed through the welding gun is used to make the weld; commonly used on steel and other metals.

Flux cored arc welding (FCAW): A welding method that uses an electric arc to melt a flux-cored wire electrode that is fed through the welding gun.

Polarity: The direction of flow of electrical current in a direct current welding circuit.

Primary current: Electrical current received from conventional power lines used to energize welding machines.

Push angle: The travel angle at which the electrode is pointing in the same direction as the welding progression.

Step-down transformer: An electrical device that uses two wire coils of different sizes to convert a high-voltage, low-current power source into a lower voltage, higher current output.

Voltage: A measurement of electromotive force or pressure that causes current to flow in a circuit.

Weld axis: A straight line drawn through the center of a weld along the length of the weld.

Weld coupon: Metal pieces to be welded together as a test or practice.

Welding procedure specification (WPS): The document containing all the detailed methods and practices required to produce a sound weld.

Work angle: An angle less than 90 degrees between a line perpendicular to the major workpiece surface and a plane determined by the electrode axis and the weld axis. In a T-joint or corner joint, the line is perpendicular to the non-butting member. The definition of work angle for a pipe weld is covered in a later module.

Additional Resources

This module presents thorough resources for task training. The following resource material is suggested further study.

Stick Electrode Welding Guide, 2004. Cleveland, OH: The Lincoln Electric Company.
www.lincolnelectric.com

Figure Credits

Topaz Publications, Inc., Module opener, Figures 6A, 6B, 7–10, 23, 28, 29, SA01B, SA02–SA04

Lincoln Electric Company, Figures 2A, 4, 5

Sumner Manufacturing Company, Inc., Figure 6C

Part #GR-85130 is available from TP Tools & Equipment @ 1-800-321-9260, **www.tptools.com**, Figure 6D

Gullco International, Inc., SA01A

©American Welding Society (AWS) *Welding Handbook* 1991, Welding Processes Volume No. 2, Edition No. 8, Miami: American Welding Society, Figure 20

NCCER CURRICULA — USER UPDATE

NCCER makes every effort to keep its textbooks up-to-date and free of technical errors. We appreciate your help in this process. If you find an error, a typographical mistake, or an inaccuracy in NCCER's curricula, please fill out this form (or a photocopy), or complete the online form at **www.nccer.org/olf**. Be sure to include the exact module ID number, page number, a detailed description, and your recommended correction. Your input will be brought to the attention of the Authoring Team. Thank you for your assistance.

Instructors – If you have an idea for improving this textbook, or have found that additional materials were necessary to teach this module effectively, please let us know so that we may present your suggestions to the Authoring Team.

NCCER Product Development and Revision

13614 Progress Blvd., Alachua, FL 32615

Email: curriculum@nccer.org
Online: www.nccer.org/olf

❑ Trainee Guide ❑ Lesson Plans ❑ Exam ❑ PowerPoints Other ____________________

Craft / Level: Copyright Date:

Module ID Number / Title:

Section Number(s):

Description:

Recommended Correction:

Your Name:

Address:

Email: Phone:

86102-14

Fire Watch

Module Seven

Trainees with successful module completions may be eligible for credentialing through NCCER's National Registry. To learn more, go to **www.nccer.org** or contact us at **1.888.622.3720**. Our website has information on the latest product releases and training, as well as online versions of our *Cornerstone* magazine and Pearson's product catalog.

Your feedback is welcome. You may email your comments to **curriculum@nccer.org**, send general comments and inquiries to **info@nccer.org**, or fill in the User Update form at the back of this module.

This information is general in nature and intended for training purposes only. Actual performance of activities described in this manual requires compliance with all applicable operating, service, maintenance, and safety procedures under the direction of qualified personnel. References in this manual to patented or proprietary devices do not constitute a recommendation of their use.

86102-14

Fire Watch

Objectives

When you have completed this module, you will be able to do the following:

1. Describe the different classes of fires and the methods used to extinguish them.
2. Explain the duties and responsibilities of a fire watch.
3. Describe the different types of fire extinguishers and their inspection requirements.

Performance Tasks

Under the supervision of the instructor, you should be able to do the following:

1. Demonstrate the RACE method in response to a simulated fire.
2. Demonstrate the PASS method on a simulated fire.
3. Complete a fire watch test.
4. Select the correct fire extinguisher for different classes of fires.
5. Inspect extinguishers and their components

Trade Terms

Combustible
Conduction
Flammable
Flash point
Incipient fire
Inerted
Radiation

Industry-Recognized Credentials

If you're training through an NCCER-accredited sponsor, you may be eligible for credentials from NCCER's Registry. The ID number for this module is 86102-14. Note that this module may have been used in other NCCER curricula and may apply to other level completions. Contact NCCER's Registry at 888.622.3720 or go to **www.nccer.org** for more information.

Contents

Topics to be presented in this module include:

Figures

1.0.0 Introduction

A fire watch is a person or persons assigned to monitor welding, torch cutting, brazing, and other types of hot work activities in order to prevent fires, to put out an **incipient fire** if one occurs, and to summon assistance if an incipient fire cannot readily be extinguished. A person performing fire watch duties must be trained for those duties and must be familiar with the different classes of fires and the specialized fire extinguishers used to combat them. Above all, a fire watch must remain alert at all times and must constantly observe the work area.

One important duty of the fire watch is to make sure **flammable** and **combustible** materials are removed from the hot work area or are otherwise protected from flames, sparks, and heat using fireproof coverings, curtains, or other shielding. Before proceeding further, the terms *flammable* and *combustible* need to be clarified. A flammable substance is one with a **flash point** below 100°F that will ignite readily and burn rapidly. A combustible substance is one with a flash point at or above 100°F that will ignite readily and burn rapidly.

OSHA has established the following eight specific conditions in which a fire watch must be posted:

1. Slag, weld, spatter, or sparks from welding or torch cutting operations (*Figure 1*) might pass through an opening and cause a fire.
2. Fire-resistant guards or curtains are not used to prevent ignition of combustible materials on or near decks, bulkheads, partitions, or overheads.
3. Combustible materials closer than 35' to the hot work in either the horizontal or vertical direction cannot be removed, protected with flameproof covers, or otherwise shielded with metal or fire-resistant guards or curtains.
4. The hot work is performed on or near insulation, combustible coatings, or sandwich-type construction that cannot be **inerted** (filled with an inert gas such as nitrogen).
5. Combustible materials adjacent to the opposite sides of bulkheads, decks, overheads, metal partitions, or sandwich-type construction may be ignited by **conduction** or **radiation**.
6. The hot work is close enough to cause ignition through heat radiation on the following:
   - Insulated pipes, bulkheads, partitions or overheads
   - Combustible materials and/or coatings
7. The work is close enough to unprotected combustible pipe or cable runs to cause ignition.
8. A marine chemist, a US Coast Guard authorized person, or a shipyard competent person, as defined in *29 CFR Part 1915 Subpart B*, requires that a fire watch be posted. These individuals have the authority to require a fire watch even if one or more of the conditions listed above is not present.

Take note of the final item in this list. The underlying message is that any area where hot work is to be performed must be inspected and certified beforehand to determine if a fire watch is necessary. A marine chemist certificate or log of inspection must be posted at the location. It must clearly state all of the safety conditions that must be satisfied before hot work can begin. If any of the conditions changes after the certificate has been posted, the area must be re-inspected and a new certificate is-

(A) OXYFUEL CUTTING

(B) WELDER AT WORK

86102-13_F01.EPS

Figure 1 Hot work in process.

sued before hot work can begin again. Fire watch certification is company-specific.

2.0.0 About Fires

Three components are required for a fire to occur: fuel, heat, and oxygen. A fire will stop burning if it is deprived of any of those three components. The fire tetrahedron shows that a fourth element – a chain reaction – is required to sustain a fire. This is represented by the fire triangle and fire tetrahedron (*Figure 2*). Although oxygen is one of the requirements, there is plenty of oxygen in the air we breathe, so the presence of air meets the requirement for oxygen. Fuel is anything that will burn, including paper, cardboard, wood, insulation, some chemicals, and certain liquids. Heat includes any heat source that will raise the temperature of a flammable material above its flash point, which is the temperature at which it will ignite. Once the flash point is reached and the material has ignited, it will continue to burn even though the heat source is removed, as long as there is oxygen and fuel.

Hot work generates heat in the form of flames, sparks, conducted heat, and radiated heat. Hot slag and sparks from oxyfuel cutting, for example, can land on nearby flammable debris and cause it to ignite. When metal such as steel is heated, the heat is conducted away from the point where the flame is applied and may cause ignition of material with which it is in contact several feet away from the immediate work area. By the same token, when metal such as steel is heated, the heat can be transferred through the air to nearby objects. If the temperature of the radiated heat exceeds the flash point of a flammable or combustible material, that material can ignite. Welding, for example, may heat the metal being welded to more than 2,000°F, which is far above the combustion point of many substances.

2.1.0 Classes of Fires

There are different classes of fires based on the material that is involved in the fire. *Figure 3* shows the standard symbols for these classes of fires. These symbols are used on extinguishers specifically designed for these classes of fires.

- *Class A* – This is a fire that involves ordinary combustibles, including paper, wood, cardboard, rubber, plastic, and fabric.

> **WARNING!**
> A fire that seems like it might be an ordinary Class A trash fire may contain objects such as spray cans that will explode when exposed to the heat of a fire.

- *Class B* – A Class B fire involves flammable liquids and gases such as oil, paint, paint thinner, diesel fuel, gasoline, fuel oil, or lubricating oil. It is important to note that it is the fumes on the surface of these liquids that is burning, rather than the liquid itself.
- *Class C* – This is a fire that involves energized electrical equipment. This type of fire can be caused by short circuits or overheating the wiring or terminations of a circuit. These fires require special extinguishing methods because

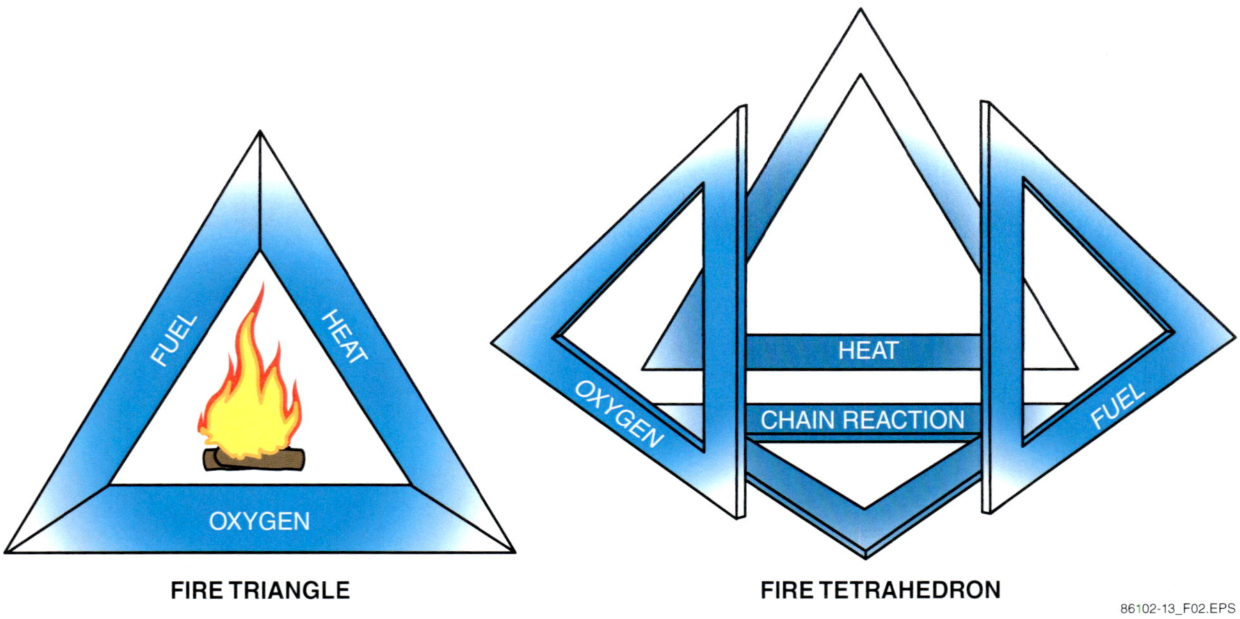

Figure 2 Fire triangle and fire tetrahedron.

Class	Materials and Proper Fire Extinguisher
Class A fires (ORDINARY COMBUSTIBLES)	These fires involve ordinary combustibles such as wood or paper. Class A fires are fought by cooling the fuel. Class A fire extinguishers contain water. Using a Class A extinguisher on any other type of fire can be very dangerous.
Class B fires (FLAMMABLE LIQUIDS)	These fires involve grease, liquids, or gases. Class B extinguishers contain carbon dioxide (CO_2) or another material that smothers fires by removing oxygen from the fire.
Class C fires (ELECTRICAL EQUIPMENT)	These fires are near or involve energized electrical equipment. Class C extinguishers are designed to protect the firefighter from electrical shock. Class C extinguishers smother fires.
Class D fires (COMBUSTIBLE METALS)	These fires involve metals. Class D extinguishers contain a powder that either forms a crust around the burning metal or gives off gases that prevent oxygen from reaching the fire. Some metals will keep burning even though they have been coated with powder from a Class D extinguisher. The best way to fight these fires is to keep using the extinguisher so the fire will not spread to other fuels.

86102-13 F03.EPS

Figure 3 Fire classifications.

water and foam will conduct electricity and are therefore a hazard to anyone nearby.

- *Class D* – A Class D fire is one that involves flammable metals such as titanium and magnesium. These fires are fought with special dry chemical extinguishing agents because water and other firefighting materials can actually cause the fire to spread.
- *Class K* – A class K fire is one that involves cooking oils or fats. This type of fire can be encountered by repair yards when working on galleys. Grease can accumulate on stoves and in kitchen vents. This type of fire is fought with a special Type K wet chemical fire extinguisher because ABC and dry chemical extinguishers are not effective for certain types of kitchen grease fires.

3.0.0 Fire Watch Duties and Responsibilities

As an entry-level fitter, you may be trained to serve as a fire watch. This is a duty with significant responsibility. As a fire watch, you will serve as a line of defense against the spread of fire. Your performance as a fire watch could be the factor that prevents injury or death to other workers and/or damage to a vessel or shipyard facility. A person assigned to fire watch duties will be held accountable for any negligence on their part.

A fire watch must be constantly vigilant and must take pains to recognize and eliminate conditions that could result in a fire. If a fire occurs despite the precautions, the fire watch must be able to quickly respond with the correct method to extinguish the fire before it spreads. If an incipient fire cannot be controlled, the fire watch must act quickly to summon help and assist in evacuating the area. There are two abbreviations that will help you recall what to do in an emergency: RACE and PASS. RACE stands for rescue, alarm, contain, extinguish (see *Figure 4*). It describes what to do if a fire occurs.

- *Rescue* – Immediately stop what you are doing and make sure everyone in the danger area is evacuated.
- *Alarm* – Activate the nearest alarm or use whatever signal method has been established (horn, siren, whistle, etc.). Call the emergency number and report the location and extent of the fire.

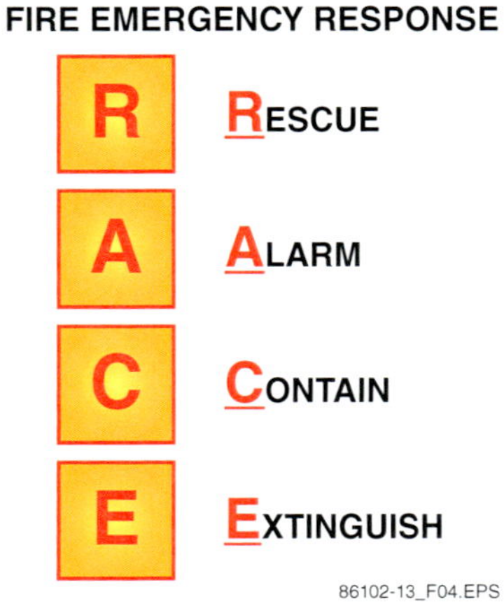

Figure 4 RACE.

- *Contain* – If possible, seal off the area by closing hatches, doors, etc. as you leave.
- *Extinguish* – If the fire is at the incipient stage (just starting), and if it safe to do so, use your extinguisher to try to put it out, as long as the extinguisher is the correct one for the situation. However, if you begin to lose control, leave the danger area and wait for help. Avoid breathing any smoke or fumes from the fire.

3.1.0 Basic Responsibilities

The primary duty of a fire watch is to safeguard life and property by closely monitoring hot work and other hazardous operations to prevent fire or explosion. In order to accomplish these duties, the worker who is assigned duty as a fire watch has the following basic responsibilities:

- Recognize and respond to different types and classes of fire.
- Inspect and set up hot work sites to reduce fire hazards.
- Use the established alert system to warn workers of safety problems.
- Follow established fire watch safety guidelines.
- Select the correct fire extinguisher for a given type of fire.
- Use the PASS (pull, aim, squeeze, sweep) method to extinguish fires (*Figure 5*).

> **NOTE**
> The specific duties and responsibilities of a fire watch may vary from one company to the next. Always become familiar with, and follow, your company's policies and procedures.

3.2.0 Preparing for Fire Watch Duty

The first step in preparing for duty is to make sure the work area has been inspected by the marine chemist or shipyard qualified person and that a hot work certificate has been posted and reflects the current conditions in the area. Make sure that a fire watch has been posted with an extinguisher in all adjoining areas that could be affected. For example, if there is an opening in a bulkhead, overhead, or deck where work is to be performed, a fire watch must be posted on the other side of the bulkhead, overhead, or deck.

Figure 5 The PASS method.

Make sure signs and barricades are posted as required by your company's policies and procedures (*Figure 6*). Be sure to remove the signs when the work is complete. If you are relieving an existing fire watch, you should re-inspect the area and either modify or replace the signs as required by your company's policies and procedures. Make sure the area has at least two evacuation routes and determine the location of the nearest emergency communication link. This can include shipboard phones, public address systems, land-lines in buildings, or wireless radios.

Inspect the work area and any adjoining areas that might be affected for flammable or combustible material, including the following:

- Pallets
- Trash bags
- Lunch bags
- Newspapers
- Coffee cups
- Rags
- Flammable fluids
- Oxygen and acetylene hoses

Dispose of any trash before work starts. Good housekeeping is one of the most effective ways of preventing a fire. If any hot work is to be done within 35 feet of any flammable or combustible material that cannot be removed, it must be covered with fireproof coverings or otherwise protected by fireproof shields or curtains.

While performing fire watch duties, it is important to follow these safe practices as prescribed by OSHA:

- Do not undertake any other duties and do not leave your post while acting as a fire watch. If you must leave the area, hot work must stop unless another fire watch relieves you.
- Wear required PPE, including hard hat, safety glasses, goggles, steel-toed boots, and long-sleeved 100-percent cotton shirt. Have hearing protection available.
- Maintain a clear, unobstructed view of the area where hot work is being performed.
- If an unsafe condition occurs while work is underway, stop the work and correct the situation.
- Make sure you have a fire extinguisher appropriate to the conditions. Be sure it is fully charged and has an up-to-date inspection tag.
- When working near welding or torch cutting, avoid breathing in smoke and fumes. Wear respiratory equipment if necessary, or if company procedure requires it. These processes can give off toxic fumes.
- Unless relieved beforehand, remain on the job for at least 30 minutes after hot work has been completed. Before leaving the site, make sure that oxygen and acetylene hoses have been removed from the space and that safety caps have been placed on the tanks. If work is finished, remove signs and barricades.
- The area and number of hot workers covered by a single fire watch can vary based on yard requirements, job conditions, and type of hazard.

Figure 6 Hot work signs.

4.0.0 Fire Extinguishers

One of the duties of a fire watch is to attempt to extinguish a fire that is in its incipient stage, that is, when it has just begun. Once a fire has begun to spread, the person on fire watch should sound the alarm, evacuate the area, and leave the firefighting to those who have the required equipment and training.

A fire watch is required to have a fire extinguisher while on duty. A portable extinguisher such as the one shown in *Figure 7* is commonly provided for that purpose.

The function of a fire extinguisher is to remove one of the three elements (oxygen, heat, fuel) required to sustain a fire. A fire extinguisher cannot remove fuel, so it is designed to remove either heat or oxygen. Heat is removed by using a coolant such as water; oxygen is removed by smothering the fire with dry chemicals or CO_2.

Fire extinguishers are rated for specific types of fires. Class A extinguishers, for example are intended for fires involving ordinary combustibles such as paper, wood, and fabric. An extinguisher that is rated only for Class A fires might contain water, which could not be used on electrical or grease fires. Using water on a Class B fire (flammable liquids, grease, or gases) could spread the fire or splash burning fuel on you. Using water on a Class C (electrical) fire could cause you to be electrocuted because water conducts electricity. Class D fire extinguishers are required for fires involving reactive metals such as sodium, potassium, magnesium, titanium, zirconium, and the metal hydrides. A Class D fire extinguisher contains a powder designed to

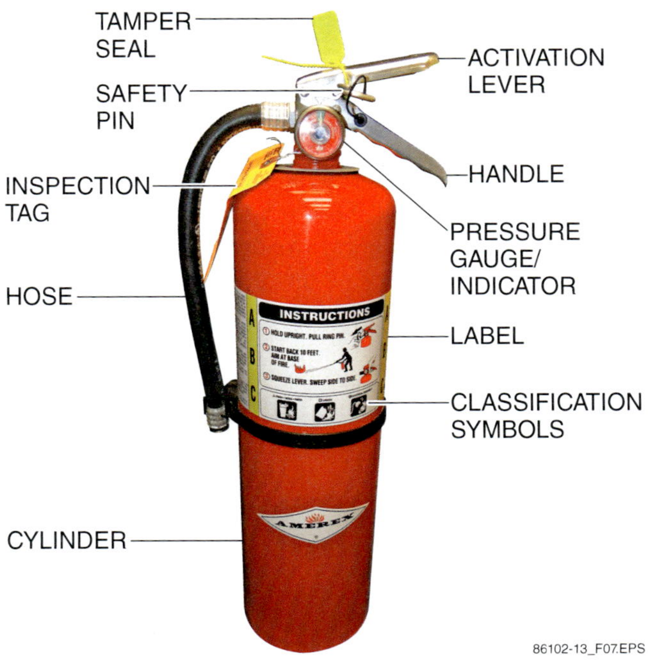

Figure 7 A portable fire extinguisher.

coat the metal to either smother the fire or keep oxygen from reaching the fire. If the powder coat fails to extinguish the fire, the best solution is to reapply the powder to keep the fire from spreading. A pictogram label (*Figure 8*) on the extinguisher denotes the type(s) of fire for which the extinguisher is intended. The extinguisher shown in the figure is a dry-chemical extinguisher, which can be used on a Class A, B, or C fire. This type is the most common because it can be used to fight most fires. Some CO_2 extinguishers can also be used on Class A, B, or C fires, but others are designed for use on only class B and C fires. This can be seen in *Figure 9*; some of the extinguishers have three pictograms, while others have only two.

CO_2 is heavier than oxygen, so it smothers the fire. It also cools as it expands, so it can serve as a coolant. Some companies permit only CO_2 extinguishers to be used on vessels because the dry chemical is a corrosive that can damage electrical systems. In addition, the fine powder emitted by the material in a dry-chemical extinguisher can cause eye and respiratory irritation when it becomes airborne. One advantage of the dry chemical extinguisher is that its working range is 10 to 20 feet, while the range of the CO_2 extinguisher is 3 to 8 feet. A pressurized-water extinguisher, which is used only for Class A fires, has a range of about 50 feet.

WARNING!

Do not use a CO_2 extinguisher on a Class D fire, as it can cause the fire to spread. Also, CO_2 is heavier than air, so it will concentrate in low areas, displacing oxygen. For that reason, a CO_2 extinguisher must never be taken into, or used in, a confined space. Before it can be used in a confined space, all personnel in the space must first be evacuated. After the extinguisher has been used in the confined space, the space must be cleared by a marine chemist before it can be re-entered. Finally, CO_2 is discharged from the extinguishers at a temperature of −109°F, so it can cause frostbite if it contacts the skin. Never discharge it while it is pointed at someone and do not hold it by the hood unless it is insulated.

NOTE

CO_2 smothers a fire by reducing the oxygen level to a point where combustion cannot be sustained. However, the fire will re-kindle if the source of heat is not removed.

86102-13_F08.EPS

Figure 8 Fire extinguisher pictograms.

Fire extinguishers must be periodically inspected to make sure they are properly charged. The needle on the extinguisher's pressure gauge must register in the green area (*Figure 10*). The inspection must be documented on the tag attached to the extinguisher. Whenever you check out an extinguisher, make sure its inspection is up to date and that the pressure gauge registers the correct charge.

NOTE

Not all extinguishers have gauges. Extinguishers without gauges must be weighed to determine if they are fully charged.

4.1.0 Using a Fire Extinguisher

The fire extinguisher must remain within arm's length during your time on fire watch. A CO_2 extinguisher should rest on its side when not in use. This will prevent the possibility of the valve breaking off if the extinguisher falls over while standing upright. The contents are under pressure, so the extinguisher can become an unguided missile if the valve breaks off. If it becomes necessary to use the extinguisher, stand it upright before breaking the seal and pulling the pin.

There are several steps to take before using a fire extinguisher to attack an incipient fire:

- Sound the alarm and make sure the area has been evacuated.
- Make sure the fire department or other designated authority has been notified.
- Make sure the fire is small and is not spreading.

86102-13_F09.EPS

Figure 9 CO_2 fire extinguishers.

- Make sure the exit is at your back. Never let a fire get between you and the exit.
- Verify that the extinguisher is rated for the type of fire.
- Only use the extinguisher if you have been trained to use it.

The handle and activation lever on the fire extinguisher are secured by a pin, which is in turn secured by a tamper seal that will break when the pin is pulled (*Figure 11*). The extinguisher cannot be activated until the pin is removed. When you check out an extinguisher, make sure the tamper seal is intact and that the pin is not bent. Keep the extinguisher upright when it is in use. Do not cradle it or carry it at an angle. If it is too heavy, stand it upright on the deck.

86102-13_F10.EPS

Figure 10 Fire extinguisher pressure gauge.

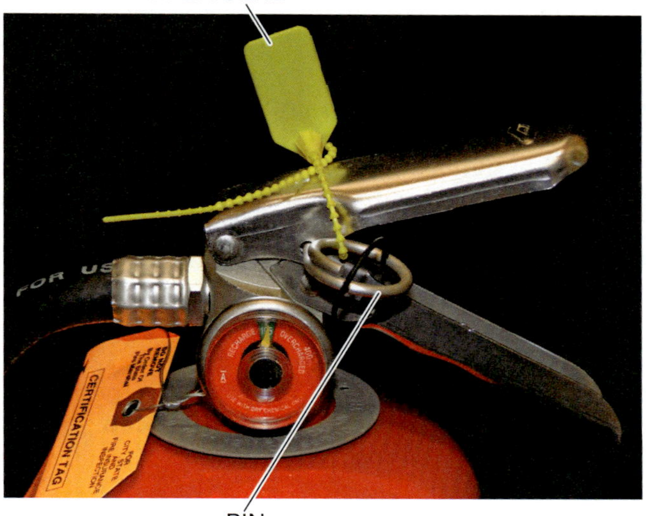

86102-13_F11.EPS

Figure 11 Pin and tamper seal.

Once the pin has been removed, squeezing the discharge handle will cause the contents of the extinguisher to be discharged. Be sure that the hose is pointed at the base of the fire before squeezing the handle. It takes 10 to 25 seconds to empty a tank containing 10 to 20 pounds of dry chemical. Remember, once an extinguisher has been used, it must be turned in to be re-charged.

WARNING!

When using a CO_2 extinguisher do not hold the horn with your bare hand, as it will become extremely cold.

Use the PASS method when attacking a fire:

- Pull the pin from the handle.
- Aim the nozzle at the base of the fire.
- Squeeze the discharge handle.
- Sweep the nozzle back and forth at the base of the fire.

As the fire is extinguished, move in closer until it is completely out. Stop shooting every three or four sweeps to check progress. Try not to use all of the contents on the initial attack, in case the fire flares up.

NOTE

Once the extinguisher has been used, it must be re-charged.

SUMMARY

The fire watch is an important line of defense against injury to personnel and damage to a vessel or structure. The person performing fire watch duties is held accountable for any negligence on their part. A fire watch is required in many instances when hot work such as welding, torch cutting, or brazing is being done and there is danger of material near the work being ignited. A marine chemist or shipyard competent person will determine when a fire watch is needed, which means that every location where hot work is being done must be inspected before the hot work can begin. Personnel assigned to fire watch must be able to recognize the different classes of fire, know the type of extinguisher to be used with each class, and be trained in the use of these extinguishers. Dry-chemical extinguishers are the most common because they can be used on most types of fires. Some CO_2 extinguishers can also be used on three different classes of fires. However, CO_2 displaces air, so it must never be used in confined spaces while they are occupied. Once CO_2 has been used in a confined space, the space must be cleared by a marine chemist before it can be re-entered.

Fire extinguishers must be periodically inspected and that inspection must be recorded on a tag attached to the extinguisher. Upon checking out an extinguisher, the fire watch should inspect it to make sure it is properly charged, that the discharge mechanism is in good condition, and that the tamper seal is intact. If a fire occurs, the fire watch must alert personnel to evacuate and contact the fire department before attempting to extinguish the fire. The PASS method should be used to attack the fire, but only if the fire is small and has not begun to spread.

Review Questions

1. A combustible material is one that has a flash point below 100°F.
   a. True
   b. False

2. The four elements required to start and sustain a fire are _____.
   a. oil, oxygen, gasoline, and fuel
   b. tetrahedron, oxygen, fuel, and heat
   c. heat, chain reaction, oxygen, and fuel
   d. heat, oxygen, oil, and fuel

3. Which of these fire classifications involves flammable liquids?
   a. Class A
   b. Class B
   c. Class C
   d. Class D

4. The A in RACE stands for _____.
   a. sound the alarm
   b. attack the fire
   c. alert the fire department
   d. activate the extinguisher

5. Which of the following is *not* a responsibility of the fire watch?
   a. Place hot work signs and barricades.
   b. Inspect the work area to identify potential fire sources.
   c. Determine if a fire watch is required.
   d. Sound the alarm in case of a fire.

6. How long must the fire watch remain on duty once the hot work is finished?
   a. 10 minutes
   b. 30 minutes
   c. 45 minutes
   d. 1 hour

7. A Class D fire extinguisher is intended for use on _____.
   a. burning metals
   b. flammable liquids
   c. trash fires
   d. electrical fires

8. The function of a dry chemical fire extinguisher is to _____.
   a. remove fuel from the fire
   b. remove heat from the fire
   c. remove oxygen from the fire
   d. drown the fire

9. Which of these extinguishers *cannot* be used on a Class C fire?
   a. Class A
   b. Class B
   c. Class C
   b. Class D

10. The A in the PASS system stands for _____.
    a. attack the fire
    b. alert other personnel
    c. sound the alarm
    d. aim at the base of the fire

Trade Terms Quiz

Fill in the blank with the correct term that you learned from your study of this module.

1. The process in which heat is transferred by direct contact is known as ________.
2. A substance with a flash point below 100°F is considered ________.
3. An area filled with a gas such as nitrogen is said to be ________.
4. A fire in its beginning stage is called a(n) ________.
5. Heat transferred in the form of invisible rays is called ________.
6. A substance with a flash point above 100°F is classified as ________.
7. The temperature at which as substance will begin to burn is its ________.

Trade Terms

Combustible
Conduction
Flammable
Flash point
Incipient fire
Inerted
Radiation

Trade Terms Introduced in This Module

Combustible: Capable of easily igniting and rapidly burning. Describes a fuel with a flash point above 100°F.

Conduction: A heat transfer process in which heat is moved from one material to another by contact.

Flammable: Capable of easily igniting and rapidly burning. Describes a fuel with a flash point below 100°F.

Flash point: The temperature at which fuel gives off enough vapors to burn.

Incipient fire: A fire in its beginning stage.

Inerted: Filled with an inert gas such as nitrogen.

Radiation: The movement of heat in the form of invisible rays or waves, such as the heat felt from a campfire.

Figure Credits

Bollinger Shipyards, Inc., Figure 1A
Trinity Yachts, Figure 1B
Topaz Publications, Inc., Figures 5, 7, 8, 10, 11
Courtesy of Amerex Corp., Figure 9

NCCER CURRICULA — USER UPDATE

NCCER makes every effort to keep its textbooks up-to-date and free of technical errors. We appreciate your help in this process. If you find an error, a typographical mistake, or an inaccuracy in NCCER's curricula, please fill out this form (or a photocopy), or complete the on-line form at **www.nccer.org/olf**. Be sure to include the exact module ID number, page number, a detailed description, and your recommended correction. Your input will be brought to the attention of the Authoring Team. Thank you for your assistance.

Instructors – If you have an idea for improving this textbook, or have found that additional materials were necessary to teach this module effectively, please let us know so that we may present your suggestions to the Authoring Team.

NCCER Product Development and Revision

13614 Progress Blvd., Alachua, FL 32615

Email: curriculum@nccer.org
Online: www.nccer.org/olf

❑ Trainee Guide ❑ Lesson Plans ❑ Exam ❑ PowerPoints Other ________________________

Craft / Level: Copyright Date:

Module ID Number / Title:

Section Number(s):

Description:

Recommended Correction:

Your Name:

Address:

Email: Phone:

86103-14

Introduction to Structural Fitter Drawings

Module Eight

Trainees with successful module completions may be eligible for credentialing through NCCER's National Registry. To learn more, go to **www.nccer.org** or contact us at **1.888.622.3720**. Our website has information on the latest product releases and training, as well as online versions of our *Cornerstone* magazine and Pearson's product catalog.

Your feedback is welcome. You may email your comments to **curriculum@nccer.org**, send general comments and inquiries to **info@nccer.org**, or fill in the User Update form at the back of this module.

This information is general in nature and intended for training purposes only. Actual performance of activities described in this manual requires compliance with all applicable operating, service, maintenance, and safety procedures under the direction of qualified personnel. References in this manual to patented or proprietary devices do not constitute a recommendation of their use.

86103-14
Introduction to Structural Fitter Drawings

Objectives

When you have completed this module, you will be able to do the following:

1. Recognize and identify basic fabrication drawing terms, components, basic lines, symbols, and bills of material.
2. Identify various drawing views such as plan, elevation, section, and details.

Performance Task

Under the supervision of the instructor, you should be able to do the following:

1. Identify components on a bill of materials and their location on a fabrication drawing.

Trade Terms

Base line
Bill of materials
Buttock line (BTK)
Center line
Detail view
Elevation view
Freeboard
General arrangement plan
Graphic scale
Isometric drawing
Legend
Longitudinal frame line
Longitudinal frames
Load water line (LWL)
Midship line
Mold line (ML)
Orthographic drawing
Plan view
Section view
Transverse frame line
Transverse frames
Water line

Industry-Recognized Credentials

If you're training through an NCCER-accredited sponsor, you may be eligible for credentials from NCCER's Registry. The ID number for this module is 86103-14. Note that this module may have been used in other NCCER curricula and may apply to other level completions. Contact NCCER's Registry at 888.622.3720 or go to **www.nccer.org** for more information.

Contents

Topics to be presented in this module include:

Figures and Tables

1.0.0 Introduction

When you are given a job to do, you will be given drawings that define how the job is to be performed and the materials that are to be used. If you are unable to correctly interpret the drawings, you will either be unable to do the job or you might do it incorrectly. The ability to interpret drawings is therefore a key factor in your development as a craft professional.

Drawings are a graphic method of representing equipment. A complete set of drawings for a vessel defines in detail everything from the complete system to the smallest piece-part. Drawings are often called blueprints because, at one time, the ammonia process used to copy drawings turned the paper blue.

A drawing is a detailed plan of a part, assembly, subassembly, or system. Drawings are used by designers to provide the information needed to build or install equipment. The structural fitter uses drawings to fabricate assemblies and install them correctly.

> **NOTE**
> It is important to remember that workers must never deviate from a drawing. If something on the drawing appears incorrect, check with a supervisor before making a change.

2.0.0 Common Drawing Elements

Certain elements are common to most drawings. These elements include the following:

- Title block
- Scales and measurements
- Symbol and abbreviation **legends**
- Notes
- Revision block

Other elements that may be found on drawings include drawing indexes and line lists. The drawing index, provided on the first page of the drawing set, is a complete list of all drawings included in the set. It is developed from the **general arrangement plan**, which is usually the first drawing created for the vessel. The pages within each category are designated by a letter followed by a number. The drawings are usually organized into the following major categories, with the drawing category in parenthesis:

- Interior/graphics (D)
- Structural (S)
- Mechanical equipment (M)
- Piping and plumbing (P)
- Electrical (E)

Drawings may be organized into work packages that contain the information necessary for a particular craft to do their work in a selected space.

2.1.0 Title Blocks

The title block (*Figure 1*) contains information used to identify the project or any part of the project. Title blocks are almost always located at the bottom right-hand corner of a drawing. The title block identifies the equipment represented on the drawing or the work task it represents and the project to which it applies. The following key information is found in the title block:

- *Drawing title* – This explains what is covered by the drawing. This may be the type of work, the work locations, or the name of the assembly or part covered by the drawing.
- *Sheet size* – There are standard sheet sizes (see *Table 1*).
- *Scale* – This shows the relationship between the size of the drawing and the size of the actual object. If the drawing is the same size as the object, the scale is full size.
- *Project number* – This is the code number assigned to the project.
- *Sheet number* – This indicates the sequence of this sheet in the series of drawings.
- *Revision number (Rev.)* – This indicates the current drawing. The number in this block can be referenced to the revision block for information about each revision.

A reference block may adjoin the title block. The reference block contains references to supporting information such as related drawing and applicable detail drawings.

Drawing sheets are zoned like road maps. Alpha characters start at the bottom and go up; numerical characters start at the right and go to the left. The title block in *Figure 1* would therefore be located in zone A1.

2.2.0 Scales and Measurements

Scales on drawings using the US or customary system are usually written in inches or in feet and inches. Measurements of less than 1 foot are written as 0 feet and the number of inches. The scale of a drawing is written in the title block. It shows the size of the drawing as compared to the actual size of the object represented.

The scale of a drawing varies according to the size of the object being drawn and the size of the paper used. The drawing scale is usually indicated by a ratio between the size of the object and the size

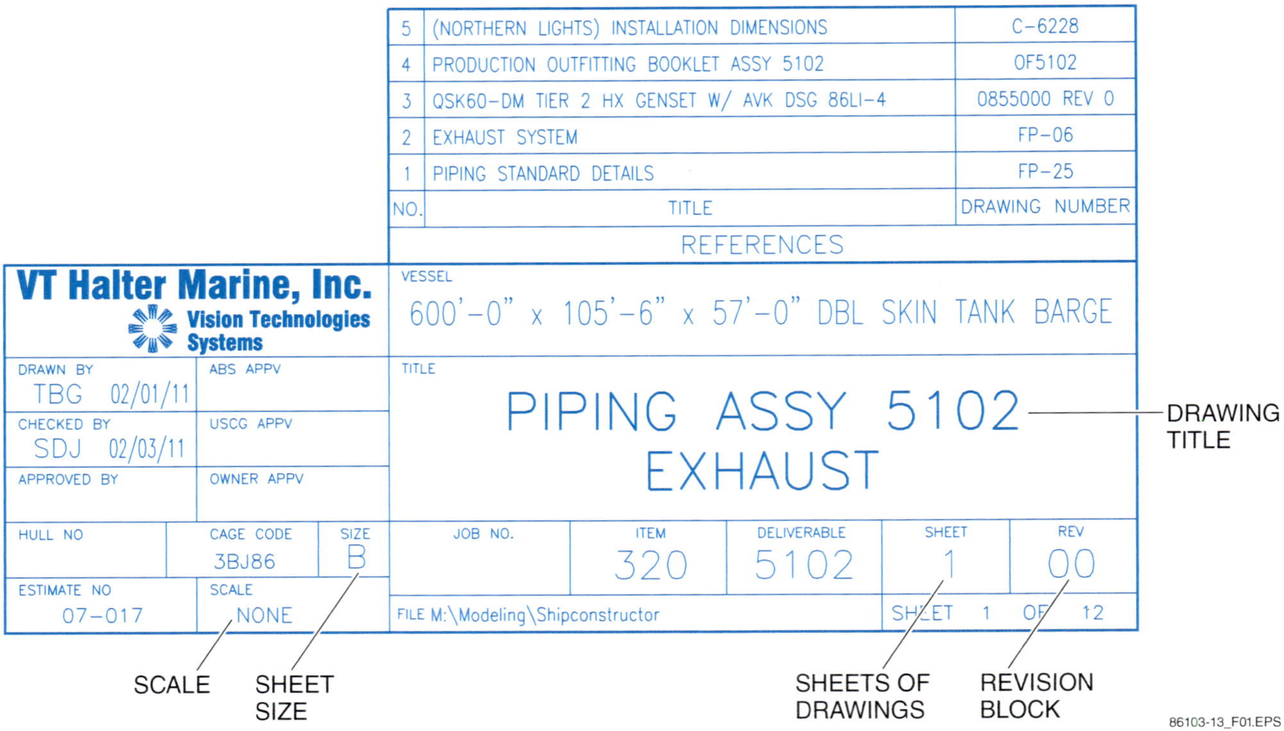

Figure 1 Example of a title block.

Table 1 Engineering Drawing Sizes

Sheet Size	Width (in)	Length (in)
A Horizontal	8.5	11.0
A Vertical	11.0	8.5
B	11.0	17.0
C	17.0	22.0
D	22.0	34.0
E	34.0	44.0
F	28.0	40.0
G	11.0	22.5 Min 90 Max
H	28.0	44.0 Min 143.0 Max
J	34.0	55.0 Min 176.0 Max
K	40.0	55.0 Min 143.0 Max

86103-13_T01.EPS

of the drawing. For example, a scale of 1 inch = 1 foot means that for every inch shown on the drawing, the actual size of the object is 1 foot. The standard scale of piping drawings is ⅜ inch = 1 foot.

Some drawings use the metric system. Since the metric system is based on units of ten, metric drawing scales are usually based on ratios. For instance, a 1:10 ratio means that one meter on the drawing equals 10 meters on the ship. If needed, conversions can be worked out using readily available conversion charts. The letters U.N. stand for "unless noted otherwise on drawing." This indicates that some parts of the drawing may use a different scale.

Graphic scales on drawings (*Figure 2*) show distances that correspond with a unit of length used on the object in the drawing. They are usually placed in or near the title block of the drawing. When a drawing has not been held to a particular scale, and cannot usefully be measured, the scale block will have the initials N.T.S., for "not to scale."

2.3.0 Lines

Drawings are made up of lines. There are both solid and broken lines, as well as thick and thin lines. Each type of line has a specific meaning. Lines can define the actual physical dimensions and shapes of objects, or can be used with arrows to indicate callouts, instructions, and notes. *Figure 3* shows some typical lines used in preparing drawings used by shipyard workers. Line usage may vary from one company to another, so it is

Figure 2 Graphic scale.

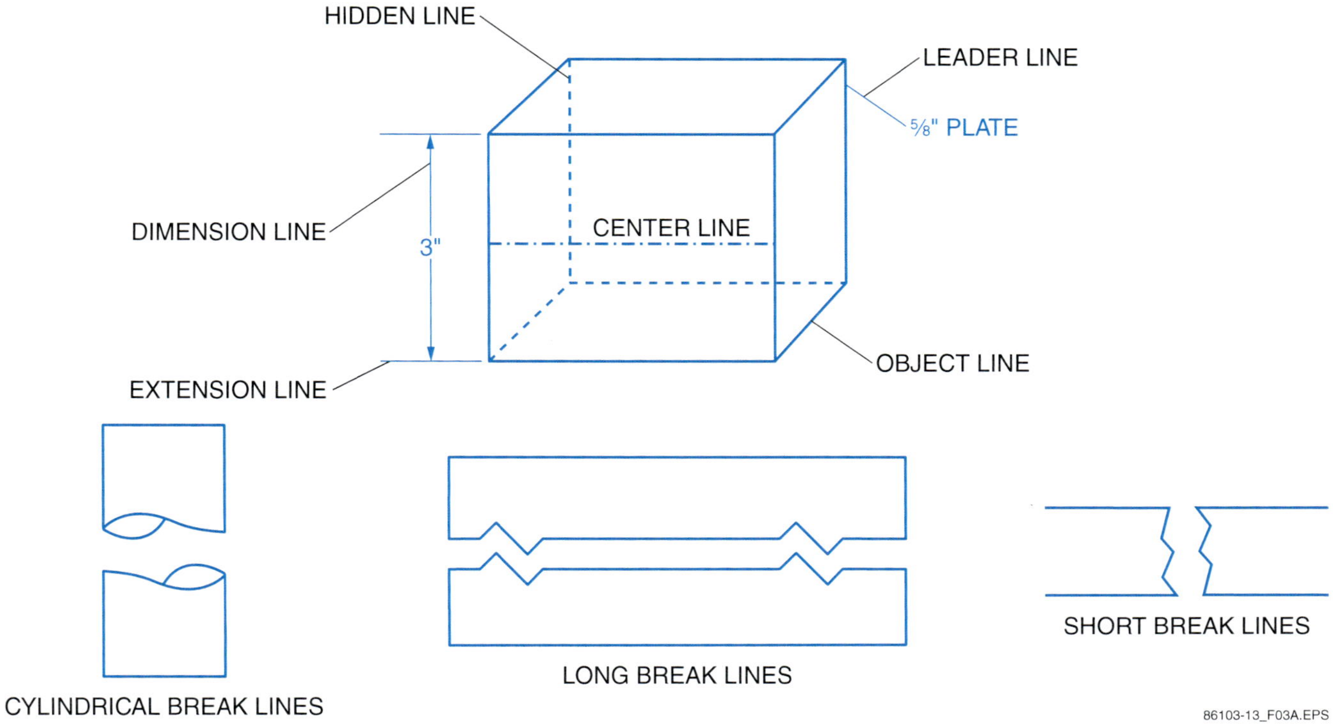

Figure 3A Lines commonly used on drawings. (1 of 2)

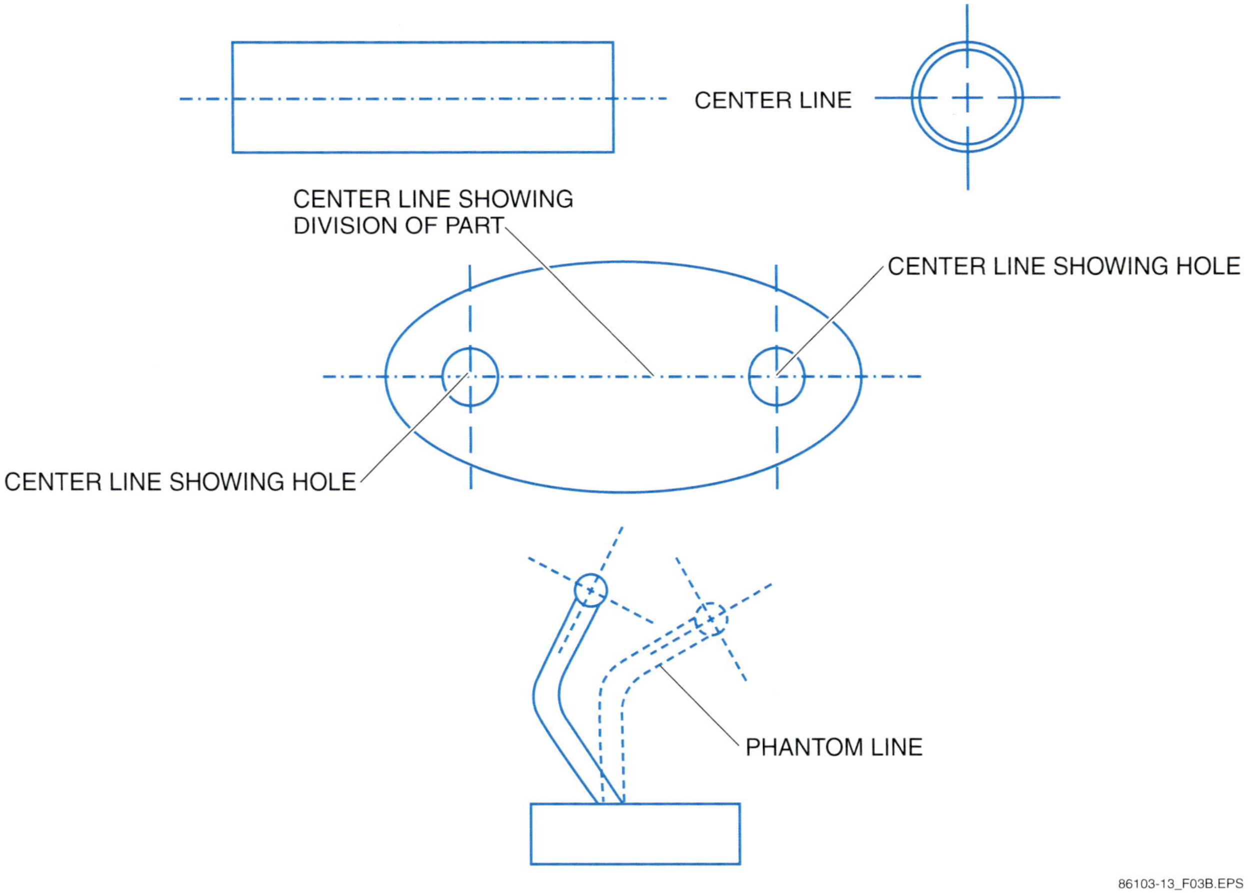

Figure 3B Lines commonly used on drawings. (2 of 2)

important to understand the convention used in your company.

- Object (visible) lines are thick, continuous lines used to represent the outline of an object.
- Hidden (invisible) lines are thin lines made up of short dashes. They are used to represent features that would not be visible from the perspective in which the object is being viewed.
- **Center lines** consist of alternately long and short thin lines. They mark the center of an object. In the case of circles, both horizontal and vertical center lines are used. The intersection of the lines marks the exact center of the circle.
- Cutting plane lines are thick lines used to identify an alternate view of the area defined by the lines. Cutting plane lines usually terminate in arrows and are labeled with letters or a letter and number that identify a **section view**, usually located nearby. The arrows show the direction in which the view is taken.
- Phantom lines are thin dashed lines used to show an alternate position of an object.
- Dimension lines are thin, solid lines used to indicate the distance between two points on the drawing. Dimension lines are bounded by extension lines.
- Leader lines are thin lines used to indicate a point at which a note, number, or other reference applies. They sometimes end in an arrowhead or dot. A leader line with a number might mean that the object being pointed to can be found at that item number listing on a **bill of materials.**
- Break lines are commonly used to shorten a long object so that it takes up less space on the drawing. Another use is to create a break in an object so that an object beneath it can be viewed.

2.4.0 Symbol and Abbreviation Legends

Symbols and abbreviations are used to simplify the preparation of drawings and to help convey a lot of information in a relatively small area. Although there are industry standards for symbols and abbreviations, companies may use their own conventions. Drawings sometimes contain legends that identify the symbols and abbreviations they have used.

2.4.1 Symbols

Symbols are a form of shorthand that allows designers to simplify drawing preparation. Symbols are used to show objects such as I-beams, channels, seams, welds. These objects are represented on drawings by small, simple line drawings. Symbols can vary from job to job. *Appendix A* contains examples of symbols used on maritime structural drawings. If non-standard symbols are used, a legend is used to explain them.

2.4.2 Abbreviations

Shipyard workers, including structural fitters, use industry-specific terms and phrases. To avoid having to write out each word or phrase every time it is used, standard abbreviations have been developed. The structural fitter should recognize these abbreviations and the meaning of each. Note that there may be some abbreviations specific to a particular site. When in doubt, the fitter should consult someone who is more familiar with the job abbreviations. *Appendix B* contains a list of abbreviations commonly used on structural drawings.

2.5.0 Notes

Notes are provided on drawings to provide information that cannot be easily represented in graphic form. Notes may be used to emphasize certain factors or to define specifications for materials or fabrication processes. Such notes may include general requirements, quality assurance criteria, calibration tolerances, and similar information. Notes may be located in a corner of a drawing, or they may be listed on a separate sheet and included with a set of drawings. Notes can be general or specific. General notes (*Figure 4*) can appear anywhere on a drawing, but usually appear at the upper left. General notes apply to the overall drawing. They may contain such in-

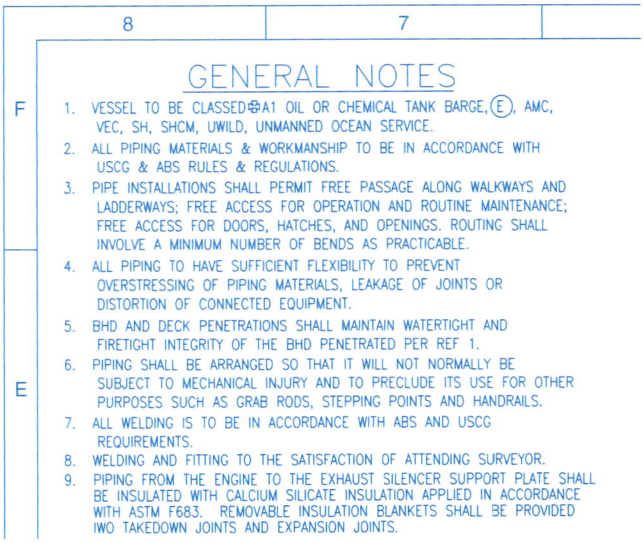

86103-13_F04.EPS

Figure 4 Example of general notes.

formation as material requirements, dimensions, requirements for hangers, or welding/brazing requirements. The important thing to know is that the general notes contain important information. Specific notes pertain to a specific location or item on the drawing and are generally tied to that location with a leader line.

WARNING!

Carefully read all notes before beginning any fabrication based on the drawing. Notes contain information that may directly affect personal protection and job quality.

2.6.0 Revision Block

The revision block is often located in the upper right hand corner of a drawing. It may also be located in the lower center of the drawing, or can be incorporated into the title block. The revision block is used to record changes that have been made to the drawing. All revisions are noted in this block, as shown in *Figure 5*. They are dated and are identified by a letter or number and a description of the revision. A drawing revision may be shown by the addition of a letter or number next to the original drawing number in the title block.

2.7.0 Proper Care of Drawings

The project drawings are as important as your tools. They must be handled carefully and kept clean and intact despite rough working conditions. Here are some guidelines to follow when handling and storing drawings:

- Keep them clean and dry.
- Fold the drawings so that the title block is visible.
- Be careful when folding and unfolding drawings to prevent tearing.
- Do not write on drawings unless authorized to do so.
- Be careful not to lay tools or pointed objects on drawings.
- Follow company policy when disposing of old drawings.
- Keep drawings out of hot-work areas where they could be damaged by sparks.

3.0.0 Reference Lines

One important function of drawings is to locate objects in relation to other objects or reference points. This information helps workers determine exactly where to install an object, drill a hole, or perform some other activity. Therefore, all measurements on drawings are stated in relation to something. For example, the location of a hole on a metal plate may be measured from the edge of the plate. The location of a bulkhead may be referenced to the center line of the ship or to a frame line. Common reference lines on a ship drawing are shown in *Figure 6*.

- The ship center line is a horizontal line running from bow to stern of the ship.
- The **base line** is a horizontal line at the bottom of the ship. This line is level and does not curve with the ship hull.
- The **midship line** is a vertical line halfway between the bow and stern. The midship line may be called the center line on some drawings.
- The **transverse frame line** represents frame members running across the ship from port to starboard (athwartship) at specified intervals. **Transverse frames** are often used as measurement reference points for distances and loca-

PROJ	NO	REVISION	RVSD	CHKD	APPD	DATE
3483	01	RELEASED FOR CONSTRUCTION		APD	NWS	JULY 06
3483	02	DELETED PART OF LINE 12037		APD	NWS	AUG 06
3483	03	⚠ ADDED WELDING SYMBOL		APD	NWS	AUG 06

86103-13_F05.EPS

Figure 5 Revision block.

tions forward and aft. Locations of objects are often referenced in terms of their relationship to transverse frames, which are numbered.

- A **longitudinal frame line** (not shown) represents structural members running fore to aft along the length of the vessel. A **longitudinal frame** can be a shell, deck, or bulkhead stiffener running fore and aft.

> **NOTE**
> The terms *transverse* and *longitudinal* can be used in different ways. Sometimes they simply refer to an axis, while in other cases they are referring to specific structural members, such as transverse frames.

Other reference lines are coordinated with one of the above mentioned horizontal or vertical reference lines. The following are examples of commonly used reference lines.

- Buttock – A given distance from a center line.
- **Buttock line (BTK)** – A line running parallel to and at specific distances from the center line of the ship. These lines are used to measure widths when it is inconvenient to use the center line as a reference point.
- **Freeboard** – The distance between the **water line** and the deck or weather deck of a ship.
- **Load water line (LWL)** – The line at which a ship will float when loaded to its carrying capacity.
- **Mold line (ML)** – The exact location at which a structure is installed.
- Water line – A line parallel to the base line and at various distances depending on the displacement of the ship.

3.1.0 Ship's Compartment Identification

There are conventions for identifying compartments on a ship so that they can be located. Compartment identifications will vary by type of ship. This section describes the convention used to identify compartments on US Navy ships. The identification number assigned locates each compartment specifically, and generally indicates the function and use of the compartment. Compartment numbers consist of four parts, separated by hyphens, for example 5-165-0-E, in the following sequence:

1. Deck number
2. Frame number
3. Position in relation to center line of ship
4. Compartment use

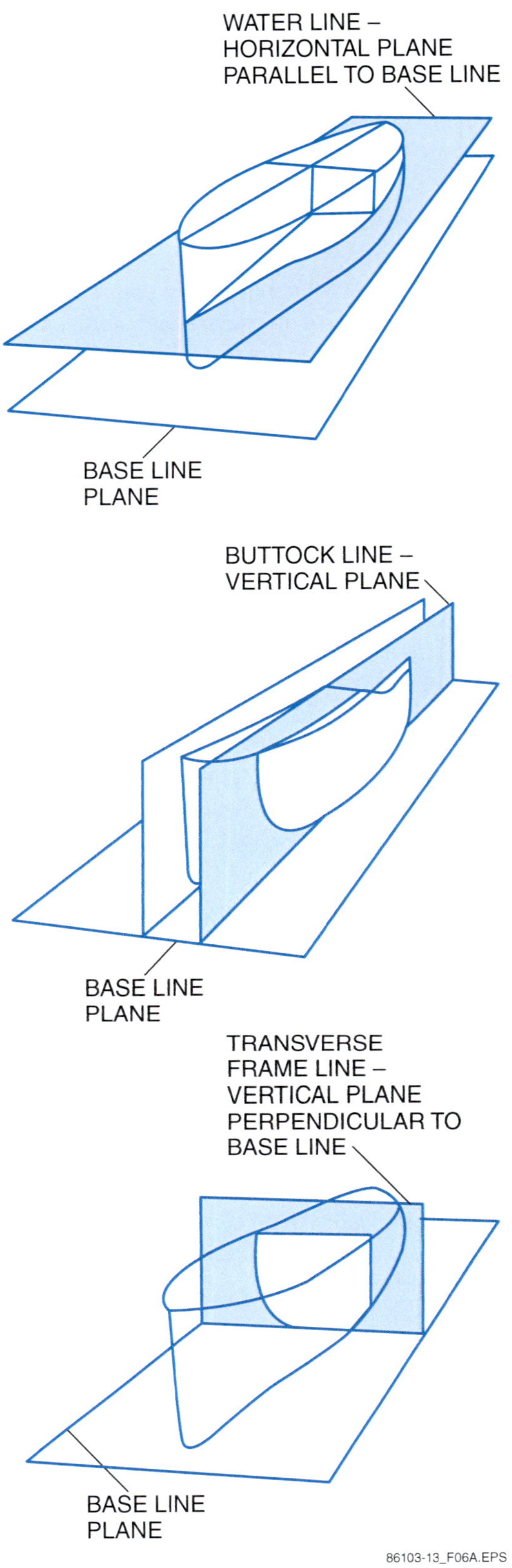

Figure 6A Reference lines in relation to the hull (1 of 2).

- *Deck number* – Numbering starts with the main deck, which is numbered 1. Lower decks are numbered consecutively, 2 through n. The level above the main deck is numbered 01. These numbers continue consecutively, as well.

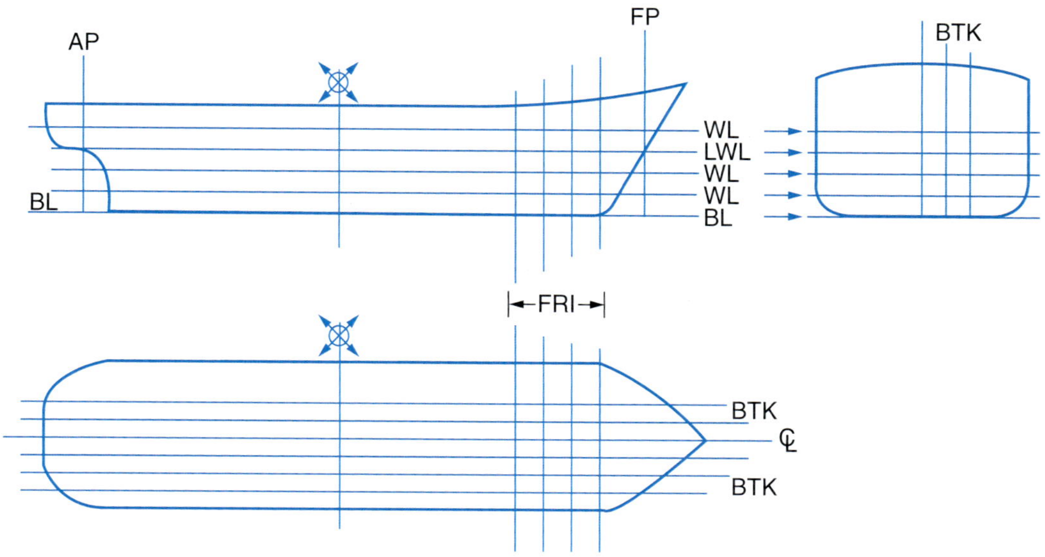

ABBREVIATIONS LEGEND	
BASE LINE	BL
WATERLINE	WL
CENTER LINE	℄
BUTTOCK LINE	BTK
FRAME LINE	FR
MIDSHIP LINE	⊗
LOAD WATER LINE	LWL
FORWARD PERPENDICULAR	FP
AFTER PERPENDICULAR	AP

86103-13_F06B.EPS

Figure 6B Reference lines in relation to the hull (2 of 2).

NOTE

Decks above the main decks are referred to as levels.

- *Frame number* – Transverse frames are consecutively numbered, based on frame spacing, until the aft transverse frame is reached. Forward of the forward perpendicular, frames are lettered starting from the perpendicular to the bull nose (A, B, C, etc.) while frames aft of the after perpendicular are double lettered to the transom (AA, BB, CC, etc.). Compartments are numbered by the frame number of the foremost bulkhead of the compartment. If this bulkhead is located between frames, the number of the foremost frame within the compartment is used. Fractional numbers are not used except where frame spacing exceeds four feet.
- *Position in relation to ship center line* – If a compartment is located so that the center line of the ship passes through it, a 0 is assigned to it. Compartments located completely to starboard of the center line are given odd numbers starting with 1, and those to port of center line are given even numbers starting with 2. If the center line of the ship passes through more than one compartment with the same forward bulkhead number, the compartment that has the portion of the forward bulkhead through which the center line of the ship passes is assigned the number 0 and the others carry numbers 01, 02, 03 etc.
- *Use of the compartment* – A capital letter is used to identify the assigned primary use of the compartment. Only one capital letter is assigned, except that on dry and liquid cargo ships a double letter identification is used to designate compartments assigned to carry cargo. Examples of compartment use are storage areas, various tanks, and living quarters.

Examples of these identifiers are as follows:

- A – Storage area
- C – Ship and Fire Control operating spaces normally manned
- E – Machinery (engineering) spaces that are normally manned
- F – Fuel or fuel oil tanks
- J – JP-5 tank
- L – Living quarters
- M – Ammunition stowage and handling
- Q – Areas not otherwise covered
- T – Vertical access trunk
- V – Void

Recalling the compartment identifier 5–165–0–E, it can be determined that it is located:

- Four decks below the Main Deck
- The foremost bulkhead is at frame 165
- It is centered on the center line of the ship
- It is used as an engineering (E) space

4.0.0 Drawing Views

Most of the drawings you use are **orthographic drawings**, which depict three-dimensional objects in two-dimensional form. Three different views are needed in order to clearly visualize a three-dimensional object in two dimensions. These are the top, front, and side views. In marine work, these views are referred to as plan (top), section (front), and elevation (side) views. These views will appear on most drawings. *Figure 7* shows how a three-dimensional object presented as an **isometric drawing** is converted to orthographic form.

In addition to plan, section, and elevation views, many drawings also require **detail views** and some have exploded views and isometric views. All these types of views are discussed in this section.

4.1.0 Plan Views

A **plan view** is a drawing of an object or area from above it, looking down on the part or system from the top. *Figure 8* is an example of a plan view. A plan views gives two dimensions, length and width. Its primary reference lines are the **frame lines** and center line. Transverse section lines are references for section views and longitudinal section lines are used for **elevation views**.

The plan view shows the exact locations of the equipment in relation to established reference points, objects, or surfaces. These reference points may be bulkheads, reference lines, or other equipment. The plan view is the view from which the elevation and section views are derived.

4.2.0 Elevation Views

An elevation view looks from port to starboard and gives the height and length of an object. Its primary references are the frame lines, molded lines of the deck, and the base line of the ship. Elevations are usually drawn looking toward the top of the left side of the drawing. Sometimes elevation drawings are called front, rear, or side elevations. *Figure 9* is an example of an elevation drawing.

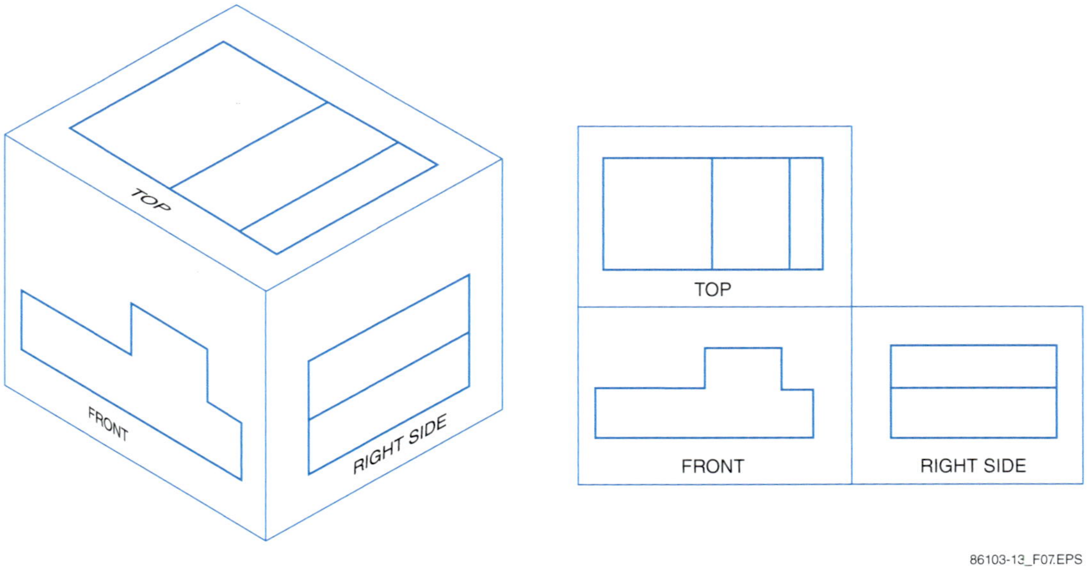

Figure 7 How plan, section, and elevation views are derived.

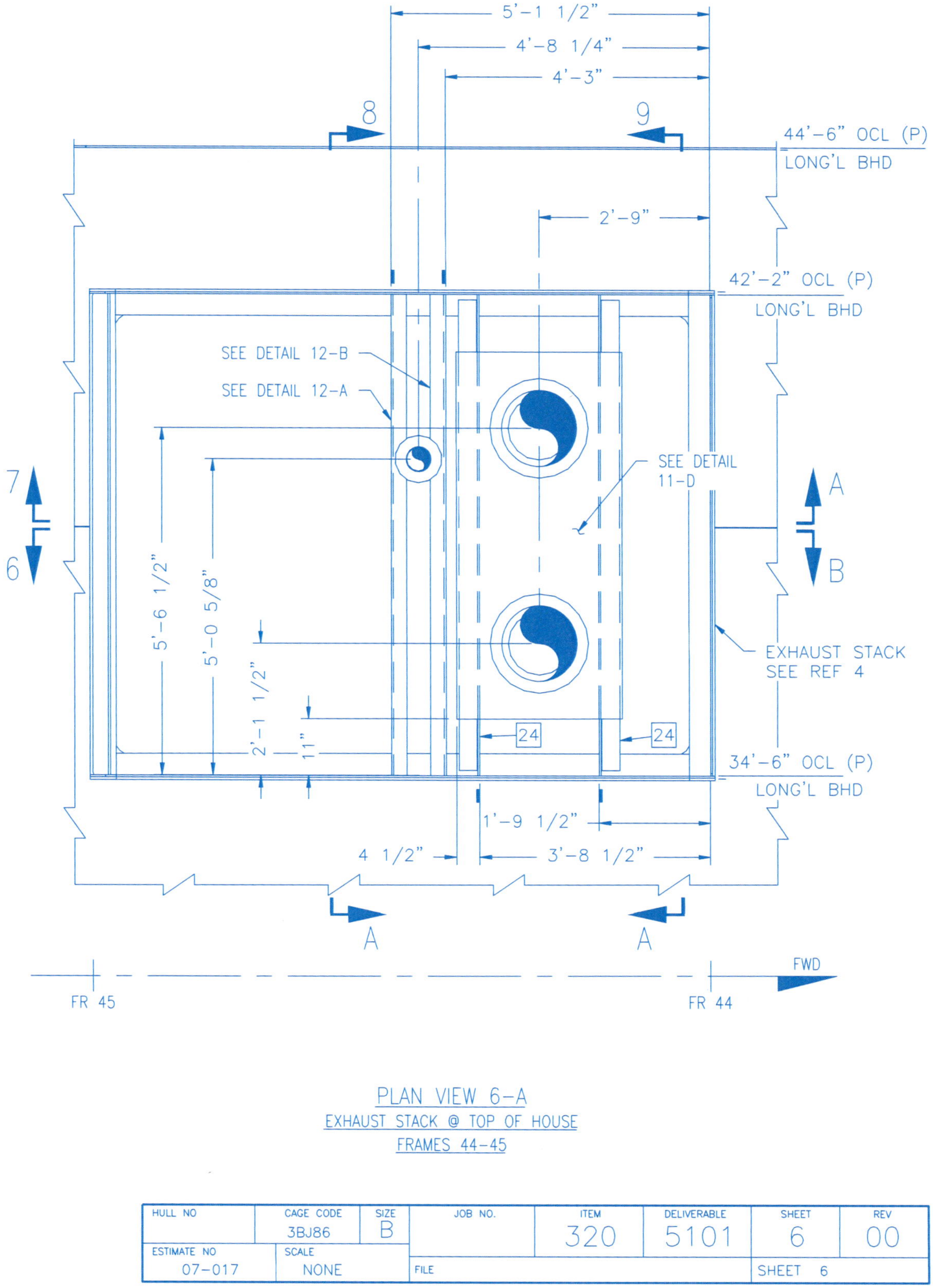

86103-13_F08.EPS

Figure 8 Plan view.

4.3.0 Section Views

It often happens that a plan or elevation view is not able to show all the necessary information. In such cases, a section view is used provide the necessary additional information. A section view is a view of an object designed to show its interior construction. Most section views are vertical, but horizontal or plan sections may be included.

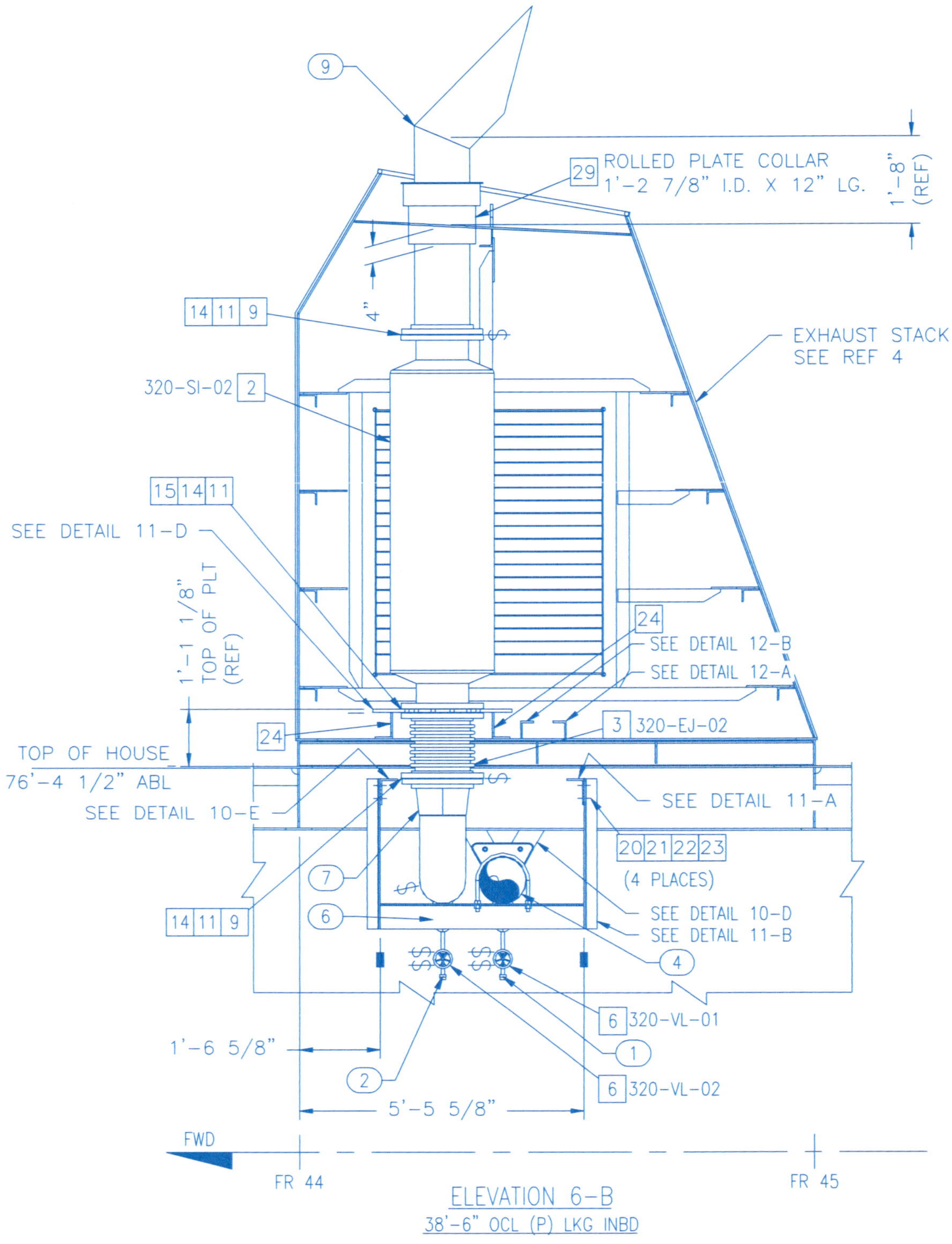

HULL NO	CAGE CODE 3BJ86	SIZE B	JOB NO.	ITEM 320	DELIVERABLE 5101	SHEET 6	REV 00
ESTIMATE NO 07-017	SCALE NONE		FILE			SHEET 6	

86103-12_F09.EPS

Figure 9 Elevation view.

In section views, the parts are usually labeled. A section view shows the width and height dimensions. Its primary reference lines are the molded lines of the decks, the ship center line, and the ship base line.

The section view shows how the equipment or parts fit together in the system. See *Figure 10*. From this drawing, the height, width, and location of the equipment can be seen. This is accomplished by drawing what the component looks like on the inside. Section drawings are usually drawn on a larger scale than plans and elevations.

4.4.0 Detail Views

There are many cases where the plan, section, and elevation views do not provide enough detail about a component. In those cases, a **detail view** is used to provide the necessary information. In the lower part of *Figure 10*, you can see a reference to Detail 11-B. That detail is shown in *Figure 11*. The detail may appear on the same sheet as its reference, or on another sheet. If the same detail is used in many locations, there may be separate drawings that contain nothing but details. In such cases, the reference will refer to a specific detail by drawing number.

4.5.0 Exploded Views

An exploded view is one in which the parts are separated from each other to show how the pieces fit together. Exploded views are uncommon and are used only when necessary to show the assembly or disassembly of complex structures.

4.6.0 Isometric Views

An isometric drawing shows two or more sides of the object in a single view. It is a method of showing an object in three-dimensional form. An isometric sketch is developed around three major lines called isometric base lines or axes. The left-side and the right-side isometric base lines each form a 30-degree angle to the horizontal base line. The vertical base line forms a 90-degree angle to the horizontal base line. Some drawings will contain an isometric view of the object depicted in the plan, elevation, and section views. *Figure 12* is an example; it shows an isometric view of the structure shown in orthographic form in the preceding figures.

5.0.0 Using Drawings

In the Core module *Introduction to Construction Drawings*, you learned that drawings define in detail how something is to be built. A complete set of drawings covers everything from the finished structure to the individual parts.

Based on your previous introductions to ship construction methods, you know that that the process begins with individual parts that are formed into subassemblies, which are in turn formed into assemblies. The assemblies are combined into blocks and units, which are then assembled to erect the ship. Every level of construction is governed by drawings that define in detail how the part or assembly is to be fabricated.

The highest level drawing is the general arrangement drawing, which shows a profile of the vessel and locates all the major subsystems, decks, bulkhead and frame locations, compartments, berths, and equipment. The lowest level drawing might represent a piece of sheet steel of specific dimensions, with openings and edge cutouts that will allow it to fit in a specific location. The next level, or subassembly drawing, might show this piece of sheet steel with brackets, pad eyes, or other hardware welded to it. This drawing would identify each part by number, show the precise location of each part, and define the method to be used to weld it.

Drawings are often assembled into work packages that cover the work to be performed in specific locations by a particular trade.

The bill of materials, sometimes called the list of materials, is the key to identifying parts on a drawing. A bill of materials is used primarily on assembly drawings that show more than one part. *Figure 13* shows a very simple example of a bill of materials and its related drawing. Note the letters in the Item column. They relate to the four circled letters on the drawing. It is more common to use numbers, rather than letters, in the Item column.

The remaining parts of the bill of materials identify the parts by their part number and description and show the quantity required for each part. In general, only drawings for piece parts and small subassemblies will have a bill of materials on the same page as the image. For large drawing sets that cover hundreds of parts, the bill of materials may take several sheets of the drawing.

In addition to part number, description, and quantity, a bill of materials may contain refer-

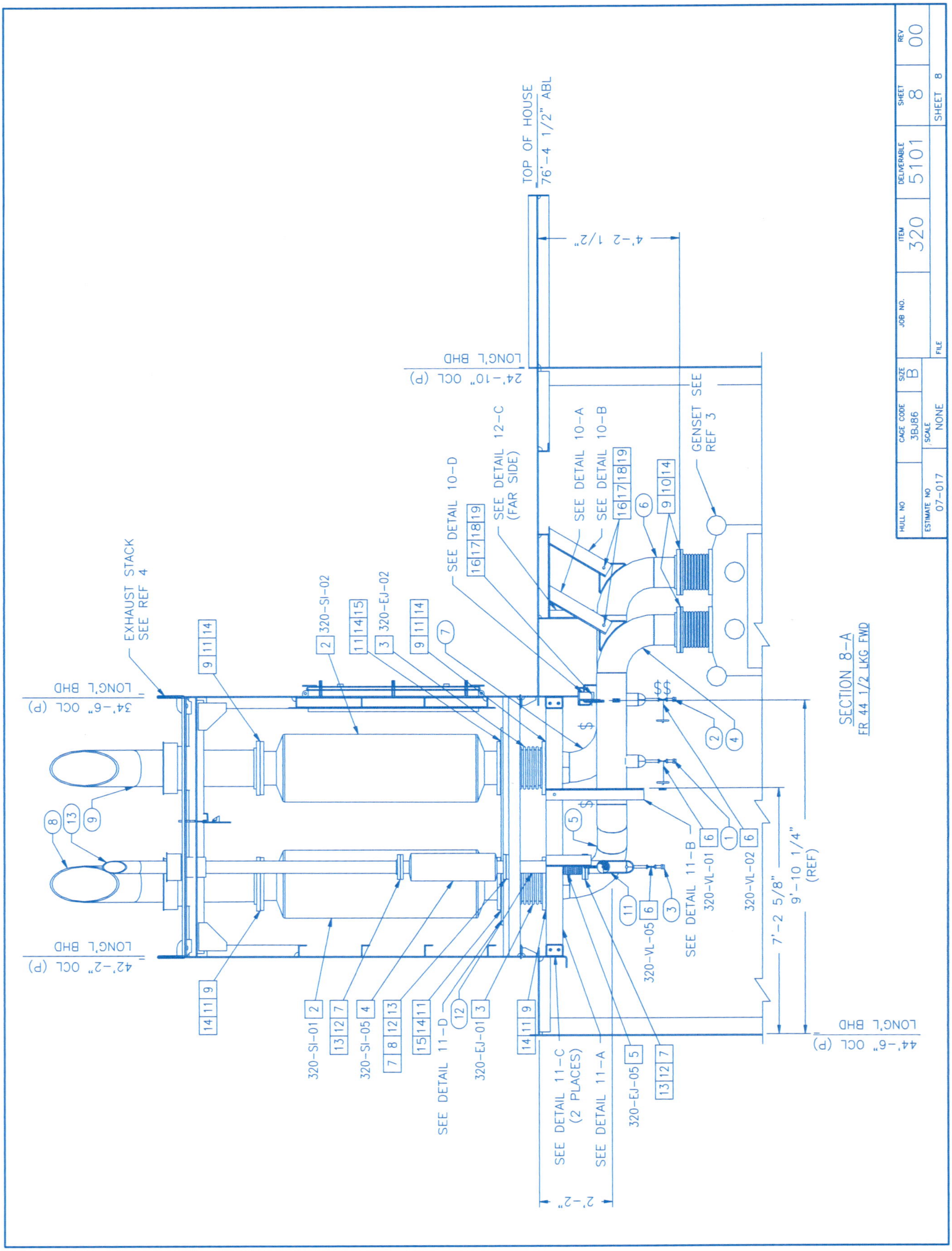

86103-13_F10.EPS

Figure 10 Example of a section view.

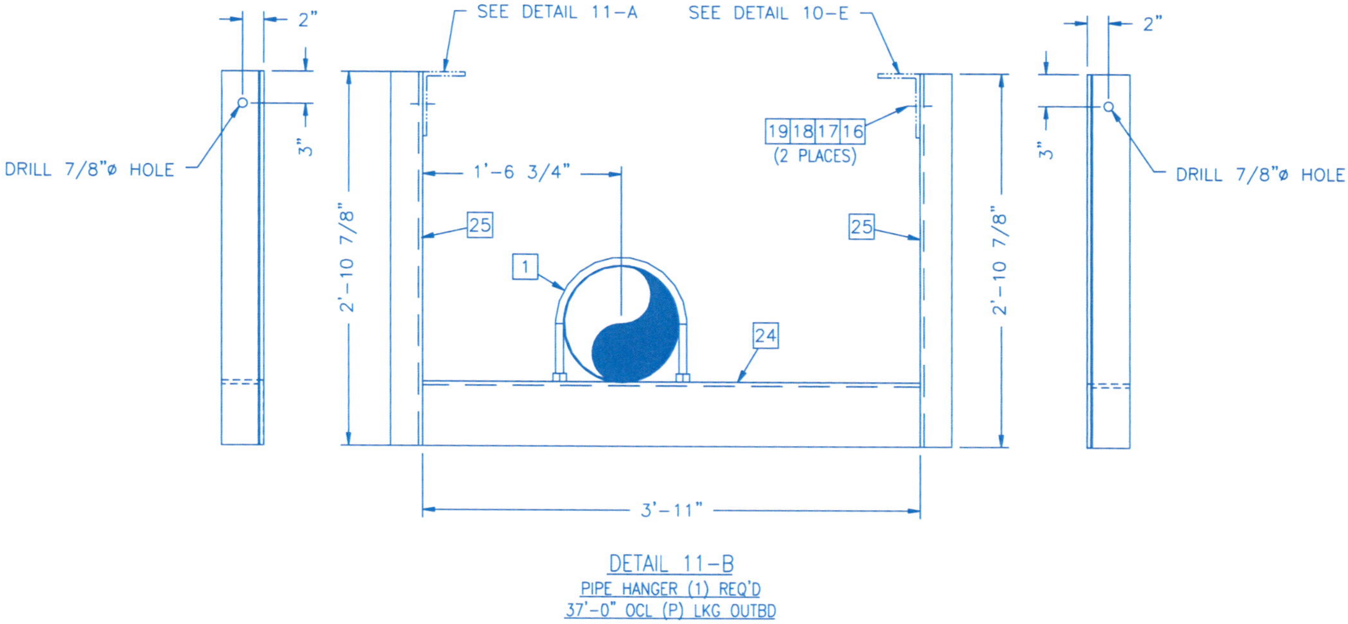

Figure 11 Example of a detail view.

ences to the sheets of the drawing on which the parts appear. Part descriptions can range from one or two words to a very detailed description that includes the dimensions of the part (see *Figure 14*).

NOTE

Appendix C contains examples of ship construction drawings. This is not a complete set of drawings, but simply selected sheets of a work package for metal joiners. These drawings are identified as *Figures A1* and *A2*, and are referred to in that manner in this text.

Figure A1 contains a portion of the list of materials, also called a bill of materials, for a shipboard crew head (restroom and shower facility). Note the "Part No." column on the far left. On other drawings, this column may also be titled "Find No." or "Key No.," depending on company practice. The numbers in this column refer to numbers in the body of the drawing. Take a look at *Figure A2*, for example. In the lower left-hand corner of elevation 20-D, there is a number 82 in a circle. In the list of materials on *Figure A1*, part number 82 is identified as a stile support angle. Take note of "REF 1" in the upper right corner of Elevation 15-D. Ref 1 is a book containing many standard details. Rather than duplicate the detail on every drawing on which it is referenced, the designers created a separate book to cover assembly details that are used in many locations.

Figure A2 contains both plan and elevation views. It also provides a good example of a detail. Detail views provide specific information about a segment of the drawing, usually details on how it is assembled. See Detail 24-E at the upper left corner of *Figure A2*, for example. It shows how the end panel support assembly is put together. The views just below the detail show how the mounting plate (147) and backing plate (148) are to be fabricated. These parts will be made in the shop, then brought to the job.

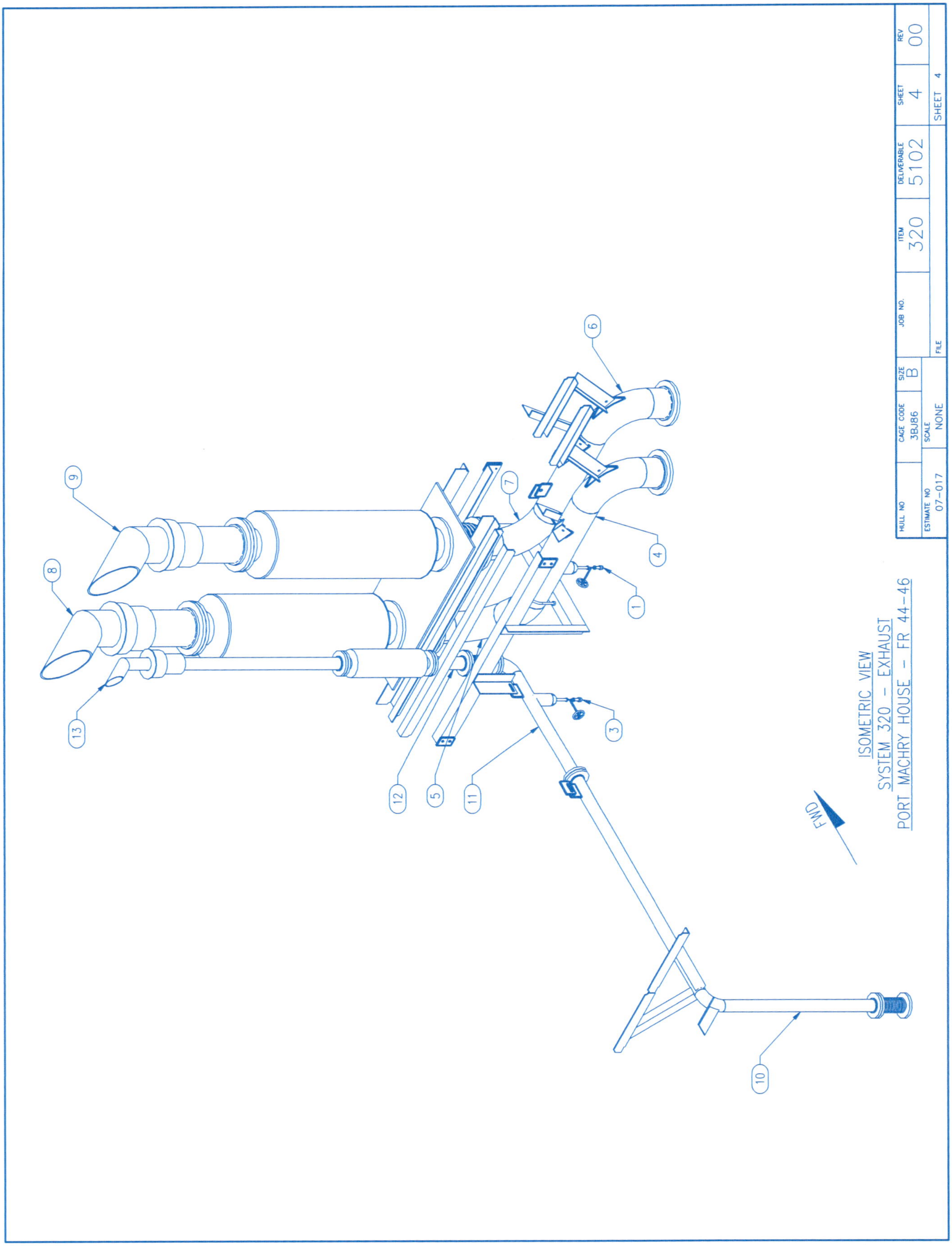

86103-13_F12.EPS

Figure 12 Example of an isometric view.

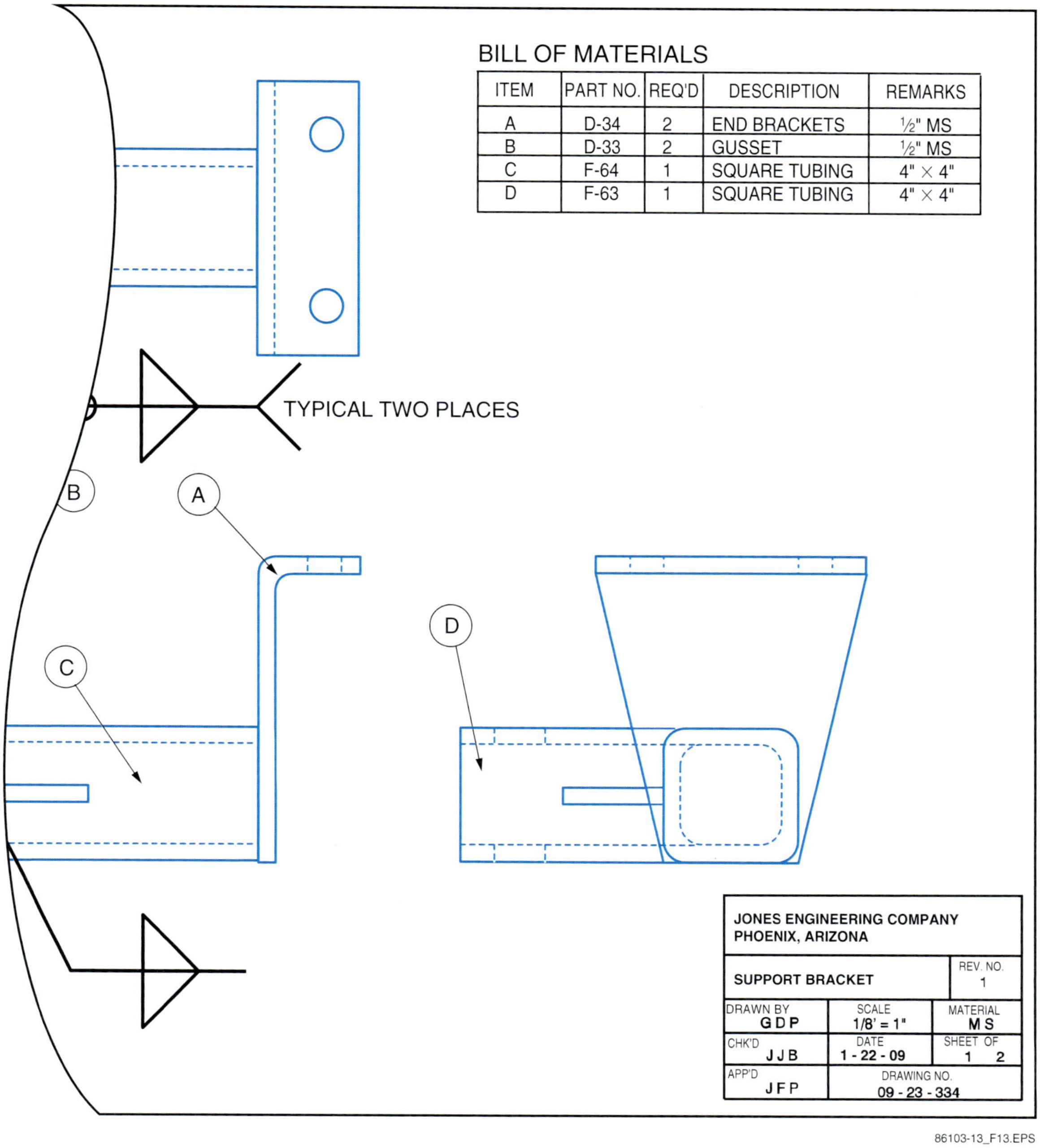

Figure 13 Bill of materials.

54 53 52 51 50 49

LIST OF MATERIAL CONTINUED

PART No.	QTY	SIZE	DESCRIPTION OF MATERIAL & USE	REMARKS
41	1 EA	90 13/16"" L	SHOWER TUBE, 2" X 1" X 11 GA TUBE, CRES 316L (INBD SHOWER ASSY)	SEE SHEET 5
41A	1 EA	44 1/16"" L	SHOWER TUBE, 2" X 1" X 11 GA TUBE, CRES 316L (USE OUTBD SHOWER SUB ASSY "B")	SEE SHEET 6
41B	1 EA	74 1/2" L	SHOWER TUBE, 2" X 1" X 11 GA TUBE, CRES 316L (USE OUTBD SHOWER SUB ASSY "A")	SEE SHEET 6
42	4 EA	NA	3" SHOWER STILE (STANDARD)	IAW DET 240-B, REF 1
43	5 EA	NA	6" SHOWER STILE (STANDARD)	IAW DET 237-B, REF 1
45	5 EA	29" L	SHOWER DIVIDER PANEL (STANDARD SIZE PANEL)	IAW 230-A, REF 1
54	2 EA	1" L	BHD CLIP FOR TUBE, 3" X 2" X 1/8" THK, CRES 316L	SEE DET 224-A, USE WITH PC# 130
56	36 EA	NA	W/C & SHWR PARTITION SUPPORT CLIP, 14 GA SHT, CRES 316L	IAW DET 128-C, REF 1
58	45 EA	NA	WC & SHWR PARTITION SUPPORT CLIP, 14 GA SHT, CRES 316L	IAW DET 128-C, REF 1
67	9 EA	NA	END PANEL CLIP, 14 GA SHEET, CRES 316L	IAW DETAIL 126-A, REF 1
68A	6 EA	NA	W/C FRONT END STILE BRKT (FOR 3" STILE) 3/16" PL, CRES 316L (SEE SHT 6 FOR OVHD SPRT FAB)	IAW DET 131-A, REF 1
68B	6 EA	NA	W/C REAR END STILE BRKT (FOR 3" STILE) 3/16" PL, CRES 316L (SEE SHT 6 FOR OVHD SPRT FAB)	IAW DET 131-A, REF 1
69A	7 EA	NA	W/C FRONT MID STILE BRKT (FOR 6" SITLE) 3/16" PL, CRES 316L (SEE SHT 6 FOR OVHD SPRT FAB)	IAW DET 131-C, REF 1
69B	7 EA	NA	W/C REAR MID STILE BRKT (FOR 6" STILE) 3/16" PL, CRES 316L (SEE SHT 6 FOR OVHD SPRT FAB)	IAW DET 131-C, REF 1
80/1	1 EA	68-1/2" LG	WC STILE MOUNTED CLOSURE PLATE, W/3 EA 1" FLANGES, 14 GA SHT	SEE FAB ON SHT 4
80/1A	1 EA	68-1/2" LG	WC STILE MOUNTED CLOSURE PLATE, W/3 EA 1" FLANGES, 14 GA SHT	SEE FAB ON SHT 4
80/1B	1 EA	68-1/2" LG	WC STILE MOUNTED CLOSURE PLATE, W/3 EA 1" FLANGES, 14 GA SHT	SEE FAB ON SHT 4
80/2	6 EA	12" LG	CLOSURE PLATE FILLER PANELS 12" L X 6" TALL, 14 GA SHT, CRES 316L (ADDED MATERIAL)	IAW DET 163-A, REF 1
80/3	3 EA	71" LG	MARGIN PLATE, FLAT BAR 1/8" X 2", CRES 316L, ADDED MATERIAL	IAW DET 163-A, REF 1
81	4 EA	3" L	WC END PANEL SPRT ANGLE, 1" X 1" X 3/16" ANGLE, CRES 316L	SIM DET 120-C, SHOP TO ASSEMBLE W/PC 146 & 147, SEE FAB SHT 3
81A	10 EA	8" L	WC DIVIDER PANEL SPRT ANGLE, 1" X 1" X 3/16" ANGLE, CRES 316L	SIM DET 120-C, SHOP TO ASSY W/PC 146 & 147
81B	2 EA	3" L	WC END PANEL SPRT ANGLE, 1" X 1" X 3/16" ANGLE, CRES 316L, USE WITH 146 (ADDED MATERIAL)	DETAIL 120-C, SHOP TO ASSY W/PC 146
81C	4 EA	9" L	WC DIVIDER PANEL SPRT ANGLE, 1" X 1" X 3/16" ANGLE, CRES 316L, USE WITH 146 (ADDED MATERIAL)	DETAIL 120-C, SHOP TO ASSY W/PC 146
82	2 EA	14" L	WC STILE SPRT ANGLE, 1 1/2" X 1 1/2" X 3/16" ANGLE, CRES 316L (ADDED MATERIAL)	IAW DETAIL 130-C
82A	1 EA	14 3/4" L	WC STILE SPRT ANGLE, 1 1/2" X 1 1/2" X 3/16" ANGLE, CRES 316L (ADDED MATERIAL)	IAW DETAIL 130-C
83	2 EA	19" L	SHOWER DIVIDER PANEL SPRT ANGLE, 1" X 1" X 3/16" ANGLE, CRES 316L, USE WITH 146 (ADDED MATERIAL)	SIM ELEV 215-B, SHOP TO ASSEMBLE W/PC 146
83A	2 EA	17" L	SHOWER DIVIDER PANEL SPRT ANGLE, 1" X 1" X 3/16" ANGLE, CRES 316L, USE WITH 146 (ADDED MATERIAL)	SIM ELEV 215-B, SHOP TO ASSEMBLE W/PC 146
83B	2 EA	15" L	SHOWER DIVIDER PANEL SPRT ANGLE, 1" X 1" X 3/16" ANGLE, CRES 316L, USE WITH 146 (ADDED MATERIAL)	SIM ELEV 215-B, SHOP TO ASSEMBLE W/PC 146
91A	4 EA	NA	SHOWER FRONT END STILE BRKT (FOR 3" STILE) 11 GA. SHT, CRES 316L	SEE SHEET 5 & 6 FOR OVHD SPRT FAB
91B	4 EA	NA	SHOWER REAR END STILE BRKT (FOR 3" STILE) 11 GA. SHT, CRES 316L	SEE SHEET 5 & 6 FOR OVHD SPRT FAB
92A	5 EA	NA	SHOWER FRONT MID STILE BRKT (FOR 6" STILE) 11 GA. SHT, CRES 316L	SEE SHEET 5 & 6 FOR OVHD SPRT FAB
92B	5 EA	NA	SHOWER REAR MID STILE BRKT (FOR 6" STILE) 11 GA. SHT, CRES 316L	SEE SHEET 5 & 6 FOR OVHD SPRT FAB
101	3 EA	NA	W/C TUBE END CAP, 11 GA. SHT, CRES 316L (FOR USE W/ AFT, MID & FWD WC SUPPORT ASSY)	SEE SHEET 4
104	2 EA	NA	SHOWER TUBE END CAP, 1 7/8" X 7/8" SQ , 11 GA. SHT, CRES 316L (FOR USE W/INBD ASSY & OUTBD SUB ASSY "B")	SEE SHEET 5 & 6 FOR OVHD SPRT FAB, IAW DET 224-A
115/1	1 EA	63 1/4" LG	SHOWER DIVIDER PANEL CLOSURE PLATE, W/3 EA 1/2" FLANGES, 14 GA SHT	SEE FAB ON SHT 7
115/1A	1 EA	63 1/4" LG	SHOWER DIVIDER PANEL CLOSURE PLATE, W/3 EA 1/2" FLANGES, 14 GA SHT	SEE FAB ON SHT 7
115/1B	1 EA	63 1/4" LG	SHOWER DIVIDER PANEL CLOSURE PLATE, W/3 EA 1/2" FLANGES, 14 GA SHT	SEE FAB ON SHT 7
115/2	6 EA	12" LG	CLOSURE PLATE FILLER PANELS 12" L X 6" TALL, 14 GA SHT, CRES 316L (ADDED MATERIAL)	IAW DET 250-A, REF 1
115/4	3 EA	65" LG	MARGIN PLATE, FLAT BAR 1/8" X 2", CRES 316L, ADDED MATERIAL	IAW DET 250-A, REF 1
127	20 EA	1" L	W/C STILE SPRT ANGLE, 2 1/2" X 2 1/2" X 3/16" ANGLE, CRES 316L	IAW DET 130-A, REF 1
130	2 EA	3" L	BACKING PLATE, 1" WIDE X 3" L X 1/8" THK, CRES 316L	DETAIL 224-A, USE WITH PC# 54
132	4 EA	NA	FLANGED 3" STILE BOTTOM CLIP, 14 GA SHT, CRES 316L	IAW DET 217-B, REF 1
133	5 EA	NA	FLANGED 6" STILE BOTTOM CLIP, 14 GA SHT, CRES 316L	IAW DET 217-C, REF 1
141	6 EA	1 1/16" L	W/C STILE BRKT END PLATE, 3" X 3/16 THK FLAT BAR OR EQL, CRES 316L	IAW DET 131-A, REF 1
146	30 EA	NA	FORMED CLIP, 14 GA SHT, CRES 316L (USE WITH 81 ,81A, 81B, 81C, 83, 83A, 83B & 83C)	IAW DET 149-C, REF 1
147	18 EA	NA	PANEL SUPPORT MOUNTING PLATE, 11 GA SHT, CRES 316L (USE WITH 81 ,81A & 83C)	SEE FAB ON SHT 3
148	18 EA	NA	PANEL SUPPORT BACKING PLATE, 11 GA SHT, CRES 316L (USE WITH 81 ,81A & 83C)	SEE FAB ON SHT 3

F E D C

86103-13_F14.EPS

Figure 14 Expanded list of material.

SUMMARY

A drawing is a detailed plan of a part or system. The drawing provides the information needed to build or install part of a system or the entire system. Fitters use drawings to install, troubleshoot, and repair piping.

The major parts of a drawing include title blocks, revision blocks, notes on the scale used, and drawing indexes. Drawings use several shortcuts, including symbols and abbreviations, which the fitter must be familiar with. Drawings contain notes and callouts that further inform the fitter as to how the assembly should be performed. Coordinates locate the equipment on an outline of the ship, referenced to several imaginary lines on the hull and superstructure.

Most drawings are made in the orthographic style. Orthographic drawings show the shape and size of a three-dimensional object through a series of related two-dimensional views. Each view shows the object from a different perspective—top, side, or front. The views appear as though the observer is looking at the object at a right angle to the surface of the object.

The bill of materials, also called the list of materials, is a key part of a drawing set. Every item called out on the drawing is listed on the bill of materials, which includes, at minimum, the item part number, quantity, and description.

Review Questions

1. If a screw or other basic hardware item specified by a drawing is not available, it is okay to substitute a different screw of the same size.
   a. True
   b. False

2. Which engineering drawing size is 17" by 22"?
   a. A
   b. B
   c. C
   d. D

3. The drawing scale is a ratio between the size of the drawing and the size of the _____.
   a. object being drawn
   b. vessel
   c. drawing paper
   d. vendor-supplied equipment

4. Interpretation of non-standard symbols used on a drawing is usually provided in a _____.
   a. revision block
   b. title block
   c. legend
   d. bill of materials

5. A revision block is used to record any changes that have been made to the drawing.
   a. True
   b. False

6. A vertical line halfway between the bow and stern of a ship is called a _____.
   a. midship line
   b. transverse line
   c. base line
   d. buttock line

7. A plan view looks at an object from the _____.
   a. top
   b. side
   c. bottom
   d. front

8. A drawing resulting from an imaginary line cut through the ship is a(n) _____.
   a. plan view
   b. elevation view
   c. side view
   d. section view

9. An isometric drawing shows, at minimum, how many sides of the object being drawn?
   a. One
   b. Two
   c. Three
   d. Four

10. Which of the following pieces of information is *not* likely to be found in a bill of materials?
   a. Item part number
   b. The price of the item
   c. The dimensions of the item
   d. A description of the item

Trade Terms Quiz

Fill in the blank with the correct term that you learned from your study of this module.

1. A single projected view of an object is called a(n) ________.
2. A drawing that shows the layout of a complete unit to be constructed is referred to as a(n) ________.
3. The ________ is the horizontal mark used to identify the bottom of the ship.
4. A(n) ________ defines the drawing's abbreviations and symbols.
5. The ________ is dependent on the ship's displacement, and is routed parallel to the base line.
6. The ________ is comprised of shell, deck, or bulkhead stiffeners that run fore and aft.
7. The scale in a drawing depicting the size relationship between the drawn version and the actual object is known as the ________.
8. A parts and materials list containing numbered entries related to items on the drawing is called a(n) ________.
9. A drawing of a single component or part is called a(n) ________.
10. A piping system drawing illustrated in three dimensions is a(n) ________.
11. A line signifying frame members running port to starboard at specified intervals is known as the ________.
12. The view depicted by an imaginary line cut through the ship or other area is called a(n) ________.
13. The ________ identifies an exact location where a structure is to be installed.
14. The ________ is routed parallel to and at defined distances from the ship's center line.
15. An ________ depicts an object's height and length, generally created from a view looking towards the upper left side of the object.
16. The water line of a ship when it is loaded to its maximum capacity is called the ________.
17. The ship's ribs formed by athwartship members are known as ________.
18. The drawing view that looks down on an object or area is known as the ________.
19. The ________ runs horizontally from the ship's bow to its stern.
20. A line representing frame members that run fore and aft along the vessel's length is called the ________.
21. The distance between the ship's deck or weather deck and its water line is called the ________.
22. The ________ is typically a vertical line positioned halfway between the bow and stern; it may be called the center line on some drawings.

Trade Terms

Base line
Bill of materials
Buttock line (BTK)
Center line
Detail view
Elevation view
Freeboard
General arrangement plan
Graphic scale
Isometric drawing
Legend
Longitudinal frame line
Longitudinal frames
Load water line (LWL)
Midship line
Mold line (ML)
Orthographic drawing
Plan view
Section view
Transverse frame line
Transverse frames
Water line

Trade Terms Introduced in This Module

Base line: A horizontal line identifying the bottom of the ship.

Bill of materials: A list of parts and materials shown on a drawing. Numbered entries on the bill of materials correspond with item (key) numbers found in the body of the drawing.

Buttock line: A line running parallel to and at specific distances from the center line of the ship.

Center line: A horizontal line running from bow to stern of the ship.

Detail view: A drawing that shows a single component or part.

Elevation view: A drawing giving the height and length of an object, usually drawn looking toward the top left side of the object.

Freeboard: The distance between the water line and the deck or weather deck of a ship.

General arrangement plan: The master drawing that shows the layout of the complete unit to be constructed. It usually contains the title block and a list of detailed drawings.

Graphic scale: A scale that shows the size relationship between the object represented in the drawing and the actual object.

Isometric drawing: A three-dimensional drawing of a piping system drawn to provide clarity.

Legend: An interpretation of symbols and abbreviations used on a drawing.

Load water line: The water line of a ship when it is loaded to its carrying capacity.

Longitudinal frame: A shell, deck, or bulkhead stiffener running fore and aft.

Longitudinal frame line: A line representing frame members running fore and aft along the length of the vessel.

Midship line: A vertical line halfway between the bow and stern; may be called the center line on some drawings.

Mold line: The exact location at which a structure is installed.

Orthographic drawing: The projection of a single view of an object.

Plan view: Drawing of an object or area from above it, looking down on the system.

Section view: A drawing of the view resulting from an imaginary line cut through the ship or area.

Transverse frame line: The line representing frame members running across the ship from port to starboard at specified intervals.

Transverse frames: Athwartships members forming the ship's ribs.

Water line: A line parallel to the ship's base line and at various distances depending on the displacement of the ship.

Appendix A

Common Drawing Symbols

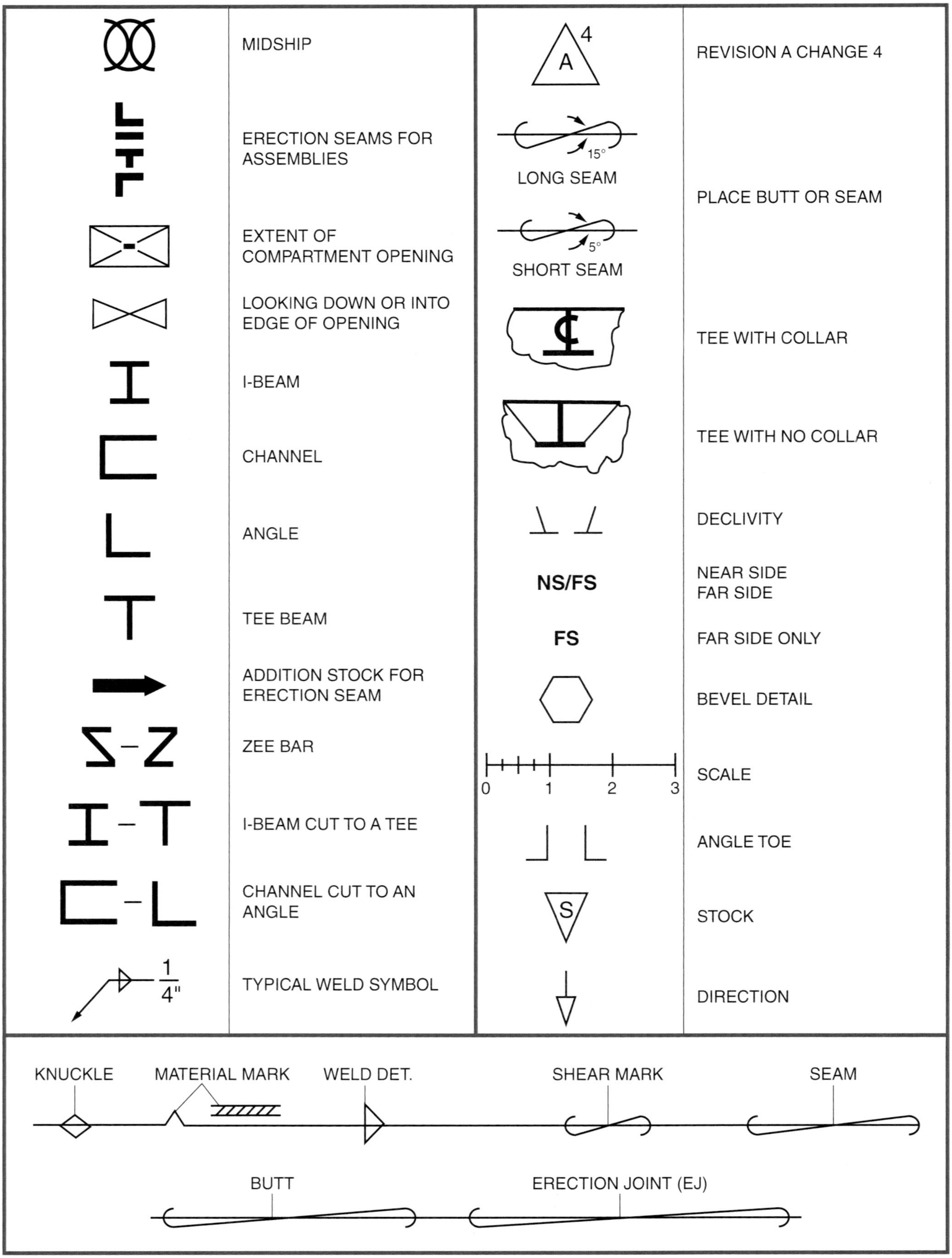

86103-13_A01.EPS

Appendix B

Common Abbreviations

Adapter	ADPT
Air preheating	A
American Iron and Steel Institute	AISI
American National Standards Institute	ANSI
American Petroleum Institute	API
American Society for Testing and Materials International	ASTM
American Society of Mechanical Engineers	ASME
Ash removal water, sluice or jet	AW
Aspirating air	AA
Auxiliary steam	AS
Bench mark	BM
Beveled	B
Beveled end	BE
Beveled, both ends	BBE
Beveled, large end	BLE
Beveled, one end	BOE
Beveled, small end	BSE
Bill of materials	BOM
Blind flange	BF
Blowoff	BO
Bottom of pipe	BOP
Butt weld	BW
Carbon steel or cold spring	CS
Cast iron	CI
Ceiling	CLG
Chain operated	CO
Chemical feed	CF
Circulating water	CW
Cold reheat steam	CR
Compressed air	CA
Concentric or concrete	CONC
Condensate	C
Condenser air removal	AR
Continue, continuation	CONT
Coupling	CPLG
Detail	DET
Diameter	DIA
Dimension	DIM
Discharge	DISCH
Double extra-strong	XX STRG
Drain	DR
Drain funnel	DF
Drawing	DWG
Drip leg or dummy leg	DL
Ductile iron	DI
Dust collector	DC
Each	EA
Eccentric	ECC
Elbolet	EOL
Elbow	ELB
Electric resistance weld	ERW
Elevation	EL
Equipment	EQUIP
Evaporator vapor	EV
Exhaust steam	E
Expansion	EXP
Expansion joint	EXP JT
Extraction steam	ES
Fabrication (dimension)	FAB
Face of flange	FOF
Faced and drilled	F&D
Factory Mutual	FM
Far side	FS
Feed pump balancing line	FB
Feed pump discharge	FD
Feed pump recirculating	FR
Feed pump suction	FS
Female	F
Female	FM
Female pipe thread	FPT
Field support or forged steel	FS

86103-13_A02A.EPS

Field weld .... FW
Figure .... FIG
Fillet weld .... W
Finish floor .... F/F
Finish floor .... FIN FL
Finish grade .... FIN GR
Fitting .... FTG
Fitting makeup .... FMU
Fitting to fitting .... FTF
Flange .... FLG
Flat face .... FF
Flat on bottom .... FOB
Flat on top .... FOT
Floor drain .... FD
Foundation .... FDN
Fuel gas .... FG
Fuel oil .... FO

Gauge .... GA
Galvanized .... GALV
Gasket .... GSKT
Grating .... GRTG

Hanger .... HGR
Hanger rod .... HR
Hardware .... HDW
Header .... HDR
Heat traced, heat tracing .... HT
Heater drains .... HD
Heating system .... HS
Heating, ventilating, and air conditioning .... HVAC
Hexagon .... HEX
Hexagon head .... HEX HD
High point .... HPT
High pressure .... HP
Horizontal .... HORIZ
Hot reheat steam .... HR
Hydraulic .... HYDR

Increaser .... INCR
Input/output .... I/O
Inside diameter .... ID
Insulation .... INS
Invert elevation .... IE
Iron pipe size .... IPS
Isometric .... ISO
Issued for construction .... IFC

Lap weld .... LW
Large male .... LM
Length .... LG
Long radius .... LR
Long tangent .... LT
Long weld neck .... LWN
Low pressure .... LP
Low-pressure drains .... DR
Low-pressure steam .... LPS
Lubricating oil .... LO

Main steam .... MS
Main system blowouts .... BL
Makeup water .... MU
Male .... M
Malleable iron .... MI
Manufacturer .... MFR
Manufacturer's Standard Society .... MSS

National Pipe Thread .... NPT
Nipolet .... NOL
Nipple .... NIP
Nominal .... NOM
Not to scale .... NTS
Nozzle .... NOZ

Outside battery limits .... OSBL
Outside diameter .... OD
Outside screw and yoke .... OS&Y
Overflow .... OF

Pipe support .... PS
Pipe tap .... PT
Piping and instrumentation diagram .... P&ID
Plain .... P
Plain end .... PE
Plain, both ends .... PBE
Plain, one end .... POE
Point of intersection .... PI
Point of tangent .... PT
Pounds per square inch .... PSI
Process flow diagram .... PFD
Purchase order .... PO

Radius .... RAD
Raised face .... RF
Raised face slip-on .... RFSO

86103-13 A02B.EPS

Raised face smooth finish	RFSF
Raised face weld neck	RFWN
Raw water	RW
Reducer, reducing	RED
Relief valve	PRV-PSV
Ring-type joint	RTJ
Rod hanger or right hand	RH
Safety valve vents	SV
Sanitary	SAN
Saturated steam	SS
Schedule	SCH
Screwed	SCRD
Seamless	SMLS
Section	SECT
Service and cooling water	SW
Sheet	SH
Short radius	SR
Slip-on	SO
Socket weld	SW
Sockolet	SOL
Stainless steel	SS
Standard weight	STD WT
Steel	STL
Suction	SUCT
Superheater drains	SD
Swage	SWG
Swaged nipple	SN
Temperature or temporary	TEMP
That is	I.E.
Thick	THK
Thousand	M
Thread, threaded	THRD
Threaded	T
Threaded end	TE
Threaded, both ends	TBE
Threaded, large end	TLE
Threaded, one end	TOE
Threaded, small end	TSE
Threadolet	TOL
Top of pipe or top of platform	TOP
Top of steel or top of support	TOS
Treated water	TW
Turbine	TURB
Typical	TYP
Underwriters Laboratories	UL
Vacuum	VAC
Vacuum cleaning	VC
Vents	V
Vitrified tile	VT
Wall thickness or weight	WT
Weld neck	WN
Weldolet	WOL
Well water	WW
Wide flange	WF

86103-13_A02C.EPS

Appendix C

Drawings

This appendix contains drawings that are referenced in Section 5.0.0 as *Figures A1* and *A2*. These drawings are examples of actual drawings used in ship construction.

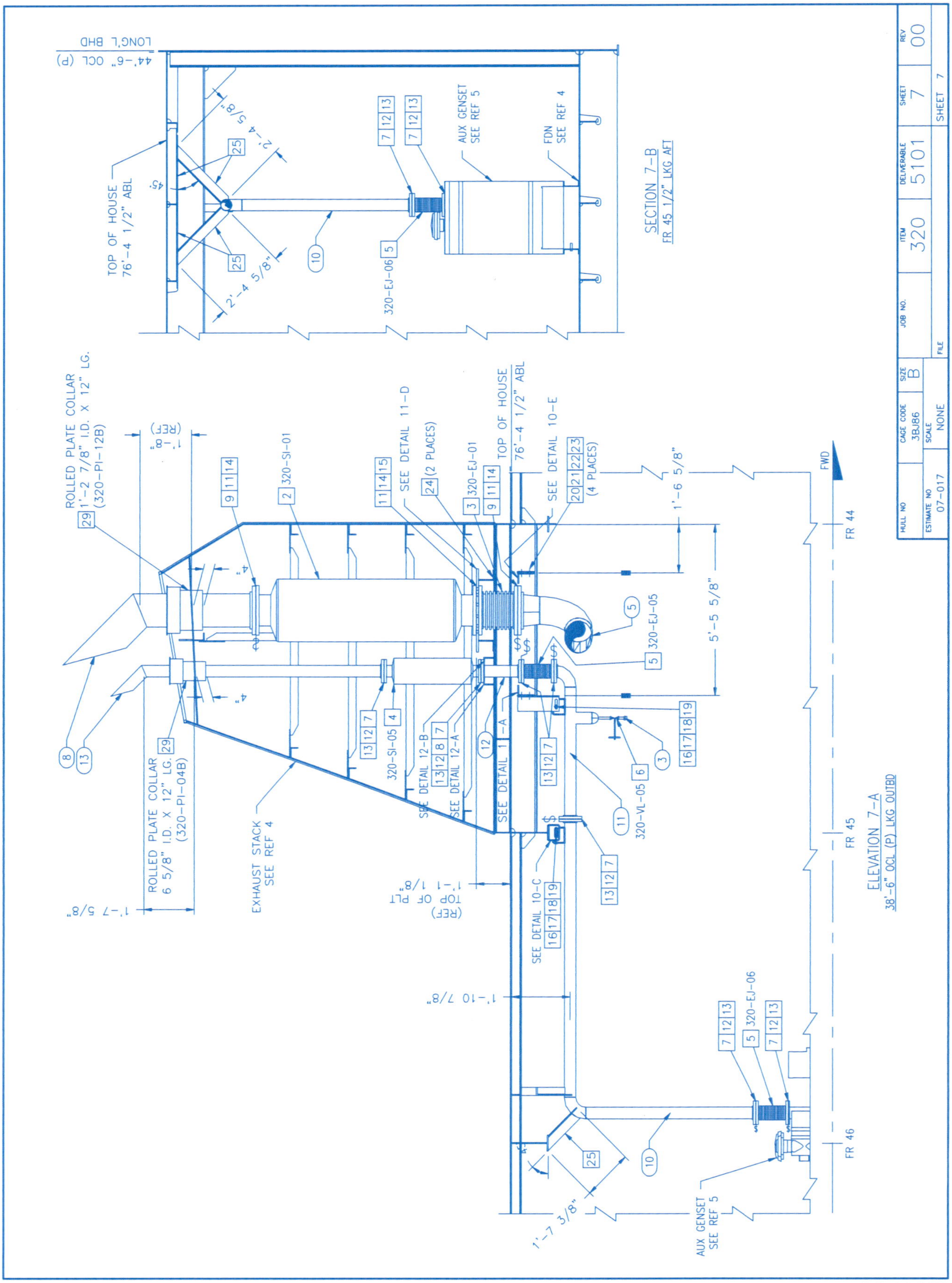

86103-13_A03A.EPS

Figure A1

SHOP BILL OF MATERIALS

Item #	Quantity	Nominal Size	Stock Description	Total Length	Weight	MARS	Length
1	1	10"	PIPE SMLS SCH 40 CS ASTM A106 ANSI B36.10	120 1/4"	Varies	10003188	Varies
2	2	4"	CAP BW SCH40 CS ASTM A234 ANSI B16.9		2.8	10004634	2 1/2"
3	1	4"	CAP BW SCH40S SS 316L ASTM A403 ANSI B16.9		2.8	10004636	2 1/2"
4	3	1/2"	CAPTHD 3000# CS ASTM A105 ANSI B16.11		0.5	10004111	7/8"
5	1	10"	ELBOW 30 BW LR SCH STD CS ASTM A234 ANSI B16.9		27.8		7 7/8"
6	4	10"	ELBOW 90 BW LR SCH STD CS ASTM A234 ANSI B16.9		83.4	10002004	23 5/8"
7	2	4"	ELBOW 90 BW LR SCH40S SS316L ASTM A403 ANSI B16.9		9	10002116	9 1/2"
8	7	4"	FLANGE SO FF 150LB CS ASTM A105 ANSI B16.5		13	10002549	1 3/8"
9	4	12"	FLANGE SO FF 150LB CS ASTM A105 ANSI B16.5		64	10002246	2 1/4"
10	2	10"	FLANGE SO FF 150LB CS ASTM A105 ANSI B16.5		40	10002231	2"
11	3	1/2"	PIPE NIPPLE 1/2" SMLS CS A106 SCH 80 TOE 3" LONG		6.9	10003172	3"
12	2	4"	PIPE SMLS CS ASTM A106-B ANSI B36.10	9"	4	10004500	4 1/2"
13	1	12"	PIPE SMLS SCH40S SS304L ASTM A312 ANSI B36.19	161"	359.5	10003213	80 1/2"
14	1	4"	PIPE SMLS SCH40S SS316L ASTM A312 ANSI B36.19	358 1/4"	Varies	10004504	Varies
15	1	1/2"	PIPE SMLS SCH80S SS316L ASTM A312 ANSI B36.19	15 1/8"	0.5	10003176	5 1/8"
16	2	12" – 10"	REDUCER CONC BW SCH STD CS ASTM A234 ANSI B16.9 12x10		32.4	10006028	8"
17		1/4"	PLATE STAINLESS STEEL	9.1 sqft			
18		3/8"	PLATE STEEL A-36	2.82 sqft			

HULL NO	CAGE CODE	SIZE	JOB NO.	ITEM	DELIVERABLE	SHEET	REV
	3BJ86	B		320	5102	3	00
ESTIMATE NO 07-017	SCALE NONE		FILE			SHEET 3	

86103-13_A03B.EPS

Figure A2

Additional Resources

This module presents thorough resources for task training. The following resource material is suggested for further study.

Process Piping Drafting. Rip Weaver. Houston, TX: Gulf Publishing Company.

Figure Credits

Topaz Publications, Inc., Module opener
Courtesy of VT Halter Marine, Inc., Figures 1, 4, 5, 8–14, Appendix C

NCCER CURRICULA — USER UPDATE

NCCER makes every effort to keep its textbooks up-to-date and free of technical errors. We appreciate your help in this process. If you find an error, a typographical mistake, or an inaccuracy in NCCER's curricula, please fill out this form (or a photocopy), or complete the online form at **www.nccer.org/olf**. Be sure to include the exact module ID number, page number, a detailed description, and your recommended correction. Your input will be brought to the attention of the Authoring Team. Thank you for your assistance.

Instructors – If you have an idea for improving this textbook, or have found that additional materials were necessary to teach this module effectively, please let us know so that we may present your suggestions to the Authoring Team.

NCCER Product Development and Revision

13614 Progress Blvd., Alachua, FL 32615

Email: curriculum@nccer.org

Online: www.nccer.org/olf

❑ Trainee Guide ❑ Lesson Plans ❑ Exam ❑ PowerPoints Other ______________________

Craft / Level: Copyright Date:

Module ID Number / Title:

Section Number(s):

Description:

Recommended Correction:

Your Name:

Address:

Email: Phone:

86104-14

Fitting I

Module Nine

Trainees with successful module completions may be eligible for credentialing through NCCER's National Registry. To learn more, go to **www.nccer.org** or contact us at **1.888.622.3720**. Our website has information on the latest product releases and training, as well as online versions of our *Cornerstone* magazine and Pearson's product catalog.

Your feedback is welcome. You may email your comments to **curriculum@nccer.org**, send general comments and inquiries to **info@nccer.org**, or fill in the User Update form at the back of this module.

This information is general in nature and intended for training purposes only. Actual performance of activities described in this manual requires compliance with all applicable operating, service, maintenance, and safety procedures under the direction of qualified personnel. References in this manual to patented or proprietary devices do not constitute a recommendation of their use.

86104-14
Fitting I

Objectives

When you have completed this module, you will be able to do the following:

1. Explain the role the fitter plays in quality assurance and safety.
2. Recognize and identify structural members and calculate their thicknesses.
3. Identify layout tools, fitting tools, and fitting aids used to fit up, align, and check plate joints.

Performance Tasks

Under the supervision of the instructor, you should be able to do the following:

1. Remove welded fittings aids properly.
2. Interpret drawings and verbal instructions for fitting/assembling.
3. Apply math calculations and use standard/metric forms of measurements to lay out and fit parts.
4. Square and mark work surfaces using the 3-4-5 method.
5. Perform proper fit procedures related to fitting tasks in the downhand (flat) position, including:
   - Longitudinals
   - Bulkheads
   - Frames
   - Miscellaneous structural items (chocks, headers, collars, brackets, clips)
6. Fit up plate joints using alignment tools and check for misalignment and poor fit-up.

Trade Terms

Arbor
Gouging
Plate stiffeners
Plumb

Industry-Recognized Credentials

If you're training through an NCCER-accredited sponsor, you may be eligible for credentials from NCCER's Registry. The ID number for this module is 86104-14. Note that this module may have been used in other NCCER curricula and may apply to other level completions. Contact NCCER's Registry at 888.622.3720 or go to **www.nccer.org** for more information.

Contents

Topics to be presented in this module include:

Figures

1.0.0 INTRODUCTION

The term *fitting* in the maritime world refers to the temporary attachment of parts such as **plate stiffeners** to prepare them for permanent attachment. One of the main functions of a structural fitter, also known as a fitter, is assembling and tack-welding the structural steel sub-assemblies and assemblies that make up the ship's hull and interior compartments. Structural fitters also build interior ship structures such as hatches, doors, and bulkheads. Among the structural fitter's duties are preparing metal surfaces for welding, and grinding tack welds to remove burrs and other imperfections. *Figure 1* shows a fitter using a track burner to cut steel plate.

The following is a list of tasks that structural fitters usually perform:

- Lay out and fabricate metal structural parts, such as plates, bulkheads, and frames, and brace them in position within the hull of a vessel for riveting or welding.
- Tack-weld fitted assemblies.
- Use oxyfuel and plasma arc cutting equipment to cut metal into specific shapes.
- Lay out the position of parts on metal, working from blueprints or templates and using a scriber and hand tools.
- Locate and mark reference lines, such as center, buttock, and frame lines.
- Position parts in the hull of a vessel.
- Align parts in relation to each other, using jacks, come-alongs, turnbuckles, clips, wedges, and mauls.
- Mark locations of holes to be drilled and install temporary fasteners to hold parts in place for welding or riveting.
- Install packings, gaskets, liners, and structural members such as doors, hatches, brackets, and clips.
- Tack-weld clips and brackets in place prior to permanent welding.
- Bend, cut, and shape plates, beams, and other heavy metal parts.

2.0.0 FITTING SAFETY

Basic safety requirements were covered in a previous module. In addition to those general safety requirements, there are job-specific safety concerns related to fitting work. Those concerns are covered in this section.

2.1.0 Personal Protective Equipment (PPE)

Due to the nature of the work, fitters must be concerned with noise; airborne dust and debris from grinding; heat and flame from cutting and welding; toxic fumes from torch-cutting certain

86104-13_F01.EPS

Figure 1 Structural fitter cutting plate.

metals; and chemical vapors. For that reason, it is especially important that the fitter wear PPE that is suitable to the work (*Figure 2*). The minimum required PPE (*Figure 2*) includes the following:

- Leather, steel-toed work boots with at least a 6" top (8" at some yards)
- Safety glasses with OSHA-approved hard impact-resistant side shields
- Ear protection
- Hard hat
- Reflective vest
- Leather gloves
- Long-sleeve cotton shirt and long pants (flame-retardant coveralls may be required by some yards)

There are other items of PPE that the fitter will need to wear in certain circumstances. These items include a full face mask, a fall protection harness, a respirator, and personal flotation device.

A full face mask (*Figure 3*) is needed whenever the fitter is using a grinder or is working around others who are using grinders. Like the worker in the figure, a fitter who is at risk of falling from an elevated position must wear a personal fall arrest system and be properly tied off to an approved anchorage point.

There are a number of situations in which a fitter might be exposed to airborne contaminants and will be required to wear respiratory protection (*Figure 4*). These situations include exposure to the following:

- Metal fumes from hot work
- Metal particles from grinding
- Paint and solvent vapors
- Adhesive vapors
- Chemical vapors and mists
- Grit from pressure blasting

Any time a fitter is working above water, a personal flotation device (PFD) must be worn (*Figure 5*). A Coast Guard-approved throwing ring with at least 90' of line must also be available nearby.

2.2.0 Safety Practices

A lot of effort goes into making a job site safe, so being safe simply means using common sense, observing safety signs and tags, and following the site safety procedures. The following are some general common-sense rules for being safe on a job site.

- Be alert for safety-related tags. Scaffolding should have a "Safe To Use" tag. If a scaffold has a yellow tag, special precautions are

86104-13_F02.EPS

Figure 2 A fitter's PPE.

86104-13_F03.EPS

Figure 3 Face protection.

86104-13_F04.EPS

Figure 4 Supplied-air respirator.

needed when using it. A red tag means it is under construction and must not be used. A scaffold must have a safe ladder or other safe means of access. It also must have proper walking surfaces, along with safety handrails and kickboards.

- Be careful around any device that can release mechanical, hydraulic, pneumatic, or electrical energy. Observe all lockout/tagout warnings and procedures.
- Note the locations of fire extinguishers in your work area. Keep in mind, however, that extinguishers are only to be used on incipient (early stage) fires, and only after the area has been evacuated and the alarm sounded.
- Make sure you know the safe evacuation route in case of an emergency.
- Be alert for the crane warning horn, which indicates a moving load.
- Do not use a forklift, pendant crane, manlift, or other rolling stock unless you have an up-to-date license for the particular equipment.
- If you are working with any chemicals, be familiar with the MSDS/SDS for each chemical.
- Use proper lifting procedures to prevent back injury.

86104-13_F05A.EPS

Figure 5A Water safety equipment. (1 of 2)

86104-13_F05B.EPS

Figure 5B Water safety equipment. (2 of 2)

WARNING!

Never enter a tank without first checking for warning signs. Confined spaces must be certified before entry.

3.0.0 Structural Members

Most of the material a fitter works with are structural members made from medium carbon steel. These structural members come in a variety of prefabricated shapes and sizes. They are used to form the various structural assemblies that make up a vessel.

Plates – Plates are sheet of steel rolled to a uniform thickness of 1¼" or greater. Plates are used to form inner bottoms, bulkheads, decks, shells, trunks, coaming, and other structural members. *Figure 6* shows plates and T-bars formed into an assembly as part of a training exercise. Also note the piece of flat bar at the bottom right of the image.

The shapes of other structural members are shown in *Figure 7* and are discussed in the text that follows.

Angles – Angles are used to join two pieces of metal that meet at approximate right angles. The legs may be equal or unequal.

Square bar – A square bar is a solid steel bar that has similar dimensions on all four sides. It is used for dock traction or ladder rungs.

Channel bar – A channel bar is a rolled bar with two parallel flanges at right angles to the web. Channels are used to make side frames, deck beams, bulkhead stiffeners, and pillars.

H and I beams – An H-beam is symmetrical; that is, it is as wide as it is tall. It has a cross-section that looks like the letter H. An I beam has a web that is taller than the flange width. In other words, it is asymmetrical.

Z-bar – A Z-bar is a rolled bar that looks like the letter Z. It is used to join surfaces that are on a different level.

T-bar – A T-bar is made from a flat bar centered on another flat bar.

Bulb bar – A bulb bar is an angle bar with a small bulb of material added to the toe of the angle at the point of maximum stress. This design provides greater strength than the angle bar.

Solid round half – This is a solid steel bar in a half-round shape.

Hollow round half – The hollow round half is a steel bar in the shape of a half cylinder.

Flat bar – A flat bar is a steel, aluminum, or stainless steel plate with a width of 6" or less. It is identified by its width and thickness.

Square tubing – Square tubing is identified by its outside dimensions and wall thickness.

The parts of the various structural members have their own identifiers, such as flanges and webs. These identifiers are shown in *Figure 8*. Fitters need to be familiar with these terms.

4.0.0 Fitting Tools

Structural fitters use a number of tools that are specific to their craft. The tools commonly used by fitters—tools they will use every day—are described in this section.

4.1.0 Common Hand Tools

Hammers – The most common hammers used by fitters are the ballpeen hammer, chipping hammer, and maul (*Figure 9*). Ballpeen hammers can weigh up to 4 pounds. The hammer face is used to strike rivets, punches, and chisels. The peen end is used for shaping metal. Chipping hammers are used to remove slag and residue from metal after welding. A maul, commonly called a sledgehammer, is used when significant driving

86104-13_F06.EPS

Figure 6 An assembly made from structural members.

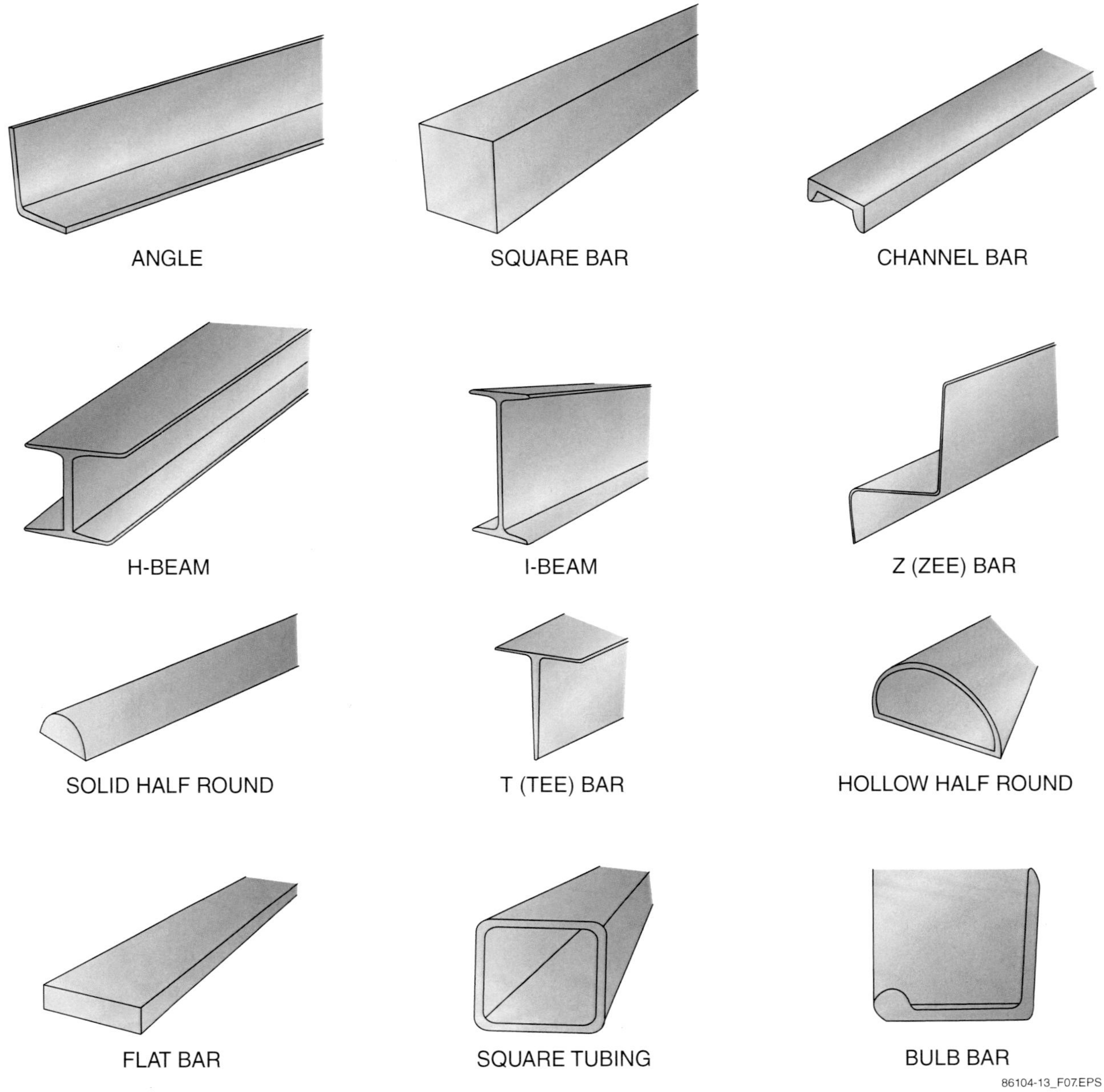

Figure 7 Structural members.

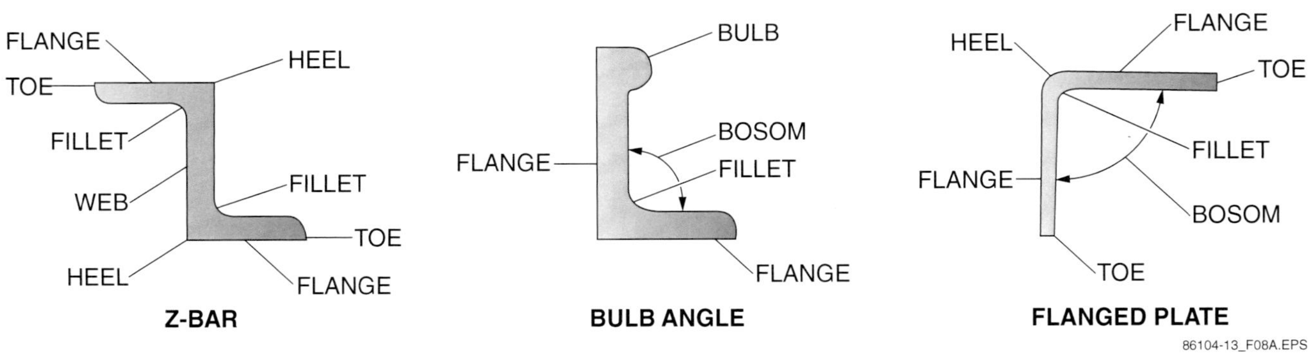

Figure 8 Structural member identifiers. (1 of 2)

force is needed. Mauls can weigh from 2 pounds to 20 pounds.

Chisels – Fitters generally use what are known as cold chisels. Of these, the flat chisel (*Figure 10*) is the most common. The chisel is held at about a 40-degree angle to the work piece and is struck with a hammer repeatedly until the metal is penetrated. Always wear eye protection when using

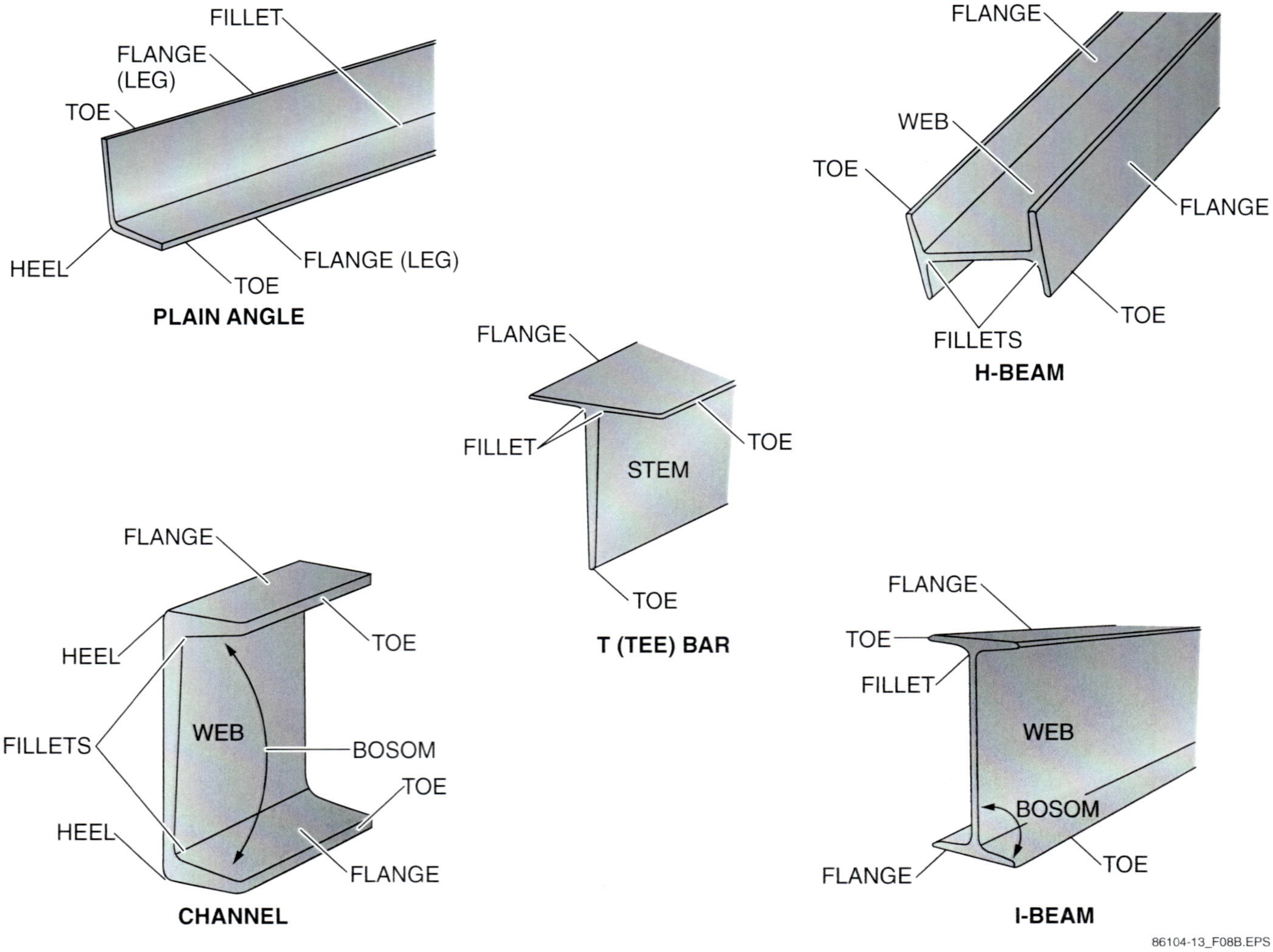

Figure 8 Structural member identifiers. (2 of 2)

a cold chisel, as it is possible for metal particles to break off. If the head of a chisel begins to mushroom, it must be replaced because pieces can easily dislodge.

Other types of cold chisels have differently shaped heads. They are the cape chisel used to cut keyways; the diamond point chisel used to cut v-grooves and clean out inside corners; and the round nose chisel used to cut grooves for oil ways in bearings.

Punches – Most punching of holes in metal parts is done by machine, but there are occasions when it is done by hand. *Figure 11* shows some of the punches used for this purpose. The center punch is used to make a dimple (indentation) in the metal so it can be drilled. The dimple keeps the drill from sliding on the metal. A prick punch has a sharper, more angled point. Its function is to leave a precisely placed small hole for use as a reference for measurements and markings.

The tapered punch or drift pin, is used to align holes that are not quite lined up so that a fastener can be inserted. Striking the drift pin with a hammer will cause the tapered pin to bring the two holes into alignment.

Pry bars – A pry bar (*Figure 12*) can be used to raise or move heavy items. In fitting work, it is used primarily to align parts.

WARNING!

Be careful when using a pry bar because it can slip. Keep hands and feet clear when using a pry bar.

Clamping tools – Clamps are used to hold objects in place, usually on a temporary basis while they are more permanently secured. C-clamps (*Figure 13*) are among the most common clamps. Welding clamps (*Figure 14A*) are also common. Other types of clamps that may be used by fitters include locking pliers (*Figure 14B*); locking sheet metal clamps (*Figure 14C*); locking chain clamps (*Figure 14D*); locking C-clamps (*Figure 14E*); and sliding bar clamps (*Figure 14F*).

Figure 9 Hammers used by fitters.

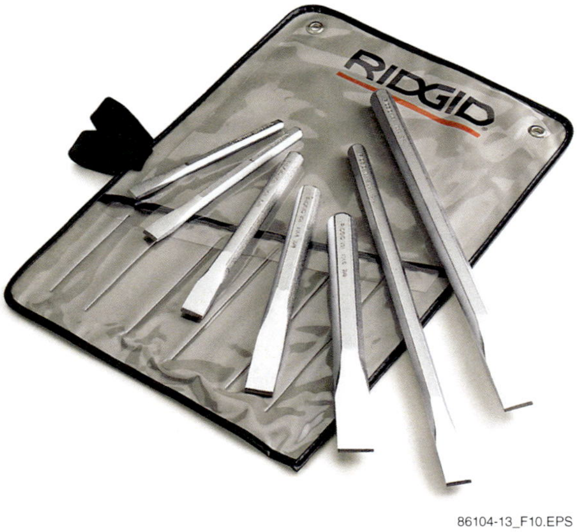

Figure 10 Cold chisels.

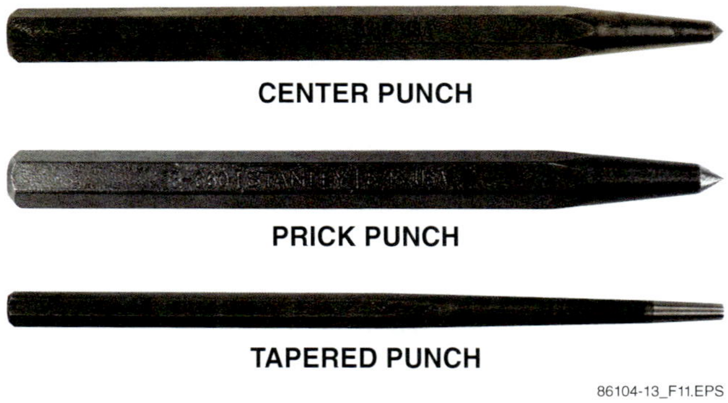

Figure 11 Punches.

Figure 12 A pry bar.

4.2.0 Hoisting and Pulling Tools

Steel is heavy, so hoisting devices are needed to lift and move it. A hoist is a device that provides a mechanical advantage to allow lifting of a load that cannot be lifted manually. The two types of hoisting devices commonly used by fitters are the chain fall (*Figure 15*), sometimes called a chain hoist, and the ratchet lever hoist (*Figure 16*), commonly called a come-along. A chain fall consists of a gear system attached to a sprocket, with a chain and hook attached to it. A chain fall can be operated in one of three ways: manually, electrically, or pneumatically, and is used only for lifting.

The ratchet-lever hoist is commonly used to lift loads. Unlike the chain fall, the ratchet-lever hoist can be used not only to lift loads, but to pull them horizontally or on an angle. It is sometimes used to align heavy assemblies.

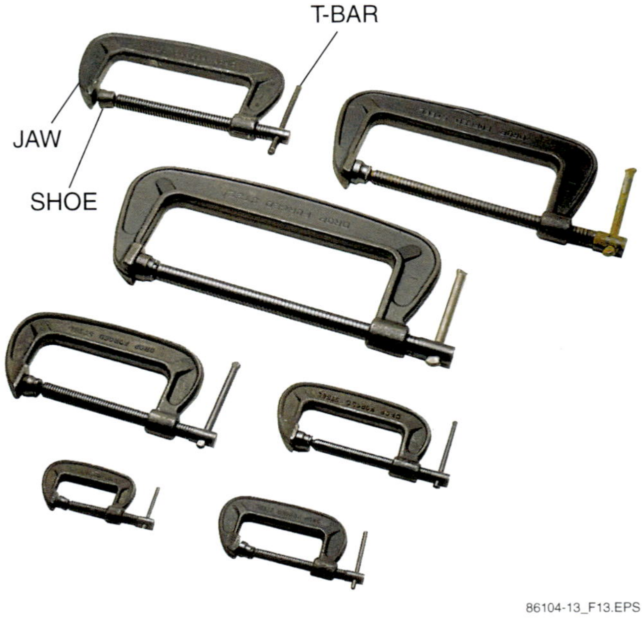

Figure 13 C-clamps.

> **NOTE**
> Some yards require special certification for anyone using a chain hoist to lift material.

Regardless of the type of hoisting device used, there are two important things to remember. One is to never exceed the rated load, which must be imprinted on the device. The other is to never engage the free-chaining handwheel or knob while the device is being used to lift a load. Free-chaining is the feature that allows the chain to be extended.

4.3.0 Levels and Squares

Levels and squares are among the layout tools commonly used by structural fitters. Levels are used to determine whether a surface is level or **plumb**. A spirit level (*Figure 17*) uses a bubble floating in alcohol. Alcohol is used rather than water because it will not freeze.

> **NOTE**
> A level cannot be used on a ship afloat.

Most spirit levels have leveling vials for both horizontal and vertical. Some also have vials for angled surfaces. There are also magnetized levels

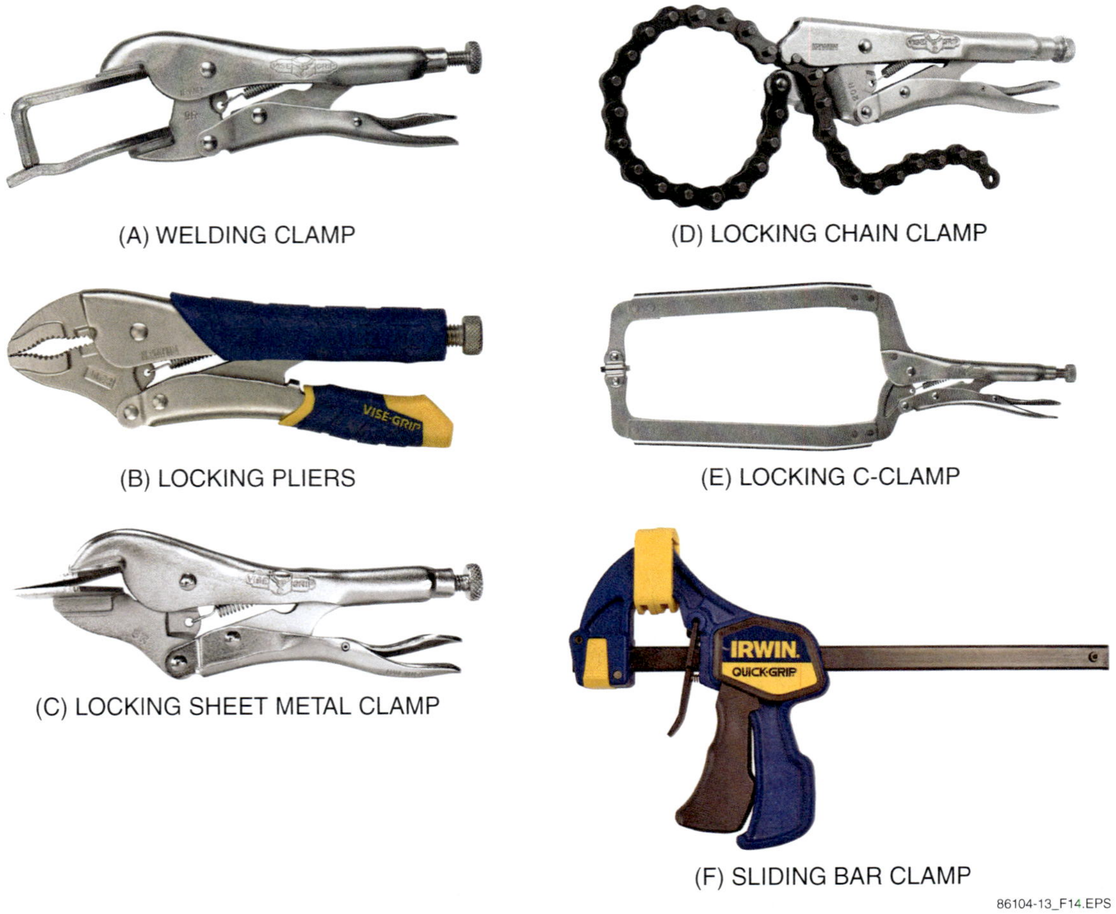

Figure 14 Common clamping devices.

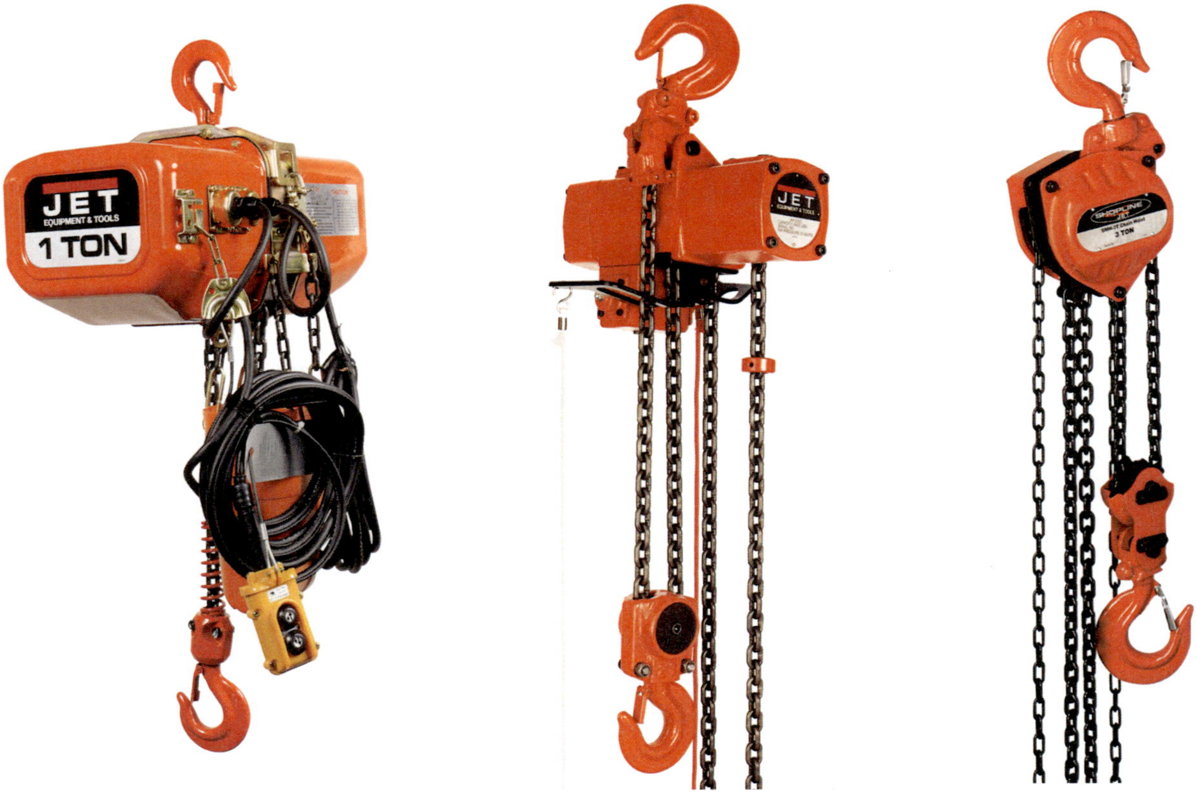

Figure 15 Chain hoists.

Figure 16 Ratchet-lever hoist.

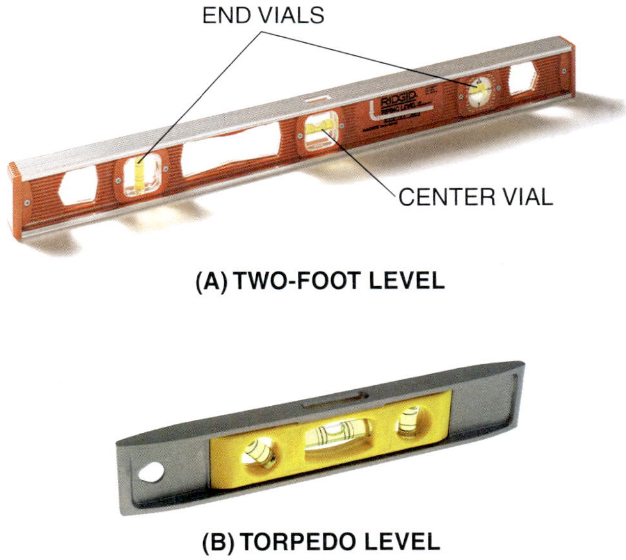

Figure 17 Spirit level.

Figure 18 Magnetic level.

that will adhere to steel, and are good for checking plumb on vertical steel structures (*Figure 18*).

While the spirit level is common, it is not the only type of level. In fact, laser levels (*Figure 19*) have become popular because they provide a high degree of accuracy. A laser level emits a high-intensity light beam, which is detected by an electronic laser detector (target). The laser level can be used to establish the correct elevation or to establish a straight line. Both fixed and rotating models of tripod-mounted laser levels are available. Spirit levels are also available with a laser beam that can be turned on to extend the level's reference line.

Squares (*Figure 20*) are used for a number of purposes, the most important of which is squaring-up parts that are perpendicular to each other. There are several types of squares:

- The framing square, often called a carpenter's square, is used for measuring, setting corners, setting bevels, and marking 45-degree angles. The shorter, narrow leg of a framing square is known as the tongue. The longer, wider leg is known as the blade or body. When one leg is placed against the edge of a workpiece, the other can be used to mark a straight line. To mark 45-degree angles, the framing square is placed on the workpiece so that the blade and tongue measurements are equal. *Figure 21* shows an example of a fitter using a framing square.
- A combination square (*Figure 22*) contains a graduated steel rule, known as the blade, and a movable head that is at a right angle to the blade. The blade can be used as a straightedge with the head removed. The movable head has both 45- and 90-degree angles. It also has a level vial and a removable scriber. Protractor heads and center heads are available for the combination square. The center head can be used to locate the exact center of round stock. The protractor head can be used to measure and mark off any angle from 0 to 180 degrees.
- A try square is used to check the accuracy of right angles. It can also be used to measure short distances, check for flatness, and lay out perpendicular lines. A try square has a calibrated brass blade and a wooden stock.
- A sliding T-bevel is an adjustable try square. The blade can be set at any angle from 0 to 180 degrees. A locking screw is used to secure the blade at the selected angle. The sliding T-bevel is used to lay out angles and to measure angles such as bevels. The angle can be set using a protractor.

Another tool used for layout is the chalk line (*Figure 23*), which is used to transfer a straight line onto a flat surface. Most straightedge devices are 3' long or less. Any line length greater than that of the nearest straightedge must be made with a chalk line. The chalk line must be stretched tightly between the marks on the workpiece. The line is then lifted at the center and allowed to snap back into place. This will leave a clear, defined straight line on the workpiece. A line can be snapped by one person using the hook on the end of the line or a clamp to secure it on one end.

86104-13_F19.EPS

Figure 19 Laser level.

4.4.0 Grinders

The grinder is one of the most common tools used by a fitter. Grinders are used to remove excess material such as burrs from metal. Grinders are also used to clean and cut metal. They can be either electrically or pneumatically powered. *Figure 24* shows a fitter using a grinder.

Fitters use grinders for such tasks as cleaning metal in preparation for welding; removing weld spatter after welding; removing paint; and removing the burrs and other residue of tack welds when fit-up aids are removed following welding. A variety of attachments are available for grinders, as shown in *Figure 25*.

Typical cutting discs used with grinders can range from 4½" to 7" in diameter, but smaller sizes are available. Grinder discs rotate at speeds ranging from 5,000 to 12,000 rpm, so it is important to handle them correctly. Applying too much pressure can cause the grinder to kick back, possibly causing injury. When using a grinder, start it before applying it to the surface. When using a grinder for surface preparation, keep the grinder as flat as possible to avoid **gouging** the base metal. If the metal is accidently gouged, the gouge must be filled with filler metal during the welding process. If you are doing a lot of grinding, it is more efficient to use a larger grinder.

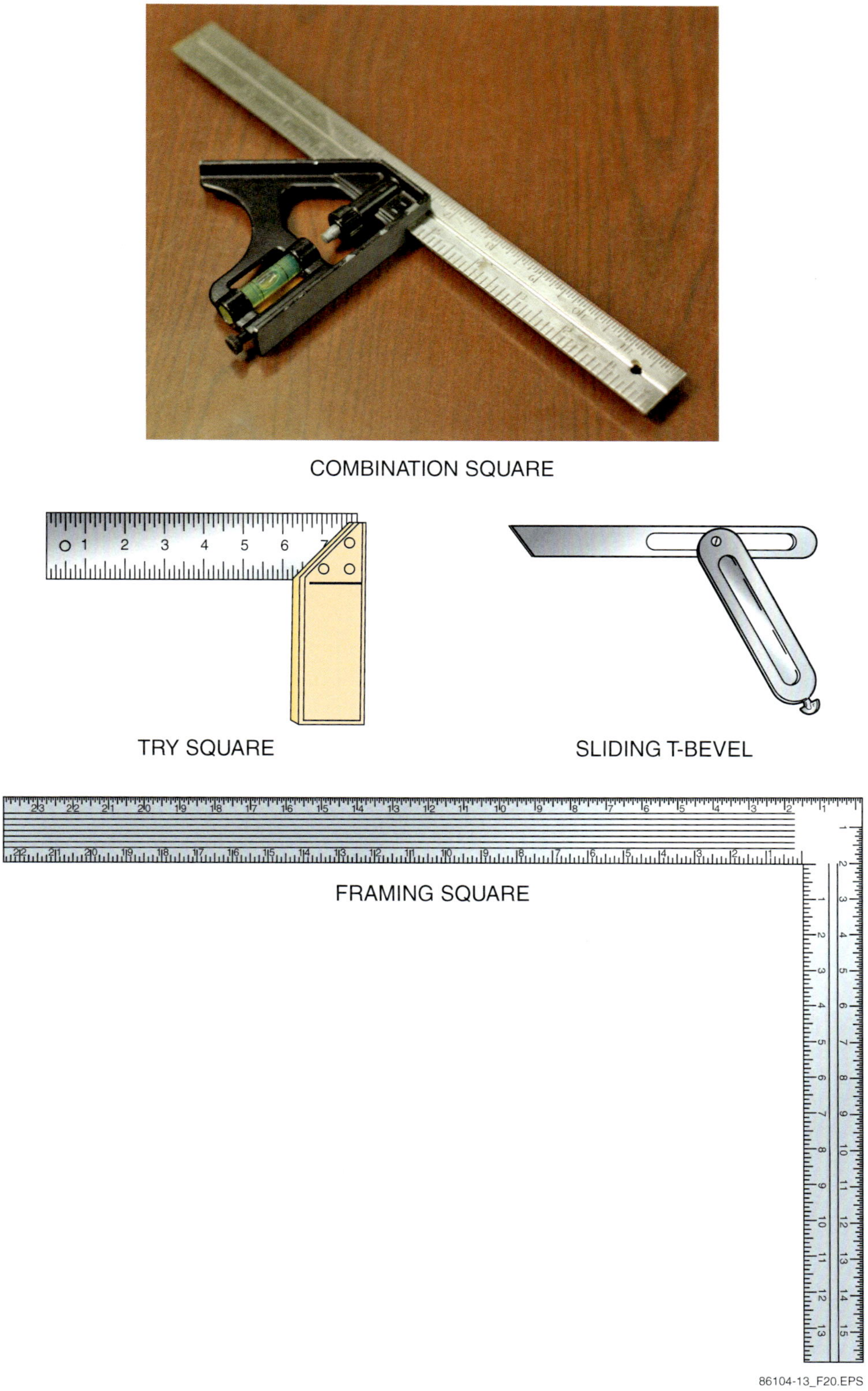

Figure 20 Squares.

WARNING!

An aluminum cutting saw is a hand-held grinder with a carbide-tipped cutting blade. These saws are called suicide saws because they can have a violent kick. When using such a tool, it should be pulled rather than pushed.

Some safety tips to observe when using a portable grinder include the following:

- Wear required PPE, including eye and face protection.
- Make sure the guard is in place.

86104-13_F21.EPS

Figure 21 Framing square in use.

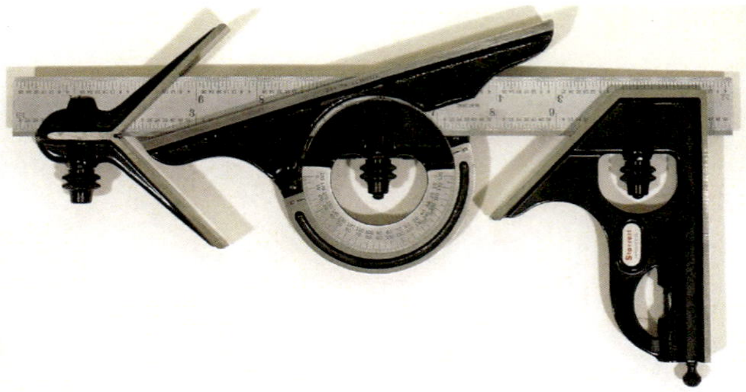

86104-13_F22.EPS

Figure 22 Combination square with multiple heads.

- Check the grinding wheel for damage. A damaged wheel can fly apart while in use.
- Make sure the grinding wheel is installed correctly and fits properly in the **arbor.**
- Make sure the wheel you are using is the correct one for the job.
- Check the electrical cord for damage.
- Inspect the work area to make sure there are no loose items that the grinder wheel could contact.
- Test the grinder at full speed with no load before using it.

> **NOTE**
> To extend the life of wire brushes, occasionally reverse direction by removing the brush and turning it over. This provides a self-sharpening action. Be sure the grinder is not plugged in while this is being done.

Keep in mind that it matters which brush you use. Crimped-wire brushes are used for light- to medium-duty work, especially for parts that could be damaged by aggressive brushing. Knot-wire brushes produce a much harder brushing

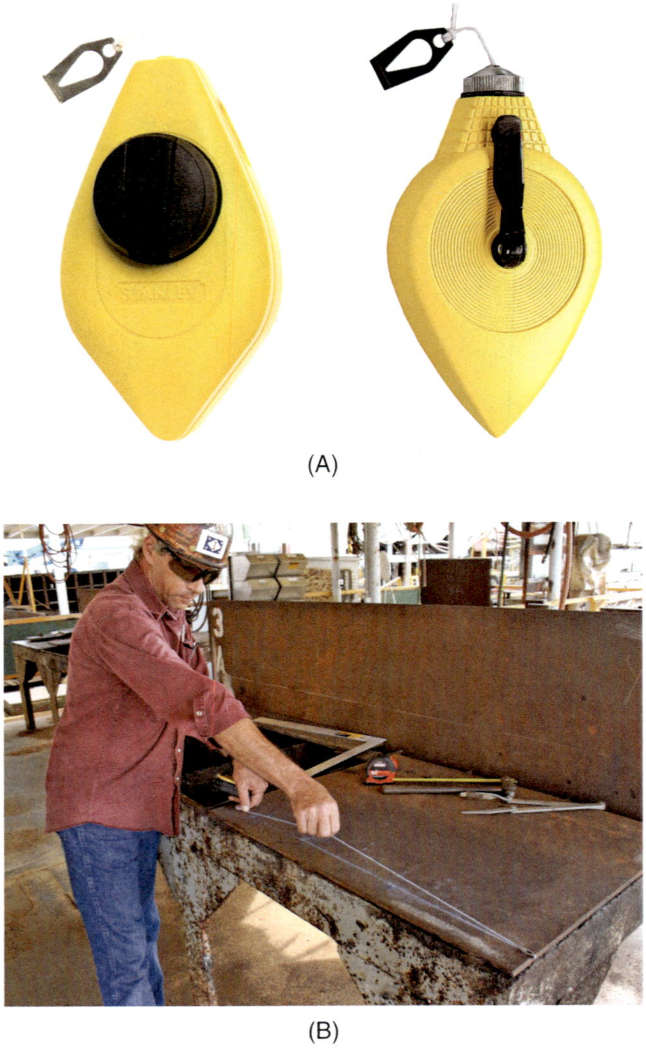

86104-13_F23.EPS

Figure 23 Chalk lines.

action and are generally used only on parts that require high-impact action. They are commonly used to remove large burrs and heavy rust, paint, or scale.

> **CAUTION**
> A wire brush that has been used on carbon steel must not be used on stainless steel or aluminum. Doing so can contaminate the metal.

4.5.0 Jacks

Jacks are used to raise and align heavy objects using force applied with a lever or screw. Jacks can be powered by hydraulic, mechanical, electrical, or manual means. Hydraulic jacks rely on the pressure created by hydraulic fluid. *Figure 26* shows examples of manually operated hydraulic jacks. Other types of jacks, such as the hydraulic cylinders show in *Figure 27* are operated by hy-

86104-13_F24.EPS

Figure 24 Grinder in use.

draulic pumps (*Figure 28*). Various types of pumps and jacks can be combined into a jacking system, as shown in *Figure 29*.

When selecting jacks, the weight, distance to be lifted, and points of contact are used to determine the type of jack or jacking system to be used. A general rule is to select devices with a capacity 50 percent greater than the load. This method prevents overloading and possible damage to the jack. It also compensates in case the load weight is underestimated. When performing jacking operations, make sure the jacks have a firm support base. When raising a load, use hardwood or other approved blocking and chase the load up with the blocking.

The Porta-power (*Figure 30*) is another type of jacking system. It is used to straighten, bend, or modify metal. In shipbuilding work, the Porta-power is used to pull shell plating and align plates on lighter structures (5⁄16" and below).

Screw jacks rely on a screw mechanism rather than hydraulic power. The screw jack consists of a screw shaft that extends from a strong support base. Screw jacks can be powered manually or with an electric motor (*Figure 31*). Screw jacks are self-locking. That is, they will remain in place when the rotating force is removed. Hydraulic jacks will release if the hydraulic pressure used to raise them is lost.

The ratchet jack (*Figure 32*), commonly known as a steamboat jack or push and pull jack, is used to pull parts together or push them apart. This device contains a handle and a ratcheting screw mechanism and uses the ratcheting action to draw parts together or push them apart.

5.0.0 Fitting Aids

Fitting aids are devices used to help bring parts into alignment and keep them aligned until they are tack-welded. Some of the fitting aids are themselves tack-welded to the parts being aligned and must be removed after the parts are tack-welded. *Figure 33* shows the fitting aids that are commonly used by structural fitters.

Pad eye – Pad eyes are attached to plates and are used to attach lifting or jacking devices.

Clip – Clips are metal pieces used to position and align plates. Landing clips are clips used in pairs to guide material into place. One common use is to guide plates that are being placed perpendicular to a horizontal surface.

Saddle – Saddles, also called yokes, are used with strongbacks for fairing parts and moving components.

Strongback – A strongback is a metal device used with a saddle to align and straighten material.

Clip plate – A clip plate, commonly known as a dog, is attached to a workpiece to serve as a restraining device for a wedge.

Adjustable clamp – An adjustable clamp can be used in place of a dog and wedge to apply force for aligning and fairing.

Eye lug – An eye lug is a larger version of a pad eye and is used to lift large pieces of metal. Eye lugs also serve as an attachment point for devices that are used to pull pieces together or push them apart.

Wedge – A wedge is a tapered part used in combination with dogs, saddles, and strongbacks to align parts.

5.1.0 Using Fitting Aids

It is common to use a dog and wedge together to fair-up plates. When attaching a dog, the tack weld should be placed as close to the wedge as possible and on the side of the dog from which the wedge will be driven. *Figure 34* shows how a wedge is used with a dog to fair-up two plates. Note the location of the tack weld.

A strongback is used with a saddle and wedge to align or move parts (*Figure 35*).

Figure 25 Grinders and attachments.

Figure 26 Hydraulic jacks.

Note the location of the tack welds. Welds are not needed along the side because the wedge will drive the flange down tightly against the plate. However, the wedge must be driven from the side on which the welds are placed; otherwise, the saddle is likely to fail. A saddle can be made of flat plate for light work, as shown in *Figure 36*. However, the flat plate saddle may not stand up to heavier work. In such cases, the saddle can be made of angle and bar stock or channel as shown in *Figure 37*.

A saddle can also be used with a jack. *Figure 38* shows the saddle and jack being used to fair-up a T-bar stiffener and a plate.

Pad eyes are fitting aids that are welded to parts in order to lift or move them. The eye lug

Figure 27 Hydraulic cylinder jacks.

Figure 28 Hydraulic pumps.

is larger than the pad eye, and is used to move heavier parts. The placement of welds depends on the direction of the lift or pull. For example, if a push and pull jack is being used to pull two plates together as shown in *Figure 39*, the sides of the lug would be welded and the weld would wrap on the side opposite from the pulling direction. The opposite would be true in the case where the plates were being pushed apart. When welding temporary attachments, the weld should be concentrated on the base metal. Welding in this manner will make it easier to break off the attachment, and will reduce damage to the base metal.

> **NOTE**
> Some pulls require the entire pad eye or eye lug to be welded. If you are not sure, check with your supervisor.

Landing clips are clips used in pairs to guide the placement of web plates. An example is shown in *Figure 40*.

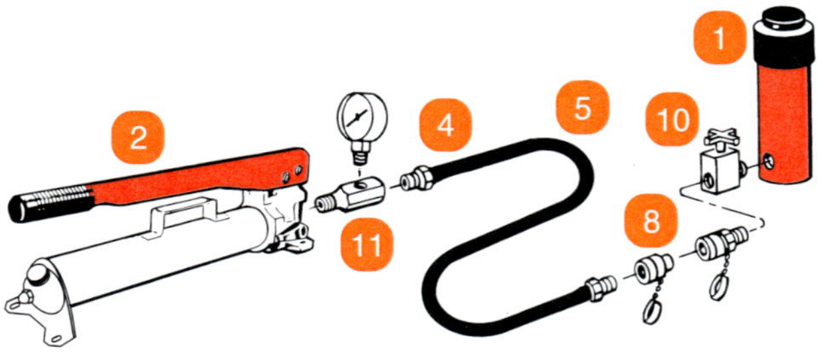

Basic single-acting system with a handpump, gauge, hose and single-acting cylinder

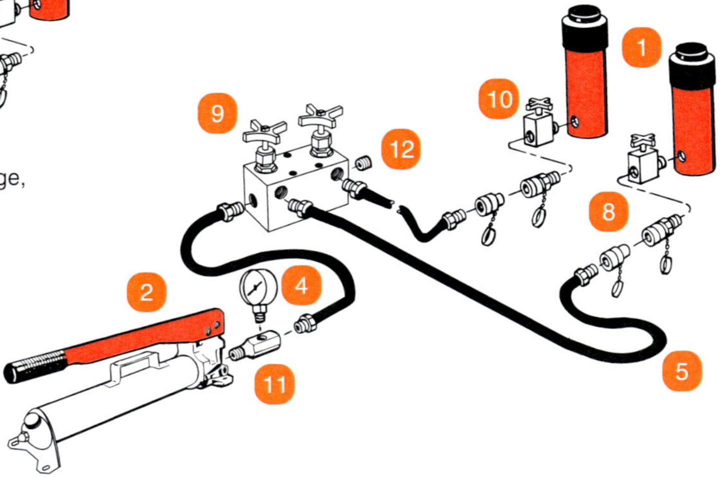

Basic single-acting system with a handpump, gauge, hose, multiple shut-off valve, load-lowering valves and muliple cylinders.

1. Cylinder – applies hydraulic force.
2. Pump – a device for converting mechanical energy to fluid energy.
3. Directional valve – controls the direction of hydraulic fluid in the system.
4. Gauge – measures psi pressure (Pounds per Square Inch) and/or force.
5. Hose – transports hydraulic fluid.
6. Manifold – allows distribution of hydraulic fluid from one source to several cylinders. (No. 9617)
7. Swivel Connector – allows proper alignment of valves and/or gauges. Used when units being connected cannot be rotated. (No. 10469)
8. Quick Coupling – "hose half" and "cylinder half" couplings are used for quick connection and fluid flow check when separated. (No. 9797 and 9798)
9. Shut-Off Valve – regulates the flow of hydraulic fluid to or from cylinders. (No. 9642 or 9644)
10. Load-Lowering Valve – allows metered lowering of cylinder and provides safety when prolonged load holding is required. (No. 9596)
11. Tee Gauge Adapter – allows for installation of pressure/tonnage gauge anywhere in the hydraulic system. (No. 9670)
12. Pipe Plug – for blocking unused ports within the system. (No. 10909)

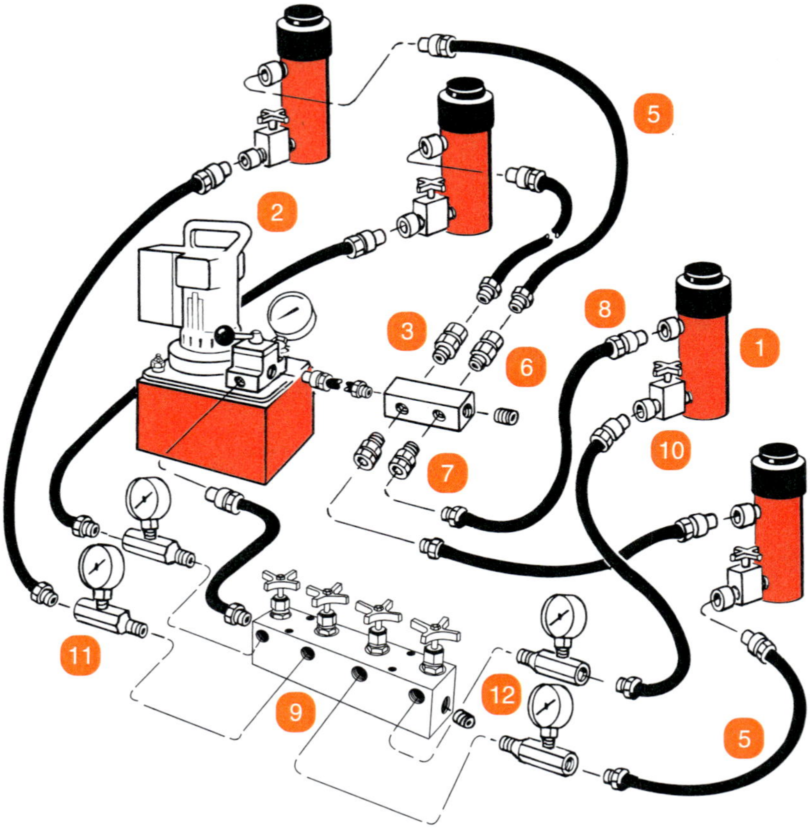

Basic double-acting system with an electric/hydraulic pump, shut-off valve, load-lowering valves and muliple double-acting cylinders.

86104-13_F29.EPS

Figure 29 Jacking systems.

Figure 30 Porta-power.

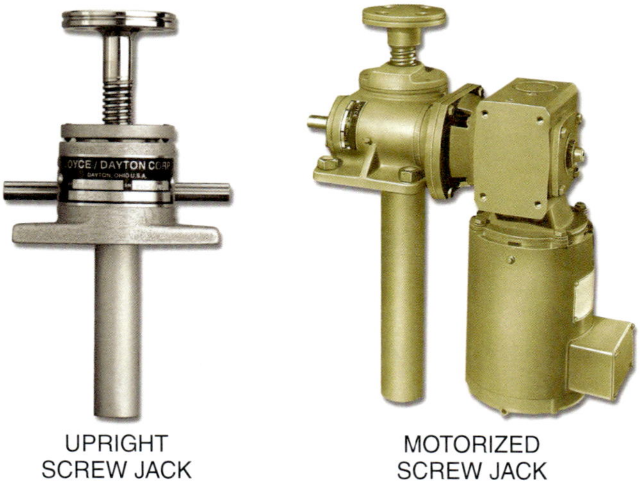

Figure 31 Screw jacks.

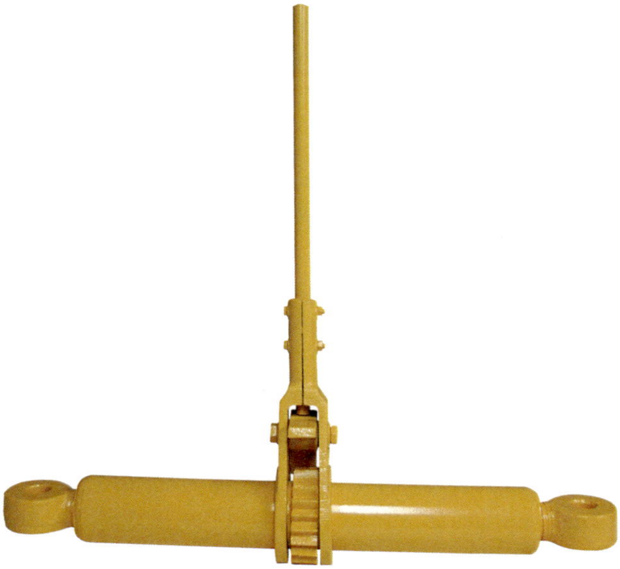

Figure 32 Steamboat jack.

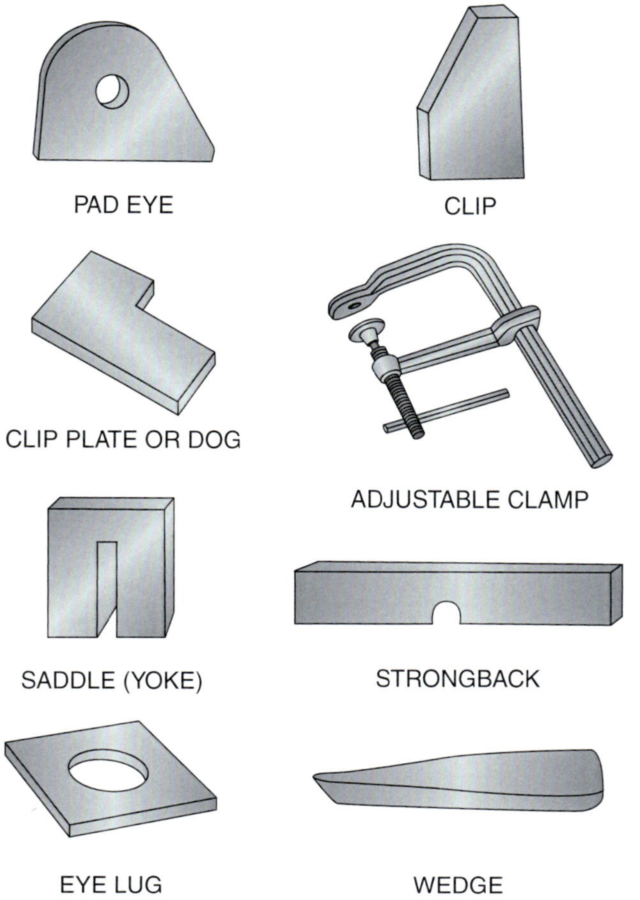

Figure 33 Fitting aids.

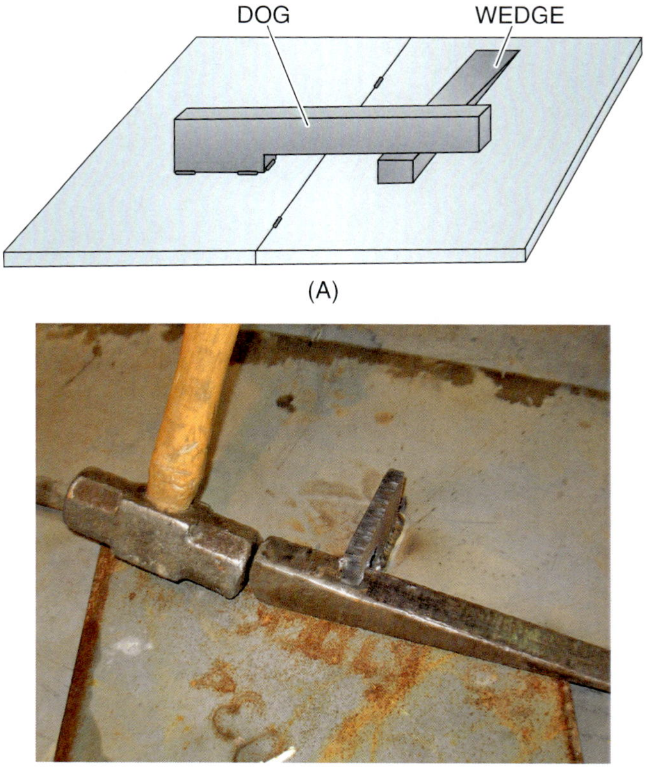

Figure 34 A wedge used with a dog to align plates.

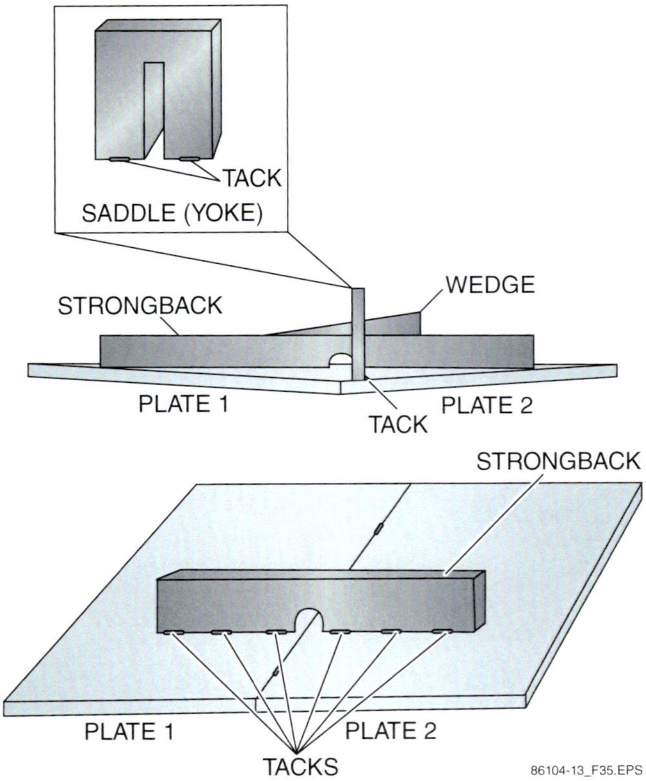

Figure 35 Using a strongback.

Figure 36 Saddle made from flat plate.

5.2.0 Removing Fitting Aids

Dogs, strongbacks, saddles and other fitting aids are temporary attachments. They must be removed once the fit-up is done. Welds in most material are removed by chipping, oxyfuel or plasma cutting, gouging, grinding, prying, or hammering. The exception occurs in the case of certain low carbon alloy materials. The following rules apply to the removal of fitting aids:

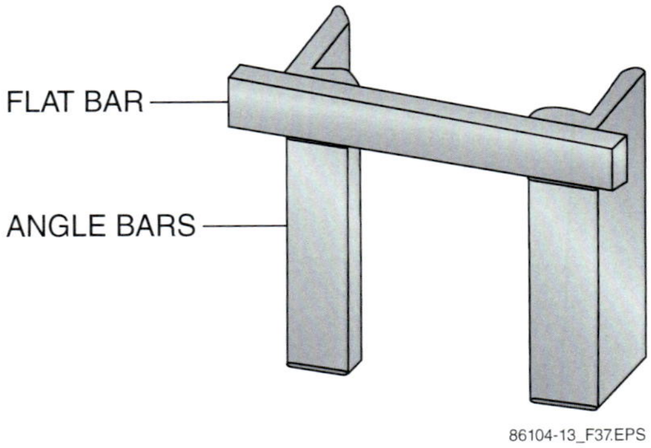

Figure 37 Saddle made from channel and bar stock.

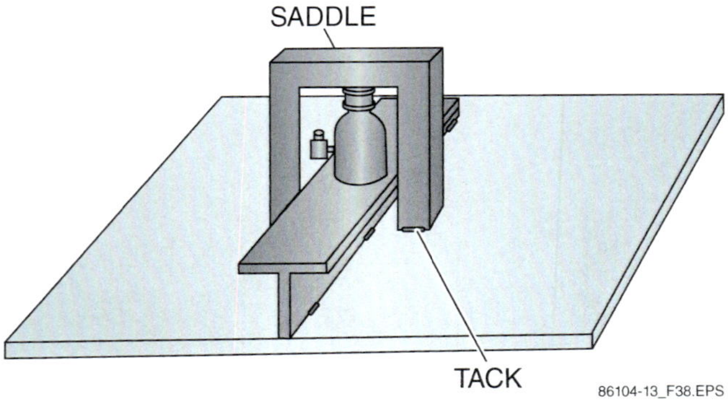

Figure 38 Bottle jack and saddle used to fair-up a T-bar stiffener and plate.

- Hammering and prying are not permitted on some ships.
- Hammering or prying may not be used on attachments that are welded on both sides.
- With the exception of the two rules above, hammering and prying should be used when possible, because applying heat to some metals can distort them.

> **NOTE**
> Most attachments are made of mild steel, so heat will cause an attachment to fail before the base metal does.

- Surface defects in the base metal must be repaired after temporary attachments are removed.

6.0.0 The Importance of Quality

Craftsmanship is a critical element of a fitter's work. If a fitter's measurements are inaccurate, or if parts are not correctly aligned, the work may have to be scrapped. If parts are not properly cleaned following removal of fitting aids, it will cause delays in the project while someone does the work. A company will not long tolerate a fitter who does poor-quality work. Fitters can be held accountable for errors.

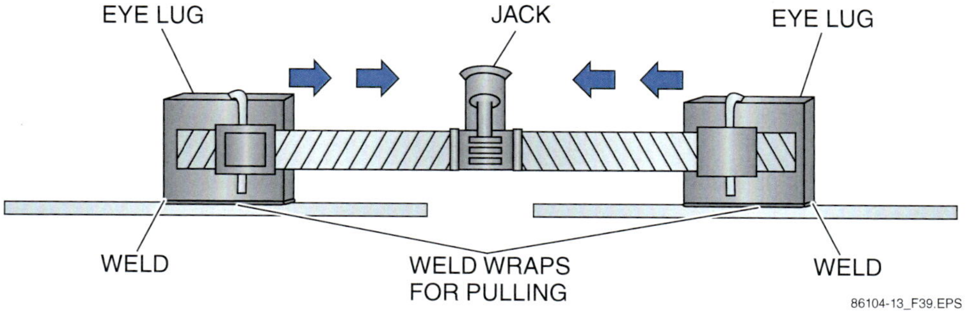

Figure 39 Weld placement for steamboat jack pull.

In some yards, fitters making tack welds are required to mark the weld symbol they used and their name or badge number in the area next to the work.

Structural fitting work requires a high degree of precision. Measurements must be accurate and work must be squared-up. There is little margin for error because errors in subassemblies can affect subsequent assembly work. Small errors in placing parts can affect the ability to install other parts.

The fitter must first learn to interpret drawings in order to identify the correct measurements. In some cases, it will be necessary to perform basic math—including adding and subtracting of fractions—to extract the measurements from a drawing.

Once you know the correct measurements, you must be able to transfer them to the workpiece. This requires the use of measuring tools such as rulers, tape measures, and folding rules (*Figure 41*). A fitter needs to be able to measure properly in order to do the following:

- Measure parts for fitting
- Measure work surfaces for squaring
- Make measurements to determine the exact locations of holes and openings in a part

The following are some tips that will help ensure accurate measurements:

- Be very careful when preparing the layout. Errors at that level are multiplied as the work progresses.
- Always measure twice.
- Make sure the hook of your measuring tape has a little play. This will allow it to be used to make accurate inside and outside measurements. Use a tape for measuring whenever possible.

> **NOTE**
> It is a good idea when measuring to start at the 1-foot or 1-inch mark, then add that amount to the end of the measurement.

- If using a ruler, make sure you account for the gap between the end of the ruler and the first measurement increment. Butt the ruler against a solid, square surface if possible.
- Be careful when marking the measurement. It is a good idea to make a V pointing to the spot to be marked (*Figure 42*).
- A Sharpie-type felt pen or soapstone marker (*Figure 43*) can be used to mark steel. Make sure the point is sharp. When using soapstone, hold the longer sharp end against a straightedge to avoid making too wide a line. Sharpen the soapstone by shaving one side with a file. This will allow you to draw a line close to the straightedge.
- A scratch awl (*Figure 44*) is a good tool for marking metal because it make a permanent mark on the surface. It also makes a fine line so that measurements are accurate.

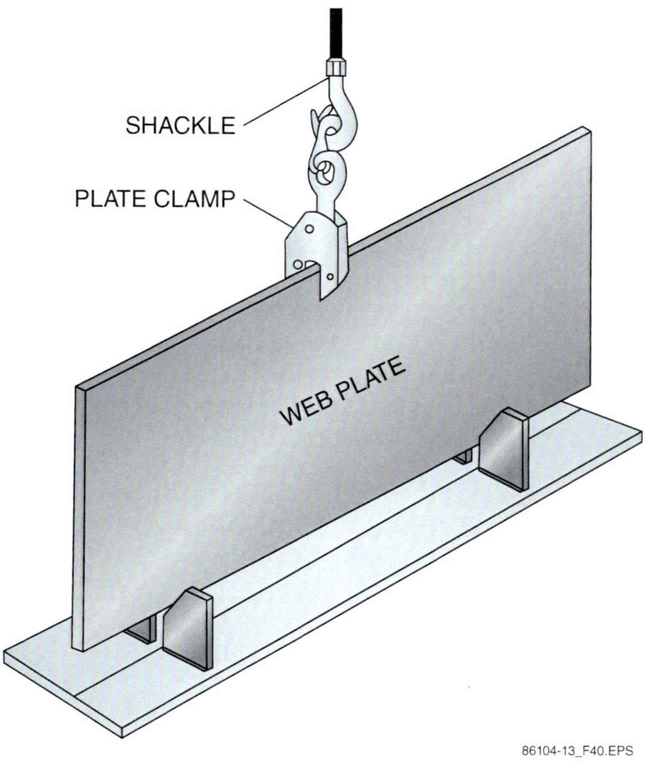

Figure 40 Using landing clips to guide a plate.

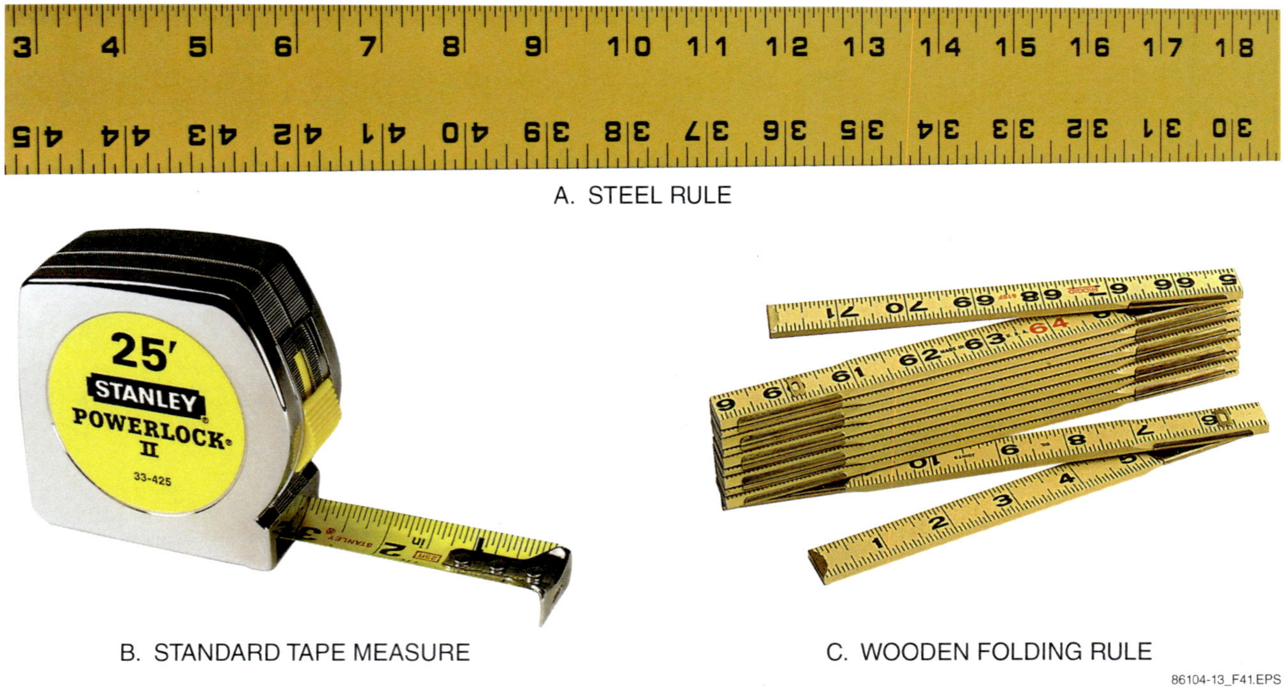

Figure 41 Measuring tools.

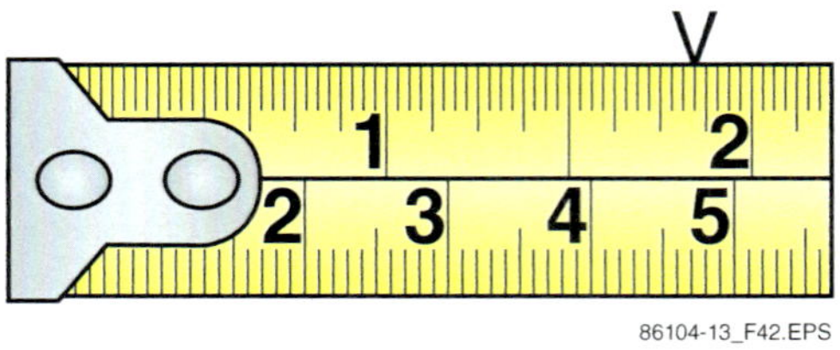

Figure 42 Make a V when marking the measurement point.

Figure 43 Sharpened soapstone marker.

Figure 44 Scratch awl.

6.1.0 Determining the Thickness of Plate and Angles

On ship construction drawings, the thickness of plate is expressed in pounds. The fitter must be able to convert that metric into inches. The rule is that 1" thick steel weighs 40.8 pounds per square foot and increases 5.1 pounds per square foot for every ⅛" increase in thickness. For example, a plate that is 1½" thick ($1\frac{4}{8}$") would weigh 61.2 pounds per square foot:

$$1" = 40.8 \text{ pounds}$$

$$\tfrac{1}{2}" = 20.4\ (4 \times 5.1)$$

$$40.8 + 20.4 = 61.2$$

Therefore, 1½" = 61.24 pounds per square foot.

A different method is used to determine the thickness of angle. It is common to use a table to determine the thickness of an angle. However, in the absence of a table, a formula can be used. This formula takes into account the weight of the material and the dimensions of the flanges.

$$T = 5W \div F1 + F2, \text{ where:}$$

T = Thickness of the angle in sixteenths of an inch

W = The weight of the angle

F1 = Width of one flange

F2 = Width of the other flange

For example the thickness of an angle with 6"and 4" flanges and a weight of 16.2 pounds per square foot is:

$$T = 5W \div F1 + F2$$

$$T = 5 \times 16.2 \div 6 + 4$$

$$T = 81 \div 10$$

$$T = 8.1 \text{ or } \tfrac{8}{16} (\tfrac{1}{2}")$$

6.2.0 Surface Preparation

An important part of doing a quality job is properly preparing the work surface. Joints to be welded must be cleaned ahead of welding in order to remove dirt, grease, paint, scale, rust, gouges, and irregularities within 4" of the area to be welded. Failure to do this could result in a failed weld. Scars and gouges should be marked so that the welder can repair them.

It is also the fitter's responsibility to make sure the surface of the workpiece is left clean when the fit-up is complete. This means removing weld spatter, paint, and grease. It also means removing burrs left when fitting aids are removed after welding.

6.3.0 The 3-4-5 Method for Checking Squareness

When preparing to fit parts, it is important to verify that the parts are square. The 3-4-5 method is an easy way to do this. If its sides are perfectly square, the corners of a rectangular or square object will form a perfect right angle (90-degree angle). In a right triangle, the sides are always in 3, 4, and 5 units. That is, one side will be 3 units, another side 4 units, and the diagonal side (the hypotenuse) will be 5 units (see *Figure 45*). The units can be inches, feet, centimeters, or millimeters. When using the 3-4-5 method, it is best to use the largest practical dimensions.

To use the 3-4-5 method, measure and mark 3 units on one side and 4 units on the other side. If the piece is square, the measurement between the two marks will be exactly 5 units. The greatest accuracy is achieved by using the largest practical dimensions.

Another method of checking squareness is to measure the diagonals. This is shown in *Figure 46*. If the piece is perfectly square, the measurements of the two diagonals will be exactly the same.

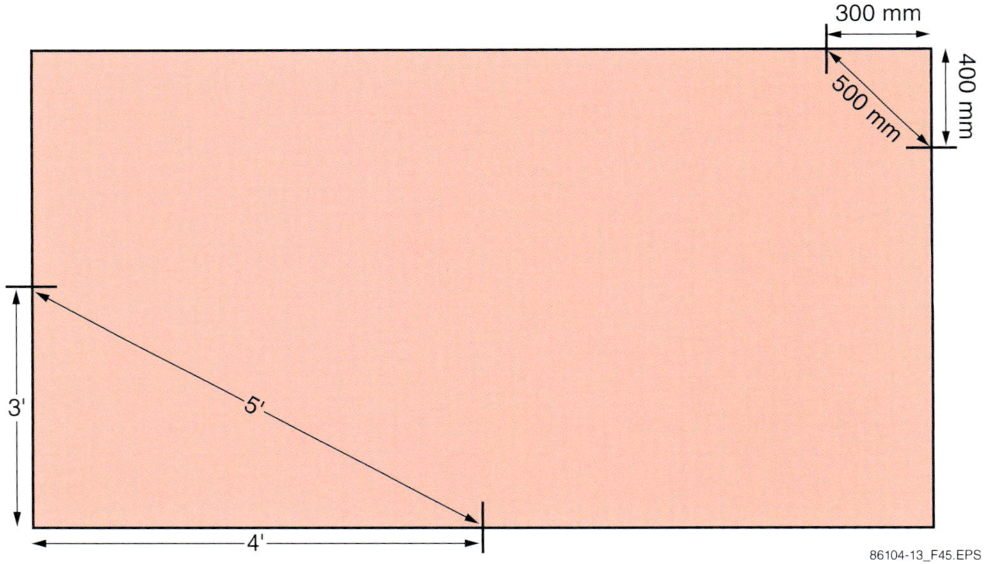

Figure 45 Example of the 3-4-5 method.

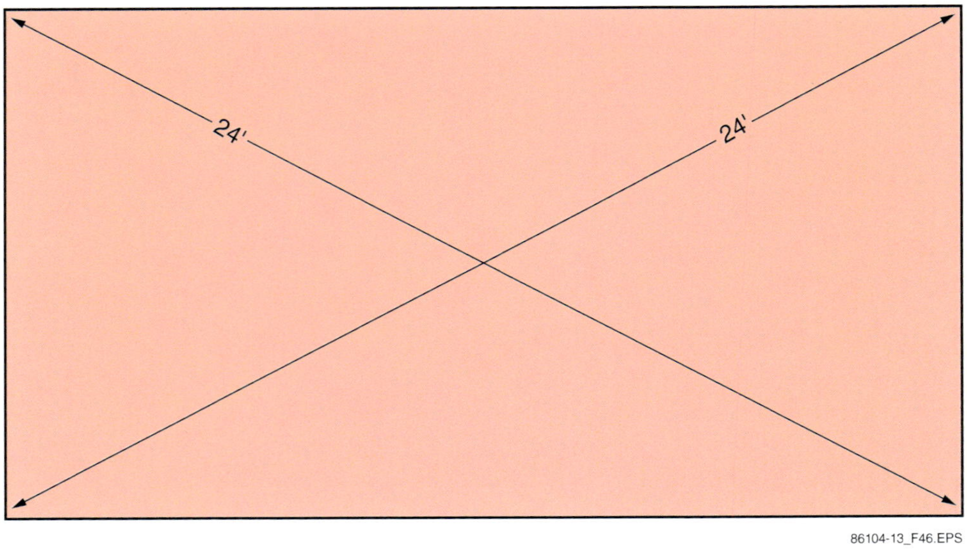

Figure 46 Measuring diagonals.

Summary

Structural fitters, sometimes called ship fitters, are responsible for laying out, aligning, and temporarily attaching parts in preparation for permanent attachment. For example, a fitter might lay out, align, and tack-weld deck plates so that they can be permanently welded together. An important part of the fitter's work is preparing surfaces for welding by using tools and cleaning aids to remove rust, oil, grease, paint, and burrs from the material to be welded. Fitters must be conscious of the unique safety hazards related to fitting work, such as the hazards that result from grinding and grit-blasting of metal that require the use of full-face protection and sometimes respiratory equipment.

Fitters must learn to recognize and work with a variety of tools such as grinders, chipping hammers, hoisting devices, jacks, levels, and squares. Fitters also must learn to use fitting aids such as dogs, saddles, strongbacks, and wedges to align parts. Fitters are responsible for the quality of their work and are held accountable for work that is incorrectly done. Quality work involves thorough preparation of material, accurate measurement, and properly aligned parts.

Review Questions

1. Which of the following is a task commonly performed by structural fitters?
   a. Joining piping systems
   b. Finish welding of faired-up assemblies
   c. Marking and drilling hole locations
   d. Painting decks and bulkheads

2. Lockout/tagout requirements apply to _____.
   a. only electrical systems and equipment
   b. any system or device that can release energy
   c. only hydraulic systems and equipment
   d. any system or equipment that contains chemicals

3. The web of an H-beam is taller than the width of the flange.
   a. True
   b. False

4. A rolled bar with two parallel flanges at right angles to the web is a _____.
   a. T-bar
   b. flat bar
   c. square bar
   d. channel bar

5. A cape chisel is used to _____.
   a. cut grooves for oilways
   b. cut keyways
   c. cut V-grooves
   d. clean out inside corners

6. A drift pin is used to _____.
   a. mark metal
   b. punch holes in metal
   c. align holes
   d. check angles

7. Which of these tools can be fitted with a protractor head?
   a. Spirit level
   b. Framing square
   c. Try square
   d. Combination square

8. Which of these tools is pulled toward the operator rather than pushed because of its potential for violent kickback?
   a. Grinder
   b. Chain fall
   c. Aluminum cutting saw
   d. Cold chisel

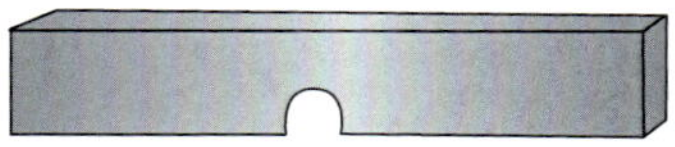

86104-13_RQ01.EPS

Figure 1

9. *Review Question Figure 1* shows a _____.
   a. side saddle
   b. strongarm
   c. dog
   d. strongback

10. What is the thickness of a plate that weighs 68 pounds per square foot based on a weight of 40.8 pounds per square foot equaling 1" in thickness?
    a. 1¼"
    b. 1½"
    c. 1⅔"
    d. 1¾"

Trade Terms Quiz

Fill in the blank with the correct term that you learned from your study of this module.

1. If a structural member is perfectly vertical, standing at a right angle to the floor, it is _______.
2. If you are cutting a groove in a steel plate, you are _______.
3. A _______ may be added to a bulkhead to make it stronger.
4. Drill bits and grinding wheels are examples of tools that are secured in a power tool's _______.

Trade Terms

Arbor
Gouging
Plate stiffeners
Plumb

Trade Terms Introduced in This Module

Arbor: In a tool, a spindle used to secure an accessory such as a drill bit or grinding wheel.

Gouging: The process of cutting a groove into metal with a torch.

Plate stiffener: An angle, T-bar, or channel used to stiffen plating of a bulkhead or other member.

Plumb: Perfectly vertical; that is at a right angle to a horizontal surface.

Additional Resources

This module presents thorough resources for task training. The following resource material is suggested further study.

Ship Construction. D.J. Eyres. Burlington, MA: Elsevier.

IPT's Metal Trades and Welding Handbook. Ronald G. Garby and Bruce J. Ashton. Edmonton, Alberta, Canada: IPT Publishing and Training Ltd.

Figure Credits

Courtesy of Bollinger Shipyards, Inc., Module opener, Figures 1, 34B, 36

Topaz Publications, Inc., Figures 3, 6, 9B, 12, 18, 19, 24, 25, 41A, 43, 44

MSA The Safety Company, Figure 4

Photo courtesy of Mustang Survival **www.mustangsurvival.com**, Figure 5A

Photo courtesy of Dave Richmond, Figure 5B

The Stanley Works, Figures 9A, 9C–9E, 11, 17B, 23A, 41B

RIDGID®, Figure 10, 17A, 41C

JET brand of WMH Tool Group, Figures 13, 15, 16

Courtesy of Irwin Tools, Figure 14

Courtesy of The LS Starrett Company, Figure 22

SPX Power Team, Figures 26–29

Joyce/Dayton Corp., Figure 31

Photo supplied by Simplex, a division of Actuant Corporation, Figure 32

NCCER CURRICULA — USER UPDATE

NCCER makes every effort to keep its textbooks up-to-date and free of technical errors. We appreciate your help in this process. If you find an error, a typographical mistake, or an inaccuracy in NCCER's curricula, please fill out this form (or a photocopy), or complete the online form at **www.nccer.org/olf**. Be sure to include the exact module ID number, page number, a detailed description, and your recommended correction. Your input will be brought to the attention of the Authoring Team. Thank you for your assistance.

Instructors – If you have an idea for improving this textbook, or have found that additional materials were necessary to teach this module effectively, please let us know so that we may present your suggestions to the Authoring Team.

NCCER Product Development and Revision

13614 Progress Blvd., Alachua, FL 32615

Email: curriculum@nccer.org

Online: www.nccer.org/olf

❑ Trainee Guide ❑ Lesson Plans ❑ Exam ❑ PowerPoints Other ______________________

Craft / Level: Copyright Date:

Module ID Number / Title:

Section Number(s):

Description:

Recommended Correction:

Your Name:

Address:

Email: Phone:

Glossary

Alloy: A metal that has had other elements added to it that substantially change its mechanical properties.

Alternating current (AC): Electrical current that reverses its flow at set intervals.

Amperage: A measurement of the rate of flow of electric current.

Arbor: In a tool, a spindle used to secure an accessory such as a drill bit or grinding wheel.

Arc: The flow of electricity through an air gap or gaseous space.

Arc blow: The deflection of the arc from its intended path by magnetic forces.

Arc strike: A discontinuity consisting of localized melting of the base metal or finished weld caused by the arc. It occurs as a result of striking the arc outside the area to be welded.

Backfire: A loud snap or pop as a torch flame is extinguished.

Backing: A weldable or nonweldable material used behind a root opening to allow defect-free welding at the open root of a joint.

Base line: A horizontal line identifying the bottom of the ship.

Base metal: Metal to be welded, cut, or brazed.

Bill of materials: A list of parts and materials shown on a drawing. Numbered entries on the bill of materials correspond with item (key) numbers found in the body of the drawing.

Bonded: The permanent joining of metallic parts to form an electrically conductive path that will assure electrical continuity and the capacity to safely conduct any current likely to be imposed on it.

Buttock line: A line running parallel to and at specific distances from the center line of the ship.

Carburizing flame: A flame burning with an excess amount of fuel; also called a reducing flame.

Center line: A horizontal line running from bow to stern of the ship.

Code: A set of regulations covering permissible materials, service limitations, fabrication, inspection, testing procedures, and qualifications of welders.

Combustible: Capable of easily igniting and rapidly burning. Describes a fuel with a flash point above 100°F.

Condensation: The process in which water from the atmosphere forms on a cool surface.

Conduction: A heat transfer process in which heat is moved from one material to another by contact.

Defect: A discontinuity or imperfection that renders a part of the product or the entire product unable to meet minimum acceptable standards or specifications.

Degauss: Neutralize a magnetic field by using equipment that produces an opposing magnetic field.

Detail view: A drawing that shows a single component or part.

Direct current (DC): Electrical current that flows in one direction only.

Discontinuity: A change or break in the shape or structure of a part that may or may not be considered a defect, depending on the code.

Distortion: The expansion and contraction of welded parts caused by the heating and subsequent cooling of the weld joint.

Drag angle: The travel angle when the electrode is pointing in a direction opposite to the welding progression.

Drag lines: The lines on the edge of the material that result from the travel of the cutting oxygen stream into, through, and out of the metal.

Dross: The material (oxidized and molten metal) that is expelled from the kerf when cutting using a thermal process. It is sometimes called slag.

Ductile: The mechanical property of a material that allows it to be bent or shaped without breaking.

Electrically grounded: Connected to earth or to some conducting body that serves in place of the earth.

Elevation view: A drawing giving the height and length of an object, usually drawn looking toward the top left side of the object.

Embrittled: Metal that has been made brittle and that will tend to crack with little bending.

Ferrous metals: Metals containing iron.

Flammable: Capable of easily igniting and rapidly burning. Describes a fuel with a flash point below 100°F.

Flash point: The temperature at which fuel gives off enough vapors to burn.

Flashback: The flame burning back into the tip, torch, hose, or regulator, causing a high-pitched whistling or hissing sound.

Flux: Material used to prevent, dissolve, or facilitate the removal of oxides and other undesirable substances on a weld.

Flux cored arc welding (FCAW): A welding method that uses an electric arc to melt a flux-cored wire electrode that is fed through the welding gun.

Freeboard: The distance between the water line and the deck or weather deck of a ship.

Fume plume: The fumes, gases, and particles from the consumables, base metal, and base metal coating during the welding process.

Gas metal arc welding (GMAW): In gas metal arc welding, sometimes called MIG, a consumable wire fed through the welding gun is used to make the weld; commonly used on steel and other metals.

General arrangement plan: The master drawing that shows the layout of the complete unit to be constructed. It usually contains the title block and a list of detailed drawings.

Gouging: The process of cutting a groove into metal with a torch.

Graphic scale: A scale that shows the size relationship between the object represented in the drawing and the actual object.

Hardenable materials: Metals that have the ability to be made harder by heating and then cooling.

Heat-affected zone: The part of the base metal that has been altered but not melted by the heat.

Hermetically sealed: Having an airtight seal.

Homogeneity: The quality or state of having a uniform structure or composition throughout.

Immediate danger to life and health (IDLH): As defined by OSHA, this term refers to any condition that poses an immediate or delayed threat to life or that would cause irreversible adverse health effects or that would interfere with an individual's ability to escape unaided from a permit space.

Incipient fire: A fire in its beginning stage.

Inclusion: Foreign matter introduced into and remaining in a weld.

Inerted: Filled with an inert gas such as nitrogen.

Isometric drawing: A three-dimensional drawing of a piping system drawn to provide clarity.

Kerf: The gap produced by a cutting process.

Laminations: Cracks in the base metal formed when layers separate.

Legend: An interpretation of symbols and abbreviations used on a drawing.

Load: The amount of force applied to a material or a structure.

Load water line: The water line of a ship when it is loaded to its carrying capacity.

Longitudinal frame: A shell, deck, or bulkhead stiffener running fore and aft.

Longitudinal frame line: A line representing frame members running fore and aft along the length of the vessel.

Low-hydrogen electrode: An electrode specially manufactured to contain little or no moisture.

Melt-through: Complete joint penetration.

Midship line: A vertical line halfway between the bow and stern; may be called the center line on some drawings.

Mold line: The exact location at which a structure is installed.

Neutral flame: A flame burning with correct proportions of fuel gas and oxygen.

Nonferrous metal: A metal, such as aluminum, copper, or brass, lacking sufficient quantities of iron to have any effect on its properties.

Notch toughness: The ability of a material to resist breaking at points where stress is concentrated.

Orthographic drawing: The projection of a single view of an object.

Oxide: The scale that forms on metal surfaces when they are exposed to air.

Oxidizing flame: A flame burning with an excess amount of oxygen.

Peen: The rounded end of a ball peen hammer used to strike a weld as it cools to reduce stress on the weld.

Pierce: To penetrate through metal plate with an oxyfuel cutting torch.

Piping porosity: A form of porosity having a length greater than its width and that is approximately perpendicular to the weld face.

Plan view: Drawing of an object or area from above it, looking down on the system.

Plate stiffener: An angle, T-bar, or channel used to stiffen plating of a bulkhead or other member.

Plumb: Perfectly vertical; that is at a right angle to a horizontal surface.

Polarity: The direction of flow of electrical current in a direct current welding circuit.

Porosity: Gas pockets, inclusions, or voids in the weld metal.

Primary current: Electrical current received from conventional power lines used to energize welding machines.

Procedure qualification: The demonstration that welds made following a specific process can meet prescribed standards.

Procedure qualification record (PQR): The document containing the results of the nondestructive and destructive testing required to qualify a welding procedure specification (WPS).

Purging gas: A gas such as nitrogen or argon that is classified as inert because it will not react with other material.

Push angle: The travel angle at which the electrode is pointing in the same direction as the welding progression.

Radiation: The movement of heat in the form of invisible rays or waves, such as the heat felt from a campfire.

Radiographic: Describes images made by passing X-rays or gamma rays through an object and recording the variations in density on photographic film.

Root face: A small facing on the end of a bevel for a groove weld.

Root opening: The space between the base metal pieces at the bottom or root of the joint.

Section view: A drawing of the view resulting from an imaginary line cut through the ship or area.

Soapstone: Soft, white stone used to mark metal.

Step-down transformer: An electrical device that uses two wire coils of different sizes to convert a high-voltage, low-current power source into a lower voltage, higher current output.

Surfacing: The application by welding, brazing, or thermal spraying of a layer of material to a surface to obtain desired properties or dimensions.

Traceability: The ability to verify that a procedure has been followed by reviewing the documentation step by step.

Transverse frame line: The line representing frame members running across the ship from port to starboard at specified intervals.

Transverse frames: Athwartships members forming the ship's ribs.

Transverse: Placed or running crosswise.

Underbead cracking: Cracking in the base metal near the weld, but under the surface.

Ventricular fibrillation: Irregular heart rhythm characterized by a rapid fluttering heartbeat.

Vertical welding: Welding with a downward progression from top to bottom.

Voltage: A measurement of electromotive force or pressure that causes current to flow in a circuit.

Water line: A line parallel to the ship's base line and at various distances depending on the displacement of the ship.

Weathering steel: Steel alloy that, under specific conditions, is designed to form a very dense oxide layer on its outer surfaces, which retards further corrosion.

Weld axis: A straight line drawn through the center of a weld along the length of the weld.

Weld coupon: Metal pieces to be welded together as a test or practice.

Welding procedure specification (WPS): The document containing all the detailed methods and practices required to produce a sound weld.

Weldment: An assembly that is fastened together by welded joints.

Work angle: An angle less than 90 degrees between a line perpendicular to the major work-piece surface and a plane determined by the electrode axis and the weld axis. In a T-joint or corner joint, the line is perpendicular to the non-butting member. The definition of work angle for a pipe weld is covered in a later module.

Index

A

B

C

D

E

F

O

P

Q

R

S

T

Z